# Lecture Notes in Computer Science    1976

Edited by G. Goos, J. Hartmanis and J. van Leeuwen

**Springer**
*Berlin*
*Heidelberg*
*New York*
*Barcelona*
*Hong Kong*
*London*
*Milan*
*Paris*
*Singapore*
*Tokyo*

Tatsuaki Okamoto (Ed.)

# Advances in Cryptology – ASIACRYPT 2000

6th International Conference on the Theory
and Application of Cryptology and Information Security
Kyoto, Japan, December 3-7, 2000
Proceedings

 Springer

Series Editors

Gerhard Goos, Karlsruhe University, Germany
Juris Hartmanis, Cornell University, NY, USA
Jan van Leeuwen, Utrecht University, The Netherlands

Volume Editor

Tatsuaki Okamoto
Nippon Telegraph and Telephone Corporation
NTT Laboratories
1-1, Hikarinooka, Yokosuka-shi, Kanagawa-ken, 239-0847 Japan
E-mail: okamoto@sucaba.isl.ntt.co.jp

Cataloging-in-Publication Data applied for


Die Deutsche Bibliothek - CIP-Einheitsaufnahme

Advances in cryptology : proceedings / ASIACRYPT 2000, 6th
International Conference on the Theory and Application of Cryptology
and Information Security, Kyoto, Japan, December 3 - 7, 2000. Tatsuaki
Okamoto (ed.). - Berlin ; Heidelberg ; New York ; Barcelona ; Hong
Kong ; London ; Milan ; Paris ; Singapore ; Tokyo : Springer, 2000
    (Lecture notes in computer science ; Vol. 1976)
    ISBN 3-540-41404-5




CR Subject Classification (1998): E.3, G.2.2, D.4.6, K.6.5, F.2.1-2, C.2, J.1

ISSN 0302-9743
ISBN 3-540-41404-5 Springer-Verlag Berlin Heidelberg New York

Springer-Verlag Berlin Heidelberg New York
a member of BertelsmannSpringer Science+Business Media GmbH
© Springer-Verlag Berlin Heidelberg 2000
Printed in Germany

Typesetting: Camera-ready by author, data conversion by Boller Mediendesign
Printed on acid-free paper      SPIN 10781195      06/3142      5 4 3 2 1 0

# Preface

ASIACRYPT 2000 was the sixth annual ASIACRYPT conference. It was sponsored by the International Association for Cryptologic Research (IACR) in cooperation with the Institute of Electronics, Information, and Communication Engineers (IEICE).

The first conference with the name ASIACRYPT took place in 1991, and the series of ASIACRYPT conferences were held in 1994, 1996, 1998, and 1999, in cooperation with IACR. ASIACRYPT 2000 was the first conference in the series to be sponsored by IACR.

The conference received 140 submissions (1 submission was withdrawn by the authors later), and the program committee selected 45 of these for presentation. Extended abstracts of the revised versions of these papers are included in these proceedings. The program also included two invited lectures by Thomas Berson (Cryptography Everywhere: IACR Distinguished Lecture) and Hideki Imai (CRYPTREC Project – Cryptographic Evaluation Project for the Japanese Electronic Government). Abstracts of these talks are included in these proceedings.

The conference program also included its traditional "rump session" of short, informal or impromptu presentations, kindly chaired by Moti Yung. Those presentations are not reflected in these proceedings.

The selection of the program was a challenging task as many high quality submissions were received. The program committee worked very hard to evaluate the papers with respect to quality, originality, and relevance to cryptography.

I am extremely grateful to the program committee members for their enormous investment of time and effort in the difficult and delicate process of review and selection.

I gratefully acknowledge the help of a large member of colleagues who reviewed submissions in their area of expertise: Masayuki Abe, Harald Baier, Olivier Baudron, Mihir Bellare, John Black, Michelle Boivin, Seong-Taek Chee, Ronald Cramer, Claude Crepeau, Pierre-Alain Fouque, Louis Granboulan, Safuat Hamdy, Goichiro Hanaoka, Birgit Henhapl, Mike Jacobson, Masayuki Kanda, Jonathan Katz, Dennis Kuegler, Dong-Hoon Lee, Markus Maurer, Bodo Moeller, Phong Nguyen, Satoshi Obana, Thomas Pfahler, John O. Pliam, David Pointch, Guillaume Poupard, Junji Shikata, Holger Vogt, Ullrich Vollmer, Yuji Watanabe, Annegret Weng, and Seiji Yoshimoto.

An electronic submission process was available and recommended. I would like to thank Kazumaro Aoki, who did an excellent job in running the electronic submission system of the ACM SIGACT group and in making a support system for the review process of the PC members. Special thanks to many people who supported him: Seiichiro Hangai and Christian Cachin for their web page supports, Joe Kilian for giving him a MIME parser, Steve Tate for supporting the SIGACT package, Wim Moreau for consulting their electronic review system, and Masayuki Abe for scanning non-electronic submissions. Special thanks go to Mami Yamaguchi and Junko Taneda for their support in arranging review reports and editing these proceedings.

I would like to thank Tsutomu Matsumoto, general chair, and the members of organizing committee: Seiichiro Hangai, Shouichi Hirose, Daisuke Inoue, Keiichi Iwamura, Masayuki Kanda, Toshinobu Kaneko, Shinichi Kawamura, Michiharu Kudo, Hidenori Kuwakado, Masahiro Mambo, Mitsuru Matsui, Natsume Matsuzaki, Atsuko Miyaji, Shiho Moriai, Eiji Okamoto, Kouichi Sakurai, Fumihiko Sano, Atsushi Shimbo, Takeshi Shimoyama, Hiroki Shizuya, Nobuhiro Tagashira, Kazuo Takaragi, Makoto Tatebayashi, Toshio Tokita, Naoya Torii. We are especially grateful to Shigeo Tsujii and Hideki Imai for their great support of the organizing committee.

The organizing committee gratefully acknowledges the financial contributions of the two organizations, Initiatives in Research of Information Security (IRIS) and the Telecommunications Advancement Organization (TAF), as well as many companies.

I wish to thank all the authors who by submitting papers made this conference possible, and the authors of accepted papers for their cooperation.

Finally, I would like to dedicate these proceedings to the memory of Kenji Koyama, who passed away in March 2000. He was 50 years old. He was one of the main organizers of the first ASIACRYPT conference held in Japan in 1991, and devoted himself to make IACR the sponsor of ASIACRYPT. He was looking forward to ASIACRYPT 2000 very much, since it was the first of the ASIACRYPT conference series sponsored by IACR. May he rest in peace.


September 2000                                        Tatsuaki Okamoto

# ASIACRYPT 2000
## 3–7 December 2000, Kyoto, Japan

Sponsored by the
*International Association for Cryptologic Research* (IACR)
in cooperation with the
*Institute of Electronics, Information and Communication Engineers* (IEICE)

**General Chair**
Tsutomu Matusmoto, Yokohama National University, Japan

**Program Chair**
Tatsuaki Okamoto, NTT Labs, Japan

**Program Committee**

Ross Anderson ................................... Cambridge University, UK
Dan Boneh ....................................... Stanford University, USA
Johannes Buchmann ........... Technical University of Darmstadt, Germany
Ivan Damgård .................................. Århus University, Denmark
Yvo Desmedt ................................ Florida State University, USA
Yongfei Han ........................................ SecurEworld, Singapore
Ueli Maurer ....................................... ETH Zurich, Switzerland
Alfred Menezes ............................ University of Waterloo, Canada
Moni Naor ...................................... Weizmann Institute, Israel
Choonsik Park ................................................ ETRI, Korea
Dingyi Pei .............................. Chinese Academy of Science, China
Phillip Rogaway ..................... University of California at Davis, USA
Kazue Sako .................................................... NEC, Japan
Kouichi Sakurai .................................. Kyushu University, Japan
Jacques Stern ................................................. ENS, France
Serge Vaudenay ........................... EPF Lausanne, Switzerland
Chung-Huang Yang ............. National Kaohsiung First University, Taiwan
Moti Yung ................................................... CertCo, USA
Yuliang Zheng ................................. Monash University, Australia

*Advisory Members*

Kazumaro Aoki (Electronic submissions) ................... NTT Labs, Japan
Eiji Okamoto (ASIACRYPT'99 program co-chair) University of Wisconsin, USA

# Table of Contents

## Number Theoretic Algorithms

## Symmetric-Key Schemes I

# Protocols II

# Invited Lecture

# Fingerprinting

# Zero-Knowledge and Provable Security

# Boolean Functions

## Cryptanalysis II

## Pseudorandomness

## Symmetric-Key Schemes II

## Public-Key Encryption and Key Distribution

## Author Index

# Cryptanalytic Time/Memory/Data Tradeoffs for Stream Ciphers

Alex Biryukov and Adi Shamir

Computer Science Department
The Weizmann Institute
Rehovot 76100, Israel.

**Abstract.** In 1980 Hellman introduced a general technique for breaking arbitrary block ciphers with $N$ possible keys in time $T$ and memory $M$ related by the tradeoff curve $TM^2 = N^2$ for $1 \leq T \leq N$. Recently, Babbage and Golic pointed out that a different $TM = N$ tradeoff attack for $1 \leq T \leq D$ is applicable to stream ciphers, where $D$ is the amount of output data available to the attacker. In this paper we show that a combination of the two approaches has an improved time/memory/data tradeoff for stream ciphers of the form $TM^2D^2 = N^2$ for any $D^2 \leq T \leq N$. In addition, we show that stream ciphers with low sampling resistance have tradeoff attacks with fewer table lookups and a wider choice of parameters.

**Keywords:** Cryptanalysis, stream ciphers, time/memory tradeoff attacks.

## 1 Introduction

There are two major types of symmetric cryptosystems: Block ciphers (which encrypt a plaintext block into a ciphertext block by mixing it in an invertible way with a fixed key), and stream ciphers (which use a finite state machine initialized with the key to produce a long pseudo random bit string, which is XOR'ed with the plaintext to obtain the ciphertext).

Block and stream ciphers have different design principles, different attacks, and different measures of security. The open cryptanalytic literature contains many papers on the resistance of block ciphers to differential and linear attacks, on their avalanche properties, on the properties of Feistel or S-P structures, on the design of S-boxes and key schedules, etc. The relatively few papers on stream ciphers tend to concentrate on particular ciphers and on particular attacks against them. Among the few unifying ideas in this area are the use of linear feedback shift registers as bit generators, and the study of the linear complexity and correlation immunity of the ciphers.

In this paper we concentrate on a general type of cryptanalytic attack known as a time/memory tradeoff attack. Such an attack has two phases: During the preprocessing phase (which can take a very long time) the attacker explores the general structure of the cryptosystem, and summarizes his findings in large tables (which are not tied to particular keys). During the realtime phase, the attacker

T. Okamoto (Ed.): ASIACRYPT 2000, LNCS 1976, pp. 1–13, 2000.

is given actual data produced from a particular unknown key, and his goal is to use the precomputed tables in order to find the key as quickly as possible.

In any time-memory tradeoff attack there are five key parameters:

- $N$ represents the size of the search space.
- $P$ represents the time required by the preprocessing phase of the attack.
- $M$ represents the amount of random access memory (in the form of hard disks or DVD's) available to the attacker.
- $T$ represents the time required by the realtime phase of the attack.
- $D$ represents the amount of realtime data available to the attacker.

## 2   Tradeoff Attacks on Block and Stream Ciphers

In the case of block ciphers, the size $N$ of the search space is the number of possible keys. We assume that the number of possible plaintexts and ciphertexts is also $N$, and that the given data is a single ciphertext block produced from a fixed chosen plaintext block. The best known time/memory tradeoff attack is due to Hellman [5]. It uses any combination of parameters which satisfy the following relationships: $TM^2 = N^2$, $P = N$, $D = 1$ (see Section 3 for further details). The optimal choice of $T$ and $M$ depends on the relative cost of these computational resources. By choosing $T = M$, Hellman gets the particular tradeoff point $T = N^{2/3}$ and $M = N^{2/3}$.

Hellman's attack is applicable to any block cipher whose key to ciphertext mapping (for a fixed plaintext) behaves as a random function $f$ over a space of $N$ points. If this function happens to be an invertible permutation, the tradeoff relation becomes $TM = N$, which is even better. An interesting property of Hellman's attack is that even if the attacker is given a large number $D$ of chosen plaintext/ciphertext pairs, it is not clear how to use them in order to improve the attack.

Stream ciphers have a very different behavior with respect to time/memory tradeoff attacks. The size $N$ of the search space is determined by the number of internal states of the bit generator, which can be different from the number of keys. The realtime data typically consists of the first $D$ pseudorandom bits produced by the generator, which are computed by XOR'ing a known plaintext header and the corresponding ciphertext bits (there is no difference between a known and a chosen plaintext attack in this case). The goal of the attacker is to find at least one of the actual states of the generator during the generation of this output, after which he can run the generator forwards an unlimited number of steps, produce all the later pseudorandom bits, and derive the rest of the plaintext. Note that in this case there is no need to run the generator backwards or to find the original key, even though this is doable in many practical cases.

The simplest time/memory tradeoff attack on stream ciphers was independently described by Babbage [2] and Golic [4], and will be referred to as the **BG attack**. It associates with each one of the $N$ possible states of the generator the string consisting of the first $log(N)$ bits produced by the generator from that state. This mapping $f(x) = y$ from states $x$ to output prefixes $y$ can be viewed as a random function over a common space of $N$ points, which is easy to evaluate but hard to invert. The goal of the attacker is to invert it on some substring of the given output, in order to recover the corresponding internal state. The preprocessing phase of the attack picks $M$ random $x_i$ states, computes their corresponding $y_i$ output prefixes, and stores all the $(x_i, y_i)$ pairs in a random access memory, sorted into increasing order of $y_i$. The realtime phase of the attack is given a prefix of $D + log(N) - 1$ generated bits, and derives from it all the $D$ possible windows $y_1, y_2, ..., y_D$ of $log(N)$ consecutive bits (with overlaps). It lookups each $y_j$ from the data in logarithmic time in the sorted table. If at least one $y_j$ is found in the table, its corresponding $x_j$ makes it possible to derive the rest of the plaintext by running the generator forwards from this known state[1]. The threshold of success for this attack can be derived from the birthday paradox, which states that two random subsets of a space with $N$ points are likely to intersect when the product of their sizes exceeds $N$. If we ignore logarithmic factors, this condition becomes $DM = N$ where the preprocessing time is $P = M$ and the attack time is $T = D$. This represents one particular point on the time/memory tradeoff curve $TM = N$. By ignoring some of the available data during the actual attack, we can reduce $T$ from $D$ towards 1, and thus generalize the tradeoff to $TM = N$ and $P = M$ for any $1 \leq T \leq D$.

This $TM = N$ tradeoff is similar to Hellman's $TM = N$ tradeoff for random permutations and better than Hellman's $TM^2 = N^2$ tradeoff for random functions (when $T = M$ we get $T = M = N^{1/2}$ instead of $T = M = N^{2/3}$). However, this formal comparison is misleading since the two tradeoffs are completely different: they are applicable to different types of cryptosystems (stream vs. block ciphers), are valid in different parameter ranges ($1 \leq T \leq D$ vs. $1 \leq T \leq N$), and require different amounts of data (about $D$ bits vs. a single chosen plaintext/ciphertext pair).

To understand the fundamental difference between tradeoff attacks on block ciphers and on stream ciphers, consider the problem of using a large value of $D$ to speed up the attack. The mapping defined by a block cipher has two inputs (key and plaintext block) and one output (ciphertext block). Since each precomputed table in Hellman's attack on block ciphers is associated with a particular plaintext block, we cannot use a common table to simultaneously analyse different ciphertext blocks (which are necessarily derived from different plaintext blocks during the lifetime of a single key). The mapping defined by a stream cipher, on the other hand, has one input (state) and one output (an ouput prefix), and thus has a single "flavour": When we try to invert it on multiple output prefixes, we can use the same precomputed tables in all the attempts. As a result, tradeoff attacks on stream ciphers can be much more efficient than tradeoff attacks on block ciphers when $D$ is large, but this possibility had not been explored so far in the research literature.

---

[1] Note that $y_j$ may have multiple predecessors, and thus $x_j$ may be different from the state we look for. However, it can be shown that these "false alarms" increase the complexity of the attack by only a small constant factor.

## 3      Combining the Two Tradeoff Attacks

In this section we show that it is possible to combine the two types of tradeoff attacks to obtain a new attack on stream ciphers whose parameters satisfy the relation $P = N/D$ and $TM^2D^2 = N^2$ for any $D^2 \leq T \leq N$. A typical point on this tradeoff relation is $P = N^{2/3}$ preprocessing time, $T = N^{2/3}$ attack time, $M = N^{1/3}$ disk space, and $D = N^{1/3}$ available data. For $N = 2^{100}$ the parameters $P = T = 2^{66}$ and $M = D = 2^{33}$ are all (barely) feasible, whereas the Hellman attack with $T = M = N^{2/3} = 2^{66}$ requires an unrealistic amount of disk space $M$, and the BG attack with $T = D = N^{2/3} = 2^{66}$ and $M = N^{1/3} = 2^{33}$ requires an unrealistic amount of data $D$.

### 3.1      Hellman's Time/Memory Tradeoff Attack on Block Ciphers

The starting point of the new attack on stream ciphers is Hellman's original tradeoff attack on block ciphers, which considers the random function $f$ that maps the key $x$ to the ciphertext block $y$ for some fixed chosen plaintext. This $f$ is easy to evaluate but hard to invert, since the problem of computing $x = f^{-1}(y)$ is exactly the cryptanalytic problem of deriving the key $x$ from the given ciphertext block $y$.

To perform this difficult inversion of $f$ with an algorithm which is faster than exhaustive search, Hellman uses a preprocessing stage which tries to cover the $N$ points of the space with a rectangular $m \times t$ matrix whose rows are long paths obtained by iterating the function $f$ $t$ times on $m$ randomly chosen starting points. The startpoints are described by the leftmost column of the matrix, and the corresponding endpoints are described by the rightmost column of the matrix (see Fig. 1). The output of the preprocessing stage is the collection of (startpoint, endpoint) pairs of all the chosen paths, sorted into increasing endpoint values. During the actual attack, we are given a value $y$ and are asked to find its predecessor $x$ under $f$. If this $x$ is covered by one of the precomputed paths, the algorithm repeatedly applies $f$ to $y$ until it reaches the stored endpoint, jumps to its associated startpoint, and repeatedly applies $f$ to the startpoint until it reaches $y$ again. The previous point it visits is the desired $x$.

A single matrix cannot efficiently cover all the $N$ points, (in particular, the only way we can cover the approximately $N/e$ leaves of a random directed graph is to choose them as starting points). As we add more rows to the matrix, we reach a situation in which we start to re-cover points which are already covered, which makes the coverage increasingly wasteful. To find this critical value of $m$, assume that the first $m$ paths are all disjoint, but the next path has a common point with one of the previous paths. The first $m$ paths contain exactly $mt$ distinct points (since they are assumed to have no repetitions), and the additional path is likely to contain exactly $t$ distinct points (assuming that $t$ is less than $\sqrt{N}$). By the birthday paradox, the two sets are likely to be disjoint as long as $t \cdot mt \leq N$, and thus we choose $m$ and $t$ which satisfy the relation $mt^2 = N$, which we call **the matrix stopping rule**.

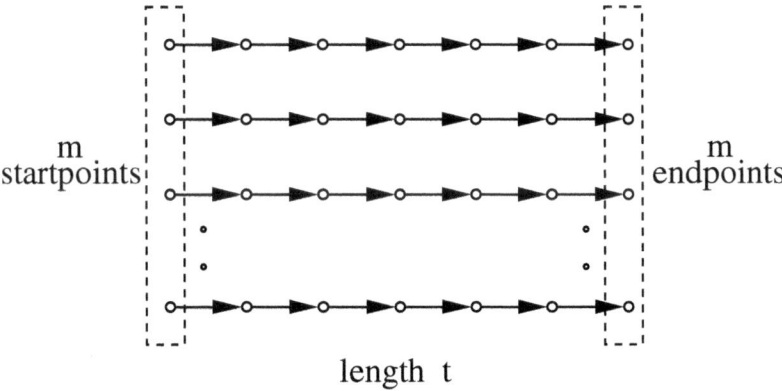

**Fig. 1.** Hellman's Matrix

A single $m \times t$ matrix with $mt^2 = N$ covers only a fraction of $mt/N = 1/t$ of the space, and thus we need $t$ "unrelated" matrices to cover the whole space. Hellman's great insight was the observation that we can use variants $f_i$ of the original $f$ defined by $f_i(x) = h_i(f(x))$ where $h_i$ is some simple output modification (e.g., reordering the bits of $f(x)$). These modified variants of $f$ have the following properties:

1. The points in the matrices of $f_i$ and $f_j$ for $i \neq j$ are essentially independent, since the existence of a common point in two different matrices does not imply that subsequent points on the two paths must also be equal. Consequently, the union of $t$ matrices (each covering $mt$ points) is likely to contain a fixed fraction of the space.
2. The problem of computing $x$ from the given $y = f(x)$ can be solved by inverting any one of the modified functions $f_i$ over the modified point $y_i = f_i(x) = h_i(f(x)$.
3. The value of $y_i = f_i(x)$ can be computed even when we do not know $x$ by applying $h_i$ to the given $y = f(x)$.

The total precomputation requires $P \approx N$ time, since we have to cover a fixed fraction of the space in all the precomputed paths. Each matrix covers $mt$ points, but can be stored in $m$ memory locations since we only keep the startpoint and endpoint of each path. The total memory required to store the $t$ matrices is thus $M = mt$. The given $y$ is likely to be covered by only one of the precomputed matrices, but since we do not know where it is located we have to perform $t$ inversion attempts, each requiring $t$ evaluations of some $f_i$. The total time complexity of the actual attack is thus $T = t^2$. To find the tradeoff curve between $T$ and $M$, we use the matrix stopping rule $mt^2 = N$ to conclude that $TM^2 = t^2 \cdot m^2t^2 = N^2$. Note that in this tradeoff formula the time $T$ can be anywhere in the range $1 \leq T \leq N$, but the space $M$ should be restricted to $N^{1/2} \leq M \leq N$, since otherwise $T > N$ and thus the attack is slower than exhaustive search.

## 3.2  An Improved Attack on Stream Ciphers

As explained earlier in this paper, the main difference between tradeoff attacks on block ciphers and on stream ciphers is that in a block cipher each given ciphertext requires the inversion of a different function, whereas in a stream cipher all the given output prefixes can be inverted with respect to the same function by using the same precomputed tables.

To adapt Hellman's attack from block ciphers to stream ciphers, we use the same basic approach of covering the $N$ points by matrices defined by multiple variants $f_i$ of the function $f$ which represents the state to prefix mapping. Note that partially overlapping prefixes do not necessarily represent neighboring points in the graph defined by the iterations of $f$, and thus they can be viewed as unrelated random points in the graph. The attack is successful if any one of the $D$ given output values is found in any one of the matrices, since we can then find some actual state of the generator which can be run forward beyond the known prefix of output bits. We can thus reduce the total number of points covered by all the matrices from about $N$ to $N/D$ points, and still get (with high probability) a collision between the stored and actual states.

There are two possible ways to reduce the number of states covered by the matrices: By making each matrix smaller, or by choosing fewer matrices. Since each evaluation step of $f_i$ adds $m$ states to the coverage, it is wasteful to choose $m$ or $t$ which are smaller than the maximum values allowed by the matrix stopping rule $mt^2 = N$. Our new tradeoff thus keeps each matrix as large as possible, and reduces the number of matrices from $t$ to $t/D$ in order to decrease the total coverage of all the matrices by a factor of $D$. However, this is possible only when $t \geq D$, since if we try to reduce the number of tables to less than 1, we are forced to use suboptimal values of $m$ and $t$, and thus enter a less efficient region of the tradeoff curve.

Each matrix in the new attack requires the same storage size $m$ as before, but the total memory required to store all the matrices is reduced from $M = mt$ to $M = mt/D$. The total preprocessing time is similarly reduced from $P = N$ to $P = N/D$, since we have to evaluate only $1/D$ of the previous number of paths. The attack time $T$ is the product of the number of matrices, the length of each path, and the number of available data points, since we have to iterate each one of the $t/D$ functions $f_i$ on each one of the $D$ given output prefixes up to $t$ times. This product is $T = t^2$, which is the same as in Hellman's original attack.

To find the time/memory/data tradeoff in this attack, we again use the matrix stopping rule $mt^2 = N$ in order to eliminate the parameters $m$ and $t$ from the various expressions. The preprocessing time is $P = N/D$, which is already free from these parameters. The time $T = t^2$, memory $M = mt/D$, and data $D$ clearly satisfy the invariant relationship:

$$TM^2D^2 = t^2 \cdot (m^2t^2/D^2) \cdot D^2 = m^2t^4 = N^2$$

This relationship is valid for any $t \geq D$, and thus for any $D^2 \leq T \leq N$. In particular, we can use the parameters $P = T = N^{2/3}$, $M = D = N^{1/3}$, which seems to be practical for $N$ up to about 100.

## 4    Time/Memory/Data Tradeoff Attacks with Sampling

One practical problem with tradeoff attacks is that random access to a hard disk requires about 8 milliseconds, whereas a computational step on a fast PC requires less than 2 nanoseconds. This speed ratio of four million makes it crucial to minimize the number of disk operations we perform, in addition to reducing the number of evaluations of $f_i$. An old idea due to Ron Rivest was to reduce the number of table lookups in Hellman's attack by defining a subset of **special points** whose names start with a fixed pattern such as $k$ zero bits.

Special points are easy to generate and to recognize. During the preprocessing stage of Hellman's attack, we start each path from a randomly chosen point, and stop it only when we encounter another special point (or enter a loop, which is unlikely when $t \leq \sqrt{N}$). Consequently, we know that the disk contains only special endpoints. If we choose $k = log(t)$, the expected length of each path remains $t$ (with some variability), and the set of $mt$ endpoints we store in all the $t$ tables contains a large fraction of the $N/t$ possible special points.

The main advantage of this approach is that during the actual attack, we have to perform only one expensive disk operation per path (when we encounter the first special point on it). The number of evaluations of $f_i$ remains $T = t^2$, but the number of disk operations is reduced from $t^2$ to $t$, which makes a huge practical difference.

Can we use a similar sampling of special points in tradeoff attacks on stream ciphers? Consider first the case of the BG tradeoff with $TM = N$, $P = M$, and $1 \leq T \leq D$. We say that an output prefix is special if it starts with a certain number of zero bits, and that a state of the stream cipher is special if it generates a special output prefix. We would like to store in the disk during preprocessing only special pairs of (state, output prefix). Unlike the case of Hellman's attack (where special states appeared on sufficiently long paths with reasonable probability, and acted as natural path terminators), in the BG attack we deal with degenerate paths of length 1 (from a state to its immediate output prefix), and thus we have to use trial and error in order to find special states.

Assume that the ratio between the number of special states and all the states is $R$, where $0 < R < 1$. Then to find the $M$ special states we would like to store during preprocessing, we have to try a much larger number $M/R$ of random states, which increases the preprocessing time from $P = M$ to $P = M/R$. The attack time reduces from $T = D$ to $T = DR$, since only the special points in the given data (which are very easy to spot) have to be looked up in the disk. To make it likely to have a collision between the $M$ special states stored in the disk and the $DR$ special states in the data, we have to apply the birthday paradox to the smaller set of $NR$ special states to obtain $MDR = NR$. The invariant satisfied for all the possible values of $R$ is thus

$$TP = MD = N \ \text{ for } \ 1 \leq T \leq D$$

An interesting consequence of this tradeoff formula is that the sampling technique had turned the original BG time/memory tradeoff ($TM = N$) into two independent time/preprocessing ($TP = N$) and memory/data ($MD = N$) tradeoffs, which are controlled by the three parameters $m$, $t$, and $R$. For $N = 2^{100}$ the first condition is easy to satisfy, since both the preprocessing time $P$ and the actual time $T$ can be chosen as $2^{50}$. However, the second condition is completely unrealistic, since neither the memory $M$ nor the data $D$ can exceed $2^{40}$.

We now describe the effect of this sampling technique on the new tradeoff $TM^2D^2 = N^2$ described in the previous subsection. The main difference between Hellman's original attack on block ciphers and the modified attack on stream ciphers is that we use a smaller number $t/D$ of tables, and force $T$ to satisfy $T \geq D^2$. Unlike the case of the BG attack, the preprocessing complexity remains unchanged as $N/D$, since we do not need any trial and error to pick the random startpoints, and simply wait for the special endpoints to occur randomly during our path evaluation. The total memory required to store the special points remains unchanged at $M = mt/D$. The total time $T$ consists of $t^2$ evaluations of the $f_i$ functions but only $t$ disk operations. We can thus conclude that the resultant time/memory/data tradeoff remains unchanged as $TM^2D^2 = N^2$ for $T \geq D^2$, but we gain by reducing the number of expensive disk operations by a factor of $t$. Rivest's sampling idea thus has no asymptotic effect on Hellman-like tradeoff curves for block and stream ciphers, but drastically changes the BG tradeoff curve for stream ciphers.

## 5    Tradeoff Attacks on Stream Ciphers with Low Sampling Resistance

The $TM^2D^2 = N^2$ tradeoff attack has feasible time, memory and data requirements even for $N = 2^{100}$. However, values of $D \geq 2^{25}$ make each inversion attack very time consuming, since small values of $T$ are not allowed by the $T \geq D^2$ condition, while large values of $T$ do not benefit in practice from the Rivest sampling idea (since the $T =$ evaluations of $f_i$ functions dominate the $\sqrt{T}$ disk operations).

At FSE 2000, Biryukov, Shamir and Wagner [3] introduced a different notion of sampling, which will be called **BSW sampling**. It was used in [3] to attack the specific stream cipher A5/1, but that paper did not analyse its general impact on the various tradeoff formulas. In this paper we show that by using BSW sampling, we can make the new $TM^2D^2 = N^2$ tradeoff applicable with a larger choice of possible $T$ values and a smaller number of disk operations.

The basic idea behind BSW sampling is that in many stream ciphers, the state undergoes only a limited number of simple transformations before emitting its next output bit, and thus it is possible to enumerate all the special states which generate $k$ zero bits for a small value of $k$ without expensive trial and error (especially when each output bit is determined by few state bits). This is almost always possible for $k = 1$, but gets increasingly more difficult when we try to force a larger number of output bits to have specific values. The sampling resistance of a stream cipher is defined as $R = 2^{-k}$ where $k$ is the maximum value for which this direct enumeration is possible. Stream ciphers were never designed to resist this new kind of sampling, and their sampling resistance can serve as a new quantifiable design-sensitive security measure. In the case of A5/1, Biryukov Shamir and Wagner show that it is easy to directly enumerate the $2^{48}$ out of the $2^{64}$ states whose outputs start with 16 zeroes, and thus the sampling resistance of A5/1 is at most $2^{-16}$. Note that BSW sampling is not applicable at all to block ciphers, since their thorough mixing of keys and plaintexts makes it very difficult to enumerate without trial and error all the keys which lead to ciphertexts with a particular pattern of $k$ bits during the encryption of some fixed plaintext.

An obvious advantage of BSW sampling over Rivest sampling is that in the BG attack we can reduce the attack time $T$ by a factor of $R$ without increasing the preprocessing time $P$. We now describe how to apply the BSW sampling idea to the improved tradeoff attack $TM^2D^2 = N^2$.

Consider a stream cipher with $N = 2^n$ states. Each state has a **full name** of $n$ bits, and an **output name** which consists of the first $n$ bits in its output sequence. If the cipher has sampling resistance $R = 2^{-k}$, we can associate with each special state a **short name** of $n - k$ bits (which is used by the efficient enumeration procedure to define this special state), and a **short output** of $n - k$ bits (which is the output name of the special state without the k leading zeroes). We can thus define a new random mapping over a reduced space of $NR = 2^{n-k}$ points, where each point can be viewed as either a short name or a short output. The mapping from short names to short outputs is easy to evaluate (by expanding the short names of special states to full names, running the generator, and discarding the $k$ leading zeroes), and its inversion is equivalent to the original cryptanalytic problem restricted to special states.

We assume that $DR \geq 1$, and thus the available data contains at least one output which corresponds to some special state (if this is not the case we simply relax the definition of special states). We try to find the short name of any one of these $DR$ special states by applying our $TM^2D^2 = N^2$ inversion attack to the reduced space with the modified parameters of $DR$ and $NR$ instead of $D$ and $N$. The factor $R^2$ is canceled out from the expression $TM^2(DR)^2 = (NR)^2$, and thus the tradeoff relation remains unchanged. However, we gain in two other ways:

1. The original range of allowed values of $T$ was lower bounded by $D^2$, which could be problematic for large values of $D$. This lower bound is now reduced to $(DR)^2$, which can be as small as 1. This makes it possible to use a wider range of $T$ parameters, and speed up actual attacks.
2. The number of expensive disk operations is reduced from $t$ to $tR$, since only the $DR$ special points in the data have to be searched in the $t/D$ matrices at a cost of one disk operation per matrix. This can greatly speed up attacks

with moderate values of $t$ in which the $t$ disk operations dominate the $t^2$ function evaluations.

Table 1 summarizes the behavior of the three types of tradeoff attacks under the two types of sampling techniques discussed in this paper. It explains why BSW sampling can greatly reduce the time $T$, even though it has no effect on the asymptotic tradeoff relation itself. Only this type of sampling enabled [3] to attack A5/1 and find its 64 bit key in a few minutes of computation on a single PC using only 4,000 disk operations, given the data contained in the first two seconds of an encrypted GSM conversation.

| Sampling type | BG attack on stream ciphers | Hellman's attack on block ciphers | Our attack on stream ciphers |
|---|---|---|---|
| Rivest | new tradeoffs: $TP = MD = N$ for $1 \leq T \leq D$ increased $P$ | unmodified tradeoff: $TM^2 = N^2$ for $1 \leq T \leq N$ fewer disk operations | unmodified tradeoff: $TM^2D^2 = N^2$ for $D^2 \leq T \leq N$ fewer disk operations |
| BSW | unmodified tradeoff: $TM = N, 1 \leq T \leq D$ | inapplicable to block ciphers | unmodified tradeoff: $TM^2D^2 = N^2$, wider range, $(RD)^2 \leq T \leq N$ even fewer disk operations |

**Table 1.** The effect of sampling on tradeoff attacks.

# References

1. D. Coppersmith, H. Krawczyk, Y. Mansour, *The Shrinking Generator*, Proceedings of Crypto'93, pp.22–39, Springer-Verlag, 1993.
2. S. Babbage, *A Space/Time Tradeoff in Exhaustive Search Attacks on Stream Ciphers*, European Convention on Security and Detection, IEE Conference Publication No. 408, May 1995.
3. A. Biryukov, A. Shamir, and D. Wagner, *Real Time Cryptanalysis of A5/1 on a PC*, Proceedings of Fast Software Encryption 2000.
4. J. Golic, *Cryptanalysis of Alleged A5 Stream Cipher*, Proceedings of Eurocrypt'97, LNCS 1233, pp. 239–255, Springer-Verlag 1997.
5. M. E. Hellman, *A Cryptanalytic Time-Memory Trade-Off*, IEEE Transactions on Information Theory, Vol. IT-26, N 4, pp.401–406, July 1980.
6. W. Meier, O. Staffelbach, *The Self-Shrinking Generator*, Proceedings of Eurocrypt'94, pp.205–214, Springer-Verlag, 1994.

# A    The Sampling Resistance of Various Stream Cipher Constructions

As we have seen in the main part of the paper low sampling resistance of a stream cipher allows for more flexible tradeoff attacks. In this appendix we briefly review several popular constructions and discuss their sampling resistance.

## A.1    Non-linear Filter Generators

In many proposed constructions a single linear feedback shift register (LFSR) is tapped in several locations, and a non-linear function $f$ of these taps produces the output stream. Such stream ciphers are called *non-linear filter generators*, and the non-linear function is called a *filter*. The sampling resistance of such constructions depends on the location of the taps and on the properties of the function $f$. A crucial factor in determining the sampling resistance of such constructions is how many bits of the function's input must be fixed so that the function of the remaining bits is linear.

Multiplexor is a boolean function, which takes $s = \log t + t$ bits of the output, and treats the first $\log t$ bits as an address of the bit in the next $t$ bits. This bit becomes the output of the function. In order to linearize the output of the multiplexor one needs to fix only $\log t$ bits. Multiplexor is thus a weak function in terms of linearization. The actual sampling resistance of the multiplexor is influenced by the minimal distance between the address taps and the minimal distance from the address taps to the output tap.

As a second example, consider the filter function

$$f(x_1, \ldots, x_s) = g(x_1, \ldots, x_{s-1}) \oplus x_s.$$

If there is a gap of length $l$ between tap $x_s$ and the other taps $x_1, \ldots, x_{s-1}$, then the sampling resistance is at most $2^{-l}$, since by proper choice of the $s-1$ bits we can linearize the output of the function $f$. Suppose that our aim is to efficiently enumerate all the $2^{n-l}$ states that produce a prefix of $l$ zeroes. We can do this by setting the $n-l$ non-gap bits to an arbitrary value, and then at each clock we choose the $x_s$ bit in a way that zeroes the function $f$ (assuming that feedback taps are not present in the gap of $l$ bits).

**Sum of Products**    A sum of products is the following boolean function: Pick a set of disjoint pairs of variables from the stream cipher's state: $(x_{i_1}, x_{i_2})$, $\ldots (x_{i_{s-1}}, x_{i_s})$. Define the filter function as:

$$f(x_1, \ldots, x_s) = \bigoplus_{j=1}^{s-1} x_{i_j} \cdot x_{i_{j+1}}.$$

A sum of products becomes a linear function if $s/2$ of its variables (one for each pair) are fixed. If these variables are all set equal to zero then $f$ becomes the constant function $f = 0$. We can thus expect this function to have a moderate resistance to sampling. The non-linear order of this function is only 2 and thus by controlling any pair $x_{i_j} x_{i_{j+1}}$ we can create any desired value of the filter function. For example if the target pair is $(x_{i_1}, x_{i_2})$ then the function $f$ can be decomposed into:

$$f(x_1, \ldots, x_s) = x_{i_1} x_{i_2} \oplus g(x_{i_3}, \ldots, x_{i_s}).$$

At each step if the value of $g$ is zero, the values of the target pair can be chosen arbitrarily out of $(0,0), (0,1), (1,0)$. If however $g = 1$, then the value of the target pair must be $(1,1)$. Thus if the control pair is in a tap-less region of size $2l$ with a gap $l$ between the controlling taps, the sampling resistance of this cipher is at most $2^{-l}$.

As another example, suppose that a consecutive pair of bits is used as a target pair. It seems problematic to use a consecutive pair for product linearization, since sometimes we have to set both bits to 1. This is however not the case if we relax our requirements, and use output prefixes with non-consecutive bits forced to have particular values. For example, prefixes in which every second bit is set to zero (and with arbitrary bits in between) can be easily generated in this sum of adjacent products.

Suppose now that in each pair the first element is from the first half of the register and the second element comes from the second half. Suppose also that the feedback function taps the most significant bit and some taps from the lower half of the register. In this case the sampling resistance is only $2^{-n/2}$. We set to arbitrary values the $n/2$ bits of the lower half of the register and guess the most significant tap bit. This way we know the input to the feedback function and linearize the output function. Forcing the output of the filter function at each step yields a linear equation (whose coefficients come from the lower half of the register and whose variables come from the upper half). After $n/2$ steps we have $n/2$ linear equations in $n/2$ variables which can be easily solved. This way we perform enumeration of all the states that produce the desired output.

Moreover, if all pairs in the product are consecutive, then even a more interesting property holds. We can linearize the function just by fixing a subset of $n/2$ even (or odd) bits of the register, and thus linearization is preserved even after shifting the register (with possible interference of the feedback function).

## A.2   Shrinking and Self-Shrinking Generators

The shrinking generator is a simple construction suggested by [1] which is not based on the filter idea. This generator uses two regularly clocked LFSRs and the output of the first one decides whether the output of the second will appear in the output stream or will be discarded. This generator has good statistical properties like long periods and high linear complexity. A year later a self-shrinking generator (which used one LFSR clocked twice) was proposed by [6]. The output of the LFSR is determined by a pair of most significant bits $a_{n-1}, a_n$ of the LFSR state: If $a_{n-1} = 1$ the output is $a_n$, and if $a_{n-1} = 0$ there is no output in this clock cycle. This construction has the following sampling algorithm: pick arbitrary value for $n/2$ decision bits, and for each pair with a decision bit equal to 1 set the corresponding output bit to 0. If the decision bit is 0 then we have freedom of choice and we enumerate both possibilities. The sampling resistance of this construction is thus $2^{-n/4}$.

# Cryptanalysis of the RSA Schemes with Short Secret Exponent from Asiacrypt '99

Glenn Durfee[1] and Phong Q. Nguyen[2]

[1] Stanford University, Computer Science Department
Stanford, CA 94305, USA
gdurf@theory.stanford.edu
http://theory.stanford.edu/~gdurf/

[2] École Normale Supérieure, Département d'Informatique,
45 rue d'Ulm, 75005 Paris, France
pnguyen@ens.fr
http://www.di.ens.fr/~pnguyen/

**Abstract.** At Asiacrypt '99, Sun, Yang and Laih proposed three RSA variants with short secret exponent that resisted all known attacks, including the recent Boneh-Durfee attack from Eurocrypt '99 that improved Wiener's attack on RSA with short secret exponent. The resistance comes from the use of unbalanced primes $p$ and $q$. In this paper, we extend the Boneh-Durfee attack to break two out of the three proposed variants. While the Boneh-Durfee attack was based on Coppersmith's lattice-based technique for finding small roots to bivariate modular polynomial equations, our attack is based on its generalization to trivariate modular polynomial equations. The attack is heuristic but works well in practice, as the Boneh-Durfee attack. In particular, we were able to break in a few minutes the numerical examples proposed by Sun, Yang and Laih. The results illustrate once again the fact that one should be very cautious when using short secret exponent with RSA.

## 1    Introduction

The RSA [13] cryptosystem is the most widely used public-key cryptosystem. However, RSA is computationally expensive, as it requires exponentiations modulo $N$, where $N$ is a large integer (at least 1024 bits due to recent progress in integer factorization [4]) product of two primes $p$ and $q$. Consequently, speeding up RSA has been a stimulating area of research since the invention of RSA. Perhaps the simplest method to speed up RSA consists of shortening the exponents of the modular exponentiations. If $e$ is the RSA public exponent and $d$ is the RSA secret exponent, one can either choose a small $e$ or a small $d$. The choice of a small $d$ is especially interesting when the device performing secret operations (signature generation or decryption) has limited computed power, such as smartcards. Unfortunately, Wiener [20] showed over 10 years ago that if $d \leq N^{0.25}$, then one could (easily) recover $d$ (and hence, the secret primes $p$ and $q$) in polynomial time from $e$ and $N$ using the continued fractions algorithm.

T. Okamoto (Ed.): ASIACRYPT 2000, LNCS 1976, pp. 14–29, 2000.

Verheul and van Tilborg [19] slightly improved the bound in 1997, by showing that Wiener's attack could be applied to larger $d$, provided an exhaustive search on about $2\log_2(d/N^{0.25})$ bits. At Eurocrypt '99, Boneh and Durfee [3] presented the first substantial improvement over Wiener's bound. Their attack can (heuristically) recover $p$ and $q$ in polynomial time if $d \leq N^{0.292}$. The attack is heuristic because it is based on the seminal lattice-based work by Coppersmith [5] on finding small roots to low-degree modular polynomial equations, in the bivariate case.[1] However, it should be emphasized that the attack works very well in practice.

At Asiacrypt '99, Sun, Yang and Laih [18] noticed that all those attacks on RSA with short secret exponent required some (natural) assumptions on the public modulus $N$. For instance, the Wiener's bound $N^{0.25}$ only holds if $p + q = O(\sqrt{N})$, and $e$ is not too large. Similar restrictions apply to the extension to Wiener's attack by Verheul-van Tilborg [19], and to the Boneh-Durfee attack [3]. This led Sun, Yang and Laih to propose in [18] simple variants of RSA using a short secret exponent that, a priori, foiled all such attacks due to the previous restrictions. More precisely, they proposed three RSA schemes, in which only the (usual) RSA key generation is modified. In the first scheme, one chooses $p$ and $q$ of greatly different size, and a small exponent $d$ in such a way that the previous attacks cannot apply. In particular, $d$ can even be smaller than $N^{0.25}$ if $p$ and $q$ are unbalanced enough. The second scheme consists of a tricky construction that selects slightly unbalanced $p$ and $q$ in such a way that both $e$ and $d$ are small, roughly around $\sqrt{N}$. The third scheme is a mix of the first two schemes, which allows a trade-off between the sizes of $e$ and $d$. Sakai, Morii and Kasahara [14] earlier proposed a different key generation scheme which achieves similar results to the third scheme, but that scheme can easily been shown insecure (see [18]).

In this paper, we show that the first and third schemes of [18] are insecure, by extending the Boneh-Durfee attack. Our attack can also break the second scheme, but only if the parameters are carelessly chosen. Boneh and Durfee reduced the problem of recovering the factors $p$ and $q$ to finding small roots of a particular bivariate modular polynomial equation derived from the basic equation $ed \equiv 1 \pmod{\phi(N)}$. Next, they applied an optimized version (for that particular equation) of Coppersmith's generic technique [5] for such problems. However, when $p$ and $q$ are unbalanced, the particular equation used by Boneh and Durfee is not enough, because it has no longer any "small" root. Our attack extends the Boneh-Durfee method by taking into account the equation $N = pq$. We work with a system of two modular equations with three unknowns; interestingly, when $p$ and $q$ are imbalanced, this approach leads to an attack on systems with $d$ even larger than the $N^{0.292}$ bound of Boneh and Durfee. The attack is extremely efficient in practice: for typical instances of two of the schemes of [18], this approach breaks the schemes within several minutes. Also, our "triviariate" version of Coppersmith's technique we use may be of independent interest.

---

[1] The bivariate case is only heuristic for now, as opposed to the (simpler) univariate case, for which the method can be proved rigorously. For more information, see [5,2,12].

The remainder of this paper is organized as follows. In Section 2, we briefly review former attacks on RSA with short secret exponents, recalling necessary background on lattice theory and Coppersmith's method to find small roots of low-degree modular polynomial equations. This is useful to explain our attacks. In Section 3, we describe the RSA schemes with short secret exponent of [18]. In Section 4, we present the new attack using the trivariate approach. We discuss an implementation of the attack and its running time on typical instances of the RSA variants in Section 5.

## 2  Former Attacks on RSA with Short Secret Exponent

All known attacks on RSA with short secret exponent focus on the equation $ed \equiv 1 \bmod \phi(N)$ (where $\phi(N) = N - (p + q) + 1$) rewritten as:

$$ed = 1 + k \left( \frac{N+1}{2} - s \right) \tag{1}$$

where $k$ is an unknown integer and $s = (p + q)/2$. The primes $p$ and $q$ can be recovered from either $d$ or $s$. Note that $k$ and $d$ are coprime.

### 2.1  The Wiener Attack

Wiener's attack [20] is based on the continued fractions algorithm. Recall that if two (unknown) coprime integers $A$ and $B$ satisfy $|x - \frac{B}{A}| < \frac{1}{2A^2}$ where $x$ is a known rational, then $\frac{B}{A}$ can be obtained in polynomial time as a convergent of the continued fraction expansion of $x$. Here, (1) implies that

$$\left| \frac{2e}{N} - \frac{k}{d} \right| = \frac{|2 + k(1 - 2s)|}{Nd}.$$

Therefore, if $\frac{k(2s-1)-2}{N} < \frac{1}{2d}$, $d$ can be recovered in polynomial time from $e$ and $N$, as $k/d$ is a convergent of the continued fraction expansion of $2e/N$. That condition can roughly be simplified to $ksd = O(N)$, and is therefore satisfied if $k$, $s$ and $d$ are all sufficiently small. In the usual RSA key generation, $s = O(\sqrt{N})$ and $k = O(d)$, which leads to the approximate condition $d = O(N^{0.25})$. But the condition gets worse if $p$ and $q$ are unbalanced, making $s$ much larger than $\sqrt{N}$. For instance, if $p = O(N^{0.25})$, the condition becomes $d = O(N^{0.125})$.

The extension of Wiener's attack by Verheul and van Tilborg [19] applies to $d > N^{0.25}$ provided exhaustive search on $O(\log_2(d/N^{0.25}))$ bits if $p$ and $q$ are balanced. Naturally, the attack requires much more exhaustive search if $p$ and $q$ are unbalanced.

### 2.2  The Boneh-Durfee Attack

**The Small Inverse Problem.** The Boneh-Durfee attack [3] looks at the equation (1) modulo $e$:

$$-k \left( \frac{N+1}{2} - s \right) \equiv 1 \; (\bmod \, e). \tag{2}$$

Assume that the usual RSA key generation is used, so that $|s| < \sqrt{e}$ and $|k| < d$ (ignoring small constants). The problem of finding such a small root $(s, k)$ of that bivariate modular equation was called the *small inverse problem* in [3], since one is looking for a number $(N + 1)/2 - s$ close to $(N + 1)/2$ such that its inverse $-k$ modulo $e$ is rather small. Note that heuristically, the small inverse problem is expected to have a unique solution whenever $|k| < d \leq N^{0.5}$. This led Boneh and Durfee to conjecture that RSA with $d \leq N^{0.5}$ is insecure.

Coppersmith [5] devised a general lattice-based technique to find sufficiently small roots of low-degree modular polynomial equations, which we will review in the next subsections, as it is the core of our attacks. By optimizing that technique to the specific polynomial of (2), Boneh and Durfee showed that one could solve the small inverse problem (and hence, break RSA) when $d \leq N^{0.292}$. This bound corresponds to the usual case of balanced $p$ and $q$. It gets worse as $p$ and $q$ are unbalanced (see [3,18]), because $s$ becomes larger.

**Lattice Theory.** Coppersmith's technique, like many public-key cryptanalyses, is based on lattice basis reduction. We only review what is strictly necessary for this paper. Additional information on lattice theory can be found in numerous textbooks, such as [6,17]. For the important topic of lattice-based cryptanalysis, we refer to the recent survey [12].

We will call *lattice* any subgroup of some $(\mathbb{Z}^n, +)$, which corresponds to the case of integer lattices in the literature. Consequently, for any integer vectors $\mathbf{b}_1, \dots, \mathbf{b}_r$, the set $L(\mathbf{b}_1, \dots, \mathbf{b}_r) = \{\sum_{i=1}^r n_i \mathbf{b}_i \mid n_i \in \mathbb{Z}\}$ of all integer linear combinations of the $\mathbf{b}_i$'s is a lattice, called the lattice *spanned* by the $\mathbf{b}_i$'s. In fact, all lattices are of that form. When $L = L(\mathbf{b}_1, \dots, \mathbf{b}_r)$ and the $\mathbf{b}_i$'s are further linearly independent (over $\mathbb{Z}$), then $(\mathbf{b}_1, \dots, \mathbf{b}_r)$ is called a *basis* of $L$. Any lattice $L$ has infinitely many bases. However, any two bases share some things in common, notably the number of elements $r$ and the Gram determinant $\det_{1 \leq i,j \leq r} \langle \mathbf{b}_i, \mathbf{b}_j \rangle$ (where $\langle, \rangle$ denotes the Euclidean dot product). The parameter $r$ is called the lattice *dimension* (or *rank*), while the square root of the Gram determinant is the lattice *volume* (or *determinant*), denoted by vol($L$). The name volume comes from the fact that the volume matches the $r$-dimensional volume of the parallelepiped spanned by the $\mathbf{b}_i$'s. In the important case of full-dimensional lattices ($r$ equal to $n$), the volume is also the absolute value of the determinant of any basis (hence the name determinant). In general, it is hard to give a "simple" expression for the lattice volume, and one contents oneself with the Hadamard's inequality to estimate the volume:

$$\text{vol}(L) \leq \prod_{i=1}^{r} \|\mathbf{b}_i\|.$$

Fortunately, sometimes, the lattice is full-dimensional and we know a specific basis which is triangular, making the volume easy to compute.

The volume is important because it enables one to estimate the size of short lattice vectors. A well-known result by Minkowski shows that in any $r$-dimensional lattice $L$, there exists a non-zero $\mathbf{x} \in L$ such that $\|\mathbf{x}\| \leq \sqrt{r} \cdot$

vol$(L)^{1/r}$, where $\|.\|$ denotes the Euclidean norm. That bound is in some (natural) sense the best possible. The LLL algorithm [9] can be viewed, from a qualitative point of view, as a constructive version of Minkowski's result. Given any basis of some lattice $L$, the LLL algorithm outputs in polynomial time a so-called *LLL-reduced* basis of $L$. The exact definition of an LLL-reduced basis is beyond the scope of this paper, we only mention the properties that are of interest here:

**Fact 1.** *Any LLL-reduced basis* $(\mathbf{b}_1, \ldots, \mathbf{b}_r)$ *of a lattice $L$ in $\mathbb{Z}^n$ satisfies:*

$$\|\mathbf{b}_1\| \leq 2^{r/2}\text{vol}(L)^{1/r} \quad and \quad \|\mathbf{b}_2\| \leq 2^{(r-1)/2}\text{vol}(L)^{1/(r-1)}.$$

**Coppersmith's Technique.** For a discussion and a general exposition of Coppersmith's technique [5], see the recent surveys [2,12]. We describe the technique in the bivariate case, following a simplified approach due to Howgrave-Graham [7].

Let $e$ be a large integer of possibly unknown factorization. Assume that one would like to find all small roots of $f(x, y) \equiv 0 \pmod{e}$, where $f(x, y)$ is an integer bivariate polynomial with at least one monomial of maximal total degree which is monic. If one could obtain two algebraically independent integral bivariate polynomial equations satisfied by all sufficiently small modular roots $(x, y)$, then one could compute (by resultant) a univariate integral polynomial equation satisfied by $x$, and hence find efficiently all small $(x, y)$. Coppersmith's method tries to obtain such equations from reasonably short vectors in a certain lattice. The lattice comes from the linearization of a set of equations of the form $x^u y^v f(x, y)^w \equiv 0 \pmod{e^w}$ for appropriate integral values of $u$, $v$ and $w$. Such equations are satisfied by any solution of $f(x, y) \equiv 0 \pmod{e}$. Small solutions $(x_0, y_0)$ give rise to unusually short solutions to the resulting linear system, hence short vectors in the lattice. To transform modular equations into integer equations, one uses the following elementary lemma, with the (natural) notation $\|h(x, y)\| = \sqrt{\sum_{i,j} a_{i,j}^2}$ for $h(x, y) = \sum_{i,j} a_{i,j} x^i y^j$ :

**Lemma 2.** *Let $h(x, y) \in \mathbb{Z}[x, y]$ be a polynomial which is a sum of at most $r$ monomials. Suppose that $h(x_0, y_0) \equiv 0 \bmod e^m$ for some positive integer $m$ where $|x_0| < X$ and $|y_0| < Y$, and $\|h(xX, yY)\| < e^m/\sqrt{r}$. Then $h(x_0, y_0) = 0$ holds over the integers.*

Now the trick is to, given a parameter $m$, consider the polynomials

$$h_{u_1, u_2, v}(x, y) = e^{m-v} x^{u_1} y^{u_2} f(x, y)^v.$$

where $u_1$, $u_2$ and $v$ are integers. Notice that any root $(x_0, y_0)$ of $f(x, y)$ modulo $e$ is a root modulo $e^m$ of $h_{u_1, u_2, v}(x, y)$, and therefore, of any integer linear combination $h(x, y)$ of the $h_{u_1, u_2, v}(x, y)$'s. If such a combination $h(x, y)$ further satisfies $\|h(xX, yY)\| < e^m/\sqrt{r}$, where $r$ is the number of monomials of $h$, then by Lemma 2, the integer equation $h(x, y) = 0$ is satisfied by all sufficiently small modular roots of $h$ modulo $e$. Thus, it suffices to find two algebraically independent such equations $h_1(x, y)$ and $h_2(x, y)$.

The use of integer linear combination suggests that we represent the polynomials as vectors in a lattice, so that finding polynomials with small norm reduces to finding short vectors in a lattice. More precisely, let $\mathcal{S}$ be a set of indices $(u_1, u_2, v)$, and choose a representation of the polynomials $h_{u_1, u_2, v}(x, y)$ with $(u_1, u_2, v) \in \mathcal{S}$ as $n$-dimensional integer vectors for some $n$. Let $L$ be the lattice in $\mathbb{Z}^n$ spanned by the vectors corresponding to $h_{u_1, u_2, v}(xX, yY)$ with $(u_1, u_2, v) \in \mathcal{S}$. Apply the LLL algorithm on the lattice, and let $h_1(xX, yY)$ and $h_2(xX, yY)$ be the polynomials corresponding to the first two vectors of the reduced basis obtained. Denoting by $r$ the dimension of $L$, one deduces from the LLL theoretical bounds that:

$$\|h_1(xX, yY)\| \le 2^{r/2} \mathrm{vol}(L)^{1/r} \text{ and } \|h_2(xX, yY)\| \le 2^{(r-1)/2} \mathrm{vol}(L)^{1/(r-1)}.$$

To apply Lemma 2, we want both of these upper bounds to be less than $e^m/\sqrt{n}$; since the factor $2^r$ is negligible with respect to $e^m$, this amounts to saying

$$\mathrm{vol}(L) \ll e^{mr}. \tag{3}$$

There are two problems. The first problem is that even if this condition is satisfied, so that Lemma 2 applies, we are not guaranteed that the integer equations $h_1(x, y) = 0$ and $h_2(x, y) = 0$ obtained are algebraically independent. In other words, $h_2$ will provide no additional information beyond $h_1$ if the two *linearly* independent short basis vectors do not also yield *algebraically* independent equations. It is still an open problem to state precisely when this can be guaranteed, although all experiments to date suggest this is an accurate heuristic assumption to make when inequality (3) holds. We note that a similar assumption is used in the work of Bleichenbacher [1] and Jutla [8].

The second problem is more down-to-earth: how can we make sure that $\mathrm{vol}(L)$ is small enough to satisfy inequality (3) ? Note that Hadamard's bound is unlikely to be useful. Indeed, in general, some of the coefficients of $f(x, y)$ are about the size of $e$, so that $\|h_{u_1, u_2, v}(xX, yY)\|$ is at least $e^m$. To address this problem, one must choose in a clever way the set of indices $\mathcal{S}$ to have a close estimate on $\mathrm{vol}(L)$. The simplest solution is to choose $\mathcal{S}$ so that $L$ is full-dimensional ($r$ equal to $n$) and the $h_{u_1, u_2, v}(xX, yY)$'s form a triangular matrix for some ordering on the polynomials and on the monomials (the vector coordinates). Since we want $\mathrm{vol}(L)$ to be small, each coefficient on the diagonal should be the smallest one of $h_{u_1, u_2, v}(xX, yY) = e^{m-v}(xX)^{u_1}(yY)^{u_2} f(xX, yY)^v$, which is likely to be the one corresponding to the monic monomial of maximal total degree of $f(x, y)$.

In the general case, $f(x, y)$ may have several monomials of maximal total degree, and the only simple choice of $\mathcal{S}$ is to cover all the monomials of total degree less than some parametrized bound. More precisely, if $\Delta$ is the total degree of $f(x, y)$, and $x^a y^{\Delta-a}$ is a monic monomial of $f(x, y)$, one defines $\mathcal{S}$ as the set of $(u_1, u_2, v)$ such that $u_1 + u_2 + \Delta v \le h\Delta$ and $u_1, u_2, v \ge 0$ with $u_1 < a$ or $u_2 < \Delta - a$. Then the volume of the corresponding lattice can be computed exactly, and it turns out that (3) is satisfied whenever $XY < e^{1/\Delta - \varepsilon}$ for and $m$ is sufficiently large.

However, depending on the shape of $f(x, y)$ (represent each monomial $x^i y^j$ by the point $(i, j)$), other choices of $\mathcal{S}$ might lead to improved bounds. Boneh and Durfee applied such tricks to the polynomial (2). In [3], they discussed several choices of $\mathcal{S}$. Using certain sets $\mathcal{S}$ for which the lattice is full-dimensional and one knows a triangular lattice basis, they obtained a first bound $d \leq N^{0.284}$ for their attack. Next, they showed that using a slightly different $\mathcal{S}$ for which the lattice is no longer full-dimensional, one ends up with the improved bound $d \leq N^{0.292}$. The latter choice of $\mathcal{S}$ is much harder to analyze. For more details, see [3].

# 3   The Sun-Yang-Laih RSA Key Generation Schemes

## 3.1   Scheme (I)

The first scheme corresponds to a simple unbalanced RSA [15] in which the parameters are chosen to foil previously known attacks:

1. Select two random primes $p < q$ such that both $p$ and $N = pq$ are sufficiently large to foil factorization algorithms such as ECM and NFS. The more unbalanced $p$ and $q$ are, the smaller $d$ can be.
2. Randomly select the secret exponent $d$ such that $\log_2 d + \log_2 p > \frac{1}{3} \log_2 N$ and $d > 2^\gamma \sqrt{p}$, where $\gamma$ is the security parameter (larger than 64).
3. If the public exponent $e$ defined by $ed \equiv 1 \pmod{\phi(N)}$ is not larger than $\phi(N)/2$, one restarts the previous step.

A choice of parameters suggested by the authors is: $p$ is a 256-bit prime, $q$ is a 768-bit prime, $d$ is a 192-bit number. Note that 192 is far below Wiener's bound (256 bits) and Boneh-Durfee's bound (299 bits).

## 3.2   Scheme (II)

The second scheme selects one of the primes in such a way that one can select $e$ and $d$ to be small at the same time:

1. Fix the bit-length of $N$.
2. Select a random prime $p$ of $\frac{1}{2} \log_2 N - 112$ bits, and a random $k$ of 112 bits.
3. Select a random $d$ of $\frac{1}{2} \log_2 N + 56$ bits coprime with $k(p - 1)$.
4. Compute the two Bézout integers $u$ and $v$ such that $du - k(p - 1)v = 1$, $0 < u < k(p - 1)$ and $0 < v < d$.
5. Return to Step 3 if $v + 1$ is not coprime with $d$.
6. Select a random $h$ of 56 bits until $q = v + hd + 1$ is prime.

The RSA parameters are $p, q, e = u + hk(p - 1)$, $d$ and $N = pq$. Notice that $e$ and $d$ satisfy the equation $ed = 1 + k\phi(N)$. They both have approximate bit-length $\frac{1}{2} \log_2 N + 56$. The primes $p$ and $q$ have approximate bit-length $\frac{1}{2} \log_2 N - 112$ and $\frac{1}{2} \log_2 N + 112$ respectively.

A possible choice of parameters for Scheme (II) might be: $p$ a 400-bit prime, $q$ a 624-bit prime, and $e$ and $d$ are each 568 bits integers.

## 3.3 Scheme (III)

The third scheme is a mix of the first two schemes, allowing a trade-off between $e$ and $d$ such that $\log_2 e + \log_2 d \approx \log_2 N + \ell_k$ where $\ell_k$ is a predetermined constant. More precisely, the scheme is a parametrized version of scheme II: $p$, $k$, $d$ and $h$ have respective bit-length $\ell_p$ (less than $\frac{1}{2} \log_2 N$), $\ell_k$, $\ell_d$, and $\log_2 N - \ell_p - \ell_d$. To resist various attacks, the following is required:

1. $\ell_k \gg \ell_p - \ell_d + 1$.
2. $4\alpha(2\beta + \alpha - 1) \gg 3(1 - \beta - \alpha)^2$, where $\alpha = \frac{\log_2 N - \ell_p}{\log_2 N + \ell_k - \ell_d}$ and $\beta = \frac{\ell_k}{\log_2 N + \ell_k - \ell_d}$.
3. $k$ must withstand an exhaustive search and $\ell_k + \ell_p > \frac{1}{3} \log_2 N$.
   A choice of parameters suggested by the authors is: $p$ is a 256-bit prime, $q$ is a 768-bit prime, $e$ is an 880-bit number, and $d$ is a 256-bit number.

## 4  The Attack Algorithm

In this section we demonstrate how to launch an attack on Schemes (I) and (III). The approach used here closely follows that taken by Boneh and Durfee [3], but differs in several crucial ways to allow it to work when the factors $p$ and $q$ of the public modulus $N$ are unbalanced. Interestingly, our attack gets better (works for larger and larger $d$) the more unbalanced the factors of the modulus become.

Recall the RSA equation

$$ed = 1 + k \left( \frac{N + 1}{2} - \frac{p + q}{2} \right).$$

We note that the Boneh-Durfee approach treats this as an equation modulo $e$ with two "small" unknowns, $k$ and $s = (p+q)/2$. This approach no longer works if $p$ and $q$ are unbalanced, since a good bound on $s$ can no longer be established. For this reason, the authors of the schemes from Section 3 hoped that these schemes would resist the lattice-based cryptanalysis outlined in Section 2.2. However, we will see that a more careful analysis of the RSA equation, namely one that does not treat $p+q$ as a single unknown quantity but instead leaves $p$ and $q$ separately as unknowns, leads to a successful attack against two of these schemes.

Writing $A = N + 1$, the RSA equation implies

$$2 + k(A - p - q) \equiv 0 \pmod{e}.$$

The critical improvement of our attack is to view this as a modular equation with *three* unknowns, $k, p, q$, with the special property that the product $pq$ of two of them is the know74n quantity $N$. We may view this problem as follows: given a polynomial $f(x, y, z) = x(A + y + z) - 2$, find $(x_0, y_0, z_0)$ satisfying:

$$f(x_0, y_0, z_0) \equiv 0 \pmod{e},$$

where

$$|x_0| < X, \quad |y_0| < Y, \quad |z_0| < Z, \quad \text{and} \quad y_0 z_0 = N.$$

Note that the bounds $X \approx ed/N$, $Y \approx p$, and $Z \approx q$ can be estimated to within a power of 2 based on the security parameters chosen for the scheme.

Following Coppersmith's method, our approach is to pick $r$ equations of the form $e^{m-v}x^{u_1}y^{u_2}z^{u_3} \cdot f^v(x, y, z)$ and to search for low-norm integer linear combinations of these polynomials. The basic idea is to start with a handful of equations of the form $y^{a+j}f^m(x, y, z)$ for $j = 0, \dots, t$ for some integers $a$ and $t$ with $t \geq 0$. Knowing $N = pq$ allows us to replace all occurrences of the monomial $yz$ with the constant $N$, reducing the number of variables in each of these equations to approximately $m^2$ instead of the expected $\frac{1}{3}m^3$. We will refer to these as the *primary polynomials*.

Since there are only $t + 1$ of these equations, this will result in a lattice that is less than full rank; we therefore include some additional equations to bring the lattice to full rank in order to compute its determinant. We refer to these as the *helper polynomials*. We have a great deal of choice in picking the helper polynomials; naturally, some choices are better than others, and it is generally a tedious but straightforward optimization problem to choose the primary and helper polynomials that are optimal. The equations we work with are the following. Fix an integer $m$, and let $a$ and $t > 0$ be integers which we will optimize later. We define

- $g_{k,i,b}(x, y, z) := e^{m-k}x^iy^az^bf^k(x, y, z)$, for $k = 0..(m-1)$, $i = 1..(m-k)$, and $b = 0, 1$; and,
- $h_{k,j}(x, y, z) := e^{m-k}y^{a+j}f^k(x, y, z)$, for $k = 0..m$ and $j = 0..t$.

The primary polynomials are $h_{m,j}(x, y, z)$ for $j = 0, \dots, t$, and the rest are helper polynomials. Following Coppersmith's technique, we form a lattice $L$ by representing $g_{k,i,b}(xX, yY, zZ)$ and $h_{k,j}(xX, yY, zZ)$ by their coefficients vectors, and use LLL to find low-norm integer linear combinations $h_1(xX, yY, zZ)$ and $h_2(xX, yY, zZ)$. The polynomials $h_1(x, y, z)$ and $h_2(x, y, z)$ have $(k, p, q)$ as a root over the integers; to remove $z$ as an unknown, we use the equality $z = N/y$, obtaining $H_1(x, y)$ and $H_2(x, y)$ which have $(k, p)$ as a solution. Taking the resultant $\text{Res}_x(H_1(x, y), H_2(x, y))$ yields a polynomial $H(y)$ which has $p$ as a root. Using standard root-finding techniques allows us to recover the factor $p$ of $N$ efficiently, completing the attack.

The running time of this algorithm is dominated by the time to run LLL on the lattice $L$, which has dimension $(m + 1)(m + t + 1)$. So it would be ideal to keep the parameters $m$ and $t$ as low as possible, limiting to a reasonable number the polynomials used to construct $L$. Surprisingly, the attack is successful even if only a handful of polynomials are used. The example given by the original authors for schemes (I) succumbs easily to this attack with $m = 3$ and $t = 1$; with these parameters, our attack generates 20 polynomials. Scheme (III) can be cryptanalyzed with parameters $m = 2$ and $t = 2$, yielding 15 polynomials. This gives lattices of dimension 20 (see Figure 1) and 15, respectively, which can be reduced *via* the LLL algorithm within a matter of seconds on a desktop computer. We discuss our implementation and the results of our experiments more in Section 5.

## 4.1   Analysis of the Attack

In order to be sure that LLL returns vectors that are "short enough" to use Lemma 2, we must derive sufficiently small bounds on the determinant of the lattice $L$ formed from the polynomials $g_{k,i,b}(xX, yY, zZ)$ and $h_{k,j}(xX, yY, zZ)$. Fortunately, this choice of polynomials makes the computation of the determinant of $L$ fairly straightforward, if somewhat tedious. We provide the details in the appendix.

**Representing the Lattice as a Triangular Matrix.** In order to compute the volume of the lattice $L$, we would like to list the polynomials $g_{k,i,b}(xX, yY, zZ)$ and $h_{k,j}(xX, yY, zZ)$ in a way that yields a triangular matrix. There is an ordering on these polynomials that leads to such a representation: we first list the $g_{k,i,b}(xX, yY, zZ)$ indexed outermost by $k = 0, \ldots, m - 1$, then $i = 0, \ldots, k$, then innermost by $b = 0, 1$. We then list $h_{k,j}(xX, yY, zZ)$ indexed outermost by $k = 0, \ldots, m$ then $j = 0, \ldots, t$. (See Figure 1 for the case of $m = 2$, $t = 1$, $a = 1$.) Each new polynomial introduces exactly one new monomial $x^{u_1} y^{u_2}$ or $x^{u_1} z^{u_3}$. Note that no monomial involving the product $yz$ appears, since $yz$ can be eliminated[2] using the identity $N = yz$.

The determinant of this matrix is simply the product of the entries on the diagonal, which for $m = 3$, $t = 1$, $a = 1$ is

$$\mathrm{vol}(L) = \det(M) = e^{40} X^{40} Y^{34} Z^4. \tag{4}$$

We expect the LLL algorithm to return vectors short enough to use Lemma 2 when

$$\mathrm{vol}(L) = e^{40} X^{40} Y^{34} Z^4 < e^{mr} = e^{60}.$$

The example given by the original authors for Scheme (I) is to use $p$ of 256 bits, $q$ of 768 bits, $d$ of 256 bits, and $e$ of 1024 bits. This gives bounds

$$X \approx ed/N \approx e^{1/4}, \quad Y \approx e^{1/4}, \quad \text{and } Z \approx e^{3/4};$$

we may then confirm

$$\det(M) = e^{40} X^{40} Y^{34} Z^4 \approx e^{59} < e^{60} = e^{mr},$$

so Lemma 2 applies.[3] Therefore, when we run the LLL algorithm on this lattice, we will get two short vectors corresponding to polynomials $h_1(x, y, z)$, $h_2(x, y, z)$; by the bound on the determinant, we know that these polynomials will have

---

[2] Caution must be taken to ensure the polynomials remain monic in the terms $x^{u_1} y^{u_2}$ and $x^{u_1} z^{u_3}$ of highest degree; if the substitution $yz \mapsto N$ causes a coefficient of such a term to be different from 1, then we multiply the polynomial by $N^{-1} \bmod e^m$ (and reduce mod $e^m$ as appropriate) before continuing.

[3] The reader may have noticed that we have suppressed the error term associated with the execution of the LLL algorithm. Interestingly, even if the LLL "fudge factor" is

| | $xy$ | $x$ | $x^2y$ | $x^2$ | $x^3y$ | $x^3$ | $x^2y^2$ | $x^2z$ | $x^3y^2$ | $x^3z$ | $x^3y^3$ | $x^3z^2$ | $y$ | $y^2$ | $xy^2$ | $xy^3$ | $x^2y^3$ | $x^2y^4$ | $x^3y^4$ | $x^3y^5$ |
|---|---|---|---|---|---|---|---|---|---|---|---|---|---|---|---|---|---|---|---|---|
| $e^3xy$ | $e^3XY$ | | | | | | | | | | | | | | | | | | | |
| $e^3xyz$ | | $e^3X$ | | | | | | | | | | | | | | | | | | |
| $e^3x^2y$ | | | $e^3X^2Y$ | | | | | | | | | | | | | | | | | |
| $e^3x^2yz$ | | | | $e^3X^2$ | | | | | | | | | | | | | | | | |
| $e^3x^3y$ | | | | | $e^3X^3Y$ | | | | | | | | | | | | | | | |
| $e^3x^3yz$ | | | | | | $e^3X^3$ | | | | | | | | | | | | | | |
| $e^2xyf$ | – | | – | – | | | $e^2X^2Y^2$ | | | | | | | | | | | | | |
| $e^2xyzf$ | | | – | – | | | | $e^2X^2Z$ | | | | | | | | | | | | |
| $e^2x^2yf$ | | | | – | | – | | | $e^2X^3Y^2$ | | | | | | | | | | | |
| $e^2x^2yzf$ | | | | – | | – | | | | $e^2X^3Z$ | | | | | | | | | | |
| $exyf^2$ | – | | – | – | – | – | | | – | – | $eX^3Y^3$ | | | | | | | | | |
| $exyzf^2$ | – | | – | – | – | – | | | – | – | | $eX^3Z^2$ | | | | | | | | |
| $e^3y$ | | | | | | | | | | | | | $e^3Y$ | | | | | | | |
| $e^3y^2$ | | | | | | | | | | | | | | $e^3Y^2$ | | | | | | |
| $e^2yf$ | – | | – | | | | | | | | | | | – | $e^2XY^2$ | | | | | |
| $e^2y^2f$ | – | | – | | | | | | | | | | | – | – | $e^2XY^3$ | | | | |
| $eyf^2$ | – | | – | – | – | | | | – | – | | | | – | | | $eX^2Y^3$ | | | |
| $ey^2f^2$ | – | | – | – | – | | | | – | | | | | – | | | – | $eX^2Y^4$ | | |
| $yf^3$ | – | | – | – | – | – | | | – | – | | | | – | | | | – | $X^3Y^4$ | |
| $y^2f^3$ | – | | – | – | – | – | | | – | – | | | | – | | | | – | – | $X^3Y^5$ |

**Fig. 1.** *Example of the lattice formed by the vectors $g_{k,i,b}(xX, yY, zZ)$ and $h_{k,j}(xX, yY, zZ)$ when $m = 2$, $t = 1$, and $a = 1$. The matrix is lower triangular. Entries marked with "–" indicate off-diagonal quantities whose values do not affect the determinant calculation. The polynomials used are listed on the left, and the monomials they introduce are listed across the top. The double line break occurs between the $g_{k,i,b}$ and the $h_{k,j}$, while the single line breaks occur between increments of $k$. The last single line break separates the helper polynomials (top) from the two primary polynomials (bottom).*

norm that is low enough to use Lemma 2. Therefore these polynomials will have $(k, p, q)$ as a solution over the integers. To turn these into bivariate equations, we use the equality $z = N/y$ to get $H_1(x, y)$ and $H_2(x, y)$ which have $(k, p)$ as a solution over the integers. We then take the resultant $Res_x(H_1(x, y), H_2(x, y))$ to obtain a univariate polynomial $H(y)$ that has $p$ as a root.

More generally, if we pick optimal values for $t$ and $a$ take $m$ sufficiently large, our attack will be successful for even larger bounds on $d$. The highest possible bound on $d$ for which our attack can work depends on the parameters chosen for the scheme. Suppose the parameter $d \approx N^\delta$ is used. The table below summarizes

---

taken into account, this bound is still good enough. We require

$$\text{vol}(L) < 2^{r^2/2}e^{59} = 2^{200}e^{59} < e^{59 + \frac{1}{5}} < e^{mr}/(\sqrt{r})^r \approx e^{60 - \frac{1}{20}}.$$

Slightly larger parameters $m$ and $t$ are required to rigorously obtain the bound for norm of the second basis vector, although in practice the LLL algorithm works well enough so that the parameters chosen here are sufficient.

the largest possible $\delta$ for which our attack can succeed. We point out the choices of parameters that give rise to the schemes of Section 3.

| | | $\log_N(e)$ | | | | | | |
|---|---|---|---|---|---|---|---|---|
| | | 1.0 | 0.9 | 0.86 | 0.8 | 0.7 | 0.6 | 0.55 |
| | 0.5 | 0.284 | 0.323 | 0.339 | 0.363 | 0.406 | 0.451 | 0.475 |
| | 0.4 | 0.296 | 0.334 | 0.350 | 0.374 | 0.415 | 0.460 | 0.483$_{\mathrm{II}}$ |
| $\log_N(p)$ | 0.3 | 0.334 | 0.369 | 0.384 | 0.406 | 0.446 | 0.487 | 0.510 |
| | 0.25 | 0.364$_{\mathrm{I}}$ | 0.398 | 0.412$_{\mathrm{III}}$ | 0.433 | 0.471 | 0.511 | 0.532 |
| | 0.2 | 0.406 | 0.437 | 0.450 | 0.470 | 0.505 | 0.542 | 0.562 |
| | 0.1 | 0.539 | 0.563 | 0.573 | 0.588 | 0.615 | 0.644 | 0.659 |

**Fig. 2.** *Largest $\delta$ (where $d < N^\delta$) for which our attack can succeed, as a function of the system parameters.*

For example, with the example for Scheme (I), where $e \approx N$ and $p \approx N^{0.25}$, our attack will be successful not only for the $\delta = 0.188$ suggested, but all the way up to $\delta < 0.364$ (assuming a large enough $m$ is used.) Similarly, our attack works in Scheme (III) up to $d < N^{0.412}$. Notice that our attack comes close to, but cannot quite reach, the $d < N^{0.55}$ required to break Scheme (II).

### 4.2   Comparison with the Bivariate Approach

Alternatively, one can consider the system of two modular equations with three unknowns as a single bivariate equation by incorporating the equation $N = pq$ into the main trivariate equation. This was independently noticed by Willi Meier [11], who also addressed the problem of breaking Schemes (I) and (III), using a bivariate approach rather than our trivariate approach. One then obtains an equation of the form $f(x,y) = x^2 y + Axy + Bx + Cy$ modulo $e$, where the unknowns are $k$ and the smallest prime among $p$ and $q$.

However, it turns out that the application of Coppersmith's technique to this particular bivariate equation yields worse bounds than with the trivariate approach previously described. For example, the bivariate approach allows one to break scheme (I) as long as $d < N^{0.135}$ (and perhaps slightly higher, if sublattices are considered as in [3]), but fails for larger $d$. One can view the bivariate approach a special case of our trivariate approach, in which one degree of freedom for optimization has been removed. One then sees that the bivariate approach constrains the choice of primary and helper polynomials in a suboptimal way, resulting in worse bounds on $d$.

## 5   Implementation

We implemented this attack using Victor Shoup's Number Theory Library [16] and the Maple Analytical Computation System [10]. The attack runs very efficiently, and in all instances of Schemes (I) and (III) we tested, it produced algebraically independent polynomials $H_1(x, y)$ and $H_2(x, y)$. These yielded a resultant $H(y) = (y - p)H_0(y)$, where $H_0(y)$ is irreducible, exposing the factor $p$ of $N$ in every instance. This strongly suggests that this "heuristic" assumption needed to complete the multivariate modular version of Coppersmith's technique is extremely reliable, and we conjecture that it always holds for suitably bounded lattices of this form. The running times of our attacks are given below.

| Scheme | size of $n$ | size of $p$ | size of $e$ | size of $d$ | $m$ | $t$ | $a$ | lattice rank | running time |
|--------|-------------|-------------|-------------|-------------|-----|-----|-----|--------------|--------------|
| I | 1024 | 256 | 1024 | 192 | 3 | 1 | 1 | 20 | 40 seconds |
| III | 1024 | 256 | 880 | 256 | 2 | 2 | 0 | 15 | 9 seconds |

These tests were run on a 500MHz Pentium III running Solaris.

## 6   Conclusions and Open Problems

We showed that unbalanced RSA [15] actually improves the attacks on short secret exponent by allowing larger exponent. This enabled us to break most of the RSA schemes [18] with short secret exponent from Asiacrypt '99. The attack extends the Boneh-Durfee attack [3] by using a "trivariate" version of Coppersmith's lattice-based technique for finding small roots of low-degree modular polynomial equations. Unfortunately, despite experimental evidence, the attack is for now only heuristic, as the Boneh-Durfee attack. It is becoming increasingly important to find sufficient conditions for which Coppersmith's technique on multivariate modular polynomials can be proved.

Our results illustrate once again the fact that one should be very cautious when using RSA with short secret exponent. To date, the best method to enjoy the computational advantage of short secret exponent is the following countermeasure proposed by Wiener [20]. When $N = pq$, the idea is to use a private exponent $d$ such that both $d_p = d \bmod (p-1)$ and $d_q = d \bmod (q-1)$ are small. Such a $d$ speeds up RSA signature generation since RSA signatures are often generated modulo $p$ and $q$ separately and then combined using the Chinese Remainder Theorem. Classical attacks do not work since $d$ is likely to be close to $\phi(N)$. It is an open problem whether there is an efficient attack on such secret exponents. The best known attack runs in time $\min(\sqrt{d_p}, \sqrt{d_q})$.

### Acknowledgements

Part of this work was done while the second author was visiting Stanford University, whose hospitality is gratefully acknowledged.

## References

1. D. Bleichenbacher. On the security of the KMOV public key cryptosystem. In *Proc. of Crypto '97*, volume 1294 of *LNCS*, pages 235–248. IACR, Springer-Verlag, 1997.

2. D. Boneh. Twenty years of attacks on the RSA cryptosystem. *Notices of the AMS*, 46(2):203–213, 1999.

3. D. Boneh and G. Durfee. Cryptanalysis of RSA with private key $d$ less than $N^{0.292}$. In *Proc. of Eurocrypt '99*, volume 1592 of *LNCS*, pages 1–11. IACR, Springer-Verlag, 1999.

4. S. Cavallar, B. Dodson, A. K. Lenstra, W. Lioen, P. L. Montgomery, B. Murphy, H. te Riele, K. Aardal, J. Gilchrist, G. Guillerm, P. Leyland, J. Marchand, F. Morain, A. Muffett, C. Putnam, C. Putnam, and P. Zimmermann. Factorization of 512-bit RSA key using the number field sieve. In *Proc. of Eurocrypt'2000*, volume 1807 of LNCS. IACR, Springer-Verlag, 2000. Factorization announced in August, 1999.

5. D. Coppersmith. Small solutions to polynomial equations, and low exponent RSA vulnerabilities. *J. of Cryptology*, 10(4):233–260, 1997. Final version of two articles from Eurocrypt '96.

6. M. Gruber and C. G. Lekkerkerker. *Geometry of Numbers*. North-Holland, 1987.

7. N. Howgrave-Graham. Finding small roots of univariate modular equations revisited. In *Cryptography and Coding*, volume 1355 of *LNCS*, pages 131–142. Springer-Verlag, 1997.

8. C. S. Jutla. On finding small solutions of modular multivariate polynomial equations. In *Proc. of Eurocrypt '98*, volume 1403 of *LNCS*, pages 158–170. IACR, Springer-Verlag, 1998.

9. A. K. Lenstra, H. W. Lenstra, Jr., and L. Lovász. Factoring polynomials with rational coefficients. *Mathematische Ann.*, 261:513–534, 1982.

10. Waterloo Maple. The Maple computational algebra system for algebra, number theory and geometry. Information available at http://www.maplesoft.com/products/Maple6/maple6info.html.

11. W. Meier. Private communication. June, 2000.

12. P. Q. Nguyen and J. Stern. Lattice reduction in cryptology: An update. In *Algorithmic Number Theory – Proc. of ANTS-IV*, volume 1838 of LNCS. Springer-Verlag, 2000.

13. R. L. Rivest, A. Shamir, and L. M. Adleman. A method for obtaining digital signatures and public-key cryptosystems. *Communications of the ACM*, 21(2):120–126, 1978.

14. R. Sakai, M. Morii, and M. Kasahara. New key generation algorithm for RSA cryptosystem. *IEICE Trans. Fundamentals*, E77-A(1):89–97, 1994.

15. A. Shamir. RSA for paranoids. *RSA Laboratories CryptoBytes*, 1(3):1–4, 1995.

16. V. Shoup. Number Theory C++ Library (NTL) version 3.6. Available at http://www.shoup.net/ntl/.

17. C. L. Siegel. *Lectures on the Geometry of Numbers*. Springer-Verlag, 1989.

18. H.-M. Sun, W-C. Yang, and C.-S. Laih. On the design of RSA with short secret exponent. In *Proc. of Asiacrypt '99*, volume 1716 of *LNCS*, pages 150–164. IACR, Springer-Verlag, 1999.

19. E. Verheul and H. van Tilborg. Cryptanalysis of less short RSA secret exponents. *Applicable Algebra in Engineering, Communication and Computing*, 8:425–435, 1997.

20. M. Wiener. Cryptanalysis of short RSA secret exponents. *IEEE Trans. Inform. Theory*, 36(3):553–558, 1990.

# A    General Calculation of the Determinant

The general formula for the determinant of the lattice we build in Section 4 is

$$\text{vol}(L) = \det(M) = e^{C_e} X^{C_x} Y^{C_y} Z^{C_z},$$

where

$$C_e = C_x = \frac{1}{6}m(m+1)(4m+3t+5),$$

$$C_y = \begin{cases} \frac{1}{6}(m^3 + 3(a+t+1)m^2 + (3t^2 + 6at + 3a^2 + 6a + 6t + 2)m \\ \qquad +(3t^2 + 6at + 3a^2 + 4a + 3t - a^3)) & \text{if } a \geq 0, \\ \frac{1}{6}(m^3 + 3(a+t+1)m^2 + (3t^2 + 6at + 3a^2 + 6a + 6t + 2)m \\ \qquad +(3t^2 + 6at + 3a^2 + 3a + 3t)) & \text{if } a < 0, \end{cases}$$

$$C_z = \begin{cases} \frac{1}{6}(m^3 - 3(a-1)m^2 + (3a^2 - 6a + 2)m + (3a^2 - 2a - a^3)) & \text{if } a \geq 0, \\ \frac{1}{6}(m^3 - 3(a-1)m^2 + (3a^2 - 6a + 2)m + (3a^2 - 3a)) & \text{if } a < 0. \end{cases}$$

We need $\det(M) < e^{mr} = e^{m(m+1)(m+t+1)}$. In order to optimize the choice of $t$ and $a$, we write $t = \tau m$ and $a = \alpha m$, and observe

$$C_e = C_x = \frac{1}{6}(3\tau + 4)m^3 + o(m^3),$$

$$C_y = \begin{cases} \frac{1}{6}(3\tau^2 + 6\alpha\tau + 3\alpha^2 + 3\alpha + 3\tau + 1 - \alpha^3)m^3 + o(m^3) & \text{if } \alpha \geq 0, \\ \frac{1}{6}(3\tau^2 + 6\alpha\tau + 3\alpha^2 + 3\alpha + 3\tau + 1)m^3 + o(m^3) & \text{if } \alpha < 0, \end{cases}$$

$$C_z = \begin{cases} \frac{1}{6}(3\alpha^2 - 3\alpha + 1 - \alpha^3)m^3 + o(m^3) & \text{if } \alpha \geq 0, \\ \frac{1}{6}(3\alpha^2 - 3\alpha + 1)m^3 + o(m^3) & \text{if } \alpha < 0. \end{cases}$$

Suppose we write $e = N^\varepsilon$, $d = N^\delta$, and $X = N^\beta$, so $Y = N^{1-\beta}$. Then $X = N^{\varepsilon\delta-1}$. So the requirement on $\det(M)$ now becomes

$$N^{\varepsilon C_e + (\varepsilon\delta-1)C_x + \beta C_y + (1-\beta)C_z} < e^{m(m+1)(m+t+1)} = N^{\varepsilon(\tau+1)m^3 + o(m^3)}.$$

The above expression holds (for large enough $m$) when

$$\varepsilon C_e + (\varepsilon\delta - 1)C_x + \beta C_y + (1 - \beta)C_z - (\tau + 1) < 0. \tag{5}$$

The left-hand-side of this expression achieves its minimum at

$$\tau_0 = (2\alpha_0\beta - \beta - \delta + 1)/(2\beta),$$

$$\alpha_0 = \begin{cases} 1 - \beta - (1 - \beta - \delta + \beta^2)^{(1/2)} & \text{if } \beta < \delta, \\ (\beta - \delta)/(2\beta - 2) & \text{if } \beta \geq \delta. \end{cases}$$

Using $\tau = \tau_0$ and $\alpha = \alpha_0$ will give us the minimum value on the left-hand-side of inequality 5, affording us the largest possible $X$ to give an attack on the largest possible $d < N^\delta$. The entries in Figure 2 were generated by plugging in $\tau_0$ and $\alpha_0$ and solving for equality in Equation 5.

It is interesting to note that formulation of the root-finding problem for RSA as a trivariate equation is strictly more powerful than its formulation as the small inverse problem. This is because the small inverse problem is not expected to have a unique solution once $\delta > 0.5$, while our attack works in many cases with $\delta > 0.5$. We note that when $\varepsilon = 1$ and $\beta = 0.5$ – as in standard RSA – our attack gives identical results to simpler Boneh-Durfee attack ($d < N^{0.284}$). Their optimization of using lattices of less than full rank to achieve the $d < N^{0.292}$ bound should also work with our approach, but we have not analyzed how much of an improvement it will provide.

# Why Textbook ElGamal and RSA Encryption Are Insecure

## (Extended Abstract)

Dan Boneh[1], Antoine Joux[2], and Phong Q. Nguyen[3]

[1] Stanford University, Computer Science Department
Stanford, CA 94305, USA
dabo@cs.stanford.edu
http://crypto.stanford.edu/~dabo/
[2] DCSSI, 18 rue du Docteur Zamenhof,
92131 Issy-les-Moulineaux Cedex, France
joux@ens.fr
[3] École Normale Supérieure, Département d'Informatique,
45 rue d'Ulm, 75005 Paris, France
pnguyen@ens.fr
http://www.di.ens.fr/~pnguyen/

**Abstract.** We present an attack on plain ElGamal and plain RSA encryption. The attack shows that without proper preprocessing of the plaintexts, both ElGamal and RSA encryption are fundamentally insecure. Namely, when one uses these systems to encrypt a (short) secret key of a symmetric cipher it is often possible to recover the secret key from the ciphertext. Our results demonstrate that preprocessing messages prior to encryption is an essential part of both systems.

## 1 Introduction

In the literature we often see a description of RSA encryption as $C = \langle M^e \rangle$ mod $N$ (the public key is $\langle N, e \rangle$) and a description of ElGamal encryption as $C = \langle My^r, g^r \rangle$ mod $p$ (the public key is $\langle p, g, y \rangle$). Similar descriptions are also given in the original papers [17,9]. It has been known for many years that this simplified description of RSA does not satisfy basic security notions, such as semantic security (see [6] for a survey of attacks). Similarly, a version of ElGamal commonly used in practice does not satisfy basic security notions (even under the Decision Diffie-Hellman assumption [5]) [1]. To obtain secure systems using RSA and ElGamal one must apply a preprocessing function to the plaintext prior to encryption,

---

[1] Implementations of ElGamal often use an element $g \in \mathbb{Z}_p^*$ of prime order $q$ where $q$ is much smaller than $p$. When the set of plaintexts is equal to the subgroup generated by $g$, the Decision Diffie Hellman assumption implies that ElGamal is semantically secure. Unfortunately, implementations of ElGamal often encrypt an $m$-bit message by viewing it as an $m$-bit integer and directly encrypting it. The resulting system is not semantically secure – the ciphertext leaks the Legendre symbol of the plaintext.

T. Okamoto (Ed.): ASIACRYPT 2000, LNCS 1976, pp. 30–43, 2000.

or a conversion to the encryption function (see [10,16,13] for instance). Recent standards for RSA [15] use Optimal Asymmetric Encryption Padding (OAEP) which is known to be secure against a chosen ciphertext attack in the random oracle model [4]. Currently, there is no equivalent preprocessing standard for El-Gamal encryption, although several proposals exist [1,10,16,13]. Unfortunately, many textbook descriptions of RSA and ElGamal do not view these preprocessing functions as an integral part of the encryption scheme. Instead, common descriptions are content with an explanation of the plain systems.

In this paper we give a simple, yet powerful, attack against both plain RSA and plain ElGamal encryption. The attack illustrates that plain RSA and plain ElGamal are fundamentally insecure systems. Hence, any description of these cryptosystems cannot ignore the preprocessing steps used in full RSA and full ElGamal. Our attack clearly demonstrates the importance of preprocessing. It can be used to motivate the need for preprocessing in introductory texts.

Our attack is based on the fact that public key encryption is typically used to encrypt session-keys. These session-keys are typically short, i.e. less than 128 bits. The attack shows that when using plain RSA or plain ElGamal to encrypt an $m$-bit key, it is often possible to recover the key in time approximately $2^{m/2}$. In environments where session-keys are limited to 64-bit keys (e.g. due to government regulations), our attack shows that both plain RSA and plain ElGamal result in a completely insecure system. We experimented with the attack and showed that it works well in practice.

## 1.1 Summary of Results

Suppose the plaintext $M$ is $m$ bits long. For illustration purposes, when $m = 64$ we obtain the following results:

- For any RSA public key $\langle N, e \rangle$, given $C = M^e \bmod N$ it is possible to recover $M$ in the time it takes to compute $2 \cdot 2^{m/2}$ modular exponentiations. The attack succeeds with probability 18% (the probability is over the choice of $M \in \{0, 1, \ldots, 2^m - 1\}$). The algorithm requires $2^{m/2}m$ bits of memory.
- Let $\langle p, g, y \rangle$ be an ElGamal public key. When the order of $g$ is at most $p/2^m$, it is possible to recover $M$ from any ElGamal ciphertext of $M$ in the time it takes to compute $2 \cdot 2^{m/2}$ modular exponentiations. The attack succeeds with probability 18% (over the choice of $M$), and requires $2^{m/2}m$ bits of memory.
- Let $\langle p, g, y \rangle$ be an ElGamal public key. Suppose $p - 1 = qs$ where $s > 2^m$ and the discrete log problem for subgroups of $\mathbb{Z}_p^*$ of order $s$ is tractable, i.e. takes time $T$ for some small $T$. When the order of $g$ is $p - 1$, it is possible to recover $M$ from any ciphertext of $M$ in time $T$ and $2 \cdot 2^{m/2}$ modular exponentiations. The attack succeeds with probability 18% (over the choice of $M$), and requires $2^{m/2}m$ bits of memory.
- Let $\langle p, g, y \rangle$ be an ElGamal public key. Suppose again $p - 1 = qs$ where $s > 2^m$ and the discrete log problem for subgroups of $\mathbb{Z}_p^*$ of order $s$ takes time $T$ for some small $T$. When the order of $g$ is either $p-1$ or at most $p/2^m$,

it is possible to recover $M$ from any ciphertext of $M$ in time $T$ plus one modular exponentiation and $2 \cdot 2^{m/2}$ additions, provided a precomputation step depending only on the public key. The success probability is 18% (over the choice of $M$). The precomputations take time $2^{m/2}T$ and $2^{m/2}$ modular exponentiations. The space requirement can optionally be decreased to $2^{m/4} \log_2 s$ bits without increasing the computation time, however with a loss in the probability of success.

All attacks can be parallelized, and offer a variety of trade-offs, with respect to the computation time, the space requirement, and the probability of success. For instance, the success probability of 18% can be raised to 35% if the computation time is quadrupled. Note that the first result applies to RSA with an arbitrary public exponent (small or large). The attack becomes slightly more efficient when the public exponent $e$ is small. The second result applies to the usual method in which ElGamal is used in practice. The third result applies when ElGamal encryption is done in the entire group, however $p-1$ has a small smooth factor (a 64-bit smooth factor). The fourth result decreases the on-line work of both the second and the third results, provided an additional precomputation stage. It can optionally improve the time/memory trade-off. The third and fourth results assume that $p - 1$ contains a smooth factor: such a property was used in other attacks against discrete-log schemes (see [2,14] for instance).

## 1.2   Splitting Probabilities for Integers

Our attacks can be viewed as a meet-in-the-middle method based on the fact that a relatively small integer (e.g., a session-key) can often be expressed as a product of much smaller integers. Note that recent attacks on padding RSA signature schemes [7] use related ideas. Roughly speaking, these attacks expect certain relatively small numbers (such as hashed messages) to be smooth. Here, we will be concerned with the size of divisors. Existing analytic results for the bounds we need are relatively weak. Hence, we mainly give experimental results obtained using the Pari/GP computer package [3].

Let $M$ be a uniformly distributed $m$-bit integer. We are interested in the probability that $M$ can be written as:

- $M = M_1 M_2$ with $M_1 \leq 2^{m_1}$ and $M_2 \leq 2^{m_2}$. See table 1 for some values.
- $M = M_1 M_2 M_3$ with $M_i \leq 2^{m_i}$. See table 2 for some values.
- $M = M_1 M_2 M_3 M_4$ with $M_i \leq 2^{m_i}$. See table 3 for some values.

The experimental results given in the tables have been obtained by factoring a large number of randomly chosen $m$-bit integers with uniform distribution. Some theoretical results can be obtained from the book [11]. More precisely, for $1/2 \leq \alpha < 1$, let $P_\alpha(m)$ be the probability that a uniformly distributed integer $M$ in $[1 \ldots 2^m - 1]$ can be written as $M = M_1 M_2$ with both $M_1$ and $M_2$ less or equal to $2^{\alpha m}$. It can be shown that $P_{1/2}(m)$ tends (slowly) to zero as $m$ grows to infinity. This follows (after a little work) from results in [11][Chapter 2] on the number $H(x, y, z)$ of integers $n \leq x$ for which there exists a divisor $d$ such that

$y \leq d < z$. More precisely, the following holds (where log denotes the neperian logarithm):

$$P_{1/2}(m) = O\left(\frac{\log\log m \cdot \sqrt{\log m}}{m^\delta}\right), \tag{1}$$

where $\delta = 1 - \frac{1+\log\log 2}{\log 2} \approx 0.086$. On the other hand, when $\alpha > 1/2$, $P_\alpha(m)$ no longer tends to zero, as one can easily obtain the following asymptotic lower bound, which corrects [8, Theorem 4, p 377]:

$$\liminf P_\alpha(m) \geq \log(2\alpha), \tag{2}$$

This is because the probability must include all numbers that are divisible by a prime in the interval $[2^{m/2}, 2^{\alpha m}]$, and the bound follows from well-known smoothness probabilities.

Our attacks offer a variety of trade-offs, due to the freedom in the factorization form, and in the choices of the $m_i$'s: the splitting probability gives the success probability of the attack, the other parameters determine the cost in terms of storage and computation time.

**Table 1.** Experimental probabilities of splitting into two factors.

| Bit-length $m$ | $m_1$ | $m_2$ | Probability |
|---|---|---|---|
| 40 | 20 | 20 | 18% |
| | 21 | 21 | 32% |
| | 22 | 22 | 39% |
| | 20 | 25 | 50% |
| 64 | 32 | 32 | 18% |
| | 33 | 33 | 29% |
| | 34 | 34 | 35% |
| | 30 | 36 | 40% |

**Table 2.** Experimental probabilities of splitting into three factors.

| Bit-length $m$ | $m_1 = m_2 = m_3$ | Probability |
|---|---|---|
| 64 | 22 | 4% |
| | 23 | 6.5% |
| | 24 | 9% |
| | 25 | 12% |

**Table 3.** Experimental probabilities of splitting into four factors.

| Bit-length $m$ | $m_1 = m_2 = m_3 = m_4$ | Probability |
|---|---|---|
| 64 | 16 | 0.5% |
| | 20 | 3% |

## 1.3   Organization of the Paper

In Section 2 we introduce the subgroup rounding problems which inspire all our attacks. In Section 3 we present rounding algorithms that break plain ElGamal encryption when $g$ generates a "small" subgroup of $\mathbb{Z}_p^*$. Using similar ideas, we present in Section 4 an attack on plain ElGamal encryption when $g$ generates all $\mathbb{Z}_p^*$, and an attack on plain RSA in Section 5.

## 2   The Subgroup Rounding Problems

Recall that the ElGamal public key system [9] encrypts messages in $\mathbb{Z}_p^*$ for some prime $p$. Let $g$ be an element of $\mathbb{Z}_p^*$ of order $q$. The private key is a number in the range $1 \leq x < q$. The public key is a tuple $\langle p, g, y \rangle$ where $y = g^x \bmod p$. To encrypt a message $M \in \mathbb{Z}_p$ the original scheme works as follows: (1) pick a random $r$ in the range $1 \leq x < q$, and (2) compute $u = M \cdot y^r \bmod p$ and $v = g^r \bmod p$. The resulting ciphertext is the pair $\langle u, v \rangle$. To speed up the encryption process one often uses an element $g$ of order much smaller than $p$. For example, $p$ may be 1024 bits long while $q$ is only 512 bits long.

For the rest of this section we assume $g \in \mathbb{Z}_p^*$ is an element of order $q$ where $q \ll p$. For concreteness one may think of $p$ as 1024 bits long and $q$ as 512 bits long. Let $G_q$ be the subgroup of $\mathbb{Z}_p^*$ generated by $g$. Observe that $G_q$ is extremely sparse in $\mathbb{Z}_p^*$. Only one in $2^{512}$ elements belongs to $G_q$. We also assume $M$ is a short message of length much smaller than $\log_2(p/q)$. For example, $M$ is a 64 bits long session-key.

To understand the intuition behind the attack it is beneficial to consider a slight modification of the ElGamal scheme. After the random $r$ is chosen one encrypts a message $M$ by computing $u = M + y^r \bmod p$. That is, we "blind" the message by *adding* $y^r$ rather than multiplying by it. The ciphertext is then $\langle u, v \rangle$ where $v$ is defined as before. Clearly $y^r$ is a random element of $G_q$. We obtain the following picture:

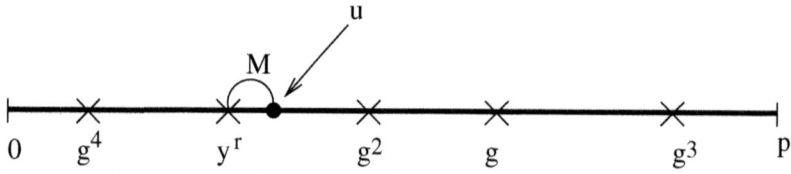

The × marks represent elements in $G_q$. Since $M$ is a relatively small number, encryption of $M$ amounts to picking a random element in $G_q$ and then slightly moving away from it. Assuming the elements of $G_q$ are uniformly distributed in $\mathbb{Z}_p^*$ the average gap between elements of $G_q$ is much larger than $M$. Hence, with high probability, there is a unique element $z \in G_q$ that is sufficiently close to $u$. More precisely, with high probability there will be a unique element $z \in G_q$ satisfying $|u - z| < 2^{64}$. If we could find $z$ given $u$ we could recover $M$. Hence, we obtain the additive version of the subgroup rounding problem:

*Additive subgroup rounding:* let $z$ be an element of $G_q$ and $\Delta$ an integer satisfying $\Delta < 2^m$. Given $u = z+\Delta \bmod p$ find $z$. When $m$ is sufficiently small, $z$ is uniquely determined (with high probability assuming $G_q$ is uniformly distributed in $\mathbb{Z}_p$).

Going back to the original multiplicative ElGamal scheme we obtain the multiplicative subgroup rounding problem.

*Multiplicative subgroup rounding:* let $z$ be an element of $G_q$ and $\Delta$ an integer satisfying $\Delta < 2^m$. Given $u = z \cdot \Delta \bmod p$ find $z$. When $m$ is sufficiently small $z$, is uniquely determined (with high probability assuming $G_q$ is uniformly distributed in $\mathbb{Z}_p$).

An efficient solution to either problem would imply that the corresponding *plain* ElGamal encryption scheme is insecure. We are interested in solutions that run in time $O(\sqrt{\Delta})$ or, even better, $O(\log \Delta)$. In the next section we show a solution to the multiplicative subgroup rounding problem.

The reason we refer to these schemes as "plain ElGamal" is that messages are encrypted *as is*. Our attacks show the danger of using the system in this way. For proper security one must pre-process the message prior to encryption or modify the encryption mechanism. For example, one could use DHAES [1] or a result due to Fujisaki and Okamoto [10], or even more recently [16,13].

# 3   Algorithms for Multiplicative Subgroup Rounding

We are given an element $u \in \mathbb{Z}_p$ of the form $u = z \cdot \Delta \bmod p$ where $z$ is a random element of $G_q$ and $|\Delta| < 2^m$. Our goal is to find $\Delta$, which we can assume to be positive. As usual, we assume that $m$, the length of the message being encrypted, is much smaller than $\log_2(p/q)$. Then with high probability $\Delta$ is unique. For example, take $p$ to be 1024 bits long, $q$ to be 512 bits long and $m$ to be 64. We first give a simple meet-in-the-middle strategy for multiplicative subgroup rounding. By reduction to a knapsack-like problem, we will then improve both the on-line computation time and the time/memory trade-off of the method, provided that $p$ satisfies an additional, yet realistic, assumption.

## 3.1   A Meet-in-the-Middle Method

Suppose $\Delta$ can be written as $\Delta = \Delta_1 \cdot \Delta_2$ where $\Delta_1 \leq 2^{m_1}$ and $\Delta_2 \leq 2^{m_2}$. For instance, one can take $m_1 = m_2 = m/2$. We show how to find $\Delta$ from $u$ in space $O(2^{m_1})$ and $2^{m_1} + 2^{m_2}$ modular exponentiations. Observe that

$$u = z \cdot \Delta = z \cdot \Delta_1 \cdot \Delta_2 \bmod p.$$

Dividing by $\Delta_2$ and raising both sides to the power of $q$ yields:

$$(u/\Delta_2)^q = z^q \cdot \Delta_1^q = \Delta_1^q \bmod p.$$

We can now build a table of size $2^{m_1}$ containing the values $\Delta_1^q \bmod p$ for all $\Delta_1 = 0, \ldots, 2^{m_1}$. Then for each $\Delta_2 = 0, \ldots, 2^{m_2}$ we check whether $u^q/\Delta_2^q \bmod p$ is present in the table. If so, then $\Delta = \Delta_1 \cdot \Delta_2$ is a candidate value for $\Delta$. Assuming $\Delta$ is unique, there will be only be one such candidate, although there will probably be several suitable pairs $(\Delta_1, \Delta_2)$.

The algorithm above requires a priori $2^{m_2} + 2^{m_1}$ modular exponentiations and $2^{m_1} \log_2 p$ bits of memory. However, we do not need to store the complete value of $\Delta_1^q \bmod p$ in the table: A sufficiently large hash value is enough, as we are only looking for "collisions". For instance, one can take the $2 \max(m_1, m_2)$ least significant bits of $\Delta_1^q \bmod p$, so that the space requirement is only $2^{m_1+1} \max(m_1, m_2)$ bits instead of $2^{m_1} \log_2 p$. Less bits are even possible, for we can check the validity of the (few) candidates obtained. Note also that the table only depends on $p$ and $q$: the same table can be used for all ciphertexts. For each ciphertext, one needs to compute at most $2^{m_2}$ modular exponentiations. For each exponentiation, one has to check whether or not it belongs to the table, which can be done with $O(m_1)$ comparisons once the table is sorted.

It is worth noting that $\Delta_1$ and $\Delta_2$ need not be prime. The probability that a random $m$-bit integer (such as $\Delta$) can be expressed as a product of two integers, one being less than $m_1$ bits and the other one being less than $m_2$ bits, is discussed in Section 1.2.

By choosing different values of $m_1$ and $m_2$ (not necessarily $m/2$), one obtains various trade-offs with respect to the computation time, the storage requirement, and the success probability. For instance, when the system is used to encrypt a 64-bit session key, if we pick $m_1 = m_2 = 32$, the algorithm succeeds with probability approximately 18% (with respect to the session key), and it requires on the order of eight billion exponentiations, far less than the time to compute discrete log in $\mathbb{Z}_p^*$.

We implemented the attack using Victor Shoup's NTL library [19]. The timings should not be considered as optimal, they are meant to give a rough idea of the attack efficiency, compared to exhaustive search attacks on the symmetric algorithm. Running times are given for a single 500 MHz 64-bit DEC Alpha/Linux. If $m = 40$ and $m_1 = m_2 = 20$, and we use a 160-bit $q$ and a 512-bit $p$, the precomputation step takes 40 minutes, and each message is recovered in less than 1 hour and 30 minutes. From Section 1.2, it also means that, given only the public key and the ciphertext, a 40-bit message can be recovered in less than 6 hours on a single workstation, with probability 39%.

## 3.2   Reduction to Knapsack-like Problems

We now show how to improve the on-line computation time ($2^{m/2}$ modular exponentiations) and the time/memory trade-off of the method. We transform the multiplicative rounding problem into a linear problem, provided that $p$ satisfies the additional assumption $p - 1 = qrs$ where $s \geq 2^m$ is such that discrete logs in subgroups of $\mathbb{Z}_p^*$ of order $s$ can be efficiently computed. For instance, if $p_1^{e_1} \cdots p_k^{e_k}$ is the prime factorization of $s$, discrete logs in a cyclic group of order $s$ can be computed with $O(\sum_{i=1}^{k} e_i(\log s + \sqrt{p_i}))$ group operations and negligible space, using Pohlig-Hellman and Pollard's $\rho$ methods (see [12]). Let $\omega$ be a generator of $\mathbb{Z}_p^*$. For all $x \in \mathbb{Z}_p^*$, $x^{qr}$ belongs to the subgroup $G_s$ of order $s$ generated by $\omega^{qr}$.

The linear problem that we will consider is known as the $k$-table problem: given $k$ tables $T_1, \ldots, T_k$ of integers and a target integer $n$, the $k$-table problem is to return all expressions (possibly zero) of $n$ of the form $n = t_1 + t_2 + \cdots + t_k$ where $t_i \in T_i$. The general $k$-table problem has been studied by Schroeppel and Shamir [18], because several NP-complete problems (e.g., the knapsack problem) can be reduced to it. We will apply (slightly modified) known solutions to the $k$-table problems, for $k = 2, 3$ and $4$.

**The Modular 2-Table Problem**  Suppose that $\Delta$ can be written as $\Delta = \Delta_1 \cdot \Delta_2$, with $0 \leq \Delta_1 \leq 2^{m_1}$ and $0 \leq \Delta_2 \leq 2^{m_2}$, as in Section 3.1. We have $u^q = \Delta_1^q \Delta_2^q \bmod p$ and therefore:

$$u^{qr} = \Delta_1^{qr} \Delta_2^{qr} \bmod p,$$

which can be rewritten as

$$\log(u^{qr}) = \log(\Delta_1^{qr}) + \log(\Delta_2^{qr}) \bmod s,$$

where the logarithms are with respect to $\omega^{qr}$.

We build a table $T_1$ consisting of $\log(\Delta_1^{qr})$ for all $\Delta_1 = 0, \ldots, 2^{m_1}$, and a table $T_2$ consisting of $\log(\Delta_2^{qr})$ for all $\Delta_2 = 0, \ldots, 2^{m_2}$. These tables are independent of $\Delta$. The problem is now to express $\log(u^{qr})$ as a modular sum $t_1 + t_2$, where $t_1 \in T_1$ and $t_2 \in T_2$. The number of targets $t_1 + t_2$ is $2^{m_1+m_2}$. Hence, we expect this problem to have very few solutions when $s \geq 2^{m_1+m_2}$. The problem involves modular sums, but it can of course be viewed as a 2-table problem with two targets $\log(u^{qr})$ and $\log(u^{qr}) + s$. The classical method to solve the 2-table problem with a target $n$ is the following:

1. Sort $T_1$ in increasing order;
2. Sort $T_2$ in decreasing order;
3. Repeat until either $T_1$ or $T_2$ becomes empty (in which case all solutions have already been found):
   (a) Compute $t = \text{first}(T_1) + \text{first}(T_2)$.
   (b) If $t = n$, output the solution which has been found, and delete $\text{first}(T_1)$ from $T_1$, and $\text{first}(T_2)$ from $T_2$;
   (c) If $t < n$ delete $\text{first}(T_1)$ from $T_1$;
   (d) If $t > n$ delete $\text{first}(T_2)$ from $T_2$;

It is easy to see that the method outputs all solutions of the 2-table problem, in time $2^{\min(m_1, m_2)+1}$. The space requirement is $O(2^{m_1} + 2^{m_2})$.

Since the original problem involves modular sums, it seems at first glance that we have to apply the previous algorithm twice (with two different targets). However, we note that a simple modification of the previous algorithm can in fact solve the modular 2-table problem (that is, the 2-table problem with modular additions instead of integer additions). The basic idea is the following. Since $T_2$ is sorted in descending order, $n - T_2$ is sorted in ascending order. The set $(n - T_2) \bmod s$ though not necessarily sorted, is almost sorted. More precisely, two adjacent numbers are always in the right order, to the exception of a single pair. This is because $n - T_2$ is contained in an interval of length $s$. The single pair of adjacent numbers in reverse order corresponds to the two elements $a$ and $b$ of $T_2$ surrounding $s - n$. These two elements can easily be found by a simple dichotomy search for $s - n$ in $T_2$. And once the elements are known, we can access $(n - T_2) \pmod s$ in ascending order by viewing $T_2$ as a circular list, starting our enumeration of $T_2$ by $b$, and stopping at $a$.

The total cost of the method is the following. The precomputation of tables $T_1$ and $T_2$ requires $2^{m_1} + 2^{m_2}$ modular exponentiations and discrete log computations in a subgroup of $\mathbb{Z}_p^*$ of order $s$, and the sort of $T_1$ and $T_2$. The space requirement is $(2^{m_1} + 2^{m_2}) \log_2 s$ bits. For each ciphertext, we require one modular exponentiation, one efficient discrete log (to compute the target), and $2^{\min(m_1, m_2)+1}$ additions. Hence, we improved the on-line work of the method of Section 3.1: loosely speaking, we replaced modular exponentiations by simple additions. We now show how to decrease the space requirement of the method.

**The Modular 3-Table Problem** The previous approach can easily be extended to an arbitrary number of factors of $\Delta$. Suppose for instance $\Delta$ can be written as $\Delta = \Delta_1 \cdot \Delta_2 \cdot \Delta_3$ where each $\Delta_i$ is less than $2^{m_i}$. We obtain

$$\log(u^{qr}) = \sum_{i=1}^{3} \log(\Delta_i^{qr}) \bmod s,$$

where the logarithms are with respect to $w^{qr}$. In a precomputation step, we compute in a table $T_i$ all the logarithms of $\Delta_i^{qr} \bmod p$ for $0 \le \Delta_i < 2^{m_i}$. We are left with a modular 3-table problem with target $\log(u^q r)$. The modular 3-table problem with target $n$ modulo $s$ can easily be solved in time $O(2^{m_1 + \min(m_2, m_3)})$ and space $O(2^{m_1} + 2^{m_2} + 2^{m_3})$. It suffices to apply the modular 2-table algorithm on tables $T_2$ and $T_3$, for all targets $(n - t_1) \bmod s$, with $t_1 \in T_1$.

Hence, we decreased the space requirement of the method of Section 3.2, by (slightly) increasing the on-line computation work and decreasing the success probability (see Section 1.2 for the probability of splitting into three factors). More precisely, if $m_1 = m_2 = m_3 = m/3$, the on-line work is one modular exponentiation, one discrete log in a group of order $s$, and $2^{2n/3}$ additions. Since an addition is very cheap, this might be useful for practical purposes.

**The Modular 4-Table Problem** Using 3 factors did not improve the time/memory trade-off of the on-line computation work. Indeed, for both modular 2-table and modular 3-table problems, our algorithms satisfy $TS = O(2^m)$, where

$T$ is the number of additions, and $S$ is the space requirement. Surprisingly, one can obtain a better time/memory tradeoff with 4 factors.

Suppose $\Delta$ can be written as $\Delta = \Delta_1 \cdot \Delta_2 \cdot \Delta_3 \cdot \Delta_4$ where each $\Delta_i$ is less than $2^{m_i}$. For instance, one can take $m_1 = m_2 = m_3 = m_4 = m/4$. We show how to find $\Delta$ from $\log(u^{qr})$ in time $O(2^{m_1+m_2} + 2^{m_3+m_4})$ and space $O(\sum_{i=1}^{4} 2^{m_i})$, provided a precomputation stage of $\sum_{i=1}^{4} 2^{m_i}$ modular exponentiations and discrete log computations in a group of order $s$.

We have $\log(u^{qr}) = \sum_{i=1}^{4} \log(\Delta_i^{qr}) \bmod s$. Again, in a precomputation step, we compute in a table $T_i$ all the logarithms of $\Delta_i^{qr} \bmod p$ for $0 \le \Delta_i < 2^{m_i}$. We are left with a modular 4-table problem, whose solutions will reveal possible choices of $\Delta_1$, $\Delta_2$, $\Delta_3$ and $\Delta_4$. Schroeppel and Shamir [18] proposed a clever solution to the basic 4-table problem, using the following idea. An obvious solution to the 4-table problem is to solve a 2-table problem by merging two tables, that is, considering sums $t_1 + t_2$ and $t_3 + t_4$ separately. However, the algorithm for the 2-table algorithm described in Section 3.2 accesses the elements of the sorted supertables sequentially, and thus there is no need to store all the possible combinations simultaneously in memory. All we need is the ability to generate them quickly (on-line, upon request) in sorted order. To implement this idea, two priority queues are used :

- $Q'$ stores pairs $(t_1, t_2)$ from $T_1 \times T_2$, enables arbitrary insertions and deletions to be done in logarithmic time, and makes the pairs with the smallest $t_1 + t_2$ sum accessible in constant time.
- $Q''$ stores pairs $(t_3, t_4)$ from $T_3 \times T_4$, enables arbitrary insertions and deletions to be done in logarithmic time, and makes the pairs with the largest $t_3 + t_4$ sum accessible in constant time.

This leads to the following algorithm for a target $n$:

1. Precomputation:
   - Sort $T_2$ into increasing order, and $T_4$ into decreasing order;
   - Insert into $Q'$ all the pairs $(t_1, \text{first}(T_2))$ for $t_1 \in T_1$;
   - Insert into $Q''$ all the pairs $(t_3, \text{first}(T_4))$ for $t_3 \in T_3$.
2. Repeat until either $Q'$ or $Q''$ becomes empty (in which case all solutions have been found):
   - Let $(t_1, t_2)$ be the pair with smallest $t_1 + t_2$ in $Q'$;
   - Let $(t_3, t_4)$ be the pair with largest $t_3 + t_4$ in $Q''$;
   - Compute $t = t_1 + t_2 + t_3 + t_4$.
   - If $t = n$, we output the solution, and apply what is planned when $t < n$ or $t > n$.
   - If $t < n$ do
     • delete $(t_1, t_2)$ from $Q'$;
     • if the successor $t_2'$ of $t_2$ in $T_2$ is defined, insert $(t_1, t_2')$ into $Q'$;
   - If $t > n$ do
     • delete $(t_3, t_4)$ from $Q''$;
     • if the successor $t_4'$ of $t_4$ in $T_4$ is defined, insert $(t_3, t_4')$ into $Q''$;

At each stage, a $t_1 \in T_1$ can participate in at most one pair in $Q'$, and a $t_3 \in T_3$ can participate in at most one pair in $Q''$. It follows that the space complexity of the priority queues is bounded by $O(|T_1| + |T_3|) = O(2^{m_1} + 2^{m_3})$. Each possible pair can be deleted from $Q'$ at most once, and the same holds for $Q''$. Since at each iteration, one pair is deleted from $Q'$ or $Q''$, the number of iterations cannot exceed the number of possible pairs, which is $O(2^{m_1+m_2} + 2^{m_3+m_4})$.

Finally, as in the 2-table case, we note that this algorithm can be adapted to modular sums, by changing the starting points in $T_2$ and $T_4$ to make sure that the modular sets are enumerated in the correct order. Hence, it is not necessary to apply the 4-table algorithm on 4 targets. If $m_1 = m_2 = m_3 = m_4 = m/4$, we obtain a time complexity of $O(2^{m/2})$ and a space complexity of only $O(2^{m/4})$, which improves the time/memory tradeoff of the methods of Sections 3.2 and 3.2. The probability that a random $m$-bit integer (such as $\Delta$) can be expressed as a product of four integers $\Delta_i$, where $\Delta_i$ has less than $m_i$ bits, is given in Section 1.2. Different values of $m_1, m_2, m_3$ and $m_4$ (not necessarily $m/4$), give rise to different trade-offs with respect to the computation time, the storage requirement, and the success probability.

Our experiments show that, as expected, the method requires much less computing power than a brute-force attack on the 64-bit key using the symmetric encryption algorithm. We implemented the attack on a PII/Linux-400 MHz. Here is a numerical example, using DSS-like parameters:

$q = 762503714763387752235260732711386742425586145191$

$p = 124452971950208973279611466845692849852574447655208586550576344180427926821830$
$\phantom{p = }386338947599247842658333549269645045449033209411448963415127034470249728887681$

The 160-bit number $q$ divides the 512-bit number $p - 1$. The smooth part of $p - 1$ is $4783 \cdot 1759 \cdot 1627 \cdot 139 \cdot 113 \cdot 41 \cdot 11 \cdot 7 \cdot 5 \cdot 2^7$, which is a 69-bit number. Our attack recovered the 64-bit secret message $14327865741237781950$ in only 2 hours and a half (we were lucky, as the maximal running time for 64 bits should be around 14 hours).

# 4    An Attack on ElGamal Using a Generator of $\mathbb{Z}_p^*$

So far, our attacks on ElGamal encryption apply when the public key $\langle p, g, y \rangle$ uses an element $g \in \mathbb{Z}_p^*$ whose order is much less than $p$. Although many implementations of ElGamal use such $g$, it is worth studying whether a "meet-in-the-middle attack" is possible when $g$ generates all of $\mathbb{Z}_p^*$. We show that the answer is positive, although we cannot directly use the algorithm for subgroup rounding.

Let $\langle p, g, y \rangle$ be an ElGamal public key where $g$ generates all of $\mathbb{Z}_p^*$. Suppose an $m$-bit message $M$ is encrypted using plain ElGamal, i.e. the ciphertext is $\langle u, v \rangle$ where $u = M \cdot y^r$ and $v = g^r$. Suppose $s$ is a factor of $p - 1$ so that in the subgroup of $\mathbb{Z}_p^*$ or order $s$ the discrete log problem is not too difficult (as in Section 3.2), i.e. takes time $T$ for some small $T$. For example, $s$ may be an integer with only small prime divisors (a smooth integer).

We show that when $s > 2^m$ it is often possible to recover the plaintext from the ciphertext in time $2^{m/2}m$ plus the time it takes to compute one discrete log in the subgroup of $\mathbb{Z}_p^*$ of order $s$. We refer to this subgroup as $G_s$. Note that when $M$ is a 64-bit session key the only constraint on $p$ is that $p - 1$ have a 64 bit smooth factor.

Let $u = M \cdot y^r$ and $v = g^r$ be an ElGamal ciphertext. As before, suppose $M = M_1 \cdot M_2$ where both $M_1$ and $M_2$ are less than $2^{m/2}$. Let $q = (p-1)/s$ then: $M_1 y^r = u/M_2 \bmod p$. Hence,

$$M_1^q (y^r)^q = u^q / M_2^q \bmod p$$

We cannot use the technique of Section 3.1 directly since we do not know the value of $y^{rq}$. Fortunately, $y^{rq}$ is contained in $G_s$. Hence, we can compute $y^{rq}$ directly using the public key $y$ and $v = g^r$. Indeed, suppose we had an integer $a$ such that $y^q = (g^q)^a$. Then $y^{rq} = g^{rqa} = v^{qa}$. Computing $a$ amounts to computing a single discrete log in $G_s$. Once $a$ is found the problem is reduced to finding $\langle M_1, M_2 \rangle$ satisfying:

$$M_1^q v^{qa} = u^q / M_2^q \bmod p \tag{3}$$

The techniques of Section 3.1 can now be used to find all such $\langle M_1, M_2 \rangle$ in the time it takes to compute $2^{m/2}$ exponentiations. Since the subgroup $G_s$ contains at least $2^m$ elements the number of solutions is bounded by $m$. The correct solution can then be easily found by other means, e.g. by trying all $m$ candidate plaintexts until one of them succeeds as a "session-key".

Note that all the techniques of Section 3.2 can also be applied. The on-line work of $2^{m/2}$ modular exponentiations is then decreased to $2^{m/2}$ additions, provided the precomputation of many discrete log in $G_s$. Indeed, by taking logarithms in (3), one is left with a modular 2-table problem. Splitting the unknown message $M$ in a different number of factors leads to other modular $k$-table problems. One can thus obtain various trade-offs with respect to the computation time, the memory space, and the probability of success, as described in Section 3.2.

To summarize, when $g$ generates all of $\mathbb{Z}_p^*$ the meet-in-the-middle attack can often be used to decrypt ElGamal ciphertexts in time $2^{m/2}$ as long as $p - 1$ contains an $m$-bit smooth factor.

## 5    A Meet-in-the-Middle Attack on Plain RSA

To conclude we remark that the same technique used for the subgroup rounding problem can be used to attack plain RSA. This was also mentioned in [8]. In its simplest form, the RSA system [17] encrypts messages in $\mathbb{Z}_N$ where $N = pq$ for some large primes $p$ and $q$. The public key is $\langle N, e \rangle$ and the private key is $d$, where $e \cdot d = 1 \bmod \phi(N)$ with $\phi(N) = (p - 1)(q - 1)$. A message $M \in \mathbb{Z}_N$ is then encrypted into $c = M^e \bmod N$. To speed up the encryption process one often uses a public exponent $e$ much smaller than $N$, such as $e = 2^{16} + 1$.

Suppose the $m$-bit message $M$ can be written as $M = M_1 M_2$ with $M_1 \leq 2^{m_1}$ and $M_2 \leq 2^{m_2}$. Then:

$$\frac{c}{M_2^e} = M_1^e \bmod N.$$

We can now build a table of size $2^{m_1}$ containing the values $M_1^e \bmod N$ for all $M_1 = 0, \ldots, 2^{m_1}$. Then for each $M_2 = 0, \ldots, 2^{m_2}$, we check whether $c/M_2^e \bmod N$ is present in the table. Any collision will reveal the message $M$. As in Section 3.1, we note that storing the complete value of $M_1^e \bmod N$ is not necessary: for instance, storing the $2\max(m_1, m_2)$ least significant bits should be enough. The attack thus requires $2^{m_1+1}\max(m_1, m_2)$ bits of memory and takes $2^{m_2}$ modular exponentiations (we can assume that the table sort is negligible, compared to exponentiations).

Using a non-optimized implementation (based on the NTL [19] library), we obtained the following results. The timings give a rough idea of the attack efficiency, compared to exhaustive search attacks on the symmetric algorithm. Running times are given for a single 500 MHz 64-bit DEC Alpha/Linux. If $m = 40$ and $m_1 = m_2 = 20$, and we use a public exponent $2^{16}+1$ with a 512-bit modulus, the precomputation step takes 3 minutes, and each message is recovered in less than 10 minutes. From Section 1.2, it also means that, given only the public key and the ciphertext, a 40-bit message can be recovered in less than 40 minutes on a single workstation, with probability at least 39%.

# 6   Summary and Open Problems

We showed that plain RSA and plain ElGamal encryption are fundamentally insecure. In particular, when they are used to encrypt an $m$-bit session-key, the key can often be recovered in time approximately $2^{m/2}$. Hence, although an $m$-bit key is used, the *effective* security provided by the system is only $m/2$ bits. Theses results demonstrate the importance of adding a preprocessing step such as OAEP to RSA and a process such as DHAES to ElGamal. The attack presented in the paper can be used to motivate the need for preprocessing in introductory descriptions of these systems.

There are a number of open problems regarding this attack:

**Problem 1:** Is there a $O(2^{m/2})$ time algorithm for the multiplicative subgroup rounding problem that works for all $\Delta$?

**Problem 2:** Is there a $O(2^{m/2})$ time algorithm for the additive subgroup rounding problem?

**Problem 3:** Can either the multiplicative or additive problems be solved in time less than $\Omega(2^{m/2})$? Is there a sub-exponential algorithm (in $2^m$)?

# Acknowledgments

We thank Paul van Oorschot and David Naccache for several conversations on this problem. We thank Adi Shamir for informing us of reference [18]. We thank Igor Shparlinski for providing us (1) and informing us of reference [11]. We thank Carl Pomerance for providing us (2) and helpful information on splitting probabilities.

# References

1. M. Abdalla, M. Bellare, P. Rogoway, "DHAES: An encryption scheme based on the Diffie-Hellman problem", manuscript, 1998.
2. R. J. Anderson, S. Vaudenay, "Minding your p's and q's", Proc of Asiacrypt '96, LNCS 1163, Springer-Verlag, pp. 26–35, 1996.
3. C. Batut, K. Belabas, D. Bernardi, H. Cohen, M. Olivier, "Pari/GP computer package version 2", available at
   `http://hasse.mathematik.tu-muenchen.de/ntsw/pari/Welcome`.
4. M. Bellare, P. Rogaway, "Optimal asymmetric encryption — how to encrypt using RSA", Proc. Eurocrypt '94, LNCS 950, Springer-Verlag, 1995.
5. D. Boneh, "The Decision Diffie-Hellman Problem", Proc. ANTS-III, LNCS 1423, Springer-Verlag, 1998.
6. D. Boneh, "Twenty Years of Attacks on the RSA cryptosystem", Notices of the AMS, 46(2):203–213, 1999.
7. J.-S. Coron, D. Naccache, J. P. Stern, "On the Security of RSA Padding", Proc. of Crypto '99, LNCS 1666, Springer-Verlag, pp. 1–18, 1999.
8. J.-S. Coron, M. Joye, D. Naccache, P. Paillier, "New Attacks on PKCS#1 v1.5 Encryption", Proc. of Eurocrypt '2000, LNCS 1807, Springer-Verlag, pp. 369–381, 2000.
9. T. ElGamal, "A public key cryptosystem and a signature scheme based on the discrete logarithm", IEEE Trans. on Information Theory, 31(4):469–472, 1985.
10. E. Fujisaki, T. Okamoto, "Secure Integration of Asymmetric and Symmetric Encryption Schemes", Proc. of Crypto '99, LNCS 1666, Springer-Verlag, pp. 537–554, 1999.
11. R. R. Hall, G. Tenenbaum, "Divisors", Cambridge University Press, 1988.
12. A. Menezes, P. v. Oorschot, S. Vanstone, "Handbook of Applied Cryptography", CRC Press, 1997.
13. T. Okamoto and D. Pointcheval, "PSEC-3: Provably Secure Elliptic Curve Encryption Scheme", Submission to IEEE P1363a, 2000.
14. P. v Oorschot, M. J. Wiener, "On Diffie-Hellman Key Agreement With Short Exponents", Proc. Eurocrypt '96, LNCS 1070, Springer-Verlag, 1996.
15. PKCS1, "Public Key Cryptography Standard No. 1 Version 2.0", RSA Labs.
16. D. Pointcheval, "Chosen-Ciphertext Security for any One-Way Cryptosystem", Proc. PKC '2000, LNCS 1751, Springer-Verlag, 2000.
17. R. L. Rivest., A. Shamir, L. M. Adleman " A method for obtaining digital signatures and public-key cryptosystems", Communications of the ACM, 21(2):120–126, 1978.
18. R. Schroeppel, A. Shamir, "A $T = O(2^{n/2})$, $S = O(2^{n/4})$ algorithm for certain NP-complete problems", SIAM J. Comput., 10(3):456–464, 1981.
19. V. Shoup, "Number Theory C++ Library (NTL) version 3.7", available at
    `http://www.shoup.net/`.

# Cryptanalysis of the TTM Cryptosystem

Louis Goubin and Nicolas T. Courtois

Bull CP8
68 route de Versailles – BP45
78431 Louveciennes Cedex
France
Louis.Goubin@bull.net, courtois@minrank.org

**Abstract.** In 1985 Fell and Diffie proposed constructing trapdoor functions with multivariate equations [11]. They used several sequentially solved stages that combine into a triangular system we call T. In the present paper, we study a more general family of TPM (for "Triangle Plus Minus") schemes: a triangular construction mixed with some $u$ random polynomials and with some $r$ of the beginning equations removed. We go beyond all previous attacks proposed on such cryptosystems using a low degree component of the inverse function. The cryptanalysis of TPM is reduced to a simple linear algebra problem called *MinRank(r)*: Find a linear combination of given matrices that has a small rank $r$.

We introduce a new attack for MinRank called 'Kernel Attack' that works for $q^r$ small. We explain that TPM schemes can be used in encryption only if $q^r$ is small and therefore they are not secure.

As an application, we showed that the TTM cryptosystem proposed by T.T. Moh at CrypTec'99 [15,16] reduces to MinRank(2). Thus, though the cleartext size is 512 bits, we break it in $\mathcal{O}(2^{52})$. The particular TTM of [15,16] can be broken in $\mathcal{O}(2^{28})$ due additional weaknesses, and we needed only few minutes to solve the challenge TTM 2.1. from the website of the TTM selling company, US Data Security.

We also studied TPM in signature, possible only if $q^u$ small. It is equally insecure: the 'Degeneracy Attack' we introduce runs in $q^{u}$ polynomial.

## 1 Introduction

The current research effort in practical public key cryptography introduced by Rivest, Shamir and Adleman, with univariate polynomials over $\mathbb{Z}_N$, is following two paths. The first is considering more complex groups, e.g. elliptic curves. The second is considering multivariate equations. Though many proposed schemes are being broken, some remain unbroken even for the simplest groups like $\mathbb{Z}_2$.

One of the paradigms for constructing multivariate trapdoor cryptosystems is the triangular construction, proposed initially in an iterated form by Fell and Diffie (1985). It uses equations that involve 1, 2, ..., $n$ variables and are solved sequentially. The special form of the equations is hidden by two linear transformations on inputs (variables) and outputs (equations). We call T this triangular construction. Let TPM (T Plus-Minus) be T with added final $u$ random (full-size) quadratic polynomials, and with $r$ of the beginning equations removed.

T. Okamoto (Ed.): ASIACRYPT 2000, LNCS 1976, pp. 44–57, 2000.

The cryptosystem TTM, proposed by T.T. Moh at CrypTec'99 is in spite of an apparent complexity, shown in 2.4 to be a subcase of TPM. The initially proposed scheme is very weak due to linear dependencies and in section 4.2, we present the solution (plaintext) to the TTM 2.1 challenge proposed by the company US Data Security, which is currently selling implementations of TTM. After this, we focus on breaking more general TPM schemes.

The general strategy to recover the secret key of TPM/TTM systems is presented in 3. It requires finding a linear combination of public equations that depends only of a subspace of variables. This gives a simple linear algebra problem called MinRank: Let us consider some $n \times n$ matrices over $GF(q)$: $M_1, \ldots, M_t$. We need to find a linear combination $M$ of the $M_i$ that has a small rank $r < n$. The name of MinRank has apparently been used first in the paper [19] that shows that MinRank is NP-complete. However the MinRank instances in TPM/TTM use very small $r$, e.g. the T.T. Moh's proposal from [16] gives $r = 2$. We note that the powerful idea of using a small rank goes back to the cryptanalysis of Shamir birational scheme [20] by Coppersmith, Stern and Vaudenay [6,7], and appears also in the Shamir-Kipnis attack on HFE [14] proposed by Patarin [17].

In 2.2 we explain how to use the TPM schemes in encryption which is possible only if $q^r$ is small. However, in the section 5 we present an attack that works precisely when $q^r$ is small, based on the small co-dimension of the kernel of the unknown matrix $M$. This 'Kernel attack' breaks in approximately $2^{52}$ a cryptosystem with 512 bit cleartexts.

Similarly in 2.2 we explain how to use the TPM schemes in signature; possible only with $q^u$ not too big. Then in section 6 we introduce the 'Degeneracy attack' on TPM based on iterative searching of degenerate polynomials. It works precisely when $q^u$ is small and the signature proposals of [15,16] are insecure.

## 2    The TPM Family of Cryptosystems

### 2.1    General Description of TPM

In the present section, we describe the general family $TPM(n, u, r, K)$, with:

- $n$, $u$, $r$ integers such that $r \leq n$. We also systematically put $m = n + u - r$.
- $K = GF(q)$ a finite field.

We first consider a function $\Psi : K^n \mapsto K^{n+u-r}$ such that $(y_1, \ldots, y_{n+u-r}) = \Psi(x_1, \ldots, x_n)$ is defined by the following system of equations:

$$
\begin{cases}
y_1 = x_1 + & g_1( & x_{n-r+1}, \ldots, x_n) \\
y_2 = x_2 + & g_2(x_1; & x_{n-r+1}, \ldots, x_n) \\
y_3 = x_3 + & g_3(x_1, x_2; & x_{n-r+1}, \ldots, x_n) \\
\quad \vdots \\
y_{n-r} = x_{n-r} + & g_{n-r}(x_1, \ldots, x_{n-r-1} ; x_{n-r+1}, \ldots, x_n) \\
y_{n-r+1} = & g_{n-r+1}(x_1, \ldots, x_n) \\
\quad \vdots \\
y_{n-r+u} = & g_{n-r+u}(x_1, \ldots, x_n)
\end{cases}
$$

with each $g_i$ $(1 \leq i \leq n+u-r)$ being a randomly chosen quadratic polynomial.

**The Public Key**

The user selects a random invertible affine transformation $s: K^n \mapsto K^n$, and a random invertible affine transformation $t: K^{n+u-r} \mapsto K^{n+u-r}$. Let $F = t \circ \Psi \circ s$. By construction, if we denote $(y'_1, \ldots, y'_{n+u-r}) = F(x'_1, \ldots, x'_n)$, we obtain an explicit set $\{P_1, \ldots, P_{n+u-r}\}$ of $(n+u-r)$ quadratic polynomials in $n$ variables, such that:

$$\begin{cases} y'_1 = P_1(x'_1, \ldots, x'_n) \\ \quad \vdots \\ y'_{n+u-r} = P_{n+u-r}(x'_1, \ldots, x'_n) \end{cases}$$

This set of $(n+u-r)$ quadratic polynomials constitute the public key of this TPM$(n, u, r, K)$ cryptosystem. Its size is $\frac{1}{8}(n+u-r)(n+1)(\frac{n}{2}+1)\log_2(q)$ bytes.

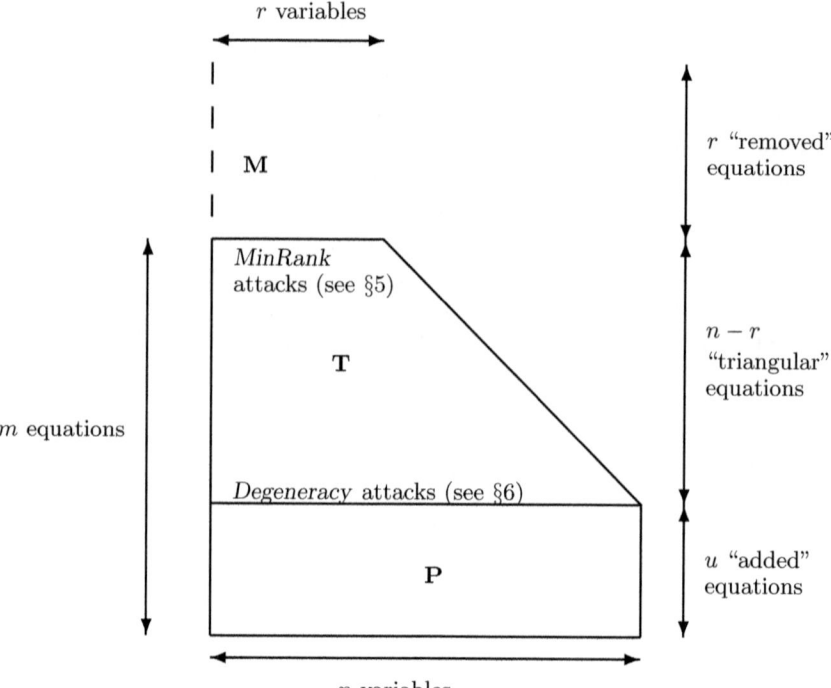

**Fig. 1.** General view of the TPM scheme – The two classes of attacks

## 2.2   Encryption Protocol (when $u \geq r$)

**Encrypting a message**

Given a plaintext $(x'_1, \ldots, x'_n) \in K^n$, the sender computes $y'_i = P_i(x'_1, \ldots, x'_n)$ for $1 \leq i \leq n+u-r$ – thanks to the public key – and sends the ciphertext $(y'_1, \ldots, y'_{n+u-r}) \in K^{n+u-r}$.

**Decrypting a message**

Given a ciphertext $(y'_1, \ldots, y'_{n+u-r}) \in K^{n+u-r}$, the legitimate receiver recovers the plaintext by the following method.

- Compute $(y_1, \ldots, y_{n+u-r}) = t^{-1}(y'_1, \ldots, y'_{n+u-r})$ ;
- Make an exhaustive search on the $r$-tuple $(x_{n-r+1}, \ldots, x_n) \in K^r$, until the $n$-tuple $(x_1, \ldots, x_n)$ obtained by $x_i = y_i - g_i(x_1, \ldots, x_{i-1}; x_{n-r+1}, \ldots, x_n)$ (for $1 \le i \le n-r$) satisfies the $u$ following equations $g_i(x_1, \ldots, x_n) = y_i$ (for $n-r+1 \le i \le n-r+u$).
- For the obtained $(x_1, \ldots, x_n)$ $n$-tuple, get $(x'_1, \ldots, x'_n) = s^{-1}(x_1, \ldots, x_n)$.

This decryption algorithm thus has a complexity essentially $\mathcal{O}(q^r)$. As a result, a TPM$(n, u, r, K)$ cryptosystem can be practically used in encryption mode only under the assumption that $q^r$ is "small enough".

The condition $u \ge r$ insures that the probability of obtaining a collision is negligible, and thus that the ciphering function $F$ can be considered as an injection from $K^n$ into $K^{n+u-r}$.

When $r = u = 0$, this kind of scheme has been considered and attacked by Fell and Diffie in [11] (in an iterated form) and by Patarin and Goubin in [18]. All these attacks explore the fact that the inverse function if of low degree in some variables, whereas the present paper cryptanalyses much more general cases with $r \ne 0$ and $u \ne 0$.

## 2.3   Signature Protocol (when $u \le r$)

**Signing a message**

Given a message $M$, we suppose that $(y'_1, \ldots, y'_{n+u-r}) = h(M) \in K^{n+u-r}$, with $h$ being a (collision-free) hash function. To sign the message $M$, the legitimate user:

- computes $(y_1, \ldots, y_{n+u-r}) = t^{-1}(y'_1, \ldots, y'_{n+u-r})$ ;
- chooses random $r$-tuples $(x_{n-r+1}, \ldots, x_n)$, until the $n$-tuple $(x_1, \ldots, x_n)$ obtained by $x_i = y_i - g_i(x_1, \ldots, x_{i-1}; x_{n-r+1}, \ldots, x_n)$ (for $1 \le i \le n-r$) satisfies the $u$ following equations $g_i(x_1, \ldots, x_n) = y_i$ (for $n-r+1 \le i \le n-r+u$).
- for the obtained $(x_1, \ldots, x_n)$ $n$-tuple, gets $(x'_1, \ldots, x'_n) = s^{-1}(x_1, \ldots, x_n)$.

This signature algorithm thus has a complexity essentially $\mathcal{O}(q^u)$. As a result, a TPM$(n, u, r, K)$ cryptosystem can be practically used in signature mode only under the assumption that $q^u$ is "small enough".

The condition $u \le r$ insures that the probability of finding no solution for $(x_1, \ldots, x_n)$ for the equation $\Psi(x_1, \ldots, x_n) = (y_1, \ldots, y_{n+u-r})$ is negligible, and thus that the ciphering function $F$ can be viewed as an surjection from $K^n$ onto $K^{n+u-r}$.

We will describe in section 6 a general attack on this signature scheme, that is also applicable when $u$ is non-zero, with $q^u$ not too large. Therefore the signature proposed by T.T. Moh in [15,16] is insecure.

## 2.4   The TTM Encryption System

In the present section, we recall the original description of the TTM cryptosystem, given by T.T. Moh in [15,16]. This definition of TTM is based on the concept of *tame automorphisms*. As we will see, TTM is a particular case of our general family TPM: it belongs to the family TPM$(64, 38, 2, \text{GF}(256))$.

### General Principle

Let $K$ be a finite field (which will be supposed "small" in real applications). We first consider two bijections $\Phi_2$ and $\Phi_3$ from $K^{n+v}$ to $K^{n+v}$, with $(z_1, \ldots, z_{n+v}) = \Phi_2(x_1, \ldots, x_{n+v})$ and $(y_1, \ldots, y_{n+v}) = \Phi_3(z_1, \ldots, z_{n+v})$ defined by the two following systems of equations :

$$
\Phi_2 : \begin{cases}
z_1 = x_1 \\
z_2 = x_2 + f_2(x_1) \\
z_3 = x_3 + f_3(x_1, x_2) \\
\vdots \\
z_n = x_n + f_n(x_1, \ldots, x_{n-1}) \\
z_{n+1} = x_{n+1} + f_{n+1}(x_1, \ldots, x_n) \\
\vdots \\
z_{n+v} = x_{n+v} + f_{n+v}(x_1, \ldots, x_{n+v-1})
\end{cases}
\qquad
\Phi_3 : \begin{cases}
y_1 = z_1 + P(z_{n+1}, \ldots, z_{n+v}) \\
y_2 = z_2 + Q(z_{n+1}, \ldots, z_{n+v}) \\
y_3 = z_3 \\
\vdots \\
y_{n+v} = z_{n+v}
\end{cases}
$$

with $f_2, \ldots, f_{n+v}$ quadratic forms over $K$, and $P$, $Q$ two polynomials of degree eight over $K$.

$\Phi_2$ and $\Phi_3$ are both "tame automorphisms" (see [15,16] for a definition) and thus are one-to-one transformations. As a result, $(x_1, \ldots, x_{n+v}) \mapsto (y_1, \ldots, y_{n+v}) = \Phi_3 \circ \Phi_2(x_1, \ldots, x_{n+v})$ is also one-to-one and can be described by the following system of equations :

$$
\begin{cases}
y_1 = x_1 + P(x_{n+1} + f_{n+1}(x_1, \ldots, x_n), \ldots, x_{n+v} + f_{n+v}(x_1, \ldots, x_{n+v-1})) \\
y_2 = x_2 + f_2(x_1) + Q(x_{n+1} + f_{n+1}(x_1, .., x_n), .., x_{n+v} + f_{n+v}(x_1, .., x_{n+v-1})) \\
y_3 = x_3 + f_3(x_1, x_2) \\
\vdots \\
y_n = x_n + f_n(x_1, \ldots, x_{n-1}) \\
y_{n+1} = x_{n+1} + f_{n+1}(x_1, \ldots, x_n) \\
\vdots \\
y_{n+v} = x_{n+v} + f_{n+v}(x_1, \ldots, x_{n+v-1})
\end{cases}
$$

T.T. Moh found a clever way of choosing $P$, $Q$ and $f_i$ such that $y_1$ and $y_2$ both become *quadratic* functions of $x_1, \ldots, x_n$ when we set $x_{n+1} = \ldots = x_{n+v} = 0$.

### Actual Parameters

This paragraph is given in the appendix. T.T. Moh chooses $n = 64$, $v = 36$ and $K = \text{GF}(256)$. As a result, TTM belongs to TPM$(64, 38, 2, \text{GF}(256))$. Applying the formula of section 2.1, the size of the public keys is 214.5 Ko.

# 3    General Strategy of TPM Attacks

In the present section, we describe a general strategy to attack a cryptosystem of the TPM Family when $r$ is "small". It will amount to solving the MinRank problem. As a result TTM, that is a TPM(64, 38, 2,GF(256)) will be broken.

## 3.1    The MinRank Problem

Let $r$ be an integer and $K$ a field. We denote by MinRank($r$) the following problem: given a set $\{M_1, \ldots, M_m\}$ of $n \times n$ matrices whose coefficients lie in $K$, find at least one $m$-tuple $(\lambda_1, \ldots, \lambda_m) \in K^m$ such that $\text{Rank}\left( \sum_{i=1}^{m} \lambda_i M_i \right) \leq r$.

The (even more) general MinRank problem has been first defined and studied by Shallit, Frandsen and Buss in [19]. It generalizes the "Rank Distance Coding" problem by Gabidulin [12], studied also in [3,22]), which itself generalizes the "Minimal Weight" problem of error correcting codes (see [1,21,2,13]). In the Shamir-Kipnis attack on the Patarin's HFE cryptosystem [14,17], the authors used an instance of MinRank($r$) with $r = \lceil \log_q n \rceil + 1$ and therefore their attack is not polynomial. In the present paper $r$ is a small constant, e.g. 2. We note that the idea of finding small ranks has first been used by Coppersmith, Stern and Vaudenay in [6,7] for breaking Shamir's birational scheme [20].

Recently Courtois proposed a new zero-knowledge scheme based on Min-Rank [10,9]. Though in the present paper only two algorithms for MinRank are introduced, another two can be found in [9].

## 3.2    Complexity of MinRank

The general MinRank problem has been proven to be NP-complete by Shallit, Frandsen and Buss (see [19]). More precisely, they prove that MinRank($r$) NP-complete when $r = n - 1$ (this corresponds to the problem of finding a linear combination of $M_1, \ldots, M_m$ that is singular). The principle of their proof consists in writing any set of multivariate equations as an instance of MinRank. It can be used in the same way to extend their result to the cases $r = n - 2$, $r = n - 3$, ... and even $r = n^\alpha$ (when $\alpha > 0$ is fixed). However, MinRank is not hard when $r$ gets smaller, indeed, in 5 we will introduce an expected polynomial time algorithm to solve the MinRank for any fixed $r$.

## 3.3    Strategy of Attack

We recall that $m = n + u - r$. We suppose $m \leq 2n$, as an encryption function with expansion rate $> 2$ is unacceptable. Moreover, if $m > \mathcal{O}(n)$, the cryptosystem is expected to be broken by Gröbner bases [8].

In each equation $y_i = x_i + g_i(x_1, \ldots, x_{i-1} ; x_{n-r+1}, \ldots, x_n)$ $(1 \leq i \leq n - r)$, the homogeneous part is given by ${}^t X A_i X$, with ${}^t X = (x_1, \ldots, x_n)$, $A_i$ being a (secret) matrix. Similarly, in each public equation $y'_i = P_i(x'_1, \ldots, x'_n)$ is given by ${}^t X' M_i X'$, with ${}^t X' = (x'_1, \ldots, x'_n)$, $M_i$ being a (public) matrix.

The fact that $(x_1, \ldots, x_n) = s(x'_1, \ldots, x'_n)$ and $(y'_1, \ldots, y'_m) = t(y_1, \ldots, y_m)$ implies that there exist an invertible $n \times n$ matrix $S$ and an invertible $m \times m$ matrix $T$ such that:

$$\begin{pmatrix} {}^t(SX')A_1(SX') \\ \vdots \\ {}^t(SX')A_m(SX') \end{pmatrix} = T^{-1} \begin{pmatrix} {}^tX'M_1X' \\ \vdots \\ {}^tX'M_mX' \end{pmatrix}.$$

Let $T^{-1} = (t_{ij})_{1 \le i,j \le m}$. We thus have, for any $X'$:

$$ {}^tX'({}^tSA_iS)X' = {}^tX'\left(\sum_{j=1}^{m} t_{ij}M_j\right)X' $$

so that:

$$\forall i, \ 1 \le i \le m, \ \sum_{j=1}^{m} t_{ij}M_j = {}^tSA_iS.$$

From the construction of $\text{TPM}(n, u, r, K)$, we have $\text{Rank}(A_1) \le r$. Since $S$ is an invertible matrix, we have $\text{Rank}(A_1) = \text{Rank}({}^tSA_1S)$ and thus Rank $\left(\sum_{j=1}^{m} t_{1j}M_j\right) \le r$, that is precisely an instance of $\text{MinRank}(r)$.

Suppose we are able to find (at least) one $m$-tuple $(\lambda_1, \ldots, \lambda_m)$ such that $\text{Rank}\left(\sum_{j=1}^{m} \lambda_j M_j\right) \le r$. With a good probability, we can suppose that:

$$\sum_{j=1}^{m} \lambda_j M_j = \mu^t SA_1S \qquad (\mu \in K^*).$$

Then we deduce the vector spaces $V_0 = S^{-1}(K^{n-r} \times \{0\}^r)$ (corresponding to $x_{n-r+1} = \ldots = x_n = 0$) and $W_0 = S^{-1}(\{0\}^{n-r} \times K^r)$ (corresponding to $x_1 = \ldots = x_{n-r} = 0$) by simply noticing that $V_0 = \text{Im}\left(\sum_{j=1}^{m} \lambda_j M_j A_1\right)$ and $W_0 = \text{Ker}\left(\sum_{j=1}^{m} \lambda_j M_j A_1\right)$.

Once we have found $V_0$ and $W_0$, we can easily deduce the vector space $V_1 = S^{-1}(\{0\} \times K^{n-r-1} \times \{0\}^r)$ of dimension 1 (corresponding to $x_1 = x_{n-r+1} = \ldots = x_n = 0$) and $W_1 = S^{-1}(K \times \{0\}^{n-r-1} \times K^r)$ (corresponding to $x_2 = \ldots = x_{n-r} = 0$): we just look for coefficients $\alpha_1, \ldots, \alpha_n, \beta_1, \ldots, \beta_m$ such that the following equation:

$$\sum_{j=1}^{m} \beta_j y'_j = \sum_{i=1}^{n} \alpha_i x_i + \delta,$$

holds for any element of $V_0$. This can be obtained by simple Gaussian reduction. We also obtain the $g_2$ quadratic function by Gaussian reduction.

By repeating these steps, we obtain two sequences of vector spaces:

$$V_0 \supseteq V_1 \supseteq V_2 \supseteq \ldots \supseteq V_{n-r-1}$$

$$W_0 \subseteq W_1 \subseteq W_2 \subseteq \ldots \subseteq W_{n-r-1}.$$

At the end, we have completely determined the secret transformations $s$ and $t$, together with the secret functions $g_i$. As a result, this algorithm completely breaks the TPM family of cryptosystems (we recovered the secret key).

## 4  Special Case Attacks on TPM

### 4.1  The 'Linearity Attack' on TTM

In this paragraph, we study the particular case of TTM, as described by T.T. Moh in [15,16]. In this case, we show that the MinRank($r$) problem is easily solved, because of the particular structure of the $Q_8$ function used in $\Phi_3$.

**Description of the Attack**

In section 3.3, we proved that an attack can be successfully performed on TTM this cryptosystem, as soon as we can find out the vector spaces $V_0 = S^{-1}(\{0\}^2 \times K^{62})$ (corresponding to $x_1 = x_2 = 0$) and $W_0 = S^{-1}(K^2 \times \{0\}^{62})$ (corresponding to $x_3 = \ldots = x_{64} = 0$). At first sight, the equations giving $y_1$ and $y_2$ seem to be quadratic in $(x_1, \ldots, x_{64})$. This leads *a priori* to an instance of MinRank(2).

However, note that the function $x \mapsto x^2$ is linear on $K = \mathrm{GF}(256)$, considered as a vector space of dimension 8 over $F = \mathrm{GF}(2)$. Therefore, considering the equations describing the (secret) $\Psi$ function of TTM[1], if we choose a basis $(\omega_1, \ldots, \omega_8)$ of $K$ over $F$ and write $x_i = x_{i,1}\omega_1 + \ldots + x_{i,8}\omega_8$ $(1 \leq i \leq 64)$, $y_1$ and $y_2$ become linear functions of $x_{1,1}, x_{1,2}, \ldots, x_{1,8}, \ldots, x_{64,1}, \ldots, x_{64,8}$. In terms of MinRank, this means that TTM leads to an instance of MinRank(0) for $8n \times 8n$ matrices (instead of an instance of MinRank(2) for $n \times n$ matrices). This leads to the following attack on TTM:

1. Let $x_i' = x_{i,1}'\omega_1 + \ldots + x_{i,8}'\omega_8$ $(1 \leq i \leq 64)$. Rewrite each public equation $y_i' = P_i(x_1', \ldots, x_{64}')$ as $y_i' = \tilde{P}_i(x_{1,1}', \ldots, x_{64,8}')$ (with $\tilde{P}_i$ a quadratic polynomial in $64 \times 8 = 512$ variables over $F = \mathrm{GF}(2)$).
2. Find the vector space of the 612-tuples $(\beta_1, \ldots, \beta_{100}, \alpha_{1,1}, \ldots, \alpha_{64,8}) \in K^{612}$ satisfying:

$$\sum_{i=1}^{100} \beta_i y_i' = \sum_{i=1}^{64} \sum_{j=1}^{8} \alpha_{i,j} x_{i,j}'.$$

This can be done by Gaussian reduction. We thus obtain the vector spaces $V_0$ and $W_0$ defined above.
3. The remaining part of the attack is exactly the same as in section 3.3.

---

[1] See $(E)$ in the appendix, in which $t_{19}$ is a linear transformation.

**Complexity of the Attack**

The main part of the algorithm consists in solving a system of linear equations on 612 variables, by Gaussian reduction. We thus obtain a complexity of approximately $2^{28}$ elementary operations to break TTM.

## 4.2   Solution to the TTM 2.1 Challenge of US Data Security

In 1997, US Data Security published on the internet 3 challenges about TTM (see [23]). On May 2nd, 2000, we managed to break the second challenge called TTM 2.1. The TTM 2.1 is a public key block cipher with plaintext block size 64 and ciphertext block size 100. It works on 8 bits finite field GF(256). The public key have been recovered with approximately 2000 queries to the "encryption oracle" available on the internet [23]. As mentioned in 2.4, its size is 214.5 Kbytes. Moreover it was broken in a simpler way that we described above. By iterative exploration of it's linearities, in 3 minutes on a PC we obtained the following plaintext which can be easily checked to be the exact solution to TTM 2.1. (note that the quotation marks are part of this plaintext):

"Tao TTP way BCKP of living hui mountain wen river moon love pt"

## 5   The 'Kernel Attack' on MinRank and TPM

In the present section we need the strategy of attack from 3.3 and use it with a new attack on MinRank($r$), which works when $q^r$ is small enough.

**Description of the Attack** (notations are as in 3.3)

1. Choose $k$ random vectors $X'^{[1]}, \ldots, X'^{[k]}$ (with $k$ an integer depending on $n$ and $m$, that we define below). Since dim Ker($^tSA_1S$) $= n-$ Rank($^tSA_1S$) $\geq n - r$, we have the simultaneous conditions $X'^{[i]} \in$ Ker($^tSA_iS$) $(1 \leq i \leq k)$ with a probability $\geq q^{-kr}$.

2. We suppose we have chosen a "good" set $\{X'^{[1]}, \ldots, X'^{[k]}\}$ of $k$ vectors (*i.e.* such that they all belong to Ker($^tSA_1S$)). Then we can find an $m$-tuple $(\lambda_1, \ldots, \lambda_m)$ such that, for all $i$, $1 \leq i \leq k$, $\left( \sum_{j=1}^{m} \lambda_j M_j \right)(X'^{[i]}) = 0$. They are solution of a system of $kn$ linear equations in $m$ indeterminates. As a result, if we let $k = \lceil \frac{m}{n} \rceil$, the solution is essentially unique and can be easily found by Gaussian reduction. We thus obtain the two vector spaces $V_0 = S^{-1}(K^{n-r} \times \{0\}^r)$ (corresponding to $x_{n-r+1} = \ldots = x_n = 0$) and $W_0 = S^{-1}(\{0\}^{n-r} \times K^r)$ (corresponding to $x_1 = \ldots = x_{n-r} = 0$).

3. The remaining part of the attack is exactly the same as in section 3.3.

**Complexity of the Attack**

The complexity of the attack is easily computed: $\mathcal{O}(q^{\lceil \frac{m}{n} \rceil r} \cdot m^3)$.

**Application to TTM**

In the particular case of TTM, we have $q = 256$, $n = 64$, $m = 100$ and $r = 2$. We thus obtain an attack on TTM with complexity $\mathcal{O}(2^{52})$.

**Note:** Compared to the $2^{28}$ of section 4.1, this attack is slower, but it does not make use of any linearity of $y_1$ and $y_2$, so that it can also be used to break possible generalizations of TTM, with more general "$Q_8$ components" (see [4] for examples of $Q_8$ which provide non linear expressions for $y_1$ and $y_2$ over $GF(2)$).

# 6   The 'Degeneracy Attack' on TPM Signature Schemes

We describe here a general attack on TMP signature schemes (recall that such schemes are possible only for $u \leq r$), when $q^u$ is not too large. From the description of the attack, its complexity is easily seen to be $\mathcal{O}(q^u \cdot n^6)$. We use the same notations as in section 3.3. In particular, $m = n + u - r$.

1. We choose a random $m$-tuple $(\beta_1, \ldots, \beta_m) \in K^m$. With a probability $q^{-u-1}$, we can suppose that $\beta_i P_i$ is a degenerate quadratic polynomial (*i.e.* a quadratic polynomial which can be rewritten with fewer variables after a linear change of variables). The fact that a quadratic polynomial is degenerate can easily be detected: for instance by using its canonical form (see [18] for some other methods).

2. Suppose we have found a "good" $m$-tuple $(\beta_1, \ldots, \beta_m)$. Considering the new set of ($< n$) variables for the quadratic form $\sum_{i=1}^{m} \beta_i P_i$, we deduce easily the vector space $W_{n-r} = S^{-1}(K^{n-r-1} \times \{0\} \times K^r)$.

3. Then we look for a $n$-tuple $(\alpha_1, \ldots, \alpha_n) \in K^n$ and a quadratic function $g^{n-r}$, such that:

$$\sum_{i=1}^{m} \beta_i y_i' = \sum_{i=1}^{n} \alpha_i x_i' + g_{n-r}(x_1', \ldots, x_n')$$

is true for any $(x_1', \ldots, x_n') \in W_{n-r}$. This can be done by Gaussian reduction. We thus obtain the vector space $V_{n-r} = S^{-1}(\{0\}^{n-r-1} \times K \times \{0\}^r)$ and the quadratic polynomial $g_{n-r}$.

4. The same principle can be repeated $n-r$ times, so as to obtain two sequences of vector spaces:

$$V_{n-r} \subseteq V_{n-r-1} \subseteq \cdots \subseteq V_0$$

$$W_{n-r} \supseteq W_{n-r-1} \supseteq \cdots \supseteq W_0.$$

At the end, as in the attack described in section 3.3, we have completely determined the secret transformations $s$ and $t$, together with the secret functions $g_i$. As a result, this algorithm completely breaks the TPM family in signature mode (we recovered the secret key).

## 7   Conclusion

We cryptanalysed a large class of cryptosystems TPM, that includes TTM as described by T.T. Moh [16]. They can be broken in polynomial time, as long as $r$ is fixed. The proposed TTM cryptosystem [16] can be broken in $2^{28}$ due to linearities. Thus we could easily break the "TTM 2.1" challenge proposed by US Data Security in October 1997. Even if $Q_8$ was nonlinear, and since $r = 2$, it is still broken in $2^{52}$ elementary operations for a 512-bit cryptosystem.

We also showed that signature schemes using TPM are insecure. There is very little hope that a secure triangular system will ever be proposed.

## References

1. E.R. Berlekamp, R.J. McEliece, H.C.A. Van Tilborg, *On the inherent intractability of certain coding problems*, IEEE Transactions on Information Theory, IT-24(3), pp. 384-386, May 1978.
2. F. Chabaud, *Asymptotic analysis of probabilistic algorithms for finding short code-words*, in Proceedings of Eurocode'92, Udine, Italy, CISM Courses and lectures n° 339, Springer-Verlag, 1993, pp. 217-228.
3. K. Chen, *A new identification algorithm*, Cryptography Policy and Algorithms Conference, LNCS n° 1029, Springer-Verlag, 1996.
4. C. Y. Chou, D. J. Guan, J. M. Chen, *A systematic construction of a $Q_{2^k}$-module in TTM*, Preprint, October 1999. Available at http://www.usdsi.com/chou.ps
5. D. Coppersmith, S. Winograd, *Matrix multiplication via arithmetic progressions*, J. Symbolic Computation (1990), **9**, pp. 251-280.
6. D. Coppersmith, J. Stern, S. Vaudenay, *Attacks on the Birational Permutation Signature Schemes*, in Advances in Cryptology, Proceedings of Crypto'93, LNCS n° 773, Springer-Verlag, 1993, pp. 435-443.
7. D. Coppersmith, J. Stern, S. Vaudenay, *The Security of the Birational Permutation Signature Schemes*, in Journal of Cryptology, 10(3), pp. 207-221, 1997.
8. N. Courtois, A. Shamir, J. Patarin, A. Klimov, *Efficient Algorithms for solving Overdefined Systems of Multivariate Polynomial Equations*, in Advances in Cryptology, Proceedings of EUROCRYPT"2000, LNCS n° 1807, Springer, 2000, pp. 392-407.
9. N. Courtois: *La sécurité des primitives cryptographiques basées sur les problèmes algébriques multivariables MQ, IP, MinRank, et HFE*, PhD thesis, Paris 6 University, 26 September 2000, partly in English.
10. N. Courtois: *The Minrank problem*. MinRank, a new Zero-knowledge scheme based on the NP-complete problem. Presented at the rump session of Crypto 2000, available at http://www.minrank.org
11. H. Fell, W. Diffie, *Analysis of a public key approach based on polynomial substitutions*, in Advances in Cryptology, Proceedings of CRYPTO'85, LNCS n° 218, Springer-Verlag, 1985, pp. 340-349.
12. E.M. Gabidulin, *Theory of codes with maximum rank distance*, Problems of Information Transmission, 21:1-12, 1985.
13. S. Harari, *A new authentication algorithm*, in Coding Theory and Applications, LNCS n° 388, Springer, 1989, pp. 204-211.

14. A. Kipnis, A. Shamir, *Cryptanalysis of the HFE public key cryptosystem*, in Advances in Cryptology, Proceedings of Crypto'99, LNCS n° 1666, Springer, 1999, pp. 19-30.

15. T.T. Moh, *A public key system with signature and master key functions*, Communications in Algebra, 27(5), pp. 2207-2222, 1999. Available at http://www.usdsi.com/public.ps

16. T.T. Moh, *A fast public key system with signature and master key functions*, in Proceedings of CrypTEC'99, International Workshop on Cryptographic Techniques and E-commerce, Hong-Kong City University Press, pp. 63-69, July 1999. Available at http://www.usdsi.com/cryptec.ps

17. J. Patarin, *Hidden Fields Equations (HFE) and Isomorphisms of Polynomials (IP): two new families of asymmetric algorithms*, in Advances in Cryptology, Proceedings of EUROCRYPT'96, LNCS n° 1070, Springer Verlag, 1996, pp. 33-48.

18. J. Patarin, L. Goubin, *Asymmetric cryptography with S-Boxes*, in Proceedings of ICICS'97, LNCS n° 1334, Springer, 1997, pp. 369-380.

19. J.O. Shallit, G.S. Frandsen, J.F. Buss, *The computational complexity of some problems of linear algebra*, BRICS series report, Aarhus, Denmark, RS-96-33. Available at http://www.brics.dk/RS/96/33

20. A. Shamir, *Efficient Signature Schemes based on Birational Permutations*, in Advances in Cryptology, Proceedings of Crypto'93, LNCS n° 773, Springer-Verlag, 1993, pp. 1-12.

21. J. Stern, *A new identification scheme based on syndrome decoding*, in Advances in Cryptology, Proceedings of CRYPTO'93, LNCS n° 773, Springer-Verlag, 1993, pp. 13-21.

22. J. Stern, F. Chabaud, *The cryptographic security of the Syndrome Decoding problem for rank distance codes*, in Advances in Cryptology, Proceedings of ASIACRYPT'96, LNCS n° 1163, Springer-Verlag, 1985, pp. 368-381.

23. *The US Data Security Public-Key Contest*, available at http://www.usdsi.com/contests.html

# Appendix: Actual Parameters for the TTM Cryptosystem

Let $Q_8$ be the function defined by

$$Q_8(q_1, \ldots, q_{30}) = q_1^8 + q_{29}^4 + q_{30}^2 + [q_2^4 + q_3^2 q_8^2 + q_4^2 q_5^2 + q_6^2 q_{12}^2 + q_7^2 q_{13}^2]$$

$$\times [q_9^4 + (q_{10}^2 + q_{14}q_{15} + q_{18}q_{19} + q_{20}q_{21} + q_{22}q_{24})(q_{11}^2 + q_{16}q_{17} + q_{23}q_{28} + q_{25}q_{26} + q_{13}q_{27})].$$

However we obtain $Q_8(q_1, \ldots, q_{30}) = t_{19}^2$ as soon as we substitute the $q_{1..30}$ with:

| $q_1 = t_1 + t_2 t_6$ | $q_2 = t_2^2 + t_3 t_7$ | $q_3 = t_3^2 + t_4 t_{10}$ | $q_4 = t_3 t_5$ |
|---|---|---|---|
| $q_5 = t_3 t_{11}$ | $q_6 = t_4 t_7$ | $q_7 = t_4 t_5$ | $q_8 = t_7^2 + t_5 t_{11}$ |
| $q_9 = t_6^2 + t_8 t_9$ | $q_{10} = t_8^2 + t_{12} t_{13}$ | $q_{11} = t_9^2 + t_{14} t_{15}$ | $q_{12} = t_7 t_{10}$ |
| $q_{13} = t_{10} t_{11}$ | $q_{14} = t_{12}^2 + t_7 t_8$ | $q_{15} = t_{13}^2 + t_{11} t_{16}$ | $q_{16} = t_{14}^2 + t_{10} t_{12}$ |
| $q_{17} = t_{15}^2 + t_{11} t_{17}$ | $q_{18} = t_{12} t_{16}$ | $q_{19} = t_{11} t_{12}$ | $q_{20} = t_8 t_{13}$ |
| $q_{21} = t_7 t_{13}$ | $q_{22} = t_8 t_{16}$ | $q_{23} = t_{14} t_{17}$ | $q_{24} = t_7 t_{11}$ |
| $q_{25} = t_{12} t_{15}$ | $q_{26} = t_{10} t_{15}$ | $q_{27} = t_{12} t_{17}$ | $q_{28} = t_{11} t_{14}$ |
| $q_{29} = t_{18} + t_1^2$ | $q_{30} = t_{19} + t_{18}^2$ | | |

We put $n = 64$, $v = 36$, and we consider the $t_i = t_i(u_1, \ldots, u_{19})$ $(1 \leq i \leq 19)$ as randomly chosen linear forms (*i.e.* homogeneous polynomials of degree one in $u_1, \ldots, u_{19}$), satisfying the following conditions:

- $t_1(u_1, \ldots, u_{19}) = u_1$ ;

- $t_{18}(u_1, \ldots, u_{19}) = u_{18}$ ;

- $t_{19}(u_1, \ldots, u_{19}) = u_{19}$ ;

- $t_6(u_1, \ldots, u_{19})$, $t_7(u_1, \ldots, u_{19})$, $t_{18}(u_1, \ldots, u_{19})$ and $t_{19}(u_1, \ldots, u_{19})$ depend only on the variables $u_6, u_7, \ldots, u_{17}$,

We thus obtain polynomials $q_i = q_i(u_1, \ldots, u_{19})$ $(1 \leq i \leq 30)$ of degree two in $u_1, \ldots, u_{19}$. Finally, we choose:

$$
\begin{cases}
P(z_{65}, \ldots, z_{100}) = Q_8(z_{93}, \ldots, z_{100}, z_{73}, \ldots, z_{92}, z_{63}, z_{64}) \\
Q(z_{65}, \ldots, z_{100}) = Q_8(z_{65}, \ldots, z_{92}, z_{61}, z_{62}) \\
f_{61}(x_1, \ldots, x_{60}) = q_{29}(x_9, x_{11}, \ldots, x_{16}, x_{51}, \ldots, x_{62}) - x_{61} \\
f_{62}(x_1, \ldots, x_{61}) = q_{30}(x_9, x_{11}, \ldots, x_{16}, x_{51}, \ldots, x_{62}) - x_{62} \\
f_{63}(x_1, \ldots, x_{62}) = q_{29}(x_{10}, x_{17}, \ldots, x_{20}, x_{15}, x_{16}, x_{51}, \ldots, x_{60}, x_{63}, x_{64}) - x_{63} \\
f_{64}(x_1, \ldots, x_{63}) = q_{30}(x_{10}, x_{17}, \ldots, x_{20}, x_{15}, x_{16}, x_{51}, \ldots, x_{60}, x_{63}, x_{64}) - x_{64} \\
f_{65}(x_1, \ldots, x_{64}) = q_1(x_9, x_{11}, \ldots, x_{16}, x_{51}, \ldots, x_{62}) \\
\quad \vdots \\
f_{92}(x_1, \ldots, x_{91}) = q_{28}(x_9, x_{11}, \ldots, x_{16}, x_{51}, \ldots, x_{62}) \\
f_{93}(x_1, \ldots, x_{92}) = q_1(x_{10}, x_{17}, \ldots, x_{20}, x_{15}, x_{16}, x_{51}, \ldots, x_{60}, x_{63}, x_{64}) \\
\quad \vdots \\
f_{100}(x_1, ! \ldots, x_{99}) = q_8(x_{10}, x_{17}, \ldots, x_{20}, x_{15}, x_{16}, x_{51}, \ldots, x_{60}, x_{63}, x_{64})
\end{cases}
$$

and randomly chosen quadratic forms for $f_i$ $(2 \leq i \leq 60)$.

Let us denote $\theta : K^{64} \to K^{100}$ the function defined by

$$
\theta(x_1, \ldots, x_{64}) = (x_1, \ldots, x_{64}, 0, \ldots, 0).
$$

Hence $(x_1, \ldots, x_{64}) \mapsto (y_1, \ldots, y_{100}) = \Phi_3 \circ \Phi_2 \circ \theta(x_1, \ldots, x_{64})$ is given by the following system:

$$(E) \begin{cases} y_1 = x_1 + [t_{19}(x_9, x_{11}, \ldots, x_{16}, x_{51}, \ldots, x_{62})]^2 \quad (= x_1 + x_{62}^2) \\ y_2 = x_2 + f_2(x_1) + [t_{19}(x_{10}, x_{17}, \ldots, x_{20}, x_{15}, x_{16}, x_{51}, \ldots, x_{60}, x_{63}, x_{64})]^2 \\ \quad (= x_2 + f_2(x_1) + x_{64}^2) \\ y_3 = x_3 + f_3(x_1, x_2) \\ \quad \vdots \\ y_{60} = x_{60} + f_{60}(x_1, \ldots, x_{59}) \\ y_{61} = q_{29}(x_9, x_{11}, \ldots, x_{16}, x_{51}, \ldots, x_{62}) \quad (= x_{61} + x_9^2) \\ y_{62} = q_{30}(x_9, x_{11}, \ldots, x_{16}, x_{51}, \ldots, x_{62}) \quad (= x_{62} + x_{61}^2) \\ y_{63} = q_{29}(x_{10}, x_{17}, \ldots, x_{20}, x_{15}, x_{16}, x_{51}, \ldots, x_{60}, x_{63}, x_{64}) \quad (= x_{63} + x_{10}^2) \\ y_{64} = q_{30}(x_{10}, x_{17}, \ldots, x_{20}, x_{15}, x_{16}, x_{51}, \ldots, x_{60}, x_{63}, x_{64}) \quad (= x_{64} + x_{63}^2) \\ y_{65} = q_1(x_9, x_{11}, \ldots, x_{16}, x_{51}, \ldots, x_{62}) \\ \quad \vdots \\ y_{92} = q_{28}(x_9, x_{11}, \ldots, x_{16}, x_{51}, \ldots, x_{62}) \\ y_{93} = q_1(x_{10}, x_{17}, \ldots, x_{20}, x_{15}, !x_{16}, x_{51}, \ldots, x_{60}, x_{63}, x_{64}) \\ \quad \vdots \\ y_{100} = q_8(x_{10}, x_{17}, \ldots, x_{20}, x_{15}, x_{16}, x_{51}, \ldots, x_{60}, x_{63}, x_{64}) \end{cases}$$

## The Public Key

The user selects a random invertible affine transformation $\Phi_1 : K^{64} \to K^{64}$, and a random invertible affine transformation $\Phi_4 : K^{100} \to K^{100}$, such that the function $F = \Phi_4 \circ \Phi_3 \circ \Phi_2 \circ \theta \circ \Phi_1$ satisfies

$$F(0, \ldots, 0) = (0, \ldots, 0).$$

By construction of $F$, if we denote $(y_1', \ldots, y_{100}') = F(x_1', \ldots, x_{64}')$, then we have an explicit set $\{P_1, \ldots, P_{100}\}$ of 100 quadratic polynomials in 64 variables, such that:

$$\begin{cases} y_1' = P_1(x_1', \ldots, x_{64}') \\ \quad \vdots \\ y_{100}' = P_{100}(x_1', \ldots, x_{64}') \end{cases}$$

This set of 100 polynomials constitutes the public key of the TTM cryptosystem.

## Encrypting a Message

Given a plaintext $(x_1', \ldots, x_{64}') \in K^{64}$, the sender computes $y_i' = P_i(x_1', \ldots, x_{64}')$ for $1 \leq i \leq 100$ (thanks to the public key) and sends the ciphertext $(y_1', \ldots, y_{100}')$.

## Decrypting a Message

Given a ciphertext $(y_1', \ldots, y_{100}') \in K^{100}$, the legitimate receiver recovers the plaintext by:

$$(x_1', \ldots, x_{64}') = {\Phi_1}^{-1} \circ \pi \circ {\Phi_2}^{-1} \circ {\Phi_3}^{-1} \circ {\Phi_3}^{-1} \circ {\Phi_4}^{-1}(y_1', \ldots, y_{100}')$$

with $\pi : K^{100} \mapsto K^{64}$ defined by $\pi(x_1, \ldots, x_{100}) = (x_1, \ldots, x_{64})$ and thus satisfies $\pi \circ \theta = \mathrm{Id}$.

# Attacking and Repairing Batch Verification Schemes

Colin Boyd and Chris Pavlovski

Information Security Research Centre
School of Data Communications
Queensland University of Technology
Brisbane, Australia
boyd@fit.qut.edu.au, chripavl@au1.ibm.com

**Abstract.** Batch verification can provide large computational savings when several signatures, or other constructs, are verified together. Several batch verification algorithms have been published in recent years, in particular for both DSA-type and RSA signatures. We describe new attacks on several of these published schemes. A general weakness is explained which applies to almost all known batch verifiers for discrete logarithm based signature schemes. It is shown how this weakness can be eliminated given extra properties about the underlying group structure. A new general batch verifier for exponentiation in any cyclic group is also described as well as a batch verifier for modified RSA signatures.

## 1   Introduction

Modular exponentiation is a fundamental operation for most practical digital signature schemes. The computational expense of both signing and verifying signatures is mainly due to the modular exponentiation required. Several techniques have been proposed in the literature to reduce this expense, including use of small exponents, and multi-exponentiation techniques [21]. An alternative way to realize a computational reduction is through use of batch cryptography.

Batch cryptography is relevant in settings where many signatures (or other primitives) need to be generated and/or verified together. Electronic commerce applications are prime examples, as typically many customers interact with the same merchant or banking server. Although techniques have been developed to improve signature generation [6,16], the majority of the recent work in the area has focused on the batch *verification* of signatures. These techniques all exploit the homomorphic properties of exponentiation in various groups to combine a set of exponentiations into one equation whose computational effort is effectively divided amongst all the individual exponentiations required.

The purpose of this paper is to illustrate flaws in a number of published batch verifiers; in some cases they are broken whilst in others we show that they do not provide the strength of verification claimed. We show that an observation of Bellare *et al.* [1], regarding the restrictions on use of certain batch verifiers, has much more serious consequences than they imply; in most applications this

T. Okamoto (Ed.): ASIACRYPT 2000, LNCS 1976, pp. 58–71, 2000.

makes the tests ineffective. Through stronger assumptions on the group structure we show how these tests may be repaired.

## 1.1    Background

The idea of batch cryptography was introduced by Fiat [6,7]; his scheme amortized the private key operations for RSA and so was designed to assist in the signing and decryption operations. His idea was to batch a number of messages together, perform one full-scale modular exponentiation to sign the messages simultaneously, and then split apart the batch into individually signed messages. This is achievable due to the homomorphic property of RSA and the use of multiple, relatively prime, public exponents, an idea introduced by Chaum [4].

Batch verification for DSA signatures was introduced by Naccache, M'Raïhi, Raphaeli and Vaudenay [15]. Their scheme is designed to verify several DSA signatures at once by checking that a batch criterion holds and is much more efficient than sequential verification of individual DSA signatures[1]. Harn subsequently proposed a new method for DSA signatures requiring interaction between signer and verifier [10] and later devised a non-interactive version [11].

Early work concerning (non-interactive) batch verification was also published by Yen and Laih [22]. Their verification techniques are proposed for batch verification of a modification of the Schnorr or Brickell-McCurley signature schemes as well as for RSA. The principle, once again, is based upon the homomorphic properties of the respective scheme. Yen and Laih also note that to remain secure from attack, the verifier must choose random exponent values and apply these during batch verification. These values prevent the signer from attempting to introduce false signatures that would otherwise satisfy the batch verification criterion (the properties of this test are discussed in more detail in section 1.2).

Recently, Bellare, Garay, and Rabin [1,2] described several techniques for conducting batch verification of exponentiation with high confidence that false values have not been mixed into the batch. The technique which they refer to as the *small exponents test*, is very similar to the algorithms of Naccache *et al.* [15] and Yen and Laih [22], while their more sophisticated *bucket test* turns out to be more efficient for larger batch instances.

## 1.2    Batch Verification of Exponentiation

First we give a general idea of how batch verification of exponentiation works in a group. Consider the situation where we are given $n$ elements $y_1, y_2, \ldots, y_n$, all in a multiplicative group $G$, and $n$ exponents $x_1, x_2, \ldots, x_n$, all integers up to some size (we will become more specific shortly). A fixed element $g \in G$ is known. The idea of batch verification is to check that $y_i = g^{x_i}$ for each $i$ without having to make this explicit calculation $n$ times. In the case that the $x_i$ values are indeed the discrete logarithms of the respective $y_i$ values we will say that

---

[1] An earlier version of the paper of Naccache *et al.* included an additional interactive batch verifier. Lim and Lee [12] showed that this version is not secure.

the batch is *correct*. A good batch verification algorithm should identify, at least with high probability, whenever one or more of the $x_i$ values is not the discrete logarithm of the respective $y_i$.

All the known batch verification techniques are based on the multiplicative property of the group. Specifically, if the batch is correct then the following equation holds.

$$\prod_{i=1}^{n} y_i = g^{\sum_{i=1}^{n} x_i} \tag{1}$$

It is easily checked that the converse is false: if equation 1 holds then it need not be the case that the batch is correct. For example, adding a constant to one $x_i$ value and subtracting the same constant from a different $x_i$ value does not change equation 1 but invalidates the batch. Another example is where the correct $x_i$ values are randomly permuted.

Various authors [15,22] have noticed this and suggested that, to turn equation 1 into a useful batch verifier, randomisation should be introduced. This is done by multiplying the $x_i$ values by small random values which must also be introduced as small exponents for the $y_i$ values. An attacker who wishes to have an incorrect batch accepted has to anticipate which random values will be used. We follow Bellare *et al.* [1] and call this idea the *small exponents test*. The algorithm is shown in table 1. Bellare *et al.* prove that the small exponents test is a good batch verifier with error bounded by $2^{-l}$ as long as $q$, the order of the group $G$, is prime. It can be seen that the algorithm uses one full exponentiation in $G$ plus $n$ multiplications to obtain $x$ and finally the cost of the $n$ small exponentiations to find $y$. Bellare *et al.* use a multi-exponentiation algorithm to show that the total average cost is $l+n(1+l/2)$ multiplications in addition to the full exponentiation.

---

GIVEN:   $g$   a   generator   of   the   group   $G$   of   prime   order   $q$,   and $(x_1, y_1), (x_2, y_2), \ldots, (x_n, y_n)$ with $x_i \in \mathbb{Z}_q$ and $y_i \in G$. Also a security parameter $l$.
CHECK: That $\forall i \in \{1, \ldots, n\} : y_i = g^{x_i}$.

1. Pick $s_1, \ldots, s_n \in \{0,1\}^l$ at random.
2. Compute $x = \sum_{i=1}^{n} x_i s_i \bmod q$ and $y = \prod_{i=1}^{n} y_i^{s_i}$.
3. If $g^x = y$ then accept, else reject.

---

**Table 1.** Small exponents test for batch verification of exponentiation [1]

We will concentrate on the small exponents test in this paper. Bellare *et al.* also propose a variation which they call the *bucket test* which can be more efficient for large batches. Our general results apply also to the bucket test and we discuss the difference further in section 3.2.

A critical assumption in the small exponents test is that the $y_i$ values lie in the group of prime order, $G$. This rules out the case where $G$ is the multiplicative group $\mathbb{Z}_n^*$ for $n$ composite as used in RSA and related algorithms. Nevertheless, Bellare *et al.* have shown that there is a simpler form of verification, which they called *screening*, that applies to RSA signatures[2]. Screening shows that the signatures must have, at some time, been formed by the true owner of the private key even though none of the individual claimed signatures might actually be correct. Screening is sufficient in applications where it is not necessary to possess the signatures, but only to know that the messages were signed; an example might be bulk verification of certificates.

## 1.3   Central Observation and Contribution

As mentioned above, it is a requirement in the proof of correctness of the small exponents test that all operations are performed within a group $G$ of prime order. Bellare *et al.* suggest that in practice this is not really a restriction as this setting is commonplace in many modern cryptographic schemes.

They observe that when the order of $G$ is not prime the small exponents test will not work. For an example they use $G = \mathbb{Z}_p^*$, which has non-prime order $p - 1$. Let $g$ be a generator of $\mathbb{Z}_p^*$, and suppose $y = g^x \bmod p$. Under these assumptions the small exponents test will not detect the invalid batch with two pairs $(x, -y \bmod p), (x, y)$ when the small exponent for the first pair is even, which occurs with probability $1/2$. Notice that if $y$ lies in some prime order subgroup $G$ then $-y$ cannot lie in $G$.

The theme of this paper revolves around the requirement of working in a prime order group, and can be summarised in two significant observations.

1. Several authors have ignored this requirement. We give explicit attacks to show that their proposed batch verifiers do not work as advertised.
2. Even when this requirement is stated, it is not usually possible to check efficiently that it actually holds in a batch presented for verification. This makes most applications, including batch verification of DSA signatures [2,15], inappropriate unless additional properties hold.

The remainder of this paper is structured as follows. In the next section we show that the claimed strong RSA batch verifiers proposed by Yen and Laih [22] actually provide only the weaker screening property. We also present an explicit attack on the batch DSA verifiers of Harn [11], showing that an outsider can forge a batch signature for messages of his choosing. In the following section we outline a general attack that is applicable to verifiers of signatures in batches, illustrating how this may be applied to the small exponents test for batch verification of DSA signatures [2,15]. The attack allows the true signer to have false signatures accepted by the verifier. We then demonstrate how this general attack may be avoided by careful choice of the prime modulus used and give a generalised small exponents test for any cyclic group. We finally present a batch verifier for modified RSA signatures.

---

[2] Coron and Naccache [5] pointed out that screening can fail if duplicate messages are present. A modified version of screening was later proven correct by Bellare *et al.* [2].

## 2    Specific Attacks on Batch Verification Schemes

In this section we look at two schemes for batch verification which do not operate
in prime order groups. The first works with a composite modulus, while the
second performs a modular reduction before verification which destroys the group
structure. We show that in both cases the verification does not provide the
assurances claimed.

### 2.1    Yen and Laih's RSA Batch Verification

Yen and Laih [22] proposed a variation of ElGamal signatures suitable for batch
signature verification. Here we consider the RSA batch verification technique
that they devised as a performance comparison with their proposed scheme.
They have essentially proposed to use the small exponents test in the RSA multiplicative group. Specifically, suppose that $S_1, \ldots, S_n$ are claimed RSA signatures
[18] on messages $m_1, \ldots, m_n$ (where these messages have been pre-processed by
any chosen hashing and redundancy functions). If the signatures are correct then
$S_i = m_i^d \bmod N$ where $d$ is the RSA private exponent and $N$ the modulus. Small
exponents $s_1, \ldots, s_n$ are chosen randomly of length $l$. The batch verification is
then to test if the following equation holds, where $e$ is the RSA public exponent.

$$\left( \prod_{i=1}^{n} S_i^{s_i} \right)^e \equiv \prod_{i=1}^{n} m_i^{s_i} \bmod N \qquad (2)$$

Notice that this test is not as efficient as the small exponents test described
in table 1 because it is not possible for the verifier to add the exponents on
the left hand side modulo the group order. Furthermore, in practice a small
value of $e$ is often used which severely limits the benefit of batch verification.
For example, if $e = 3$ then the batch verification can never be as efficient as
individual verification of the signatures with any reasonable failure probability.
But regardless of the test's efficiency it is wrong to assume that is provides more
than screening; this means that use of the small exponents is redundant since
Bellare et al. showed that equation 2 provides screening with all $s_i = 1$, at least
in the case of full domain hashing.

The simplest attack is to replace some $S_i$ values by $-S_i$ and some $m_i$ values
by $-m_i$ (all modulo $N$). Then the test will still succeed with probability $1/2$
depending on the parity of the $s_i$ values chosen. This attack can be launched
by any party. It can be compounded by the signer who can choose an element
$\alpha$ of small order $t$ in the multiplicative group ($t$ should be smaller than $2^l$).
Any $S_i$ value can then be replaced by $\alpha S_i \bmod N$ and the test will succeed with
probability $1/t$. Note that it is easy to find such an $\alpha$ if the factorisation of $N$ is
known.

### 2.2    Harn's DSA Batch Verification

Harn [11] proposed an algorithm which is essentially a direct application of
equation 1 to variants of DSA signatures. Specifically he considers the following

signature algorithm. Primes $p$ and $q$ are chosen with $q|p-1$ and a generator $g$ of the group $G$ of order $q$ is published. A user's private key is a number $x$ in $\mathbb{Z}_q$ and the corresponding public key is $y = g^x \bmod p$. A signature of a message $m$ (again pre-processed by hashing) is a pair $(r, s)$ where both $r$ and $s$ lie in $\mathbb{Z}_q$. A claimed signature pair is correct if the following verification equation holds.

$$r = (g^{sr^{-1} \bmod q} y^{mr^{-1} \bmod q} \bmod p) \bmod q$$

Now suppose that $m_1, \ldots, m_n$ is a batch of messages with corresponding set of claimed signatures $(r_1, s_1), \ldots, (r_n, s_n)$. Applying the multiplicative property, the following equation holds, which is also the proposed batch verification test.

$$\prod_{i=1}^{n} r_i \bmod q = (g^{\sum_{i=1}^{n} s_i r_i^{-1} \bmod q} y^{\sum_{i=1}^{n} m_i r_i^{-1} \bmod q} \bmod p) \bmod q \qquad (3)$$

Our first observation is that this test can provide no more than screening. For suppose that a batch of correct signatures is known. Keep the $r_i$ values the same and then choose the $n-1$ values $s_1', \ldots, s_{n-1}'$ randomly and finally solve the equation

$$\sum_{i=1}^{n} s_i' r_i^{-1} \bmod q = \sum_{i=1}^{n} s_i r_i^{-1} \bmod q$$

to obtain the value $s_n'$. Then the batch $(r_1, s_1'), \ldots, (r_n, s_n')$ satisfies the test but almost certainly none of the signatures is correct.

Now we show that the situation is compromised even further by an explicit attack. With high probability it is possible for an attacker who is not the signer to find signatures for any chosen message set. We only need to assume that the attacker has any known signature for this scheme: this gives values $A$, $B$ and $C$ with $A = (g^B y^C \bmod p) \bmod q$. We suppose that the attacker has chosen two messages for signing, say $m_1$ and $m_2$ (the attack is easily generalised to any number of messages). The attack works by making verification equation 3 the same as for the known signature. This is done in two steps.

1. Solve for $r_1$ and $r_2$ to ensure that
$$r_1 r_2 \equiv A \bmod q$$
$$m_1 r_1^{-1} + m_2 r_2^{-1} \equiv C \bmod q.$$

2. Solve for $s_1$ and $s_2$ to ensure that

$$s_1 r_1^{-1} + s_2 r_2^{-1} \equiv B \bmod q.$$

The simultaneous equations in step 1 can be reduced to the quadratic equation $(m_2/A)r_1^2 - Cr_1 + m_1 \bmod q$ which can be solved by completion of the square as long as the discriminant $C^2 - 4m_1 m_2/A$ is a quadratic residue modulo $q$. On the assumption that $m_1$ and $m_2$ are random (they are the result of hashing) this will be the case with probability $1/2$. Step 2 can then be completed by choosing $s_1$ randomly and solving for $s_2$.

The attack can be generalised for any number of messages to be forged. In step 1 all but two of the $r_i$ values can be chosen randomly and then the remaining two found by solving a quadratic equation as described above. Step 2 proceeds as above with all but one of the $s_i$ values chosen at random. It is interesting to note that this attack will not work if random small exponents are added to the verification equation. However, since there is no security proof it would be dangerous to rely on such a test.

## 3    General Attack on the Small Exponents Test

In this section we show that the small exponents test described in table 1 is much less useful that it at first appears. We will show that many of the proposed applications for the test are, in fact, not appropriate at all.

### 3.1    Attacking Batch Verification of DSA

In order to explain the weakness we first describe the batch DSA verification proposed by Bellare $et$ $al.$ [2]. (Note that this application was not included in the shortened version of the paper published at Eurocrypt'98 [1]). As previously suggested by Naccache $et$ $al.$ [15] the verification algorithm is applied not to the original DSA signature scheme but to a slightly altered version.

The setting is again in a subgroup $G$ of $\mathbb{Z}_p^*$ of prime order $q$ where a user's private key is $x \in \mathbb{Z}_q$ with public key $y = g^x \in G$. The signature of a (pre-processed) message $m$ is a triple $(\lambda, s, m)$ which satisfies the following verification equation, where $r = \lambda \bmod q$.

$$\lambda = g^{ms^{-1} \bmod q} y^{rs^{-1} \bmod q} \bmod p$$

The difference in original DSA is that $\lambda$ is replaced by $r$, and the verification equation is reduced modulo $q$. This means that the original DSA signature is only twice the size of $q$ instead of the size of $q$ plus the size of $p$ in the revised version. Since typical sizes of $p$ and $q$ would be 1024 and 160 bits respectively, this is a significant extra overhead which might be worthwhile for the computational gains of batch verification. Note that the modified version can easily be converted into an original DSA version at any time by replacing $\lambda$ with $r$. Bellare $et$ $al.$ applied the small exponents test to a batch of modified DSA signatures as shown in table 2.

We now apply our main observation to the algorithm: at no time in the algorithm is it checked that the $\lambda_i$ values are actually within the group $G$ as they should be. Once this is observed it is straightforward to develop an attack. (In contrast to the attack in section 2.2 only the true signer can carry out this attack.) Similar to the attack on Yen-Laih's algorithm in section 2.1, the idea of the attack is to replace one or more $\lambda_i$ values by $-\lambda_i$ and the signatures will be accepted with probability $1/2$. Because the $b_i$ values in the test depend on $\lambda_i$ the attacking signer needs to choose $\lambda_i$ first and then find $s_i$. Specifically the signer proceeds as follows to run the attack with one or more of the messages $m_i$.

---

GIVEN: Public parameters $p, q, g$ a public key $y$ and a batch of claimed signatures: $(\lambda_1, s_1, m_1), \ldots, (\lambda_n, s_n, m_n)$ with $s_i \in \mathbb{Z}_q$ and $\lambda_i \in G$. Also a security parameter $l$.

CHECK: That $\forall i \in \{1, \ldots, n\} : \lambda_i = g^{m_i s_i^{-1} \bmod q} y^{r_i s_i^{-1} \bmod q} \bmod p$.

1. For $i = 1, \ldots, n$ set $a_i = s_i^{-1} m_i \bmod q$ and $b_i = s_i^{-1} \lambda_i \bmod q$.
2. Pick $w_1, \ldots, w_n \in \{0, 1\}^l$ at random.
3. Compute $A = \sum_{i=1}^{n} a_i w_i \bmod q$, $B = \sum_{i=1}^{n} b_i w_i \bmod q$, and $R = \prod_{i=1}^{n} \lambda_i^{w_i}$.
4. If $g^A y^B = R$ then accept, else reject.

---

**Table 2.** Small exponents test for batch verification of modified DSA [2]

1. Choose $k_i$ randomly in $\mathbb{Z}_q$ and set $L_i = g^{k_i} \bmod p$.
2. Set $\lambda_i = -L_i \bmod p$, $r_i = \lambda_i \bmod q$ and $s_i = k_i^{-1}(m_i + x r_i) \bmod q$.
3. Present $(\lambda_i, s_i, m_i)$ to the verifier as part of the batch.

It follows that

$$g^{m_i s_i^{-1} \bmod q} y^{r_i s_i^{-1} \bmod q} \bmod p = g^{k_i} \bmod p = L_i$$

and since $L_i^2 \equiv \lambda_i^2 \bmod p$ this will go undetected if the verifier chooses this $w_i$ to be even which happens with probability $1/2$.

As with the attack on Yen-Laih, it can be generalised by substituting $\lambda_i = \alpha L_i \bmod p$ for an element $\alpha$ with any order $t$ where $t | p - 1$ and $t \le 2^l$. Usually there will be many such $t$ values that can be chosen. Then the signature will be accepted with probability $1/t$.

We would like to emphasise that this does not invalidate the theorem proven by Bellare *et al.* regarding the security of their small exponents test since it is an assumption in table 1 that the $y_i$ values are in the group $G$. Furthermore, strictly the application is correct as long as the $\lambda_i$ values are in $G$, but this is not a reasonable assumption in practice.

### 3.2   Other Schemes Susceptible to the Attack

Several other published schemes make essentially the same unjustified assumption. An attack on the earlier DSA batch verification scheme of Naccache *et al.* [15] is identical to that proposed above. A similar attack on the batch verifier for a Schnorr signature variant proposed by Yen and Laih [22] is possible. Note that if all (or most) of the small exponents will be chosen to be odd, such as is suggested by Naccache *et al.*, the substitution should be made on an even number of $\lambda_i$ values for the attack to succeed.

Another application that is vulnerable is a recent proposal for batch verification of coins in Brands' cash scheme [17]. In this proposal the merchant essentially uses the small exponents test during the payment protocol to verify a batch of coins together; a batch test is also used by the bank at deposit time. A possible consequence of the above attack is that a customer can frame a merchant since there is a high probability that the customer can have a bad coin accepted at payment time but that it will be rejected by the bank during deposit.

The alternative *bucket test* of Bellare *et al.* [1] is also vulnerable to the same attack, since it basically consists of a series of small exponent tests run on random partitions of the batch. However, in many instances it will detect the attack with much higher probability than the small exponents test. The bucket test uses an additional parameter $m$, repeats the partitioning $\lceil l/(m-1) \rceil$ times, and runs the small exponents test with parameter $m$ in place of $l$. So a value $\lambda_i$ replaced by $-\lambda_i$ will be detected with probability $1/2$ for every repetition, or $2^{-\lceil l/(m-1) \rceil}$ overall. This is still much worse than the claimed probability of failure of $2^{-l}$.

## 4   Repairing the Small Exponents Test

An obvious way to prevent the attack is to check that the $\lambda_i$ values in table 2 are indeed in $G$, as required by the small exponents test. However, there does not appear to be any way to do this that does not totally negate the computational savings of the test. For example, to test directly that $\lambda_i^q \mod p = 1$ would require $n$ extra exponentiations. Note that it is not sufficient to check, for example, that the product of the $\lambda_i$ values are in $G$.

The main problem in ensuring that the proof still holds is to avoid elements of low order in the 'large group'. The element of order 2 is always present in $\mathbb{Z}_p^*$ so we have to accept that there may sign changes in a batch that passes the test. In this section we show that through judicious choice of $p$ it is possible to avoid any other problems.

### 4.1   Dealing with Prime Order Subgroups

First of all we assume that $p$ is chosen to be of the form $p-1 = 2rq$ where $r$ and $q$ are both primes. The modified form of the small exponents test is shown in table 3; the differences from that in table 1 are small but significant. In particular there is no assumption that the $y_i$ values lie in $G$. A consequence of this difference is that exponentiations are only known to be correct up to a possible multiple of -1. This should be acceptable in most applications since it can always be corrected if a particular value is later found to be incorrect.

The computational cost of the modified test is identical to that of table 1. Using an improved algorithm for multiexponentiation, Bellare *et al.* [1] calculated the total cost of the test as $l + n(1 + l/2)$ multiplications plus the cost of the exponentiation. The exact cost will depend on the size of the values of $p$, $q$ and $l$ (as well as the algorithms used for exponentiation and multi-exponentiation). Reasonable values today might be $|p| = 1024$, $|q| = 160$ and $l = 60$.

**Theorem 1.** *Suppose $p$ is a prime and $G$ a subgroup of $\mathbb{Z}_p^*$ of prime order $q$. If $p-1 = 2qr$ where $r$ is prime and $\min(q,r) > 2^l$ then the algorithm in table 3 is a batch verifier which fails with probability at most $2^{-l}$.*

---

GIVEN: $g$ a generator of the subgroup $G$ of $\mathbb{Z}_p^*$ and $(x_1, y_1), (x_2, y_2), \ldots, (x_n, y_n)$ with $x_i \in \mathbb{Z}_q$ and $y_i \in \mathbb{Z}_p^*$. Also a security parameter $l$.
CHECK: That $\forall i \in \{1, \ldots, n\} : \pm y_i = g^{x_i}$.

1. Pick $s_1, \ldots, s_n \in \{0, 1\}^l$ at random.
2. Compute $x = \sum_{i=1}^n x_i s_i \bmod q$ and $y = \prod_{i=1}^n y_i^{s_i}$.
3. If $g^x = y$ then accept, else reject.

---

**Table 3.** Modified small exponents test for batch verification of exponentiation in $\mathbb{Z}_p^*$

*Proof* The proof is basically similar to that of Bellare *et al.* for their small exponents test but there are a few extra problems to consider. Suppose that $g_0$ is a generator of $\mathbb{Z}_p^*$ and suppose, without loss of generality, that $g = g_0^{2r}$. We can then write $y_i = g_0^{x_i'}$ for some $x_i'$ with $1 \leq x_i' \leq p - 1$. Suppose that the test passes; then the following equation holds.

$$g_0^{2rx \bmod p-1} = g_0^{\sum_{i=1}^n x_i' s_i \bmod p-1}$$

Because $g_0$ is a generator of $\mathbb{Z}_p^*$ we have

$$2r(s_1 x_1 + \ldots + s_n x_n) \equiv x_1' s_1 + \ldots + x_n' s_n \bmod (p-1)$$

which we may re-write as the following.

$$s_1(x_1' - 2r x_1) + \ldots + s_n(x_n' - 2r x_n) \bmod (p-1) = 0 \qquad (4)$$

Suppose that for at least one value of $i$ we have $\pm y_i \neq g^{x_i}$. Without loss of generality let us assume that $i = 1$. If we suppose that the values of $s_2, \ldots, s_n$ have been chosen, then equation 4 is a linear equation in $s_1$ and the number of solutions for $s_1$ is either 0 or $\nu = (p - 1, 2r x_1 - x_1')$. Because $p - 1 = 2qr$, $\nu$ can take any of the eight values $\{1, 2, q, r, 2r, 2q, qr, 2qr\}$. But the case $\nu = 2qr$ means that $2r x_1 \equiv x_1' \bmod p - 1$ so $y_i = g^{x_i}$ which we have assumed is not true.

The next largest case is $\nu = qr$, so that we have either $2r x_1 \equiv x_1' \bmod p - 1$ or $2r x_1 + qr \equiv x_1' \bmod p - 1$. The former possibility is ruled out and the latter possibility means that $y_1 = g_0^{x_1'} = g_0^{2r x_1 + qr} = -g^{x_1}$ which is also assumed not to hold.

The remaining cases do not satisfy the check so we need to show that they occur with small probability. The next largest case is $\nu = 2r$. Although in this case there are many solutions to equation 4, these solutions are evenly distributed in the sense that if $X$ is any solution for $s_1$ then $X + q$ is also a solution. This means that there is at most one solution for $s_1$ in the range $0 \leq s_1 \leq 2^l$ since $q > 2^l$. A similar argument holds for all other possible value of $\nu$. Since $s_1$ is chosen randomly the probability that equation 4 holds when $\pm y_1 \neq g^{x_1}$ is thus at most $2^{-l}$. The same is then true if all $s_1, \ldots, s_n$ are drawn independently and randomly. $\qquad \square$

It can be seen from the proof that the requirement that $p-1 = 2rq$ is stronger than necessary. In fact it is necessary only that $p - 1$ has no factors smaller than $q$ apart from 2. Efficient methods to generate primes satisfying either of these conditions have been described by Lim and Lee [13]. They suggest that to satisfy $p-1 = 2rq$, the prime $r$ should be chosen first and then random primes $q$ of the desired size chosen until $p$ is prime. For $|p| = 1024$ only around 710 trials for $q$ will be required which is a very practical requirement.

## 4.2   Generalisation and Applications

There are a number of ways that the modified small exponents test can be extended. We give a generalised form in table 4 which applies to any cyclic group. This algorithm can only give assurance of the correctness of the batch up to multiplication by an element of order less than $2^l$. Therefore, in applications it will be useful to ensure that the group order has as few small factors as possible. The following theorem shows that the algorithm is a correct batch verifier. The proof, which is a generalisation of the proof of Theorem 1, is omitted due to space restrictions.

**Theorem 2.** *The algorithm in table 4 is a batch verifier which fails with probability at most $2^{-l}$.*

---

GIVEN: $g$ a generator of a cyclic group $H$ of order $w$ and $(x_1, y_1), (x_2, y_2), \ldots, (x_n, y_n)$ with $x_i \in \mathbb{Z}_w$ and $y_i \in H$. Also a security parameter $l$.
CHECK: That $\forall i \in \{1, \ldots, n\} : \alpha y_i = g^{x_i}$ for some element $\alpha \in H$ of order less than $2^l$.

   1. Pick $s_1, \ldots, s_n \in \{0, 1\}^l$ at random.
   2. Compute $x = \sum_{i=1}^{n} x_i s_i \bmod w$ and $y = \prod_{i=1}^{n} y_i^{s_i}$.
   3. If $g^x = y$ then accept, else reject.

---

**Table 4.** Generalised small exponents test for batch verification of exponentiation in any cyclic group

There are a number of useful applications of our modified small exponents tests.

- DSA batch verification can be achieved by adapting the algorithm of table 2 to verify the signature up to multiplication of each $\lambda_i$ by -1. The algorithm is identical except
  - we require that $p - 1$ has no factors smaller than $2^l$ apart from 2.
  - we do not assume that $\lambda_i$ is in a prime order subgroup.
  Of course the attack in section 3.1 still holds so if this verification is to be used it is necessary to adapt the DSS algorithm so that $(r, s, m)$ will be a correct signature if either of the following checks passes.

$$r = (g^{ms^{-1} \bmod q} y^{rs^{-1} \bmod q} \bmod p) \bmod q \qquad (5)$$

$$(p - r) \bmod q = (g^{ms^{-1} \bmod q} y^{rs^{-1} \bmod q} \bmod p) \bmod q \qquad (6)$$

Although it seems intuitively reasonable that this extension to DSA signatures is as secure as original DSA we do not offer any proof.

- Bellare *et al.* [1] have asked whether a batch verifier for exponentiation can be found for $\mathbb{Z}_p^*$ rather than in a prime order subgroup. The algorithm in table 4 answers this question in the affirmative (up to multiplication by -1) with the condition that $p - 1$ has no small factors apart from 2.
- The *bucket test* of Bellare *et al.* [1] is an extension of the small exponents test and it is immediate to extend our test in the same way. The computational cost will be the same as that of the original bucket test.

## 5 Batch Verification of RSA Signatures

As mentioned previously, Bellare *et al.* introduced screening as a weaker form of verification for RSA signatures. In this section we will use the ideas from the previous section to derive batch verification of a slightly modified definition of RSA signatures. This variation was already used by Gennaro *et al.* [8] in a different context. Specifically, the set of signatures on a message $m$, randomised appropriately, is defined as

$$\mathrm{SIG}(m) = \{S_m : S_m = \alpha m^d, ord(\alpha) \leq 2\}.$$

For an RSA modulus, $N = pq$, there are four possible signatures of every message. In addition to 1 and -1 there are two 'non-trivial' square roots of unity and knowledge of either of these allows $N$ to be factorised. Consequently an oracle to forge a signature in $\mathrm{SIG}(m)$ can be used to forge an ordinary RSA signature either directly or by allowing factorisation of $N$.

The next restriction we need is that $N$ should be the product of two safe primes: $N = pq$ where $(p - 1)/2$ and $(q - 1)/2$ are also prime. Since there is an efficient method to prove that $N$ is of this form [3,14] this property can be checked when the public key is certified or, if necessary, prior to the batch verification. Our batch verification algorithm for RSA is given in table 5. The proof of the following result is omitted due to space restrictions.

**Theorem 3.** *The algorithm in table 5 is a batch verifier which fails with probability at most $2^{-l}$. Its cost is approximately $l(n+2)+1.5|e|+n-1$ multiplications modulo $N$.*

Batch verification of RSA is counter-productive when $e$ is small, as the cost of conventional sequential verification of the $n$ signatures will be $1.5n|e|$ multiplications. The algorithm is worthwhile when $e$ satisfies

$$|e| \gg \frac{l(n+2)}{1.5(n-1)} + \frac{2}{3}.$$

---

GIVEN: A modulus $N$ which is the product of two primes and $(S_1, m_1), (S_2, m_2), \ldots, (S_n, m_n)$ with $S_i, m_i \in \mathbb{Z}_N$. Also a security parameter $l$ with $2^l < \min(p', q')$.

CHECK: That $\forall i \in \{1, \ldots, n\} : \alpha S_i^e = m_i$ for some element $\alpha \in \mathbb{Z}_n^*$ of order not more than 2.

1. Check that $(S_i, N) = 1$ for all $i$.
2. Pick $s_1, \ldots, s_n \in \{0,1\}^l$ at random.
3. Compute $x = \left(\prod_{i=1}^n S_i^{s_i}\right)^e \bmod N$ and $y = \prod_{i=1}^n m_i^{s_i} \bmod N$.
4. If $x = y$ then accept, else reject.

---

**Table 5.** Small exponents test for batch verification of RSA signatures

Thus for large $n$ we require that $|e| \approx 2l/3$ before the test becomes useful. Small values of $e$ such as 3 or $2^{16} + 1$ will never benefit from our batch verification. There are certain situations where a random, or large, $e$ is desirable [19]. For a random $e$ our test provides immediate gains for any reasonable size of $N$.

## 6   Conclusion

In this paper we have outlined several new attacks on batch verification techniques in the literature including a general attack on batch verification which affects most of the prominent schemes. We have shown how this attack may be avoided by careful choice of the modulus and weakening the acceptance condition. We have also provided a new batch verifier for exponentiation in any cyclic group and a batch verifier for modified RSA signatures. These results answer many of the open questions posed by Bellare *et al.* [1].

## Acknowledgements

We are grateful to Wenbo Mao for a number of helpful comments, and to the anonymous referees for constructive advice on presentation of the results.

## References

1. M. Bellare, J. A. Garay, T. Rabin, Fast Batch Verification for Modular Exponentiation and Digital Signatures, Proceedings of Eurocrypt'98, LNCS, Vol. 1403, pp.236-250, Springer-Verlag, 1998.
2. M. Bellare, J. A. Garay, T. Rabin, Fast Batch Verification for Modular Exponentiation and Digital Signatures, available online at
http://www-cse.ucsd.edu/users/mihir.
3. J.Camenisch and M.Michels, Proving in Zero-Knowledge that a Number is the Product of Two Safe Primes, Advances in Cryptology – Eurocrypt'99, Springer-Verlag, pp.107-122, 1999.

4. D. Chaum, A. Fiat, M. Naor, Untraceable electronic cash, Proceedings of Crypto '88, pp.319-227, 1988.
5. J.-S. Coron, D. Naccache, On The Security Of RSA Screening, Proceedings of PKC'99, Springer-Verlag, 1999.
6. A. Fiat, Batch RSA, Advances in Cryptology – Crypto '89, LNCS, Vol. 435, Springer-Verlag, pp.175-185, 1990.
7. A. Fiat, Batch RSA, Journal of Cryptology, 10, 2, pp.75-88, 1997.
8. R. Gennaro, H. Krawczyk and T. Rabin, RSA-Based Undeniable Signatures, Advances in Cryptology – Crypto'97, Springer-Verlag, 1997, pp.132-149.
9. A.J.Menezes, P.C. van Oorschot and S.A. Vanstone, Handbook of Applied Cryptography, CRC Press, 1996.
10. L. Harn, DSA Type Secure Interactive Batch Verification Protocols, Electronics Letters, 31, 4, pp.257-258, 16th February 1995.
11. L. Harn, Batch Verifying Multiple DSA-type Digital Signatures, Electronics Letters, 34, 9, pp.870-871, 30th April 1998.
12. C.H. Lim and P.J. Lee, Security of Interactive DSA Batch Verification, Electronics Letters, 30, 19, pp.1592-1593, 15th September, 1994.
13. C.H. Lim and P.J. Lee, A Key Recovery Attack on Discrete Log-based Schemes Using a Prime Order Subgroup, Advances in Cryptology – Crypto'97, Springer-Verlag, 1997, pp. 249-262.
14. T.V. Le, K.Q. Nguyen and V. Varadharajan, How to Prove That a Committed Number is Prime, Advances in Cryptology – Asiacrypt'99, Springer-Verlag, 1999, pp.208-218.
15. D. Naccache, D. M'Raïhi, D. Raphaeli, S. Vaudenay, Can DSA be improved: complexity trade-offs with the digital signature standard, Proceedings of Eurocrypt'94, pp.85-94, 1994.
16. D. M'Raïhi and D. Naccache, Batch Exponentiation – A Fast DLP-Based Signature Generation Strategy, 3rd ACM Conference on Computer and Communications Security, pp.58-61, 1996.
17. C. Pavlovski, C. Boyd, E. Foo, Detachable Electronic Coins, Proceedings of ICICS '99, LNCS, Vol. 1726, Springer-Verlag, pp.54-70, 1999.
18. R. Rivest, A. Shamir and L. Adleman, A method for obtaining digital signatures and public key cryptosystems, Comm. ACM, pp.120-126, Vol. 21, No. 2, 1978.
19. H.-M. Sun, W.-C. Yang and C.-S. Laih, On the Design of RSA with Short Secret Exponent, Advances in Cryptology - Asiacrypt'99, Springer-Verlag, 1999, pp.150-164.
20. Y. Yacobi, M. Beller, Batch Diffie-Hellman Key Agreement Systems and their Application to Portable Communications, Proceedings of Eurocrypt 92, Vol. 658, pp.208-217, 1992.
21. S.M.Yen, C.S.Laih and A.K.Lenstra, Multi-exponentiation, IEE Proceedings, Part E: Computers and Digital Techniques, 141, 6, pp.325-326, 1994.
22. S. Yen, C. Laih, Improved Digital Signature Suitable for Batch Verification, IEEE Transactions on Computers, Vol. 44, No. 7, pp.957-959, July 1995.

# Cryptography Everywhere
## (IACR Distinguished Lecture)

Thomas A. Berson

Anagram Laboratories
P.O. Box 791
Palo Alto, CA 94302, USA
Xerox Palo Alto Research Center
3333 Coyote Hill Rd, Palo Alto, CA 94304 USA
berson@anagram.com

# 1   Abstract

The past twenty years have seen cryptography move from arcane to common-place, from difficult to easy, from expensive to cheap. Many influences are at work. These include: the professionalization of cryptographers, in which the IACR has played a significant role; the creation of textbooks and of courses; the steady growth of computational power delivered by the operation of Moore's Law; the algorithmic advances made by cryptographic researchers and engineers; the rise of e-commerce and wireless infrastructures which have a seemingly endless appetite for cryptographic services; the entry of many young people into the field; and the easing of government export controls. We envisage a near future where cryptographic operations will be as pervasive, cheap and unremarkable as IP protocol operations have become today.

Some things about this future are already clear. Cryptographic operations will disappear into the infrastructure. The complexities of cryptography and of cryptographic key management will be hidden from users. New sorts of protocols will become practical. New sorts of businesses will be possible. We will describe several such protocols and businesses. Other important aspects of this future are less clear, such as the social, economic, and political implications. We will hazard guesses at these and other impacts of cryptography everywhere.

# 2   Pointer to Further Detail

Further materials may be found at
http://www.anagram.com/berson/ac2000.html.

T. Okamoto (Ed.): ASIACRYPT 2000, LNCS 1976, pp. 72–72, 2000.

# Security of Signed ElGamal Encryption

Claus Peter Schnorr[1] and Markus Jakobsson[2]

[1] Fachbereich Mathematik/Informatik,
Universität Frankfurt, PSF 111932,
D-60054 Frankfurt am Main, Germany.
schnorr@cs.uni-frankfurt.de
[2] Information Sciences Laboratory,
Bell Laboratories Murray Hill, New Jersey 07974.
markusj@research.bell-labs.com

**Abstract.** Assuming a cryptographically strong cyclic group $G$ of prime order $q$ and a random hash function $H$, we show that ElGamal encryption with an added Schnorr signature is secure against the *adaptive chosen ciphertext attack*, in which an attacker can freely use a decryption oracle except for the target ciphertext. We also prove security against the novel *one-more-decryption attack*. Our security proofs are in a new model, corresponding to a combination of two previously introduced models, the Random Oracle model and the Generic model. The security extends to the distributed threshold version of the scheme. Moreover, we propose a very practical scheme for private information retrieval that is based on blind decryption of ElGamal ciphertexts.

## 1 Introduction and Summary

We analyse a very practical public key cryptosystem in terms of its security against the strong *adaptive chosen ciphertext attack* (CCA) of [RS92], in which an attacker can access a decryption oracle on arbitrary ciphertexts (except for the target ciphertext.) Let a *signed ElGamal encryption* of a message be an ElGamal ciphertext together with a Schnorr signature of that ciphertext — the public signature key is given by the ElGamal ciphertext. We prove that this scheme is secure against generic attacks where both the group $G$ and the random hash function $H$ are black boxes.

*The traditional versus the new security model.* Assuming a strong cyclic group $G$ and a random hash function $H$ we prove tight bounds on the success probability of a generic attacker performing some $t$ generic steps. Our approach has practical consequences. It yields very practical cryptographic schemes that are provably secure in a reasonable, new security model, the *random oracle and generic model* (ROM+GM). The ROM goes back to FIAT AND SHAMIR [FS86] and has been further enhanced by BELLARE AND ROGAWAY [BR93], while the generic model (GM) goes back to NECHAEV [Ne94] and SHOUP [Sh97]. We introduce the combination of these two models, the result of which seems to cover all practical attacks at hand. Namely, security in ROM+GM allows a separation of

T. Okamoto (Ed.): ASIACRYPT 2000, LNCS 1976, pp. 73–89, 2000.

potential weaknesses of the group $G$, the hash function $H$ and the cryptographic protocols using $G$ and $H$. It allows a modular replacement of weak hash functions or groups without forcing changes to the cryptographic protocols. Whereas the security guarantees of most efficient groups and hash functions are merely heuristics based on the absence of known attacks, we obtain tight bounds on the success of arbitrary generic attacks. While we do not have to rely on any unproven assumption, it is the case that our security guarantees hinge on the existence of strong hash functions $H$ and groups $G$ for which the combination $(G, H)$ has no weaknesses. On the other hand, we do *not* assume that the discrete logarithm (DL) problem or to the Diffie-Hellman problem is hard — our security proof contains a hardness proof of the DL-problem in the generic model.

The new ROM+GM is a powerful tool for proving security against interactive attacks. In this paper we merely consider encryption. For security in ROM+GM of Schnorr signatures — in particular security of blind signatures against the one-more signature forgery — see [SJ99]. Recently, it has been shown [Sc00] that the generation of secret DL-keys from short random seeds through a strong hash function is secure in GM.

*Notions of security.* Let $G$ be a cyclic group of prime order $q$ with generator $g$, and let $\mathbf{Z}_q$ be the field of integers modulo $q$. A Diffie-Hellman key pair consists of a random secret key $x \in \mathbf{Z}_q$ and the corresponding public key $h = g^x \in G$. Diffie-Hellman keys give rise to many cryptographic schemes, for example ELGAMAL *encryption* [E85]. An ElGamal ciphertext of message $m \in G$ is a pair $(g^r, mh^r) \in G^2$ for random $r \in \mathbf{Z}_q$. ElGamal encryption is *indistinguishable* [GM84] — it is secure against a passive, merely eavesdroping adversary. Formally, an attacker, given distinct messages $m_0, m_1$ and a corresponding target ciphertext $cip_b$ for random $b \in \{0, 1\}$, cannot guess $b$ better than with probability $\frac{1}{2}$. However, ElGamal encryption is completely insecure against various active attacks, where a decryption oracle can be used under appropriate conditions.

A powerful active attack is the CCA-attack of RACKOFF AND SIMON [RS92]. CCA-security means indistinguishability against an adversary that can freely use a decryption oracle except for the target ciphertext. DOLEV, DWORK AND NAOR [DDN91] propose another notion of security against active attacks, called *non-malleability*. Here the adversary — which is given a decryption oracle — tries to create another ciphertext that is related in an interesting way to the target ciphertext. Non-malleability and CCA-security have been shown to be equivalent [DDN98].

*Previous work.* The public key encryption schemes of SHOUP, GENNARO [SG98], CRAMER, SHOUP [CS98], ABDALLA, BELLARE, ROGAWAY [ABR98], FUJISAKI, OKAMOTO [FO99], SHOUP [Sh00] and ZHENG, SEBERRY [ZS92] all extend variants of ElGamal encryption by an added signature or tag. This idea first appears in [ZS92] without a security proof. CCA-security has been proved in [SG98, CS98, ABR98, FO99, Sh00]. The schemes in [SG98, CS98, ABR98, Sh00] either use an involved tag construction or key generation to simplify the reduction to the discrete log or to the Diffie-Hellman problem, the tag in [ABR98] uses symmetric encryption. We consider the very practical, signed extension of ElGamal encryption, which was independently proposed by TSIOUNIS AND YUNG [TY98] and JAKOBSSON [J98]. Herein, an ElGamal ciphertext $(g^r, mh^r)$ is completed by a SCHNORR *signature* [Sc91] providing a proof of knowledge of the plaintext $m$ and of the secret $r$ — the public signature key $g^r$ is given by the ciphertext. CCA-security of this *signed ElGamal encryption* has been shown in [TY98] under the assumption that the signer really "knowns" the secret signature key $r$. That assumption holds in the ROM if there is only a logarithmic number of interactions with the decryption oracle.[1]

*Our results.* We "validate" the [J98,TY98]-assumption that the signer really "knows" the secret key $r$ in the ROM+GM. We give a plaintext extractor, and we prove security against a generic CCA-attacker performing some number $t = o(\sqrt{q})$ of interactions and generic group steps. A CCA-attacker can freely use a decryption oracle except for the target ciphertext. We show that a generic CCA-attacker using $t$ generic steps, and given distinct messages $m_0, m_1$, a target ciphertext $cip_b$ for random $b \in_R \{0,1\}$, cannot predict $b$ with probability better than $\frac{1}{2} + t^2/q$. This probability is over the random hash function $H$, the random public encryption key $h$, the coin tosses of the encipherer, and the random bit $b$. This bound is almost tight, as a generic attacker, given the public key $h$, can compute the secret decryption key with probability $\binom{t}{2}/q$ in $t$ generic steps. This result improves the known security guarantees for signed ElGamal encryption. Moreover, our security proofs extend to a straightforward distributed threshold version of signed ElGamal encryption, see [SG98] for the threshold setting.

Furthermore, we introduce the *one-more decryption attack* and we show that signed ElGamal encryption is secure against this attack. In the one-more decryption attack the adversary attempts to partially decrypt $\ell + 1$ ciphertexts by asking a decryption oracle some $\ell$ times. The new attack is not covered by the adaptive chosen ciphertext attack, as the latter relates to a single target ciphertext. Interestingly, security against the one-more attack follows from plaintext awareness (PA) as defined in [BR94]. Proving PA is the core of the proof of Theorem 1 and 2. [2] For motivation of the one-more decryption attack, we propose a practical scheme for private information retrieval. It is based on blind decryption and security against the *random one-more attack* — which is a weak version of the one-more decryption attack.

*Generalized (signed) ElGamal encryption.* Finally, we propose a more general variant of (signed) ElGamal encryption with two major advantages. *Firstly,* for long messages our generalized encryption is very fast and its data expansion rate approaches 1. *Secondly,* the generalized encryption does not require messages to

---

[1] The FFS-extractor of Feige-Fiat-Shamir, in the oracle replay mode of POINTCHEVAL AND STERN [PS96], extracts the secret signature key from signed ElGamal encryptions. The FFS-extractor has a constant delay factor, and thus can in polynomial time at most be iterated a logarithmical number of times.

[2] It seems that PA is the most important security notion for encryption. E.g., [BDPR98] show that PA and IND-CPA imply CCA-security while the converse does not hold. PA requires the ROM, security proofs without assuming the ROM do not prove PA.

be encoded into the group generated by the public key $h$.[3] Let the message space be an arbitrary additive group $M$, e.g., $M = \mathbf{Z}_q^n$ for some $n$. Let a generalized ElGamal ciphertext be a pair $(g^r, m + H_M(h^r))$ for random $r \in \mathbf{Z}_q$, where $H_M : G \to M$ is a random hash function. We then add a Schnorr signature (using the public signature key $g^r$) to the ciphertext $(g^r, m + H_M(h^r)) \in G \times M$. This signed generalized ElGamal encryption has provably the same security as signed ElGamal encryption, without any further assumptions.

*The structure of the paper.* In Section 2, we introduce the generic model for interactive algorithms that use a hash oracle and an oracle for decryption. We propose a setup for the GM that slightly differs from the [Sh97] proposal in that we do not assume a random binary encoding of group elements. We examplify the difference of the two setups for the baby-step-giant-step DL-algorithm. While our generic algorithms do not allow for efficient sorting of group elements this does not affect the number of generic steps as equality tests of group elements are free of charge.

In Section 3, we review signed ElGamal encryption, which is based on the original ElGamal encryption. Moreover, we generalize the original and the signed ElGamal encryption. Then we introduce the main tools for proving security in the GM. We show in Lemma 1 and 2 that a collision-free, non-interactive generic attacker $\mathcal{A}$ gets no information on the secret random data — the secret key, the random number $r$, etc. — except that $\mathcal{A}$ observes the absence of collisions. Lemma 1 bounds the probability for non-trivial collisions. This bound also covers the leakage of secret information through the absence of collisions.

Section 4 presents the proof of CCA-security of signed ElGamal encryption in the ROM+GM. It gives a generic extractor that extracts the signature key $\bar{r} = \log_g \bar{h}$ from a signed ElGamal ciphertext $(\bar{h}, \bar{f}, c, z)$, produced by the attacker. We also prove security against the one-more decryption attack. We motivate this novel attack by interesting services for trading encrypted data.

## 2    The Random Oracle and the Generic Model

**The Random Oracle Model (ROM).** Let $G$ be a group of prime order $q$ with generator $g$, a range $M$ of messages, and let $\mathbf{Z}_q$ denote the field of integers modulo $q$. Let $H$ be an *ideal* hash function with range $\mathbf{Z}_q$, modelled as an oracle that given an input (query) in $G \times M$, outputs a random number in $\mathbf{Z}_q$. Formally, $H$ is a random function $H : G \times M \to \mathbf{Z}_q$ chosen at random over all functions of that type with uniform probability distribution. There is an ongoing debate on whether the assumption of a random hash function is realistic or too generous. The problem is that random functions can in principle not be implemented by

---

[3] Encoding of arbitrary bit sequences into sequences of group elements is easy for particular groups such as $\mathbf{Z}_q^*$ that correspond to an interval of integers. For general groups, even for subgroups of $\mathbf{Z}_N^*$ or subgroups of elliptic curves, an encoding into group elements is impractical. Known extensions of ElGamal encryption — see e.g., [MOV] section 8.26 — do not solve this encoding problem.

public algorithms. CANETTI, GOLDREICH, HALEVI [CGH98] present an artificial "counter-example" that is provably secure in the ROM but which cannot be implemented in a secure way replacing the random oracle by a computable function family.[4] Nevertheless, the security achievable in the ROM seems to in practice eliminate all attacks at hand.

**The Generic Model (GM).** Generic algorithms for $G$ do not use the binary encodings of the group elements, as they access group elements only for group operations and equality tests. NECHAEV [Ne94] proves that the discrete logarithm problem is hard in such a model. The generic model of algorithms was further elaborated on by SHOUP [Sh97]. We present the Shoup model in a slightly different setup[5] and we extend it to algorithms that interact with a decryption oracle. Encryptions are for the private/public key pair $(x, h)$, where $x$ is random in $\mathbf{Z}_q$ and $h = g^x$. We describe the extended generic model in detail, first focusing on non-interactive algorithms and thereafter on algorithms interacting with oracles for hashing and decryption.

The *data of a generic algorithm* is partitioned into group elements in $G$ and non-group data. The *generic steps* for group elements are multivariate exponentiations:

- mex: $\mathbf{Z}_q^d \times G^d \to G$, $(a_1, ..., a_d, g_1, ..., g_d) \mapsto \prod_i g_i^{a_i}$ with $d \geq 0$.

The cases $d = 2, a_1 = 1, a_2 = \pm 1$ present multiplication/division. The case $d = 0$ presents *inputs* in $G$ — e.g., $g$, $h$ are inputs for the DL-computation.

**Def.** A (non-interactive) *generic algorithm* is a sequence of $t$ generic steps[6]

- $f_1, \ldots, f_{t'} \in G$ (inputs) $1 \leq t' < t$,

- $f_i = \prod_{j=1}^{i-1} f_j^{a_j}$ for $i = t' + 1, \ldots, t$, where $(a_1, \ldots, a_{i-1}) \in \mathbf{Z}_q^{i-1}$ depends arbitrarily on $i$, the non-group input and the set $\mathcal{CO}_{i-1} := \{(j, k) \mid f_j = f_k, 1 \leq j < k \leq i - 1\}$ of previous *collisions* of group elements.

Typical non-group inputs are represented by elements in $\mathbf{Z}_q$ — which we assume to be given — various integers in $\mathbf{Z}_q$ contained in given ciphertexts or signatures. $\mathcal{CO}_t$ is the set of all collisions of the algorithm.

---

[4] In [CGH98] a mechanism for the implementation of random hash functions has been added to the ROM. The artificial "counter-example" is defined relative to that mechanism using the function ensemble that implements the random oracle.

[5] We count the same generic steps as in [Sh97]; however, we allow arbitrary multivariate exponentiations while Shoup merely uses multiplication and division. The technical setup in [Sh97] looks different as groups $G$ are *additive* and associated with a random injective encoding $\sigma : G \to S$ of the group $G$ into a set $S$ of bit strings — the generic algorithm performs arbitrary computations on these bit strings. Addition/subtraction is done by an oracle that computes $\sigma(f_i \pm f_j)$ when given $\sigma(f_i), \sigma(f_j)$ and the specified sign bit. As the encoding $\sigma$ is random it contains only the information about which group elements coincide — this is what we call the set of *collisions*.

[6] We can allow a generic algorithm to perform a number $t$ of generic steps, where $t$ varies with the input. We can let the algorithm decide after each step whether to terminate depending arbitrarily on the given non-group data. Then the number $t$ of generic steps depends on the computed non-group data.

Some group inputs $f_i$ depend on random coin flips, e.g., the random public key $h = g^x$ depends on the random secret key $x \in_R \mathbf{Z}_q$. The *probability space* consists of the random group elements of the input. The logarithms $\log_g f_i$ of the random inputs $f_i$ play the role of *secret parameters*. Information about the secret parameters can only be revealed by collisions. E.g., $g^a = f_i^b$ implies $\log_g f_i = a/b$. We let the non-group input and the generator $g$ not depend on random bits.

The *output* of a generic algorithm consists of

- non-group data that depend arbitrarily on the non-group input and on the set $\mathcal{CO}_t$ of all collisions,

- group elements $f_{\sigma_1}, \ldots, f_{\sigma_d}$ where the integers $\sigma_1, \ldots, \sigma_d \in \{1, \ldots, t\}$ depend arbitrarily on the non-group input and on $\mathcal{CO}_t$.

For the sake of clairifying the GM, we give an example of a generic algorithm:

**The baby-step-giant-step DL-algorithm.** This algorithm is given $q$ and $g, h \in G$ and computes $\log_g h \in \mathbf{Z}_q$ in $2\sqrt{q}$ generic steps.

**1.** Compute $k := \lceil \sqrt{q} \rceil$, $l := \lceil q/k \rceil$ so that $lk - k < q \leq lk$. The computation of the non-group data $k, l$ is for free.

**2.** Form the lists $L_1 := \{g^i \mid 0 \leq i < k\}$ in $k-1$ multiplications and $L_2 := \{hg^{jk} \mid 0 \leq j < l\}$ in $l$ multiplications. Clearly, $L_1 \cap L_2 \neq \emptyset$.

**3.** Find a collision by testing all equalities $g^i = hg^{jk}$. Note that the detection of the collision is for free. An equality implies $\log_g h = i - jk \bmod q$.

While this algorithm performs $\#L_1 \times \#L_2$ "free" equality tests, the corresponding Turing machine — in the [Sh97]-setup — constructs a collision differently, using only $O(\sqrt{q} \log_2 q)$ equality tests. It sorts the binary encodings of the $g^i$ and inserts the encodings of $hg^{jk}$ into the sorted list.

Going back to the description of the model we work in, we now elaborate on **interactive, generic algorithms**. We count the following generic steps :

- group operations,     $\mathrm{mex}: \mathbf{Z}_q^d \times G^d \to G$,     $(a_1, ..., a_d, g_1, ..., g_d) \mapsto \prod_i g_i^{a_i}$,

- queries to the hash oracle $H$,

- interactions with a decryption oracle (*decryptor* for short) — see 3.1 [7].

A *generic adversary* $\mathcal{A}$ — attacking an encryption scheme — is an interactive algorithm that interacts with a decryptor. It performs some $t$ generic steps resulting in $t' \leq t$ group elements $f_1, ..., f_{t'}$. $\mathcal{A}$ iteratively selects the next generic step — a group operation, a query to $H$, an interaction with the decryptor — depending arbitrarily on the non-group input and on previous collisions of group elements.

The *input* consists of the generator $g$, the public key $h \in G$, the group order $q$, a collection of messages and ciphertexts and so on, all of which can be broken down into group elements and non-group data.

The computed *group elements* $f_1, ..., f_{t'} \in G$ are the group elements contained in the input, such as $g, h$. When counting the number of group operations, we

---

[7] Other types of interactions are possible for other signature/encryption schemes, other cryptographic protocols using groups of non-prime order, groups of unknown order or using several distinct groups.

count each input as one operation. As a decryptor interaction is counted as a generic step the number $t'$ of group elements is bounded by the number $t$ of generic steps, $t' \leq t$. We have $t = t'$ for a non-interactive $\mathcal{A}$.

The given *non-group data* consists of the non-group data contained in the input, the previous hash replies $H(Q)$ of queries $Q$, and the set of previous collisions of group elements.

A *decryptor interaction* (defined in subsection 3.1) is a two round deterministic protocol. A claimed ciphertext is send to the decryptor, which performs a generic group operation using the secret decryption key $x$, verifies the Schnorr signature using the public key $g^r$ contained in the ciphertext, and — in case that this signature is correct — outputs the decrypted message. If the signature is invalid the decryptor outputs a random element of $G$. $\mathcal{A}$'s interactions with the decryptor are sequential as the interleaving of these two-round interactions is necessarily trivial.

$\mathcal{A}$'s *output* and *transmission* to the decryptor consists of non-group data $NG$ and previously computed group elements $f_\sigma$, where $NG$ and $\sigma$, $1 \leq \sigma \leq t'$, depend arbitrarily on given non-group data.

$\mathcal{A}$'s *transmission* to the hash oracle $H$ depends arbitrarily on given group elements and given non-group data. The *probability space* consists of the random $H$ and the random input group elements.

The *restriction of the generic model* is that $\mathcal{A}$ can use group elements only for generic group operations, equality tests and for queries to the hash oracle, whereas non-group data can be arbitrarily used without charge. The computed group elements $f_1, ..., f_{t'}$ are given as explicit multiplicative combinations of group elements in the input and from decryptor interactions. Let the group elements in the input and from decryptor interactions be $g_1, ..., g_\ell$. By induction on $j$, a computed $f_j \in G$ is of the form $f_j = g_1^{a_{j,1}} ... g_\ell^{a_{j,\ell}}$, where the exponents $a_{j,1}, ..., a_{j,\ell} \in \mathbf{Z}_q$ depend arbitrarily on given non-group data. $\mathcal{A}$ can arbitrarily use the coefficients $a_{j,1}, ..., a_{j,\ell}$ from this explicit representation of $f_j$. A generic adversary is deterministic, which is not a restriction as its coin flips would be useless.[8]

*Trivial collisions.* We call a collision $(i, j) \in \mathcal{CO}_t$ trivial if $f_i = f_j$ holds with probability 1, i.e., if it holds for all choices of the secret data such as the secret key $x$ and the random bits $r$ of the encipherer. We write $f_i \equiv f_j$ for a trivial collision.[9] Trivial collisions do not release any information about the secret data while non-trivial collisions can completely release some secret data.

---

[8]  $\mathcal{A}$ could select interior coin flips that maximize the probability of success — there is always a choice for the internal coin flips that does not decrease $\mathcal{A}$'s probability of success. It is useless for $\mathcal{A}$ to generate random group elements — in particular ones with unknown DL. Using one generic step, $\mathcal{A}$ could replace random elements in $G$ by some deterministic $g^a$ where $a \in \mathbf{Z}_q$ is chosen as to maximize the probability of success.

[9]  Trivial collisions occur in testing correctness of an ElGamal ciphertext $(g^r, mh^r)$ and its message $m$. In case of a correct message-ciphertext pair the test results in a trivial collision. Also, identical repetitions of a group operation yield a trivial collision.

Trivial collisions can be ignored, and so, we can exclude them from $\mathcal{CO}_t$ so that $\mathcal{CO}_t$ consists only of non-trivial collisions.

# 3   Signed ElGamal Encryption, Non-interactive Attacks

We define Schnorr signatures, based on an ideal hash function $H : G \times M \to \mathbf{Z}_q$, where $M$ is the set of messages. Hereafter we define signed ElGamal encryption as well as the generalized concepts of the original and of signed ElGamal encryption.

   Lemma 1 and 2 are our main tools for proving security in GM. These show — for a collision-free attacker — that the secret data $x, r$, etc. are stat. indep. of all non-group data. There is, however, a minor leakage of secret information as the secret data are not perfectly random in the absence of collisions. We show in Prop. 2 that ElGamal encryption is indistinguishable (or semantically secure) against generic non-interactive attacks. Prop. 2 is part of the CCA-security proof of Theorem 1.

   *Private/public key for signatures.* The *private key* $x$ is random in $\mathbf{Z}_q$. The corresponding *public key* $h = g^x \in G$ is random in $G$, $x = \log_g h$.

   A SCHNORR *signature* on a message $m$ is a triple $(m, c, z) \in M \times \mathbf{Z}_q^2$ such that $H(g^z h^{-c}, m) = c$. In order to *sign* a message $m \in M$, pick a random $r \in_R \mathbf{Z}_q$, compute $g^r$, $c := H(g^r, m)$ and $z := r + cx$. Output the *signature* $(m, c, z)$.

   In order to *verify* a signature $(m, c, z)$ check that $H(g^z h^{-c}, m) = c$. The signing protocol produces a correct signature since $g^z h^{-c} = g^{r+cx} h^{-c} = g^r$.

## 3.1   Definition of Signed ElGamal Encryption

The private/public key pair for encryption is $x$, $h = g^x$ where $x$ is random in $\mathbf{Z}_q$. The basic encryption scheme is for messages in $M = G$, ElGamal ciphertexts are in $G \times M$, the added Schnorr signature signs pairs in $G \times M$ and uses a random hash function $H : G^2 \times M \to \mathbf{Z}_q$. We also propose a generalized scheme, where the message space $M$ is an arbitrary additive group.

   In order to *encipher* a message $m \in G$, we pick random $r, s \in_R \mathbf{Z}_q$, compute $g^r$, $m\,h^r$, $c := H(g^s, g^r, m h^r)$ and $z := s + cr$ and output the *ciphertext* $(g^r, m h^r, c, z) \in G^2 \times \mathbf{Z}_q^2$.

   A *decryption oracle (decryptor)* is a function that decrypts valid ciphertexts: The user sends a claimed ciphertext $(\bar{h}, \bar{f}, c, z)$ to the decryptor. The decryptor checks that $H(g^z \bar{h}^{-c}, \bar{h}, \bar{f}) = c$ and sends, if that test succeeds, $m := \bar{f}/\bar{h}^x$ to the user. If the test fails the decryptor sends a random message in $G$. For simplicity, we disregard the impact of that random message to the probability.

   The decryption is correct as $\bar{h} = g^r$, $\bar{f} = m\,h^r$ yields $\bar{f}/\bar{h}^x = m\,g^{rx}g^{-rx} = m$.

**Remarks 1.** A signed ciphertext $(g^r, m h^r, c, z)$ consists of an ElGamal ciphertext $(g^r, m h^r)$ and a Schnorr signature $(c, z)$ of the "message" $(g^r, m h^r)$ for the public signature key $g^r$. The signature $(c, z)$ does not contain any information about $m$ as $(c, z)$ depends on $m$ exclusively via some hash value that is statistically independent of $m$.

**2. Threshold Distributed Version.** The validity of the ciphertext $(\bar{h}, \bar{f}, c, z)$ is tested prior to and separate from decryption. Hence, the security properties of the scheme are preserved in the more general setting of threshold cryptography, see [SG98]. It is possible for a distributed entity to perform the decryption in a controlled manner after each server first having verified that indeed the decryption is allowed i.e., that the signature in the ciphertext is valid. If this were not locally verifiable, it would make a threshold decryption severely more complex.

**3. Comparison with other secure DL-cryptosystems.** We count the number of exponentiations per encryption/decryption and the number of on-line exp. per enc. (exponentiations not depending on the message). [10]

|                       | exp./enc. | on-line/enc. | exp./dec. |
|-----------------------|:---------:|:------------:|:---------:|
| Signed ElGamal enc.   | 3         | 0            | 2         |
| [FO99] El Gamal       | 2         | 2            | 2         |
| [ABR 98]              | 2         | 0            | 1         |
| [CS98], [Sh00]        | 4         | 1            | 2         |
| [SG98], TDH1, TDH2    | 5         | 2            | 5         |

The relative efficiency of [FO99], [ABR98] is due to the usage of further cryptographic primitives. [FO99] uses private encryption, [ABR98] uses private encryption and message authentication code. Signed ElGamal encryption and TDH1, TDH2 of [SG98] are amenable to a secure distributed threshold decryption. Signed EG-encryption and the [FO99] EG-scheme are plaintext aware. Signed ElGamal encryption virtually combines all the good properties.

**Generalized (signed) ElGamal encryption.** Let the message space $M$ be an arbitrary additive group, e.g., $M = \mathbf{Z}_q^n$. Let $H : G^2 \times M \to \mathbf{Z}_q$ be a random hash function and let $H_M : G \to M$ be a second random hash function that is statistically independent of $H$. Then replace in the basic encryption scheme $mh^r \in G$ by $m + H_M(h^r) \in M$.

The generalized ElGamal ciphertext is $(g^r, \bar{f})$, where $\bar{f} = m + H_M(h^r)$, the generalized signed ElGamal ciphertext is $(g^r, \bar{f}, c, z)$, and $c = H(g^s, g^r, \bar{f})$, $z = s + cr$. Decrypt a signed ciphertext $(\bar{h}, \bar{f}, c, z)$ into $\bar{f} - H_M(\bar{h}^x)$ provided that the signature $(c, z)$ of $(\bar{h}, \bar{f})$ is correct, i.e., $H(g^z \bar{h}^{-c}, \bar{h}, \bar{f}) = c$.

For $M = \mathbf{Z}_q^n$ the bit length of the ciphertext is $\log_2 \|G\| + (n+2)\log_2 q$, the message is $n \log_2 q$ bits long and $\|G\|$ is the bit length of the group elements. The data expansion rate is $1 + \frac{2}{n} + \frac{\log_2 \|G\|}{n \log_2 q}$ which is near to 1 for large $n$.

The short generalized ciphertexts are as secure as the original ones. Encryption requires only a long[11] and a short hash as well as a long and a short addition. The three exponentiations $g^r, h^r, g^s$ can be done beforehand.

---

[10] We count an expression $g^z \bar{h}^{-c}$ as 1 exponentiation even though it is slightly more expensive than a full exponentiation.

[11] Long hash values in $\mathbf{Z}_q^n$ can be generated using a random hash function $H_M : G \to \mathbf{Z}_q^n$ according to the following, or some related, approach: $(H_M(f, 1), \ldots, H_M(f, n))$.

## 3.2    Basic Tools for Proving Security in GM

This subsection studies to a generic, non-interactive adversary $\mathcal{A}$ that performs some $t$ generic steps in attacking the indistinguishability of ElGamal encryption. Given $q$, the public key $h = g^x$, two messages $m_0, m_1 \in G$ and an ElGamal ciphertext $cip_b = (g^r, m_b h^r)$ for random $r, x \in_R \mathbf{Z}_q$ and $b \in_R \{0, 1\}$, $\mathcal{A}$ guesses $b$. We show that $\mathcal{A}$ does not succeed better than with probability $\frac{1}{2} + 2\binom{t}{2}/q$.

The probability space consists of the random group elements $g^r, g^x, m_b g^{rx}$, or equivalently of the random $r, x \in_R \mathbf{Z}_q$ and $b \in_R \{0, 1\}$. Let $\mathcal{A}$ compute the group elements $f_1, ..., f_t$. We let the *Main Case* be the part of the probability space where there are no non-trivial collisions among $f_1, ..., f_t$, i.e., $\mathcal{CO}_t = \emptyset$.

**Lemma 1.** *Non-trivial collisions among* $f_1, ..., f_t$ *occur at most with probability* $2\binom{t}{2}/q$. *The probability refers to the random* $b, r, x$.

*Proof.* In order to prove the claim we show for $i < j$, $f_i \not\equiv f_j$, for constant $b$ and random $r, x \in \mathbf{Z}_q$ that $\Pr_{r,x}[f_i = f_j] \leq \frac{2}{q}$. This implies

$$\Pr_{r,x}[\mathcal{CO}_t \neq \emptyset] \leq \sum_{1 \leq i < j \leq t} \Pr_{r,x}[f_i = f_j] \leq 2\binom{t}{2}/q.$$

The input group elements are $g, g^r, h, m_b h^r, m_0, m_1$. Let $\log_g m_0, \log_g m_1$ be given, then all computed group elements are explicit combinations of $(g_1, g_2, g_3, g_4) = (g, g^r, h, m_b h^r)$, thus $f_j = \prod_{\nu=1}^4 g_\nu^{a_{j,\nu}}$ where the exponents $a_{j,1}, ..., a_{j,4} \in \mathbf{Z}_q$ depend arbitrarily on given non-group data, but not on $b, r, x$. Consider $r, x$ as formal variables over $\mathbf{Z}_q$. Then $\log_g f_j$ is a polynomial in $\mathbf{Z}_q[r, x]$ of the form $a_{j,1} + a_{j,2}r + a_{j,3}x + a_{j,4}(\log_g m_b + rx)$. The *difference polynomial* $\log_g f_i - \log_g f_j \in \mathbf{Z}_q[r, x]$ has total degree $d \geq 1$ as we assume that trivial collisions have been eliminated. Importantly, trivial collisions do not depend on $b$. [12] As $1 \leq d \leq 2$, the probability that $f_i(r, x) = f_j(r, x)$ for random $r, x$ is[13] at most $\frac{2}{q}$, thus proving the claim. Here we use a Lemma attributed to SCHWARTZ [Sch80] [14]  □

*The leakage of secret information through the absence of collisions.* Here we pay attention to the fact that $b, r, x$ are not perfectly random if $\mathcal{CO}_t = \emptyset$. By Lemma 1 a $2\binom{t}{2}/q$-fraction of the probability space is excluded in the Main Case. The SHANNON entropy of the secret parameters $b, r, x$ decreases accordingly. We can neglect this minor leakage of secret information through the absence of collisions. Thus, for a "collision-free" attacker the secret data are statistically indepependent of the computed non-group data:

---

[12] The formal polynomial $\log_g f_i - \log_g f_j \in \mathbf{Z}_q[r, x]$ is of the form $c_1 + c_2 r + c_3 x + c_4(\log_g m_b + rx)$. The coefficients $c_1, ..., c_4 \in \mathbf{Z}_q$ only depend on $q$ and previous non-trivial collisions. If $f_i \equiv f_j$ holds for some $b \in \{0, 1\}$ then $c_4 = 0$ and $f_i \equiv f_j$ holds for all $b \in \{0, 1\}$. Hence the identity $f_i \equiv f_j$ does not depend on $b$.

[13] The factor 2 disappears if $m_b h^r$ is removed from the input — then the difference polynomial has total degree at most 1.

[14] Lemma [Sch80] A multivariate polynomial $F \in \mathbf{Z}_q[X_1, ..., X_k]$ of total degree $d$ satisfies for random $x_1, ..., x_k \in \mathbf{Z}_q$ that $\Pr_{x_1,...,x_k}[F(x_1, ..., x_k) = 0] \leq d/q$.

**Lemma 2.** *In the Main Case the random $b, r, x$ are stat. indep. of the computed non-group data except that the $b, r, x$ leading to collisions are excluded.*

*Proof.* The random $b, r, x$, enter into the generic computation only via the group elements $g^r, g^x, m_b g^{rx}$. Therefore, $b, r, x$ enter into non-group data only via nontrivial collisions of group elements. □

**Proposition 1.** Generic DL-Complexity Lower Bound [Ne94,Sh97]. *Let $\mathcal{A}$, upon input $g$ and $h = g^x \in_R G$, output $y \in \mathbf{Z}_q$. Then $\Pr_x[y = \log_g h] \leq \binom{t}{2}/q + \frac{1}{q}$.*

*Proof.* We use Lemma 1 and 2 for a generic $\mathcal{A}$ with input $g, h$ — without inputs $g^r, m_b h^r$. The factor 2 in Lemma 1 disappears as the polynomials $\log_g f_j$ have total degree $\leq 1$. For a collision-free $\mathcal{A}$, $x$ is statistically independent of the non-group output $y$, and thus $\Pr_h[y = \log_g h] = \frac{1}{q}$. By Lemma 1, non-trivial collisions occur at most with probability $\binom{t}{2}/q$. □

**Proposition 2.** Indistinguishability. *Let a generic, non-interactive $\mathcal{A}$ be given $g, h$, two messages $m_0, m_1 \in G$ and a ciphertext $(g^r, m_b h^r)$ for random $r \in_R \mathbf{Z}_q$ and $b \in_R \{0, 1\}$. Let $\mathcal{A}$ output a guess $b'$ for $b$. Then $\Pr_{b,x,r}[b' = b] \leq \frac{1}{2} + 2\binom{t}{2}/q$.*

*Proof.* In the Main Case $b, r, x$ are stat. indep. of the non-group output $b'$, thus $\Pr_{b,x,r}[b' = b] = \frac{1}{2}$. The Main Case occurs except with probability $2\binom{t}{2}/q$. □

**Extension 1.** Obviously Prop.2 extends to generalized ElGamal ciphertexts $(g^r, m + H_M(h^r))$ for a random function $H_M : G \to M$. Whereas $\mathcal{A}$ can arbitrarily use the hash values $H_M(f_1), ..., H_M(f_t)$ of the computed group elements these hash values are statistically independent random numbers except for collisions $f_i = f_j$.

**Extension 2.** Prop. 2 extends to signed ElGamal encrytion and to generalized signed ElGamal encrytion. This is because the added Schnorr signature does not contain any information about the plaintext.

## 4    Security Against Interactive Attacks

We study the security of signed ElGamal encryption in ROM+GM. Signed El-Gamal encryption was independently proposed by TSIOUNIS AND YUNG [TY98] and JAKOBSSON [J98]. We show in Theorem 1 that this scheme is indistinguishable against the adaptive chosen ciphertext attack (CCA). This is equivalent to non-malleability against CCA [DDN98]. We refer to non-malleability as defined in [DDN98] and to the strong chosen ciphertext attack proposed by RACKOFF AND SIMON [RS92]. The adversary has access to a decryption oracle which can be used arbitrarily except for the target ciphertext.

Moreover, we introduce the one-more-decryption attack and we show in Theorem 2 that signed ElGamal encryption is secure against this attack. An adversary can — after some $\ell$ interactions with the decryption oracle — not decrypt more than $\ell$ ciphertexts. More precisely, he gets non-negligible information about at most $\ell$ encrypted plaintexts. The core of the proof of Theorems 1 and 2 shows that signed ElGamal encryption is plaintext aware. Therefore, the attackers decryption requests for self-constructed ciphertexts can be eliminated.

Theorem 1 proves indistinguishability against a CCA-adversary $\mathcal{A}$. The adversary is given a target ciphertext $cip_b$ and a decryption oracle for the decryption of arbitrary ciphertexts except for $cip_b$. The attack is called adaptive because the queries to the decryption oracle may depend on the challenges and their corresponding answers. We let the generic adversary $\mathcal{A}$ perform some $t$ generic steps: group operations, inputs in $G$, queries to the oracle $H$, and queries to the decryption oracle not including the target ciphertext.

**Theorem 1.** *Let the attacker $\mathcal{A}$ be given $g, h$, distinct messages $m_0, m_1$, a target ciphertext $cip_b$ corresponding to $m_b$ for a random bit $b \in_R \{0,1\}$, and oracles for $H$ and for decryption. Then a generic $\mathcal{A}$ using $t$ generic steps cannot predict $b$ with a better probability than $\frac{1}{2} + t^2/q$. The probability space consists of the random $x, H, b$ and the coin tosses $r$ of the encipherer.*

*Proof.* We present a generic extractor $\mathcal{E}$ that extracts the secret key $\bar{r} = \log_g \bar{h}$ from a signed ciphertext $(\bar{h}, \bar{f}, c, z)$ produced by $\mathcal{A}$. Given $\bar{r}, \bar{h}, \bar{f} = \bar{m}\bar{h}^{\bar{r}}$ the plaintext $\bar{m}$ can be extracted in one generic step. Thus, signed ElGamal encryption is — in a generic way — plaintext aware as defined in [BR94].

Let $(\bar{h}, \bar{f}, c, z)$ be the first claimed ciphertext that $\mathcal{A}$ transmits to the decryptor. $\mathcal{A}$ has produced it without interacting with the decryptor. Let this non-interactive, generic computation compute group elements $f_1, ..., f_{t'}$, $t' \leq t$. By Lemma 1 non-trivial collisions among $f_1, ..., f_{t'}$ occur with probability no more than $2\binom{t}{2}/q$. By Lemma 2 the secret $b, r, x$ are statistically independent of the non-group data of a collision-free computation of $(\bar{h}, \bar{f}, c, z)$.

In the ROM the equation $c = H(g^z \bar{h}^{-c}, \bar{h}, \bar{f})$, required for a valid signature, necessitates that $\mathcal{A}$ selects $c$ from given hash values $H(f_\sigma, f_j, \bar{f})$ for given group elements $f_\sigma$, $f_j = \bar{h}$, $\bar{f}$. Otherwise, the equation $c = H(g^z \bar{h}^{-c}, \bar{h}, \bar{f})$ holds with probability $\frac{1}{q}$ as $H$ is random. $\mathcal{A}$ gets $c = H(f_\sigma, f_j, \bar{f})$ from the hash oracle[15] and must compute $z$ so that $g^z \bar{h}^{-c} = f_\sigma$, i.e., $\mathcal{A}$ must compute $z = \log_g(f_\sigma f_j^c)$.

The computed $z$ does not depend on $x, r$ whereas $\log_g(f_\sigma f_j^c)$ may depend. We distinguish between the two values as follows: We let $z' := \log_g(f_\sigma f_j^c)$ denote the value required for a signature, whereas the computed $z$ is from $\mathcal{A}$'s transmission $(f_j, \bar{f}, c, z)$.

Let the target ciphertext be $cip_b = (g^r, m_b h^r, c_b, z_b)$, where the random $r, x \in_R \mathbf{Z}_q$, $b \in_R \{0,1\}$ are secret and $h = g^x$. Let $\log_g m_0, \log_g m_1$ be given, then $\mathcal{A}$'s group steps refer to the given group elements $(g_1, g_2, g_3, g_4) := (g, g^r, h, m_b h^r)$. $\mathcal{A}$ computes $f_i := \prod_{\nu=1}^4 g_\nu^{a_{i,\nu}}$ for $i = 1, ..., t'$ using exponents $a_{i,1}, ..., a_{i,4} \in \mathbf{Z}_q$ that arbitrarily depend on given non-group data, but not on $b, r, x$. Hence $z'$ is of the form

---

[15] $\mathcal{A}$'s choice of $c, \sigma$ is determined by the claimed ciphertext $(\bar{h}, \bar{f}, c, z)$ via $f_\sigma = g^z \bar{h}^{-c}$.

$$z' = \log_g(f_\sigma f_j^c)$$
$$= a_{\sigma,1} + c\,a_{j,1} + (a_{\sigma,2} + c\,a_{j,2})\,r + (a_{\sigma,3} + c\,a_{j,3})\,x + (a_{\sigma,4} + c\,a_{j,4})(\log_g m_b + rx).$$

Considering $r, x$ as formal variables over $\mathbf{Z}_q$, $z'$ is a polynomial in $\mathbf{Z}_q[r, x]$. The random $b, c, r, x$ are statistically independent of $a_{\sigma,1}, ..., a_{\sigma,4}, a_{j,1}, ..., a_{j,4}$.

Obviously $z'$ has total degree $d = 0$ if and only if $a_{\sigma,k} + c\,a_{j,k} = 0$ for $k = 2, 3, 4$. If the total degree $d$ is non-zero then $1 \le d \le 2$, and thus $z' = z(c)$ holds with probability at most $\frac{2}{q}$ for random $r, x$ and arbitrary functions $z(c)$. There are two subcases of the case $d = 0$: either $f_j = g^{a_{j,1}}$, $f_\sigma = g^{a_{\sigma,1}}$ or $a_{\sigma,k} = -c\,a_{j,k}$ for $k = 2, 3, 4$. The second case occurs with probabilty $\le \frac{1}{q}$ as the hash value $c$ is statistically indepependent of $a_{\sigma,k}, a_{j,k}$.

Thus, a collision-free $\mathcal{A}$ succeeds not better than with probability $\frac{3}{q}$ in generating a correct signature $(c, z)$ except that $\mathcal{A}$ sets $f_j = g^{a_{j,1}}$, $f_\sigma = g^{a_{\sigma,1}}$. So, let the extractor $\mathcal{E}$ compute $\bar{r} := a_{j,1}$ by mimicking $\mathcal{A}$'s computation of $\bar{h} = f_j$.

*Eliminating all interactions with the decryptor.* The plaintext corresponding to $(\bar{h}, \bar{f}, c, z)$ is $\bar{f}/h^{\log_g \bar{h}} = \bar{f}/h^{a_{j,1}}$ except for a probability $\frac{3}{q}$. This eliminates the first interaction with the decryptor and the call for $H(f_\sigma, f_j, \bar{f})^{16}$ using one generic step for computing $\bar{f}/h^{a_{j,1}}$. This decreases the number of generic steps and reduces $\mathcal{A}$'s probability of success by at most $\frac{3}{q}$. Let there be $\ell$ interactions with the decryptor. We iteratively eliminate them by the above method.[17] This transforms $\mathcal{A}$ into a non-interactive generic $\mathcal{A}'$ that performs $t - \ell$ generic steps. Proposition 2 applies to the non-interactive $\mathcal{A}'$, because the Schnorr signature in $cip_b$ is useless for decryption.[18] Also, the oracle $H$ is useless without a decryptor. Thus, the non-interactive $\mathcal{A}'$ predicts $b$ with a probability not exceeding $(t - \ell)^2/q + \frac{1}{2}$. This proves Theorem 1 as $(t - \ell)^2 + 3\ell \le (t - \ell + \ell)^2 = t^2$ for $t - \ell \ge 3$. Note that $t - \ell \ge 4$ due to the input group elements $g, g^r, h, m_b h^r$. $\square$

Theorem 1 can easily be extended to the one-more decryption attack.

**Theorem 2.** *Let the attacker $\mathcal{A}$ be given $g, h$, ciphertexts $cip_1, ..., cip_d$, the corresponding messages $m_1, ..., m_d$ in random order and oracles for $H$ and for decryption. Let the generic $\mathcal{A}$ perform $t$ generic steps including some $\ell < d$ arbitrary queries to the decryption oracle. Then $\mathcal{A}$ cannot produce $\ell + 1$ message-ciphertext pairs with a probability better than $\frac{1}{d-\ell} + t^2/q$. The probability space consists of the random $x, H$, the coin tosses of the encipherer and the random ordering of the messages.*

---

[16] The transformed $\mathcal{A}$ gets the plaintext $\bar{f}/h^{a_{j,1}}$ of the first decryptor interaction without using the signature and its hash value required for the decryption request. If $\mathcal{A}$ does not get $c$ from the oracle $H$ we remove the call for decrypting $(\bar{h}, \bar{f}, c, z)$ decreasing the number of generic steps and decreasing $\mathcal{A}$'s probability of success by at most $\frac{1}{q}$.

[17] This iterative elimination is impossible in the ROM without assuming the GM, see footnote 1.

[18] The signature contained in a ciphertext does not reveal any information about the message $m$. The signature depends on $m$ exclusively via the hash value $c$ that is statistically independent of $m$.

*Proof.* We have shown that signed ElGamal encryption is plaintext aware, and the attacker can only construct ciphertexts corresponding to known plaintexts. In particular, the adversary $\mathcal{A}$ can be transformed into a generic adversary $\mathcal{A}'$ that does not query the decryptor about any self-constructed ciphertext, performs $t$ generic steps and succeeds essentially with the same probability as $\mathcal{A}$. $\mathcal{A}'$ can only query the decryption oracle about $\ell$ of the input ciphertexts. These $\ell$ decryptions give no information about the $d - \ell$ remaining input ciphertexts. This is because the random bits of the ciphertexts are stat. indep. We can therefore eliminate the $\ell$ decryptions and the resulting $\ell$ message-ciphertext pairs. This transforms $\mathcal{A}'$ into a non-interactive adversary where the argument of Lemma 1 applies. Consider the impact of a random permutation of the remaining $d - \ell$ messages for a collision-free attacker. By Lemma 2 the random permutation is statistically independent of $\mathcal{A}'$'s guess of a correct message-ciphertext pair. Therefore, $\mathcal{A}'$ cannot guess a correct pair with a probability better than $\frac{1}{d-\ell}$. By Lemma 1 non-trivial collisions occur with probability at most $2\binom{t}{2}/q$, hence the claim.  $\square$

**Trading encrypted information.** Suppose a user wants to buy sensitive digital information, e.g., digital music, videos, pictures, stock market analysis, etc. Let the digital information be freely accessible in encrypted form in a public data bank. For simplicity, let each encrypted package cost \$1. Let the users have access to a public decryption oracle that charges \$1 per decryption. For the security of such trade of encrypted information the encryption scheme must be secure against the one-more decryption attack.

This type of service does not require CCA-security. However, it would be nice to have an encryption that allows for blind decryption so that no information is revealed in a decryptor interaction. Blind decryption guarantees anonymity of the buyer of digital information. It is well known that the original ElGamal ciphertexts allow for blind decryption.[19] Even though, ElGamal encryption is insecure against the one-more decryption attack we show below that it is secure against the weaker *random one-more attack*, where the enciphered plaintexts are statistically independent messages — e.g. secret keys that are unknown to the attacker.

**Efficient scheme for private information retrieval (PIR).** Let the information packages $m_i$ of the public data bank be each encrypted under a private key $k_i$ of a secure symmetric encryption scheme. Let $m_i$ contain a content description $descr_i$ of $m_i$ and a signed ElGamal ciphertext $cip(k_i) = (g^r, k_i h^r, c, z)$ of the key $k_i \in G$. Let $(c, z)$ be a signature of $(g^r, k_i h^r, descr_i)$ with public key $g^r$. Suppose a user wants to anonymously buy $\ell$ packages $m_i$ of his choice. He checks the Schnorr signature $(c, z)$ of $cip(k_i) = (g^r, k_i h^r, c, z)$ in package $m_i$ and stops if the signature is invalid. Otherwise, he blinds the ElGamal ciphertext $(g^r, k_i h^r)$ into $(g^{r+s}, uk_i h^{r+s})$ for random $s \in \mathbf{Z}_q$, $u \in G$, and asks the decryption oracle to decrypt $(g^{r+s}, uk_i h^{r+s})$. As the blinded ciphertext is statistically

---

[19]  Blind decryption of the ElGamal ciphertext $(g^r, mh^r)$: The user picks random $u \in G$ and $s \in \mathbf{Z}_q$ and asks for decryption of $(g^{r+s}, umh^{r+s})$. He gets $m$ from the plaintext $um$ transmitted by the decryptor by multiplication with $u^{-1}$.

independent of $(g^r, k_i h^r)$ no information is revealed about which $k_i$ he gets. As the user pays for $\ell$ decryptions it is important that he cannot get $\ell + 1$ keys $k_i$.

**Security against the random one-more attack.** Consider the above PIR for random $k_i \in G$. Clearly, $\ell + 1$ keys $k_i \in_R G$ have SHANNON *entropy* $(\ell + 1) \log_2 q$. But each decryption reveals no more than $\log_2 q$ bits of a plaintext in $G$, $|G| = q$. Thus, $\ell$ decryptions cannot reveal $\ell + 1$ statistically independent keys $k_i$.

Another application would be an electronic service for delivering sensitive, possibly unpleasant messages like court orders, summons, admonitions and so on. Such messages can be sent in encrypted form, given access to a decryption oracle that combines the decryption with an acknowledgement of the receipt of the decrypted message. This makes sure that a recipient can only read the message by acknowledging receipt. For such a service it would be important that the encryption is CCA-secure, so that the receipt correctly specifies the revealed message. However, we also need security against the one-more decryption attack as users may want to decrypt several ciphertexts. Signed ElGamal encryption can be used for such a service.

**Security of Schnorr signatures with short hash values.** Let the hash values of $H$ be random in an interval $[0, 2^k[ \subset [0, q[ \cong \mathbf{Z}_q$. The size of that interval enters into the proof of Theorem 1 merely at the point, where we argue that the case $a_{\sigma,k} = -ca_{j,k}$ for $k = 2, 3, 4$ has probability $\leq \frac{1}{q}$. For random hash values $c \in_R [0, 2^k[$ that case has probability $\leq 2^{-k}$.

Consequently, in the case of Theorem 1 a CCA-attacker does not succeed better than with probability $\frac{1}{2} + t^2/q + \ell(2^{-k} - \frac{1}{q})$, where $\ell$ is the number of decryptor interactions. This shows that random hash values can securely range over a set of $\sqrt{q}$ values.

**Security of Schnorr Signatures in the ROM+GM.** The proof of Theorem 1 contains a security proof for Schnorr signatures in the ROM+GM:

**Corollary 1.** *Let $A$ be a generic algorithm that is given $g$, the public signature key $h \in_R G$ and a random hash oracle. Using $t$ generic steps — group operations and hash queries — $A$ cannot produce a Schnorr signature with a probability better than $\frac{3}{q} + \binom{t}{2}/q$. The probability space consists of the random $h, H$.*

**Security against the chosen message attack.** Corollary 1 extends to the case that the adversary $A$ has a signature oracle and can ask the oracle for signatures on messages of its choice. An interaction with the signature oracle is counted as generic step. The goal of the attack is to generate a new signature which is not produced by the signature oracle. The proof of the extension is straightforward.

Unlike the case of Theorems 1 and 2, Corollary 1 and its extension have a counterpart in the ROM without assuming the GM, see POINTCHEVAL AND STERN [PS96]. Howewer, the security theorems and their proofs in the ROM use completely different arguments — the probability bounds are less tight.

# References

[ABR98]    *M. Abdalla, M. Bellare and P. Rogaway*: DHES: An Encryption Scheme Based on the Diffie-Hellman Problem. Contributions to P1363, ftp: //stdgbbs.ieee.org/pub/p1363/contributions/aes-uhf.ps

[BDPR98]   *M. Bellare, A. Desai, D. Pointcheval and P. Rogaway*: Plaintext Awareness, Non-Malleability, and Chosen Ciphertext Security: Implications and Separations. Crypto'98, LNCS 1462, pp. 26–45, 1998.

[BL96]     *D. Boneh and R.J. Lipton*: Algorithms for black-box fields and their application in cryptography. Crypto'96, LNCS 1109, pp. 283–297, 1996.

[BR93]     *M. Bellare and P. Rogaway*: Random Oracles are Practical: a Paradigms for Designing Efficient Protocols. 1st ACM Conference on Computer Communication Security, pp. 62–73, 1993.

[BR94]     *M. Bellare and P. Rogaway*: Optimal Asymmetric Encryption. Eurocrypt'94, LNCS 950, pp. 92–111, 1995.

[CGH98]    *R. Canetti, O. Goldreich and S. Halevi*: The Random Oracle Methodology, Revisited. STOC'98, ACM Press, pp. 209–218, 1998.

[CS98]     *R. Cramer and V. Shoup*: A Practical Public Key Cryptosystem Provably Secure against Adaptive Chosen Ciphertext Attack. Crypto'98, LNCS 1462, pp. 13–25, 1998.

[DDN91]    *D. Dolev, C. Dwork and M. Naor*: Non-Malleable Cryptography. STOC'91, ACM Press pp. 542–552, 1991.

[DDN98]    *D. Dolev, C. Dwork and M. Naor*: Non-Malleable Cryptography. Manuscript (updated, full length version of STOC paper), 1998.

[E85]      *T. ElGamal*: A Public Key Cryptosystem and a Signature Scheme Based on Discrete Logarithms. IEEE Trans. Inform. Theory, 31, pp. 469–472, 1985.

[FO99]     *E. Fujisaki and T. Okamoto*: Secure Integration of Asymmetric and Symmetric Encryption Schemes. Crypto'99, LNCS 1666, pp. 537–554, 1999.

[FFS88]    *U. Feige, A. Fiat and A. Shamir*: Zero-knowledge proofs of identity. J. Cryptology, 1 , pp. 77–94, 1988.

[FS87]     *A. Fiat and A. Shamir*: How to Prove Yourself: Practical Solutions of Identification and Signature Problems. Proc. Crypto'86, LNCS 263, pp. 186–194, 1987.

[GM84]     *S. Goldwasser and S. Micali*: Probabilistic Encryption. J. Computer and System Sciences, 28, pp. 270–299,1984.

[J98]      *M. Jakobsson*: A Practical Mix. Eurocrypt'98, LNCS 1403, pp. 448–461, 1998.

[MOV96]    *A. Menezes, P. van Oorschot and S. Vanstone*: Handbook of Applied Cryptography. CRC Press, Inc., 1996.

[Ne94]     *V.I. Nechaev*: Complexity of a Determinate Algorithm for the Discrete Logarithm. Mathematical Notes 55, pp. 165-172, 1994.

[RS92]     *C. Rackoff and D.R. Simon*: Non-Interactive Zero-Knowledge Proof of Knowledge and Chosen Ciphertext Attack. Crypto'91, LNCS 576, pp. 433–444, 1992.

[Sch80]    *J. Schwartz*: Fast probabilistic algorithms for verification of polynomial identities. J. ACM, 27(4), pp. 701–717, 1980.

[Sc91]     *C.P. Schnorr*: Efficient Signature Generation for Smart Cards. Journal of Cryptology 4 (1991), pp. 161–174.

[SJ99]     *C.P. Schnorr and M. Jakobsson*: Security of Discrete Log Cryptosystems in the Random Oracle and Generic Model. TR report University Frankfurt and Bell Laboratories 1999.

[Sc00]    *C.P. Schnorr* : Small Generic Hardcore Subsets for the Discrete Logarithm: Short Secret DL-Keys. Presented at rump session of Eurocrypt'2000.

[Sh97]    *V. Shoup* : Lower Bounds for Discrete Logarithms and Related Problems. Eurocrypt'97, LNCS 1233, pp. 256-266, 1997.

[Sh00]    *V. Shoup* : Using Hash Functions as a Hedge against Chosen Ciphertext Attack. Eurocrypt'2000, LNCS 1807, pp. 275–288, 2000.

[SG98]    *V. Shoup and R. Gennaro* : Securing Threshold Cryptosystems against Chosen Ciphertext Attacks. Eurocrypt'98, LNCS 1404, pp. 1–16, 1998.

[TY98]    *Y. Tsiounis and M. Yung*, On the Security of ElGamal Based Encryption. PKS'98, LNCS 1431, pp. 117-134, 1998.

[ZS92]    *Y. Zheng and J. Seberry*, Practical Approaches to Attaining Security against Adaptively Chosen Ciphertext Attacks. Crypto'92, LNCS 740, pp. 292-304, 1992.

# From Fixed-Length to Arbitrary-Length RSA Padding Schemes

Jean-Sébastien Coron[1], Francois Koeune[2], and David Naccache[3]

[1] Ecole Normale Supérieure
45 rue d'Ulm
Paris, F-75005, France
coron@clipper.ens.fr
[2] UCL Crypto Group
Bâtiment Maxwell, place du Levant 3
Louvain-la-Neuve, B-1348, Belgium
fkoeune@dice.ucl.ac.be
[3] Gemplus Card International
34 rue Guynemer
Issy-les-Moulineaux, F-92447, France
david.naccache@gemplus.com

**Abstract.** A common practice for signing with RSA is to first apply a hash function or a redundancy function to the message, add some padding and exponentiate the resulting padded message using the decryption exponent. This is the basis of several existing standards.

In this paper we show how to build a secure padding scheme for signing arbitrarily long messages with a secure padding scheme for fixed-size messages. This focuses more sharply the question of finding a secure encoding for RSA signatures, by showing that the difficulty is not in handling messages of arbitrary length, but rather in finding a secure redundancy function for short messages, which remains an open problem.

**Key words :** Signature scheme, provable security, padding scheme.

## 1 Introduction

Since the discovery of public-key cryptography by Diffie and Hellman [4], one of the most important research topics has been the design of practical and provably secure cryptosystems. A proof of security is usually a computational reduction between breaking the cryptosystem and solving a well established problem such as factoring large integers, computing the discrete logarithm modulo a prime $p$ or extracting a root modulo a composite integer. RSA [10] is based on this last problem.

A common practice for signing with RSA is to first apply a hash (or a redundancy) function to the message $m$, add some padding and raise the padded message to the decryption exponent. This is the basis of numerous standards such as ISO/IEC-9796-1 [6], ISO 9796-2 [7] and PKCS#1 v2.0 [8].

T. Okamoto (Ed.): ASIACRYPT 2000, LNCS 1976, pp. 90–96, 2000.

Many padding schemes have been designed and many have been broken (see [9] for a survey). The Full Domain Hash (FDH) scheme and the Probabilistic Signature Scheme (PSS) [2] were among the first practical and provably secure signature schemes. Those schemes are provably secure in the random oracle model [1], in which the hash function is assumed to behave as a truly random function.

However, security proofs in the random oracle model are not "real" proofs, and can be only considered as heuristic, since in the real world the random oracle is replaced by a function which can be computed by all parties. A recent result by Canneti, Goldreich and Halevi [3] shows that a security proof in the random oracle does not necessarily imply security in the "real world".

In this paper we do not model hash functions as random oracles nor assume the existence of collision-resistant hash-functions. Instead, we assume the existence of a secure deterministic padding function $\mu$ for signing fixed-length message and show how to build a secure padding scheme for signing arbitrarily long messages. This focuses more sharply the question of finding a secure encoding for RSA signatures, by showing that the difficulty is not in handling messages of arbitrary length, but rather in finding a secure redundancy function for short messages, which remains an open problem.

# 2 Definitions

## 2.1 Signature Schemes

The digital signature of a message $m$ is a string which depends on $m$ and on some secret known only to the signer, in such a way that anyone can check the validity of the signature. The following definitions are based on [5].

**Definition 1 (signature scheme).** *A signature scheme is defined by the following :*

*- The key generation algorithm* Generate *is a probabilistic algorithm which given* $1^k$, *outputs a pair of matching public and secret keys,* $\{pk, sk\}$.

*- The signing algorithm* Sign *takes the message $M$ to be signed and the secret key* sk *and returns a signature $x = $* $\mathsf{Sign}_{sk}(M)$. *The signing algorithm may be probabilistic.*

*- The verification algorithm* Verify *takes a message $M$, a candidate signature $x'$ and the public key* pk. *It returns a bit* $\mathsf{Verify}_{pk}(M, x')$, *equal to 1 if the signature is accepted, and 0 otherwise. We require that if $x \leftarrow$* $\mathsf{Sign}_{sk}(M)$, *then* $\mathsf{Verify}_{pk}(M, x) = 1$.

## 2.2 Security of Signature Schemes

The security of signature schemes was formalized in an asymptotic setting by Goldwasser, Micali and Rivest [5]. Here we use the definitions of [2] which provide a framework for the concrete security analysis of digital signatures. Resistance against adaptive chosen-message attacks is considered : a forger $\mathcal{F}$ can dynamically obtain signatures of messages of its choice and attempts to output a valid forgery. A *valid forgery* is a message/signature pair $\{M, x\}$ such that $\mathsf{Verify}_{\mathsf{pk}}(M, x) = 1$ whilst the signature of $M$ was never requested by $\mathcal{F}$.

**Definition 2.** *A forger $\mathcal{F}$ is said to $(t, q_{sig}, \epsilon)$-break the signature scheme* $\{\mathsf{Generate}, \mathsf{Sign}, \mathsf{Verify}\}$ *if after at most $q_{sig}$ signature queries and $t$ processing time, it outputs a valid forgery with probability at least $\epsilon$.*

**Definition 3.** *A signature scheme $\{\mathsf{Generate}, \mathsf{Sign}, \mathsf{Verify}\}$ is $(t, q_{sig}, \epsilon)$-secure if there is no forger who $(t, q_{sig}, \epsilon)$-breaks the scheme.*

## 2.3   The RSA Cryptosystem

RSA [10] is the most widely used public-key cryptosytem. It may be used to provide both secrecy and digital signatures.

**Definition 4 (The RSA cryptosystem).** *RSA is a family of trapdoor permutations. It is specified by :*

*- The RSA generator $\mathcal{RSA}$, which on input $1^k$, randomly selects 2 distinct $k/2$-bit primes $p$ and $q$ and computes the modulus $N = p \cdot q$. It randomly picks an encryption exponent $e \in \mathbb{Z}^*_{\phi(N)}$ and computes the corresponding decryption exponent $d$ such that $e \cdot d = 1 \bmod \phi(N)$. The generator returns $\{N, e, d\}$.*

*- The encryption function $f : \mathbb{Z}^*_N \to \mathbb{Z}^*_N$ defined by $f(x) = x^e \bmod N$.*

*- The decryption function $f^{-1} : \mathbb{Z}^*_N \to \mathbb{Z}^*_N$ defined by $f^{-1}(y) = y^d \bmod N$.*

## 2.4   The Standard RSA Signature Scheme

Let $\mu$ be a padding function taking as input a message of size $k + 1$ bits and returning an integer of size $k$ bits. We consider in figure 1 the classical RSA signature scheme $\{\mathsf{Generate}, \mathsf{Sign}, \mathsf{Verify}\}$ which signs fixed-length $k + 1$-bits messages.

# 3   The New Construction

We construct in figure 2 a new signature scheme $\{\mathsf{Generate'}, \mathsf{Sign'}, \mathsf{Verify'}\}$ using function $\mu$. The new construction enables to sign messages of size $2^a \cdot (k - a)$ bits where $a$ is comprised between 0 and $k - 1$ and $k$ is the size of the modulus in bits. The maximum length that can be handled is then $2^{k-1}$ bits for $a = k - 1$ or $a = k - 2$. The construction can be recursively iterated to sign messages of arbitrary length. For bit strings $m_1$ and $m_2$, we let $m_1 \| m_2$ denote the concatenation of $m_1$ and $m_2$.

This construction preserves the resistance against adaptive chosen message attack of the signature scheme :

```
System parameters
    an integer k > 0
    a function μ : {0, 1}^{k+1} → {0, 1}^k
Key generation : Generate
    {N, e, d} ← RSA(1^k)
    public key : {N, e}
    private key : {N, d}
Signature generation : Sign
    y ← μ(m)
    return y^d mod N
Signature verification : Verify
    y ← x^e mod N
    y' ← μ(m)
    if y = y' then return 1 else return 0.
```

**Fig. 1.** The classical RSA scheme using function $\mu$ for signing fixed-length messages.

**Theorem 1.** *If the signature scheme* {Generate, Sign, Verify} *is* $(t, q_{sig}, \epsilon)$ *secure, then the signature scheme* {Generate', Sign', Verify'} *which signs messages of length* $2^a \cdot (k - a)$ *bits is* $(t', q'_{sig}, \epsilon')$ *secure, where :*

$$t' = t - 2^a \cdot q_{sig} \cdot \mathcal{O}(k^2) \tag{1}$$

$$q'_{sig} = q_{sig} - 2^{a+1} \tag{2}$$

$$\epsilon' = \epsilon \tag{3}$$

*Proof.* Let $\mathcal{F}'$ be a forger that $(t', q'_{sig}, \epsilon')$-breaks the signature scheme {Generate', Sign', Verify'} . We construct a forger $\mathcal{F}$ that $(t, q_{sig}, \epsilon)$-breaks the signature scheme {Generate, Sign, Verify} using $\mathcal{F}'$. The forger $\mathcal{F}$ has oracle access to a signer $\mathcal{S}$ for the signature scheme {Generate, Sign, Verify} and its goal is to produce a forgery for {Generate, Sign, Verify} . The forger $\mathcal{F}$ will answer the signature queries of $\mathcal{F}'$ itself.

The forger $\mathcal{F}$ is given as input $\{N, e\}$ where $N, e$ were obtained by running Generate. It starts running $\mathcal{F}'$ with the public key $\{N, e\}$.

When $\mathcal{F}'$ asks the signature of the $j$-th message $m_j$ with $m_j = m_j[1]||\dots||m_j[r_j]$, $\mathcal{F}$ computes :

$$\alpha_j = \prod_{i=1}^{r_j} \mu(0||i||m_j[i]) \bmod N$$

and requests from $\mathcal{S}$ the signature $s_j = \mu(1||\alpha_j)^d \bmod N$ of the message $1||\alpha_j$, and returns $s_j$ to $\mathcal{F}'$. Let $q$ be the total number of signatures requested by $\mathcal{F}'$.

```
System parameters
    an integer k > 0
    an integer a ∈ [0, k − 1]
    a function μ : {0,1}^{k+1} → {0,1}^k
Key generation : Generate'
    {N, e, d} ← RSA(1^k)
    public key : {N, e}
    private key : {N, d}
Signature generation : Sign'
    Split the message m into blocks of size k − a bits
    such that m = m[1]|| . . . ||m[r].
    let α = ∏_{i=1}^{r} μ(0||i||m[i]) mod N
    where i in 0||i||m[i] is the a-bit string representing i.
    let y ← μ(1||α)
    return y^d mod N
Verification : Verify'
    y ← x^e mod N
    let α = ∏_{i=1}^{r} μ(0||i||m[i]) mod N
    let y' ← μ(1||α)
    if y = y' then return 1 else return 0.
```

**Fig. 2.** The new construction using function $\mu$ for signing long messages.

Eventually $\mathcal{F}'$ outputs a forgery $\{m', s'\}$ for the signature scheme $\{$Generate', Sign', Verify'$\}$ with $m' = m'[1]|| \ldots ||m'[r']$, from which $\mathcal{F}$ computes :

$$\alpha' = \prod_{i=1}^{r'} \mu(0||i||m'[i]) \mod N$$

We distinguish two cases :

**First case :** $\alpha' \notin \{\alpha_1, \ldots, \alpha_q\}$. In this case $\mathcal{F}$ outputs the forgery $\{1||\alpha', s'\}$ and halts. This is a valid forgery for the signature scheme $\{$Generate, Sign, Verify$\}$ since $s' = \mu(1||\alpha')^d$ and the signature of $1||\alpha'$ was never asked to the signer $\mathcal{S}$.

**Second case :** $\alpha' \in \{\alpha_1, \ldots, \alpha_q\}$, so there exist $c$ such that $\alpha' = \alpha_c$. Let denote $m = m_c$, $\alpha = \alpha_c$ and $r = r_c$. We have :

$$\prod_{i=1}^{r'} \mu(0||i||m'[i]) \mod N = \prod_{i=1}^{r} \mu(0||i||m[i]) \mod N \qquad (4)$$

The message $m'$ is distinct from the message $m$ because the signature of $m$ has been requested by $\mathcal{F}'$ whereas the signature of $m'$ was never requested by $\mathcal{F}$, since $m'$ is the message of the forgery. Consequently there exist an integer $j$ such that :

$$0||j||m'[j] \notin \{0||1||m[1], \ldots, 0||r||m[r]\} \tag{5}$$

or

$$0||j||m[j] \notin \{0||1||m'[1], \ldots, 0||r'||m'[r']\} \tag{6}$$

We assume that condition (5) is satisfied (condition (6) leads to the same result). In this case $\mathcal{F}$ asks $\mathcal{S}$ for the signatures $x'_i$ of the messages $0||i||m'[i]$ for $i \in [1, r']$ and $i \neq j$, and the signatures $x_i$ of the messages $0||i||m[i]$ for $i \in [1, r]$. Since from (4) :

$$\mu(0||j||m'[j]) = \left( \prod_i \mu(0||i||m[i]) \right) \left( \prod_{i \neq j} \mu(0||j||m'[j]) \right)^{-1} \bmod N$$

the forger $\mathcal{F}$ can compute the signature of $0||j||m'[j]$ from the other signatures :

$$x'_j = \mu(0||j||m'[j])^d = \left( \prod_i x_i \right) \left( \prod_{i \neq j} x'_j \right)^{-1} \bmod N$$

and $\mathcal{F}$ finally outputs the forgery $\{0||j||m'[j], x'_j\}$. This is a valid forgery for the signature scheme $\{\mathsf{Generate},\ \mathsf{Sign},\ \mathsf{Verify}\}$ since the signature of $0||j||m'[j]$ was never asked from the signer $\mathcal{S}$.

We assume that $\mu$ can be computed in time linear in $k$, as is the case for most padding functions. The running time of $\mathcal{F}$ is then the running time of $\mathcal{F}'$ plus the time necessary for the multiplications modulo $N$, which is quadratic.

$\square$

Note that $q_{sig}$ must me greater than $2^{a+1}$ so that equation (2) holds. The security reduction is tight : the probability of success of $\mathcal{F}$ is exactly the probability of success of $\mathcal{F}'$.

## 4    Conclusion and Further Research

We have reduced the problem of designing a secure deterministic general-purpose RSA padding scheme to the problem of designing a one block secure padding scheme, by providing an efficient and secure tool to extend the latter into the former. As stated previously, this focuses more sharply the question of finding a secure encoding for RSA signatures, by showing that the difficulty is not in handling messages of arbitrary length, but rather in finding a secure redundancy function for short messages, which remains an open problem.

Our construction assumes that the padding function $\mu$ takes as input messages larger than the modulus; padding schemes such as ISO/IEC 9697-1 are consequently uncovered. A possible line of research could be a construction similar to ours, using a small (1024-bit) inner modulus and a larger (2048-bit) outer modulus.

## 5    Acknowledgements

We thank Jean-Marc Robert and Geneviève Arboit for useful discussions and the anonymous referees for their comments.

## References

1. M. Bellare and P. Rogaway, *Random oracles are practical : a paradigm for designing efficient protocols*, proceedings of the First Annual Conference on Computer and Commmunications Security, ACM, 1993.

2. M. Bellare and P. Rogaway, *The exact security of digital signatures - How to sign with RSA and Rabin*, proceedings of Eurocrypt'96, LNCS vol. 1070, Springer-Verlag, 1996, pp. 399-416.

3. R. Canetti, O. Goldreich and S. Halevi, *The Random Oracle Methodology, Revisited*, STOC '98, ACM, 1998.

4. W. Diffie and M. Hellman, *New directions in cryptography*, IEEE Transactions on Information Theory, IT-22, 6, pp. 644-654, 1976.

5. S. Goldwasser, S. Micali and R. Rivest, *A digital signature scheme secure against adaptive chosen-message attacks*, SIAM Journal of computing, 17(2):281-308, april 1988.

6. ISO/IEC 9796, *Information technology - Security techniques - Digital signature scheme giving message recovery, Part 1 : Mechanisms using redundancy*, 1999.

7. ISO/IEC 9796-2, *Information technology - Security techniques - Digital signature scheme giving message recovery, Part 2 : Mechanisms using a hash-function*, 1997

8. RSA Laboratories, PKCS #1 : *RSA cryptography specifications*, version 2.0, September 1998.

9. J.F. Misarsky, *How (not) to design signature schemes*, proceedings of PKC'98, Lecture Notes in Computer Science vol. 1431, Springer Verlag, 1998.

10. R. Rivest, A. Shamir and L. Adleman, *A method for obtaining digital signatures and public key cryptosystems*, CACM 21, 1978.

# Towards Signature-Only Signature Schemes

Adam Young[1] and Moti Yung[2]

[1] Columbia University, New York, NY, USA.
`ayoung@cs.columbia.edu`
[2] CertCo, New York, NY, USA.
`moti@cs.columbia.edu`

**Abstract.** We consider a problem which was stated in a request for comments made by NIST in the FIPS97 document. The question is the following: Can we have a digital signature public key infrastructure where the public (signature verification) keys cannot be abused for performing encryption? This may be applicable in the context of, say, exportable/escrow cryptography. The basic dilemma is that on the one hand, (1) to avoid framing by potentially misbehaving authorities we do not want them to ever learn the "signing keys" (e.g., Japan at some point declared a policy where signature keys may be required to be escrowed), and on the other hand (2) if we allow separate inaccessible public signature verification keys, these keys (based on trapdoor functions) can be used as "shadow public-keys," and hence can be used to encrypt data in an unrecoverable manner. Any solution within the "trapdoor function" paradigm of Diffie and Hellman does not seem to lead to a solution which will simultaneously satisfy (1) and (2).

The cryptographic community so far has paid very limited attention to the problem. In this work, we present the basic issues and suggest a possible methodology and the first scheme that may be used to solve much of the problem. Our solution takes the following steps: (1) it develops the notion of a *nested trapdoor* which our methodology is based on, (2) we implement this notion based on a novel composite "double-decker" exponentiation technique which embeds the RSA problem within it (the technique may be of independent interest), (3) we analyze carefully what can be and what cannot be achieved regarding the open problem by NIST (our analysis is balanced and points out possibilities as well as impossibilities), and (4) we give a secure signature scheme within a public key infrastructure, wherein the published public key can be used for signature verification only (if it is used for encryptions, then the authorities can decrypt the data). The security of our scheme is based on RSA. We then argue how the scheme's key cannot be abused (statically) based on an additional assumption. We also show that further leakages and subliminal leakages when the scheme is in (dynamic) use are not added substantially beyond what is always possible by a simple adversary; we call this notion *competitive leakage*. We also demonstrate such simple leaking adversary.

We hope that our initial work will stimulate further thoughts on the non-trivial issue of signature-only signatures.

T. Okamoto (Ed.): ASIACRYPT 2000, LNCS 1976, pp. 97–115, 2000.

**Key words:** Public Key Cryptosystems (PKCS), Public Key Infrastructure (PKI), digital signature, decryption, nested trapdoor, abuse freeness, (subliminal) leakage, design validation proofs, NIST, FIPS.

# 1   Introduction

Traditionally, implementations of efficient digital signature algorithms have been constructed such that the public verification keys can be made to function as encryption keys. The separation of an encryption capability from that of signature verification seems interesting in general, especially in the contexts of exportable cryptography. In fact, in any formulated policy for encryption, it seems that bodies like recovery agents have no legitimate need to be able to obtain the signing private keys of other users. But in all of the traditional systems, signing infrastructures can immediately be used as a shadow public key encryption infrastructures (a notion put forth by Kilian and Leighton [KL95]). Thus, the natural question posed by NIST, given the needs of the US federal agencies, was: "How can a public-key system incorporating both privacy as well as signatures which does not support unrecoverable encryption be designed?" This problem (the "NIST problem" in the sequel) was specifically stated in detail in a request for comments made by NIST in a revision of the DSA FIPS [FIPS97]. It read as follows:

*"The Administration policy is that cryptographic keys used by Federal agencies for encryption (i.e., to protect the confidentiality of information) shall be recoverable through an agency or third-party process and that keys used for digital signature (i.e., for integrity and authentication of information) shall not be recoverable. Agencies must be able to ensure that signature keys cannot be used for encryption. Any algorithms proposed for digital signature must be able to be implemented such that they do not support encryption unless keys used for encryption are distinct from those used for signature and are recoverable."*

Comments were received from various computer security companies. Yet **none** of the comments attempted to give a technical solution to the problem. The problem as stated seems quite nontrivial. It implies the existence of a public key scheme and infrastructure that can only be used for signatures and not for public key encryptions. The problem as a whole is a purely technical challenge (the need to separate escrowed encryption from signature was noted in [FY95, An97]). In this paper we attempt a solution to this problem. Note that minimizing the abuse and direct leakage of a signature scheme will make such a scheme more easily exportable, and may help prevent that which almost happened in Japan where policy makers a number of years ago attempted to escrow signature keys for law enforcement purposes.

Obviously in any on-going communication system, if one allows leakage using timing channels and other side information like correct/incorrect message structure (or other side subliminal channels) one cannot prevent leakage of information, and this information can be used cryptographically (say, for key exchanges). For such covert channels, no crypto at all is in fact needed, and just a marking scheme (e.g. error correcting code) or a timing convention is needed. In fact, against such attacks, communication itself has to be halted forever, which is unrealistic. This is what Rivest has pointed out as the "Chaffing and Winnowing" method [R98] using authentication channels for marking. Knowing that we cannot eliminate covert channels, the harder question is "what can be done to prevent 'direct use' of the published signature verification key for encryption?" as asked by NIST. Here we also add (perhaps beyond NIST's requirements) the issue of preventing (as much as possible) **additional** direct leakages when the scheme is dynamically used. We seek to eliminate leakage channels with much larger bandwidth (and amount of work) when compared to trivial covert channels; we call this property "competitive leakage," borrowing terminology from the field of on-line algorithms.

The above issues together try to eliminate the added advantage for "abusers" resulting from the direct introduction (publishing and using) of a signature scheme. One has to understand what can and cannot be done in this area, and the answer has to be scientific and forthright. We take a first step in this direction by attempting to answer the publicly posed NIST problem.

## 2   Analysis of the NIST Problem

We want to prevent published keys from being abused for encryption and further prevent direct and easy use of a small number of published signatures for encryption. We note that the difficulty of completely formalizing abuse-freeness (especially in the context of signature schemes) has been noted in the literature (see Desmedt et al. [D89, BD-etal96]). We add a key registration and publication stage to a signature scheme, and we define a digital signature infrastructure that is "signature-only" as following:

**Definition 1.** *A Signature-Only Digital Signature Infrastructure is a digital signature algorithm (which is secure against existential forgeries under adaptive chosen plaintext attacks [GMRi88]) with the following additional properties:*

1. **Security:** *It is not possible for the CA's (together with hypothetical other authorities) to forge a signature of any user in the system (even an existential forgery attack).*
2. **Published Key misuse freeness:** *The public verification key cannot be used to encrypt data in an unrecoverable manner.*
3. **Published Key leakage resistance:** *The public verification key has a very small subliminal channel (e.g., $< 32$ bits or so), and thus cannot be used to effectively display a complete shadow public key.*
4. **Signature leakage resistance:** *The signatures created using the private signing key are shadow public key resistant. Namely, it is not possible to use the properties of a given signature per se, to derive a shadow public-key. In addition we can strengthen the requirement and demand competitive leakage: the number of signatures (and complexity of derivation) required to get a shadow public key is approximately the same as in a trivially leaking channel (available by the communication).*

The first property above strengthens the traditional security and requires it to hold against the certification authorities (CA) which may hold extra information. The second property deals with abuse as required by NIST, whereas the other properties deal with add-on leakage of the system (compared to a non cryptographic system), which we have added to NIST's original requirement. We note that we do not attempt to eliminate subliminal leakage, but rather try to minimize add-on leakage as compared to existing trivial channels. In fact, we will show that every on-going signing leads to trivial (unavoidable) leakage.

**Additional related work:** The notion of a subliminal channel has been suggested by Gus Simmons [Si85]. A subliminal channel in a cryptosystem is a channel that allows a user of the system to leak information of his or her choosing through that channel. Simmons showed that information can be leaked in digital signatures, including DSA signatures [Si93]. It is known that in many operational environments, covert channels inherently exist. Within the context of key recovery systems, Kilian and Leighton showed that subliminal channels can be exploited to display unescrowed public keys within channels that exist in escrowed public keys, which are displayed by the CA [KL95]. They call this form of attack which easily adds a public key directory, shadow public key abuse. Abuses of cryptosystems have been discussed by Desmedt [D89]. It was originally believed (by some) that the NSA designed DSA [DSS91] could not be used for public key encryption or key exchange operations. Yet it was shown how the DSA verification key can be used for encryption (overtly [NR94] and even covertly within a single signature [YY97]). Such capabilities cannot be permitted if one is to design a scheme solving the NIST problem.

**Methodology:** In our methodology, we first discuss the security of the scheme (reducing from a known problem), we then show (under a new assumption) how the public keys in the scheme we present cannot be used as trapdoors in known systems in the sense that, if they are used, then the authorities are able to recover data encrypted with them. We then present methods that limit additional subliminal leakage of information. We do not employ cumbersome interactive techniques (e.g., the schemes in [D89, KL95]), but rather we employ non-interactive methods that assure that near-random choices of public parameters are made (making random choices via the use of one-way hash functions which are regarded as random oracles– an idea rooted in the DSA parameter generation procedure). We also guarantee leakage-freeness in the signatures themselves by using deterministic (and pseudorandomized) signing algorithms rather than randomized signing algorithms.

To claim security, we reduce the ability to forge (by the CA which holds more information than the signature verifying users) to the ability to break RSA as a one-way function (in the random oracle model). We assume RSA is a one-way function:

**Security Assumption:** Given $n \in Z^+$ s.t. $n = pq$ where $p$ and $q$ are distinct random primes of size $k$, and a fixed $e \in Z^+$ s.t. $gcd(e, \phi(n)) = 1$, and given a random $c \in Z_n^*$, finding $m \in Z_n^*$ s.t. $m^e = c \pmod{n}$ is hard (i.e., for any polynomial it takes more than polynomial in $k$ time, for all $k$ large enough).

Naturally, we need to claim that a public key cannot be used for public key encryptions. Since public key information is available, we **must** assume or show the non-existence of a (probabilistic) poly-time algorithm that uses the public key as it was created to encrypt data without the CA being able to decrypt that data. This is how the notion of a "nested trapdoor" helps: on one hand the CA cannot sign (since the CA does not know the innermost trapdoor or otherwise is in violation of the security assumption), but on the other hand, the CA can decrypt based on the information it holds (using the middle trapdoor).

For the outermost trapdoor, we now give the new assumption about the public key information that forms the basis for our proposed scheme. It is needed, given the state of the art of subliminal freeness in signature schemes. It basically states (based on the state of the art) that once the factorization of a randomly generated specially structured number is known (and its structure proven), it cannot serve as a public encryption key. More precisely:

**No-Abuse Assumption:** There does not exist a secure public key encryption algorithm that takes $n$ as the input public key, where $n$ (of size $k$) is the product of two large known (to the CA only, in our case) primes $p$ and $q$ with $p = 2tp_1q_1 + 1$ and $q = 2q_2 + 1$, such that $p_1$ and $q_1$ are secret primes, $t$ is a small public safe prime, and $q_2$ is prime (and given $e$ a fixed number of size polynomial in $k$, such that $gcd(e, \phi(\phi(n))) = 1$).

Note that even though $p_1q_1$ is known (to the CA only), the encryption algorithm does not know it since it only takes $n$ (and $e$) as input. Given the state of the art, the above assumption is mathematically plausible (its validation is an interesting issue, and knowing whether a weaker assumption suffices is left open). The plausibility of the assumption can be supported as follows:

- The assumption's public refutation can be done by proposing an encryption scheme which bypasses the above factorization knowledge (of the CA in our case) by using either (1) an old trapdoor, or (2) a new one. The latter, in fact, would be an interesting new public key suggestion.
- If a more general version of the assumption (not restricting to the exact structure of $n$ above but using generic $n$) does not hold, then it implies the existence of a shadow public key attack on **any** factoring based key escrow system where w.l.o.g. $p - 1$ is hard to factor.

**Main Idea:** Now that we discussed the assumptions, let us describe briefly the main idea behind the scheme. The value $\phi(\phi((n)))$ effectively constitutes the private signing key of the user, $Z_{\phi(n)}$ is the domain in which the signatures are computed, and $Z_n$ is the domain in which signature verification is performed. Our scheme is setup such that neither $\phi(n)$ nor $\phi(\phi(n))$ is known to the verifier, and such that only knowledge of $n$ is needed to verify. Also, we insure that $\phi(n)$ *is* known to the CA. Hence, any PKCS predicated on the difficulty of factoring $n$ provides no security against the CA. So, in our scheme there is "double-decker" exponentiation. The upper most deck is known only to the signer, the middle deck is known to the CA and the signer, and the bottom deck is known to everyone. We call this a nested trapdoor construction.

Note that double-decker techniques based on the discrete log problem have been utilized before [St96, CaSt97, YY98]. However, the novel double-decker application we present requires the use of composites and is predicated on the problem of factoring.

**Remark:** Naturally, any infrastructure that provides authenticity can be abused via leakage by criminals to permit *authenticated* communications which are untappable (since there are non cryptographic means to leak data). This is irrelevant since the problem posted by NIST appears to deal with direct abuse of signatures for encryption (obviously, NIST did not overlook the possibility of subliminal/covert channel use and direct "key exchanges"). For instance, the certified signature keys can be used to conduct an *authenticated* key exchange. When criminals are given the ability to authenticate themselves then this type of abuse is entirely unavoidable. **Hence, the property that we require is that criminals be forced to go outside of the provided key recovery/authentication mechanisms to conduct the untappable authenticated communications** (e.g., force criminals to do a key exchange, which they can authenticate if they so choose). What is achieved here, in fact, is that under our assumptions and in our proposed system, a user *cannot* abuse keys directly, and usage of the system competitively, to conduct secure communication. Recall that by 'competitively' we mean that we do not allow substantial increase in leakage via the keys or the signatures (requirements beyond NIST's original problem). The scheme can therefore be viewed as the first unescrowed digital signature scheme which, as much as possible, does not defeat the purpose of a recoverable PKI when added to a recoverable PKI on its outset (but requires extra work from abusers). We note that in general, in the area of abusing escrow encryption, the malicious adversaries can always bypass the system and the remaining scientific challenges are to protect against more benign adversaries which do not invest much extra effort (e.g., they use the available keys) or when this bypassing is to be revealed by other means in the communication system.

## 3    Re–examination of Existing Signature Schemes

In this section we will analyze the shortcomings of various digital signature algorithms in light of the NIST requirement. None of the schemes were designed with NIST's problem in mind, so these are not weaknesses but rather a reflection of the state of the art.

**RSA:** We take the opportunity to look at an RSA variant which makes the RSA signing function a uniform trapdoor permutation via pre-hashing. In this system, $n$ is the product of two large primes $p$ and $q$, and $e$ is a public value such that $gcd(e, (p-1)(q-1)) = 1$. $e$ and $n$ are the user's public verification keys, and the inverse of $e$ mod $\phi(n)$ is $d$, the user's private signing key. The signature on $m$ is $s = H(i||m)^d \bmod n$, where $i$ is the smallest positive integer making $H(i||m) \bmod n \in Z_n^*$ (as in [BR93]). To verify a signature we check that $s^e \bmod n = H(i||m)$. Clearly $(e, n)$ can be used to encrypt data. If $d$, $p$, or $q$ is given to the authorities, it follows that the authorities (CA) can forge signatures.

This applies to Rabin [Ra78] and every factoring based scheme, including Esign where $n = p^2 q$.

**ElGamal and Relatives:** In ElGamal [ElG85], the public key is $y = g^x \bmod p$, where $x$ is the private signing key. Here $g$ is a public generator modulo the public prime $p$. If the private key $x$ is not given to the authorities, $y$ can be used for encryption (e.g., the ElGamal encryption scheme). If $x$ is given to the authorities, obviously they can forge signatures. The same holds for DSA, Elliptic Curve DSA, Schnorr, undeniable signatures, Nyberg and Rueppel, etc.

**Fiat-Shamir:** In Fiat-Shamir, $n$ is the product of two primes and none of the users know the factorization of $n$. To generate a public key, a user generates $k$ different quadratic residues $v_1, v_2,...,v_k$ modulo $n$. This vector is the public key. The scheme therefore succumbs to the following shadow public key attack. The user chooses a random $x$ and squares it to get $v_1$. Let $g = v_1$ be a generator of hopefully a large subgroup containing all quadratic residues in $Z_n^*$. To generate $v_2$, a malicious user chooses $w$ at random and sets $v_2 = g^{2w} \bmod n$. Thus, $v_1$ and $v_2$ are quadratic residues, and constitute a shadow public key for generalized ElGamal mod $n$. The shadow private key is $2w$. (Even, if we only have $v_2$, $v_1$ can be derived from $n$ and the user's name, or be a constant)

**Okamoto '92:** To date, no large subliminal channel in this scheme is known. Recall that in this scheme, the verification key is $v = g_1^{-s_1} g_2^{-s_2} \bmod p$. Here $g_1$ and $g_2$ have order $q$ modulo the public prime $p$. The values for $g_1$, $g_2$, and $q$ are public. The private key is $(s_1, s_2)$. Both $s_1$ and $s_2$ are chosen randomly modulo $q$. Okamoto is based on the representation problem modulo $p$ (and so are fail-stop signatures).

To sign a message $m$ in Okamoto, we choose $r_1, r_2 \in_R Z_q$. We then compute $e = H(g_1^{r_1} g_2^{r_2} \bmod p, m)$. Here $H$ is a one-way hash function. We then compute $y_1 = r_1 + es_1 \bmod q$ and $y_2 = r_2 + es_2 \bmod q$. The signature on $m$ is the triple $(e, y_1, y_2)$. To verify the signature we check that $e = H(g_1^{y_1} g_2^{y_2} v^e \bmod p, m)$.

At first sight it seems as if the scheme is a good candidate for a solution. For, suppose that we don't give to the authorities $s_1$ and $s_2$. Then the authorities can't forge signatures. But, then we need to insure that $v$ cannot be used as a shadow public key. Suppose that $v$ can be used as a public key in a public key encryption algorithm. Since Okamoto is extendible to three or more bases, maybe there is no encryption algorithm if the representation uses three bases, or four bases, etc. This line of reasoning begs the question as to whether or not there exists a 'generalized Okamoto' public key encryption algorithm.

We will now answer this question in the affirmative. Indeed, there is a public key algorithm that uses public keys based on the representation problem with any number of bases. We will demonstrate this algorithm where two bases are used. To public key encrypt a message $m$ using $v$ as in Okamoto's scheme, we do the following:

1. $k \in_R Z_q$, $a = g_1^k \bmod p$, $b = v^k \bmod p$, $c = g_2^k m \bmod p$
2. The ciphertext of $m$ is $(a, b, c)$

To decrypt we compute:

1. $a' = a^{-s_1} \bmod p$ which equals $g_1^{-s_1 k} \bmod p$
2. $b' = b/a' \bmod p$ which equals $g_2^{-s_2 k} \bmod p$
3. $m = c/(b'^{-s_2^{-1}}) \bmod p$

This algorithm can be easily extended to handle representations using more bases. The ciphertext is an $(m+1)$-tuple if $m$ bases are used in the representation of $v$. Thus, this scheme and it's extentions using more bases does not meet the requirements.

**Naor Yung signature scheme:** A digital signature algorithm (which is presented as an important plausibility result, not as an efficient scheme) was presented in [NY89] that is based on the existence of one-way permutations (Rompel uses this exact setting but with any one-way function). The system is provably secure against adaptive chosen message attacks. It operates by having each user publish a row of windows, where each window is a pair of values that form commitments of the user via a one-way permutation value. Since the scheme (and its follow up work) is not based on trapdoor functions it departs from the Diffie-Hellman paradigm. Does this mean it is shadow public key resistant? No, the problem is that even though the system *itself* does not employ a trapdoor function, it could display one subliminally assuming trapdoor functions exist. The problem is that each window can leak several (say, 20) subliminal bits, thus allowing the system to directly leak a shadow public key (e.g., an RSA public key which is not held by the authorities).

# 4   The Signature Scheme

In this section we describe our construction which is a first answer to the NIST problem. We present the scheme and the user registration procedure.

## 4.1   Key Generation

Let $e$ be an odd value (fixed and bounded as discussed below). Let $M$ be a security parameter (which is say, a power of 2). Let $p_1$ and $q_1$ be $M/2$ bit primes. Let $q = 2q_2 + 1$ be a safe prime. These primes adhere to the following constraints for proper system operation:

1. Each of $p_1 - 1$, $q_1 - 1$, and $q_2 - 1$ have a large prime in their factorization.
2. $\gcd(e, (p_1 - 1)(q_1 - 1)(q_2 - 1)) = 1$.
3. There exists an $M_1$-bit (e.g., $M_1 = 63$) safe prime $t$ s.t. $p = 2tp_1q_1 + 1$ is prime[1] and s.t. $\gcd(e, \phi(t)) = 1$.

---

[1] An alternate implementation chooses $(q - 1)/2$ to have the same form as $(p - 1)/2$, but $(q - 1)/2$ has a different factorization.

If we want $|pq|$ to be a power of 2, we can choose $q$ to be $M - M_1 - 1$ bits long. We incorporate $t$ into $p - 1$ to make key generation fast. To provide protection against shadow public key abuse (see [KL95] for the attack on composites), the following additional constraints are needed to reduce subliminal leakage:

1. $s$ and $s'$ are chosen (randomly).
2. $H_1(s)$ (or $H_1(s) + 1$, see below) is the same as the upper half of the bits in the bit representation of $p_1 q_1$.
3. $q$ is chosen by computing $q = H(s')$ and testing for primality and strong primality (more elaborate methods are possible, e.g., the method for generating parameters for DSA).

To accomplish step 2, an algorithm similar to [L98, YY96] can be employed. Thus, either $H_1(s)$ is the upper order bits, or $H_1(s) + 1$ is the upper order bits due to a borrow bit being taken. Here $H_1$ and $H$ are suitable (ideal) one-way hash functions. This step is to avoid the leakage of $M/2$ bits in the composite $p_1 q_1$. The values $p, q$, and $t$ are found to satisfy the above. The key generation algorithm then performs the following computations:

1. $n_1 = 2tp_1 q_1 q_2$
2. Compute the smallest value $s''$ that makes $g' = H_2(s, s', s'', n)$ a generator mod $p$, $g'' = H_3(s, s', s'', n)$ a generator mod $q$, and $\gcd(x_i, n_1) = 1$, where $x_i = h'_i(s, s', s'', n, i)$ for $i = 1, ..., K$ ($K$ odd), and $H_2, H_3, h'_i$ being appropriate ideal hash functions. We insist that $s''$ is, say, at most 24 bits in length.
3. Chinese remainder $g \equiv g' \mod p$ with $g \equiv g'' \mod q$ to get $g \mod pq$ ($g$ then has order $\lambda(pq)$).
4. $n = pq$
5. Compute $g_i = g^{x_i} \mod n$ for $i = 1, ..., K$
6. $d = e^{-1} \mod \phi(\phi(n))$ (We will use $\phi^2$ to denote $\phi(\phi())$.)
7. Compute $T$ to be a NIZK proof of knowledge of the factorization of $(p-1)/2t$ into two distinct primes (in the random oracle model). A modification of the techniques of [GHY89] and [BFL91] enables this (the short proof of [PS00] may apply as well). $T$ can also be an interactive ZK proof if we allow interaction.

Note that $n_1 = \lambda(n)$ is the Carmichael function $\lambda$ of $n$. The public verification key is $(g, g_1, ..., g_K, e, n)$. The private signing key of the user is $(x_1, x_2, ..., x_K, d, n_1)$. To register the public verification key with the CA, the user sends to the CA the tuple $(s, s', s'', p, q, t, e, T)$. The value for $s''$ must be sent, to assure the CA with very high probability that $g'$ generates $Z_p$. In practice $n$ is about twice the size of a regular RSA modulus.

It is obviously imperative that the $g, g_i$'s and $n$, when taken together, cannot easily allow $n$ to be factored. Otherwise, the composite $\lambda(n)$ would be available to all users and could be used as a shadow public key. Since random $g, g_i$'s are samplable, the following fact holds:

**Fact 1** *The values for the $g, g_i$'s and $n$, when taken together, cannot be used to factor $n$.*

## 4.2   CA Registration

The CA receives $(s, s', s'', p, q, t, e, T)$. The CA computes $v_1$ to be $(p-1)/2t$. The CA then sets $z$ to be the upper half of the bit representation of $v_1$. The CA computes $n$, $g'$, $g''$, $g$, and the $g_i$'s in the same way as the user. The CA verifies all of the following things:

1. $p$ is prime, $t$ is prime, $t \mid p-1$, $|p-q|$ is large, $g''$ generates $Z_q$, etc.
2. $H_1(s)$ or $H_1(s) + 1$ equals $z$.
3. $g^{(p-1)/2}, g^{(p-1)/t}, g^{(p-1)/v_1} \neq 1 \bmod p$.
4. checks that $T$ is valid.

$T$ convinces the CA that the user knows the signing private key. $T$ and step 1 proves that there are 16 possible orders for $g'$. $T$ and step 3 proves that $ord_p(g') \in \{2tp_1, 2tq_1, 2tp_1q_1\}$, i.e., it can be only 3 of those 16. Recall that if $p$ is prime and $w \mid p-1$, then the number of least residues mod $p$ with order $w$ is $\phi(w)$. Since $g'$ is chosen by an oracle, it has order $2tp_1$ with probability $(t-1)(p_1-1)/p$, which is guaranteed to be negligible since $v_1$ must be made difficult to factor. By making $p$ large enough, this quantity is still negligible even after a polynomial number of oracle queries by a cheating prover. The same holds for $2tq_1$. So, the order of $g'$ mod $p$ is $2tp_1q_1$ with overwhelming probability. Anyway, recently an efficient method to prove in ZK that a number is maximal order was shown [JY00]. Clearly $g$ has order $\lambda(n)$ if $g'$ and $g''$ are generators. Similarly, the CA verifies that the order of $g_i$'s (for $i = 1, ..., K$) are the same as the order of $g$ (i.e., that $gcd(x_i, n_1) = 1$).

If all of the verifications pass then the CA publishes (and certifies) the string $(g, g_1, ..., g_K, e, n)$ as the user's public verification key.

## 4.3   Signing and Verifying Messages

Let $a \| b$ denote the concatenation of string $a$ with string $b$, let $h_i$'s be ideal full domain hash functions into $\{1, 3, 5, ..., (n+1)/2 - 1\}$ (a random oracle answer concatenated with a 1 as the least significant bit). To sign an arbitrary message $m$, the user computes $c = (x_1 h_1(j \| m) + x_2 h_2(j \| m) + ... + x_K h_k(j \| m))^d \bmod n_1$, where $j > 0$ is the smallest integer making all $h_j(j \| m) \bmod n_1 \in Z_{n_1}^*$ (as required by the full domain hash method [BR93], and will be 1 almost always, thus for brevity we omit the notation $j \| m$ and use $m$ instead). The signature on $m$ is $c$. Note that the signer can use the Chinese Remaindering method to make signing faster. The verifier checks that:

1. $c < n/2$

2. $g_1^{h_1(m)} * g_2^{h_2(m)} * ... * g_K^{h_K(m)} \ (mod \ n) = g^{c^e} \ (mod \ n)$

If (1) and (2) are not satisfied then the signature is rejected. To substantiate verification (1), we need to show that $\lambda(n) < n/2$ for all properly chosen $n$'s. This can be seen from the following.

$$\lambda(n) = lcm(p-1, q-1) = lcm(2tp_1q_1, 2q_2) = 2tp_1q_1q_2$$
$$n/2 = (1/2)(4tp_1q_1q_2 + 2tp_1q_1 + 2q_2 + 1) > 2tp_1q_1q_2$$

So, the claim holds. The reason we insist on making signatures less than $n/2$ is to avoid subliminal leakage of information in signatures. Below we elaborate more on this.

Note that since $e > 2$ the only way to verify the signature is to determine $g^{c^e} \bmod n$ by exponentiating $e$ times:

$$((g^c \bmod n)^c \bmod n)^c ... \bmod n$$

This is necessary because the modulus in the exponent is unknown to the verifier, thus the verifier cannot first compute $c^e \bmod n_1$ and then use this as the exponent for $g$. Note that signing is done in the upper most deck with the signature value in the middle-deck, and verification is performed in the bottom deck. The system therefore utilizes composite-based double-decker exponentiation. For efficiency we assume $e$ to be small (theoretically, $e$ which is polynomial in the size of $n$ is still feasible).

**Proposition 2.** *Signature verification is complete (i.e., properly generated keys and signatures will always verify).*

*Proof.* If $n$ is generated properly, then verification (1) will always pass. Now consider verification (2). One can verify that $c^e = x_1h_1(m) + x_2h_2(m) + ... + x_Kh_K(m) \bmod \lambda(n)$. Since $g$ and the $g_i$'s have order $\lambda(n)$, by the construction of the $g_i$'s the claim is proved. □

# 5   Security Against Forgeries: Validation Proof

Recall that the user sends to the CA the tuple $(s, s', s'', p, q, t, e, T)$. We need to show that given this information, and given a polynomial number of message/signature pairs, it is not possible for the CA to forge messages (i.e., that the CA can't choose messages and sign them, or even produce existential forgeries). We will do this in two steps. First we will show that the tuple that is given to the CA during certification does not help the CA forge, then we will show that a slight modification of our system is identical in security to RSA, and that this modification implies that our scheme is also secure. We will then show even further that in the random oracle model, our system is secure against adaptive chosen message attacks (we are aware of the advantages of having such proofs, e.g. [BR93], as well as the existence of limitations for certain implementation constructs which are not applied here [CGH98]).

Note that the CA knows $\lambda(n)$, since $\lambda(n) = n_1 = (1/2)(p-1)(q-1)$. The CA does not know $\phi(\phi(n))$, since the CA doesn't know the factorization of $p_1q_1$. We need to show that no information about the factorization of $p_1q_1$ is leaked to the CA by $(s, s', s'', p, q, t, e, T)$, excluding deliberate leakage (i.e., assuming honest user). First, nothing is leaked by $T$ since this is a zero-knowledge proof. Also, $t$ need not even be sent by the user, since it can be found easily by the CA. The seeds $s$, $s'$, and $s''$ contribute the information constituting the upper order bits of $p_1 q_1$, the prime $q$, and the generators $g'$, $g''$, and the exponents bringing $g$ to $g_i$ where $i = 1, ..., K$, and nothing else, since they constitute the preimages under the random oracle hash functions $H_1$, $H_2$, and $H_3$, and $h'_1, ..., h'_K$. To see this, note that under the random oracle assumption, each hash value is chosen uniformly at random from its respective set, and the result is therefore independent (as viewed by all poly–time algorithms) of the random hash function's input. It follows that the only information leaked by $s$, $s'$, and $s''$ is the upper order bits of $p_1 q_1$ (which is already known to the CA from $p$), and also $q$, $g'$, $g''$, $g$, and $g_1, ..., g_K$. If we can show that no information about the factorization of $p_1 q_1$ is leaked by $g'$ then we will be done.

Suppose that given a generator $g'$ of $Z_p$, and given $p$, an oracle A can be used to factor $p-1$, where $p$ is prime. In other words, $A(g', p)$ outputs the factorization of $p - 1$. Then A can be used to factor any composite $n$ as follows. Choose a prime $k$ randomly s.t. $2kn+1$ is a prime number $p'$. Choose a value $h \in_R Z_{p'}$ and run $A(h, p')$. If A fails, choose $(k, h)$ over and repeat. Since there are $\phi(p' - 1)$ generators of $Z_{p'}$, we expect to factor $n$ with probability $\phi(p' - 1)/(p' - 1)$ each time we invoke A. Thus, $p - 1$ will be factored in expected poly–time. We have shown via random reduction that factoring $n$ is no harder than implementing the oracle A. So, implementing the oracle A which uses $g'$ and $p$ to factor $p - 1$ is no easier than factoring $n$. Since $g'$ is chosen (accessed) almost randomly , it follows that $g'$ contributes in no way to the CA's ability to factor $(p - 1)/2t$. Thus, we have shown that:

**Lemma 3.** *No information about the factorization of $p_1 q_1$ (and hence $\phi^2(n)$) is leaked to the CA by $(s, s', s'', p, q, t, e, T)$.*

To forge, the CA need not necessarily output a $c$ satisfying $c < \lambda(n)$, since the users who will perform verifications do not know $\lambda(n) = n_1$. The CA therefore needs only to output a $c < n/2$ s.t. $g^{x_1 h_1(m) + x_2 h_2(m) + ... + x_K h_K(m)} \pmod{n} = g^{c^e} \pmod{n}$ to forge, since users can be certain that $c$ must be less than $n/2$. But, if the CA knows a valid $c > n_1$, then the CA knows a valid signature in the correct range, since $c \bmod n_1 < n_1$. Thus, a CA can output an acceptable forgery implies that it knows a forgery in the correct range. The following proves that forgeries are possible in our system by the CA iff the CA can break RSA.

**Theorem 4.** *The CA can forge signatures of users in our system (without hashing) using exponent $e$ iff the CA can forge signatures in the RSA scheme (without hashing) using exponent $e$.*

*Proof.* Let $n = pq$, and let $e$ be a small number s.t. $gcd(e, \phi(\phi(n))) = 1$. Furthermore, let $t$ be a safe prime, and let $p = 2tp_1 q_1 + 1$, $q = 2q_2 + 1$, and $q_2$ be prime. Suppose that the CA has an oracle $A_e$ that uses an exponent $e$ such that when given $(n, m)$ produces $c < n/2$ s.t. $g^{x_1 m + ... + x_K m} \bmod n = g^{c^e} \bmod n$.

In other words, $c = A_e(n, m) < n/2$ is a valid signature on $m$ in our algorithm (without hashing), hence the CA can forge signatures in our system. Let $n'$ be an RSA modulus, i.e., $n' = p_3 q_3$ where $p_3$ and $q_3$ are prime. Let the RSA exponent be the same $e$ as in our scheme. Thus, $gcd(e, \phi(n')) = 1$. Let $d'$ be s.t. $ed' + w\phi(n') = 1$ for some integer $w$. The CA can use $A_e$ to forge RSA signatures for the public key $(e, n')$ by choosing a small safe prime $t$ randomly s.t. $p' = 2tn' + 1$ is prime, and by choosing a random safe prime $q'$ where $q' = 2q'' + 1$. The CA makes sure that $gcd(e, \phi(t)\phi(q'')) = 1$. Let $d''$ be s.t. $ed'' = 1 \ mod \ \phi(2tp_3q_3q'')$. To forge an RSA signature, the CA chooses $r \in_R Z^*_{\lambda(p'q')}$ and computes $c' = r^{-1}A_e(p'q', r^e m/(x_1 + ... + x_K)) \ mod \ 2tn'q''$. It follows that $c' = m^{d''} \ mod \ 2tn'q''$. The CA then sets $c = c' \ mod \ n'$.

$$c' = m^{d''} + k2tp_3q_3q'' \text{ for some integer } k$$

$$c = c' \ mod \ p_3q_3 = m^{d'' \ mod \ \lambda(p_3q_3)} \ mod \ p_3q_3$$

But, $d'' \ mod \ \lambda(p_3q_3)$ is the inverse of $e$ modulo $\lambda(p_3q_3)$. To see this note that,

$$ed'' + v\lambda(2tp_3q_3q'') = 1 \text{ for some integer } v$$

But, $\lambda(p_3q_3) \mid \lambda(2tp_3q_3q'')$. So, $e(d'' \ mod \ \lambda(p_3q_3)) = 1 \ mod \ \lambda(p_3q_3)$. Thus, $c$ is a valid signature on $m$ in RSA mod $p_3q_3$.

Conversely, suppose that the CA can forge signatures in the RSA scheme using the exponent $e$. So, assume that the CA has an oracle $B_e$ that uses $e$, a composite of two different primes $n_1$ and a message $m$, that produces a valid RSA signature $c \ mod \ n_1$ on $m$. That is, $c = B_e(n_1, m)$ is a valid RSA signature on $m$. A CA knowing the composite $p_1q_1$ and the prime $q_2$ of a user can use $B_e$ as an oracle to forge in our system as follows. The CA computes $c' = B_e(p_1q_1, r^e m(x_1 + ... + x_K))/r \ mod \ p_1q_1$ where $r$ is chosen randomly from its respective message space, and computes $c'' = (m(x_1 + ... + x_K))^{e^{-1}} \ mod \ 2tq_2$. Note that it was verified by the CA that $gcd(e, \phi(t)) = 1$, $p = 2tp_1q_1 + 1$ is prime, and that $q = 2q_2 + 1$ is a safe prime. The CA Chinese remainders $c' \ mod \ p_1q_1$ with $c'' \ mod \ 2tq_2$ to produce $c \ mod \ 2tp_1q_1q_2$. □

Note that the above proof constitutes a randomized reduction where if oracle $A_e$ exists, then we show the existence of only one other oracle, namely oracle $B_e$, and vice-versa (i.e., existence of $A_e$ does not necessarily imply the existence of $B_{e'}$ for all $e'$). However, we need only assume that every oracle succeeds on a fixed fraction of the message space. Thus, our system is no more or less secure than RSA. Since the above holds for the CA, clearly *the same holds for all of the users* in the system, who have access to even less information than the CA.

We will now prove that our signature scheme is secure against adaptive chosen message attacks under the random oracle model, assuming that RSA is secure. Since $t$ and $q_2$ are known to the CA, the CA can always compute the partial signature $c_1 = (x_1 h_1(m) + x_2 h_2(m) + ... + x_K h_K(m))^{e^{-1} \ mod \ \phi(tq_2)} \ mod \ 2tq_2$ on a message $m$. Therefore, when the user sends the CA a signature $c$ on message $m$, the user can send the signatures $c_1$ and $c_2 = (x_1 h_1(m) + x_2 h_2(m) + $

$... + x_K h_K(m)^{e^{-1} \ mod \ \lambda(p_1 q_1)} \ mod \ p_1 q_1$ instead, since the CA can just Chinese Remainder $c_1$ and $c_2$ to get $c$. For the purposes of this proof, we will assume a digital signature scheme in which everyone knows $p$, $q$, $t$, $g'$, $g''$, and the $g_i$'s and in which the two signatures $(c_1, c_2)$ constitute the actual signature on $m$. The generation and verification of $c_1$ is straightforward, and the generation and verification of $c_2$ is as in RSA with hashing. Though it is important to note that this proof holds for the specific full domain hash function $Hash(m) = x_1 h_1(m) + x_2 h_2(m) + ... + x_K h_K(m)$. Now, since the basic $h_i$ are random oracles over their domain (of odd numbers), and adding mod $n_1$ assures that we stay within the domain (provided $K$ is odd), we can use $Hash$ as an ideal (random oracle) hash function (over a certain domain), and the arguments of the security of the full domain hash hold.

**Theorem 5.** *Assuming the random oracle assumption holds on the $h_i$'s, and assuming that our security assumption holds (i.e., RSA being a one-way function), the signatures $c_2$ are secure against adaptive chosen message attacks with respect to the CA.*

*Proof.* We apply the Full-Domain Hash analyzed in [BR93, C00] based on a random oracle over a sub-domain of $Z^*_{n_1} = Z^*_{2t p_1 q_1 q_2}$ (explained below), and get a full domain hash $Hash$, which takes messages $m$ and gives a random element from that full domain. Clearly, $c_1$ provides no assurance that $m$ is authentic, since $c_1$ can be produced by anyone with the CA's knowledge. It follows that our digital signature scheme is secure against adaptive chosen message attacks iff the key generation algorithm for $c_2$, the signing algorithm for $c_2$, and the verification algorithm for $c_2$ is secure against adaptive chosen message attacks. It was shown in [BR93, C00] that indeed these algorithms (using a full-domain hash) are secure against adaptive chosen message attacks. Now, using the Chinese Remainder Theorem, $c_2 = (x_1 h_1(m) + x_2 h_2(m) + ... + x_K h_K(m))^{e^{-1}} \ mod \ p_1 q_1$ In the security proof the value of $c_2$ before the exponentiation (namely, $x_1 h_1(m) + x_2 h_2(m) + ... + x_K h_K(m)$ is in fact assumed to be chosen by a random oracle drawing elements from $Z^*_{p_1 q_1}$, which is the domain over which we want to prove security. This can be assumed due to the 1-1 mapping between the elements assured to be (almost always) in $Z^*_{n_1}$ and the set of their remainders: $c_1, c_2$. Furthermore, due to the random oracle assumption which gives us a degree of freedom in fixing polynomially many values (of $c_2$) in the attack protocol, when arguing about a value, we can always solve for the randomly chosen $c_2$ element of our choice (by "playing" with the actual value of one of the basic hash values $h_i$) and then, in the simulation, having a desired challenged value or probed element to be a random element of our choice in $Z^*_{p_1 q_1}$ (which we already know the result of applying the exponentiation to). By doing so we determined $c_1$ with an arbitrary value, due to the algebraic relation. Under the above proper random oracle assumption, we can turn the signature values ($c_2$) into desired (random) challenges whose inverses are actually accessible anyway (by our choice), and then "chosen plaintext" which is the input to the RSA operation is reduced to a "random plaintext" attack on the RSA operation (namely, it reduces the adaptively chosen plaintext forgery attack to the security of RSA as a one-way function into the domain of the $c_2$ elements). The full domain hash assumption enables us to argue that the "altered random oracle" (which is modified in only a few polynomially-many places) is statistically indistinguishable from the original one. A more sophisticated oracle answering strategy is in [C00].     □

## 6   Arguing Published Key Abuse Freeness

To show the resistance of $(g, g_1, ..., g_K, e, n)$ to abuses, it is necessary to show two things:

1. Key misuse freeness: the values in $\{g, g_1, ..., g_K, e, n\}$ cannot be used directly as a shadow public key in any PKCS.
2. Key leakage resistance: the information in $(g, g_1, ..., g_K, e, n)$ cannot display enough subliminal bits to constitute a secure shadow public key.

To show (1), first notice that we claim that $n$ cannot be used for unrecoverable RSA/factoring-based encryptions using $n$ as the composite modulus, simply because the CA knows the factorization of $n$. From the No-Abuse assumption there is no other way to exploit $n$ (here is where we strongly employ this assumption).

Next observe that the way $g$ is generated, makes it random maximal order element. Assuming discrete logarithm is a hard problem on random instances, even when the factorization is known (this is a regular discrete log assumption), will prevent the user from using it as a base for a public key. As for $g_i$'s, the CA is aware of the same information about them as the user does. Finally, since small enough $e$ can be fixed among all the users, it cannot be or contribute to a shadow public key.

We will now observe a simple yet strong result which proves about the system containing $(n, e)$ the following:which uses a fixed $e$ (without the $g$ and the $g_i$'s):

**Theorem 6.** *If there exists a shadow public key attack on $pq$ where $pq$ "is" the shadow public key then there exists a shadow public key attack on any composite $pq$ based key escrow system where the user chooses $p$ and $q$ at the users own discretion and where $p$ (or $q$) is escrowed.*

This directly follows from the fact that what is published in this modified system is exactly what is published in any composite $pq$ based key escrow system (where users can choose the form of, w.l.o.g., $p - 1$).

Next we argue (2), namely that there is no key leakage (based on the current state of the art). We need to show that there is no subliminal channel in $n$ of significant bandwidth. Recall that $\lambda(n)$ has $p_1 q_1$ and $q_2$ in its factorization. The value $p_1 q_1$ was generated in such a way as to minimize subliminal channels in it. This was done by making its upper order bits a one-way hash (behaving like a random oracle) of a random number $s$, and is precisely the method used to foil high bandwidth shadow public key attacks in RSA moduli (by filling the subliminal channel). By forming $p_1q_1$ in this way, there can only be a small (logarithmic length) subliminal channel in $p_1q_1$ (we may also insist that the values chosen via the one way function have a fixed number of bits). Similarly, $q_2$ can only have a similarly small subliminal channel since it is chosen by a random oracle. Since $t$ is a small odd value, it provides no significant subliminal channel either. It follows that $\lambda(n)$ is mostly random and has no substantial subliminal channel, so neither does $n$ (which is built from it in a given format).

Now consider $g$. The value $g$ is computed based on $g'$ and $g''$. But, $g'$ is chosen by a random oracle whose input are the values $s$, $s'$, $n$, and $s_1$. So, $g'$ has no substantial subliminal channel. We make $s$, $s'$, and $n$ inputs to a random oracle and increment $s''$ from zero to find $g'$ as a precautionary measure to limit the amount of subliminal leakage in $g$. This precaution works because $s''$ is small, and thus a malicious user is forced to regenerate $n$ if the user fails to leak enough information in $g$ by the time $s_1$ reaches its maximum allowable value. Thus, the key generation time is inhibited when key generation is modified to leak subliminal information via $g$. The same argument applies to $g''$, so $g$ is mostly random and has no substantial subliminal channel. In fact, $g$ is effectively chosen almost uniformly at random (by a random oracle). Furthermore $g_i$ is chosen via $x_i$ and $g$, and $x_i$ is formed from the same information (under a different random oracle) as $g'$ and $g''$. We note that using one-way hashing to generate trap-door free parameters was first proposed by NIST with respect to DSA. This concludes the arguments suggesting that $(g, g_1, ..., g_K, e, n)$ does not leak enough subliminal information. In practice it may be possible to leak on the order of $O(\log \log(n))$ bits or so if an attacker works hard (searching for a pattern), but again, we are protecting only in the relative sense: it may be that so little amount of information can be subliminally sent outside the cryptographic system.

## 7 Arguing Signature Leakage Resistance

Can a single signature leak information? Clearly, if there were a way to output more than one signature for a given $m$, it may be possible to leak bits of subliminal information. So, we would like to show that for each message $m$, with overwhelming probability there is only one valid signature $c$ on $m$. Then, since the signing algorithm is deterministic, we will be well on our way to showing that our signatures are leakage resistant from this aspect.

**Proposition 7.** *With overwhelming probability, it is not possible to find an $m$ s.t. $c, c' < n/2$ will both be accepted as valid signatures on $m$.*

*Proof.* From signature verification it is clear that all the valid signatures on $m$ are specified by

$$c_j = j\lambda(n) + (x_1h_1(m) + ... + x_Kh_K(m))^d \bmod \lambda(n)) < n/2, \text{ for } j \geq 0$$

Clearly $c_0$ will always be a valid signature. If $h_j$ is a random oracle, then for each $m$, the probability that $c_j$ is a valid signature is negligible if $j \geq 1$. We will prove this by first proving that it holds for $j = 1$, and the fact that it holds for all $j > 1$ will be immediate. Note that if $j = 1$ then $c_j$ is a valid signature iff $(x_1 h_1(m) + ... + x_K h_K(m))^d \mod \lambda(n)) < n/2 - \lambda(n) = tp_1 q_1 + q_2 + 1/2$ Since the $h_j$'s are random oracles, then $(x_1 h_1(m) + ... + x_K h_K(m))^d \mod \lambda(n)$ is chosen randomly from $Z_{\lambda(n)}$, independent of $m$. Thus, $(x_1 h_1(m) + ... + x_K h_K(m))^d \mod \lambda(n) < n/2 - \lambda(n)$ with probability $(tp_1 q_1 + q_2 + 1/2)/(2tp_1 q_1 q_2)$, which is negligible. $\qquad\square$

Next, we analyze the leakage of a sequence of signatures, which we want to show to be competitive, namely that there are trivial channels in the usage of signatures that allow the leakage of a public key. Next we present a leakage design, based on ordering of signatures only (and not assuming further timing, errors or error correcting based leakage). This shows that in on-going usage there are plenty of opportunities to leak information (which is expected).

**Universal attack:** this attack works as follows. The user encodes the shadow public key 15 bits at a time, by successively taking 8 unique signatures at a time, and ordering them in a specific way in the post to the bulletin board, say (assuming that scheduling the signatures has this degree of freedom). There are 8! ways to order the signatures (which are distinguished and ordered by their numeric value), so 15 bits can be encoded. This allows us to encode efficiently. For example in 168 signatures we can encode a 314 bit shadow public key (e.g. based on an elliptic curve over $GF(2^{157})$). This is a "non-cryptographic leakage" which is trivial (and with additional covert information and better encoding we can do with much fewer signatures).

What leakage can we produce in our system? We can, for example leak $\lambda(n)$ over the course of using the signature system, exploiting the structure of our system. In fact, it works as following: When signing a message $m_i$ we produce signature $c_i$. the attacker gets information (over the integers) of the form $(c_i)^e = x_1 h_1(m) + ... + x_K h_K(m) + r_i \lambda(n)$ where the $x_i$'s are fixed and unknown and the unknown last (free) term $(r_i \lambda(n))$ a multiple of $\lambda(n)$ ($r_i$ is associated with $c_i$). The fixed unknown $x_i$'s can be eliminated from the expressions as additional signatures are added, after $K + 1$ equations, we are left with a large integer representing a multiple of $\lambda(n)$, which can then be used as a public key (and if the process is continued we can get smaller multiples via gcd's and get to isolate $\lambda(n)$). This type of leakage was pointed first by Bleichenbacher. Notice that by choosing $K$ large enough and choosing $e$ large enough we force the attacker to collect a large enough number of signatures and to work over large integers, thus leakage in the scheme more competitive.

There is an obvious tradeoff between efficiency and the competitive factor of this specific leakage scheme. Small $K$ gives a signature whose signing is as expensive as RSA (of double the size) and whose verification takes $e + K$ exponentiations (we implemented the scheme for small values). Finally, we remark that naturally, other leakages may still exist and their presence is an open problem. It is very hard to quantify leakage of this form, and it may be that moderate complexity measures are enough since we deal with prevention in light of existing trivial channels (achieving "competitive leakage").

## 8  Conclusion

In reaction to NIST's problem, we have presented a framework and an implementation for a digital signature PKI where the public verification keys cannot be abused safely for encryption. We proved its security, argued its no-abuse property, and further argued concretely, based on the sate of the art, why the leakage with on-going use is competitive (compared with universally available means of leakage, which we also pointed out). Numerous issues remain to be investigated like the validation of our no-abuse assumption, whether our assumptions are necessary, and whether or not weaker assumptions can be used. We employed the random oracle proof methodology numerous times regardless of its possible weaknesses. A question one might ask is whether or not there are solutions which do not employ random oracles, yet are efficient, etc. (there may exist a situation where a random oracle is used for arguing that some elements are chosen almost at random, but where standard complexity-theoretic proofs are used for the security argument). The notion of "non-leakage" (or reduced leakage) and its reduction and implementation is another unexplored area: formalization and characterization of the issues, notions, and solutions in the area are left open.

## References

[FIPS97]  Announcing Plans to Revise Federal Information Processing Standard 186, Digital Signature Standard. In volume 62, n. 92 of *Federal Register*, pages 26293–26294, May 13, 1997.

[An97]  R. Anderson. The GCHQ Protocol and Its Problems. In *Advances in Cryptology—Eurocrypt'97*, pages 134-148, 1997. Springer-Verlag.

[BFL91]  J. Boyar, K. Friedl, C. Lund. Practical zero-knowledge proofs: Giving hints and using Deficiencies. In *Journal of Cryptology*, 4(3), pages 185–206, 1991.

[BR93]  M. Bellare, P. Rogaway. Random Oracles are Practical: A Paradigm for Designing Efficient Protocols. newblock In *1st Conf. Computer & Comm. Security*, ACM, pages 62–73, 1993.

[BD-etal96]  Burmester, Desmedt, Itoh, Sakurai, Shizuya, Yung. A progress report on Subliminal-Free Channels. In *Workshop on Information Hiding*, Cambridge U.K., LNCS, pages 157–168, 1996.

[CGH98]  R. Canetti, O. Goldreich, S. Halevi. The Random Oracle Methodology, Revisited. In *ACM STOC '98*.

[C00]  J.-S. Coron. On the Exact Security of Full Domain Hash In *Advances in Cryptology—CRYPTO '00*, pages 229–235, 2000. Springer-Verlag.

[D89]  Y. Desmedt. Abuses in Cryptography and How to Fight Them. In *Advances in Cryptology—CRYPTO '89*, pages 375–389, 1990. Springer-Verlag.

[DH76]  W. Diffie, M. Hellman. New Directions in Cryptography. In volume IT-22, n. 6 of *IEEE Transactions on Information Theory*, pages 644–654, Nov. 1976.

[DSS91]  Proposed Federal Information Processing Standard for Digital Signature Standard (DSS). In volume 56, n. 169 of *Federal Register*, pages 42980–42982, 1991.

[ElG85]  T. ElGamal. A Public-Key Cryptosystem and a Signature Scheme Based on Discrete Logarithms. In *Advances in Cryptology—CRYPTO '84*, pages 10–18, 1985. Springer-Verlag.

[FY95]    Y. Frankel, M. Yung. Escrow Encryption Systems Visited: Attacks, Analysis and Designs. In *Advances in Cryptology—CRYPTO '95*, pages 222–235, 1987. Springer-Verlag.

[GHY89]  Z. Galil, S. Haber and M. Yung. Minimum-Knowledge Interactive Proof for Decision Problems. In *SIAM Journal on Computing*, 1988.

[GMRi88] S. Goldwasser, S. Micali and R. Rivest. Digital Signature Scheme Secure against Adaptive Chosen Plaintext Attack. In *SIAM Journal on Computing* 1988.

[JY00]    A. Juels and M. Yung. Manuscript.

[KL95]    J. Kilian and F.T. Leighton. Fair Cryptosystems Revisited. In *Advances in Cryptology—CRYPTO '95*, pages 208–221, 1995. Springer-Verlag.

[L98]     A. Lenstra. Generating RSA Moduli with Predetermined Portion. In *Advances in Cryptology—Asiacrypt '98*, pages 1–10, 1998. Springer-Verlag.

[LS]      M. Liskov, R. D. Silverman. A Statistical Limited-Knowledge Proof for Secure RSA Keys. Submitted to the IEEE P1363 Working Group. Available at http://grouper.ieee.org/groups/1363/contrib.htm.

[NR94]    K. Nyberg, R. Rueppel. Message Recovery for Signature Schemes Based on the Discrete Logarithm Problem. In *Advances in Cryptology—Eurocrypt '94*, pages 182–193, 1994. Springer-Verlag.

[NY89]    M. Naor, M. Yung. Universal One-Way Hash Functions and their Cryptographic Applications. STOC '89, pages 33–43, 19889. ACM.

[Ok92]    T. Okamoto. Provably Secure and Practical Identification Schemes and Corresponding Signature Schemes. In *Advances in Cryptology—Crypto '92*, pages 31–53, 1993. Springer-Verlag.

[PS00]    G. Poupard and J. Stern,  Short Proofs of Knowledge of Factoring.  In *PKC'2000*, LNCS 1751 Springer Verlag, pages 147–166, 2000.

[Ra78]    M. Rabin. A Public-key and Signature Scheme as Secure as Factoring, MIT Tech. Report, 1978.

[R98]     R. Rivest,  Chaffing and Winnowing: confidentiality without encryption. CryptoBytes 4 (1), 1998, pages 12–17.

[RSA78]   R. Rivest, A. Shamir, L. Adleman. A method for obtaining Digital Signatures and Public-Key Cryptosystems. In *Communications of the ACM*, volume 21, n. 2, pages 120–126, 1978.

[Si85]    G. Simmons. The Subliminal Channels and Digital Signatures. In *Advances in Cryptology—Eurocrypt '84*, pages 51–57, Springer-Verlag.

[Si93]    G. Simmons. Subliminal Communication is Easy Using the DSA. In *Advances in Cryptology—Eurocrypt '93*, Springer-Verlag.

[St96]    M. Stadler. Publicly Verifiable Secret Sharing. In *Advances in Cryptology—Eurocrypt '96*, pages 190–199.

[CaSt97]  J. Camenisch, M. Stadler. Efficient Group Signature Schemes for Large Groups. In *Advances in Cryptology—Crypto '97*, pages 410–424.

[YY96]    A. Young, M. Yung. The Dark Side of 'Black-Box' Cryptography, or: Should We Trust Capstone? In *Advances in Cryptology—CRYPTO '96*,pages 89–103, Springer-Verlag.

[YY97]    A. Young, M. Yung. Kleptography: Using Cryptography Against Cryptography. In *Advances in Cryptology—Eurocrypt '97*, pages 62–74, Springer-Verlag.

[YY98]    A. Young, M. Yung. Auto-Recoverable and Auto-Certifiable Cryptosystems In *Advances in Cryptology—Eurocrypt '98*, Springer-Verlag.

# A New Forward-Secure Digital Signature Scheme

Michel Abdalla[1] and Leonid Reyzin[2]

[1] Department of Computer Science & Engineering
University of California at San Diego
La Jolla, California 92093
mabdalla@cs.ucsd.edu
http://www-cse.ucsd.edu/users/mabdalla
[2] Laboratory for Computer Science
Massachusetts Institute of Technology
Cambridge, MA 02139
reyzin@theory.lcs.mit.edu
http://theory.lcs.mit.edu/~reyzin

**Abstract.** We improve the Bellare-Miner (Crypto '99) construction of signature schemes with forward security in the random oracle model. Our scheme has significantly shorter keys and is, therefore, more practical. By using a direct proof technique not used for forward-secure schemes before, we are able to provide better security bounds for the original construction as well as for our scheme.

Bellare and Miner also presented a method for constructing such schemes without the use of the random oracle. We conclude by proposing an improvement to their method and an additional, new method for accomplishing this.

**Keywords:** forward security, digital signatures, proven security, concrete security.

## 1 Introduction

### 1.1 The Problem

Many cryptographic techniques today, whether only available in the literature or actually used in practice, are believed to be quite secure. Several, in fact, can be proven secure (with appropriate definitions) under very reasonable assumptions. In a vast majority of solutions, however, security guarantees last only as long as secrets remain unrevealed. If a secret is revealed (either accidentally or via an attack), security is often compromised not only for subsequent uses of the secret, but also for prior ones. For example, if a secret signing key becomes known to an adversary, one cannot trust any signature produced with that key, regardless of when; if a secret decryption key becomes known to an adversary, then any encrypted message, even if sent long before, is not guaranteed to remain private.

To address this problem, several different approaches have been suggested. Many attempt to lower the chance of exposure of secrets by distributing them across several systems, usually via secret sharing. As pointed out in [3], this

T. Okamoto (Ed.): ASIACRYPT 2000, LNCS 1976, pp. 116–129, 2000.
© Springer-Verlag Berlin Heidelberg 2000

method is usually quite costly, and may, in fact, be too expensive to be implemented by a typical individual user. Moreover, since each of the systems may be susceptible to the same attack, the actual risk may not decrease.

A complementary approach is to reduce the potential damage in case secrets are exposed. In what is often called *forward security*, the main idea is to ensure that secrets are used only for short time periods, and that compromise of a secret does not affect anything based on secrets from prior time periods. One of the challenges in designing such a system is to be able to change secret information without the inconvenience of changing public information, such as the public key.

This approach has been known in the context of key agreement as *forward secrecy* [14,8]. In the context of digital signatures, it was first proposed, together with a few simple solutions, by Anderson in [2]. Bellare and Miner formalized Anderson's approach and provided more solutions in [3]

The specific problem addressed in this paper is that of designing a forward-secure signature scheme.

## 1.2   Forward-Secure Signature Schemes

Informally, a *key-evolving signature scheme* is one whose operation is divided into time periods, with a different secret key for each time period. Each secret key is used to sign messages only during a particular time-period, and to compute a new secret key at the end of that time period. It is then erased. As in ordinary signature schemes, however, there is only one public key, which remains the same through all the time periods. The verification algorithm checks not only that a signature is valid, but also that it was generated during a specific time period.

Such a scheme is *forward-secure* if it is infeasible for an adaptive chosen-message adversary to forge signatures for past time periods, even if it discovers the secret key for the current time period. Note that, in particular, this implies that past secret keys cannot be recovered from the current one. In a forward-secure signature scheme, even if the current secret key is compromised, signatures from past time periods can still be trusted.

Anderson [2] proposed a construction of forward-secure signature schemes in which the size of secret key (but not the public key) grows linearly with the number of time periods. The first forward-secure signature schemes in which key sizes do not grow linearly were proposed by Bellare and Miner in [3]. Their most efficient scheme, forward-secure in the random oracle model of [4] (assuming factoring is hard), uses ideas from the Fiat-Shamir [10] and Ong-Schnorr [16] identification and signature schemes.

As mentioned in [3], although still practical, their scheme requires very large keys, mainly because the original Fiat-Shamir scheme required very large keys (in fact, the forward-secure scheme of [3] does not add much to the already large key).

### 1.3   Our Contributions

MAIN RESULT. We propose a new forward-secure digital signature scheme, with much shorter keys than those in the scheme of [3]. In fact, our keys are comparable in size to those used in similar ordinary signature schemes.

Similarly to the scheme of [3], our scheme is based on signature schemes that are derived from three-round identification protocols. Specifically, the scheme is based on a generalized version of Micali's signature scheme [17], which is in many ways similar to the schemes of Ong-Schnorr [16], Guillou-Quisquater [13] and Ohta-Okamoto [19]. It is quite simple and efficient, although the computational efficiency of some components is less than that of the scheme of [3]. Our scheme can also be proven forward secure in the random oracle model, assuming factoring is hard.

OTHER CONTRIBUTIONS. While [3] use reduction to identification schemes to prove security, we use a direct proof technique. This enables us to provide a tighter exact security analysis for our scheme than the indirect technique of [3]. In fact, our technique can also be applied to the scheme of [3] to obtain a tighter security analysis for that scheme (which we present in Section 3.5).

We also present methods of achieving forward security in signature schemes without relying on random oracles. In general, they are less efficient than our main construction, and are not practical. However, they are still of interest, and can be viewed as an improvement on the tree-based construction of [3].

## 2   Definitions

All definitions provided here are based on those given in [3], which in turn are based on those given in [12] and [5]. Due to space constraints, we provide little discussion of our formal definitions; more discussion can be found in [3] and in the full version of our paper [1].

### 2.1   Forward-Secure Digital Signature Schemes

A forward-secure digital signature scheme is, first of all, a key-evolving digital signature scheme. A key-evolving signature scheme is very similar to a standard one, except that its operation is divided into time periods, each of which uses a different secret key to sign a message. The keys are updated by an algorithm that computes the secret key for the new time period based on the current secret key. Note that the public key stays the same.

**Definition 1.** *A key-evolving digital signature scheme is a quadruple of algorithms,* FSIG = (FSIG.key, FSIG.update, FSIG.sign, FSIG.vf), *where:*

- FSIG.key, *the* key generation *algorithm, takes as input a security parameter* $k \in \mathsf{N}$ *(given in unary as* $1^k$*) and the total number of periods* $T$ *and returns a pair* $(SK_0, PK)$*, the initial secret key and the public key;*

- FSIG.sign, *the* signing *algorithm, takes as input the secret key $SK_j$ for the current time period $j$ and the message $M$ to be signed and returns a pair $\langle j, sign \rangle$, the signature of $M$ for time period $j$;*
- FSIG.update, *the* secret key update *algorithm, takes as input the secret key for the current period $SK_j$ and returns the new secret key $SK_{j+1}$ for the next period.*
- FSIG.vf, *the* verification *algorithm, takes as input the public key $PK$, a message $M$, and a candidate signature $\langle j, sign \rangle$, and returns 1 if $\langle j, sign \rangle$ is a valid signature of $M$ or 0, otherwise.*

*It is required that* $\mathsf{FSIG.vf}_{PK}(M, \mathsf{FSIG.sign}_{SK_j}(M)) = 1$ *for every message $M$ and time period $j$. We also assume that the secret key $SK_j$ for time period $j \leq T$ always contains both the value $j$ itself and the value $T$ of the total number of periods. Finally, we adopt the convention that $SK_{T+1}$ is the empty string and that* $\mathsf{FSIG.update}_{SK_T}$ *returns* $SK_{T+1}$. ∎

When we work in the random oracle model, all the above-mentioned algorithms would additionally have oracle access to a public hash function $H$, which is assumed to be random in the security analysis.

SECURITY. Forward-security for key-evolving signature schemes is defined similarly to the way security is defined for classical signature schemes in [12], except that the adversary is allowed, in addition to the usual adaptive chosen-message attack, to "break-in" and learn the secret key for a given time period. Its task is then to forge a signature on a new message for a time-period *prior* to the one whose secret key it learned. Formally, this adversary is modeled via the following experiment (in the random-oracle model). In this experiment, the adversary is denoted by $F$, and works in either the chosen-message attack stage (cma) or the forgery stage (forge). It indicates its desire to switch from cma to forge by outputing the string breakin. Its state is preserved between invocations.

**Experiment** F-Forge-RO(FSIG, $F$)
    Select $H: \{0,1\}^* \to \{0,1\}^l$ at random
    $(PK, SK_0) \xleftarrow{R} \mathsf{FSIG.key}^H(k, \ldots, T)$
    $j \leftarrow 0$
    Repeat
        $j \leftarrow j + 1$
        $SK_j \leftarrow \mathsf{FSIG.update}^H(SK_{j-1})$ ; $d \leftarrow F^{H, \mathsf{FSIG.sign}^H_{SK_j}(\cdot)}(\mathsf{cma}, PK)$
    Until $(d = \mathsf{breakin})$ or $(j = T)$
    If $d \neq \mathsf{breakin}$ and $j = T$ then $j \leftarrow T + 1$
    $(M, \langle b, sign \rangle) \leftarrow F^H(\mathsf{forge}, SK_j)$
    If $\mathsf{FSIG.vf}^H_{PK}(M, \langle b, sign \rangle) = 1$ and $1 \leq b < j$
        and $M$ was not queried of $\mathsf{FSIG.sign}^H_{SK_b}(\cdot)$ in period $b$
        then return 1 else return 0

**Definition 2.** *Let* FSIG *be a key-evolving signature scheme, and $F$ an adversary. We let $\mathbf{Succ}^{\text{fwsig}}(\text{FSIG}[k, \ldots, T], F)$ denote the probability that the experiment $F\text{-}Forge\text{-}RO(\text{FSIG}[k, \ldots, T], F)$ returns 1. Then the insecurity of* FSIG *is the function*

$$\mathbf{InSec}^{\text{fwsig}}(\text{FSIG}[k, \ldots, T]; t, q_{\text{sig}}, q_{\text{hash}}) = \max_{F} \left\{ \mathbf{Succ}^{\text{fwsig}}(\text{FSIG}[k, \ldots, T], F) \right\},$$

*where the maximum here is taken over all adversaries $F$ making a total of at most $q_{\text{sig}}$ queries to the signing oracles across all the stages and for which the running time of the above experiment (including the time needed to answer the adversary's queries) is at most $t$ and at most $q_{\text{hash}}$ queries are made to the random oracle $H$.* ∎

The insecurity function above follows the concrete security paradigm and gives us a measure of how secure or insecure the scheme really is. Therefore, we want its value to be as small as possible.

## 2.2   Factoring

Let $A$ be an adversary for the problem of factoring Blum integers. That is, $A$ gets as input an integer $N$ that is the product of two primes, each congruent to 3 modulo 4, and tries to compute these prime factors. We define the following experiment using notation from [3].

**Experiment** Factor$(k, A)$
    Randomly choose two primes $p$ and $q$, such that:
        $p \equiv q \equiv 3 \pmod 4$, $2^{k-1} \le (p-1)(q-1)$, and $pq < 2^k$
    $N \leftarrow pq$
    $(p', q') \leftarrow A(N)$
    If $p'q' = N$ and $p' \ne 1$ and $q' \ne 1$ then return 1 else return 0

**Definition 3.** *[Factoring] Let $A$ be an adversary for the problem of factoring Blum integers and let $\mathbf{Succ}^{\text{fac}}(A, k)$ denote the probability that experiment* Factor$(k, A)$ *returns 1. The insecurity of factoring Blum integers is the function*

$$\mathbf{InSec}^{\text{fac}}(k, t) = \max_{A} \left\{ \mathbf{Succ}^{\text{fac}}(A, k) \right\},$$

*where the maximum here is taken over all adversaries $A$ for which the above experiment runs in time at most $t$.* ∎

## 3   Our Scheme

We start by explaining some number theory that provides intuition for our construction. We then present a slight variation of a signature scheme due to Micali [17]. The scheme has similarities to the schemes of Ong-Schnorr [16], Guillou-Quisquater [13] and Ohta-Okamoto [19] and, like they, is based on the idea of Fiat and Shamir [10] for converting identification schemes into signature schemes.

We then modify the signature scheme to make it forward-secure, and prove its security.

The schemes in this section are in the random oracle model. We will call the oracle $H : \{0,1\}^* \rightarrow \{0,1\}^l$.

## 3.1  Number Theory

Let $k$ and $l$ be two security parameters. Let $p_1 \equiv p_2 \equiv 3 \pmod{4}$ be two primes of approximately equal size and $N = p_1 p_2$ be a $k$-bit integer (such $N$ is called a *Blum* integer). To simplify further computations, we will assume not only that $N > 2^{k-1}$, but also that $|Z_N^*| = N - p_1 - p_2 + 1 \geq 2^{k-1}$. Let $Q$ denote the set of non-zero quadratic residues modulo $N$. Note that $|Q| \geq 2^{k-3}$. Note also that for $x \in Q$, exactly one of its four square roots is also in $Q$ (this follows from the fact that $-1$ is a non-square modulo $p_1$ and $p_2$ and the Chinese remainder theorem). Thus, squaring is a permutation over $Q$. From now on, when we speak of "the square root of $x$," we mean the single square root in $Q$; by $x^{2^{-k}}$ we will denote the single $y \in Q$ such that $x = y^{2^k}$.

Let $U \in Q$. Following [12], define $F_0(Z) = Z^2 \bmod N$, $F_1(Z) = UZ^2 \bmod N$, and, for an $l$-bit binary string $\sigma = b_1 \ldots b_l$, define $F_\sigma : Q \rightarrow Q$ as $F_\sigma(Z) = F_{b_l}(\ldots (F_{b_2}(F_{b_1}(Z)))\ldots) = Z^{2^l} U^\sigma \bmod N$ (note that $U^\sigma$ is a slight abuse of notation, because $\sigma$ is a binary string, rather than an integer; what is really meant here is $U$ raised to the power of the integer represented in binary by $\sigma$). Because squaring is a permutation over $Q$ and $U \in Q$, $F_\sigma$ is a permutation over $Q$.

Note that $F_\sigma(Z)$ can be efficiently computed by anybody who knows $N$ and $U$. Also, if one knows $p_1$ and $p_2$, one can efficiently compute $Z = F_\sigma^{-1}(Y)$ for a given $Y$ (as shown by Goldreich in [11]) by computing $S = 1/U^{2^{-l}} \bmod N$ and then letting $Z = Y^{2^{-l}} S^\sigma \bmod N$ (these calculations can be done modulo $p_1$ and $p_2$ separately, and the results combined using the Chinese remainder theorem). However, if one does not know the square root of $U$, then $F_\sigma^{-1}$ is hard to compute, as shown in the Lemma below (due to [12]).

**Lemma 1.** *Given $Y \in Q$, two different strings $\sigma$ and $\tau$ of equal length, $Z_1 = F_\sigma^{-1}(Y)$ and $Z_2 = F_\tau^{-1}(Y)$, one can compute $V \in Q$ such that $V^2 \equiv U \bmod N$.*

*Proof.* The proof is by induction on the length of the strings $\sigma$ and $\tau$.

If $|\sigma| = |\tau| = 1$, then assume, without loss of generality, that $\sigma = 0$ and $\tau = 1$. Then $F_0(Z_1) = F_1(Z_2) = Y$, i.e., $Z_1^2 \equiv UZ_2^2 \pmod{N}$, so we can set $V = Z_1/Z_2 \bmod N$.

For the inductive case, let $\sigma$ and $\tau$ be two strings of length $m + 1$. Let $\sigma'$ and $\tau'$ be their $m$-bit prefixes, respectively. If $F_{\sigma'}(Z_1) = F_{\tau'}(Z_2)$, we are done by the inductive hypothesis. Otherwise, the last bit of $\sigma$ must be different from the last bit of $\tau$, so, without loss of generality, assume the last bit of $\sigma$ is 0 and the last bit of $\tau$ is 1. Then $F_0(F_{\sigma'}(Z_1)) = F_1(F_{\tau'}(Z_2))$, and the same proof as for the base case works here.

We will now provide a geometric interpretation of the discussion above. Consider a complete binary tree of depth $l$ where each node stores a value in $Q$. The root (at the top of the tree) stores $Y$. The values at the children of a node that stores $A$ are $F_0^{-1}(A)$ at the left child and $F_1^{-1}(A)$ at the right child. Then computing $F_\sigma^{-1}(Y)$ means finding the value at the leaf for which the path from the root is given by $\sigma$ (where right-to-left in $\sigma$ corresponds to top-to-bottom in the tree).

It is clearly easy to compute the values "up" the tree from a given node. What the lemma says is that it is hard to compute the values "down" the tree without the ability to take square roots: in fact, if one knows two paths from the bottom of the tree, then one can get the square root of $U$ by looking at the children of the point where the two paths join together.

Finally, note that the value $R$ stored at the bottom-left leaf of the tree is $F_{00...0}^{-1}(Y) = Y^{2^{-l}}$, so if one knows $S = 1/U^{2^{-l}}$ and $R$, then one can compute the value at any leaf (given by $\sigma$) by computing $RS^\sigma \bmod n$.

## 3.2   The $2^l$-th Root Signature Scheme

The discussion above suggests the following signature scheme, which is similar to the schemes of [16] and [17] (an interactive three-round identification scheme can be designed similarly).

The signer generates a modulus $N$, picks a random $S \in Q$ to keep as its secret key, computes $U = 1/S^{2^l}$ and outputs $(N, U)$ as its public key.

To sign a message $M$, it first generates a random $R \in Q$ and computes $Y = R^{2^l}$. Note that this gives it the ability to find any leaf of the binary tree described above, rooted at $Y$. It therefore computes $\sigma = H(Y, M)$ and $Z = F_\sigma^{-1}(Y) = RS^\sigma \bmod N$ which it outputs as the signature.

The verifier checks that $Z \not\equiv 0 \pmod{N}$ and computes $Y' = F_\sigma(Z) = Z^{2^l}U^\sigma \pmod{N}$. It then verifies that $\sigma = H(Y', M)$.

We will not prove the security of this scheme here. The intuition, however, is the following: the verifier believes the signature because the signer was able to go down a random (given by $H$) path in the tree rooted at $Y$. Because the ability to go down two different paths implies the knowledge of the square root of $U$, the ability to go down a random path out of $2^l$ probably also implies that knowledge.

One point worth mentioning is that the verifier does not know if $U, Z \in Q$. All it knows is that $U, Z \not\equiv 0 \pmod{N}$, so either $U, Z \in Z_N^*$ or else one of the gcd's $(U, N)$, $(Z, N)$ gives a factorization of $N$. We therefore need the following reformulation of Lemma 1.

**Lemma 2.** *Given $Z_1, Z_2, U \in Z_N^*$ and two different strings $\sigma$ and $\tau$ of equal length such that $Z_1^{2^l}U^\sigma \equiv Z_2^{2^l}U^\tau \pmod{N}$, one can compute $V \in Z_N^*$ such that $V^2 \equiv U \pmod{N}$.*

*Proof.* The proof is the same as for Lemma 1.

In fact, now that we have this lemma, $S$ and $R$ picked by the signer need not be in $Q$: they can come from $Z_N^*$.

## 3.3   The Forward-Secure Signature Scheme

Note that the security of the above scheme hinges on the value $S$ and the number $l$ of squaring operations that separates it from $U$. It is $S$ that allows the signer to go from the leftmost leaf of the tree to any leaf and it is $l$ that determines the maximum depth of the tree.

Thus, a reasonable way of making the scheme forward-secure is to start out with a deep tree, and to use smaller and smaller depths for subsequent time periods. Then new values of $S$ can be obtained from old values of $S$ simply by squaring. Old values of $S$ cannot be recovered from new ones.

While making the tree deeper, however, there is no need to make it any wider. The width of the tree is only used to ensure that $\sigma$ is sufficiently random, so the adversary cannot guess what $\sigma$ will be and thus forge a signature. Therefore, the tree will remain complete to a certain sufficient depth, and from that point, each node will only have the left child (given by $F_0^{-1}$). The length of $\sigma$ will remain the same ($l$). This will make the scheme more efficient.

Now there is a question of how much up the tree we should go with each time period (that is, by how many squarings the current value of $S$ should be separated from the previous value $S'$). Note that, in order to compute a signature with respect to $S'$, one only needs $S'^\sigma$, not $S'$ itself. Thus, if $S \equiv S'^{2^x} \pmod{N}$ and the last $x$ bits of $\sigma$ are 0, then $S$ will allow one to compute the signature. Therefore, we should separate $S$ from $S'$ by $|\sigma|$ squarings, so that a forgery is possible for exactly one value of $\sigma$, as before. A smaller separation makes no sense without the corresponding reduction in the length of $\sigma$ and, therefore, the width of tree.

Having given the intuition, we refer the reader to Figure 1 for the complete description of our forward-secure scheme.

## 3.4   Security Analysis

We state the following theorem that will allow us to upper-bound the insecurity function for this signature scheme. Its proof combines ideas from [20], [3] and [18]. The proof technique used here can also be used to improve the bound on the insecurity function of the forward-secure scheme of [3] (see Section 3.5 for more details).

**Theorem 1.** *If there exists a forger $F$ for $\mathsf{FSIG}[k, l, T]$ that runs in time at most $t$, asking at most $q_{\mathrm{hash}}$ hash queries and $q_{\mathrm{sig}}$ signing queries, such that $\mathbf{Succ}^{\mathrm{fwsig}}(\mathsf{FSIG}[k, l, T], F) \geq \varepsilon$, then there exists an algorithm $A$ that factors Blum integers generated by $\mathsf{FSIG.key}(l, T)$ in expected time at most $t'$ with probability at least $\varepsilon'$, where*

```
algorithm FSIG.key(k, T)                    algorithm FSIG.update(SK)
begin                                       begin
   Generate random primes p₁, p₂ such that:    parse SK as (N, T, j, Sⱼ)
      p₁ ≡ p₂ ≡ 3  (mod 4)                      if  j = T then
      2^{k-1} ≤ (p₁ - 1)(p₂ - 1)                   SK ← ε
      p₁p₂ < 2^k                                else
   N ← p₁p₂                                        SK ← (N, T, j + 1, Sⱼ^{2^l} mod N)
   S₀ ←ᴿ Z*_N                                   return SK
   U ← 1/S₀^{2^{l(T+1)}} mod N              end
   SK ← (N, T, 0, S₀)
   PK ← (N, U, T)
   return (S, U)
end

algorithm FSIG.sign^H(M, SK)                algorithm FSIG.vf^H(M, PK, sign)
begin                                       begin
   parse SK as (N, T, j, Sⱼ)                   parse PK as (N, U, T)
   R ←ᴿ Z*_N                                   parse sign as (j, (Z, σ))
   Y ← R^{2^{l(T+1-j)}} mod N                  if Z ≡ 0  (mod N)
   σ ← H(j, Y, M)                                 return 0
   Z ← RSⱼ^σ mod N                             else
   return (j, (Z, σ))                            Y' ← Z^{2^{l(T+1-j)}} U^σ mod N
end                                             if σ = H(j, Y', M) then
                                                   return 1
                                                else
                                                   return 0
                                            end
```

**Fig. 1.** *Our forward-secure digital signature scheme*

$$t' = 2t + O(k^2 lT + k^3)$$

$$\varepsilon' = \frac{\left(\varepsilon - 2^{3-k} q_{\text{sig}}(q_{\text{hash}} + 1)\right)^2}{2T^2(q_{\text{hash}} + 1)} - \frac{\varepsilon - 2^{3-k} q_{\text{sig}}(q_{\text{hash}} + 1)}{2^{l+1}T}.$$

PROOF IDEA. To factor its input $N$, $A$ will select a random $x \in Z*_N$, compute $v = x^2 \bmod N$, and attempt to use the adversary to find a square root $y$ of $v$. Because $v$ has four square roots and $x$ is random, with probability $1/2$ we have that $x \not\equiv \pm y \pmod{N}$ and, hence $A$ will be able to find a factor of $N$ by computing the gcd of $x - y$ and $N$.

So, the task of $A$ is to find a square root of $v$ without using $x$. Note that $A$ gets to provide the public key for $F$ and to answer its signing and hashing queries. The idea, then, is to base to the public key $U$ on $v$ and run $F$ once to get a signature $(b, (Z, \sigma))$. Note that $F$ had to ask a hash query on $(b, Y, M)$ where $Y = Z^{2^{k(T+1-b)}} U^\sigma$—otherwise, the probability of its correctly guessing $\sigma$ is at most $2^{-l}$. Then, run $F$ the second time with the same random tape, giving the same answers to all the oracle queries before the query $(b, Y, M)$. For $(b, Y, M)$ give a new answer $\tau$. Then, if $F$ again forges a signature $(b, (Z', \tau))$ using $Y$ and $M$, we will have a condition similar to that of Lemma 2, and will be able to compute a square root of $v$. Please refer to the full version of this paper [1] for the actual proof.

**Theorem 2.** *Let* $\mathsf{FSIG}[k, l, T]$ *represent our key evolving signature scheme with modulus size* $k$, *challenge length* $l$, *and number of time periods* $T$. *Then for any* $t$, $q_{\mathrm{sig}}$, *and* $q_{\mathrm{hash}}$,

$$\mathbf{InSec}^{\mathrm{fwsig}}(\mathsf{FSIG}[k, l, T]; t, q_{\mathrm{sig}}, q_{\mathrm{hash}}) \leq$$
$$T\sqrt{2(q_{\mathrm{hash}} + 1)\mathbf{InSec}^{\mathrm{fac}}(k, t')} + 2^{-l}T(q_{\mathrm{hash}} + 1) + 2^{3-k}q_{\mathrm{sig}}(q_{\mathrm{hash}} + 1) \,,$$

*where* $t' = 2t + O(k^3 + k^2 l T)$.

*Proof.* The value for the insecurity function can be computed simply by solving for $(\varepsilon - 2^{3-k}q_{\mathrm{sig}}(q_{\mathrm{hash}} + 1))/T$ the quadratic equation in Theorem 1 that expresses $\varepsilon'$ in terms of $\varepsilon$ to get

$$(\varepsilon - 2^{3-k}q_{\mathrm{sig}}(q_{\mathrm{hash}} + 1))/T$$
$$= 2^{-l-1}(q_{\mathrm{hash}} + 1) + \sqrt{2^{-2l-2}(q_{\mathrm{hash}} + 1)^2 + 2\varepsilon'(q_{\mathrm{hash}} + 1)}$$
$$\leq 2^{-l-1}(q_{\mathrm{hash}} + 1) + \sqrt{2^{-2l-2}(q_{\mathrm{hash}} + 1)^2} + \sqrt{2\varepsilon'(q_{\mathrm{hash}} + 1)}$$
$$= 2^{-l}(q_{\mathrm{hash}} + 1) + \sqrt{2\varepsilon'(q_{\mathrm{hash}} + 1)},$$

and then solving the resulting inequality for $\varepsilon$.

## 3.5   Discussion

Note that, for any reasonable choices of $q_{\mathrm{sig}}$ and $q_{\mathrm{hash}}$, the minimally secure value for the modulus size $k$ (which should be greater than 512) makes the term $2^{3-k}q_{\mathrm{sig}}(q_{\mathrm{hash}} + 1)$ negligible. The term $2^{-l}T(q_{\mathrm{hash}} + 1)$ allows one to find a value for $l$ (the size of the hash values) that depends, mainly, on $q_{\mathrm{hash}}$ (which is the number of hash values an adversary is believed to be capable of computing).

Finally, the term $T\sqrt{2(q_{\mathrm{hash}} + 1)\mathbf{InSec}^{\mathrm{fac}}(k, t')}$ allows one to find the value for $k$ that depends, mainly, on the assumed insecurity of factoring and on $q_{\mathrm{hash}}$ (because $T$, which is related to the efficiency of the scheme, is probably much less than $q_{\mathrm{hash}}$).

Using our direct proof technique, the bound on the insecurity of the scheme of [3] can be improved by a factor of almost $\sqrt{T q_{\mathrm{hash}}}$ ([3] lose this factor by using

an indirect proof, which first reduces the security of the signature scheme to the
security of the corresponding identification scheme). The resulting bound is

$$T\sqrt{2l(q_{\text{hash}} + 1)\mathbf{InSec}^{\text{fac}}(k, t')} + 2^{-l}T(q_{\text{hash}} + 1) + 2^{3-k}q_{\text{sig}}(q_{\text{hash}} + 1) \, ,$$

which is worse than that of our scheme by a factor of at most $\sqrt{l}$. Thus, the two
schemes have almost the same security for the same parameters $l, k, q_{\text{sig}}, q_{\text{hash}}$.

The size of both the public and the private keys in the scheme of [3] is about
$k(l + 1)$ bits, while the size of the keys is in our scheme is about $2k$ bits. So the
keys in our scheme are about $(l + 1)/2$ times shorter.

The efficiency of key generation and update algorithms is about the same for
both schemes.

Signing for both scheme can be decomposed into two components: off-line
(before the message is known) and on-line (once the message is available). The
off-line component for time period $j$ for the scheme of [3] takes time $T - j + 1$
modular squarings, while for our scheme it takes $l$ times more. The on-line component takes about $l/2$ multiplications for [3] and $3l/2$ for our scheme. However,
because the on-line signing component in our scheme involves exponentiation of
a fixed based, precomputation techniques are available. Specifically, if the signer,
using a variation of the technique of Lim and Lee [15], precomputes 3 additional
powers of $S_j$ at the cost of increasing the secret key size by a factor of 2.5,
the on-line component will take about $l/2$ multiplications—as long as in the [3]
scheme. Precomputation of more values will reduce the on-line component of
signing even further, at the expense of the secret key length and the efficiency
of the update algorithm.

Finally, verification for time period $j$ for the scheme of [3] takes about $T +
1 - j + l/2$ modular multiplications, while in our scheme about $l(T + 1 - j) + 3l/2$
modular multiplications are needed. Again, precomputing powers of the public
key may be used to reduce the $3l/2$ term, but this term is not very significant
unless $j$ is close to $T$.

Thus, our scheme has slightly better security, much shorter keys, and comparable efficiency for the on-line component of signing. The efficiency of the off-line
component of signing and that of verifying is worse, however. Because each secret
key needs to be separated by $l$ squarings from the previous one (Section 3.3), we
believe that the efficiency of off-line signing and verifying cannot be improved
without a significant change in the design idea.

## 4    Schemes in the Standard Model

Both our scheme above and the Bellare-Miner's scheme were proven secure based
on the hardness of factoring and on the assumption that the hash function $H$
behaves like a random function. The main reason for this is that, when converting an identification scheme to a signature scheme (á la Fiat-Shamir [10]), the
challenge produced by the hash function should be as random as that produced
by an honest verifier, so as to maintain the security of this transformation.

One way of avoiding random oracles in the design of forward-secure signature schemes is to use the binary certification tree method suggested by Bellare and Miner [3]. It works as follows. Each node of the tree represents a pair of keys, a secret key and the related public key, used for an (ordinary) signature scheme. At the leaf level, each key is associated to a certain time period. Thus, the total number of leaves equals the total number of time periods. Each key at an internal node is used to certify the keys of its two children. The public key for the forward-secure scheme is the public key at the root of the tree. To sign a message in a certain time period, we use the secret key of the corresponding leaf and attach to the signature a certification chain based on the path from the root to that leaf so that the verifier can check the validity of the key itself. To maintain forward security, nodes are created dynamically. The secret key of an internal node is deleted as soon as it certifies the keys of its children. At any time, we only keep those keys on the path from the root to the leaf associated to the current time period, plus the right sibling of those nodes which are the left child of their parents. Consequently, as Bellare and Miner already pointed out, the lengths of both the secret key and signature are logarithmic in the total number of time slots.

Clearly, the scheme obtained via the binary tree certification method is less efficient than our scheme above and the random-oracle scheme of [3]. However, by properly instantiating the scheme, one can reduce its key length while maintaining its efficiency. The key observation for doing so is that we do not need the full power of ordinary signature schemes at the internal nodes, since they only need to certify two other nodes. Hence, we can use more "light-weight" schemes at these nodes, such as one-time signature schemes [9]. These are schemes which can only withstand single-message attacks, i.e. the signing key can be used only once. They are usually very efficient and have the potential for using smaller keys due to the restriction they impose on the attack. By using such schemes, we were actually able to achieve some improvements (see the full version of our paper [1]), but, unfortunately, given what is currently known, this still does not seem to give us a practical implementation without random oracles.

Another way of avoiding the use random oracles in the design of forward-secure signature schemes is by using ideas of Cramer and Damgård [6]. They show how to convert a secure identification scheme of the type commit-challenge-respond (which they refer to as *signature protocols*) into a secure signature scheme without relying on random oracles. The transformation is based on the idea of authentication trees. In this model, each message has a leaf associated to it. Signing a message is simply a matter of computing the path, which they call authentication path, from the leaf associated with that message to the root. To avoid having to precompute and store the whole tree, nodes are created dynamically in a way very similar to that of the GMR scheme. And like the GMR scheme, the resulting scheme is not memoryless and needs to remember the signature of the previous message to be able to compute the next signature. The length of each signature also grows logarithmically with the number of signed messages. This can, however, be improved to give a memoryless scheme, using the same modifications that Goldreich [11] suggested for the GMR scheme. The length of each signature will now be the same, although still logarithmic in the total number of messages ever to be signed.

In the case of forward security, we would have to start with a forward-secure identification scheme (such as the one given in [3]), and then apply to it the same type of transformation described above with one main difference: we also have to account for the index of the current time period. But we can easily do so by simply replacing a message in the original case by a pair message-index in our case. Although we do not prove this result, our claim is that forward security will be preserved. The main advantage of such an approach is that we can obtain a signature scheme which is forward secure based solely on the security of the corresponding identification scheme (and thus, if we use the scheme of [3], solely on the hardness of factoring). Moreover, the lengths of both the secret and public keys are independent of the total number of time periods. Its main disadvantages are that the resulting signature scheme would be far less efficient than the one we suggest in Section 3, and would have signatures whose length is a function of the total number of signed messages (and, therefore, related to the total number of time periods).

## 5   Acknowledgments

We are grateful to Mihir Bellare for encouraging us to work together and for advice along the way. The first author is supported by CAPES under Grant BEX3019/95-2. The second author is supported by the National Science Foundation Graduate Research Fellowship and a grant from the NTT corporation.

## References

1. M. ABDALLA AND L. REYZIN, "A New Forward-Secure Digital Signature Scheme," *Cryptology ePrint Archive Report 2000/002* at http://eprint.iacr.org/ (full version of this paper). Also available from authors' websites.
2. R. ANDERSON, Invited lecture, *Fourth Annual Conference on Computer and Communications Security*, ACM, 1997.
3. M. BELLARE AND S. MINER, "A forward-secure digital signature scheme," *Advances in Cryptology – Crypto 99 Proceedings*, Lecture Notes in Computer Science Vol. 1666, M. Wiener ed., Springer-Verlag, 1999.
4. M. BELLARE AND P. ROGAWAY, "Random oracles are practical: a paradigm for designing efficient protocols," *Proceedings of the First Annual Conference on Computer and Communications Security*, ACM, 1993.
5. M. BELLARE AND P. ROGAWAY, "The exact security of digital signatures: How to sign with RSA and Rabin," *Advances in Cryptology – Eurocrypt 96 Proceedings*, Lecture Notes in Computer Science Vol. 1070, U. Maurer ed., Springer-Verlag, 1996.
6. R. CRAMER AND I. DAMGÅRD, "Secure signature schemes based on interactive protocols," *Advances in Cryptology – Crypto 95 Proceedings*, Lecture Notes in Computer Science Vol. 963, D. Coppersmith ed., Springer-Verlag, 1995.

7. R. CRAMER AND V. SHOUP, "Signature schemes based on the Strong RSA Assumption," *Sixth Annual Conference on Computer and Communications Security*, ACM, 1999.

8. W. DIFFIE, P. VAN OORSCHOT, AND M. WIENER, "Authentication and authenticated key exchanges," *Designs, Codes and Cryptography*, 2, 1992, pp. 107–125.

9. S. EVEN, O. GOLDREICH, AND S. MICALI, "On-line/Off-line digital signatures," *Jounal of Cryptology*, Vol. 9, 1996, pp. 35–67.

10. A. FIAT AND A. SHAMIR, "How to prove yourself: Practical solutions to identification and signature problems," *Advances in Cryptology – Crypto 86 Proceedings*, Lecture Notes in Computer Science Vol. 263, A. Odlyzko ed., Springer-Verlag, 1986.

11. O. GOLDREICH, "Two remarks concerning the GMR signature scheme," *Advances in Cryptology – Crypto 86 Proceedings*, Lecture Notes in Computer Science Vol. 263, A. Odlyzko ed., Springer-Verlag, 1986.

12. S. GOLDWASSER, S. MICALI AND R. RIVEST, "A digital signature scheme secure against adaptive chosen-message attacks," *SIAM Journal of Computing*, Vol. 17, No. 2, pp. 281–308, April 1988.

13. L. GUILLOU AND J. QUISQUATER, "A practical zero-knowledge protocol fitted to security microprocessor minimizing both transmission and memory," *Advances in Cryptology – Eurocrypt 88 Proceedings*, Lecture Notes in Computer Science Vol. 330, C. Gunther ed., Springer-Verlag, 1988.

14. C. GÜNTHER, "An identity-based key-exchange protocol," *Advances in Cryptology – Eurocrypt 89 Proceedings*, Lecture Notes in Computer Science Vol. 434, J-J. Quisquater, J. Vandewille ed., Springer-Verlag, 1989.

15. C. H. LIM AND P .J. LEE, "More Flexible Exponentiation with Precomputation," *Advances in Cryptology – Crypto 94 Proceedings*, Lecture Notes in Computer Science Vol. 839, Y. Desmedt ed., Springer-Verlag, 1994

16. H. ONG AND C. SCHNORR, "Fast signature generation with a Fiat-Shamir like scheme," *Advances in Cryptology – Eurocrypt 90 Proceedings*, Lecture Notes in Computer Science Vol. 473, I. Damgård ed., Springer-Verlag, 1990.

17. S. MICALI, "A secure and efficient digital signature algorithm," *Technical Report MIT/LCS/TM-501*, Massachusetts Institute of Technology, Cambridge, MA, March 1994.

18. S. MICALI AND L. REYZIN, "Improving the exact security of Fiat-Shamir signature schemes." In R. Baumgart, editor, *Secure Networking — CQRE [Secure] '99*, volume 1740 of *Lecture Notes in Computer Science*, pages 167–182, Springer-Verlag, 1999.

19. K. OHTA AND T. OKAMOTO. "A Modification of the Fiat-Shamir Scheme," *Advances in Cryptology – Crypto 88 Proceedings*, Lecture Notes in Computer Science Vol. 403, S. Goldwasser ed., Springer-Verlag, 1988, pp. 232-243.

20. D. POINTCHEVAL AND J. STERN, "Security proofs for signature schemes," *Advances in Cryptology – Eurocrypt 96 Proceedings*, Lecture Notes in Computer Science Vol. 1070, U. Maurer ed., Springer-Verlag, 1996.

# Unconditionally Secure Digital Signature Schemes Admitting Transferability

Goichiro Hanaoka[1], Junji Shikata[1], Yuliang Zheng[2], and Hideki Imai[1]

[1] The Third Department, Institute of Industrial Science, the University of Tokyo
7-22-1 Roppongi, Minato-ku, Tokyo 106-8558, Japan
Phone & Fax: +81-3-3402-7365
{hanaoka,shikata}@imailab.iis.u-tokyo.ac.jp
imai@iis.u-tokyo.ac.jp
[2] School of Network Computing
Monash University, McMahons Road, Frankston
Melbourne, VIC 3199, Australia
Phone: +61 3 9904 4196, Fax: +61 3 9904 4124
yuliang.zheng@infotech.monash.edu.au

**Abstract.** A potentially serious problem with current digital signature schemes is that their underlying hard problems from number theory may be solved by an innovative technique or a new generation of computing devices such as quantum computers. Therefore while these signature schemes represent an efficient solution to the short term integrity (unforgeability and non-repudiation) of digital data, they provide no confidence on the long term (say of 20 years) integrity of data signed by these schemes. In this work, we focus on signature schemes whose security does not rely on any unproven assumption. More specifically, we establish a model for unconditionally secure digital signatures in a group, and demonstrate practical schemes in that model. An added advantage of the schemes is that they allow unlimited transfer of signatures without compromising the security of the schemes. Our scheme represents the first unconditionally secure signature that admits provably secure transfer of signatures.

## 1 Introduction

Digital signatures are an important technology for ensuring the unforgeability and non-repudiation of digital data. While some data may only require the assurance of integrity for a relatively short period of time (say up to 5 years), some other important data, such as court records and speeches by a parliamentarian, require the assurance of integrity for a long period of time (say up to 50 years).

Currently, digital signature schemes based on number theoretic problems are the prevalent methods used in providing data integrity. These schemes rely for their security on the assumed computational difficulty of computing certain number theoretic problems, such as factoring large campsites or solving discrete logarithms in a large finite field. RSA [20], Fiat-Shamir [11], ESIGN [19]

T. Okamoto (Ed.): ASIACRYPT 2000, LNCS 1976, pp. 130–142, 2000.

and many other schemes are based on the difficulty of factoring. On the other hand, ElGamal [10], Schnorr [24], DSA [9] and others, are based on discrete logarithms. Progress in computers as well as further refinement of various algorithms has made it possible to solve the number theoretic problems of larger sizes. As an example, in August 1999, a team of researchers from around the world succeeded in cracking an 512-bit RSA composite by the use of the Number Field Sieve [3] over the Internet. One can safely predict that even larger composites will be factored in the future. In addition, one cannot rule out the possibility of the emergence of innovative algorithms that solve efficiently these number theoretic problems in the future. More importantly, in the past few years there has been significant progress in quantum computers. It has been known that quantum computers can solve both factoring and discrete logarithm problems with ease [25,1], hence advances in the design and manufacturing of quantum computers poses a real threat to the long term security of all the digital signature schemes based on number theoretic problems.

The above discussions indicate the necessity of digital signature schemes that provide assurance of long term integrity. In the past decade, several attempts by various researchers have been made to address the problem. However, schemes proposed by these researchers are essentially variants of authentication codes, and none of these schemes has addressed the transferability of signatures among recipients.

The major contribution of this work is to propose the first digital signature scheme that admits transferability, and provable unconditional security against impersonation, substitution, and transfer with a trap. A potentially useful property of our proposed scheme is that a public key of a user can be associated with the user's unique name, resulting in an identity-based signature scheme.

## 1.1   Related Work

Chaum and Roijakkers [4] made the first attempt to construct an unconditionally secure signature scheme using cryptographic protocols. Their basic scheme was impractical, as it could only sign a single bit message. Furthermore, in their scheme, the level of security of a signature decreased as the signature moved from one verifier to another. In practice, it is important for a signature scheme to have *transferability*, i.e., its security is not compromised when a signature is transferred among users. Recently an improved version of Chaum-Roijakkers scheme has been proposed in [14]. However, the author of this improved scheme has not addressed the transferability of his signature scheme.

In another development, Chaum, Heijst and Pfitmann proposed a different version of unconditionally secure signature schemes [5]. However, its unconditional security was guaranteed only for signers.

There have also been attempts to modify unconditionally secure authentication codes [12,26] with the aim of enhancing the codes with extra security properties. It is tempting to transform an unconditionally secure authentication code into a digital signature. There are, however, two technical hurdles that are hard to overcome. First, authentication codes, especially the conventional Cartesian authentication codes, do not provide the function of non-repudiation, as a receiver can easily forge a sender's message and vice versa. Second, the receiver is always designated, meaning a signature cannot be verified by another party who does not have the shared key. These two properties must be removed for an authentication code to be converted into a digital signature.

An extension of authentication codes is *authentication codes with arbitration* or $A^2$-codes [27,28,15,16,18,14]. These codes involve a trusted third party called an arbiter. The arbiter can help resolve a dispute when a receiver forges a sender's message or the sender claims that a message is forged by the receiver. $A^2$-codes have been further improved to require a less trustworthy arbiter. These codes are called $A^3$-codes [2,7,13,29,14,30]. A property shared by both codes is that the receiver of a signature has to be designated.

As yet another extension, *multi-receiver authentication codes* (MRA) [8,21,14] have been extensively studied in the literature. With a MRA scheme, a broadcast message can be verified by any of the receivers. Although earlier MRA schemes required the sender to be designated, the so-called *MRA with dynamic sender* or DMRA have been proposed [22,23] to relax the requirement of a designated sender. It is important to note that these schemes make sense only in broadcasting. If MRA or DMRA is used for point-to-point authentication, then the sender can easily generate a fraudulent message that is accepted by the receiver, but not by other participants. The situation is made complex due to the fact that the same fraudulent message may have been generated by the receiver himself. A further problem is that MRA or DMRA does not provide transferability. In particular, if an authenticated message is transferred from one verifier to another, the second verifier can forge a message that appears to be perfectly valid to the next verifier. For the above reasons, neither MRA nor DMRA satisfies the non-repudiation requirement of a digital signature.

In summary, although unconditionally secure authentication codes can be enhanced to satisfy some of the properties of a digital signature, not all of the requirements can be fulfilled. Especially, none of the enhanced authentication schemes had addressed transferability.

## 1.2   Main Results

In this paper, we present an unconditionally secure identity-based signature scheme. First, we propose a novel model of a signature system called an *Identity-based Signature Schemes for Unconditional Security in a Group* (ISSUSG). As an example implementation of the model, a concrete $(n, \omega, \psi, p_1, p_2)$-*secure* scheme in ISSUSG is demonstrated, where $n$ indicates the total number of users, $\omega$ the maximum number of "bad" users who may collude, $\psi$ is the maximum number of signatures a user is allowed to generate, and $p_1$ and $p_2$ indicate the best probabilities for an attacker to succeed.

Our approach is an information theoretic one, and the security of our scheme does not rely on any assumption on the computational power of an attacker. Therefore, when the parameters of our scheme are properly chosen, the security of the scheme will not be affected by future advancement in computing or an algorithmic breakthrough in number theory. An important property of our scheme is that it admits unlimited transfer of signatures from one user to another, without compromising the security of the signature scheme in any way. A further advantage is that the scheme can be made identity-based by associating the unique name of a user to the signature generation and verification algorithms. The scheme is shown to achieve the lower bound on the required memory size of a signature.

As a by-product, we note that our unconditionally secure digital signature scheme can be used as an $A^3$-code and also as a DMRA. In fact, one may view our scheme as one that fulfills the requirements of both an $A^3$-code and a DMRA scheme.

The organization of the remaining part of this paper is as follows: In Section 2, we present our new model of an identity-based signature scheme for unconditional security, which we call an *Identity-based Signature Scheme for Unconditionally Security in a Group* (ISSUSG). In Section 3, a concrete unconditionally secure identity-based signature scheme in the model is proposed. In Section 4, some remarks related to our scheme are discussed. Section 5 presents the system-parameter settings when practical memory devices are used. In Section 6, we discuss how to handle long messages in our scheme. Finally, Section 7 concludes the paper with some final remarks.

## 2   The Model

In the model we consider, signatures are assumed to work in a group. Namely, only members in the group can generate and/or verify signatures. New users are allowed to join the group even after the system is set up, as long as the total number of users does not exceed a pre-defined threshold (this threshold is denoted by $n$). When the threshold is sufficiently large, in practice our signature scheme can be used in many applications when conventional public key signature schemes are used. Therefore, the group orientation of our scheme should not present any difficulties in practical applications.

We assume that there is a trusted authority, denoted by TA, and $n$ users $\mathcal{U} = \{U_1, U_2, \cdots, U_n\}$. For each user $U_i \in \mathcal{U}$ $(1 \le i \le n)$, for convenience we use the same symbol $U_i$ to denote the identity of the user. The TA produces a pair of signing and verification-keys on behalf of a user. Once being given a pair of keys, a user can then generate and/or verify signatures by using his own signing-key and verification-key, respectively. A more formal definition is given below:

**Definition 1** *A scheme $\Pi$ is an Identity-based Signature Scheme for Unconditional Security in a Group (ISSUSG) if it is constructed as follows:*

**1. Notation:**
   $\Pi$ *consists of* (TA, $\mathcal{U}$, $\mathcal{M}, \mathcal{S}, \mathcal{V}, \mathcal{A}$, **Sig**, **Ver**), *where*
   − *TA is a trusted authority,*
   − $\mathcal{U}$ *is a finite set of users (to be precise, users' unique names),*

- $\mathcal{M}$ *is a finite set of possible messages,*
- $\mathcal{S}$ *is a finite set of possible signing-keys,*
- $\mathcal{V}$ *is a finite set of possible verification-keys,*
- $\mathcal{A}$ *is a finite set of possible signatures,*
- **Sig** : $\mathcal{S} \times \mathcal{M} \longrightarrow \mathcal{A}$ *is a signing-algorithm,*
- **Ver** : $\mathcal{M} \times \mathcal{A} \times \mathcal{V} \times \mathcal{U} \longrightarrow \{accept, reject\}$ *is a verification-algorithm.*

2. **Key Pair Generation and Distribution by TA:**
   *For each user $U_i \in \mathcal{U}$, the TA chooses a signing-key $s_i \in \mathcal{S}$ and a verification-key $v_i \in \mathcal{V}$, and transmits the pair $(s_i, v_i)$ to $U_i$ via a secure channel. After delivering these keys, the TA erases the pair $(s_i, v_i)$ from his memory. And each user keeps secret both his signing-key and verification-key.*

3. **Signature Generation:**
   *For a message $m \in \mathcal{M}$, $U_i$ generates a signature $\alpha = $ **Sig**$(s_i, m) \in \mathcal{A}$ by using the signing-key in conjunction with the signing-algorithm. The pair $(m, \alpha)$ is regarded as a signed message of $U_i$.*

4. **Signature Verification:**
   *On receiving $(m, \alpha)$ from $U_i$, a user $U_j$ checks whether $\alpha$ is valid by using his verification-key $v_j$. More precisely, $U_j$ accepts $(m, \alpha)$ as a valid, signed message from $U_i$ if and only if* **Ver**$(m, \alpha, v_j, U_i) = accept.$

The main difference between our definition of signature schemes and that of conventional ones based on public-key cryptography lies in the fact that in our model each user is required to keep secret both his signing-key and verification-key.

In order to discuss in a formal way the security of a signature scheme in our model, we define the probability of success of various types of attacks. We consider three broad types of attacks: *impersonation, substitution* and *transfer with a trap*. Of these attacks, the first two are usually taken into account in discussing the security of authentication codes, especially $A^2$-codes, $A^3$-codes, and MRA codes. The third type of attacks, transfer with a trap, is new, and will be formally defined later.

Consider the case where there are $n$ users among whom up to $\omega$ user may be dishonest (and hence may collude). Each user is allowed to sign up to $\psi$ signatures. We now discuss in a more formal way the three types of attacks.

1) *Impersonation:*
   $t$ users, with $t \leq \omega$, launch an attack against a pair of users $U_i$ and $U_j$ by generating a signed message with the hope that $U_j$ accepts it as being a valid signature from $U_i$. This attack may be executed after the colluders observe at most $\psi(n-1)$ signed messages generated by users other than $U_i$.

2) *Substitution:*
   $t$ users, with $t \leq \omega$, construct a fraudulent message $m'$ to replace a message genuinely signed by $U_i$, with the hope that $U_j$ will accept it as being an authentic message from $U_i$. This attack may be executed after the colluders observe at most $\psi n$ signed messages generated by any users. Among the observed messages, at least one but up to $\psi$ may be generated by $U_i$.

3) *Transfer with a trap:*

   After $U_j$ receives a valid pair $(m, \alpha)$ from $U_i$, $t$ colluders, where $t \leq \omega$, attempt to generate a new pair $(m, \alpha')$ with $\alpha \neq \alpha'$. Note that both the singer $U_i$ and the user $U_j$ could be among the colluders. The colluders hope that another user $U_k$ will accept $(m, \alpha')$ as being a valid message-signature pair from $U_i$, but no other users will. The risk with this attack is that when $U_j$ transfers such a pair $(m, \alpha')$ to $U_k$ and $U_k$ then transfers it to another user $U_l$, $U_l$ finds that the pair is invalid. When this happens, $U_k$ is in a sense trapped by the colluders.

To formally define the probabilities of success in the above three attacks, some notations are introduced first.

Let $\mathcal{W} := \{W \subset \mathcal{U} \mid |W| \leq \omega\}$. Each element of $\mathcal{W}$ represents a group of possibly colluding users. Let $s_W$ and $v_W$ be the set of signing-keys and that of verification-keys for a $W \in \mathcal{W}$, respectively.

**Definition 2** *The success probabilities of impersonation, substitution and transfer with a trap attacks, denoted by $P_I$, $P_S$ and $P_T$ respectively, are formally defined as follows:*

1) *Success probability of impersonation: for $W \in \mathcal{W}$ and $U_i, U_j \in \mathcal{U}$ with $U_i, U_j \notin W$, we define $P_I(U_i, U_j, W)$ as*

$$P_I(U_i, U_j, W) := \max_{s_W, v_W} \max_{1 \leq k \leq n, k \neq i} \max_{c_k = \{(m_{k,l}, \alpha_{k,l})\}} \max_{(m, \alpha)}$$
$$\Pr(U_j \text{ accepts } (m, \alpha) \text{ as valid from } U_i | s_W, v_W, \{c_k\}),$$

*where $c_k = \{(m_{k,l}, \alpha_{k,l})\}$ is taken over a family of possible sets of valid signed messages generated by $U_k$ $(1 \leq k \leq n, \ k \neq i)$ such that $0 \leq |c_k| \leq \psi$ $(1 \leq k \leq n, \ k \neq i)$. Note that $m_{k,l}$ are not necessarily distinct. Then, $P_I$ is given as*

$$P_I := \max_{U_i, U_j, W} \Pr(U_i, U_j, W)$$

*where $W \in \mathcal{W}$ and $U_i, U_j \in \mathcal{U}$ with $U_i, U_j \notin W$.*

2) *Success probability of substitution: for $W \in \mathcal{W}$ and $U_i, U_j \in \mathcal{U}$ with $U_i, U_j \notin W$, we define $P_S(U_i, U_j, W)$ as*

$$P_S(U_i, U_j, W) := \max_{s_W, v_W} \max_{1 \leq k \leq n} \max_{c_k = \{(m_{k,l}, \alpha_{k,l})\}} \max_{(m, \alpha)}$$
$$\Pr(U_j \text{ accepts } (m, \alpha) \text{ as valid from } U_i | s_W, v_W, \{c_k\})$$

*where $c_k = \{(m_{k,l}, \alpha_{k,l})\}$ is taken over a family of possible sets of valid signed messages generated by $U_k$ $(1 \leq k \leq n)$ such that $0 < |c_i| \leq \psi$ and $0 \leq |c_k| \leq \psi$ $(1 \leq k \leq n, \ k \neq i)$, and $(m, \alpha)$ is taken such that $m \neq m_{i,l}$ for any $l$. Note that $m_{k,l}$ are not necessarily distinct. Then, $P_S$ is given as*

$$P_S := \max_{U_i, U_j, W} \Pr(U_i, U_j, W)$$

*where $W \in \mathcal{W}$ and $U_i, U_j \in \mathcal{U}$ with $U_i, U_j \notin W$.*

*3) Success probability of transfer with a trap: for $W \in \mathcal{W}$ and $U_i, U_j \in \mathcal{U}$ with $U_j \notin W$ we define $P_T(U_i, U_j, W)$ as*

$$P_T(U_i, U_j, W) := \max_{s_W, v_W} \max_{1 \leq k \leq n, k \neq i} \max_{c_k = \{(m_{k,l}, \alpha_{k,l})\}} \max_{(m,\alpha)} \max_{(m,\alpha')}$$
$$\Pr(U_j \text{ accepts } (m, \alpha') \text{ as valid from } U_i | s_W, v_W, \{c_k\}, (m, \alpha))$$

*where $c_k = \{(m_{k,l}, \alpha_{k,l})\}$ is taken over a family of possible sets of valid signed messages generated by $U_k$ ($1 \leq k \leq n$, $k \neq i$) such that $0 \leq |c_k| \leq \psi$ ($1 \leq k \leq n$, $k \neq i$), $(m, \alpha)$ is taken over the set of possible signed messages generated by $U_i$, and $\alpha'$ is taken such that $\alpha \neq \alpha'$. Then, $P_T$ is given as*

$$P_T := \max_{U_i, U_j, W} \Pr(U_i, U_j, W)$$

*where $W \in \mathcal{W}$ and $U_i, U_j \in \mathcal{U}$ with $U_j \notin W$.*

Now we are ready to define the concept of an $(n, \omega, \psi, p_1, p_2)$-secure ISSUSG signature scheme. Here both $p_1$ and $p_2$ are security parameters whose meanings will be made precise in the following definition.

**Definition 3** *Let $\Pi$ be an ISSUSG with $n$ users. Then, $\Pi$ is $(n, \omega, \psi, p_1, p_2)$-secure if the following conditions are satisfied: as long as there exist at most $\omega$ colluders and each user is allowed to generate at most $\psi$ signatures, the following inequalities hold:*

$$\max\{P_I, \ P_S\} \leq p_1$$
$$P_T \leq p_2$$

*where $P_I$, $P_S$ and $P_T$ are the probabilities of success in impersonation, substitution and transfer with a trap attacks, respectively.*

We note that there is an alternative definition of security in which one may use a single security parameter $p$ instead and define the success probability as

$$\max\{P_I, \ P_S, \ P_T\} \leq p.$$

In practice, however, some applications may attach more weight to strength against impersonation and substitution than against transfer with a trap, while some other applications may have an emphasis on robustness against transfer with a trap. By introducing two separate parameters $p_1$ and $p_2$, we have an opportunity to design a signature scheme with fine-tuned level of security.

## 3    Implementation

### 3.1    Protocol

In this section, an implementation of the ISSUSG will be presented. It is constructed by the use of a polynomial with $\omega + 2$ variables over a finite field.

As before, let $\mathcal{U} := \{U_1, U_2, \cdots, U_n\}$ be the set of $n$ users and TA the trusted authority.

## 1. Key Pair Generation and Distribution by TA:

Let $F_q$ be the finite field with $q$ elements such that $q \geq n$. The TA picks uniformly at random $n$ elements $v_1, v_2, \ldots, v_n$ in $F_q^\omega$ for users $U_1, U_2, \ldots, U_n$ respectively, and constructs a polynomial $F(x, y_1, \ldots, y_\omega, z)$ as follows:

$$F(x, y_1, \ldots, y_\omega, z) = \sum_{i=0}^{n-1} \sum_{k=0}^{\psi} a_{i0k} x^i z^k + \sum_{i=0}^{n-1} \sum_{j=1}^{\omega} \sum_{k=0}^{\psi} a_{ijk} x^i y_j z^k$$

where the coefficients $a_{ijk}$ are chosen uniformly at random from $F_q$. Moreover, we assume that a user's identity $U_l$ and a message $m$ are also from $F_q$.

For each user $U_l$ $(1 \leq l \leq n)$, the TA computes a *signing-key* $s_l :=$ $F(U_l, y_1, \ldots, y_\omega, z)$, and a *verification-key* $\tilde{v}_l := F(x, v_l, z)$. $v_l$ and $\tilde{v}_l$ together form a pair of verification-keys for user $U_l$. The TA then sends both the signing-key and the pair of verification-keys to $U_l$ over a secure channel. Once the keys are delivered, there is no need for the TA to keep the user's keys.

## 2. Signature Generation:

For a message $m \in F_q$, $U_i$ generates a signature by

$$\alpha = F(U_i, y_1, \ldots, y_\omega, z)|_{z=m} = F(U_i, y_1, \ldots, y_\omega, m)$$

using his signing-key.

## 3. Signature Verification:

On receiving $(m, \alpha)$ from $U_i$, user $U_j$ checks whether $\alpha$ is valid by the use of his verification-keys $v_j$ and $\tilde{v}_j$. More specifically, $U_j$ calculates evaluation values $r_1, r_2$ using his verification-keys $\tilde{v}_j = F(x, v_j, z)$ and $v_j := (v_{1,j}, \ldots, v_{\omega,j})$ as follows:

$$r_1 := F(x, v_j, z)|_{x=U_i, z=m},$$
$$r_2 := \alpha|_{(y_1, \ldots, y_\omega)=(v_{1,j}, \ldots, v_{\omega,j})}.$$

$U_j$ accepts $(m, \alpha)$ as being a valid message-signature pair from $U_i$ if and only if $r_1 = r_2$.

We can show that the above signature scheme is an $(n, \omega, \psi, (\frac{2}{q} - \frac{1}{q^2}), \frac{1}{q})$-secure ISSUSG scheme.

**Theorem 1** *The above scheme results in an $(n, \omega, \psi, (\frac{2}{q} - \frac{1}{q^2}), \frac{1}{q})$-secure ISSUSG scheme.*

Due to the lack of space, the proof of Theorem 1 is omitted. It will be provided in the full version of this paper.

The above scheme can be modified slightly, resulting in yet another $(n, \omega, \psi, \frac{1}{q}, \frac{1}{q-1})$-secure ISSUSG scheme.

**Theorem 2** *In the above construction, the following modification produces also an $(n, \omega, \psi, \frac{1}{q}, \frac{1}{q-1})$-secure ISSUSG scheme:*
*Instead of choosing randomly, the TA may choose $n$ elements $v_1, \ldots, v_n \in F_q^\omega$, as verification-keys, such that for any $\omega + 1$ vectors*

$$v_{i_1} = (v_{1,i_1}, \ldots, v_{\omega,i_1}), \ldots, v_{i_{\omega+1}} = (v_{1,i_{\omega+1}}, \ldots, v_{\omega,i_{\omega+1}}),$$

*the $\omega + 1$ new vectors $(1, v_{1,i_1}, \ldots, v_{\omega,i_1}), \ldots, (1, v_{1,i_{\omega+1}}, \ldots, v_{\omega,i_{\omega+1}})$ are linearly independent.*

Note that our scheme can be used in place of an authentication code, MRA or DMRA. In fact our scheme is cryptographically stronger than the authentication codes, with an added benefit of being transferable, although it requires more memory space than MRA and DMRA.

## 3.2   Memory Sizes

The following theorem states the required memory size for our construction, and its proof is obvious.

**Theorem 3** *The required memory size in the above constructions is given as follows:*

$$|\mathcal{A}| = q^{(\omega+1)}, \qquad \textit{(size of signature)}$$
$$|\mathcal{S}| = q^{(\omega+1)(\psi+1)}, \qquad \textit{(size of signing-key)}$$
$$|\mathcal{V}| = q^{\omega+n(\psi+1)}, \qquad \textit{(size of verification-key)}.$$

**Corollary 1** *The construction proposed in Theorem 2 is optimal in terms of the memory size of a signature.*

The proof follows from [23].

It is not yet clear to the authors as to whether the scheme also achieves optimality in terms of memory size for signing-keys and verification-keys.

# 4   Some Remarks on Our Scheme

This section shows useful extensions of the scheme presented above, and discusses some of the properties of the scheme. More detailed discussions will be provided in the full version of this paper.

## 4.1   Signature Scheme for $t$ Senders

In some applications, users who might sign are specified first. When there are only $t$ specified senders in the system, we can easily specialize our scheme to produce a *signature scheme for $t$ senders*. Namely, by changing the degree $n-1$ of $x$ in $F(x, y_1, \cdots, y_\omega, z)$ to $t-1$, a signature scheme for $t$ senders is obtained. Based on this restriction, the required memory for verification-keys can be reduced from $q^{\omega+n(\psi+1)}$ to $q^{\omega+t(\psi+1)}$. Note that the required memory sizes for signatures and signing-keys are still the same as in the non-restrictive scheme.

## 4.2   Arbiter

We can also introduce an *arbiter* which can resolve a dispute between a signer and a recipient. In one such implementation, the arbiter will be given a pair of verification-keys, whereas no user will. The arbiter can notify users of the result of verification of a signature. We note that any user can play the role of an arbiter for other users.

## 4.3   Reduction of Memory Size for Verification-Key

In the proposed schemes in Section 3, the degree of $x$ in $F(x, y_1, \cdots, y_\omega, z)$ is set as $n - 1$. If the degree of $x$ is $\omega + d$ instead $(d \leq n - \omega - 2)$, the system may be attacked as follows: when the same message is signed by $d + 1$ signers, $\omega$ colluders can forge a victim's signature of the same message by using their own secrets and the generated signatures. To prevent the scheme from this attack, the degree of $x$ is set to $n - 1$, which is the primary contributor to the required memory size of verification-keys.

If in a practical system it is known that the chance for the same message to be signed by $d+1$ signers is extremely small, the degree of $x$ may be set to be smaller than $n - 1$. This will reduce the required memory size for verification-keys.

## 4.4   Active Attacks against Verification-Keys

As already discussed earlier, the proposed scheme is unconditionally secure against passive attacks. In an active attack, an adversary may manage to obtain some information on verification-keys. As an example, by selecting a random element from $F_q^{\omega+1}$ as a forged signature and obtaining the verification result from a targeted victim, the adversary obtains some information on the victim's verification-key. We can show that the information obtained does not help succeed with a non-negligible probability in impersonation, substitution or transfer with a trap. Thus such an active attack is not an issue in practice. Details will be presented in the full paper.

# 5   Practical Systems Based on Memory Devices

In this section, we discuss the values of security parameters in the proposed schemes. Table 1 shows the value of $\psi$ according to the values of the number of users and memory devices which may contain users' signing-keys, assuming the worst case where $\omega = n - 1$. One can see that using commonly available memory devices, the number of signatures that can be generated by a user is sufficiently large even for a large organization that has 1,000 to 10,000 users.

Table 2 gives data on a more realistic setting. One can see that compared with the previous table, the number of signatures that can be signed by a user increases significantly.

We note that the capacity of memory devices is getting larger and larger, and their prices are dropping as fast. This helps significantly the usability of the proposed signature scheme.

**Table 1.** The number of signatures a user can generate, assuming that $|q|$ has 160 bits and $\omega = n - 1$.

| | $n = 1,000$ | $n = 10,000$ | $n = 100,000$ | $n = 1,000,000$ |
|---|---|---|---|---|
| 2HD disk(1.44MByte) | 71 | 6 | 0 | 0 |
| ZIP(100MByte) | 4,999 | 499 | 49 | 4 |
| CD-R(650MByte) | 32,499 | 3,249 | 324 | 31 |
| DVD-RAM(5.2GByte) | 259,999 | 25,999 | 2,599 | 259 |

**Table 2.** The number of signatures a user can generate, assuming that $|q|$ has 160 bits and $\omega$ is determined appropriately for each $n$.

| | $n = 1,000$ | $n = 10,000$ | $n = 100,000$ | $n = 1,000,000$ |
|---|---|---|---|---|
| | $\omega = 500$ | $\omega = 2,000$ | $\omega = 10,000$ | $\omega = 50,000$ |
| 2HD disk(1.44MByte) | 142 | 34 | 6 | 0 |
| ZIP(100MByte) | 9,979 | 2,497 | 498 | 98 |
| CD-R(650MByte) | 64,869 | 16,240 | 3,248 | 648 |
| DVD-RAM(5.2GByte) | 518,961 | 129,934 | 25,996 | 5,198 |

## 6  On Handling Long Messages

In our proposed scheme, the length of messages to be signed is restricted to be $|q|$ or less. An important question that is yet to be addressed is how to handle longer messages, without significantly increasing the size of such a message.

In practice, one may use the technique of applying a one-way hashing to a long message prior to signing it. Some examples of one-way hash algorithms are SHA-1 [17], HAVAL [31] and RIPEMD-160 [6]. Although this will lose the unconditional security property of the proposed signature scheme, we note that a good one-way hash function would remain secure even if one employed quantum computers in attacking it.

## 7  Conclusions

We have proposed unconditionally secure identity-based signature schemes. More specifically, we have established a model for unconditionally secure digital signatures in a group, and demonstrated practical schemes in that model. An added advantage of the scheme is that it allows unlimited transfer of signatures without compromising the security of the scheme. Although there is a limit on the number of signatures a user can generate, this limitation is not an issue in practice thanks to the development in inexpensive memory devices with a huge capacity. Specifically, by using a DVD-RAM, 25,999 signatures can be generated by a user in an organization of 10,000 employees.

We are currently working on other possible implementations of ISSUSG, as well as the problem on how to sign long message without losing unconditional security.

## Acknowledgments

Part of this work was supported by Research for the Future Program (RFTF), Japan Society for the Promotion of Science (JSPS), under contract number JSPS-RETF 96P00604. The first author was supported by a Research Fellowship from JSPS.

The authors would like to thank Tsutomu Matsumoto and Tsuyoshi Nishioka for their valuable comments. The first author would like also to thank Yumiko C. Hanaoka for her help in preparing this paper.

## References

1. D. Boneh and R. J. Lipton, "Quantum cryptanalysis of hidden linear functions," Proc. of CRYPTO'95, LNCS 963, Springer-Verlag, pp.424-437, 1995.
2. E. F. Brickell and D. R. Stinson, "Authentication codes with multiple arbiters," Proc. of Eurocrypt'88, LNCS 330, Springer-Verlag, pp.51-55, 1988.
3. S. Cavallar, B. Dodson, A. K. Lenstra, et al., "Factorization of a 512-bit RSA modulus," Proc. of Eurocrypt'00, LNCS 1807, Springer-Verlag, pp.1-18, 2000.
4. D. Chaum and S. Roijakkers, "Unconditionally secure digital signatures," Proc. of CRYPTO'90, LNCS 537, Springer-Verlag, pp.206-215, 1990.
5. D. Chaum, E. Heijst and B. Pfitzmann, "Cryptographically strong undeniable signatures, unconditionally secure for the signer," Proc. of CRYPTO'91, LNCS 576, Springer-Verlag, pp.470-484, 1991.
6. H. Dobbertin, A. Bosselaers and B. Preneel, "RIPEMD160: strengthened version of RIPEMD," Proc. of FSE'96, LNCS 1039, Springer-Verlag, pp.71-82, 1996.
7. Y. Desmedt and M. Yung, "Arbitrated unconditionally secure authentication can be unconditionally protected against arbiter's attack," Proc. of CRYPTO'90, LNCS 537, Springer-Verlag, pp.177-188, 1990.
8. Y. Desmedt, Y. Frankel and M. Yung, "Multi-receiver/Multi-sender network security: efficient authenticated multicast/feedback," Proc. of IEEE Infocom'92, pp.2045-2054, 1992.
9. "Proposed federal information processing standard for digital signature standard (DSS)," Federal Register, vol. 56, no. 169, 30, pp.42980-42982, 1991.
10. T. ElGamal, "A public key cryptosystem and a signature scheme based on discrete logarithms," IEEE Trans. on Inform. Theory, IT-31, 4, pp.469-472, 1985.
11. A. Fiat and A. Shamir, "How to prove yourself: practical solutions to identification and signature problems," Proc. of CRYPTO'86, LNCS 263, Springer-Verlag, pp.186-194, 1986.
12. E. N. Gilbert, F. J. MacWilliams and N. J. A. Sloane, "Codes which detect deception," Bell System Technical Journal, 53, pp.405-425, 1974.
13. T. Johansson, "Lower bounds on the probability of deception in authentication with arbitration", IEEE Trans. Inform. Theory, IT-40, 5, pp.1573-1585, 1994.
14. T. Johansson, "Further results on asymmetric authentication schemes," Information and Computation, 151, pp.100-133, 1999.

15. K. Kurosawa, "New bound on authentication code with arbitration," Proc. of CRYPTO'94, LNCS 839, Springer-Verlag, pp.140-149, 1994.
16. K. Kurosawa and S. Obana, "Combinatorial bounds for authentication codes with arbitration," Proc. of Eurocrypt'95, LNCS 921, Springer-Verlag, pp.289-300, 1995.
17. NIST, "Secure hash standard," *FIPS PUB 180-1*, Department of Commerce, Washington D.C., 1995.
18. S. Obana and K. Kurosawa, "$A^2$-code = affine resolvable + BIBD," Proc. of ICICS'97, LNCS 1334, Springer-Verlag, pp.118-129, 1997.
19. T. Okamoto, "A fast signature scheme based on congruential polynomial operations," IEEE Trans. on Inform. Theory, IT-36, 1, pp.47-53, 1990.
20. R. Rivest, A. Shamir and L. Adleman, "A method for obtaining digital signature and public-key cryptosystems," Communication of the ACM, vol.21, no.2, pp.120-126, 1978.
21. R. Safavi-Naini and H. Wang, "New results on multi-receiver authentication codes," Proc. of Eurocrypt'98, LNCS1403, pp.527-541, 1998.
22. R. Safavi-Naini and H. Wang, "Broadcast authentication in group communication," Proc. of Asiacrypt'99, LNCS1716, Springer-Verlag, pp.399-411, 1999.
23. R. Safavi-Naini and H. Wang, "Multireceiver authentication codes: models, bounds, constructions and extensions," Information and Computation, 151, pp.148-172, 1999.
24. C. Schnorr, "Efficient signature generation by smart cards," Journal of Cryptology, 4, pp.161-174, 1991.
25. P. W. Shor, "Polynomial-time algorithms for prime factorization and discrete logarithms on a quantum computer," SIAM J. Comp., 26, no.5, pp.1484-1509, 1997.
26. G. J. Simmons, "Authentication theory/coding theory," Proc. of CRYPTO'84, LNCS 196, Springer-Verlag, pp.411-431, 1984.
27. G. J. Simmons, "Message authentication with arbitration of transmitter/ receiver disputes," Proc. of Eurocyrpt'87, Springer-Verlag, pp.151-165, 1987.
28. G. J. Simmons, "A Cartesian construction for unconditionally secure authentication codes that permit arbitration," Journal of Cryptology, 2, pp.77-104, 1990.
29. R. Taylor, "Near optimal unconditionally secure authentication," Proc. of Eurocyrpt'94, LNCS 950, Springer-Verlag, pp.244-253, 1994.
30. Y. Wang and R. Safavi-Naini, "$A^3$-codes under collusion attacks," Proc. of Asiacrypt'99, LNCS 1716, Springer-Verlag, pp.390-398, 1999.
31. Y. Zheng, J. Pieprzyk and J. Seberry, "HAVAL - A one-way hashing algorithm with variable length of output," Proc. of Auscrypt'92, LNCS 718, Springer-Verlag, pp.83-104, 1993.

# Efficient Secure Multi-party Computation[*]
## (Extended Abstract)

Martin Hirt[1], Ueli Maurer[1], and Bartosz Przydatek[2][**]

[1] ETH Zurich, Switzerland,
{hirt,maurer}@inf.ethz.ch
[2] Carnegie Mellon University, USA,
bartosz@cs.cmu.edu

**Abstract.** Since the introduction of secure multi-party computation, all proposed protocols that provide security against cheating players suffer from very high communication complexities. The most efficient unconditionally secure protocols among $n$ players, tolerating cheating by up to $t < n/3$ of them, require communicating $\mathcal{O}(n^6)$ field elements for each multiplication of two elements, even if only one player cheats.

In this paper, we propose a perfectly secure multi-party protocol which requires communicating $\mathcal{O}(n^3)$ field elements per multiplication. In this protocol, the number of invocations of the broadcast primitive is independent of the size of the circuit to be computed. The proposed techniques are generic and apply to other protocols for robust distributed computations.

Furthermore, we show that a sub-protocol proposed in [GRR98] for improving the efficiency of unconditionally secure multi-party computation is insecure.

## 1 Introduction

### 1.1 Secure Multi-party Computation

The goal of secure multi-party computation, as introduced by Yao [Yao82], is to enable a set of $n$ players to compute an arbitrary agreed function of their private inputs. The computation must guarantee the correctness of the outputs while preserving the secrecy of the players' inputs, even if some of the players are corrupted by an active adversary and misbehave maliciously.

As the first general solution to this problem, Goldreich, Micali, and Wigderson [GMW87] presented a protocol, based on cryptographic intractability assumptions, which allows $n$ players to securely compute an arbitrary function even if an adversary corrupts any $t < n/2$ of the players. In the secure-channels model, where bilateral secure channels between every pair of players are assumed, Ben-Or, Goldwasser, and Wigderson [BGW88] and independently Chaum, Crépeau,

---

[*] Research supported by the Swiss National Science Foundation (SNF), SPP project no. 5003-045293. Full version at http://www.inf.ethz.ch/department/TI/um/ .
[**] Research done at ETH Zurich, Switzerland.

T. Okamoto (Ed.): ASIACRYPT 2000, LNCS 1976, pp. 143–161, 2000.

and Damgård [CCD88] proved that unconditional security is possible if at most $t < n/3$ of the players are corrupted. In a model where additionally physical broadcast channels are available, unconditional security is achievable if at most $t < n/2$ players are corrupted [RB89, Bea91b, CDD+99].

## 1.2   Efficiency Considerations

All proposed multi-party protocols that provide security against misbehaving players suffer from high communication complexities. This is in sharp contrast to their private (but non-resilient) counterparts, for which reasonably efficient solutions are known [BGW88]. The communication overhead of resilient multi-party protocols over private protocols is due mainly to the sophisticated techniques for achieving resilience against faults. Specifically, these techniques make extensive use of a broadcast primitive, which must be realized with a protocol for Byzantine agreement (e.g., [PSL80, DFF+82, FM88, BGP89, CW89]). Such protocols are very communication-intensive. The necessity of the broadcast channel is independent of whether or not actual faults occur: often broadcast is used to complain about an inconsistency, but when no inconsistency is detected, the players must nevertheless broadcast a confirmation message (the inherent information of the message is one bit). Many researchers take a broadcast channel for granted, neglecting the fact that this primitive does not exist in most realistic scenarios for distributed computing, and hence must be simulated. Broadcast is an efficiency bottleneck, in both information-theoretic and cryptographic settings; reducing the number of broadcast invocations is therefore crucial for reducing the overall communication complexity of distributed protocols.

There is a line of research that focused on reducing the communication complexity of multi-party protocols. First, several works [BB89, BMR90, BFKR90] concentrated on reducing the round complexity of such protocols. However, the price for the low round complexity is a substantially increased message complexity. With the current results, namely $\mathcal{O}(n^6)$ field elements per multiplication, the main efficiency bottleneck seems to be the message complexity rather than the round complexity. First steps towards lower message complexities were taken in [BFKR90]. The proposed protocol is very efficient, but it only tolerates adversaries corrupting up to $t = \mathcal{O}(\log n)$ players. Protocols with optimal resilience (i.e., $t < n/3$) were proposed in [FY92] and in [GRR98]. Their approach is to first perform a private protocol with fault-detection (for the whole protocol in [FY92], and for a part of the protocol in [GRR98]), and only in case of faults to repeat the computation with a slow but resilient protocol. Although this approach can improve the best-case complexity of the protocol (when no adversary is present), it cannot speed up the protocol in the presence of a malicious adversary: a single corrupted player can persistently enforce the robust but slow execution, annihilating (and even inverting) any efficiency gain.

## 1.3  Contributions

This paper significantly improves the message complexity of unconditionally secure multi-party computations, without increasing the round complexity in a relevant manner. We consider a set of $n$ players, where up to $t < n/3$ of them can be corrupted by a computationally unbounded, adaptive, active adversary. We present a protocol that allows the players to securely compute an agreed function specified as an arithmetic circuit over a finite field $\mathbb{F}$, requiring communication of $\mathcal{O}(mn^3)$ field elements, where $m$ denotes the number of multiplication gates in the circuit. The total number of invocations of the broadcast primitive in the whole protocol is only $\mathcal{O}(n^2)$, independent of the circuit size.

This is to be compared with the most efficient unconditionally secure protocol known so far, namely the protocol of Beaver [Bea91a], which requires $\mathcal{O}(mn^6)$ field elements. Other protocols whose goal is to improve the message complexity of unconditionally secure multi-party protocols [FY92, GRR98] fail to do so in the presence of faults. The new protocol improves even on the cryptographically secure protocol [GRR98], which communicates $\mathcal{O}(mn^4)$ field elements[1] (but tolerates up to $t < n/2$ corruptions). Recently, a protocol with cryptographic security for evaluating Boolean circuits was proposed in which $\mathcal{O}(mn^3k)$ bits are communicated, where $k$ is a security parameter [CDN00]. The round complexities of all considered protocols are essentially equal. All stated complexities include the costs of simulating the broadcast channels by a protocol for Byzantine agreement.

The techniques that allow this speed-up are generic and apply to many protocols for general multi-party computation as well as to special-purpose protocols, in both the cryptographic model and the information-theoretic model. One key technique is *player elimination*. In contrast to previous protocols where only evident misbehavior leads to elimination and where slowing down the protocol is still possible without being detected, we proceed more rigorously: Whenever a fault occurs (and slows down the protocol execution), a set of players which contains at least a certain number of corrupted players (but possibly also some honest ones) is identified and eliminated from the further protocol execution. This ensures that faults occur only rarely, namely at most $t$ times during the entire computation, which in turn allows to reduce the number of consistency checks performed in the protocol: Rather than after each gate, the consistency checks are performed only after a sequence of gates, a so-called *segment*. During the entire computation, up to $t$ segments can fail and require re-computation, but with an appropriate size of the segments, the total cost of re-computation will be much smaller than the savings due to the reduced number of the checks.

Furthermore, we show that the very efficient protocol of [GRR98] for the verification of equality of shared values is insecure (cf. App. A), thus invalidating previously stated efficiency improvements.

---

[1] In this protocol, the field must be large for security reasons.

## 1.4   Outline

In Sect. 2 we introduce the general framework for efficient resilient protocols. This framework is not specific for multi-party computation. The new multi-party computation protocol is described in Sect. 3, and its efficiency is analyzed and compared with known protocols in Sect. 4. Finally, some conclusions and open problems are mentioned in Sect. 5.

# 2   Framework for Efficient Resilient Protocols

## 2.1   Introduction

Distributed protocols resilient against misbehavior of some of the players require in general much more communication than their private (but non-resilient) counterparts, even when no cheating occurs. The reasons for this contrast are two-fold: First, in a model where players might deviate from the protocol, expensive consistency checks must be performed frequently, and agreement must be reached on whether or not faults occurred. Second, if indeed at least one player misbehaves, then inconsistencies will occur, and costly fault-recovery procedures must be applied. Note that the consistency checks are necessary even when no cheating occurs, whereas fault recovery is necessary only when at least one player misbehaves.

   In this section, we describe a framework for efficient resilient protocols that overcomes these disadvantages. The key idea is to eliminate at least one malicious player (and potentially some honest players) each time a fault is detected. Hence the number of fault-recovery invocations is bounded by the maximal number of corrupted players and is independent of the length of the protocol. Furthermore, the resulting seldom occurrence of faults allows to reduce the frequency of consistency checks and thereby to significantly reduce the communication-overhead caused by them.

   The techniques presented in this section apply to many applications in several models, including those relying on intractability assumptions. The adversary can be static or adaptive, but not mobile: A mobile adversary [OY91, CH94] may release some of the corrupted players during the protocol execution and thereby regain the capability of corrupting new players, which contradicts the idea of elimination of corrupted players.

## 2.2   Incorporating Resilience into a Private Protocol

We consider a private protocol that proceeds in rounds (e.g., in each round one gate is evaluated) and wish to execute this protocol in a resilient manner. In contrast to the classical approach to resilient protocols, where after each round some consistency checks are performed and agreement on whether or not a fault occurred is reached, we divide the protocol into *segments*, each consisting of a sequence of rounds, and only at the end of each segment the consistency of the data held by the players is checked and the players agree on whether or not a fault occurred (*fault detection*). If a fault is detected, then a set of players is identified which contains at least a certain number of cheaters (*fault localization*), the players in the set are eliminated from the further protocol execution (*player elimination*), and the failed segment is repeated (*fault correction*). If privacy is an issue, then after each round some checks must be performed, but no agreement must be reached on the fact whether or not a fault occurred (*weak fault detection*).

During a protocol consisting of $m$ rounds, the classical approach invokes $m$ times fault detection and, if at least one player misbehaves permanently, $m$ times fault-recovery. In our approach, where the protocol is divided into segments of $m_s$ rounds, only the weak fault detection is invoked $m$ times. Fault detection is performed $m/m_s$ times, and fault localization, player elimination, and fault correction are invoked at most $t$ times. By selecting $m_s$ appropriately, the overhead for the (in total up to $t$) repetitions of a segment will not dominate the total complexity of the protocol, and the costs of fault detection and fault localization are independent of $m$ (and polynomial in $n$). In many applications, this will significantly reduce the overall complexity of the protocol.

We now describe the steps in more detail:

1. **Private computation with weak fault detection.** All rounds of the segment are computed according to the private computation. The computation of this step must be *verifiable*, i.e. it must be possible to check later (see below) whether or not any faults occurred. However, *robustness* is not required, i.e. if faults occur, then the computation may fail (in such a case it must be possible to perform an appropriate fault localization, see below). In order to preserve privacy even in case of faults, consistency checks are performed after each round, and every player sends to every other player one bit indicating whether or not he observed an inconsistency. A player who observed or was informed about an inconsistency will use default (random) dummy values unrelated to the actual values in all further rounds of the segment.

2. **Fault detection.** The goal of fault detection is to reach agreement on whether or not a fault occurred during the current segment. Typically, fault detection is achieved by having every player broadcast (with a protocol for Byzantine agreement) a binary message according to whether or not he observed or was informed about an inconsistency in any round of the current segment, and a fault is detected if at least one player complains. The following steps 3. to 5. are performed if and only if a fault is detected.

3. **Fault localization.** The purpose of fault localization is to find out which players are corrupted or, because agreement about this can usually not be reached, at least to narrow down the set of players containing the cheaters. The output of fault localization is a set $\mathcal{D}$ with $|\mathcal{D}| = p$ players, guaranteed to contain at least $r$ cheaters, denoted as $(r, p)$-*localization*.

4. **Player elimination.** The set $\mathcal{D}$ agreed upon during fault localization is eliminated from the further computation. In general, after eliminating the players in $\mathcal{D}$, the protocol cannot be continued immediately, but it must be

transformed to capture the new setting with $n - p$ players and at most $t - r$ cheaters.

5. **Fault correction.** Since some players are eliminated whenever a fault is detected, faults can be corrected simply by repeating the current segment of the protocol.

# 3    Constructing Efficient Multi-party Computation Protocols

In this section we present a construction of efficient multi-party computation protocols in the secure-channels model, based on the framework with player-elimination from the previous section. We first formally define the considered model, then we describe the main (top-level) protocol and finally all required sub-protocols.

## 3.1    Model

We consider the well-known secure-channels model as used in [BGW88, CCD88]: The set $\mathcal{P} = \{P_1, \ldots, P_n\}$ of $n$ players is connected by bilateral synchronous reliable secure channels. Broadcast channels are not assumed to be available. The goal of the protocol is to compute an agreed function, specified as an arithmetic circuit over a finite field $\mathbb{F}$ with $|\mathbb{F}| > n$. The number of inputs to the circuit is denoted by $n_I$, the total number of outputs by $n_O$,[2] the number of multiplication gates in the circuit by $m$, and the multiplicative depth by $d$ (i.e., the maximal number of multiplication gates in any path of the circuit). To each player $P_i$ a unique public value $\alpha_i \in \mathbb{F} \setminus \{0\}$ is assigned. There are no further assumptions about the field.[3] The computation of the function must be secure with respect to a computationally unbounded adaptive active adversary who can corrupt up to $t$ of the players, where $t$ is a given threshold with $t < n/3$. Once a player is corrupted, the adversary can read all his information and can make the player misbehave arbitrarily. The security of our protocol is perfect, i.e. unconditional with zero failure probability. Formal definitions of security can be found in [Can00] and in [MR98], and our protocol is secure for any of these definitions.

To simplify the presentation, we adopt the following convention throughout the description of the protocols: Unless otherwise stated, whenever a player does not receive an expected message, or receives a malformed message, then a default value for this message is taken.

---

[2] $n_O$ specifies the total number of outputs — if the same value is given as output to several players, then this value is counted several times.

[3] This is in contrast to the protocol in [BGW88], where the existence of an $n$-th root of unity in $\mathbb{F}$ is assumed.

## 3.2   Main Protocol

The protocol follows the classical approach for secure multi-party computation: First, each player secret-shares his input(s) among the players. Second, the circuit is evaluated with the shared values. Third, the output value(s) are reconstructed towards the authorized players.

According to the framework from Sect. 2, the circuit will be divided into segments. If the evaluation of a segment fails, then some players are eliminated and the segment is repeated. Clearly, *all* players must be able to provide input and receive output, including players that are eliminated in the protocol evaluation (also honest players can be eliminated). This is achieved by using a resilient protocol (which does not make use of the player-elimination technique) for sharing input values. No special measures are necessary for receiving output, because the secret-reconstruction protocol can also be performed towards an eliminated player (this player only receives values and cannot cause inconsistencies).

**Sharing.** The sharing is based on Shamir's secret-sharing scheme [Sha79], extended to a two-dimensional sharing [GHY87, BGW88, CCD88, RB89, FHM98]. Each value is shared among the players with a polynomial of degree $t$, and each share is again shared among the players with a polynomial of degree $t$. Formally, a value $s$ is $t$-*shared* among the players if there exist degree-$t$ polynomials $f$ and $f_1, \ldots, f_n$ with $s = f(0)$ and $f_i(0) = f(\alpha_i)$. The information held by player $P_i$ is the share $s_i = f(\alpha_i)$, the polynomial $f_i$, and the share-shares $s_{ji} = f_j(\alpha_i)$ (for $j = 1, \ldots, n$). The polynomials in the sharing must be randomly chosen such that any set of $t$ players does not obtain any information about the secret.

**Segmentation.** Due to the linearity of the secret-sharing scheme, linear functions of shared values can be computed non-interactively, and hence only multiplication gates are relevant for the communication complexity. In order to partition the circuit with $m$ multiplication gates and multiplicative depth $d$ into segments, we select an ordering of the gates which satisfies the partial order defined by the circuit (i.e., the inputs of the $i$-th gate must be provided by gates with index smaller than $i$). Every segment consists of a number of consecutive gates, subject to the following bounds:

- the number $m_s$ of multiplication gates in each segment is at most $\lceil m/n \rceil$,
- the multiplicative depth $d_s$ of each segment is at most $\lceil d/n \rceil$.

Furthermore, in every segment (except the last) at least one of the above bounds is satisfied with equality, hence the total number of segments is smaller than $2n$.

At the end of every segment, fault detection is performed and agreement is reached on whether or not a fault occurred within the segment. If no fault occurred, then the computation of this segment is completed, and the next segment is started. If a fault is detected, then a $(1, 2)$-localization $\mathcal{D} \subset \mathcal{P}$ will be found and eliminated (we will not consider other types of localizations), and the evaluation of the segment is repeated. During the whole circuit evaluation, at most $t$ segments fail. The described segmentation guarantees that the repeated computation will not dominate the overall protocol complexity, neither in terms of the number of communicated bits nor in terms of the number of communication rounds.

**Protocol Overview.** Let $\mathcal{P}$ denote the set of players, where $n = |\mathcal{P}|$, and $t < n/3$ the upper bound on the number of cheaters. During the computation, players can be eliminated, and then $\mathcal{P}'$ will denote the set of remaining players, $n' = |\mathcal{P}'|$, and $t'$ the upper bound on the number of cheaters in this set.

0. Set $\mathcal{P}' := \mathcal{P}$, $n' := n$, $t' := t$.

1. Input stage: Every player $P$ providing input secret-shares his input value (Sect. 3.3).

2. Computation stage (Sect. 3.4): For each segment of the circuit:

   2.1 For each gate in the segment (all gates at the same level can be evaluated in parallel):
   - If the gate is linear: Call the sub-protocol for the evaluation of linear functions.
   - If the gate is a multiplication gate: Call the multiplication sub-protocol. Players that have detected (or were notified about) a fault earlier in this segment use default shares.

   2.2 For each $P_i \in \mathcal{P}'$, broadcast one bit according to whether or not a fault was observed (or notified) in the segment. If at least one player reports a fault, then the segment fault-localization procedure is invoked to find a $(1, 2)$-localization $\mathcal{D}$, and $\mathcal{P}'$ is set to $\mathcal{P}' \setminus \mathcal{D}$, $t'$ is set to $t' - 1$, and step 2. is restarted (for the same segment).

3. Output stage: For every player $P$ that is to receive output: Call the sub-protocol for receiving output (Sect. 3.5).

## 3.3   Input Stage

In the input stage, every player secret-shares his input(s). Let $\mathcal{P}$ be the set of players, at most $t$ of which are corrupted, and let $P$ be a designated dealer holding a secret input $s$. The protocol for providing $s$ as input is a variation of the verifiable secret-sharing (VSS) protocol of Ben-Or, Goldwasser and Wigderson [BGW88]:

1. DISTRIBUTION. The dealer $P$ selects at random a polynomial $p(x, y) = \sum_{i,j=0}^{t} r_{ij} x^i y^j$ of degree $t$ in both variables, where $p(0, 0) = s$, and sends the polynomials $f_i(x) = p(x, \alpha_i)$ and $\widetilde{f}_i(y) = p(\alpha_i, y)$ to player $P_i$ (for $i = 1, \ldots, n$).[4] This implicitly defines the polynomial $f(x) = p(0, x)$.

2. CONSISTENCY CHECKS. Each pair of players $P_i, P_j$ (for $1 \leq i, j \leq n$) checks whether $f_i(\alpha_j) \overset{?}{=} \widetilde{f}_j(\alpha_i)$. For this, $P_i$ sends $f_i(\alpha_j)$ to $P_j$, and $P_j$ checks whether the received value is equal to $\widetilde{f}_j(\alpha_i)$.

---

[4] An efficiency gain of a factor 2 can be achieved by setting $r_{ij} = r_{ji}$, and hence $f_i(x) = \widetilde{f}_i(x)$. One can prove that privacy is not violated by this technique. See [CDM00] for more details.

3. COMPLAINT STAGE. Every player broadcasts a message (containing one bit) indicating whether all consistency checks were successful or at least one test failed. In case of a complaint, the player afterwards broadcasts a bit-vector, where the $j$-th bit indicates whether or not the player has observed an inconsistency with player $P_j$. The dealer answers the complaints by broadcasting the corresponding correct values.

4. ACCUSATION STAGE. If a player $P_j$ observes more than $t$ inconsistencies or discovers that the dealer's answers contradict his own values, he broadcasts an accusation. In such a case the dealer broadcasts both polynomials $f_j(x)$ and $\tilde{f}_j(y)$. The published polynomials can cause some new inconsistencies with the values held by some other players, who react again with accusations, and so on.[5] If more than $t$ players have accused, or if the dealer did not answer all the complaints and accusations, a default sharing (e.g., the constant sharing of 0) is taken.

In the protocol of [BGW88], the share of player $P_i$ is $s_i = f(\alpha_i) = f_i(0)$, and the second dimension of the sharing is not used. In our scheme, the share of player $P_i$ is the polynomial $f_i$ (and in particular $s_i = f_i(0)$), as well as the share-shares $s_{ji} = \tilde{f}_i(\alpha_j) = p(\alpha_i, \alpha_j)$ (for $j = 1, \ldots, n$).

In order to analyze the security of this secret-sharing protocol we distinguish two cases: (a) If the dealer is honest, all shares and share-shares of honest players will be consistent, and only values held by corrupted players can be published. No honest player will accuse the dealer, hence there will be at most $t$ accusations. Clearly, in this case the outcome will be a proper $t$-sharing. (b) If the dealer is corrupted, then at the end of the protocol (if there were not more than $t$ accusations) the cross-over points of all honest players are consistent, and their share-shares uniquely define a two-dimensional polynomial $p'(x, y)$, satisfying the conditions for a proper $t$-sharing. If there were more than $t$ accusations, then at least one of the accusations origins from an honest player, and indeed the dealer is cheating. In this case it is legitimate to take some default value as the dealer's secret.

## 3.4   Computation Stage

The computation of the circuit proceeds segment by segment. We denote the current set of players with $\mathcal{P}'$, where $n' = |\mathcal{P}'|$, and the current upper bound on the number of cheaters in $\mathcal{P}'$ with $t'$. Without loss of generality, we assume that $\mathcal{P}' = \{P_1, \ldots, P_{n'}\}$. A segment is computed as follows: First, the gates of the segment are computed. Linear functions can be computed robustly (as no communication is needed). In contrast, the computation of multiplication gates is private and verifiable, but not robust. At the end of each multiplication sub-protocol, the (honest) players inform each other in a weak fault detection

---

[5] One can show that two rounds of accusations are sufficient to reach agreement. After two rounds of accusations, either the total number of accusations exceeds $t$, or all accusations in the second round originate from corrupted players.

procedure whether or not they observed an inconsistency. If a player observed such an inconsistency, or was informed about one in weak fault detection, then he continues the computation of the segment with default values independent of the actual shares. At the end of each segment, fault detection is performed and, if necessary, fault localization, player elimination and fault correction.

**Linear Functions.** Let $\mathcal{L}$ be a linear function, and assume that the values $a, b, \ldots$ are $t$-shared with polynomials $f, f_1, \ldots, f_{n'}, g, g_1, \ldots, g_{n'}, \ldots$, respectively. Due to the linearity of $\mathcal{L}$, the polynomials $h = \mathcal{L}(f, g, \ldots)$ and $h_i = \mathcal{L}(f_i, g_i, \ldots)$ define a $t$-sharing of $c = \mathcal{L}(a, b, \ldots)$. Hence, player $P_i$ can compute his share of $c$ as $h_i = \mathcal{L}(f_i, g_i, \ldots)$ and $c_{ji} = \mathcal{L}(a_{ji}, b_{ji}, \ldots)$ (for $j = 1, \ldots, n'$). The privacy of this protocol is trivial (there is no communication), and the correctness is due to the linearity of the sharing.

**Multiplication.** The crucial sub-protocol for multiplication is a *re-sharing* protocol. A re-sharing protocol is a protocol that takes a degree-$\gamma$ sharing of a value $s$ and generates an independent degree-$\delta$ sharing of $s$. This re-sharing is possible in a verifiable (but non-robust) manner if $t' < n' - \gamma$. Privacy can be guaranteed if $t' \leq \gamma$ and $t' \leq \delta$.

The protocol for computing the $t$-shared product $c$ of two $t$-shared values $a$ and $b$ proceeds in three steps: First, both inputs $a$ and $b$ are re-shared with degree $t'$. Second, every player locally multiplies his respective shares and share-shares of $a$ and $b$, resulting in a degree-$2t'$ sharing of $c$. And third, this degree-$2t'$ sharing of $c$ is re-shared to a degree-$t$ sharing.

We have to show that the necessary (and sufficient) conditions for all re-sharings are satisfied: After a sequence of $k$ $(1, 2)$-localizations and eliminations, we have $n' = n - 2k$ and $t' = t - k$. The requirements for the re-sharing are $t' < n' - t$ and $t' < n' - 2t'$, and both are satisfied for $3t < n$.

*Re-sharing protocol.* The goal of re-sharing is to transform a $\gamma$-sharing of a value $s$ into a proper and independent $\delta$-sharing of $s$, where $t' < n' - \gamma$, $t' \leq \gamma$ and $t' \leq \delta$. The re-sharing sub-protocol can fail in the presence of malicious players. However, if it fails, all (honest) players will learn so, and at the end of the segment, agreement on whether or not such a fault occurred will be reached and the segment will be repeated if necessary.

Roughly speaking, our re-sharing protocol works along the lines of degree reduction of [BGW88, GRR98], but it is significantly more efficient, due to various techniques in the spirit of the player-elimination framework (cf. Sect. 2).

Assume that $s$ is $\gamma$-shared with the polynomials $f$ and $f_1, \ldots, f_{n'}$, and player $P_i$ holds the polynomial $f_i(x)$ (hence his share $s_i = f_i(0)$), and his share-shares $s_{ji} = f_j(\alpha_i)$ (for $j = 1, \ldots, n'$). The value $s$ can be expressed as a linear combination (Lagrange interpolation) of the values $s_1, \ldots, s_{n'}$ [BGW88, GRR98]. Therefore, once the values $s_1, \ldots, s_{n'}$ are $\delta$-shared, the required $\delta$-sharing of $s$ can be computed by a distributed evaluation of the appropriate linear function (as described in Sect. 3.4). Thus, the re-sharing can be performed as follows:

Every player $\delta$-shares his share $s_i$, proves that the shared value is indeed $s_i$, and computes his degree-$\delta$ share of $s$ as a linear combination of the received shares of $s_1, \ldots, s_{n'}$.

We describe the steps in more detail:

1. NON-ROBUST VSS. Every player $P_i$ shares his share $s_i$ with the degree-$\delta$ polynomials $h^{(i)}, h_1^{(i)}, \ldots, h_{n'}^{(i)}$ in a non-robust but verifiable manner. The protocol works like the first two steps of the VSS in the input stage (Sect. 3.3):

   a) $P_i$ selects at random a polynomial $p^{(i)}(x, y)$ of degree $\delta$ in both variables, where $p^{(i)}(0, 0) = s_i$, and sends the polynomials $h_j^{(i)}(x) = p^{(i)}(x, \alpha_j)$ and $\tilde{h}_j^{(i)}(y) = p^{(i)}(\alpha_j, y)$ to player $P_j$ (for $j = 1, \ldots, n'$). This implicitly defines the polynomial $h^{(i)}(x) = p^{(i)}(0, x)$.

   b) Each pair of players $P_j, P_k$ (for $1 \le j, k \le n'$) verifies the equality of their common shares. For this, $P_j$ sends $h_j^{(i)}(\alpha_k)$ to $P_k$, who then checks whether the received value is equal to $\tilde{h}_k^{(i)}(\alpha_j)$.

2. PROVING CORRECTNESS. Every player $P_i$ proves that $h^{(i)}(0) = f_i(0)$ by showing that the free coefficient of the polynomial $h^{(i)}(x) - f_i(x)$ is equal to zero. This is done in two steps:

   a) Let $\mu = \max(\gamma, \delta)$. $P_i$ computes the polynomial $g^{(i)}(x) := (h^{(i)}(x) - f_i(x))/x$ (whose degree is at most $\mu - 1$), and distributes the shares on $g^{(i)}$ among the players. For this purpose the non-robust VSS protocol from Step 1 is used, where the corresponding two-dimensional polynomial, say $q^{(i)}(x, y)$, is chosen randomly, but such that $q^{(i)}(0, x) = g^{(i)}(x)$.

   b) Every player $P_k$ checks whether $\alpha_k g^{(i)}(\alpha_k) = h^{(i)}(\alpha_k) - f_i(\alpha_k)$.

3. WEAK FAULT DETECTION. Every player sends to every other player one bit indicating whether or not any of his consistency checks in Steps 1, 2a and 2b, have failed.

4. LAGRANGE INTERPOLATION. Every player $P_i$ who has neither detected nor was informed about any inconsistencies computes his degree-$\delta$ share of $s$ as a linear combination of his shares of $s_1, \ldots, s_{n'}$.

It is easy to see (using basic algebra), that if no player has reported inconsistencies during the weak fault detection, then the result of re-sharing is a proper $\delta$-sharing of $s$. Otherwise, if at least one (honest) player has sent or received a bit indicating inconsistencies, it will be possible to identify a $(1, 2)$-localization.

**Fault Detection.** At the end of the segment, every player $P_i$ *broadcasts* one bit indicating whether or not an inconsistency was observed by or reported to $P_i$ in one of the re-sharing protocols in the segment. If all players broadcast a confirmation (i.e., no inconsistency was observed), then the computation of the segment is completed and the next segment can be started. If at least one player broadcasts a complaint, then fault localization is invoked.

**Fault Localization.** The goal of fault-localization is to identify a $(1, 2)$-localization $\mathcal{D}$, i.e. a set $\mathcal{D} \subset \mathcal{P}$ containing two players, at least one of them being corrupted. These players will then be eliminated from the protocol, and hence fault localization is invoked at most $t$ times.

The two players to be eliminated are selected from the players involved in the first fault that occurred in the current segment. In order to determine the first fault, every player who complained during fault detection broadcasts the index (relative to the segment) of the re-sharing protocol, in which for the first time an inconsistency occurred, together with a number denoting the step of the re-sharing protocol in which the fault was detected (Step 1, 2a or 2b), or reported (Step 3). Among all the broadcast indices the smallest one is selected. Let $P_k$ denote the player who complained about the selected re-sharing protocol.[6] The method of determining the $(1,2)$-localization $\mathcal{D}$ depends on the step of the re-sharing protocol in which the first fault appeared. Four cases must be distinguished:

(i) The first fault is in Step 1, i.e. for some $i$ and $j$, the value $h_j^{(i)}(\alpha_k)$ sent by $P_j$ differs from $\widetilde{h}_k^{(i)}(\alpha_j)$:

 $P_k$ broadcasts $i$, $j$, and $\widetilde{h}_k^{(i)}(\alpha_j)$. On this request, $P_j$ broadcasts $\widetilde{h}_j^{(i)}(\alpha_k)$, and $P_i$ broadcasts $p^{(i)}(\alpha_k, \alpha_j)$. Given these three values, the set $\mathcal{D}$ is determined as follows:
 - If $\widetilde{h}_k^{(i)}(\alpha_j) = h_j^{(i)}(\alpha_k)$, then $\mathcal{D} := \{P_j, P_k\}$, else
 - if $p^{(i)}(\alpha_k, \alpha_j) \neq \widetilde{h}_k^{(i)}(\alpha_j)$, then $\mathcal{D} := \{P_i, P_k\}$, else
 - $p^{(i)}(\alpha_k, \alpha_j) \neq h_j^{(i)}(\alpha_k)$, and $\mathcal{D} := \{P_i, P_j\}$.

(ii) The first fault is in Step 2a: analogously to the case (i).

(iii) The first fault is in Step 2b, i.e., for some $i$ the check $\alpha_k g^{(i)}(\alpha_k) \overset{?}{=} h^{(i)}(\alpha_k) - f_i(\alpha_k)$ failed:

 According to $P_k$, player $P_i$ is cheating, so $P_k$ broadcasts the index $i$, and $\mathcal{D}$ is set to $\{P_i, P_k\}$.

(iv) The first fault is in Step 3, i.e., $P_k$ claims that in Step 3 some player reported a fault to him:

 Since no player admits the discovery of an inconsistency (as follows from the rule for choosing $P_k$), obviously either $P_k$ is lying or the player who reported the fault to him was malicious. $P_k$ broadcasts the index $i$ of the player $P_i$ who in Step 3 reported the fault to him, and $\mathcal{D}$ is set to $\{P_i, P_k\}$.

It is obvious that all players find the same set $\mathcal{D}$, and that in each case at least one player in $\mathcal{D}$ is corrupted, hence $\mathcal{D}$ is a $(1,2)$-localization.

**Player Elimination.** All players set $\mathcal{P}'$ to $\mathcal{P}' \setminus \mathcal{D}$, and reduce $t'$ to $t' - 1$.

**Fault Correction.** Fault correction is achieved by repeating the failed segment. Since after each failure at least one malicious player is eliminated, at most $t$ segments will be repeated in a complete protocol run.

---

[6] If there are several such players, we consider those who have broadcast the smallest step-number, and from that group the player with the smallest index $k$ is chosen.

## 3.5   Output Stage

Let $P$ be the designated player supposed to receive a value $s$ that is $t$-shared among the players in $\mathcal{P}'$ with the polynomials $f$ and $f_1, \ldots, f_{n'}$. First, every player $P_i \in \mathcal{P}'$ sends the polynomial $f_i(x)$ and the share-shares $s_{1i}, \ldots s_{n'i}$ to $P$. Then, $P$ interpolates the secret $s$ from the shares $s_i = f_i(0)$ for all $i$ where $f_i(x)$ is consistent with all but (at most) $t'$ share-shares $s_{ij}$. Note that this protocol needs neither error correction nor broadcast.

The privacy of this protocol is obvious. The correctness can be proven as follows: At most $t'$ players send a bad polynomial $f_i' \neq f_i$, and they will be inconsistent with at least $n' - t - t' > t'$ share-shares. Hence, $P$ will ignore bad polynomials and interpolate the correct secret $s$.

# 4   Complexity Analysis

In this section we analyze the communication complexity of the proposed multi-party computation protocol and compare it with the most efficient protocols known before. We focus on the case when an adversary is present and neglect the efficiency gain that some protocols (e.g., [FY92]) achieve when no fault at all occurs.

The communication complexity of a protocol is characterized by two quantities: the *message complexity* (MC, the total number of bits transmitted by all players during the protocol), and the *round complexity* (RC, the number of communication rounds of the protocol).

When analyzing the communication complexity of a multi-party protocol, one must also include the communication costs for simulating the broadcast channels. For most protocols in the literature (but not for ours), these costs are dominating the overall complexity of the protocol. We consider two different types of broadcast sub-protocols: Protocols with optimal message complexity ($\mathcal{O}(n^2)$, but $\mathcal{O}(n)$ rounds), e.g., [BGP89, CW89, DR85, HH91], and protocols with optimal round complexity ($\mathcal{O}(1)$, but $\mathcal{O}(n^4)$ messages), e.g., [FM88]. So far, no broadcast protocol with $\mathcal{O}(1)$ rounds and $\mathcal{O}(n^2)$ messages is known. In the cryptographic setting, such a protocol is known for a model where a trusted dealer is available in the set-up phase [CKS00], but this requirement contradicts the main purpose of of secure multi-party computation, namely getting rid of the need for a trusted party. There exist also various techniques which improve the efficiency of (stand-alone) protocols for Byzantine agreement, e.g. "early stopping" [DRS82]. However, they lead to "staggered termination", and it is unclear how and whether at all they are applicable for multi-party computation protocols.

## 4.1   Complexity of the New Protocol

The communication complexity of the proposed MPC protocol (cf. Sect. 3) is stated in the following theorem. This result is achieved by employing a Byzantine agreement protocol with optimal message complexity [BGP89, CW89].

**Theorem 1.** The protocol of Sect. 3 allows a set of $n$ players, with at most $t < n/3$ of them being corrupted, to securely compute a function over a finite field $\mathbb{F}$, using $\mathcal{O}(d+n^2)$ communication rounds and with total communication complexity $\mathcal{O}(n_I n^4 + mn^3 + n_O n^2)$ field elements, where $n_I$ and $n_O$ denote the number of inputs and outputs, respectively, $m$ denotes the number of multiplications and $d$ the multiplicative depth of the circuit computing the function.

The detailed analysis of this protocol is omitted from this extended abstract. We give only a very brief overview. The VSS protocol for providing one input requires in the worst case $\mathcal{O}(n^2)$ field elements to be broadcast, which results in $\mathcal{O}(n^4)$ field elements per input when using the most efficient broadcast protocols [BGP89, CW89]. Each multiplication requires each player to secret-share (with the non-robust VSS protocol) one element, which adds up to $\mathcal{O}(n^3)$ field elements per multiplication, and hence $\mathcal{O}(m_s n^3)$ elements per segment with $m_s$ multiplication gates. Fault-detection requires $\mathcal{O}(n)$ bits to be broadcast per segment, and fault-localization requires $\mathcal{O}(n \log m_s + \log |\mathbb{F}|)$ bits to be broadcast at the end of up to $t$ segments. For the proposed segmentation with $m_s = \lceil m/n \rceil$ and $d_s = \lceil d/n \rceil$, at most $2n$ segments are computed, which results in a total message complexity of $\mathcal{O}(mn^3)$ field elements. Only $\mathcal{O}(n^2)$ field elements must be broadcast in total (independently of the circuit size!), which does not dominate the overall costs when $m \geq n$. The message complexity of secret reconstruction is $\mathcal{O}(n^2)$ elements per output (broadcast is not needed).

## 4.2 Comparison with Other Protocols

The complexity of the new protocol is compared with the most efficient multiparty computation protocols for the unconditional model known before. In the sequel, we summarize the most important results. A more detailed complexity analysis can be found in [Prz99]. For simplicity we focus on the complexity of the evaluation of the circuit, and ignore the complexities of providing inputs and receiving inputs. The following table lists the message complexity (MC) and the round complexity (RC) of the most efficient protocols for the unconditional model, once when a broadcast protocol with optimal bit complexity is applied, and once when a broadcast protocol with optimal round complexity is applied. The second last row in the table refers to the protocol of [BGW88], where the "Rabin's trick" [GRR98] for simpler multiplication is used. Note that the other technique for increasing the efficiency of [BGW88] suggested in the same paper, namely the efficient proof that a shared secret is indeed the product of two shared factors, is shown to be insecure (see App. A), and hence its impact on the complexity is not analyzed.

For completeness, in Table 2 we also state the complexities of the best protocol for the cryptographic model [GRR98], in which up to $t < n/2$ of the players can be corrupted, but the security of the protocol relies on unproven assumptions. Subsequently to our work, a new protocol with cryptographic security was proposed in [CDN00], and its complexity is also listed in the table (where $k$ denotes the security parameter). In contrast to other protocols, here the function must be specified as a Boolean circuit, and the complexity is indicated in bits.

| MPC protocol | Broadcast protocol | MC | RC |
|---|---|---|---|
| [BGW88] | [FM88]<br>[BGP89, CW89] | $\mathcal{O}(mn^8)$<br>$\mathcal{O}(mn^6)$ | $\mathcal{O}(d)$<br>$\mathcal{O}(dn)$ |
| [CCD88] | [FM88]<br>[BGP89, CW89] | $\mathcal{O}(mn^9)$<br>$\mathcal{O}(mn^7)$ | $\mathcal{O}(dn)$<br>$\mathcal{O}(dn^2)$ |
| [Bea91a] | [FM88]<br>[BGP89, CW89] | $\mathcal{O}(mn^8)$<br>$\mathcal{O}(mn^6)$ | $\mathcal{O}(d)$<br>$\mathcal{O}(d+n)$ |
| [FY92] | [FM88]<br>[BGP89, CW89] | $\mathcal{O}(mn^8)$<br>$\mathcal{O}(mn^6)$ | $\mathcal{O}(d)$<br>$\mathcal{O}(dn)$ |
| [BGW88, GRR98] | [FM88]<br>[BGP89, CW89] | $\mathcal{O}(mn^8)$<br>$\mathcal{O}(mn^6)$ | $\mathcal{O}(d)$<br>$\mathcal{O}(dn)$ |
| this paper | [BGP89, CW89] | $\mathcal{O}(mn^3)$ | $\mathcal{O}(d+n^2)$ |

**Table 1.** Worst-case communication complexities of unconditional MPC protocols.

| MPC protocol | Broadcast protocol | MC | RC |
|---|---|---|---|
| [GRR98] | [BGP89, CW89]<br>[FM88] | $\mathcal{O}(mn^4)$<br>$\mathcal{O}(mn^6)$ | $\mathcal{O}(dn)$<br>$\mathcal{O}(d)$ |
| [CDN00] | [BGP89, CW89]<br>[FM88] | $\mathcal{O}(mn^3k)$<br>$\mathcal{O}(mn^5k)$ | $\mathcal{O}(dn)$<br>$\mathcal{O}(d)$ |

**Table 2.** Worst-case communication complexities of cryptographic MPC protocols.

## 5   Conclusions and Open Problems

General secure multi-party computation protocols for evaluating an algebraic circuit will have important applications in distributed information systems. One major reason why such protocols are not yet widely used in practical applications is their hopeless inefficiency. In particular, they all make extensive use of a reliable broadcast channel, which in any reasonable application scenario is not available, and hence must be simulated by an expensive protocol among the players.

In this paper we proposed a new framework for communication-efficient distributed protocols, applied it to secure multi-party computations, resulting in a very efficient protocol. We stress that the message complexity (and possibly the round complexity), but not the computation complexity, are the bottlenecks in most distributed applications.

There are several open problems to be solved to make general multi-party protocols applicable in distributed systems. The main issue is definitely the model:

It is an open problem to generalize the framework to the asynchronous model, and to convert the used techniques accordingly. Furthermore, it might be interesting to generalize the results to non-threshold adversary structures [HM00]. Finally, it is questionable whether comparable efficiency improvements can be achieved in a model with mobile adversaries, where player elimination seems not to be applicable.

# References

[BB89]      J. Bar-Ilan and D. Beaver. Non-cryptographic fault-tolerant computing in a constant number of rounds of interaction. In *Proc. 8th ACM Symposium on Principles of Distributed Computing (PODC)*, pp. 201–210, 1989.

[Bea91a]    D. Beaver. Efficient multiparty protocols using circuit randomization. In *Advances in Cryptology — CRYPTO '91*, vol. 576 of *LNCS*, pp. 420–432, 1991.

[Bea91b]    D. Beaver. Secure multiparty protocols and zero-knowledge proof systems tolerating a faulty minority. *Journal of Cryptology*, pp. 75–122, 1991.

[BFKR90]    D. Beaver, J. Feigenbaum, J. Kilian, and P. Rogaway. Security with low communication overhead (extended abstract). In *Advances in Cryptology — CRYPTO '90*, pp. 62–76, 1990.

[BGP89]     P. Berman, J. A. Garay, and K. J. Perry. Towards optimal distributed consensus (extended abstract). In *Proc. 21st IEEE Symposium on the Foundations of Computer Science (FOCS)*, pp. 410–415, 1989. Expanded version: Bit optimal distributed consensus. In *Computer Science Research*, 1992.

[BGW88]     M. Ben-Or, S. Goldwasser, and A. Wigderson. Completeness theorems for non-cryptographic fault-tolerant distributed computation. In *Proc. 20th ACM Symposium on the Theory of Computing (STOC)*, pp. 1–10, 1988.

[BMR90]     D. Beaver, S. Micali, and P. Rogaway. The round complexity of secure protocols (extended abstract). In *Proc. 22nd ACM Symposium on the Theory of Computing (STOC)*, pp. 503–513, 1990.

[Can00]     R. Canetti. Security and composition of multi-party cryptographic protocols. *Journal of Cryptology*, 13(1):143–202, 2000.

[CCD88]     D. Chaum, C. Crépeau, and I. Damgård. Multiparty unconditionally secure protocols (extended abstract). In *Proc. 20th ACM Symposium on the Theory of Computing (STOC)*, pp. 11–19, 1988.

[CDD+99]    R. Cramer, I. Damgård, S. Dziembowski, M. Hirt, and T. Rabin. Efficient multiparty computations secure against an adaptive adversary. In *Advances in Cryptology — EUROCRYPT '99*, vol. 1592 of *LNCS*, pp. 311–326, 1999.

[CDM00]     R. Cramer, I. Damgård, and U. Maurer. General secure multi-party computation from any linear secret sharing scheme. In *Advances in Cryptology — EUROCRYPT '00*, vol. 1807 of *LNCS*, pp. 316–334, 2000.

[CDN00]     R. Cramer, I. Damgård, and J. B. Nielsen. Multiparty computation from threshold homomorphic encryption. Manuscript, 2000.

[CH94]      R. Canetti and A. Herzberg. Maintaining security in the presence of transient faults. In *Advances in Cryptology — CRYPTO '94*, vol. 839 of *LNCS*, pp. 425–438, 1994.

[CKS00]   C. Cachin, K. Kursawe, and V. Shoup. Random oracles in Constantino-
          ple: Practical asynchronous Byzantine agreement using cryptography.
          In *Proc. 19th ACM Symposium on Principles of Distributed Computing
          (PODC)*, pp. 123–132, 2000.

[CW89]    B. A. Coan and J. L. Welch. Modular construction of nearly optimal
          Byzantine agreement protocols. In *Proc. 8th ACM Symposium on Prin-
          ciples of Distributed Computing (PODC)*, pp. 295–305, 1989. Expanded
          version: Modular construction of a Byzantine agreement protocol with op-
          timal message bit complexity. In *Information and Computation*, 97(1):61-
          85, 1992.

[DFF+82]  D. Dolev, M. J. Fischer, R. Fowler, N. A. Lynch, and H. R. Strong. An
          efficient algorithm for Byzantine agreement without authentication. *Infor-
          mation and Control*, 52(3):257–274, 1982.

[DR85]    D. Dolev and R. Reischuk. Bounds on information exchange for Byzantine
          agreement. *Journal of the ACM*, 32(1):191–204, 1985.

[DRS82]   D. Dolev, R. Reischuk, and H. R. Strong. 'Eventual' is earlier than 'Im-
          mediate'. In *Proc. 23rd IEEE Symposium on the Foundations of Com-
          puter Science (FOCS)*, pp. 196–203, 1982. Final version: Early Stopping
          in Byzantine Agreement. In *Journal of the ACM*, 37(4):720-741, October
          1990.

[FHM98]   M. Fitzi, M. Hirt, and U. Maurer. Trading correctness for privacy in
          unconditional multi-party computation. In *Advances in Cryptology —
          CRYPTO '98*, vol. 1462 of *LNCS*, pp. 121–136, 1998.

[FM88]    P. Feldman and S. Micali. Optimal algorithms for Byzantine agreement.
          In *Proc. 20th ACM Symposium on the Theory of Computing (STOC)*,
          pp. 148–161, 1988. Expanded version in *SIAM Journal on Computing*
          26(4):873–933, August 1997.

[FY92]    M. K. Franklin and M. Yung. Communication complexity of secure com-
          putation. In *Proc. 24th ACM Symposium on the Theory of Computing
          (STOC)*, pp. 699–710, 1992.

[GHY87]   Z. Galil, S. Haber, and M. Yung. Cryptographic computation: Secure fault-
          tolerant protocols and the public-key model. In *Advances in Cryptology
          — CRYPTO '87*, vol. 293 of *LNCS*, pp. 135–155. Springer-Verlag, 1987.

[GMW87]   O. Goldreich, S. Micali, and A. Wigderson. How to play any mental
          game — a completeness theorem for protocols with honest majority. In
          *Proc. 19th ACM Symposium on the Theory of Computing (STOC)*, pp.
          218–229, 1987.

[GRR98]   R. Gennaro, M. O. Rabin, and T. Rabin. Simplified VSS and fast-
          track multiparty computations with applications to threshold cryptogra-
          phy. In *Proc. 17th ACM Symposium on Principles of Distributed Comput-
          ing (PODC)*, 1998.

[HH91]    V. Hadzilacos and J. Y. Halpern. Message-optimal protocols for byzantine
          agreement. In *Proc. 10th ACM Symposium on Principles of Distributed
          Computing (PODC)*, pp. 309–324, 1991. Final version in *Mathematical
          Systems Theory*, 26:41-102, October 1993.

[HM00]    M. Hirt and U. Maurer. Player simulation and general adversary structures
          in perfect multiparty computation. *Journal of Cryptology*, 13(1):31–60,
          2000. Extended abstract in *Proc. 16th of ACM PODC '97*.

[MR98]    S. Micali and P. Rogaway. Secure computation: The information theo-
          retic case. Manuscript, 1998. Former version: Secure computation, In *Ad-*

vances in Cryptology — CRYPTO '91, volume 576 of LNCS, pp. 392–404, Springer-Verlag, 1991.

[OY91]    R. Ostrovsky and M. Yung. How to withstand mobile virus attacks (extended abstract). In Proc. 10th ACM Symposium on Principles of Distributed Computing (PODC), pp. 51–59, 1991.

[Prz99]   B. Przydatek. Efficiency in multi-party computation. Master's thesis, ETH Zurich, 1999.

[PSL80]   M. Pease, R. Shostak, and L. Lamport. Reaching agreement in the presence of faults. Journal of the ACM, 27(2):228–234, 1980.

[RB89]    T. Rabin and M. Ben-Or. Verifiable secret sharing and multiparty protocols with honest majority. In Proc. 21st ACM Symposium on the Theory of Computing (STOC), pp. 73–85, 1989.

[Sha79]   A. Shamir. How to share a secret. Communications of the ACM, 22:612–613, 1979.

[Yao82]   A. C. Yao. Protocols for secure computations. In Proc. 23rd IEEE Symposium on the Foundations of Computer Science (FOCS), pp. 160–164. IEEE, 1982.

# A    Security Flaw in [GRR98]

In Appendix B ("Computing Multiplication with Faults") of [GRR98],[7] a very efficient sub-protocol was proposed for proving that for three shared values $a$, $b$, and $c$, the equation $c = ab$ holds. This sub-protocol was intended to replace the (rather inefficient) verification sub-protocol ("tool (II)") of [BGW88]. We show in the sequel that this new sub-protocol of [GRR98] is insecure. First we briefly summarize the protocol and then demonstrate the security flaw.

Assume that player $P$ has shared the values $a$, $b$, and $c$ with polynomials $f(x)$, $g(x)$, and $h(x)$ respectively, all of degree at most $t$. Let $a_i$, $b_i$, and $c_i$ denote the corresponding shares of player $P_i$, $i = 1, \ldots, n$. The protocol of [GRR98] works as follows:

1. The dealer $P$ shares (using "normal" secret sharing, not VSS) a random value with a polynomial $r(x)$ of degree $2t - 1$. The share $r_i$ of player $P_i$ is $r_i = r(\alpha_i)$. Furthermore, $P$ computes and broadcasts the polynomial $R(x) = x \cdot r(x) + f(x) \cdot g(x) - h(x)$. $R(x)$ is a random polynomial of degree $2t$, and if $c = ab$ holds then $R(0) = 0$.

2. Every player $P_i$ verifies that $R(0) = 0$ and $R(\alpha_i) = \alpha_i \cdot r_i + a_i \cdot b_i - c_i$. $P_i$ broadcasts either "OK", if both checks were successful, or otherwise a request to make his values public.

3. If in the previous step some requests occurred (at most $t$), $P$ broadcasts all the requested data. If there were more than $t$ requests, $P$ is clearly cheating.

This protocol does not guarantee correctness, in contrast to what is claimed in the paper and was believed before. The dealer $P$ can pass this verification even if $c = ab$ does not hold:

---

[7] After the security problem was discovered, this appendix was deleted from the version available online.

1. Instead of selecting a random polynomial $r(x)$ of degree $2t - 1$, the dealer first selects a (random) polynomial $R(x)$ of degree $2t$ with $R(0) = 0$, then computes and distributes the "shares" $r_1, \ldots, r_n$ as $r_i = \alpha_i^{-1}(R(\alpha_i) - a_i \cdot b_i + c_i)$. The dealer can do so because the degree of the polynomial $r(x)$ cannot be verified. Finally, $P$ broadcasts the polynomial $R(x)$.

Clearly, the checks in Step 2 of all players will succeed, and no (honest) player will complain.

# Mix and Match:
# Secure Function Evaluation via Ciphertexts
## (Extended Abstract)

Markus Jakobsson[1] and Ari Juels[2]

[1] Information Sciences Research Center
Bell Labs
Murray Hill, New Jersey 07974
`markusj@research.bell-labs.com`
[2] RSA Laboratories
RSA Security Inc.
Bedford, MA 01730, USA
`ajuels@rsasecurity.com`

**Abstract.** We introduce a novel approach to general secure multiparty computation that avoids the intensive use of verifiable secret sharing characterizing nearly all previous protocols in the literature. Instead, our scheme involves manipulation of ciphertexts for which the underlying private key is shared by participants in the computation. The benefits of this protocol include a high degree of conceptual and structural simplicity, low message complexity, and substantial flexibility with respect to input and output value formats. We refer to this new approach as *mix and match*.

While the atomic operations in mix and match are logical operations, rather than full field operations as in previous approaches, the techniques we introduce are nonetheless highly practical for computations involving intensive bitwise manipulation. One application for which mix and match is particularly well suited is that of sealed-bid auctions. Thus, as another contribution in this paper, we present a practical, mix-and-match-based auction protocol that is fully private and non-interactive and may be readily adapted to a wide range of auction strategies.

**Key words:** auction, general secure multiplayer computation, millionaires' problem, secure function evaluation

## 1 Introduction

Consider the following scenario. Alice and Bob have respective fortunes $A$ and $B$. They wish to determine who is richer, i.e., whether $A > B$, but do not wish to reveal any additional information about their fortunes. This task is known as the *millionaires' problem* [44]. It is a special instance of the more general setting in which Alice and Bob, or indeed a larger number of players, wish to compute

T. Okamoto (Ed.): ASIACRYPT 2000, LNCS 1976, pp. 162–177, 2000.

the output of a function $f$ on secret inputs without revealing any additional information.

Alice and Bob can take one of several approaches. They might confide their fortunes to a trusted third party charged with the task of determining honestly whether $A > B$ and not leaking any information to either party. Alternatively, they might construct a piece of trusted hardware for the same purpose. It has been known for some time, however, that Alice and Bob can in fact *simulate* a trusted party or device in such a way as to enable secure computation of general functions [44,27]. Computation of this sort involving two or more players is known as *secure multiparty computation*. For general functions $f$ it is known as *general secure multiparty computation* or *secure function evaluation*.

All of the current approaches to general secure multiparty computation rely on the simulation of a circuit $C_f$ for the function $f$ of interest. This circuit is typically viewed as being composed of gates implementing two operators, such as + (modular addition) and × (modular multiplication), that together allow for the realization of an arbitrary computable function. In nearly every protocol in the literature with robustness against active adversaries, the lynchpin is a cryptographic primitive known as *verifiable secret sharing* (VSS), introduced in [14]. Players distribute their inputs to $C_f$ by dealing shares to other players through a VSS protocol. At any stage in the computation, concealed values are held distributively. To simulate a + gate, players perform local addition of their shares. To simulate a × gate, they perform an interactive protocol involving multiplication of pairs of shares held by different players.

In this paper, we investigate a different approach to secure function evaluation. Rather than employing multi-player sharing of individual inputs or intermediate computational results, we consider a representation of these values as ciphertexts. We concentrate in particular in this paper on use of the El Gamal cryptosystem [24], although use of other semantically secure cryptosystems, such as Cramer-Shoup [18], is possible. Distribution of trust among the players in our scheme relies on sharing of a single, underlying private decryption key. Players perform the operations required by the computation using well established techniques for distributed manipulation of El Gamal ciphertexts.

A brief sketch of our approach is as follows. Having agreed upon a function $f$ and a circuit representation $C_f$, the players provide El Gamal ciphertexts of their input bits. Gates in $C_f$ are each represented by a boolean function, such as $AND$ or $NOT$ (although others are possible). For each gate, the players construct a logical table corresponding to the function computed by the gate, the entries in this table consisting of El Gamal ciphertexts. In an initial blinding phase, the players use a primitive known as a *mix network* to blind and perform row-wise permutation of these tables in a distributed fashion. The basis of the subsequent computation phase, which we refer to as *matching*, is a primitive called a *plaintext equality test* ($\mathcal{PET}$). The $\mathcal{PET}$ primitive enables players to determine in a distributed fashion whether two given ciphertexts represent the same plaintext. Players evaluate the circuit $C_f$ iteratively, using $\mathcal{PET}$ to perform table lookups. For each gate, they compare ciphertext input values to ciphertext values in the corresponding blinded logical table. When the correct row in the table is found, the players obtain an output ciphertext from the third column. Due to the use of blinded permutation, they do not learn the plaintext corresponding to the output value. The output ciphertext is used as input to the next gate (table). We refer to this approach as *mix-and-match* computation.

## 1.1  Previous Work

The idea of performing secure computation by means of blinded table lookups was essentially the basis for the original proposal of Yao [44], whose two-player scheme was predicated on the hardness of factoring. Goldreich, Micali, and Wigderson [27] generalized the basis of Yao's scheme to use of any one-way trapdoor permutation. The idea behind both approaches in their two-party instantiations is as follows. Alice constructs a circuit $C_f$ using boolean gates represented as randomly permuted, blinded logical tables. Inputs to a gate (table) are randomly generated tags representing different bit values. Each set of tags, representing a given set of inputs to a table, serves as a decryption key for a particular row of the table, and thus a particular output tag for the gate. By means of a 1-2 oblivious transfer protocol, Alice blindly transfers to Bob the tags representing his input values for the circuit, and also sends Bob her table representation of $C_f$. For each gate in $C_f$, Bob uses the input tags to decrypt output tags representing the corresponding gate output. He is thereby able to evaluate $C_f$ without further interaction with Alice. See [27] for further details.

Chaum, Damgård, and van de Graaf [12] extend the notion of blinded table mixing and lookup to a multiparty scenario. In their scheme, each player in turn blinds the logical table for a given gate. The basis of this scheme is a homomorphic commitment scheme that enables one player to alter the commitment of another without knowing the correct decommitment.[1] Players provide cut-and-choose proofs of correct behavior. The security of the scheme is unconditional for one player, and for the others is based on the quadradic residuosity problem.

The Chaum *et al.* scheme is not robust against an active adversary, in the sense that such an adversary may corrupt the computation irretrievably or force it to halt. To achieve robustness, the authors recommend incorporation of VSS to enable reconstruction of the commitments in their scheme. Similarly, since the introduction of secure multiparty computation in [27], such protocols have generally employed VSS as a means of enforcing robustness for the computation on each gate. Ben-Or, Goldwasser, and Widgerson [4] and Chaum, Crépeau, and Damgård [11] introduced the first protocols enabling security against an active adversary in the non-cryptographic model, that is, one in which players are assumed to have unbounded computing power, but cannot eavesdrop on honest players. Their approach has loosely formed the basis of the majority of subsequent work, even some of the most recent and efficient constructions such as [15,26].

---

[1] Manipulation of homomorphic commitments by $n$ players here in fact yields a kind of $(n, n)$-VSS protocol in this scheme.

Other work related to our own is that of Franklin and Haber [22]. As we do here, they propose a secure multiparty computation method dependent on manipulation of ciphertexts. In their scheme, the underlying encryption scheme is a special variant of El Gamal. The Franklin and Haber system, however, is only secure against passive adversaries.

While drawing on table-based approaches to secure function evaluation, and particularly on the frameworks presented in [12,22], mix and match offers robustness without the use of VSS on input values or sharing of intermediate values. The mix-and-match approach consequently achieves several benefits unavailable in conventional VSS-based approaches:

1. Mix and match is conceptually very simple.
2. The message complexity for mix and match is quite low. In the random oracle model, the broadcast message complexity is $\mathcal{O}(nN)$ group elements, where $n$ is the number of players and $N$, the number of gates in the circuit.
3. Sharing in mix and match occurs only at the level of a decryption key, and not for input or intermediate computational values.

This last property means that mix and match has the advantage of natural flexibility in terms of both input and output formats and player participation. For example, players contributing inputs need not even know what players or how many will be performing the computation, and need not themselves participate. Outputs may be made to take the form of ciphertexts under an arbitrary key, with no additional protocol overhead. We note that a similar property emerges in independent work by Cramer, Damgård, and Nielsen [16].

A drawback to mix and match is the fact that the atomic computational unit is the boolean formula, rather than the field operations employed in secure multiparty computation schemes based on [4]. For many functions, therefore, such as threshold signature computation [25], it is probably substantially less efficient than, e.g., [30]. For functions involving intensive bitwise manipulations, however, mix and match is highly competitive. A good example is the millionaires' problem, or natural multi-player extensions such as auction protocols. To highlight this strength, we present a flexible non-interactive auction protocol in this paper based on mix and match. This protocol, as we show, has several practical advantages over state-of-the-art proposals for non-interactive secure auctions.

### 1.2  Organization

We present model details and definitions in section 2. We describe our mix-and-match scheme in section 3. In section 4, we briefly discuss the literature on auction protocols and outline a non-interactive, fully private auction protocol based on mix and match.

## 2   Model and Building Blocks

In elaborating mix and match, we consider the cryptographic model of secure multiparty computation. This involves $n$ players, $P_1, P_2, \ldots, P_n$, who are as-

sumed to share an authenticated broadcast channel, and an *adversary* with resources polynomially bounded in all security parameters. We consider an adversary who may corrupt up to $t < n/2$ of these players in an active fashion, i.e.,
the adversary gains access to their private information, and may govern their
behavior in an arbitrary fashion. We assume that the adversary is *static*, that
is, she must choose in advance which players she wishes to corrupt. Our results
can be extended straightforwardly to more complex adversarial structures. We
can achieve security in the mix-and-match protocol reducible to the *Decision
Diffie-Hellman* (DDH) assumption (see, e.g., [35]) on the group $\mathcal{G}$ over which
the computation takes place. To achieve the best possible efficiency and simplicity here, however, we additionally invoke the random oracle model (see, e.g.,
[3]). The asymptotic costs presented in this paper assume malicious adversarial
behavior.

## 2.1   Building Blocks

*El Gamal cryptosystem:* We employ the El Gamal cryptosystem [24] as the
basis for our constructions. Encryption in the El Gamal cipher takes place over
a group[2] $\mathcal{G}_q$ of prime order $q$.

Let $g$ be a generator of $\mathcal{G}_q$. This generator is typically regarded as a system
parameter, as it may correspond to multiple key pairs. A private encryption key
consists of an integer $x \in_U Z_q$, where $\in_U$ denotes uniform random selection.
The corresponding public key is defined to be $y = g^x$. To encrypt a message
$m \in \mathcal{G}_q$ under public key $y$, we select $a \in_U Z_q$, and compute the ciphertext
$(\alpha, \beta) = (my^a, g^a)$. To decrypt this ciphertext using the private key $x$, we compute $\alpha/\beta^x = my^a/(g^a)^x = m$.

The El Gamal cryptosystem is *semantically secure* [28] under the Decision
Diffie-Hellman (DDH) assumption over $\mathcal{G}_q$. Informally, this means that an attacker who selects message pair $(m_0, m_1)$ is unable to distinguish between encryptions of these two messages with probability significantly greater than $1/2$,
i.e., than a random guess. See [43] for details.

Let $(\alpha_0\alpha_1, \beta_0\beta_1) = (\alpha_0, \beta_0) \otimes (\alpha_1, \beta_1)$. Another useful property of the El
Gamal cryptosystem is the fact that it possesses a *homomorphism* under the
operator $\otimes$. In particular, observe that if $(\alpha_0, \beta_0)$ and $(\alpha_1, \beta_1)$ represent ciphertexts corresponding to plaintexts $m_0$ and $m_1$ respectively, then $(\alpha_0, \beta_0) \otimes (\alpha_1, \beta_1)$
represents an encryption of the plaintext $m_0 m_1$. A consequence of this homo-
morphic property is that it is possible, using knowledge of the public key alone,
to derive a random *re-encryption* $(\alpha', \beta')$ of a given ciphertext $(\alpha, \beta)$. This is
accomplished by computing $(\alpha', \beta') = (\alpha, \beta) \otimes (\gamma, \delta)$, where $(\gamma, \delta)$ represents an
encryption of the plaintext value 1. It is possible to prove quite efficiently in
zero-knowledge that $(\alpha', \beta')$ represents a valid re-encryption of $(\alpha, \beta)$ using, e.g.,

---

[2] Most commonly, we let $p = 2q + 1$, and we let $\mathcal{G}_q$ be the set of quadratic residues
in $Z_p^*$. In this setting, plaintexts not in $\mathcal{G}_q$ can be mapped onto $\mathcal{G}_q$ by appropriate
forcing of the LeGendre symbol, e.g., through multiplication by a predetermined
non-residue.

a variant of the Schnorr proof of knowledge protocol [17,42]. This proof may also be made non-interactive. See [6] for an overview. We let $(\alpha, \beta) \equiv (\alpha', \beta')$ denote equivalence of underlying plaintexts for $(\alpha, \beta)$ and $(\alpha', \beta')$, and $(\alpha, \beta) \not\equiv (\alpha', \beta')$ denote non-equivalence.

The plaintext 0 has only a degenerate ciphertext in the El Gamal cryptosystem. We can represent it by some other plaintext in our protocol.

*Distributed decryption for El Gamal:* A final useful property of the El Gamal cipher is that it may easily be converted into a threshold protocol. With use of a *distributed key generation* protocol such as that in [8] or [9], players generate a private key $x$ that is held according to a $t$-out-of-$n$ sharing scheme for some $t < n/2$. In particular, each player $P_i$ obtains a private share $x_i$ consisting of the evaluation of a $(t - 1)$-degree polynomial on, e.g., the value $i$ over an appropriate field. To decrypt a ciphertext $(\alpha, \beta)$, each player $P_i$ publishes the value $\beta_i = \beta^{x_i}$, along with a ZK proof of correct exponentiation, that is, a ZK proof of knowledge of $z$ such that $\log_g y_i = \log_\beta \beta_i$ using, e.g., an appropriate variant on the protocol described in [13]. Players then compute $\beta^x = \prod_{i=1}^{t} \beta_{a_i}^{\lambda_{a_i}}$ on $t$ correct shares $\{\beta_{a_i}\}_{i=1}^{t}$, where $\lambda_{a_i}$ is the LaGrange coefficient for the $a_i^{th}$ share. Assuming use of non-interactive proofs, with security consequently relying on the random oracle model, the broadcast round complexity of the protocol is $\mathcal{O}(1)$, the message complexity is $\mathcal{O}(n)$ group elements, and the per-player computational costs are $\mathcal{O}(n)$ exponentiations. We do not provide further details, but instead refer the reader to, e.g., [25], which describes a threshold DSS scheme with similar properties.

*Proof of knowledge of El Gamal plaintext:* Given knowledge of the encryption exponent $a$ of an El Gamal ciphertext $(\alpha, \beta) = (my^a, g^a)$, a player can prove knowledge of $m$ in zero knowledge. This may be accomplished simply by means of an (honest-verifier) zero-knowledge proof of knowledge of $a$ such that $\beta = g^a$ using, e.g., a variant of the Schnorr identification protocol [42] with a challenge carefully generated jointly by all servers. Soundness may then be based on the discrete log problem. The proof of knowledge may also be replaced with a non-interactive protocol through use of Fiat-Shamir techniques [20]. In this case, the ciphertext and accompanying proof may be regarded as a *plaintext-aware* encryption, and security depends additionally on use of the random oracle model. In this case, the broadcast round complexity of this protocol is $\mathcal{O}(1)$, the message complexity is $\mathcal{O}(1)$ group elements, and the per-player computational costs are $\mathcal{O}(1)$ exponentiations.

*Mix network* $(\mathcal{MN})$: The second key tool in our construction is known as a *mix network*. This primitive for privacy was introduced by Chaum [10], and has recently received considerable attention, both in terms of implementation improvements [1,32,33,34,37] and a wide variety of ideas for applications, of which some examples may be found in [23,31,38,41]. Intuitively, a mix network is a multi-party protocol that takes as input a list of ciphertext items and from this produces a new, random list of ciphertext items such that there is a one-to-one correspondence between the underlying plaintexts of input and output items. In other words, the underlying output plaintexts represent a random permutation of the underlying input plaintexts. The security of a mix network is characterized by the infeasibility for an adversary of determining which output items correspond to which input items.

While there are many flavors of mix network, the type we employ is based on the El Gamal cipher, and works as follows. An El Gamal public key $y$ is published for use by all players, who are assumed to share an authenticated broadcast channel (or, equivalently, a bulletin board). Some subset of $n$ players known as *mix servers* share the corresponding private key according to a $(t, n)$-threshold scheme as described above. Input to the mix network consists of a sequence $(\alpha_1, \beta_1), (\alpha_2, \beta_2), \ldots, (\alpha_k, \beta_k)$ of El Gamal ciphertexts posted by the players to the bulletin board.[3] The mix servers perform a sequence of distributed operations on these inputs. The output of the mix network is a random permutation and re-encryption of the inputs, namely a sequence $(\alpha'_{\sigma(1)}, \beta'_{\sigma(1)}), (\alpha'_{\sigma(2)}, \beta'_{\sigma(2)}), \ldots, (\alpha'_{\sigma(k)}, \beta'_{\sigma(k)})$ where $(\alpha'_i, \beta'_i)$ represents a random re-encryption of $(\alpha_i, \beta_i)$, and $\sigma$ is a random permutation on $k$ elements.

There are a number of variants on this basic primitive. For example, inputs to the mix network may be plaintexts, rather than ciphertexts; alternatively, the converse is possible. Note that these are really just special cases of what is described here: a plaintext is a degenerate form of ciphertext with encryption factor 0. Another important variant that we employ here is a mix network in which the input consists of a two-dimensional matrix of ciphertexts $\{(\alpha_{i,j}, \beta_{i,j})\}$ for $1 \leq i \leq k$ and $1 \leq j \leq v$. The output consists of a random, blinded permutation of the rows of this matrix, with no alteration of the order of underlying plaintexts within rows. In other words the output is $\{(\alpha'_{\sigma(i),j}, \beta'_{\sigma(i),j})\}$ for a random permutation $\sigma$ on $k$ elements, where $(\alpha'_{i,j}, \beta'_{i,j})$ represents a re-encryption of $(\alpha_{i,j}, \beta_{i,j})$. We do not provide details on this extension of the basic mix network primitive, but simply note that it may be implemented with overhead linear in the number $v$ of input columns.

We can base the mix network $\mathcal{MN}$ for the mix-and-match protocol on any of several constructions proposed in the literature. For small input sizes, the construction of Abe [2] or the similar construction of Jakobsson and Juels [33] is most efficient. Given that these schemes are publicly verifiable and possess easily provable security properties, we adopt either construction where we must make explicit reference to the properties of the mix network in mix and match.

The mix networks described in [2] and [34] have the following properties. They are secure, that is, both private and robust, against a static, active adversary that corrupts $t < n/2$ mix servers. Let $k$ be the number of input elements. With underlying interactive zero-knowledge proof protocols, i.e., those involving challenges carefully generated jointly by the servers, privacy for this protocol

---

[3] To prevent attacks involving one player posting a re-encryption of the ciphertext of another player, it is sometimes necessary for ciphertexts to be encrypted in a manner that is *plaintext aware*. This may be accomplished through, e.g., a ZK proof of knowledge of the discrete log of $\beta$ for a ciphertext $(\alpha, \beta)$ (see [32,43]).

may be reduced to the DDH assumption, and protocol robustness to the discrete log problem. By using non-interactive proof protocols under the Fiat-Shamir heuristic [20], we introduce additional dependence on the random oracle model for security. In this latter case, the asymptotic broadcast round complexity is $\mathcal{O}(n)$. The message complexity is $\mathcal{O}(nk \log k)$, while the total computational cost per server is $\mathcal{O}(nk \log k)$ exponentiations.

*Distributed plaintext equality test* ($\mathcal{PET}$): Let $(\alpha, \beta)$ and $(\alpha', \beta')$ be El Gamal ciphertexts with respective underlying plaintexts $m_1$ and $m_2$. In the $\mathcal{PET}$ protocol, players jointly determine whether $m_1 = m_2$, i.e., whether $(\alpha, \beta) \equiv (\alpha', \beta')$.

Consider the ciphertext $(\epsilon, \zeta) = (\alpha/\alpha', \beta/\beta')$. If $(\alpha, \beta) \equiv (\alpha', \beta')$, then $(\epsilon, \zeta)$ represents an encryption of the plaintext integer 1; otherwise, it represents an encryption of the quotient $m_1/m_2$. The idea behind the $\mathcal{PET}$ protocol, therefore, is to have the parties blind and then decrypt $(\epsilon, \zeta)$ in such a way that the resulting output is 1 if $(\alpha, \beta) \equiv (\alpha', \beta')$, and a random integer otherwise. Player $P_i$ blinds $(\epsilon, \zeta)$ by raising each element in the pair to a random exponent $z_i \in Z_q$. This form of blinding leaves the plaintext intact if it is equal to 1 and randomizes it otherwise. The players then combine their blinded shares and perform a distributed decryption on the resulting ciphertext. The protocol is as follows:

1. Each player $P_i$ selects $z_i \in_U Z_q$. She publishes a Pedersen commitment [39] $C_i = g^{z_i} h^{r_i}$ to $z_i$, where $h$ is a generator such that $\log_g h$ is unknown to any coalition of servers and $r_i \in_U Z_q$ is selected by $P_i$.
2. Each player computes $(\epsilon_i, \zeta_i) = (\epsilon^{z_i}, \zeta^{z_i})$ and broadcasts it.
3. Each player $P_i$ proves to the other players that $(\epsilon_i, \zeta_i)$ is well formed relative to the commitment $C_i$. In particular, she provides a zero-knowledge proof of knowledge of a pair $(z_i, r_i) \in Z_q^2$ such that $C_i = g^{z_i} h^{r_i}$ and $\epsilon_i = \epsilon^{z_i}$ and $\zeta_i = \zeta^{z_i}$. This may be accomplished efficiently using appropriate variants on protocols elaborated in [13,17].
4. The players jointly decrypt $(\gamma, \delta) = (\prod_{i=1}^n \epsilon_i, \prod_{i=1}^n \zeta_i)$.
5. If the resulting plaintext is 1, then the players conclude that $(\alpha, \beta) \equiv (\alpha', \beta')$. Otherwise, they conclude that $(\alpha, \beta) \not\equiv (\alpha', \beta')$.

If any player is found to be deviating from the protocol, that player is excluded from further participation.

Like the other building blocks presented here, $\mathcal{PET}$ is minimal knowledge under the DDH assumption. By "minimal knowledge", we mean that players learn nothing beyond whether or not $(\alpha, \beta) \equiv (\alpha', \beta')$. To be more precise, even given malicious adversarial behavior (such as refusal of servers to participate in step 3), the distribution of protocol transcripts may be simulated by any entity that knows whether or not $(\alpha, \beta) \equiv (\alpha', \beta')$. The simulated transcript is indistinguishable from a correct one under the DDH assumption. Under the discrete log problem, it is infeasible for any adversary controlling $t < n/2$ servers to cause a server to deviate from the protocol without detection. Assuming non-interactive proof protocols with security in the random oracle model, the protocol may be executed with $\mathcal{O}(1)$ broadcast rounds, with a message complexity of $\mathcal{O}(n)$ group elements. The computational costs per player are $\mathcal{O}(n)$ exponentiations.[4]

The $\mathcal{PET}$ algorithm is the tool that enables us to perform comparisons between encrypted values input to gates and encrypted values in blinded tables. In other words, it is the basic tool for lookups in blinded tables.

## 3   The Mix and Match Protocol

We are now ready to describe in detail our main protocol, the mix-and-match scheme. Recall that players must agree in advance on a representation of the target function $f$ as a circuit $C_f$. Let us suppose that this circuit consists of $N$ gates, denoted by $G_1, G_2, \ldots, G_N$. We may assume, without loss of generality, that the numbering of gates is such that every gate $G_{i+1}$ has circuit depth at least that of $G_i$. Thus, evaluation of gate values may proceed in order of index number. For simplicity of presentation, we assume that all gates $G_i$ are binary, i.e., each gates has two inputs and one output, all of which are bit values. We also assume that the function $f$ is binary, i.e., the output is a single bit. We let gate $G_N$ be the output gate for $f$. We later give a brief description of how to extend the described scheme to non-binary gates and functions $f$ quite straightforwardly.

Let us denote the sequence of input bits of player $i$ by $B_i = \{b_1, b_2, \ldots, b_k\}$. Thus, the aim of the protocol is for players to compute $f(B_1, B_2, \ldots, B_n)$ without revealing any additional information about any of $B_1, B_2, \ldots, B_n$. Let us denote the lookup table corresponding to gate $G_i$ by $T_i$. As we assume that gates are binary, table $T_i$ contains three columns and four rows; the first two columns represent input bit values, and the third, the corresponding output bit. Table 1, for example, depicts the logical table corresponding to an AND gate. This is, of course, just a standard truth table.

| left | right | output |
|------|-------|--------|
| 0 | 0 | 0 |
| 0 | 1 | 0 |
| 1 | 0 | 0 |
| 1 | 1 | 1 |

Figure 1. Logical table representing an AND gate

We let $T_i[u, v]$ represent the value in row $u$ and column $v$ of table $T_i$. We denote by $\overline{T}_i$ the blinded, permuted table yielded by application of $\mathcal{MN}$ to $T_i$.

---

[4] Pedersen commitments are included here for technical reasons, namely to aid in security proofs for the protocol. Under the random oracle assumption on hash function $h$, we can have each player $P_i$ instead publish a commitment $C_i = h(\epsilon_i, \zeta_i)$, and subsequently have all players decommit. This reduces protocol costs by a significant constant factor.

We present our mix-and-match protocol in terms of four steps:

1. **Input protocol:** Each player contributing input to $f$ broadcasts El Gamal encryptions of her input bits $B_i$ under the public key $y$. (Note that as the integer 0 has only a degenerate ciphertext in the El Gamal cryptosystem, it is convenient to represent a '0' bit by the plaintext $g^{-1}$ and a '1' bit by the plaintext $g$.) Players prove knowledge of the associated plaintexts.
2. **Mixing:** Players apply the mix network $\mathcal{MN}$ to the tables $\{T_i\}$ under public key $y$. Each player in turn mixes all tables. The output is the set of blinded tables $\overline{T}_1, \overline{T}_2, \ldots, \overline{T}_N$.
3. **Matching:** For gates $G_1, G_2, \ldots, G_N$ in order, the players do the following. Let $l_i$ (left) and $r_i$ (right) be the ciphertext input values to gate $G_i$. The players use $\mathcal{PET}$ in order to compare the pair $(l_i, r_i)$ with each row $u$ in $\overline{T}_i$ until a match is found. For each row $u$, the players check whether $\mathcal{PET}(l_i, \overline{T}_i[u, 1]) = 1$ and $\mathcal{PET}(r_i, \overline{T}_i[u, 2]) = 1$. If both checks hold, then the players determine that the encrypted output value $o_i$ of gate $G_i$ is $\overline{T}_i[u, 3]$. The players do this for $u = 1, 2, 3$, and then 4 until a match is found.
4. **Output:** After evaluating the last gate, $G_N$, the players obtain $o_N$, a ciphertext encrypting $f(B_1, B_2, \ldots, B_n)$. They jointly decrypt this ciphertext value to reveal the output of the function $f$.

If a player has provided an invalid input ciphertext $(\alpha, \beta)$, i.e., a ciphertext whose plaintext does not represent a bit, then the matching step as applied to that ciphertext will fail. In other words, no matching row will be found. This will reveal the invalidity of the input to participating players. An alternative strategy to identify invalid inputs is for players to provide validity proofs along with their inputs. Let $(\alpha, \beta)$ be a ciphertext input in which a '0' bit is represented by the plaintext value $g^{-1}$, and a '1' bit is represented by the plaintext value $g$. Such a proof then takes the form of a proof of knowledge of $(z \mid \alpha/g = y^z, \beta = g^z)$ or $(z' \mid \alpha g = y^{z'}, \beta = g^{z'})$. See [6,17] for descriptions of how to construct disjunctive and conjunctive proofs of knowledge efficiently.

If players determine that a player $P_i$ participating in the function evaluation protocol has cheated or failed, they expel him from the protocol, according to standard practice in the literature for threshold security protocols. They then rewind and recompute as necessary. Due to space limitations, we do not prove security results, but simply state that our mix-and-match construction meets the security requirements formalized by Canetti [7] for secure multiplayer protocols. The interactive variant does this in a computational sense, i.e., there is an *ideal process adversary* capable of producing a simulation indistinguishable from a real one under the DDH assumption. The non-interactive variant depends additionally on use of the random oracle model.

**Extensions:** As explained above, it is easy to extend the mix-and-match protocol to non-binary gates $G_i$. For $G_i$ to take as input a $j$-tuple of values, we construct $T_i$ such that columns $1, 2, \ldots, j$ contain input values. (We increase the number of rows correspondingly to $2^j$.) On evaluating $G_i$, we compare the input tuple with the first $j$ columns of a given row in $\overline{T}_i$. For $G_i$ to yield multiple output values, it suffices to have the output column in $T_i$ carry multiple values. Input or output values can be made non-binary simply by formulating the entries in $T_i$ appropriately. Additionally, the circuit for $f$ can have multiple output gates, requiring simply that players perform multiple decryptions in the final step of the mix-and-match protocol.

The set of players providing inputs to $f$ in step 1 may be distinct or arbitrarily overlapping with the set of players performing the secure computation. We see this principle at work in our auction protocol in section 4. Similarly, since the set of players performing the mixing operation need have knowledge only of $y$ and not $x$, this set may be disjoint from the set of players performing the matching.

**Remarks:**

- The multiplicative homomorphism property of the El Gamal cipher allows for secure multiplication of plaintext values at the cost of a single modular multiplication. This observation may often be exploited to reduce the cost of the protocol. For example, if the left and right input bits to an AND gate are ciphertexts $l_i$ and $r_i$ respectively, each with plaintext value $g^{-1}$ (representing a '0' bit) or $g$ (representing a '1' bit), then the product $l_i r_i$ will have one of three corresponding plaintexts, $g^{-2}$, 1, or $g^2$. We can thus condense the AND gate to include only three rows.
- One means of reducing the number of gates is to use the El Gamal variant proposed by Franklin and Haber [22] for use in secure multiparty computation. This is an El Gamal cryptosystem in which -1 and 1 are both valid plaintexts. In consequence of the multiplicative homomorphism of El Gamal, it is possible for players to compute XOR non-interactively with this scheme.
- It may easily be seen that the transcripts of all players' proofs constitute a publicly verifiable proof of the correctness of the computation.

### 3.1    Performance

The full protocol in the random oracle model, i.e., with non-interactive proofs, may be achieved in $\mathcal{O}(n + d)$ broadcast rounds, where $d$ is the depth of the circuit $C_f$. As players invoke $\mathcal{MN}$ once per gate and $\mathcal{PET}$ a constant number of times per gate, the overall message complexity is $\mathcal{O}(nN)$ group elements, while the computational complexity per player is likewise $\mathcal{O}(nN)$ exponentiations. As noted above, mix and match can be implemented with use of interactive proofs using, e.g., techniques in [16], thereby eliminating the random oracle assumption as a security requirement at the expense of slightly higher protocol costs. Asymptotic costs in the non-interactive mix-and-match protocol are on a par with the best contemporaneous results, such as [16,30]. It is important to note, though, that these latter two results achieve full field operations for each gate. (The result in [16] is in the computational model with an assumed broadcast channel, and tolerates an adversary in control of any minority coalition. The scheme in [30] is in the private channels model with security for $t < n/3$.)

# 4   Auctions

We now show how to apply mix and match to the construction of an auction protocol. We consider two, possibly overlapping sets of participants, $m$ bidders, denoted by $A_1, A_2, \ldots, A_m$, and $n$ servers, denoted by $P_1, P_2, \ldots, P_n$. All participants are assumed to share an authenticated broadcast channel. We achieve the following properties in our proposed scheme:

1. *Non-interactivity:* Bidders submit bids in a non-interactive fashion. That is, they broadcast their bids to the servers, but need not participate subsequently in the auction protocol except to learn the outcome.
2. *Auction adaptability:* Our auction protocol is readily adaptable with little overhead to a range of auction types, such as highest-price auctions and Vickrey auctions, as well as to related non-auction procedures, such as polling.
3. *Full privacy:* The only information revealed at the conclusion of the auction is that essential to public execution of resulting transactions. In a highest-price auction, for example, only the winning bid and the identity of the winning bidder are revealed.
4. *Robustness:* An adversary consisting of a coalition of bidders or a minority coalition of servers cannot disrupt the auction protocol or undermine the privacy guarantees. (The servers are simply players as described in the mix-and-match scheme above, so that the computation achieves the same robustness and security characteristics.)
5. *Multiple servers:* Our auction protocol accommodates an arbitrary number of servers and a range of trust models on these servers.
6. *Public verifiability:* The proof transcripts of the auction servers are publicly verifiable. That is, any player (and indeed, any external entity) can verify the correctness of the auction execution without trust in the auction servers.

The principle drawback of our scheme is the intensive communication it requires among servers. Given that this may occur in a manner that is offline from the perspective of bidders, however, it does not pose a serious practical limitation.

In principle, it is possible to achieve the above set of properties with use of any general secure multiparty computation technique and any threshold public-key cryptosystem. The most difficult property to achieve in practice is that of non-interactivity. One method is as follows. Given an appropriate circuit representation for the computation, bidders submit ciphertext bids which the the servers decompose into shares using their private shares of the ciphertext key. This approach, however, is rather inefficient, as the circuit must be very large.

Another, more efficient approach to building a non-interactive protocol is for a bidder to make the sharing implicit in her bid. The bidder submits verifiable secret sharings of the component bits of her bid, along with ciphertexts of the shares. These ciphertexts may be encrypted under the keys of the individual servers, or under a shared key. In the latter case, the ciphertexts may be *directively decrypted*, i.e., decrypted for a unique recipient. This approach, while fairly practical, is still cumbersome. The bidder must, at a minimum, know how many servers are participating, and, to achieve a practical scheme, must submit $nk$ ciphertexts, where $k$ is the number of bits composing her bid. Additionally, as noted above, we believe that mix and match is quite competitive with other secure function evaluation protocols for applications, like auctions, involving intensive bitwise manipulation.

In consequence of the difficulties involved in deploying standard general secure function evaluation techniques, a number of secure protocols have been proposed in the literature that are specially tailored for auctions. One of the earliest of these is the scheme of Franklin and Reiter [21]. This scheme is not fully private, in the sense that it only ensures the confidentiality of bids until the end of the protocol (although the authors mention a fully private variant). Some more recent schemes include those of Harkavy, Tygar, and Kikuchi [29], Cachin [5], Sako [40], Di Crescenzo [19], and Naor, Pinkas, and Sumner [36]. The Harkavy *et al.* scheme is fully privacy preserving, but involves intensive bidder involvement [29], and is not easily adaptable to different auction types or to related protocols. The scheme of Cachin involves two servers, and requires some communication among bidders. At the end of the protocol, a list of bidders is obtained, but not the bid amounts. The scheme of Di Crescenzo [19] requires no communication between bidders, and has low round complexity, but involves the participation of only a single server. The scheme of Sako [40] works on a different principle from these others, involving opening of bids in what is effectively a privacy-preserving Dutch-style auction. While efficient for small auctions, it involves costs linear in the range of possible bids, and does not allow for extension to second-price and other auction types. The scheme of Naor *et al.* [36] is the first to seek to achieve the first four auction properties enumerated above.[5]

### 4.1    A Mix-and-Match Auction Protocol

We now present our auction protocol, achieving all six of the properties enumerated above. We describe an architecture for executing highest-bid auctions, although variants such as Vickrey auctions may be achieved through simple modifications imposing minimal additional overhead. Let $B_i = b_i^{(k)}, b_i^{(k-1)}, \ldots, b_i^{(1)}$ be a bitwise representation of the bid $B_i$ of bidder $A_i$. Let $E[b]$ represent the El Gamal encryption of a given bit $b$. To avoid cumbersome details, we use somewhat loose notation here, and also do not consider the association of user identities with bids. Also for the sake of simplicity, we assume that there are no ties between bids. The protocol is as follows.

1.  Each bidder $A_i$ submits her bid consisting of El Gamal ciphertexts $E[b_i^{(k)}]$, $E[b_i^{(k-1)}], \ldots, E[b_i^{(1)}]$ along with proofs of knowledge of the associated plaintexts. Let $E[B_i]$ denote the $k$-tuple of ciphertexts representing the submitted bid of $A_i$. (This submission $E[B_i]$ may be also digitally signed by $A_i$.)

---

[5] A security flaw which we do not have space to describe here, however, allows one of the servers to tamper with bids in this protocol.

2. For each pair of encrypted bids, players do the following:
   - Servers apply a mix-and-match-based millionaires' problem protocol to pairs of encrypted bids $(E[B_i], E[B_j])$. Let $E[w]$ denote the ciphertext outcome of the comparison on the bid pair $(\overline{E}_i, \overline{E}_j)$. Here $w = 1$ if bid $i$ is higher and $w = -1$ if bid $j$ is higher.[6]
   - Servers construct a two-row table $T$ in which the first row contains the pair $(-1, E[B_i])$, and the second row contains the pair $(1, E[B_j])$. Thus each row contains $k + 1$ columns. Players mix $T$ to obtain blinded table $\overline{T}$. Thus $\overline{T}$ consists of rows $(E[-1], \overline{E}[B_i])$ and $(E[1], \overline{E}[B_j])$ in a random order, where $\overline{E}[B_i]$ and $\overline{E}[B_j]$ denote re-encryptions of ciphertexts $E[B_i]$ and $E[B_j]$.
   - Servers match $w$ against the first column entries of $\overline{T}$. When they find a match, they output the ciphertext bid in the corresponding row. This will be $\overline{E}[B_i]$ if $w = -1$ and $\overline{E}[B_j]$ if $w = 1$.
3. Servers repeat the previous step following a tennis tournament format until only the ciphertext $E[B_t]$ of a winning bid remains.
4. Servers jointly decrypt the winning bid $E[B_t]$.

Many variants of this basic scheme are possible. For example, to handle ties, players might execute a sorting algorithm based on pairwise comparisons, rather than a tennis tournament. In this case, we must enforce a secret, random tie-breaking mechanism for comparisons between equal bids. Once an ordered list is obtained, it suffices to compare bids from highest to lowest until all highest bids are identified. We leave further details to the reader.

Assuming use of the publicly verifiable mix network proposed in [2,34] and correct server behavior, it may easily be seen that the asymptotic computational cost of the protocol described above is $\mathcal{O}(knm)$ exponentiations per server, while the message complexity is also $\mathcal{O}(knm)$. With the use of fast workstations, a crude estimate suggests that for $m = 100, n = 5$, and $k = 20$, i.e., for 100 bidders, 5 servers, and bids ranging values ranging from 1 to just over 1,000,000, an auction may be conducted in under 3 minutes on fast workstations.

## Acknowledgments

The authors wish to extend their thanks to Ivan Damgård, Moti Yung, and the anonymous reviewers of this paper.

## References

1. M. Abe. Universally verifiable mix-net with verification work independent of the number of mix-servers. In K. Nyberg, editor, *EUROCRYPT '98*, pages 437–447. Springer-Verlag, 1998. LNCS no. 1403.
2. M. Abe. A mix-network on permutation networks. In K.Y. Lam, C. Xing, and E. Okamoto, editors, *ASIACRYPT '99*, pages 258–273, 1999. LNCS no. 1716.

---

[6] The plaintext -1 may be encoded as other integer, as needed.

3. M. Bellare and P. Rogaway. Random oracles are practical: A paradigm for designing efficient protocols. In *ACM CCS '93*, pages 62–73. ACM, 1993.
4. M. Ben-Or, S. Goldwasser, and A. Wigderson. Completeness theorems for non-cryptographic fault-tolerant distributed computations. In *STOC '88*, pages 1–10. ACM, 1988.
5. C. Cachin. Efficient private bidding and auctions with an oblivious third party. In G. Tsudik, editor, *ACM CCS '99*, pages 120–127. ACM, 1999.
6. J. Camenisch and M. Michels. Proving that a number is the product of two safe primes. In J. Stern, editor, *EUROCRYPT '99*, pages 107–122. Springer-Verlag, 1999. LNCS no. 1592.
7. R. Canetti. Security and composition of multiparty cryptographic protocols. *Journal of Cryptology*, 13(1):143–202, 2000.
8. R. Canetti, R. Gennaro, S. Jarecki, H. Krawczyk, and T. Rabin. Adaptive security for threshold cryptosystems. In M. Weiner, editor, *CRYPTO '99*, pages 98 – 115. Springer-Verlag, 1999. LNCS no. 1166.
9. R. Canetti, R. Gennaro, S. Jarecki, H. Krawczyk, and T. Rabin. The (in)security of distributed key generation in dlog-based cryptosystems. In J. Stern, editor, *EUROCRYPT '99*, pages 295–310. Springer-Verlag, 1999. LNCS no. 1592.
10. D. Chaum. Untraceable electronic mail, return addresses, and digital pseudonyms. *Communications of the ACM*, 24(2):84–88, 1981.
11. D. Chaum, C. Crépeau, and I. Damgård. Multiparty unconditionally secure protocols. In *STOC '88*, pages 11–19. ACM, 1988.
12. D. Chaum, I. Damgård, and J. van de Graaf. Multiparty computations ensuring privacy of each party's input and correctness of the result. In C. Pomerance, editor, *CRYPTO '87*, pages 87–119. Springer-Verlag, 1987. LNCS no. 293.
13. D. Chaum and T.P. Pedersen. Wallet databases with observers. In E.F. Brickell, editor, *CRYPTO '92*, pages 89–105. Springer-Verlag, 1992. LNCS no. 740.
14. B. Chor, S. Goldwasser, S. Micali, and B. Awerbuch. Verifiable secret sharing and achieving simultaneity in the presence of faults. In *FOCS '85*, pages 383–395. IEEE Computer Society, 1985.
15. R. Cramer, I. Damgård, S. Dziembowski, M. Hirt, and T. Rabin. Efficient multiparty computations secure against an adaptive adversary. In J. Stern, editor, *EUROCRYPT '99*, pages 311–326. Springer-Verlag, 1999. LNCS no. 1592.
16. R. Cramer, I. Damgård, and J.B. Nielsen. Multiparty computation from threshold homomorphic encryption, 2000. IACR ePrint archive manuscript.
17. R. Cramer, I. Damgård, and B. Schoenmakers. Proofs of partial knowledge and simplified design of witness hiding protocols. In Y.G. Desmedt, editor, *CRYPTO '94*, pages 174–187. Springer-Verlag, 1994. LNCS no. 839.
18. R. Cramer and V. Shoup. A practical public-key cryptosystem provably secure against adaptive chosen ciphertext attack. In H. Krawczyk, editor, *CRYPTO '98*, pages 13–25. Springer-Verlag, 1998. LNCS no. 1462.
19. G. Di Crescenzo. Private selective payment protocols. In P. Syverson, editor, *Financial Cryptography '00*, 2000. To appear.
20. A. Fiat and A. Shamir. How to prove yourself: Practical solutions to identification and signature problems. In J. L. Massey, editor, *EUROCRYPT '86*, pages 186–194. Springer-Verlag, 1986. LNCS no. 263.
21. M. Franklin and M. Reiter. The design and implementation of a secure auction server. *IEEE Transactions on Software Engineering*, 22(5):302–312, 1996.
22. M.K. Franklin and S. Haber. Joint encryption and message-efficient secure computation. *Journal of Cryptology*, 9(4):217–232, 1996.

23. A. Fujioka, T. Okamoto, and K. Ohta. A practical secret voting scheme for large scale elections. In J. Seberry and Y. Zheng, editors, *AUSCRYPT '92*, pages 244–251. Springer-Verlag, 1992. LNCS no. 718.

24. T. El Gamal. A public key cryptosystem and a signature scheme based on discrete logarithms. *IEEE Transactions on Information Theory*, 31:469–472, 1985.

25. R. Gennaro, S. Jarecki, H. Krawczyk, and T. Rabin. Robust threshold DSS signatures. In U. Maurer, editor, *EUROCRYPT '96*, pages 354–371. Springer-Verlag, 1996. LNCS no. 1070.

26. R. Gennaro, M. Rabin, and T. Rabin. Simplified VSS and fast-track multiparty computations with applications to threshold cryptography. In *PODC '98*, pages 101–111. ACM, 1998.

27. O. Goldreich, S. Micali, and A. Wigderson. How to play any mental game. In *STOC '87*, pages 218–229. ACM, 1987.

28. S. Goldwasser and S. Micali. Probabilistic encryption. *J. Comp. Sys. Sci*, 28(1):270–299, 1984.

29. M. Harkavy, J.D. Tygar, and H. Kikuchi. Electronic auctions with private bids. In *3rd USENIX Workshop on Electronic Commerce*, pages 61–73, 1999.

30. M. Hirt, U. Maurer, and B. Przydatek. Efficient secure multi-party computation. In T. Okamoto, editor, *ASIACRYPT '00*, 2000. To appear.

31. P. Horster, M. Michels, and H. Petersen. Some remarks on a receipt free and universally verifiable mix-type voting scheme. In K. Kim and T. Matsumoto, editors, *ASIACRYPT '96*, pages 125–132. Springer-Verlag, 1996. LNCS no. 1163.

32. M. Jakobsson. A practical mix. In K. Nyberg, editor, *EUROCRYPT '98*, pages 448–461. Springer-Verlag, 1998. LNCS no. 1403.

33. M. Jakobsson. Flash mixing. In *PODC '99*, pages 83–89. ACM, 1999.

34. M. Jakobsson and A. Juels. Millimix: Mixing in small batches, June 1999. DIMACS Technical Report 99-33.

35. A.J. Menezes, P.C. van Oorschot, and S.A. Vanstone. *Handbook of Applied Cryptography*. CRC Press, 1996.

36. M. Naor, B. Pinkas, and R. Sumner. Privacy preserving auctions and mechanism design. In *1st ACM Conf. on Electronic Commerce*, pages 129–139. ACM, 1999.

37. W. Ogata, K. Kurosawa, K. Sako, and K. Takatani. Fault tolerant anonymous channel. In *ICICS '97*, pages 440–444. Springer-Verlag, 1997. LNCS no. 1334.

38. C. Park, K. Itoh, and K. Kurosawa. All/nothing election scheme and anonymous channel. In T. Helleseth, editor, *EUROCRYPT '93*. Springer-Verlag, 1993. LNCS no. 921.

39. T. Pedersen. Non-interactive and information-theoretic secure verifiable secret sharing. In J. Feigenbaum, editor, *CRYPTO '91*, pages 129–140. Springer-Verlag, 1991. LNCS no. 576.

40. K. Sako. An auction protocol which hides bids of losers. In H. Imai and Y. Zheng, editors, *PKC '00*, pages 422–432. Springer-Verlag, 2000. LNCS no. 1751.

41. K. Sako and J. Kilian. Receipt-free mix-type voting scheme - a practical solution to the implementation of a voting booth. In L.C. Guillou and J.-J. Quisquater, editors, *EUROCRYPT '95*. Springer-Verlag, 1995. LNCS no. 921.

42. C.P. Schnorr. Efficient signature generation by smart cards. *Journal of Cryptology*, 4:161–174, 1991.

43. Y. Tsiounis and M. Yung. On the security of ElGamal-based encryption. In H. Imai and Y. Zheng, editors, *PKC '98*, pages 117–134. Springer-Verlag, 1998. LNCS no. 1431.

44. A.C. Yao. Protocols for secure computations (extended abstract). In *FOCS '82*, pages 160–164. IEEE Computer Society, 1982.

# A Length-Invariant Hybrid Mix

Miyako Ohkubo and Masayuki Abe

NTT Laboratories
Nippon Telegraph and Telephone Corporation
1-1 Hikari-no-oka, Yokosuka-shi, Kanagawa-ken, 239-0847 Japan
miyako.ookubo@east.ntt.co.jp, abe@isl.ntt.co.jp

**Abstract.** This paper presents a secure and flexible Mix-net that has the following properties; it efficiently handles long plaintexts that exceed the modulus size of underlying public-key encryption as well as very short ones (*length-flexible*), input ciphertext length is not impacted by the number of mix-servers (*length-invariant*), and its security in terms of anonymity is proven in a formal way (*provably secure*). One can also add robustness i.e. it outputs correct results in the presence of corrupt servers. The security is proved in the random oracle model by showing a reduction from breaking the anonymity of our Mix-net to breaking a sort of indistinguishability of the underlying symmetric encryption scheme or solving the Decision Diffie-Hellman problem.

**Keywords:** Mix-net, Hybrid Encryption, Anonymous Channel, Hybrid Mix

## 1 Introduction

### 1.1 Background

Mix-net is a cryptographic primitive that provides anonymity to message senders. It takes a list of encrypted messages sent from a sufficient number of users and outputs a list of corresponding plaintexts sorted in random order so that it conceals the correspondence between each plaintext and user. Accordingly, it provides anonymity by hiding the individual user in the mass. Such a primitive was first introduced in [7] with a heuristic construction based on public key encryption. Since then, many works have improved its usability and security. In [18], Park, et al., constructed a scheme based on El Gamal encryption, where the encryption work and resulting ciphertext length were independent of the number of mix-servers. Robustness was addressed in [22,16,17,3,12,1,13,14,9]. Attacks are found in [20,19,16,9].

A promising application of Mix-net is electronic voting as it can convey any style of ballots, e.g., simple binary value of Yes/No voting and free-format questionnaires, without changing the protocol. It is also useful in other applications such as anonymous payments and anonymous bids.

To support wide availability, Mix-net should be able to efficiently handle messages of various lengths that differ depending on the application. Some applications where users anonymously send signatures issued by an authority (possibly

T. Okamoto (Ed.): ASIACRYPT 2000, LNCS 1976, pp. 178–191, 2000.

in a blind way) need Mix-net to convey signature-message pairs which require thousands of bits to be stored depending on the signature algorithm. Despite the need for handling long plaintexts, all previous Mix-nets limit messages to be shorter than a single block of ElGamal encryption or handle long messages in a heuristic way. For example, if ElGamal encryption is implemented over an elliptic curve for speed with typical settings, the messages are limited to just 160 bits, which greatly limits applicability. Although one can handle long messages simply by dividing each message into some blocks and repeating the atomic mix processing a sufficient number of times, such an approach results in an inefficient scheme.

Very short plaintexts are also dealt in an inefficient way as they are expanded to one block of underlying public-key encryption. For instance, ElGamal encryption with 1024 bit modulus expands 1 bit message to $1024 + 1024$ bits ciphertext. Hence, previous schemes incur higher communication costs than actually needed.

A common approach that overcomes such shortcomings would be to use hybrid encryption schemes that combine asymmetric key exchange and symmetric (common key) encryption. Although some provably secure hybrid encryption schemes are available in the literature, e.g., [10,23], applying those schemes does not immediately result in a secure and efficient Mix-net. It is not clear whether a secure hybrid encryption scheme provides security even in the context of Mix-net. Furthermore, a straightforward use of hybrid encryption in the original construction of Chaum [7] obviously extends the resulting input ciphertext linearly depending on the number of servers.

## 1.2   Our Contribution

This paper presents Mix-nets that realize, for the first time, the following properties all at the same time.

- **Length-flexibility**: The size of the public-key of Mix-net does not limit plaintext length. Plaintexts of any length are encrypted efficiently in terms of computation and resulting ciphertext length.
- **Length-invariance**: The length of input ciphertexts is independent of the number of mix-servers.
- **Provable security**: The security, in terms of anonymity, of our Mix-net can be proven in the random oracle model [5] assuming the intractability of the Decision Diffie-Hellman problem and the availability of a symmetric encryption scheme that ensures a sort of indistinguishability.

Furthermore, we show an approach to add *robustness* so that correct output is obtained even if some of the users and servers behave maliciously.

To achieve the above goals, we developed a novel hybrid encryption scheme with group decryption feature that suits Mix-net. Informally, it conceals the correspondence between inputs and outputs at each step of group decryption performed by each server.

Our scheme saves communication cost for short messages as well as long ones since the encryption only extends the message with modulus length. For instance, input ciphertext is $1024 + 1$ bits long for a 1 bit message with 1024 bit modulus. Computational cost of encryption grows linearly depending on the number of servers. We show, however, in section 7, that the cost can actually be smaller than that of previous standard schemes in many settings.

We first introduce a basic scheme that highlights our key idea. It is secure only against honest but curious users and mix-servers. We then add security to withstand distrustful users (mix-servers are still honest but curious). If needed, one can add individual verifiability to these basic schemes in a simple standard way in order to detect the deviation of servers with some probability. Such schemes would be applicable for applications where mix-servers are chosen carefully and thus are more creditable than users. Such schemes would also be used in the applications, such as anonymous donation or payment, in which each user is not concerned about the input of other users and thus individual verifiability is sufficient.

We then add robustness by following [9]. As in their scheme, the resulting Mix-net is robust in such a sense that it outputs a correct result and provides anonymity in the presence of corrupt servers, but does not provide universal verifiability. That is, only the servers can be convinced of the correctness of the results while no external parties can verify them. Such a model was also addressed in [12,13]. Accordingly, such scheme would be useful, for instance, for small scale applications where every user can act as a mix-server.

## 2 Model

### 2.1 Scenario

There are $n$ users and $m$ mix-servers. Let $\mathsf{U}_i$ and $\mathsf{M}_j$ denote user $i$ and server $j$, respectively. For simplicity, we assume that all communication between these participants are done via a bulletin board. The scenario consists of three phases.

**Preliminary phase:** The maximum length of each plaintext, say $\ell_{msg}$ is announced to all users together with other application-dependent information. It is stressed that $\ell_{msg}$ is independent of the public key size and is determined by the Mix-net application. Theoretically, $\ell_{msg}$ can be any positive integer bound by a polynomial of the security parameter.

**Casting phase:** Each user encrypts his message and sends it to the bulletin board. Appropriate padding may be applied to the message before encryption so that the length of the message equals $\ell_{msg}$.

**Mixing phase:** Let $L_0$ be a list of all ciphertexts sent from the users. The first server takes $L_0$ and outputs a list, $L_1$, to the bulletin board. Similarly, server $i$ takes $L_{i-1}$ and outputs $L_i$. The final output of the mix-net is $L_m$. If all servers work correctly, $L_m$ is a list of plaintexts sorted in random-order.

Let $\bar{L}_0$ be a list of messages obtained by correctly decrypting each ciphertext in $L_0$. The output of mix-net, $L_m$, is said to be *correct* if there exists a permutation between $\bar{L}_0$ and $L_m$. We say, informally, that the mix-net provides *anonymity* if it is intractable to distinguish two plain messages in $L_m$ that originate from two honest users.

## 2.2    Adversaries

We represent the power of an adversary by $(t_u, t_s)^{**}$ where $*$ is either A or P meaning *active* or *passive*, respectively. For instance, $(t_u, t_s)^{AP}$-adversary means that the adversary can thoroughly control up to $t_u$ users (i.e., active against users), and can obtain views from up to $t_s$ mix-servers (i.e., passive against servers). Note that adversaries that are passive against servers only attempt to violate anonymity as they can not control the servers so as to output incorrect results. We assume that the adversaries are static meaning that they decide, before the protocol begins, which users and servers they will attack.

In this paper, we deal with the following types of adversaries. The types are listed in order of increasing strength.

- $(n-2, m-1)^{PP}$-adversary; There are at least two unattacked users and an unattacked server. The adversary can obtain views from attacking users and servers but can not control either of them. Our basic scheme is safe against this type of adversary.
- $(n-2, m-1)^{AP}$-adversary; The same as above, but can control corrupt users. Since the adversary can send any ciphertexts through the corrupt users and let the servers decrypt them, it can launch chosen ciphertext attacks. Our extended scheme withstands this type of adversary.
- $(n-2, \mathcal{O}(\sqrt{m}))^{AA}$-adversary; This type of adversary, which is the strongest of the three, attempts to violate anonymity *or correctness*. Our third scheme withstands such an adversary.

## 3    The Basic Scheme

Let $\mathcal{G}$ be a discrete logarithm instance generator such that $(p, q, g) \leftarrow \mathcal{G}(1^k)$ where $k$ is a security parameter, and $p$, $q$ are primes that satisfy $q|p-1$, and $g$ is an element of $Z_p^*$ whose order is $q$. Let $\langle g \rangle$ denote a unique subgroup of $Z_p^*$ generated by $g$. All the subsequent arithmetic operations are performed in modulo $p$ unless otherwise stated.

$(\mathcal{E}, \mathcal{D}, \mathcal{K}, \mathcal{M}, \mathcal{C})$ denotes a symmetric encryption scheme where $\mathcal{E}, \mathcal{D}$ are the encryption and decryption algorithms and $\mathcal{K}, \mathcal{M}, \mathcal{C}$ are the spaces for keys, messages, and ciphertexts, respectively. $\mathcal{E}_K(x)$ denotes the result of encrypting plaintext $x$ with common key $K$. Similarly, $\mathcal{D}_K(x)$ denotes the plaintext obtained by decrypting ciphertext $x$ with key $K$. We assume that the symmetric encryption scheme is length-preserving, i.e., $\mathcal{M} = \mathcal{C} = \{0,1\}^{\ell_{msg}}$. Let $H$ be a hash function, $H : \langle g \rangle \to \mathcal{K}$.

**[Key generation]**
Server $i$ randomly selects a pair of private keys $a_i, x_i$ from $Z_q^*$ and computes

$$h_i := h_{i-1}^{a_i}, \text{ and}$$
$$y_i := h_i^{x_i},$$

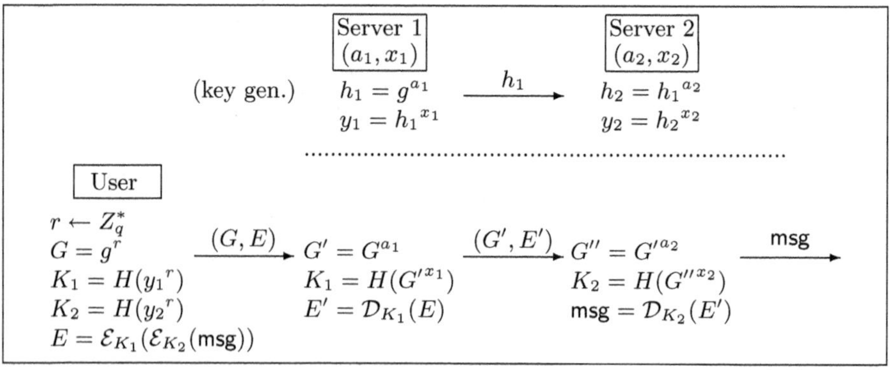

**Fig. 1.** The basic hybrid Mix-net with two servers. The input to Server 1 is a list of $(G, E)$s made by each user. Server 1 outputs a list of randomly ordered $(G', E')$s. Server 2 finally outputs a list of randomly ordered plaintexts. $\mathcal{E}, \mathcal{D}$ are the encryption and decryption algorithms. $H$ is a hash function.

for $h_{i-1}$ given from the previous server. (Let $h_0 = g$ for the first server.) It then publishes $y_i$ and $h_i$ as a pair of public keys.

**[Encryption]**
User encrypts message $msg \in \{0, 1\}^{\ell_{msg}}$ to ciphertext $C$ as

$$C := (G, E) = (g^r, \ \mathcal{E}_{K_1} \cdots \mathcal{E}_{K_m}(msg))$$

where $r$ is randomly taken from $Z_q^*$, and $K_1, \ldots, K_m$ are session keys for symmetric encryption $\mathcal{E}$ and are computed as

$$K_i := H(y_i^r).$$

**[Mix Decryption]**
For $i = 1$ to $m$, server $i$ decrypts each ciphertext $C = (G, E)$ in list $L_{i-1}$ to get $C' = (G', E')$ as

$$G' := G^{a_i},$$
$$E' := \mathcal{D}_{K_i}(E) \quad \text{where} \quad K_i = H(G'^{x_i}).$$

(For the last server, let $C' = E'$.) Server $i$ then selects a random permutation $\pi_i$ of $\{1, \ldots, n\}$ and puts the resulting ciphertexts into $L_i$ in the random order defined by $\pi_i$.

Figure 1 illustrates the above basic scheme with two servers. For each $i$, $H(G'^{x_i}) = H(h_{i-1}^{r a_i x_i}) = H(h_i^{r x_i}) = H(y_i^r) = K_i$ holds. Thus, if every server works correctly, correct session keys are retrieved by each server and the correct plaintext is obtained.

# 4   Security of the Basic Scheme

## 4.1   Definitions and Assumptions

For $\rho := (p, q, g)$ generated by $\mathcal{G}(1^k)$, we assume that all poly-time algorithms solve the following problems only with negligible (in $k$) advantage over randomly guessing.

**Definition 1. (Computational Diffie-Hellman Problem : CDHP)**

> **Input:** $(\rho, g^a, g^b)$ where $a, b \leftarrow Z_q$.

**Output:** $g^{ab}$

**Definition 2. (Decision Diffie-Hellman Problem : DDHP)**

> **Input:** $(\rho, g^a, G_0, G_1, G_b{}^a)$ where $b \leftarrow \{0, 1\}$, $a \leftarrow Z_q$, $G_0, G_1 \leftarrow \langle g \rangle$.
> **Output:** $b$

**Definition 3. (Matching Diffie-Hellman Problem : MDHP)**

> **Input:** $(\rho, g^a, G_0, G_1, G_b{}^a, G_{\bar{b}}{}^a)$ where $b \leftarrow \{0, 1\}$, $a \leftarrow Z_q$, $G_0, G_1 \leftarrow \langle g \rangle$
> **Output:** $b$

It holds that CDHP > DDHP > MDHP, i.e., CDHP is the hardest to solve. The reverse relation between CDHP and DDHP is not known. For DDHP and MDHP, we can show that MDHP = DDHP following [11] or [21].

Next we define the Matching Find-Guess problem, which is closely related to the Find-Guess problem [10] which defines a sort of indistinguishability of symmetric encryption schemes.

**Definition 4. (Matching Find-Guess Problem : MFGP)**

> **Input:** $(\mathcal{E}_{K_0}(x_0), \mathcal{E}_{K_1}(x_1), x_b, x_{\bar{b}})$ where $x_0, x_1 \leftarrow \mathcal{M}$, $K_0, K_1 \leftarrow \mathcal{K}$, $b \leftarrow \{0, 1\}$.
> **Output:** $b$

We say that a symmetric encryption is secure in the sense of MFG if for all poly-time algorithms MFGP can be solved only with negligible advantage over $1/2$. Clearly, a one-time pad provides security in the sense of MFG. A stream cipher also provide the same security if its generator produces a pseudorandom bit-stream. For the sake of efficiency, we expect that existing carefully designed symmetric encryption schemes used in an appropriate mode of operation such as OFB provide such security as well.

In our construction, session keys for symmetric encryption are derived by applying hash function $H$ to the results of the Diffie-Hellman key exchange. Namely, the Diffie-Hellman key exchange and the symmetric encryption are *connected* by hash function $H$. The security of our hybrid encryption scheme is related to the following problem.

**Definition 5. (MFG-MDH Joint Problem)**

  **Input:** $(\rho, g^a, g^{ax}, M_0, G_0, M_1, G_1, M'_b, G'_b, M'_{\bar{b}}, G'_{\bar{b}})$
  *Where $a, x \leftarrow Z^*_q$, $b \leftarrow \{0, 1\}$, and*
  *for $i = (0, 1)$, $G_i \leftarrow \langle g \rangle$, $M_i \leftarrow \mathcal{M}$, $G'_i = G^a_i$, $M'_i = \mathcal{D}_{K_i}(M_i)$*
  *where $K_i = H(G'^x_i)$.*

**Output:** $b$

It will be shown that if $H$ is an ideal hash function then solving the above joint problem is as hard as solving either MFGP or MDHP.

## 4.2    Theorems and Proofs

**Theorem 1.** *The basic scheme provides anonymity in the presence of $(n-2, m-1)^{\text{PP}}$-adversary if DDHP and the MFGP are intractable.*

To support this theorem, we will prove the following lemmas.

**Lemma 1.** *If there exists an $(n-2, m-1)^{\text{PP}}$-adversary $\mathcal{A}_{Xp}$ that breaks anonymity in our Mix-net, then there exists machine $\mathcal{A}_M$ that solves the MFG-MDH joint problem with probability non-negligibly better than $1/2$.*

**Lemma 2.** *If $\mathcal{A}_M$ exists, then there exists machine $\mathcal{A}_D$ that solves, at least, either MFGP or MDHP with probability non-negligibly better than $1/2$.*

**Lemma 3.** *If $\mathcal{A}_D$ solves MDHP, then there exists a machine $\mathcal{A}_H$ that solves DDHP with probability non-negligibly better than $1/2$.*

Here, we sketch the proof of Lemma 1 and put the proof of Lemma 2 in the Appendix. Lemma 3 can be proven in the same way as shown in [11], so its proof is omitted.

*Proof of Lemma 1* (sketch): Let $\mathsf{M}_\xi$ be the server that $\mathcal{A}_{Xp}$ does not attack, i.e, the one whose view is not given to $\mathcal{A}_{Xp}$. Given an MFG-MDH joint problem instance $(\mathbf{p}, \mathbf{q}, \mathbf{g}, \mathbf{g^a}, \mathbf{g^{ax}}, \mathbf{M_0}, \mathbf{G_0}, \mathbf{M_1}, \mathbf{G_1}, \mathbf{M'_b}, \mathbf{G'_b}, \mathbf{M'_{\bar{b}}}, \mathbf{G'_{\bar{b}}})$, $\mathcal{A}_M$ simulates the view of $\mathcal{A}_{Xp}$ as follows.

  *Simulating Keys:* For server $\mathsf{M}_\xi$, $\mathcal{A}_M$ sets $h_\xi = \mathbf{g^a}$ and $y_\xi = \mathbf{g^{ax}}$. For the keys of descending servers, $\mathsf{M}_{\xi+1}, \ldots, \mathsf{M}_m$, $\mathcal{A}_M$ follows the key generation procedure, i.e. randomly chooses private keys $a_i, x_i$ and computes corresponding public keys $h_i = h^{a_i}_{i-1}, y_i = h^{x_i}_i$. For the keys of ascending servers, a little thought is needed. Let $h_{\xi-1} = \mathbf{g}$. Then, for $i = \xi - 1$ to $1$, $\mathcal{A}_M$ chooses $a_i, x_i$ and computes $h_{i-1} = h^{1/a_i}_i, y_i = h^{x_i}_i$. It finally sets $\rho = (\mathbf{p}, \mathbf{q}, h_0)$.

  *Simulating Lists:* $\mathcal{A}_M$ puts $(\mathbf{M'_b}, \mathbf{G'_b})$ and $(\mathbf{M'_{\bar{b}}}, \mathbf{G'_{\bar{b}}})$ into random positions in $L_\xi$. It then randomly generates other entries of $L_\xi$ by taking $M$ randomly from ciphertext space $\mathcal{C}$ and computing $G$ as $h^r_\xi$ with randomly chosen $r$. Next, it selects random permutation $\pi_\xi$ and computes each entry of $L_{\xi-1}$ by encrypting the corresponding entry in $L_\xi$. (This is possible because $\mathcal{A}_M$ can retrieve the correct session key by computing $y_\xi^r$.) For the two special entries of $L_{\xi-1}$ that correspond to $(\mathbf{M}_\mathbf{b}', \mathbf{G}_\mathbf{b}')$ and $(\mathbf{M}_\mathbf{\bar{b}}', \mathbf{G}_\mathbf{\bar{b}}')$, $\mathcal{A}_M$ inserts $(\mathbf{M_0}, \mathbf{G_0})$ and $(\mathbf{M_1}, \mathbf{G_1})$.

Now, the rest of ascending lists, $L_{\xi-2}$ to $L_1$, can be computed in order by *encrypting* the previous lists with the simulated private keys and randomly generated permutations. (Note that those permutations are chosen so that the two special entries of $L_{\xi-1}$ correspond to the inputs from unattacked users $\mathsf{U}_1$ and $\mathsf{U}_2$.) Similarly, the rest of descending lists, $L_{\xi+1}$ to $L_m$, can be computed by *decrypting* from $L_\xi$ to $L_{m-1}$ in order. In the course of the above simulation, $\mathcal{A}_M$ consults random oracle $H$ to compute the session keys.

Views of attacked users and servers can be appropriately simulated by using messages in $L_m$ and random choices of above simulation. Given the perfectly simulated views and lists, and free access to $H$, $\mathcal{A}_{Xp}$ distinguishes two messages in $L_m$ originated from $\mathsf{U}_1$ and $\mathsf{U}_2$. From the result of $\mathcal{A}_{Xp}$, $\mathcal{A}_M$ can derive the correspondence between two special positions in $L_{\xi-1}$ and $L_\xi$ where the given instances were placed by using the permutation taken by the simulated servers, except for $\mathsf{M}_\xi$. The success provability of $\mathcal{A}_M$ is the same as that of $\mathcal{A}_{Xp}$.     □

## 5   Securing against Corrupt Users

The key idea to add security against corrupt users is to make the underlying encryption non-malleable so that they cannot launch chosen ciphertext attacks. Although several efficient non-malleable encryption schemes are available (e.g. [24,25,8,6,2,10,23]), few meet our requirements. For our security proof, we need the underlying encryption scheme that provides *plaintext awareness* [4] and *public verifiability*. The latter functionality allows the validity of ciphertexts to be checked without using the decryption key. Our solution is based on [24].

Overall, the protocols are unchanged except that users attach a kind of proof and the mix-servers screen ciphertexts that come with invalid proofs.

**[Encryption]**
Message $msg$ is encrypted to $C = (G, E)$ in the same way as in the basic scheme. Let $G = g^r$. A proof of knowing $r$ is defined as $P := (e, z, \bar{G}, \tilde{G}, \eta, \bar{\eta}, \tilde{\eta})$ such that

$$\bar{g} := H_2(G), \tilde{g} := H_3(G),$$
$$\bar{G} := \bar{g}^r, \tilde{G} := \tilde{g}^r,$$
$$\eta := g^\varsigma, \bar{\eta} := \bar{g}^\varsigma, \tilde{\eta} := \tilde{g}^\varsigma,$$
$$e := H_4(E, g, \bar{g}, \tilde{g}, G, \bar{G}, \tilde{G}, \eta, \bar{\eta}, \tilde{\eta}),$$
$$z := \varsigma - re \bmod q,$$

where $\varsigma \leftarrow Z_q$ and $H_2, H_3, H_4$ are hash functions. The output is $(C, P)$.

**[Mix Decryption]**
Each server first verifies that

$$e = H_4(E, g, \bar{g}, \tilde{g}, G, \bar{G}, \tilde{G}, g^z G^e, \bar{g}^z \bar{G}^e, \tilde{g}^z \tilde{G}^e)$$

holds for each input ciphertext. Here $\bar{g}$, $\tilde{g}$ are computed as $\bar{g} := H_2(G)$, $\tilde{g} := H_3(G)$, respectively. After they agree on the result of the verification, they put $C$, which came with valid $P$, into list $L_0$. The rest of the process is the same as that of the basic scheme.

The benefit of $P$ is that it makes the simulation of a server, say $\mathsf{M}_\xi$, possible without knowing its private keys. The trick is as follows. We want to derive $K_\xi$ and $G'$ for each ciphertext in $L_\xi$. Set $\bar{g} = H_2(G) = y_\xi{}^u$ and $\tilde{g} = H_3(G) = h_\xi{}^v$ taking $u$ and $v$ randomly from $Z_q$. It follows that valid $P$ should contain $\bar{G} = \bar{g}^r = y_\xi^{ru}$ and $\tilde{G} = \tilde{g}^r = h_\xi^{rv}$. Thus, we can compute $H(\bar{G}^{1/u}) = H(y_\xi^r) = K_\xi$ and $\tilde{G}^{1/v} = h_\xi^r = G'$ as expected.

**Theorem 2.** *The extended scheme provides anonymity in the presence of $(n - 2, m - 1)^{\mathrm{AP}}$-adversaries if DDHP and MFGP are intractable.*

To prove the above theorem, it is sufficient to prove the following lemma. The rest of the proof is supported by Lemma 2 and 3.

**Lemma 4.** *If there exists an $(n - 2, m - 1)^{\mathrm{AP}}$-adversary that breaks anonymity in our Mix-net, there exists machine that solves the MFG-MDH joint problem.*

*Proof (sketch)* The difference of this proof from that of Lemma 1 is twofold:

- the proof-part $P$ of the inputs from honest users has to be simulated, and
- the simulator $A_M$ has to correctly decrypt the input ciphertexts coming from a corrupt user without knowing the private keys of the unattacked server.

For the first point, we will use the standard simulation technique for the honest verifier public-coin zero-knowledge proofs by regarding $H_4$ as a random oracle.

For the second point, we exploit the plaintext awareness (PA) of the underlying encryption scheme. $A_M$ first computes $L_m$ by using the PA property, and subsequently computes $L_{m-1}, ..., L_\xi$ by encrypting each entry of the previous list with the simulated public keys. $A_M$ then computes $L_0$ to $L_{\xi-1}$ by correctly performing decryption with the simulated decryption keys. In this way, the resulting $L_{\xi-1}$ and $L_\xi$ have the same relation as the one in the real execution with regard to the public key of unattacked server $\mathsf{M}_\xi$.     □

# 6     Securing against Corrupt Servers

Robustness is added following [9]. Let us briefly introduce the main idea here and omit the details due to page restriction.

To prevent corrupt servers from behaving maliciously, we group servers in such a way that every group contains at least one honest server, and at least one group consists only of honest servers. Such grouping is easily formed by placing

$t + 1$ servers in each of $t + 1$ groups when $m = (t + 1)^2$. Then, a representative member in each group executes mix decryption and all other members are given all private information (secret keys and random choices) used in mix decryption and monitor its behavior. Thus malicious deviation in each group is always detected by an honest member. Since there is at least one perfectly honest group, the group works correctly and outputs correct result.

## 7   Efficiency Analysis

Here, the computational cost of our basic scheme is compared to that of [18], which is one of the widely known schemes based on ElGamal encryption. It provides the same level of security as our basic scheme. Their scheme is summarized as follows; Server $i$ has private key $x_i$ and public key $y_i = g^{x_i}$. For each ElGamal ciphertext $(G, M) = (g^r, msg \cdot y^r)$ where $y = \prod_{j=1}^{m} y_j$, server $i$ computes $(G', M') = (Gg^t, Mg^{-x_i}\hat{y}_i^t)$ where $\hat{y}_i = \prod_{j=i}^{m} y_j$. In this scheme, $q$ must be very large so that all potential messages are in the subgroup generated by $g$. Hence we assume $|p| \approx |q|$. On the other hand, since our scheme needs randomness sufficient for generating symmetric keys, $|q|$ can be much smaller than $|p|$.

Table 1 shows the number of modular multiplications needed for encryption assuming the use of the binary method for exponentiation. For double-base and single-base exponentiation, we assume the simple table-lookup method described in [15] which costs $\frac{7}{4}|q|$ multiplications for a double-base exponentiation, and $\frac{3}{2}|q|$ for a single-base exponentiation. Although user's computation is linear in the number of servers, for a typical setting, say $|p| = 1024$ and $|q| = 160$, our scheme enjoys lesser computation (excl. symmetric encryption) up to 11 servers.

This advantage will be lost if one considers elliptic curve implementation where $|p| \approx |q|$. However, our scheme still saves computation if messages exceeds 160 bits as symmetric encryption is 100 to 1000 times faster than scalar multiplication over an elliptic curve.

| Scheme | User | Each Server |
|---|---|---|
| El Gamal [18] | $\frac{3}{2}|p| \times 2$ | $\frac{7}{4}|p| + \frac{3}{2}|p|$ |
| Ours (basic) | $\frac{3}{2}|q| \times (m + 1)$ | $\frac{3}{2}|q| \times 2$ |

**Table 1.** Number of modular multiplications per message. $m$ is number of servers.

## 8   Open Problems

The resulting robust scheme still has some issues that must be resolved. First, it is preferable to provide public verifiability so that anyone outside of the mix can be convinced of the output. Second, optimal resiliency should be provided. At least, we need linear resiliency, i.e., $\mathcal{O}(m)$, instead of $\mathcal{O}(\sqrt{m})$.

Our security proof of Lemma 1 is for the case where honest users select messages uniformly from $\{0,1\}^{\ell_{msg}}$. However, the users might be restricted to choose messages from exponentially sparse space (with length $\ell_{msg}$). In such a case, our simulation in the proof of Lemma 1 is not suitable. That is because the plain messages obtained by decrypting the given instance of the joint problem are not likely to fall into the exponentially sparse space. Since we assume that the underlying encryption scheme provides security equally for all messages, such restriction on message space is not likely to impact security. It remains, however, as an open problem to prove this in a formal way.

## Acknowledgment

The authors thank Kazue Sako for her invaluable suggestions and comments.

## References

1. M. Abe. Mix-networks on permutation networks. *Asiacrypt '99*, LNCS 1716, pages 258–273. 1999.
2. M. Abe. Robust distributed multiplication without interaction. In M. Wiener, editor, *in Cryptology — CRYPTO '99*, LNCS 1666, pages 130–147. 1999.
3. M. Abe. Universally verifiable mix-net with verification work independent of the number of mix-servers. *IEICE Trans. Fundamentals*, E83-A(7):1431–1440, July 2000. Presented in Eurocrypt'98.
4. M. Bellare, A. Desai, D. Pointcheval, and P. Rogaway. Relations among notions of security for public-key encryption schemes. *CRYPTO '98*, LNCS 1462, pages 26–45. 1998.
5. M. Bellare and P. Rogaway. Random oracles are practical: a paradigm for designing efficient protocols. *ACM CCCS*, pages 62–73. 1993.
6. R. Canetti and S. Goldwasser. An efficient threshold public key cryptosystem secure against adaptive chosen ciphertext attack. *Eurocrypt '99*, LNCS 1592, pages 90–106. 1999.
7. D. L. Chaum. Untraceable electronic mail, return address, and digital pseudonyms. *Comm. of the ACM*, 24:84–88, 1981.
8. R. Cramer and V. Shoup. A practical public key cryptosystem provably secure against adaptive chosen ciphertext attack. *CRYPTO '98*, LNCS 1462, pages 13–25. 1998.
9. Y. Desmedt and K. Kurosawa. How to break a practical MIX and design a new one. *Eurocrypt 2000*, LNCS 1807, pages 557–572. 2000.
10. E. Fujisaki and T. Okamoto. Secure integration of asymmetric and symmetric encryption schemes. *CRYPTO '99*, LNCS 1666, pages 537–554. 1999.
11. H. Handschuh, Y. Tsiounis, and M. Yung. Decision oracles are equivalent to matching oracles. *PKC '99*, LNCS 1560, pages 276–289. 1999.
12. M. Jakobsson. A practical mix. *Eurocrypt '98*, LNCS 1403, pages 448–461. 1998.
13. M. Jakobsson. Flash mixing. *PODC99*, 1999.
14. A. Juels and M. Jakobsson. Millimix. Tech. Report 99-33, DIMACS, June 1999.

15. A. Menezes, P. Oorschot, and S. Vanstone. *Handbook of Applied Cryptography.* CRC Press, 1997.

16. M. Michels and P. Horster. Some remarks on a receipt-free and universally verifiable mix-type voting scheme. *Asiacrypt '96*, LNCS 1163, pages 125–132. 1996.

17. W. Ogata, K. Kurosawa, K. Sako, and K. Takatani. Fault tolerant anonymous channel. *ICICS '98*, LNCS 1334, pages 440–444. 1998.

18. C. Park, K. Itoh, and K. Kurosawa. Efficient anonymous channel and all/nothing election scheme. *Eurocrypt '93*, LNCS 765, pages 248–259. 1994.

19. B. Pfitzmann. Breaking an efficient anonymous channel. *Eurocrypt '94*, LNCS 950, pages 339–348. 1995.

20. B. Pfitzmann and A. Pfitzmann. How to break the direct RSA implementation of MIXes. *Eurocrypt '89*, LNCS 434, pages 373–381. 1989.

21. T. Saitoh. The efficient reductions between the decision Diffie-Hellman problem and related problems. In the joint meeting of SCI2000 and ISAS2000, July 2000. available from the author.

22. K. Sako and J. Kilian. Receipt-free mix-type voting scheme — a practical solution to the implementation of a voting booth —. *Eurocrypt '95*, LNCS 921, pages 393–403. 1995.

23. V. Shoup. Using hash functions as a hedge against chosen ciphertext attack. *Eurocrypt 2000*, LNCS 1807, pages 275–288. 2000.

24. V. Shoup and R. Gennaro. Securing threshold cryptosystems against chosen ciphertext attack. *Eurocrypt '98*, LNCS 1403, pages 1–16. 1998.

25. Y. Tsiounis and M. Yung. On the security of El Gamal based encryption. *PKC '98*, LNCS 1431, pages 117–134. 1998.

## Appendix (Proof of Lemma 2)

We first show that $\mathcal{A}_M$ can be used to solve CDHP, or else can be used to solve either MFGP or MDHP. Let $(\mathbf{M_0}, \mathbf{M_1}, \mathbf{M'_{b_F}}, \mathbf{M'_{\overline{b_F}}})$ be an instance of MFGP and $(\rho, \mathbf{h}, \mathbf{G_0}, \mathbf{G_1}, \mathbf{G^a_{b_D}}, \mathbf{G^a_{\overline{b_D}}})$ be an instance of MDHP. Let $y := \mathbf{h}^x$ for $x \leftarrow Z_q$. The input to $\mathcal{A}_M$ is a tuple such as : $Input^*_{\mathcal{A}_M} := \{\rho, y, \mathbf{h}, (\mathbf{M_0}, \mathbf{G_0}), (\mathbf{M_1}, \mathbf{G_1}),$ $(\mathbf{M'_{b_F}}, \mathbf{G^a_{b_D}}), (\mathbf{M'_{\overline{b_F}}}, \mathbf{G^a_{\overline{b_D}}})\}$. Note that this may not be a correct instance of the joint problem as $b_D$ and $b_F$ may not be the same. Now we observe the behavior of $\mathcal{A}_M$ given this input. Let $\mathsf{q}_F$ be the maximum number of queries from $\mathcal{A}_M$ to $H$. Here $\mathsf{q}_F$ is limited to a polynomial in the security parameter $\kappa$. Let $Q_i$ denote the $i$-th query to $H$. If there exists $i$ such that $Q_i = (G^{ax}_{b_D})$, then $G^{ax}_{b_D}$ is the answer of CDHP $[h, y, G^a_{b_D}]$ $([h, y, G^a_{b_D}]=[g^a, g^{ax}, g^{ar}], G^{ax}_{b_D} = g^{axr})$. Similarly, if there exists $i$ such that $Q_i = (G^{ax}_{\overline{b_D}})$, $G^{ax}_{\overline{b_D}}$ is the answer of CDHP $[h, y, G^a_{\overline{b_D}}]$. Define $\mathcal{P}_{DH}$ as $\mathcal{P}_{DH} = Pr[^\exists i \in \{1, \ldots, \mathsf{q}_F\}, ^\exists j \in \{0, 1\} ; Q_i = (G^{ax}_j)]$. As above, $\mathcal{A}_M$ can be used to solve CDHP.

Next we show that $\mathcal{A}_M$ can be used to solve either MFGP or MDHP. Suppose that no such queries exist. In this case, the symmetric keys used in the MFGP are independent of the MDHP part because of the randomness of $H$; the adversary $\mathcal{A}_D$ makes those keys randomly without asking random oracle $H$. We next consider the relation between $b_F$ and $b_D$.

1. The case of $b_F = b_D$.

   $Input^*_{\mathcal{A}_M}$ is perfectly indistinguishable from the correct input coming from the answer of random oracle $H$. So $\mathcal{A}_M$ will output $\tilde{\tilde{b}}$ as $\tilde{\tilde{b}} = b$ with probability $(> \frac{1}{2} + \mu_A)$, i.e., the same as the success probability of $\mathcal{A}_{Xp}$

2. The case of $b_F \neq b_D$.

   In this case, the distribution of $Input^*_{\mathcal{A}_M}$ is not correct.

   If $\mathcal{A}_M$ does not stop within $t_A$, $\tilde{\tilde{b}}$ is randomly taken from $\{0, 1\}$. Now, define $\mathcal{P}_F$ and $\mathcal{P}_D$ as follows. $\mathcal{P}_F = Pr[\tilde{\tilde{b}} = b_F | b_F \neq b_D]$, $\mathcal{P}_D = Pr[\tilde{\tilde{b}} = b_D | b_F \neq b_D]$.

   If $b_F \neq b_D$, $\tilde{\tilde{b}}$ must equal to either $b_F$ or $b_D$. Hence, $\mathcal{P}_F + \mathcal{P}_D = 1$.
   Since both $b_F = b_D$ and $b_F \neq b_D$ happen with probability $\frac{1}{2}$, we have

$$Pr[\tilde{\tilde{b}} = b_F] = (1 - \mathcal{P}_{DH})\{\frac{1}{2}(\frac{1}{2} + \mu_A) + \frac{\mathcal{P}_F}{2}\} = (1 - \mathcal{P}_{DH})\{\frac{1 + 2\mathcal{P}_F}{4} + \frac{\mu_A}{2}\} \quad (1)$$

$$Pr[\tilde{\tilde{b}} = b_D] = (1 - \mathcal{P}_{DH})\{\frac{1}{2}(\frac{1}{2} + \mu_A) + \frac{\mathcal{P}_D}{2}\} = (1 - \mathcal{P}_{DH})\{\frac{1 + 2\mathcal{P}_D}{4} + \frac{\mu_A}{2}\} \quad (2)$$

According to equation (8), either $\mathcal{P}_F$ or $\mathcal{P}_D$ is not less than $\frac{1}{2}$. Therefore, either Equation (1) or (2) is not negligible, or both.

The common key used in MFGP is perfectly indistinguishable from the actual common key derived from the answer of random oracle $H$ because it is decided randomly. So if the answer of the CDHP is not contained among the list of queries from $\mathcal{A}_M$ to $H$, we can say that $(M_0, M_1, M'_b, M'_{\tilde{b}})$ and $(G_0, G_1, G'_b, G'_{\tilde{b}})$ have no relation to each other, though they affect each other through the common key in the actual input. Hence MFGP and MDHP are independent, and neither provides any help in solving the other.

Based on the above observation, we construct Block T1, Block T2, Block T3 that solve CDHP, MFGP and MDHP, respectively.

## [Block T1]

1. Receive CDHP instance $(\rho, \mathbf{g}^\alpha, \mathbf{g}^\beta)$.
2. Make an MDHP instance as follows.
   - Choose $b_1 \leftarrow \{0, 1\}$.
   - $y := \mathbf{g}^\alpha$, $G'_{b_1} := \mathbf{g}^\beta$, $G'_{\bar{b}_1} \leftarrow \langle g \rangle$
   - $a \leftarrow Z^*_q$, $g := \mathbf{g}^{\frac{1}{a}}$, $h := \mathbf{g}^a$, $G_0 := G'_0{}^{\frac{1}{a}}$, $G_1 := G'_1{}^{\frac{1}{a}}$
   - $\rho = (\mathbf{p}, \mathbf{q}, g)$
3. Make an MFGP instance as $M_i \leftarrow \mathcal{M}$, $K_i \leftarrow \mathcal{K}$, $M'_i := \mathcal{D}_{K_i}(M_i)$ for $i = 0, 1$.
4. Choose $b \leftarrow \{0, 1\}$.
5. Choose $I$ randomly from $1 \leq i \leq q_F$.
6. Input the following to $\mathcal{A}_M$.
   $Input'_{\mathcal{A}_M} = \{\rho, h, y, (M_0, G_0), (M_1, G_1), (M'_b, G'_b), (M'_{\tilde{b}}, G'_{\tilde{b}})\}$
7. If $\mathcal{A}_M$ poses query to $H$, return a random value chosen from key space $\mathcal{K}$. If it is the $I$-th query, output and stop.

Observe that the simulation is perfect only if the correct answer of CDHP is $Q_I$, or it is not asked to $H$ (otherwise we have answered to the query with randomly chosen session key that may confuse $\mathcal{A}_M$).

**[Block T2]**

1. Receive MFGP instance $(\mathbf{M_0}, \mathbf{M_1}, \mathbf{M'_{b_F}}, \mathbf{M'_{\bar{b}_F}})$.
2. Make an MDHP instance as follows.
   - $(p, q, g) \leftarrow \mathcal{G}(1^\kappa)$
   - $a, x \leftarrow Z_q^*,\ y := g^{ax},\ h := g^a$
   - $r_0, r_1 \leftarrow Z_q^*,\ G_0 := g^{r_0}, G_1 := g^{r_1},\ G'_0 := h^{r_0}, G'_1 := h^{r_1}$
3. Choose $b_D \leftarrow \{0, 1\}$. Next input the following to $\mathcal{A}_M$.
   $$Input''_{\mathcal{A}_M} = \{\rho, h, y, (\mathbf{M_0}, G_0), (\mathbf{M_1}, G_1), (\mathbf{M'_{b_F}}, G'_{b_D}), (\mathbf{M'_{\bar{b}_F}}, G'_{\bar{b}_D})\}$$
4. For all queries from $\mathcal{A}_M$ to $H$, return a random value from $\mathcal{K}$.
5. Output $\tilde{b}$ that $\mathcal{A}_M$ outputs.

**[Block T3]**

1. Receive MDHP instance $(\rho, \mathbf{h}, \mathbf{G_0}, \mathbf{G_1}, \mathbf{G^a_{b_D}}, \mathbf{G^a_{\bar{b}_D}})$.
2. Make an MFGP instance as $M_i \leftarrow \mathcal{M},\ K_i \leftarrow \mathcal{K},\ M'_i := \mathcal{D}_{K_i}(M_i)$ for $i = \{0, 1\}$.
3. Choose $b_F \leftarrow \{0, 1\}$. Next input the following to $\mathcal{A}_M$.
   $$Input'''_{\mathcal{A}_M} = \{\rho, h, y, (M_0, \mathbf{G_0}), (M_1, \mathbf{G_1}), (M'_{b_F}, \mathbf{G'_{b_D}}), (M'_{\bar{b}_F}, \mathbf{G'_{\bar{b}_D}})\}$$
4. For all queries from $\mathcal{A}_M$ to $H$, return a random value from $\mathcal{K}$.
5. Output $\tilde{b}$ that $\mathcal{A}_M$ outputs.

By using the above blocks, we construct $\mathcal{A}_D$ as follows.

**[Construction of $\mathcal{A}_D$]**

1. Receive CDHP instance $(\rho, \mathbf{g^\alpha}, \mathbf{g^\beta})$, MFGP instance $(\mathbf{M_0}, \mathbf{M_1}, \mathbf{M'_{b_F}}, \mathbf{M'_{\bar{b}_F}})$, and MDHP instance $(\rho, \mathbf{h}, \mathbf{G_0}, \mathbf{G_1}, \mathbf{G^a_{b_D}}, \mathbf{G^a_{\bar{b}_D}})$.
2. Input each instance to the appropriate block.
3. Output the result provided by each block as the answer to the corresponding problem.

Now we discuss the success probability of $\mathcal{A}_D$.

**Case 1** ($\mathcal{P}_{DH}$ is not negligible.)
   The output from block T1 is a correct answer of CDHP if $Q_I = G'^{\log_h y}_{b'}$ for $b' = b_1$. This happens with probability $\frac{\mathcal{P}_{DH}}{2q_F}$ which is not negligible.

**Case 2** ($\mathcal{P}_{DH}$ is negligible.)
   If $\mathcal{P}_F \geq 1/2$, from Equation (1), we have

$$Pr[\tilde{b} = b_F] = (1 - \mathcal{P}_{DH})\{\frac{1 + 2\mathcal{P}_F}{4} + \frac{\mu_A}{2}\}$$

$$\geq (\frac{1}{2} + \frac{\mu_A}{2}) - (\frac{1}{2} + \frac{\mu_A}{2})\mathcal{P}_{DH} \geq \frac{1}{2} + \mu_{MFG}$$

for some $\mu_{MFG}$ which is not negligible. Otherwise, if $\mathcal{P}_{DH} \geq 1/2$, we have $Pr[\tilde{b} = b_D] \geq \frac{1}{2} + \mu_{MFG}$ for some $\mu_{MFG}$ which is not negligible. Thus, either MFGP or MDHP will be solved with an advantage that is not negligible. $\square$

# Attack for Flash MIX

Masashi Mitomo* and Kaoru Kurosawa

Tokyo Institute of Technology,
2-12-1 O-okayama, Meguro-ku, Tokyo 152-8552, Japan
mitomo@flab.fujitsu.co.jp, kurosawa@ss.titech.ac.jp

**Abstract.** A MIX net takes a list of ciphertexts $(c_1, \cdots, c_N)$ and outputs a permuted list of the plaintexts $(m_1, \cdots, m_N)$ without revealing the relationship between $(c_1, \cdots, c_N)$ and $(m_1, \cdots, m_N)$. This paper shows that the Jakobsson's flash MIX of PODC'99, which was believed to be the most efficient robust MIX net, is broken. The first MIX server can prevent computing the correct output with probability 1 in our attack. We also present a countermeasure for our attack.

## 1 Introduction

A MIX net takes a list of ciphertexts $(c_1, \cdots, c_N)$ of users $1, \cdots, N$ and outputs a permuted list of the plaintexts $(m_1, \cdots, m_N)$ without revealing the relationship between $(c_1, \cdots, c_N)$ and $(m_1, \cdots, m_N)$. MIX nets have found many applications in anonymous communication [4], election schemes [4,7,13,15] and payment systems [9].

The original MIX net was proposed by Chaum [4]. B.Pfitzmann and A.Pfitzmann, however, showed an attack by a sender, which is more complicated than a simple repeated ciphertext attack [14].

Another problem of Chaum's MIX net, based on RSA, is that the size of each ciphertext $c_i$ is very long proportionally to the number of MIX servers $v$. Park et al. overcame this problem by using ElGamal encryption scheme so that the size of each $c_i$ became independent of $v$ [13]. Almost all MIX nets proposed after this paper are based on ElGamal encryption scheme.

A general method to achieve verifiability is to have each MIX server to prove that he behaved correctly in zero knowledge. Sako and Kilian [15] showed such an efficient proof system for Park et al.'s MIX net. This scheme is the first *universally verifiable* MIX net.

On the other hand, Ogata et al. showed the first *robust* MIX net which is also universally verifiable [12]. In this scheme, the computational cost of each MIX server is $O(\kappa t N)$ and the external verifier's cost is also $O(\kappa t N)$, where $\kappa$ is the security parameter and $t$ denotes the number of malicious MIX servers.

At Eurocrypt'98, Abe showed a robust MIX net in which the external verifier's cost is reduced to $O(\kappa N)$ [1]. At the same time, Jakobsson showed a more efficient robust MIX net, called practical MIX [8] (but not universally verifiable).

---

* He is currently working for Fujitsu Laboratories Ltd.

T. Okamoto (Ed.): ASIACRYPT 2000, LNCS 1976, pp. 192–204, 2000.
© Springer-Verlag Berlin Heidelberg 2000

Instead of cut and choose methods, he introduced a method of so called repetition robustness. However, this scheme was recently broken by Desmedt and Kurosawa (DK attack) [5].

At PODC'99, Jakobsson proposed his second robust MIX net, called flash MIX [10]. This scheme is the most efficient robust MIX net known so far which satisfies $v = O(t)$, where $v$ is the number of MIX servers. (The MIX net recently proposed by [5] requires $v = O(t^2)$.) In flash MIX, the computational cost of each MIX server is only $O(tN)$. The DK attack [5] for practical MIX [8] does not work for flash MIX directly because two dummy elements are inserted into the input list at the beginning of the protocol in flash MIX. Actually, Jakobsson *proved* the security of flash MIX in [10, Theorem 1 and Theorem 2].

In this paper, however, we show that flash MIX is broken. In our attack, the first MIX server can prevent computing the correct output with probability 1. This means that his security proof is wrong. Our attack is a variant of the DK attack for practical MIX [5]. We also present a countermeasure for our attack. It will be a further work to study about the security of our countermeasure.

Flash MIX consists of the first re-encryption phase, the second re-encryption phase and the unblinding protocol in which each MIX server proves that he behaved correctly in the first and the second re-encryption phases. Now our malicious first MIX server executes the first re-encryption phase honestly, but cheats in the second re-encryption phase. He computes his invalid output lists from not only his input lists of the second re-encryption phase but also the input to the flash MIX itself so that no cheating is detected in the unblinding protocol.

*Other related works.* Abe showed MIX nets which are efficient for small $N$ [2,3]. In Abe's MIX nets, the cost of each MIX server is $O(tN \log N)$. Jakobsson and Juels showed a MIX net which has the same advantage in [11]. In their MIX net, the cost of each MIX server is $O(tN \log^2 N)$. Since these complexities grow faster in $N$ than the other schemes, these schemes suit small $N$.

On the other hand, Desmedt and Kurosawa showed an MIX net in which the cost of each MIX server is only $O(N)$ while $v = O(t^2)$ [5].

## 2    Model of MIX Net

### 2.1    Model and Definitions

In the model of MIX nets, there exist three types of participants: *users*, a *bulletin board*, and the *MIX servers*.

1. The *users* post encrypted messages $(c_1, \cdots, c_N)$ to the *bulletin board*.
2. After the bulletin board fills up, or after some other triggering event occurs, the *mix servers* compute a randomly permuted list of decryptions $(m_1, \cdots, m_N)$ of all valid encryptions posted on the bulletin board.

MIX nets must satisfy privacy, verifiability and robustness. Suppose that at most $t$ among $v$ MIX servers and at most $N-2$ among $N$ senders are malicious. Then we say that a MIX net satisfies :

- *t-privacy* if the relationship between $(c_1, \cdots, c_N)$ and $(m_1, \cdots, m_N)$ is kept secret.
- *t-verifiability* if an incorrect output of the MIX net is detected with overwhelming probability.
- *t-robustness* if it can output $(m_1, \cdots, m_N)$ correctly with overwhelming probability.

We say that a MIX net is *t-resilient* if it satisfies *t*-privacy, *t*-verifiability and *t*-robustness.

## 2.2   ElGamal Based Encryption Scheme for Users

Let $p$ be a safe prime, i.e., $p, q$ be primes such that $p = 2q + 1$, and $g$ be a generator of $G_q$. Let $y = g^x \bmod p$, where $x$ is a secret key. The public key of ElGamal encryption scheme is $(p, q, g, y)$.

To encrypt a value $m \in G_q$, a random number $r \in_u Z_q$ is chosen and the ciphertext $(a, b) = (g^r, my^r)$ is calculated. For decryption, $m = b/a^x$ is calculated. (To guarantee that $m \in G_q$, we should let $m = (M \mid p)M$ for an original message $M \in [1 \ldots (p-1)/2]$, where $(M \mid p)$ is the Jacobi symbol of $M$.)

The MIX servers share a secret key $x$ using a $(t+1, v)$ threshold scheme [16], where $v$ denotes the number of MIX servers.

## 3   Flash MIX

Jakobsson proposed his second *t*-resilient MIX net, called flash MIX, at PODC'99 [10]. (His first *t*-resilient MIX net [8] was broken [5].) This scheme is the most efficient robust MIX net known so far which satisfies $v = O(t)$. (The MIX net recently proposed by [5] requires $v = O(t^2)$.) In flash MIX, the computational cost of each MIX server is only $O(tN)$.

For $(a, b)$, let

$$(c, d) = (ag^\beta, by^\beta).$$

We say that $(c, d)$ is a re-encryption of $(a, b)$ and $\beta$ is the re-encryption exponent. For $(a_1, b_1)$ and $(a_2, b_2)$, we say that $(a_1 a_2, b_1 b_2)$ is the product of $(a_1, b_1)$ and $(a_2, b_2)$.

## 3.1   Functionality

The input to flash MIX is a list of ciphertexts

$$((a_1, b_1), \cdots, (a_N, b_N)),$$

where $(a_i, b_i)$ is an ElGamal encryption of a message $m_i$ with respect to the public key $(p, q, g, y)$. The output is a random permutation of

$$((a'_1, b'_1), \cdots, (a'_N, b'_N)),$$

where
$$(\acute{a}_i, \acute{b}_i) = (a_i g^{r_i}, b_i y^{r_i})$$

is a random re-encryption of $(a_i, b_i)$. $(\acute{a}_i, \acute{b}_i)$ can later be decrypted by a $(t+1, v)$-threshold decryption scheme.

Flash MIX starts with $t + 1$ MIX servers, say, MIX servers $1, \cdots t + 1$. If cheating is detected during any step of the protocol, then a cheater detection phase commences. In the cheater detection phase, a cheater is detected and replaced. Afterwards, the protocol is restarted.

It consists of two subprotocols, the blinding protocol and the unblinding protocol.

## 3.2   Blinding Protocol

Flash MIX first executes the blinding protocol as follows.

(1) Generation and insertion of dummies.
   Two dummies $(a_{N+1}, b_{N+1})$ and $(a_{N+2}, b_{N+2})$ are constructed collectively by all MIX servers such that $a_{N+1}, b_{N+1}, a_{N+2}$ and $b_{N+2}$ are random elements of $G_q$. Let
$$L_0 = ((a_1, b_1), \cdots, (a_{N+2}, b_{N+2})). \tag{1}$$

(2) Duplication.
   $\tau \geq 2$ copies of $L_0$ are created, where
$$\tau = 1 - \frac{\log_2 \epsilon}{2\log_2 N} \tag{2}$$

   for $\epsilon$ which denotes the maximum failure probability. They are denoted by $L_{1,0}, L_{2,0}, \cdots, L_{\tau,0}$.

(3) First re-encryption.
   For $j = 1, 2, \cdots, t + 1$, MIX server $j$ takes as input the lists
$$L_{1,(j-1)}, L_{2,(j-1)}, \cdots, L_{\tau,(j-1)}.$$

   He re-encrypts each element of each lists given to him, and forwards random permutations of the resulting lists to the next server. His output lists are denoted by
$$L_{1,j}, \cdots, L_{\tau,j}.$$

   The final result of this step is denoted by
$$L'_{1,0} \triangleq L_{1,t+1}, \cdots, L'_{\tau,0} \triangleq L_{\tau,t+1}.$$

   Since at least one of the $t+1$ MIX servers is assumed to be honest, they are randomly re-encrypted and permuted lists of $L_0$.

(4) Second re-encryption.
   The $t + 1$ MIX servers execute similar re-encryption on input $L'_{1,0}, \cdots, L'_{\tau,0}$. The input lists of MIX server 1 are $L'_{1,0}, \cdots, L'_{\tau,0}$ and the output lists are denoted by $L'_{1,1}, \cdots, L'_{\tau,1}$. The input lists of MIX server $j \geq 2$ are denoted by $L'_{1,(j-1)}, \cdots, L'_{\tau,(j-1)}$ and the output lists are denoted by $L'_{1,j}, \cdots, L'_{\tau,j}$.

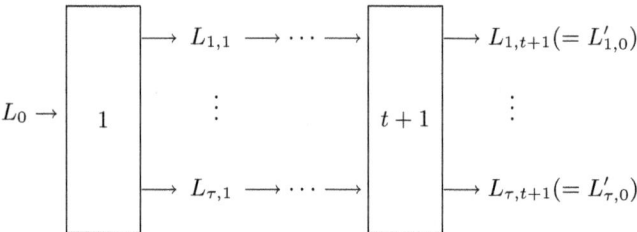

**Fig. 1.** First re-encryption

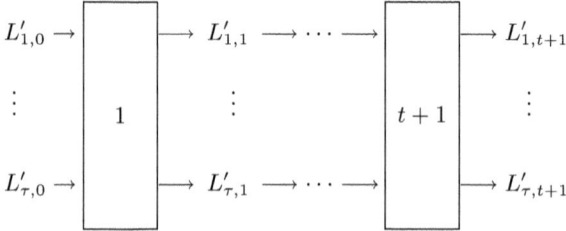

**Fig. 2.** Second re-encryption

### 3.3 Unblinding Protocol

After the blinding protocol, the unblinding protocol is executed in which each MIX server proves that he behaved honestly in the blinding protocol.

(1) Verifying the first re-encryption.

  Each MIX server reveals the re-encryption exponents and the random permutation which he used in the first re-encryption. They are checked by the other MIX servers.

(2) Aggregation.

  After the above step, everyone can compute the aggregate permutations $\Pi_1, \cdots, \Pi_\tau$ and the aggregate re-encryption exponents $\beta_{i,j}$ of the first re-encryption such that

$$L'_{1,0} = \Pi_1\left((a_1 g^{\beta_{1,1}}, b_1 y^{\beta_{1,1}}), \cdots, (a_{N+2} g^{\beta_{1,N+2}}, b_{N+2} y^{\beta_{1,N+2}})\right),$$

$$\vdots \tag{3}$$

$$L'_{\tau,0} = \Pi_\tau\left((a_1 g^{\beta_{\tau,1}}, b_1 y^{\beta_{\tau,1}}), \cdots, (a_{N+2} g^{\beta_{\tau,N+2}}, b_{N+2} y^{\beta_{\tau,N+2}})\right)$$

(3) Verification of dummy values.

  In this phase, each MIX server proves that he behaved honestly about the two dummies in the second re-encryption.

 (3.1) MIX server 1 publishes how he permuted the two dummies in $L'_{1,1}, \cdots, L'_{\tau,1}$. (Note that after the verifying the first re-encryption phase, he knows the positions of the two dummies in $L'_{1,0}, \cdots, L'_{\tau,0}$.)

Next, he reveals the re-encryption exponent he used for the second dummy. He also proves that he knows the re-encryption exponent he used for the first dummy in zero-knowledge.

They are checked by the other MIX servers.

(3.2) MIX server 2 behaves similarly to MIX server 1. (Note that from Step 3.1, he knows the positions of the two dummies in $L'_{1,1}, \cdots, L'_{\tau,1}$.)

(3.3) MIX server $3, \cdots, t+1$ behave similarly.

(4) Verification of products.

In this phase, each MIX server proves that he behaved honestly about the product of all elements except the second dummy of each list in the second re-encryption.

(4.1) MIX server 1 behaves as follows. For $i = 1, 2, \cdots, \tau$, let

$$(A_i, B_i) \triangleq \text{the product of all elements of } L'_{i,0}$$
$$\text{except the second dummy.}$$
$$(C_i, D_i) \triangleq \text{the product of all elements of } L'_{i,1}$$
$$\text{except the second dummy.}$$

Then it holds that

$$C_i = A_i g^{\mu_i} \text{ and } D_i = B_i y^{\mu_i} \tag{4}$$

for some $\mu_i$. MIX server 1 publishes such $\mu_i$, for $1 \le i \le \tau$. The other MIX servers verify that eq.(4) holds for $1 \le i \le \tau$.

(4.2) MIX servers $2, 3, \cdots, t+1$ behave similarly.

(5) Verification of relative sorting.

Each MIX server $j$ proves that $L'_{1,j}$ is a permuted and re-encrypted version of $L'_{i,j}$ for $2 \le i \le \tau$ in the second re-encryption.

Let $f$ be a keyed function that can be modelled by a random oracle. For simplicity, we assume that the range and the domain of $f$ are equal but for a negligible fraction of values.

(5.1) MIX server 1 behaves as follows. Let

$$L'_{1,1} = ((a'_1, b'_1), \cdots, (a'_{N+2}, b'_{N+2})),$$
$$L'_{i,1} = ((c'_1, d'_1), \cdots, (c'_{N+2}, d'_{N+2})),$$

Then $L'_{1,1}$ is a permuted and re-encrypted version of $L'_{i,1}$ for $2 \le i \le \tau$. That is,

$$L'_{1,1} = \Phi_i \left( (c'_1 g^{\gamma_{i,1}}, d'_1 y^{\gamma_{i,1}}), \cdots, (c'_{N+2} g^{\gamma_{i,N+2}}, d'_{N+2} y^{\gamma_{i,N+2}}) \right) \tag{5}$$

for some $\Phi_i$ and $\{\gamma_{i,j}\}$. Note that MIX server 1 can compute such $\Phi_i$ and $\{\gamma_{i,j}\}$ from $\Pi_i$, $\{\beta_{i,j}\}$ of eq.(3) and the random numbers he used in the second re-encryption.

Now MIX server 1 proves that eq.(5) holds by revealing the so called tag lists $T_{1,1}, \cdots, T_{\tau,1}$ and the so called offset lists $E_{2,1}, \cdots, E_{\tau,1}$ such that

$$
T_{1,1} = (R_1, \cdots, R_{N+2}),
$$
$$
T_{i,1} = \Phi_i^{-1}(R_1, \cdots, R_{N+2}) \text{ for } 2 \le i \le \tau,
$$
$$
E_{i,1} = \Phi_i^{-1}(\gamma_{i,1}, \cdots, \gamma_{i,N+2}) \text{ for } 2 \le i \le \tau,
$$

where $R_1, \cdots, R_{N+2}$ are unique elements in the domain of $f$. (Revealing the tag lists and the offset lists is almost equivalent to revealing $\Phi_i$ and $\{\gamma_{i,j}\}$).

The other MIX servers verify that eq.(5) holds by using the above tag lists and offset lists.

(5.2) Each MIX server $i(\ge 2)$ applies the function $f$, keyed with a secret and random key, to all the elements of his input tag lists. His tag lists are obtained by applying the permutation he used in the second re-encryption to the above updated lists. He also generates his offset lists by using his input tag lists and input offset lists. He then reveals his tag lists and offset lists. The other MIX servers verify them.

(6) Output of flash MIX.

If no cheater was found, then dummies are removed from the re-encrypted and permuted first list copy, and the resulting list is output.

## 3.4   Security

The DK attack [5] for practical MIX [8] does not work for flash MIX directly because two dummy elements are inserted into the input list at the beginning of the protocol in flash MIX. Actually, Jakobsson argued the security of flash MIX as follows [10, Proof of Theorem 1].

In order to successfully alter some elements in the final output, the adversary has to alter at least two elements of the re-encrypted and permuted first list copy, none of which are the second dummy. (See Step (4) and (6) of the previous subsection.) In order for this not to be noticed in the step where lists are relatively sorted and compared, the adversary has to select the same two elements from the remaining $\tau - 1$ list copies.

He claimed that this probability was smaller than $\epsilon$. (See eq.(2) for $\epsilon$.) In the next section, however, we show that this claim is not true.

## 4   Attack for Flash MIX

In this section, we show that MIX server 1 can prevent computing the correct output.

## 4.1    Attack for the Blinding Protocol

In the blinding protocol of flash MIX, our malicious MIX server 1 executes the first re-encryption honestly, but cheats in the second re-encryption. He computes his invalid output lists $L'_{1,1}, \cdots, L'_{\tau,1}$ from not only his input lists $L'_{1,0}, \cdots, L'_{\tau,0}$ of the second re-encryption phase but also the input to the flash MIX itself ($L_0$ of eq.(1)). See Fig.3 and Fig.4.

Now our malicious MIX server 1 executes the second re-encryption phase as follows.

(1) MIX server 1 first chooses random numbers $\alpha_1, \cdots, \alpha_N$ such that

$$\alpha_1 + \cdots + \alpha_N = 1 \bmod q. \tag{6}$$

(2) For $1 \leq i \leq \tau$, $L'_{i,0}$ is written as follows.

$$L'_{i,0} = \Pi_i \left( (a'_{i,1}, b'_{i,1}), \ldots, (a'_{i,N+2}, b'_{i,N+2}) \right)$$

where $\Pi_i$ is the aggregate permutation,

$$a'_{i,k} \triangleq a_k g^{\beta_{i,k}} \quad, \quad b'_{i,k} \triangleq b_k y^{\beta_{i,k}}. \tag{7}$$

and $\beta_{i,k}$ is the aggregate re-encryption exponent. (See eq.(3).)

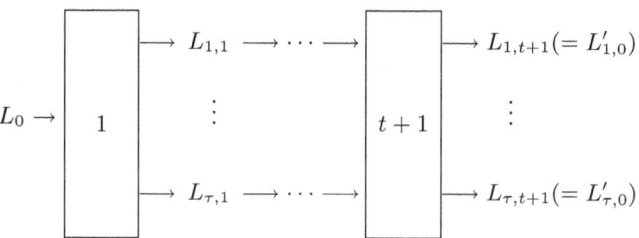

**Fig. 3.** First re-encryption of our attack

MIX server 1 does not know $\Pi_i$. However, note that he can compute the products $a'_{i,1} \cdots a'_{i,N+2}$ and $b'_{i,1} \cdots b'_{i,N+2}$. Now MIX server 1 computes

$$\widetilde{A_i} \triangleq a'_{i,1} \cdots a'_{i,N+2} / a_{N+1} a_{N+2} \quad \text{and} \quad \widetilde{B_i} \triangleq b'_{i,1} \cdots b'_{i,N+2} / b_{N+1} b_{N+2}, \tag{8}$$

where $(a_{N+1}, b_{N+1})$ and $(a_{N+2}, b_{N+2})$ are the two dummy elements which are inserted into the input list at the beginning of the protocol (see eq.(1)).
(3) Next for $i = 1, \ldots, \tau$, MIX server 1 publishes

$$L'_{i,1} \triangleq \theta_i \left( (\widetilde{A_i}^{\alpha_1} g^{t_{i,1}}, \widetilde{B_i}^{\alpha_1} y^{t_{i,1}}), \cdots, (\widetilde{A_i}^{\alpha_N} g^{t_{i,N}}, \widetilde{B_i}^{\alpha_N} y^{t_{i,N}}), \right.$$

$$\left. (a_{N+1} g^{t_{i,N+1}}, b_{N+1} y^{t_{i,N+1}}), (a_{N+2} g^{t_{i,N+2}}, b_{N+2} y^{t_{i,N+2}}) \right) \tag{9}$$

where $\theta_i$ and $t_{i,1}, \cdots, t_{i,N+2}$ are randomly chosen by MIX sever 1.

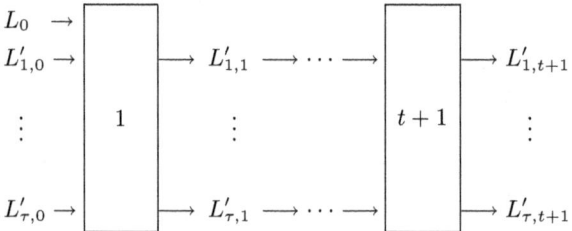

**Fig. 4.** Second re-encryption of our attack

Let

$$\widetilde{A} \overset{\triangle}{=} a_1 \cdots a_N, \quad \widetilde{B} \overset{\triangle}{=} b_1 \cdots b_N. \tag{10}$$

Then note that for $i = 1, \ldots, \tau$, $L'_{i,1}$ is a randomly re-encrypted and permuted list of $(\widetilde{a}_1, \widetilde{b}_1), \cdots, (\widetilde{a}_N, \widetilde{b}_N), (a_{N+1}, b_{N+1}), (a_{N+2}, b_{N+2})$, where

$$(\widetilde{a}_1, \widetilde{b}_1) = (\widetilde{A}^{\alpha_1}, \widetilde{B}^{\alpha_1})$$

$$\vdots$$

$$(\widetilde{a}_N, \widetilde{b}_N) = (\widetilde{A}^{\alpha_N}, \widetilde{B}^{\alpha_N}).$$

## 4.2   Attack for the Unblinding Protocol

We next show that MIX server 1 can behave properly in each phase of the unblinding protocol so that no cheating is detected.

**Theorem 1.** *MIX server 1 can behave properly so that the verifying the first re-encryption phase ends successfully.*

*Proof.* Our MIX server 1 executed the first re-encryption phase honestly. Therefore, he can execute the *verifying* the first re-encryption phase correctly.   □

**Theorem 2.** *MIX server 1 can behave properly so that the verification of dummy values phase ends successfully.*

*Proof.* Everyone knows $\Pi_i$ of eq.(3). MIX server 1 knows $\theta_i$ of eq.(9). Therefore, MIX server 1 knows how the two dummies are permuted from $L'_{i,0}$ to $L'_{i,1}$. Hence, MIX server 1 can publish a description of how the two dummies are permuted from $L'_{i,0}$ to $L'_{i,1}$ for $1 \le i \le \tau$.
  Next let

$$z_{i,N+1} = t_{i,N+1} - \beta_{i,N+1}$$
$$z_{i,N+2} = t_{i,N+2} - \beta_{i,N+2}$$

where $\beta_{i,j}$ are defined in eq.(3) and $t_{i,j}$ are defined in eq.(9). Then $z_{i,N+1}$ and $z_{i,N+2}$ are the re-encryption exponents of the two dummies from $L'_{i,0}$ to $L'_{i,1}$.

MIX server 1 can compute $z_{i,N+1}$ and $z_{i,N+2}$ because he know $\{t_{i,j}\}$ and $\{\beta_{i,j}\}$. Therefore, he can reveal $z_{i,N+2}$. He also proves that he knows $z_{i,N+1}$ in zero-knowledge.

Then the verification of dummy values phase ends successfully.     □

**Theorem 3.** *MIX server 1 can behave properly so that the verification of products phase ends successfully.*

*Proof.* From eq.(3), it holds that

$$A_i = a_1 \cdots a_{N+1} \cdot g^{\beta_{i,1}+\cdots+\beta_{i,N+1}}$$
$$B_i = b_1 \cdots b_{N+1} \cdot y^{\beta_{i,1}+\cdots+\beta_{i,N+1}}$$

On the other hand, from eq.(9), it holds that

$$C_i = (\widetilde{A_i}^{\alpha_1} g^{t_{i,1}}) \cdots (\widetilde{A_i}^{\alpha_N} g^{t_{i,N}}) \cdot (a_{N+1} g^{t_{i,N+1}})$$
$$= \widetilde{A_i}^{\alpha_1+\cdots+\alpha_N} \cdot a_{N+1} \cdot g^{t_{i,1}+\cdots+t_{i,N+1}}$$
$$= \widetilde{A_i} \cdot a_{N+1} \cdot g^{t_{i,1}+\cdots+t_{i,N+1}} \tag{11}$$

from eq.(6). Substitute eq.(7) into eq.(8). Then we have

$$\widetilde{A_i} = a'_{i,1} \cdots a'_{i,N+2}/a_{N+1}a_{N+2}$$
$$= a_1 g^{\beta_{i,1}} \cdots a_{N+2}g^{\beta_{i,N+2}}/a_{N+1}a_{N+2}$$
$$= a_1 \cdots a_N \cdot g^{\beta_{i,1}+\cdots+\beta_{i,N+2}} \tag{12}$$

Further, substitute eq.(12) into eq.(11). Then we have

$$C_i = a_1 \cdots a_N \cdot g^{\beta_{i,1}+\cdots+\beta_{i,N+2}} \cdot a_{N+1} \cdot g^{t_{i,1}+\cdots+t_{i,N+1}}$$
$$= a_1 \cdots a_{N+1} \cdot g^{(\beta_{i,1}+\cdots+\beta_{i,N+2})+(t_{i,1}+\cdots+t_{i,N+1})}$$

Similarly, we have

$$D_i = b_1 \cdots b_{N+1} \cdot y^{(\beta_{i,1}+\cdots+\beta_{i,N+2})+(t_{i,1}+\cdots+t_{i,N+1})}$$

Now let

$$\mu_i = t_{i,1} + \cdots + t_{i,N+1} + \beta_{i,N+2}. \tag{13}$$

Then it is clear that eq.(4) is satisfied.

MIX server 1 can compute the above $\mu_i$ because everyone knows $\beta_{i,N+2}$ and $\{t_{i,j}\}$ is chosen by MIX server 1. Note that $\beta_{i,N+2}$ is computable for everyone by aggregating the re-encryption exponents of the second dummy, which are published in the verification of dummy values. He reveals this $\mu_i$. Then the verification of products phase ends successfully.     □

**Theorem 4.** *MIX server 1 can behave properly so that the verification of relative sorting phase ends successfully.*

*Proof.* Substitute eq.(12) into eq.(9). Then we have

$$L'_{i,1} \triangleq \theta_i \left( \left( \widetilde{A}^{\alpha_1} g^{t_{i,1} + \alpha_1 \cdot (\beta_{i,1} + \cdots + \beta_{i,N+2})}, \widetilde{B}^{\alpha_1} y^{t_{i,1} + \alpha_1 \cdot (\beta_{1,1} + \cdots + \beta_{1,N+2})} \right), \right.$$

$$\vdots$$

$$\left( \widetilde{A}^{\alpha_N} g^{t_{i,1} + \alpha_N \cdot (\beta_{1,1} + \cdots + \beta_{1,N+2})}, \widetilde{B}^{\alpha_N} y^{t_{i,1} + \alpha_N \cdot (\beta_{i,1} + \cdots + \beta_{i,N+2})} \right),$$

$$\left( a_{N+1} g^{t_{i,N+1}}, b_{N+1} y^{t_{i,N+1}} \right),$$

$$\left. \left( a_{N+2} g^{t_{i,N+2}}, b_{N+2} y^{t_{i,N+2}} \right) \right)$$

where $\widetilde{A}$ and $\widetilde{B}$ is defined in eq.(10). Let

$$\gamma_{i,1} \triangleq t_{1,1} - t_{i,1} + \alpha_1 \cdot (\beta_{1,1} + \cdots + \beta_{1,N+2}) - \alpha_1 \cdot (\beta_{i,1} + \cdots + \beta_{i,N+2}),$$

$$\vdots$$

$$\gamma_{i,N} \triangleq t_{1,N} - t_{i,N} + \alpha_N \cdot (\beta_{1,1} + \cdots + \beta_{1,N+2}) - \alpha_N \cdot (\beta_{i,1} + \cdots + \beta_{i,N+2}),$$

where $t_{i,j}$ is defined in eq.(9). Note that MIX server 1 can compute $\gamma_{i,1}, \cdots, \gamma_{i,N}$ for $2 \leq i \leq \tau$.

Now MIX server 1 reveals the tag lists $T_{1,1}, \cdots, T_{\tau,1}$ and the offset lists $E_{2,1}, \cdots, E_{\tau,1}$ such that

$$T_{i,1} = \theta_i(R_1, \cdots, R_{N+2}) \text{ for } 1 \leq i \leq \tau,$$
$$E_{i,1} = \theta_i(\gamma_{i,1}, \cdots, \gamma_{i,N+2}) \text{ for } 2 \leq i \leq \tau,$$

where $\theta_i$ is defined in eq.(9) and $R_1, \cdots, R_{N+2}$ are unique elements in the domain of $f$. It is easy to see that eq.(5) is satisfied with $\Phi_i = \theta_1 \theta_i^{-1}$.

Therefore, the verification of relative sorting phase ends successfully.     □

Theorem 1 $\sim$ 4 show that each phase of the unblinding protocol ends successfully and no cheating is detected.

## 4.3     Output of Flash MIX

Let the input to flash MIX be a list of ciphertexts

$$((a_1, b_1), \cdots, (a_N, b_N)),$$

where $(a_i, b_i)$ is an ElGamal encryption of a message $m_i$ with respect to the public key $(p, q, g, y)$.

Then after threshold decryption, flash MIX must output a random permutation of

$$(m_1, \cdots, m_N).$$

However, in our attack, flash MIX outputs

$$((m_1 \cdots m_N)^{\alpha_1}, \cdots, (m_1 \cdots m_N)^{\alpha_N})$$

which is clearly different from $(m_1, \cdots, m_N)$. Therefore, flash MIX does not compute the correct output without being detected.

## 5  Countermeasure

In this section, we show a countermeasure for our attack. The blinding protocol is unchanged. The new unblinding protocol is as follows.

(1) Open dummies of the first re-encryption.
    Each MIX server publishes how he permuted the two dummies in the first re-encryption. He next proves that he knows the re-encryption exponents of the two dummies in zero-knowledge.
(2) Verification of dummy values in the second re-encryption.
    Unchanged.
(3) Verification of products in the second re-encryption.
    Unchanged.
(4) Verifying the first re-encryption.
    Unchanged.
(5) Aggregation.
(6) Verification of relative sorting in the second re-encryption.
    Unchanged.

Note that

1. (1) is newly introduced. In (1), the re-encryption exponent of the second dummy is not revealed.
2. (4) was put at the beginning of the unblinding protocol in the original scheme.

Then our attack does not work. Theorem 1, 2 and 4 hold. However, Theorem 3 does not hold.

It will be a further work to study about the security of our countermeasure.

## References

1. M. Abe, "Universally Verifiable Mix-net with Verification Work Independent of the Number of Mix-centers," Eurocrypt '98, pp. 437–447.
2. M. Abe, "A Mix-Network on permutation networks," ISEC Technical report 99-10 (in Japanese) (May, 1999)
3. M. Abe, "Mix-Networks on permutation networks," Asiacrypt '99, pp. 258–273.
4. D. Chaum, "Untraceable electronic mail, return addresses, and digital pseudonyms," Communications of the ACM, ACM 1981, pp. 84-88 "Undeniable Signatures,"

5. Y.Desmedt and K.Kurosawa, "How to break a practical MIX and design a new one", Eurocrypt'2000.
6. T. ElGamal, "A Public-Key Cryptosystem and a Signature Scheme Based on Discrete Logarithms," Crypto '84, pp. 10-18
7. A. Fujioka, T. Okamoto and K. Ohta, "A practical secret voting scheme for large scale elections," Auscrypt '92, pp. 244-251
8. M. Jakobsson, "A practical MIX," Eurocrypt '98, pp. 448–461.
9. M. Jakobsson and D. M'Raihi, "Mix-based Electronic Payments," SAC'98, pp. 157–173.
10. M. Jakobsson, "Flash Mixing," PODC'99, pp. 83–89.
11. M. Jakobsson and A. Juels "Millimix: Mixing in small batches," DIMACS Technical report 99-33 (June 1999)
12. W. Ogata, K. Kurosawa, K. Sako, K. Takatani, "Fault Tolerant Anonymous Channel," ICICS '97, pp. 440-444
13. C. Park, K. Itoh, K. Kurosawa, "All/nothing election scheme and anonymous channel," Eurocrypt '93, pp. 248-259
14. B. Pfitzmann and A. Pfitzmann. "How to break the direct RSA-implementation of MIXes," Eurocrypt '89, pp. 373-381
15. K. Sako, J. Kilian, "Receipt-Free Mix-Type Voting Scheme," Eurocrypt '95, pp. 393-403
16. A. Shamir, "How to Share a Secret," Communications of the ACM, Vol. 22, 1979, pp. 612-613

# Distributed Oblivious Transfer

Moni Naor[1*] and Benny Pinkas[2**]

[1] Dept. of Computer Science and Applied Math, Weizmann Inst. of Science,
Rehovot 76100, Israel.
naor@wisdom.weizmann.ac.il
[2] STAR Lab, Intertrust Tech., 4750 Patrick Henry Drive, Santa Clara, CA 95054.
bpinkas@intertrust.com, benny@pinkas.net

**Abstract.** This work describes distributed protocols for oblivious transfer, in which the role of the sender is divided between several servers, and a chooser (receiver) must contact a threshold of these servers in order to run the oblivious transfer protocol. These distributed oblivious transfer protocols provide information theoretic security, and do not require the parties to compute exponentiations or any other kind of public key operations. Consequently, the protocols are very efficient computationally.

## 1   Introduction

Oblivious Transfer (abbrev. OT) refers to several types of two-party protocols where at the beginning of the protocol one party, the *sender*, has an input, and at the end of the protocol the other party, the *chooser* (sometimes called the receiver), learns some information about this input in a way that does not allow the sender to figure out what the chooser has learned. In this paper we are concerned with 1-out-of-2 OT protocols where the sender's input consists of two strings $(m_0, m_1)$ and the chooser can choose to get either one of these inputs and learn nothing about the other string.

Distributed oblivious transfer protocols distribute the task of the sender between several servers. Security is ensured as long as a limited number of these servers collude. The constructions we describe have three major advantages compared to single server based oblivious transfer: (1) They are more efficient since they only involve the evaluation of polynomials over relatively small fields (and no exponentiations). (2) They provide information theoretic security, thus making the task of composing such a protocol with other protocols easier. (3) They also provide better security guarantee when applied to the multi party protocols based on the auction architecture of of [21] (see below).

* Part of this work was done while visiting Stanford University and IBM Almaden Research Center. Partly supported by DOD Muri grant administered by ONR and DARPA contract F30602-99-1-0530.
** Most of this work was done while the author was at the Weizmann Institute of Science and the Hebrew University of Jerusalem, and was supported by an Eshkol grant of the Israeli Ministry of Science.

T. Okamoto (Ed.): ASIACRYPT 2000, LNCS 1976, pp. 205–219, 2000.

The setting of distributed oblivious transfer involves, as in the basic 1-out-of-2 protocol, a sender with two inputs $m_0, m_1$, and a chooser with an input $\sigma \in \{0, 1\}$. There are also $n$ servers $S_1, \ldots, S_n$. The sender generates for every server $S_i$ a transfer function $F_i$, which is sent to the server. Apart from this message there is no interaction between the servers and the sender, or between the servers themselves. Server $S_i$ then uses the function $F_i$ to answer a query of the chooser. The sender never interacts with the chooser and can be offline when the chooser sends his queries.

**Related Work.** The notion of 1-out-2 oblivious transfer was suggested by Even, Goldreich and Lempel [13], as a generalization of Rabin's "oblivious transfer" [23]. Further generalization to 1-out-of-$N$ oblivious transfer was introduced by Brassard, Crépeau and Robert [7] under the name ANDOS (all or nothing disclosure of secrets). For an up-to-date definition of OT and oblivious function evaluation see Goldreich [16].

Reductions between various types of oblivious transfer protocols have been investigated extensively and they all turn out to be information theoretically equivalent (See [6,8,12,11,9]). These reductions emphasize the importance of distributed oblivious transfer, since they enable other types of OT protocols to be based on the efficient constructions of distributed OT presented in this paper. In particular, a protocol for distributed 1-out-of-$N$ OT can be constructed using the (non-information theoretic) reduction of Naor and Pinkas [20] to $OT_1^2$. The protocol uses $\log N$ invocations of distributed $OT_1^2$, and $N$ invocations of a pseudo-random function. The resulting $OT_1^N$ protocol is very efficient and does not require any public key operations.

Oblivious transfer protocols are the foundation of secure distributed computation. Since its proposal by Rabin [23] OT has enjoyed a large number of applications and in particular Kilian [19] and Goldreich and Vainish [17] have shown how to use OT in order to implement general oblivious function evaluation, i.e., to enable parties to evaluate any function of their inputs without revealing more information than necessary. Oblivious transfer can be implemented under a variety of assumptions (see e.g. [6,13,5]). Essentially every known suggestion of public-key cryptography allows also to implement OT (although there is no general theorem that implies this state of affairs), and the complexity of 1-out-of-2 OT is typical of public-key operations [6,5]. OT can be based on the existence of trapdoor permutations, factoring, the Diffie-Hellman assumption and the hardness of finding short vectors in a lattice (the Ajtai-Dwork cryptosystem). On the other hand, given an OT protocol it is a simple matter to implement secret-key exchange using it. Therefore from the work of Impagliazzo and Rudich [18] it follows that there is no black-box reduction of OT from one-way functions. This result is quite discouraging if one attempts to improve the efficiency of OT protocols, since one-way functions are typically more efficient than public key operations by a few orders of magnitude.

There are many works which solve problems which are related (at least syntactically) to ours. The work of Beaver et. al. [4] on locally random reductions enables to distribute a function between many servers, such that a user can compute the function by contacting these servers. The construction guarantees that the servers cannot learn which values the users compute, but on the other hand it does not provide security against a user who attempts to compute the function in many locations. This is also the case with PIR (private information retrieval) protocols [10]. SPIR protocols [15] address the security of the sender as well, but the emphasis of both these types of protocols is different than ours: they consider communication overhead as the major resource that must be minimized (at the cost of increasing the computation overhead). In the PIR context Gertner et. al. [14] proposed a system where the database owner solicits the help of several servers which are not fully trusted. A related line of work is that of "commodity based cryptography" [3], where OT is treated as a resource, but our work puts a much more stronger emphasis on simplicity and efficiency.

Very recently Rivest has considered a model with a "trusted initializers" who (similarly to the sender in our scenario) participates only in and initial setup [24]. The difference with our setting (i) The trusted party should provide secret information to the receiver/chooser as well; this is unacceptable in application such as the privacy preserving architecture discussed below. (ii) the online sender knows the the values $m_0$ and $m_1$, whereas the servers in our scenario do not gain information about them.

**Application to the Privacy Preserving Architecture** An architecture for executing auctions, economic mechanism design and negotiations was proposed in [21]. The goal is to preserve the privacy of the inputs of the participants (so that no nonessential information about them is divulged, even a posteriori) while maintaining communication and computational efficiency. This goal is achieved by adding another party, the auction *issuer*, in addition to the bidders and the auctioneer. This party's role is to generate the programs ("garbled circuits") for computing the auctions prior to the auction and to run a variant of OT called proxy OT after the the bids have been submitted. Other than that it does not take an active part in the protocol. The auction issuer is not a trusted party, but is assumed not to collude with the auctioneer. In the original protocol of [21] the privacy of bidders is preserved as long as the auction issuer and the auctioneer do not collude.

Employing the distributed oblivious transfer protocols proposed in this paper allows splitting the role of the auction issuer into two parts (this was the motivation for our work). One of them needs a central server that acts only offline. It prepares the garbled circuits and acts as the sender preparing the inputs for the $n$ servers in the distributed OT protocol. During the execution of the auction these $n$ servers, called the online auction servers, operate after the bids are submitted. The central auction issuer can be better safeguarded than the online servers, since it operates offline. Privacy is guaranteed as long as the auctioneer does not collude with a coalition of *several* (more than the given threshold) of the online auction servers.

## 2   Definitions

A distributed $k$-out-of-$n$ $OT_1^2$ protocol involves three types of parties:

- A **sender** which has two inputs $m_0, m_1$. It is convenient to assume that both these inputs are elements in a field $\mathcal{F}$.
- A **chooser** that has an input bit $\sigma \in \{0, 1\}$.
- Additional $n$ **servers**, $S_1, \ldots, S_n$.

The protocol is composed of the following functional steps:

- The sender generates for each server $S_i$ a function $F_i$, which depends on $(m_0, m_1)$ and on random coin tosses of the sender.
- The chooser contacts $k$ different servers. She sends to server $S_i$ a query $q_i$ which is a function of $\sigma$ and of $i$, and of private random coin tosses. The server answers the query with $F_i(q_i)$.

A distributed $k$-out-of-$n$ $OT_1^2$ protocol must guarantee the following properties:

- **Reconstruction:** If the chooser receives information from $k$ servers she can compute $m_\sigma$. That is, there is an efficient algorithm for computing $m_\sigma$ from any set $\{i_j, F_{i_j}(q_{i_j})\}_{j=1}^k$.
- **Sender's privacy:** Given any $k$ values $\{i_j, F_{i_j}(q_{i_j})\}_{j=1}^k$ the chooser must gain information about a single input $m_\sigma$, and no information about the other input of the sender. (A weaker requirement is that she can compute at most a single linear combination of $m_0$ and $m_1$.)
- **Chooser's privacy:** No coalition of less than $t$ servers gains any information about $\sigma$, where $t$ is a parameter in the range $1 \le t \le k$. The parameter $t$ should ideally be as close as possible to $k$.
- **Chooser-servers collusion:** A coalition of the chooser with $\ell$ corrupt servers cannot learn about $m_0, m_1$ more than be learned by the chooser herself (where $\ell$ is a parameter).

An additional requirement is that if the chooser receives information from less than $k$ servers she gains no information about $m_0$ or $m_1$. There might be applications in which this requirement is not important, since the emphasis might be on the chooser having to contact *at most* $k$ servers. This requirement is not supported in all of the protocols that we present. Namely, in the protocol of Section 3.2 the receiver can obtain information about a single input after receiving information from less than $k$ servers. However, in this case she compromises her own privacy and risks that a coalition of fewer than $k$ servers can learn $\sigma$.

Note that the privacy of both the sender and the receiver is based on information theory and does not depend on any computational assumption. Furthermore, the protocol is very simple, the chooser simply asks server $S_i$ for a value of $F_i(\cdot)$ and receives an answer, and this process is considerably more efficient than a $OT_1^2$ protocol (since in all protocols $F_i$ is simply a polynomial).

The privacy of the sender depends on the chooser getting shares from at most $k$ servers. We discuss in Section 5 how to ensure that this is indeed the case.

The protocols use bivariate polynomials in a way which is similar to that used by the oblivious polynomial evaluation protocols of [20]: The sender defines a bivariate polynomial $Q(x, y)$ which hides his input, and the chooser defines a secret univariate polynomial $S(x)$ and interpolates $Q(x, S(x))$ which reveals to her one value of the sender's input. However, in [20] a single sender knows the polynomial $Q$ and the chooser uses $OT_1^N$ in order to learn the values of this polynomial at different locations, without revealing them to the sender. In the current work each server knows part of the polynomial, and the chooser simply asks servers to reveal to her values of the polynomial at different points. The chooser does not have to use $OT$ in order to hide these points from the servers, since as long as not too many of them collude they cannot learn her input.

**Why Secret Sharing Isn't Enough:** The first naive approach for designing a distributed $OT_1^2$ scheme is probably to suggest using simple $k$-out-of-$n$ secret sharing for sharing $m_0$ and $m_1$ between the servers. Namely, each input should be divided into $n$ shares, and each of the $n$ servers is given a share. The chooser should obtain $k$ shares of one of the schemes to reconstruct one of the inputs. The problem with this method is, of course, that the chooser must hide from the servers the identity of the input whose shares it requires. This essentially requires the chooser to run a $OT_1^2$ protocol with each of the servers.

## 3    Protocols for Distributed Oblivious Transfer

This section describes several protocols for distributed $OT_1^2$. The protocols follow the generic structure described in Table 1.

---

1. **Input:** The sender's input is a pair $m_0, m_1 \in \mathcal{F}$. The chooser's input is $\sigma \in \{0, 1\}$.
2. The sender generates a bivariate polynomial $Q(x, y)$, s.t. $Q(0, 0) = m_0$, $Q(0, 1) = m_1$.
3. The sender sends the univariate polynomial $Q(i, \cdot)$ to server $S_i$.
4. The chooser chooses a random polynomial $S$ s.t. $S(0) = \sigma$, and defines a univariate polynomial $R$ to be $R(x) = Q(x, S(x))$. The degree of $R$ is $k - 1$.
5. The chooser asks server $S_i$ for the value $R(i) = Q(i, S(i))$.
6. After receiving $k$ values of $R$ the chooser interpolates $R$ and computes $R(0)$.

---

**Fig. 1.** The basic steps of the distributed $OT_1^2$ protocol.

The main difference between the different protocols is the type of the polynomial $Q(x, y)$ that is generated by the sender. This choice affects all other parameters of the protocol. In particular, the first type of protocols uses a polynomial $Q(x, y)$ which is defined as the sum of a polynomial in $x$ and a linear polynomial in $y$, and has no monomials which include both $x$ and $y$. We denote such polynomials as *sparse*. Since the sender is only required to compute sparse polynomials, his task is greatly reduced (compared to the computation of full polynomials). This type of protocols is secure as long as there is no collaboration between the chooser and a corrupt server. It is also possible to make it immune against a collusion between the chooser and a single (or a few) servers.

We describe a different type of protocols which can protect the sender's privacy against a collusion between the chooser and a large set of servers. This type of protocols uses *full* bivariate polynomials in which the coefficients of all the monomials are non-zero (with high probability).

### 3.1   Using a Sparse Polynomial

The most basic and straightforward protocol employs a bivariate polynomial, where the degree of $y$ is 1 and there are no monomials which contain both $x$ and $y$. The protocol is described in Figure 2. It has the following properties.

- **Reconstruction:** After receiving information from $k$ servers, the chooser can learn $m_\sigma$, by interpolating the polynomial $R$.
- **Sender's privacy:** After receiving information from $k$ servers, the chooser cannot learn more than a single linear equation of $m_0$ and $m_1$ (this is proved in theorem 1). We later show in Section 4 how to ensure that the chooser learns exactly $m_0$ or $m_1$ and not any other combination of these values.
- Information from less than $k$ servers does not reveal to the chooser any information about $m_0$ and $m_1$ (since the degree of $x$ in $Q$ is $k-1$).
- **Chooser's privacy:** No coalition of at most $t = k-1$ servers can learn any information about $\sigma$ (this is proved in Theorem 2 and is based on the degree of $S$ being $k-1$).
- **No security against chooser-server collusion:** A coalition of the chooser with one corrupt server reveals to the chooser both $m_0$ and $m_1$ (after running the protocol). At the end of this Section we describe a method to address this problem if the chooser colludes with a single corrupt server (or a small number of corrupt servers). Section 3.2 describes a scheme which is secure against a collusion between the receiver and a large number of servers.
- **Overhead:** The sender has to choose $O(K)$ elements and has to send to each server $O(1)$ elements. Each server has to compute a linear polynomial a single time. The chooser should contact $k$ servers, and her total communication overhead is $O(k)$. The computation of $m_\sigma$ involves interpolation of a $k-1$ degree polynomial in order to find its free coefficient. This can be done in $O(k^2)$ multiplications using Lagrange's interpolation formula, or $O(k \log^2 k)$ multiplications using FFT (see e.g. [1] p. 299). The operations are done over the field $\mathcal{F}$ which can be rather small[1] and are therefore efficient by a few orders of magnitude compared to the public key operations required (following [18]) for non-distributed oblivious transfer.

---

[1] Typically the field should contain $m_0, m_1$. However, if these elements are large the sender can choose two random keys $k_0, k_1$ (say, 128 bits long) and use them to encrypt $m_0, m_1$, respectively. The OT protocol should be run for the inputs $k_0, k_1$, and therefore the field $\mathcal{F}$ should only be large enough to contain them.

*Initialization:* The sender generates a linear polynomial $P_y(y) = b_1 \cdot y + b_0$, s.t.

$$P_y(0) = m_0, P_y(1) = m_1. \quad (I.e., m_0 = b_0, \; m_1 = b_1 + b_0.)$$

The sender generates a random masking polynomial $P_x(x)$ of degree $k - 1$, s.t. $P_x(0) = 0$. Namely, $P_x(x) = \sum_{j=1}^{k-1} a_j x^j$. It also defines a bivariate polynomial

$$Q(x,y) = P_x(x) + P_y(y) = \sum_{j=1}^{k-1} a_j x^j + b_1 y + b_0$$

The sender provides server $S_i$ with the function $F_i(y)$ which is the result of substituting $x = i$ in the polynomial $Q$. Namely,

$$F_i(y) = Q(i,y) = \sum_{j=1}^{k-1} a_j i^j + b_1 y + b_0 = b_1 y + \left(\sum_{j=1}^{k-1} a_j i^j + b_0\right)$$

*Transfer:* The chooser generates a random polynomial $S(x)$ of degree $k - 1$, subject to the constraint $S(0) = \sigma$. I.e, $S(x) = \sum_{j=0}^{k-1} s_j x^j$ where $s_0 = \sigma$. Consider the polynomial $R(x)$ which is generated by substituting $S(x)$ instead of $y$ in $Q$,

$$R(x) = Q(x, S(x)) = \sum_{j=1}^{k-1} a_j x^j + b_1 \sum_{j=0}^{k-1} s_j x^j + b_0 = \sum_{j=1}^{k-1} (a_j + b_1 s_j) x^j + b_1 s_0 + b_0$$

The chooser's goal is to interpolate $R$ and compute $R(0) = Q(0, S(0)) = Q(0, \sigma) = m_\sigma$. The degree of $R$ is $k-1$, and therefore the chooser should obtain $k$ values of $R$ in order to interpolate it. She approaches $k$ different servers and asks server $S_i$ for the value $F_i(S(i)) = Q(i, S(i)) = R(i)$. After receiving $k$ answers she can interpolate $R$ and compute $R(0) = m_\sigma$.

**Fig. 2.** A distributed $OT_1^2$ protocol using a sparse linear polynomial.

## Proofs of Privacy

**Theorem 1 (Sender's privacy).** *After receiving information from $k$ servers, the chooser cannot learn more than a single linear combination of $m_0$ and $m_1$.*

**Proof:** When the chooser sends to server $i$ the query $y_i$, she receives the answer $F_i(y_i) = Q(i, y_i) = \sum_{j=1}^{k-1} a_j i^j + b_1 y_i + b_0$. The receiver therefore obtains the following set of $k$ equations:

$$\underbrace{\begin{pmatrix} i_1^{k-1} & i_1^{k-2} & \cdots & i_1 & y_{i_1} & 1 \\ i_2^{k-1} & i_2^{k-2} & \cdots & i_2 & y_{i_2} & 1 \\ \vdots & \vdots & \vdots & \vdots & \vdots & \vdots \\ i_k^{k-1} & i_k^{k-2} & \cdots & i_k & y_{i_k} & 1 \end{pmatrix}}_{A} \cdot \begin{pmatrix} a_{k-1} \\ \vdots \\ a_1 \\ b_1 \\ b_0 \end{pmatrix} = \begin{pmatrix} F_{i_1}(y_{i_1}) \\ F_{i_2}(y_{i_2}) \\ \vdots \\ F_{i_k}(y_{i_k}) \end{pmatrix}$$

It should be shown that no matter what values the chooser assigns to the $y_i$'s, she does not learn more than a single linear combination of $b_0, b_1$. In other words, that the rows of the matrix $A$ do not span both the vector $e_k = (0, \ldots, 0, 1, 0)$ and the vector $e_{k+1} = (0, \ldots, 0, 0, 1)$. The matrix $A$ has $k+1$ columns and $k$ rows. Consider the matrix $A'$ with $k+1$ rows which is formed by taking the first $k-1$ rows of $A$ and appending to them the vectors $e_k, e_{k+1}$. The determinant of $A'$ is different than 0 (since the sub-matrix of size $(k-1) \times (k-1)$ in the upper-left corner is Van Der Monde). Therefore, the first $k-1$ rows of $A$ do not span any of $e_k, e_{k+1}$, and the matrix $A$ which has just a single additional row cannot span both vectors. □

**Theorem 2 (Chooser's privacy).** *A coalition of $k-1$ servers does not learn any information about $\sigma$.*

**Proof:** The coalition receives $k-1$ values of $S(i)$ for $i \neq 0$. The polynomial $S$ is of degree $k$ and is random except for $S(0) = \sigma$. The information that the coalition learns could have been equally likely derived from a polynomial $S$ with $S(0) = 0$ as from a polynomial with $S(0) = 1$. □

**How to Protect against a Collusion between the Chooser and a Single Server:** The main drawback of the protocol is that a collusion between the chooser and one of the servers reveals both $m_0$ and $m_1$. This happens since each server $S_i$ knows a polynomial $F_i(y)$ which reveals $b_1 = m_1 - m_0$. We describe below a simple solution against a collusion between a chooser and a *single* server. This solution is general and is good for *any* distributed OT scheme. The aim of the rest of the paper is to deal with larger collusions.

In order to protect against a coalition of the chooser with a *single* server, the sender divides the $n$ servers into all possible $n$ subsets of $n-1$ servers. It defines $n$ random shares $\{m_{0,i}\}_{i=1}^n$ that satisfy $m_0 = \oplus_{i=1}^n m_{0,i}$, and similarly shares $\{m_{1,i}\}_{i=1}^n$ that satisfy $m_1 = \oplus_{i=1}^n m_{1,i}$. Next, it defines $n$ schemes for $(k-1)$-out-of-$(n-1)$ distributed $OT_1^2$. The $i$th scheme enables to transfer either one of $(m_{i,0}, m_{i,1})$, and is assigned to the members of the $i$th subset of servers.

The chooser should contact $k$ servers, and run the $n$ distributed $OT_1^2$ protocols, learning $\{m_{\sigma,i}\}_{i=1}^n$. She should then combine the results to compute $m_\sigma$.

This protocol ensures that a coalition of $t = k-2$ servers cannot learn which element the receiver learned, and that any $k$ servers enable the receiver to learn only a single share. A coalition of the chooser with a single server cannot learn any additional information, since this server has no information about one of the $OT_1^2$ schemes. This method can be generalized to handle a collusion of the chooser with $t$ servers, but this would require running $\binom{n}{t}$ distributed $OT_1^2$ protocols.

## 3.2    Using a Full Polynomial

In order to protect against large chooser-servers collusions, the sender should use a bivariate polynomial which includes all possible monomials, and in which the degree of $y$ is high. This approach yields a tradeoff between the number of servers that can compromise the chooser's privacy, and the size of a chooser-servers collusion that can compromise the sender's privacy. The protocol is described in Figure 3.

---

*Initialization:* The sender generates a random bivariate polynomial $Q(x, y)$ of degree $d_x$ in $x$ and degree $d_y$ in $y$, subject to the constraints

$$Q(0, 0) = m_0, \quad Q(0, 1) = m_1.$$

Namely, $Q(x, y) = \sum_{j=0}^{d_x} \sum_{l=0}^{d_y} a_{j,l} x^j y^l$, where $a_{0,0} = m_0$ and $\sum_{l=0}^{d_y} a_{0,l} = m_1$. It should also hold that $d_x = (k-1)/2$ (the parameter $k$ must be even). The sender sends to server $S_i$ the function $F_i(y)$ which is the result of substituting $x = i$ in the polynomial $Q$. Namely,

$$F_i(y) = \sum_{l=0}^{d_y} \left( \sum_{j=0}^{d_x} a_{j,l} \cdot i^j \right) \cdot y^l.$$

*Transfer:* The chooser generates a random polynomial $S(x)$ of degree $d_s$, where the degree satisfies[a] $d_y d_s = d_x = (k-1)/2$. The polynomial $S$ is random subject to the constraint $S(0) = \sigma$.
Consider the polynomial $R(x)$ which is generated by substituting $S(x)$ instead of $y$ in $Q$,

$$R(x) = Q(x, S(x))$$

The chooser should interpolate $R$ and compute $R(0) = Q(0, S(0)) = Q(0, \sigma) = m_\sigma$. The degree of $R$ is $k - 1 = d_x + d_y d_s$, and therefore the chooser should obtain $k$ values of $R$ in order to interpolate it. She approaches $k$ different servers and asks server $S_i$ for the value $F_i(S(i)) = Q(i, S(i)) = R(i)$. After receiving $k$ answers she can interpolate $R$ and compute $R(0) = m_\sigma$.

---

[a] We assume that the degrees are chosen such that this equality holds. Otherwise it must hold that $d_y d_s < d_x$.

**Fig. 3.** A distributed $OT_1^2$ protocol using a full polynomial.

The protocol has the following properties:

- **Reconstruction:** As in the previous protocol, after receiving information from $k$ servers the chooser can learn $m_\sigma$, since the degree of $R$ is $k$.
- **Sender's privacy:** After receiving information from $k$ servers, the chooser cannot learn more than a single linear equation of $m_0$ and $m_1$. This is proved in Theorem 3 in the Appendix. We show in Section 4 how to ensure that she learns exactly $m_0$ or $m_1$.
- **Chooser's privacy:** No coalition of at most $t = d_s = (k-1)/(2d_y)$ servers can learn any information about $\sigma$ (if the chooser acts according to the protocol). This follows from the degree of $S$.

- Information from less than $k$ servers might reveal to the chooser information about $m_0$ or $m_1$ (e.g., if she sets $S(x)$ to be of degree smaller than $d_x$, the degree of $R = Q(x, S(x))$ would be smaller than $k$). However, this affects the chooser's privacy, namely reveals $\sigma$ to a coalition of less than $(k-1)/(2d_y)$ servers. If the chooser receives information from less than $d_x$ servers she learns no information about either $m_0$ or $m_1$.

- **Security against chooser-servers collusion:** A coalition of the chooser with $d_x - \frac{2d_x}{d_y+1}$ corrupt servers, does not reveal to the chooser more than a single linear equation of $m_0$ and $m_1$. This is proved in Theorem 4 in the Appendix.

- **Overhead:** The sender in preparing the polynomial has to choose $O(kd_y)$ elements and send $d_y$ elements per server. Each server has to compute a polynomial of degree $d_y$ a single time. The overhead of the chooser is as in the sparse polynomial scheme.

This construction, therefore, gives a tradeoff between chooser privacy against a coalition of corrupt servers, and sender's privacy against a coalition between the chooser and corrupt servers. Once $n$ and $k$ are fixed, The tradeoff depends on a parameter $d_y$. The size of a coalition of corrupt servers against which the chooser is secure is $(k-1)/(2d_y) = d_x/d_y$, whereas the size of a coalition of corrupt servers that can help the chooser learn more than a single input is $d_x - \frac{2d_x}{d_y+1}$.

## 4    Preventing the Chooser from Learning Linear Combinations

Suppose that the chooser must be forced to learn either $m_0$ or $m_1$, and it is required to prevent her from learning linear combinations of the two inputs[2].

The following method can be used to ensure that the chooser learns either $m_0$ or $m_1$, but not any other linear combination of the two inputs. We describe it for the protocol of Section 3.1 which uses a sparse bivariate polynomial.

The protocol is run simultaneously with two polynomials $P_y^1 = (a \cdot m_1 - b \cdot m_0)y + m_0 \cdot b$, and $P_y^2 = (a-b)y + b$, and corresponding polynomials $Q^1$ and $Q^2$. (The first polynomial hides $m_1$ multiplied by $a$, and $m_0$ multiplied by $b$, whereas the second polynomial hides $a$ and $b$). The chooser sends a single value $S(i)$ to server $i$ and receives the values $Q^1(i, S(i))$ and $Q^2(i, S(i))$.

If the chooser operates according to the protocol, she learns the values $m_0 \cdot b$ and $b$ if $S(0) = 0$, and can then compute $m_0$. Similarly, she can compute $m_1$ if she sets $S(0) = 1$.

---

[2] A heuristic approach for achieving this property might encrypt the inputs $m_0$ and $m_1$ using two random keys $k_0$ and $k_1$, respectively, and run the distributed OT protocol to let the chooser learn either $k_0$ or $k_1$. If the chooser chooses to learn a linear combination of both keys then presumably she would not be able to decrypt any of the encryptions. This approach can be proved to be secure in the random oracle world, i.e. if a function $H$ which is modeled as a random oracle is used to encrypt each $m_i$ using $k_i$.

The chooser cannot learn any other linear combination of $m_0$ and $m_1$. The important property of the protocol is that the chooser learns the same linear combination of the coefficients of both $P_y^1$ and $P_y^2$. Suppose that in this combination the coefficient of $y$ is multiplied by $\alpha$ and the free coefficient is multiplied by $\beta$. The chooser therefore learns the following equations:

$$\begin{pmatrix} m_1\alpha & m_0(\beta - \alpha) \\ \alpha & \beta - \alpha \end{pmatrix} \begin{pmatrix} a \\ b \end{pmatrix}$$

If this matrix is non singular then any value of $m_0, m_1$ corresponds to a different pair $a, b$, and no information is divulged about $m_0$ or $m_1$. The matrix is singular only if $m_0 = m_1$ (but we can ensure that this does not happen if we append a different prefix to each input), or if $\alpha = 0$ or $\alpha = \beta$. These last two cases reveal to the chooser the value of $m_0$ or $m_1$, respectively, and are therefore legitimate.

# 5    Ensuring that a Chooser Does Not Obtain More than $k$ Shares

Distributed oblivious transfer prevents the chooser from learning more than a single input as long as she does not obtain information from more than $k$ servers. This property raises the following question: how should we ensure that the chooser receives information from at most $k$ servers? (note that this problem does not exist if the system implements an $n$-out-of-$n$ access structure). This issue might be regarded as orthogonal to the schemes themselves. Alternatively, there might be some centralized mechanism for limiting the number of servers that send information to the chooser. However, it might be difficult to operate such a mechanism in a distributed setting.

We now describe two solutions that are applicable for the case $k > n/2$ (or any other quorum system). The solutions can be combined with any protocol for distributed OT. Therefore there is no need to postulate any external mechanism enforcing the limit on the number of servers accessed in this case.

*A solution for $k > n/2$ (or any other quorum system):* The servers share a key $K$ for a pseudo-random function $F$ (pseudo-random functions are commonly modeled by block ciphers). The key $K$ is known to each of the servers. Denote the subset of $k$ servers that the user approaches as $\mathcal{S}$, $|\mathcal{S}| = k$. The user sends the names of all servers in $\mathcal{S}$ to each of the servers she contacts.

Each such server, $S_i$, operates as follows:

- It verifies that $\mathcal{S}$ contains the names of $k$ servers including $S_i$, and that it did not previously send an answer to the chooser for a different set $\mathcal{S}'$ which contains $S_i$ (for the same OT).
- It computes $\alpha_\mathcal{S} = \oplus_{S_i \in \mathcal{S}} F_K(\mathcal{S}, S_i)$, where $F_K$ is a pseudo-random function $F$ keyed by $K$.
- It sends to the chooser its answer, as defined in the distributed OT protocol, encrypted by $\alpha_\mathcal{S}$. In addition it sends her $F_K(\mathcal{S}, S_i)$.

After receiving answers from all servers in $S$ the chooser can compute $\alpha_S$ and decrypt the answers. Since $k > n/2$, every two different subsets of $k$ servers, $S$ and $S'$, intersect, and therefore the chooser cannot compute both $\alpha_S$ and $\alpha_{S'}$.

The above solution can be generalized to any access structure which is based on a quorum system[3]. Assume, for simplicity, that each quorum contains the same number of servers, $k$. The system should use a $k$-out-of-$n$ threshold access structure. In addition each server $S_i$ should verify that $S$ is a legitimate quorum which contains $S_i$, and encrypt its answer with $\alpha_S$ as described above. Since each two quorums intersect, the chooser can only decrypt $k$ answers of a single quorum.

*A solution for $k > n/2$ (and any other quorum system) secure against chooser-servers coalition:* The drawback of the previous solution is that even a single server cooperating with the chooser can reveal $K$ and enable the chooser to decrypt messages from more than $k$ servers. The following solution solves the problem chooser-server coalition, provided the size of the coalition is less than $2k - n$.

The sender defines in advance $n(n - 1)$ strings $\{\alpha_{i,j}\}_{1 \leq i,j \leq n, i \neq j}$ for every *ordered* pair of servers, and gives server $S_i$ the $2(n-1)$ strings $\{\alpha_{i,j}, \alpha_{j,i} \mid i \neq j\}$. The chooser sends to server $S_i$ the set $S$ of $k$ servers which she is querying. The server first verifies that $S_i \in S$ and that it was not asked to answer the chooser using a different set $S'$ of servers. It then sends its answer encrypted by $\oplus_{S_j \in S, j \neq i} \alpha_{i,j}$. It also sends to the chooser the values $\{\alpha_{j,i} \mid S_j \in S, j \neq i\}$. The chooser must receive answers from all the servers in $S$ before she can decrypt them. This method can be applied to any access structure which is based on a quorum system, provided a coalition does not cover any intersection of quorums.

# References

1. Aho A., Hopcroft J. and Ullman J., *The design and analysis of computer algorithms*, Addison-Wesley, 1974.
2. D. Beaver, *Foundation of Secure Interactive Computation*, Advances in Cryptology - Crypto '91, pp. 377–391, 1991.
3. D. Beaver, "Commodity-Based Cryptography", STOC 1997, pp. 446-455.
4. D. Beaver, J. Feigenbaum, J. Kilian, and P. Rogaway. "Locally Random Reductions: Improvements and Applications", Journal of Cryptology 10(1): 17-36 (1997).
5. M. Bellare and S. Micali, *Non-interactive oblivious transfer and applications*, Advances in Cryptology - Crypto '89, pp. 547-557, 1990.
6. G. Brassard, C. Crépeau and J.-M. Robert *Information Theoretic Reduction Among Disclosure Problems*, 27th FOCS, pp. 168–173, 1986.
7. G. Brassard, C. Crépeau and J.-M. Robert, *All-or-Nothing Disclosure of Secrets*, Advances in Cryptology - Crypto '86, LNCS 263, Springer, pp. 234–238, 1987.

---

[3] A quorum system is a collection of subsets of some ground set such that any two subsets intersect. They have been considered for cryptographic purposes previously, e.g. [22].

8. G. Brassard, C. Crépeau and M. Santha, *Oblivious Transfer and Intersecting Codes*, IEEE Trans. on Inform. Theory, Vol. 42(6), pp. 1769–1780, 1996.
9. C. Cachin, *On the foundations of oblivious transfer*, Advances in Cryptology - Eurocrypt '98, LNCS 1403, pp. 361-374. Springer, 1998.
10. B. Chor, O. Goldreich, E. Kushilevitz and M. Sudan, *Private Information Retrieval*, J. of ACM 45(6), 1998, pp. 965–981.
11. C. Crépeau, *Equivalence between two flavors of oblivious transfers*, Advances in Cryptology – Crypto '87 , LNCS 293, pp. 350–354, 1988.
12. C. Crépeau and J. Kilian, *Achieving oblivious transfer using weakened security assumptions*, FOCS '88, pp. 42–52, 1988.
13. S. Even, O. Goldreich and A. Lempel, *A Randomized Protocol for Signing Contracts*, Communications of the ACM **28**, pp. 637–647, 1985.
14. Y. Gertner, S. Goldwasser, T. Malkin, *A Random Server Model for Private Information Retrieval or How to Achieve Information Theoretic PIR Avoiding Database Replication,*. RANDOM 1998, LNCS 1518, Springer, pp. 200–217.
15. Y. Gertner, Y. Ishai, E. Kushilevitz, and T. Malkin, *Protecting Data Privacy in Private Information Retrieval Schemes*, Proc. 30th STOC 1998, pp. 151–160.
16. O. Goldreich, *Secure Multi-Party Computation* (working draft) Version 1.1, 1998.
17. O. Goldreich and R. Vainish, *How to Solve any Protocol Problem - An Efficiency Improvement*, Advances in Cryptology - Crypto '87, LNCS 293, 1988, pp. 73–86.
18. R. Impagliazzo and S. Rudich, *Limits on the Provable Consequences of One-Way Permutations,* STOC '89, pp. 44–61, 1989.
19. J. Kilian, **Use of Randomness in Algorithms and Protocols**, MIT Press, Cambridge, Massachusetts, 1990.
20. M. Naor and B. Pinkas, *Oblivious Transfer and Polynomial Evaluation*, Proc. of the 31st ACM Symp. on Theory of Computer Science, 1999, pp. 245–254.
21. M. Naor, B. Pinkas and R. Sumner, *Privacy Preserving Auctions and Mechanism Design*, Proc. of the 1st ACM conf. on Electronic Commerce, November 1999, pp. 129–139 .
22. M. Naor and A. Wool, *Access Control and Signatures via Quorum Secret Sharing*, IEEE Transactions on Parallel and Distributed Systems 9(9), 1998, pp. 909–922.
23. M. O. Rabin, *How to exchange secrets by oblivious transfer*, Tech. Memo TR-81, Aiken Computation Laboratory, 1981.
24. R. Rivest, *Unconditionally Secure Commitment and Oblivious Transfer Schemes Using Private Channels and a Trusted Initializer*, manuscript. Available: http://theory.lcs.mit.edu/~rivest/publications.html

# A   Privacy for the Protocol Which Uses Full Polynomials

## A.1   Sender's Privacy

We first prove that if $d_y = 1$ then the chooser can learn only a single linear equation of $m_0$ and $m_1$, and then prove this for any degree $d_y$.

**Lemma 1.** *Let $Q(x, y)$ be a bivariate polynomial in which $x$ is of degree $d_x$ and $y$ is linear. Denote by $P(y) = ay + b = Q(0, y)$ the polynomial which is equal to $Q$ constrained to the line $x = 0$ (i.e. to the $y$ axis). Any $2d_x + 1$ values $Q(x_i, y_i)$ where all the $x_i$-s are distinct and different from $0$ do not yield more than a single linear equation on the coefficients $a$ and $b$.*

**Proof:** Denote the polynomial as $Q(x,y) = \sum_{i=0}^{d_x} \sum_{j=0}^{1} a_{i,j} x^i y^j$ (i.e. $a = a_{0,1}$ and $b = a_{0,0}$). The $2d_x + 1$ values of $Q(x,y)$ define $2d_x + 1$ linear relations for the $2d_x + 2$ coefficients $a_{i,j}$. Assume wlog that these equations are linearly independent (otherwise Alice has made redundant queries). Note that this implies that not all $y_i$ values are the same (if all $y_i$ were the same then for all $1 \le i \le d_x + 1$ columns $i$ and $d_x + 1 + i$ would have been linearly dependent).

The equations can be represented by a matrix $A$ with $2d_x + 1$ rows and $2d_x + 2$ columns,

$$
\begin{pmatrix}
1 & x_1 & \cdots & x_1^{d_x} & y_1 & y_1 x_1 & \cdots & y_1 x_1^{d_x} \\
1 & x_2 & \cdots & x_2^{d_x} & y_2 & y_2 x_2 & \cdots & y_2 x_2^{d_x} \\
  & \vdots & & \vdots & & \vdots & & \\
1 & x_{2d_x+1} & \cdots & x_{2d_x+1}^{d_x} & y_{2d_x+1} & y_{2d_x+1} x_{2d_x+1} & \cdots & y_{2d_x+1} x_{2d_x+1}^{d_x}
\end{pmatrix}
\begin{pmatrix}
a_{0,0} \\
\vdots \\
a_{d_x,0} \\
a_{0,1} \\
\vdots \\
a_{d_x,1}
\end{pmatrix}
=
\begin{pmatrix}
P(x_1,y_1) \\
P(x_2,y_2) \\
\vdots \\
P(x_{2d_x+1}, y_{2d_x+1})
\end{pmatrix}
$$

We will prove that it cannot be the case that both $e_{0,0}$ and $e_{0,1}$ are defined by these equations. In other words, let $e_{i,j}$ be the $2d_x + 2$ entry vector in which all entries are 0 except for the $(i + 1 + j \cdot (d_x + 1))$'th entry which is 1 (i.e. only the coefficient of $a_{i,j}$ is 1). We will prove that the rows of the matrix $A$ cannot span both $e_{0,0}$ and $e_{0,1}$.

The vector space is of dimension $2d_x + 2$, the vectors $e_{0,0}$ and $e_{0,1}$ are orthogonal and the rank of $A$ is $2d_x + 1$ (all its rows are linearly independent). Therefore $A$ spans a vector in the linear subspace generated by $e_{0,0}$ and $e_{0,1}$. Assume wlog that this vector is of the form $v = (\alpha, 0, \ldots, 0, 1, 0, \ldots, 0)$, i.e. that its first entry equals $\alpha$ and its $(d_x + 2)$'th entry equals 1. The vector $v$ can be represented as a linear combination of the rows of $A$, and we can therefore replace one of the rows of $A$ (say the last row) with $v$. Wlog we prove that this revised matrix (and therefore also $A$) cannot span $e_{0,0}$ in addition to $v$. Consider the matrix $B'$ which is constructed by adding to the revised matrix the row $e_{0,0}$. It has $2d_x + 2$ rows and $2d_x + 2$ columns.

$$
B' =
\begin{pmatrix}
1 & 0 & \cdots & 0 & 0 & 0 & \cdots & 0 \\
\alpha & 0 & \cdots & 0 & 1 & 0 & \cdots & 0 \\
1 & x_1 & \cdots & x_1^{d_x} & y_1 & y_1 x_1 & \cdots & y_1 x_1^{d_x} \\
  & \vdots & & & & \vdots & & \\
1 & x_{2d_x} & \cdots & x_{2d_x}^{d_x} & y_{2d_x} & y_{2d_x} x_{2d_x} & \cdots & y_{2d_x} x_{2d_x}^{d_x}
\end{pmatrix}
$$

The lemma is proven by the following claim, which shows that all the rows of $B'$ are linearly independent. The proof appears in the full version of the paper.

**Claim:** The determinant of a matrix $B'$ in which all the $x_i$-s are distinct and different from 0 and not all $y_i$ values are equal, cannot be 0.

Following is a privacy theorem for polynomials in which the degree of $y$ is greater than linear. The proof is similar to that of Lemma 1.

**Theorem 3.** *Let $Q(x,y)$ be a bivariate polynomial in which $x$ is of degree $d_x$ and $y$ of degree $d_y$. Denote by $P(y) = \sum_{j=0}^{d_y} a_{0,j} y^j = P(0,y)$ the polynomial which is* *equal to $Q$ constrained to the line $x = 0$ (i.e. to the $y$ axis). Denote the coefficients of the elements free of $x$, i.e. $a_{0,0}, a_{0,1}, \ldots, a_{0,d_y}$, as the $y$ coefficients. Then given any $2d_x + 1$ values $Q(x_i, y_i)$ where all the $x_i$-s are distinct and different from $0$, at most a single linear relation is defined between the $y$ coefficients.*

## A.2    Chooser-Servers Collusion

The following theorem demonstrates that a collusion between the chooser and $d_x - \frac{2d_x}{d_y+1}$ servers (in addition to the $k$ servers that were contacted by the chooser), cannot learn about $m_0, m_1$ more than can be learned by the chooser herself. The proof appears in the full version of the paper.

**Theorem 4.** *Let $Q(x, y)$ be a bivariate polynomial in which $x$ is of degree $d_x$ and $y$ of degree $d_y$. Denote by $P(y) = \sum_{j=0}^{d_y} a_{0,j} y^j = P(0, y)$ the polynomial which is equal to $Q$ constrained to the line $x = 0$ (i.e. to the $y$ axis). Denote the coefficients of the elements free of $x$, i.e. $a_{0,0}, a_{0,1}, \ldots, a_{0,d_y}$, as the $y$ coefficients. Then given any $2d_x + 1$ values $Q(x_i, y_i)$ where all the $x_i$-s are distinct and different from $0$, and given the restrictions of $Q(x, y)$ to $\ell$ different $x$ values, where $\ell \leq d_x - \frac{2d_x}{d_y+1}$, at most a single linear relation is defined between the $y$ coefficients.*

# Key Improvements to XTR

Arjen K. Lenstra[1] and Eric R. Verheul[2]

[1] Citibank, N.A., Technical University Eindhoven,
1 North Gate Road, Mendham, NJ 07945-3104, U.S.A.,
arjen.lenstra@citicorp.com
[2] PricewaterhouseCoopers, GRMS Crypto Group,
Goudsbloemstraat 14, 5644 KE Eindhoven, The Netherlands,
Eric.Verheul@[nl.pwcglobal.com, pobox.com]

**Abstract.** This paper describes improved methods for XTR key representation and parameter generation (cf. [4]). If the field characteristic is properly chosen, the size of the XTR public key for signature applications can be reduced by a factor of three at the cost of a small one time computation for the recipient of the key. Furthermore, the parameter set-up for an XTR system can be simplified because the trace of a proper subgroup generator can, with very high probability, be computed directly, thus avoiding the probabilistic approach from [4]. These non-trivial extensions further enhance the practical potential of XTR.

## 1 Introduction

In [1] it was shown that conjugates of elements of a subgroup of $GF(p^6)^*$ of order dividing $\phi_6(p) = p^2 - p + 1$ can be represented using $2\log_2(p)$ bits, as opposed to the $6\log_2(p)$ bits that would be required for their traditional representation. In [4] an improved version of the method from [1] was introduced that achieves the same communication advantage at a much lower computational cost. The resulting representation method is referred to as XTR, which stands for Efficient and Compact Subgroup Trace Representation. As shown in [4], solving the XTR version of a particular discrete logarithm related problem is equivalent to solving the same problem in its traditional $GF(p^6)$ setting, which is as hard as solving the problem in the full multiplicative group $GF(p^6)^*$.

It is argued in [4] that XTR is an excellent alternative to either RSA or Elliptic Curve Cryptosystems using random curves over prime fields (ECC), because it combines most of the advantages of RSA and ECC without having any of their disadvantages. More specifically, it is shown in [4] that, with the exception of signature applications, XTR keys are much smaller than RSA keys of equivalent security, and at most twice as big as ECC keys. Furthermore, parameter and key selection for XTR is very fast compared to RSA, and thus much faster than ECC. Finally, for almost all cryptographic applications XTR is faster than ECC when random curves over prime fields are used; the exception is signature verification where ECC is slightly faster than XTR.

In this paper we describe three improvements to XTR. We present a careful analysis of Scipione del Ferro's classical method to solve cubic equations. As a

T. Okamoto (Ed.): ASIACRYPT 2000, LNCS 1976, pp. 220–233, 2000.
© Springer-Verlag Berlin Heidelberg 2000

result we are able to reduce the XTR public key size for signature applications by a factor of three if the field characteristic is not equal to 8 modulo 9. Because that is not unduly restrictive, it follows that XTR public keys are at most twice as long as ECC public keys for *all* applications of XTR. This is, in our opinion, an important enhancement of XTR. As a side result we get a method to find the trace of a proper subgroup generator that is 50% faster than the method presented in [4]. Finally, we give a much faster deterministic method for the same problem that works only if the characteristic is not equal to 8 modulo 9. None of these two improved XTR parameter selection methods is of crucial importance for practical applications of XTR, but the last method in particular makes implementation of XTR even easier. The resulting algorithms are all very practical and allow easy implementation.

In Section 2 we review XTR. In Section 3 we present Scipione del Ferro's method and the resulting improved parameter selection method. An even faster parameter selection method is given in Section 4, and the key size reduction methods are given in Section 5.

## 2    XTR

In this section we review some of the results from [4]. Let $p$ be prime and let $F(c, X)$ for $c \in \mathrm{GF}(p^2)$ be the polynomial $X^3 - cX^2 + c^p X - 1 \in \mathrm{GF}(p^2)[X]$. For $n \in \mathbf{Z}$ we denote by $c_n$ the sum of the $n^{\mathrm{th}}$ powers of the roots of $F(c, X)$, i.e., if $F(c, h_j) = 0$ for $j = 0, 1, 2$, then $c_n = h_0^n + h_1^n + h_2^n$. Notice that $c_1 = c$. It is shown in [4] that $c_n \in \mathrm{GF}(p^2)$, that $c_{-n} = c_n^p$, and that $F(c_n, h_j^n) = 0$ for $j = 0, 1, 2$. Furthermore, if $p \equiv 2 \bmod 3$, then $p^{\mathrm{th}}$ powering in $\mathrm{GF}(p^2)$ is effectively free, and $c_n$ can be computed given $c = c_1$ in $8 \log_2(n)$ multiplications in $\mathrm{GF}(p)$ using a Fibonacci-like recurrence relation (cf. [4]). The values $c_{n-1}$ and $c_{n+1}$ are obtained at no extra cost as a side result of the computation of $c_n$.

It is shown in [4] that if $F(c, X)$ is irreducible, then the roots of $F(c, X)$ take the form $h, h^{p^2}, h^{p^4}$ for some $h \in \mathrm{GF}(p^6)$ of order dividing $p^2 - p + 1$ and $> 3$. This implies that in these circumstances $c_n$ is of the form $Tr(h^n)$, where $Tr(y) = y + y^{p^2} + y^{p^4} \in \mathrm{GF}(p^2)$ is the trace over $\mathrm{GF}(p^2)$ of $y \in \mathrm{GF}(p^6)$, i.e., the sum of the conjugates over $\mathrm{GF}(p^2)$ of $y$. The trace over $\mathrm{GF}(p^2)$ is $\mathrm{GF}(p^2)$-linear. Vice versa, it is shown that the minimal polynomial of any $h \in \mathrm{GF}(p^6)$ of order dividing $p^2 - p + 1$ and $> 3$ is equal to $F(Tr(h), X)$, illustrating the fundamental idea of XTR that for such $h$ the trace value fully specifies $h$'s minimal polynomial, and thus the conjugates of $h$.

Let $g \in \mathrm{GF}(p^6)$ have order $q$ for a prime $q > 3$ dividing $p^2 - p + 1$. It follows from the results cited above that $Tr(g^n) \in \mathrm{GF}(p^2)$ and $F(Tr(g^n), g^n) = 0$ for any $n$. Furthermore, if $p \equiv 2 \bmod 3$ then $Tr(g^n)$ can be computed given $Tr(g)$ in $8 \log_2(n)$ multiplications in $\mathrm{GF}(p)$, which is almost three times faster than computing $g^n$ from $g$ using traditional exponentiation methods. Thus, in XTR we replace powers of $g$ by their traces, thereby saving a factor of three both in storage and in computing time. Note that an actual representation of $g$ is not required, and that it suffices to have its trace $Tr(g)$. Given $Tr(g)$, the order $q$ subgroup generated by (the unknown) $g$ is called the XTR group.

XTR parameter selection is the problem of finding primes $p$ and $q$ such that $q$ divides $p^2 - p + 1$, $q > 3$, $p \equiv 2 \bmod 3$, and $p \equiv 3 \bmod 4$, and the trace $Tr(g)$ of a generator of the XTR group. The primes $p$ and $q$ of appropriate sizes can be found using either of the two methods given in [4]. To find a proper $Tr(g)$ it suffices to find $c \in \mathrm{GF}(p^2) \setminus \mathrm{GF}(p)$ such that $F(c, X) \in \mathrm{GF}(p^2)[X]$ is irreducible, such that $c_{(p^2-p+1)/q} \neq 3$, and to put $Tr(g) = c_{(p^2-p+1)/q}$ (cf. [4]). The probability that $c_{(p^2-p+1)/q} = 3$ if $F(c, X)$ is irreducible is only $1/q$, so usually the first irreducible $F(c, X)$ works. In Section 3 we describe a fast way to test $F(c, X)$ for irreducibility (assuming a randomly selected $c \in \mathrm{GF}(p^2)$), and in Section 4 we show how irreducible polynomials of the form $F(c, X)$ can be written down directly if $p \not\equiv 8 \bmod 9$.

The ability to quickly compute $Tr(g^n)$ given $Tr(g)$ suffices for efficient implementation of many cryptographic protocols. But in some cryptographic applications, most notably verification of digital signatures and authentication responses, values of the form $Tr(g^{a+kb})$ have to be computed, for $a, b \in \mathbf{Z}$, given $Tr(g)$ and $Tr(g^k)$ for some secret integer $k$ (the private key). It is shown in [4] that computation of $Tr(g^{a+kb})$ can efficiently be done if additionally $Tr(g^{k-1})$ and $Tr(g^{k+1})$ are known. Thus, whereas for many applications the XTR public key data consist of just $p$, $q$, $Tr(g)$, and $Tr(g^k)$ (for unknown $k$), in some applications $Tr(g^{k-1})$ and $Tr(g^{k+1})$ must be included in the XTR public key data as well. This considerably increases the transmission overhead for the XTR public key data. In Section 4 we show how this problem can be dealt with. First we show that $Tr(g^{k-1})$ (or $Tr(g^{k+1})$) can easily be determined as a function of $Tr(g)$, $Tr(g^k)$ and $Tr(g^{k+1})$ (or $Tr(g^{k-1})$). And next we show how $Tr(g^{k+1})$ (or $Tr(g^{k-1})$) can be quickly computed based on just $Tr(g)$ and $Tr(g^k)$, assuming that $p \not\equiv 8 \bmod 9$. Both methods impose very mild restrictions on the choice of the private key $k$ and have no negative impact on the security of XTR.

## 3    Finding a Root of a Cubic Equation

We describe Scipione del Ferro's classical method (cf. [6], page 559) to compute the roots of a third-degree equation in its full generality, after which we apply it to test the third-degree polynomial $F(c, X) \in \mathrm{GF}(p^2)[X]$ as in Section 2 for irreducibility.

**Algorithm 3.1 (Scipione del Ferro, $\sim$1465-1526)** To find the roots of the third-degree polynomial $f(X) = aX^3 + bX^2 + dX + e$ in a field of characteristic $p$ unequal to 2 or 3, perform the following steps.

1. Compute the polynomial $f(X - b/(3a))/a = X^3 + f_1 X + f_0$ with $f_1 = (3ad - b^2)/(3a^2)$ and $f_0 = (27a^2e - 9abd + 2b^3)/(27a^3)$.
2. Compute the discriminant $\Delta = f_0^2 + 4f_1^3/27$ of the polynomial $X^2 + f_0 X - f_1^3/27$, and compute its roots $r_{1,2} = (-f_0 \pm \sqrt{\Delta})/2$.
3. If $r_1 = r_2 = 0$, then let $u = v = 0$. Otherwise, let $r_1 \neq 0$, compute a cube root $u$ of $r_1$, and let $v = -f_1/(3u)$. Note that $v$ is a cube root of $r_2$.

4. The roots of $f(X)$ are $u+v-b/(3a)$, $uw+vw^2-b/(3a)$, and $uw^2+vw-b/(3a)$, where $w \in \mathrm{GF}(p^2)$ is a non-trivial cube root of unity, i.e., $w^3 = 1$ and $w^2 + w + 1 = 0$.

**Theorem 3.2** *Let $f(X) \in \mathrm{GF}(p^2)[X]$ be such that $\Delta$ as in Step 2 of Algorithm 3.1 is in $\mathrm{GF}(p)$. The following four statements are equivalent.*

1. *$f(X)$ is reducible over $\mathrm{GF}(p^2)$.*
2. *$f(X)$ has a root in $\mathrm{GF}(p^2)$.*
3. *$f(X)$ has three roots in $\mathrm{GF}(p^2)$.*
4. *The roots $r_1$ and $r_2$ as in Step 2 of Algorithm 3.1 are cubes in $\mathrm{GF}(p^2)$.*

**Proof.** $1 \Leftrightarrow 2$ and $3 \Rightarrow 2$ are trivial. We prove $2 \Leftrightarrow 4$ and $4 \Rightarrow 3$.

'$4 \Rightarrow 2$'. If there is a $u$ in $\mathrm{GF}(p^2)$ such that $u^3 = r_1$, then $u - f_1/(3u) - b/(3a)$ is a root of $f(X)$ in $\mathrm{GF}(p^2)$ (cf. Step 4 of Algorithm 3.1).

'$2 \Rightarrow 4$'. If $f(X)$ has a root in $\mathrm{GF}(p^2)$, then there is a cube root $u$ of $r_1$ such that $u+v-b/(3a) \in \mathrm{GF}(p^2)$, with $v = -f_1/(3u)$, so that $u+v$ is in $\mathrm{GF}(p^2)$. Since also $uv = -f_1/3$ is in $\mathrm{GF}(p^2)$, it follows that $u \in \mathrm{GF}(p^4)$. On the other hand, $r_1, r_2 \in \mathrm{GF}(p^2)$ because $\Delta \in \mathrm{GF}(p)$. Since $u^3 = r_1$ it follows that $u \in \mathrm{GF}(p^6)$. From $u \in \mathrm{GF}(p^4) \cap \mathrm{GF}(p^6)$ it follows that $u \in \mathrm{GF}(p^2)$ so that $r_1$ is a cube in $\mathrm{GF}(p^2)$. It follows from $r_2 = (-f_1/(3u))^3$ that $r_2$ is a cube in $\mathrm{GF}(p^2)$ as well.

'$4 \Rightarrow 3$'. If $u - f_1/(3u) - b/(3a)$ is a root of $f(X)$ with $u$ in $\mathrm{GF}(p^2)$ then $uw - f_1w^2/(3u) - b/(3a)$ and $uw^2 - f_1w/(3u) - b/(3a)$, with $w \in \mathrm{GF}(p^2)$ as in Step 4 of Algorithm 3.1, are the two other roots of $f(X)$ (cf. Step 4 of Algorithm 3.1), and all three roots are in $\mathrm{GF}(p^2)$.

**Lemma 3.3** *For any $c \in \mathrm{GF}(p^2)$ the discriminant $\Delta$ as in Step 2 of Algorithm 3.1 of $f(X) = F(c, X)$ is in $\mathrm{GF}(p)$.*

**Proof.** It follows from a straightforward computation that $\Delta = 1 - 2c^{p+1}/3 - c^{2p+2}/27 + 4(c^3 + c^{3p})/27$. This implies that $\Delta^p = \Delta$ so that $\Delta \in \mathrm{GF}(p)$.

**Corollary 3.4** *The polynomial $F(c, X) \in \mathrm{GF}(p^2)[X]$ is reducible over $\mathrm{GF}(p^2)$ if and only if the $r_1$ from Step 2 of an application of Algorithm 3.1 to $f(X) = F(c, X)$ is a cube in $\mathrm{GF}(p^2)$.*

**Proof.** Immediate from Lemma 3.3 and Theorem 3.2.

An element $x \in \mathrm{GF}(p^2)$ is a cube if and only if $x^{(p^2-1)/3} = 1$, which is the case if and only if $x^{p(p+1)/3} = x^{(p+1)/3}$. Thus, testing if an element of $\mathrm{GF}(p^2)$ is a cube can be done at the cost of a $(p+1)/3^{\mathrm{th}}$ powering in $\mathrm{GF}(p^2)$ followed by a $p^{\mathrm{th}}$ powering (which is free in $\mathrm{GF}(p^2)$, cf. Section 2).

**Algorithm 3.5 (Irreducibility test)** To decide if $F(c, X) \in \mathrm{GF}(p^2)[X]$ is irreducible over $\mathrm{GF}(p^2)$, perform the following steps.

1. Compute $F(c, X + c/3) = X^3 + f_1 X + f_0 \in \mathrm{GF}(p^2)[X]$ with $f_1 = c^p - c^2/3$ and $f_0 = (-27 + 9c^{p+1} - 2c^3)/27$ (cf. $p \neq 3$).

2. If $\Delta = f_0^2 + 4f_1^3/27 \in \mathrm{GF}(p)$ (cf. $p \neq 3$) is a quadratic non-residue in $\mathrm{GF}(p)$ then $F(c, X)$ is reducible (cf. Lemma 3.6).
3. Otherwise, compute a root $r_1 \in \mathrm{GF}(p^2)$ (cf. Corollary 3.4) of $X_2 + f_0 X - (f_1/3)^3$: $r_1 = (-f_0 + \sqrt{\Delta})/2$ (cf. $p \neq 2$).
4. Compute $y = r_1^{(p+1)/3} \in \mathrm{GF}(p^2)$, then $F(c, X)$ is irreducible $\iff y \neq y^p$.

**Lemma 3.6** *The discriminant $\Delta$ as in Step 2 of Algorithm 3.5 is a quadratic residue in $\mathrm{GF}(p)$ if and only if either $F(c, X)$ is irreducible in $\mathrm{GF}(p^2)[X]$ or all roots in $\mathrm{GF}(p^2)$ of $F(c, X)$ have order dividing $p + 1$.*

**Proof.** According to Algorithm 3.1 the roots $h_0$, $h_1$, $h_2$ in $\mathrm{GF}(p^6)$ of $F(c, X) \in \mathrm{GF}(p^2)[X]$ can be written as $u + v + y$, $u\alpha + v\alpha^2 + y$, and $u\alpha^2 + v\alpha + y$ with $u$ and $v$ as in Algorithm 3.1, $y$ some element of $\mathrm{GF}(p^2)$, and $\alpha$ as in Section 4. Without loss of generality we have that $h_0 = u + v + y$, $h_1 = u\alpha + v\alpha^2 + y$, and $h_2 = u\alpha^2 + v\alpha + y$. Multiplying the three identities by 1, $\alpha^2$, and $\alpha$, respectively, we get

$$h_0 = u + v + y, \quad \alpha^2 h_1 = u + v\alpha + y\alpha^2, \quad \alpha h_2 = u + v\alpha^2 + y\alpha.$$

Adding these identities and using that $\alpha^2 + \alpha + 1 = 0$ we find that $u = U/3$ where $U = h_0 + \alpha^2 h_1 + \alpha h_2$.

According to Algorithm 3.1 we have that $U^3/27 = u^3 = r_1$ where $r_1 = (-f_0 + \sqrt{\Delta})/2$ and $f_0 = (-27 + 9c^{p+1} - 2c^3)/27$ (cf. Algorithm 3.5). Since $(-27 + 9c^{p+1})/27 \in \mathrm{GF}(p)$ we have that $\sqrt{\Delta} \in \mathrm{GF}(p)$ if and only if $U^3 - c^3 \in \mathrm{GF}(p)$. With $c_3 = Tr(g^3) = c^3 - 3c^{p+1} + 3$ (cf. Corollary 2.3.5.$i$ and $ii$ in [4]) and $c^{p+1} \in \mathrm{GF}(p)$ this is the case if and only if $U^3 - Tr(g^3) \in \mathrm{GF}(p)$. With $Tr(g^3) = h_0^3 + h_1^3 + h_2^3$ it follows from a straightforward computation that

$$U^3 - Tr(g^3) = 3(h_0^2 h_2 + h_1^2 h_0 + h_2^2 h_1 - 2)\alpha + 3(h_0^2 h_1 + h_1^2 h_2 + h_2^2 h_0 - 2)\alpha^2,$$

$$= 3(h_0/h_1 + h_1/h_2 + h_2/h_0 - 2)\alpha + 3(h_0/h_2 + h_1/h_0 + h_2/h_1 - 2)\alpha^2.$$

where the last identity follows from $h_0 h_1 h_2 = 1$. According to Lemma 2.3.2.$iv$ in [4] we have that $F(c, h_j^{-p}) = 0$ for $j = 0, 1, 2$. Thus either $h_j = h_j^{-p}$ for $j = 0, 1, 2$ (i.e., all roots have order dividing $p + 1$), or $h_0 = h_0^{-p}$, $h_1 = h_2^{-p}$, and $h_2 = h_1^{-p}$, or $h_j = h_{j+1 \bmod 3}^{-p}$ for $j = 0, 1, 2$. According to Lemma 2.3.2.$vi$ in [4], the last case is equivalent with $F(c, X)$ being irreducible in $\mathrm{GF}(p^2)[X]$. We prove that $U^3 - Tr(g^3) \in \mathrm{GF}(p)$ if and only if the first or the last case applies.

Let $w = h_0/h_1 + h_1/h_2 + h_2/h_0$ and $z = h_0/h_2 + h_1/h_0 + h_2/h_1$. If the first or the last case applies, then $w^p = z$ so that $(U^3 - Tr(g^3))^p = U^3 - Tr(g^3)$, and thus $U^3 - Tr(g^3) \in \mathrm{GF}(p)$. If the second case applies, then $w^p = w$ and $z^p = z$ so that $w, z \in \mathrm{GF}(p)$. Now, if additionally $U^3 - Tr(g^3) \in \mathrm{GF}(p)$ then $w = z$ so that the polynomial $X^3 - wX^2 + zX - 1 = X^3 - wX^2 + wX - 1$ has 1 as a root. As this polynomial has root-set $\{h_0/h_1, h_1/h_2, h_2/h_0\}$, it follows that $h_1 = h_2$, or one of $h_1, h_2$ is equal to $h_0$. As the order of $h_0$ divides $p + 1$ by assumption, it follows in each case that the same is true for $h_1$ and $h_2$. That is, the first case applies (and we are in the situation that both the first and second case applies).

**Theorem 3.7** *Finding the trace of a generator of the XTR group can be done in an expected number $\frac{q}{q-1}(7.2\log_2(p)+8\log_2((p^2-p+1)/q))$ plus a small constant number of multiplications in $GF(p)$.*

**Proof.** The correctness of Algorithm 3.5 follows from Corollary 3.4 and Lemma 3.6. Because $\Delta$ is a quadratic residue in $GF(p)$ if $F(c, X)$ is irreducible (cf. Appendix A) Step 3 of Algorithm 3.5 takes a $((p+1)/4)^{\text{th}}$ powering in $GF(p)$ (cf. $p \equiv 3 \bmod 4$). Assuming that a squaring in $GF(p)$ takes 80% of the time of a multiplication (cf. [2]), Step 3 of Algorithm 3.5 can be expected to require $1.3\log_2(p)$ multiplications in $GF(p)$. Step 4 of Algorithm 3.5 takes an expected $\log_2(p)$ squarings and $0.5\log_2(p)$ multiplications in $GF(p^2)$, for an expected total of $3.5\log_2(p)$ multiplications in $GF(p)$ (cf. Lemma 2.1.1 in [4]). Thus the total expected cost of Steps 3 and 4 of Algorithm 3.5 is $4.8\log_2(p)$ multiplications in $GF(p)$. According to Lemma 3.2.1 in [4] the probability that $F(c, X)$ is irreducible for a random $c \in GF(p^2)$ is about one third. Furthermore, it can be proved along the lines of the proof of the same lemma that for a random $c$ the $\Delta$ as in Step 2 of Algorithm 3.5 is a quadratic non-residue with probability $1/2$. The theorem now follows with Section 2 and the fact that the cost of the Jacobi sum test to test the quadratic residuosity of $\Delta$ is bounded by a small constant number of multiplications in $GF(p)$.

**Remark 3.8** It follows that a proper $Tr(g)$ can be found more than 50% faster than described in [4]. Theorem 3.7 is however just a side result of a more important consequence of Scipione del Ferro's method, namely the key size reduction method presented in Section 5. Before we can present that method we need some other results that also lead to yet another, even faster, way to find $Tr(g)$.

## 4   Improved Parameter Selection if $p \not\equiv 8 \bmod 9$

In this section we prove that if $p \not\equiv 8 \bmod 9$ (but $p \equiv 2 \bmod 3$), then an irreducible $F(c, X) \in GF(p^2)[X]$ can be written down directly. This follows from a general argument shown to us by H.W. Lenstra, Jr., that applies even to the characteristic zero case. We present a simplified description that applies just to non-zero characteristics.

So far we have considered $p \equiv 2 \bmod 3$, because this implies that the polynomial $(X^3 - 1)/(X - 1) = X^2 + X + 1 \in GF(p)[X]$ is irreducible over $GF(p)$ and $\{\alpha, \alpha^2\}$ with $\alpha^2 + \alpha + 1 = 0$ forms an optimal normal basis for $GF(p^2)$ over $GF(p)$. As shown in [4] this leads to a very efficient and convenient representation of $GF(p^2)$ in which $p^{\text{th}}$ powering is free. Here we restrict the choice of $p$ to $p \equiv 2 \bmod 9$ or $p \equiv 5 \bmod 9$, i.e., $p \equiv 2 \bmod 3$ but $p \not\equiv 8 \bmod 9$. For these $p$ the polynomial $(Z^9 - 1)/(Z^3 - 1) = Z^6 + Z^3 + 1 \in GF(p)[Z]$ is irreducible over $GF(p)$, as follows from the well known result that the $t^{\text{th}}$ cyclotomic polynomial $\phi_t(Z)$ is irreducible over $GF(p)$ if $GF(t)^*$ is cyclic and generated by $p \bmod t$. The multiplicative group $GF(t)^*$ is cyclic if and only if either $t = 2, 4$, or $t$ is a power of an odd prime, or $t$ is twice a power of an odd prime, or $t$ is four times the power of an odd prime that is $2 \bmod 3$. Applying this to $t = 9$ and

$p \equiv 2, 5 \bmod 9$ it follows that $\phi_9(Z) = Z^6 + Z^3 + 1 \in \mathrm{GF}(p)[Z]$ is irreducible over $\mathrm{GF}(p)$.

Let $\zeta$ denote a zero of $Z^6 + Z^3 + 1$. This $\zeta$ enables us to conveniently represent elements of $\mathrm{GF}(p^6)$, either using a basis over $\mathrm{GF}(p)$ or using a basis over $\mathrm{GF}(p^2)$. For the purposes of the present section we use a basis over $\mathrm{GF}(p)$ and write elements of $\mathrm{GF}(p^6)$ as $\sum_{i=0}^{5} a_i \zeta^i$ for $a_i \in \mathrm{GF}(p)$. In this representation elements of the subfield $\mathrm{GF}(p^2)$ of $\mathrm{GF}(p^6)$ correspond to elements of the form $a_3 \zeta^3 + a_0$; this follows from $3p^2 \equiv 3 \bmod 9$ and a counting argument. The element $\sum_{i=0}^{5} a_i \zeta^i$ can be written as $(a_5 \zeta^6 + a_2 \zeta^3)\zeta^{-1} + (a_4 \zeta^6 + a_1 \zeta^3)\zeta^{-2} + (a_3 \zeta^6 + a_0 \zeta^3)\zeta^{-3}$. Since $\zeta^3 = \alpha$ with $\alpha$ as above this implies that $\{\zeta^{-1}, \zeta^{-2}, \zeta^{-3}\}$ forms a basis for $\mathrm{GF}(p^6)$ over $\mathrm{GF}(p^2)$, using the representation of $\mathrm{GF}(p^2)$ as used in [4]. Obviously, the latter basis is equivalent to the basis $\{\zeta^2, \zeta, 1\}$ which we found convenient for implementation purposes. This basis simply leads to squaring and multiplication in $\mathrm{GF}(p^6)$ at the cost of 12 and 18 multiplications in $\mathrm{GF}(p)$, respectively. Note that one can move back and forth between the representations of $\mathrm{GF}(p^6)$ at the cost of a small constant number of additions in $\mathrm{GF}(p)$.

None of the above bases is optimal normal. For the calculations in this section that is not a problem, since they had to be carried out just once. For practical applications of XTR it is not a disadvantage either, because in the key recovery application (cf. Section 5) at most three multiplications in $\mathrm{GF}(p^6)$ have to be carried out per XTR key recovery. Note that if $p \bmod 7$ generates $\mathrm{GF}(7)^*$ the polynomial $(X^7 - 1)/(X - 1)$ is irreducible over $\mathrm{GF}(p)$ and leads to an optimal normal basis for $\mathrm{GF}(p^6)$ over $\mathrm{GF}(p)$ (cf. [3]). We chose not to use this representation because it imposes an additional restriction on $p$ without leading to significant advantages.

**Lemma 4.1** *The trace over* $\mathrm{GF}(p^2)$ *of* $\sum_{i=0}^{5} a_i \zeta^i \in \mathrm{GF}(p^6)$ *equals* $3(a_3 \zeta^3 + a_0) = 3(a_3\alpha + a_0) = -3a_0\alpha^2 + 3(a_3 - a_0)\alpha \in \mathrm{GF}(p^2)$.

**Proof.** Because the trace is $\mathrm{GF}(p^2)$-linear it suffices to show that the trace of $\zeta^i$ is zero for $i = 1, 2, 4, 5$ and $3\zeta^i$ for $i = 0, 3$. This follows trivially from $\zeta^9 = 1$, $\zeta^6 + \zeta^3 + 1 = 0$, and the fact that the trace of $\zeta^i$ equals $\zeta^i + \zeta^{ip^2} + \zeta^{ip^4}$.

**Lemma 4.2** *For* $x \in \mathrm{GF}(p^6)$ *the trace over* $\mathrm{GF}(p^2)$ *of* $x^p$ *equals the* $p^{\mathrm{th}}$ *power of the trace of* $x$ *over* $\mathrm{GF}(p^2)$.

**Proof.** The trace over $\mathrm{GF}(p^2)$ of $x^p$ equals $x^p + x^{p^3} + x^{p^5}$ which is the $p^{\mathrm{th}}$ power of the trace $x + x^{p^2} + x^{p^4}$ of $x$ over $\mathrm{GF}(p^2)$.

A particularly convenient property of our representation of $\mathrm{GF}(p^6)$ is that it enables us to do several calculations without using the specific value of $p$. The following result is an example.

**Proposition 4.3** *Let* $a \in \mathrm{GF}(p)$, *let* $\zeta$ *and* $\alpha = \zeta^3$ *be as above, and let* $Q = (p^6 - 1)/(p^2 - p + 1)$. *Then the trace over* $\mathrm{GF}(p^2)$ *of the element* $(\zeta + a)^Q$ *of* $\mathrm{GF}(p^6)$ *of order dividing* $p^2 - p + 1$ *equals*

$$\frac{-3}{a^6 - a^3 + 1}((a^2 - 1)^3 \alpha + a^3(a^3 - 3a + 1)\alpha^2)$$

if $p \equiv 2 \bmod 9$ and the $p^{\mathrm{th}}$ power thereof if $p \equiv 5 \bmod 9$, where $a^6 - a^3 + 1 \neq 0$.

**Proof.** If $a^6 - a^3 + 1 = 0$, then $b = a^3$ is a zero in $\mathrm{GF}(p)$ of the sixth cyclotomic polynomial $X^2 - X + 1$. It follows that $b^6 = 1$. With $b^{p-1} = 1$ and $\gcd(p-1, 6) = 2$ we find that $b^2 = 1$ so that $b = \pm 1$. But neither $+1$ nor $-1$ is a zero of $X^2 - X + 1$, and we conclude that $a^6 - a^3 + 1 \neq 0$.

From $Q = (p^6 - 1)/(p^2 - p + 1) = p^4 + p^3 - p - 1$ it follows that

$$(\zeta + a)^Q = \frac{(\zeta + a)^{p^4}(\zeta + a)^{p^3}}{(\zeta + a)^p(\zeta + a)} = \frac{(\zeta^{p^4} + a)(\zeta^{p^3} + a)}{(\zeta^p + a)(\zeta + a)}.$$

With $\zeta^9 = 1$ this reduces to

$$\frac{(\zeta^7 + a)(\zeta^8 + a)}{(\zeta^2 + a)(\zeta + a)}$$

if $p \equiv 2 \bmod 9$ and to

$$\frac{(\zeta^4 + a)(\zeta^8 + a)}{(\zeta^5 + a)(\zeta + a)}$$

if $p \equiv 5 \bmod 9$. If $p \equiv 5 \bmod 9$ the $p^{\mathrm{th}}$ power of the former expression equals the latter, so that if $p \equiv 5 \bmod 9$ the trace of $(\zeta + a)^Q$ equals the $p^{\mathrm{th}}$ power of the trace of $(\zeta + a)^Q$ when $p \equiv 2 \bmod 9$ (cf. Lemma 4.2). For the computation of the trace of $(\zeta + a)^Q$ when $p \equiv 2 \bmod 9$ one easily verifies that

$$\frac{a^6 - a^3 + 1}{\zeta + a} = (a^3 - \zeta^3 - 1)(\zeta^2 - a\zeta + a^2)$$

and that

$$\frac{a^6 - a^3 + 1}{\zeta^2 + a} = -a\zeta^5 + (a^3 - 1)\zeta^4 + a^2\zeta^3 - a^4\zeta^2 - \zeta + a^5.$$

With $\zeta^6 + \zeta^3 + 1 = 0$ the trace of

$$\frac{(\zeta^7 + a)(\zeta^8 + a)}{(\zeta^2 + a)(\zeta + a)}$$

then follows from a straightforward computation and Lemma 4.1.

**Corollary 4.4** If $a \neq 0, \pm 1$ then

$$\frac{-3}{a^6 - a^3 + 1}((a^2 - 1)^3 \alpha + a^3(a^3 - 3a + 1)\alpha^2) \in \mathrm{GF}(p^2)$$

is the trace over $\mathrm{GF}(p^2)$ of an element of $\mathrm{GF}(p^6)$ of order dividing $p^2 - p + 1$ and $> 3$.

**Proof.** If $p \equiv 2 \bmod 9$ it follows from Proposition 4.3 that there is an $x \in$ $\mathrm{GF}(p^6)^*$ of order dividing $p^2 - p + 1$ with the required trace over $\mathrm{GF}(p^2)$. If $p \equiv 5 \bmod 9$ it follows in the same way, after taking conjugates over $\mathrm{GF}(p)$ and using Lemma 4.2. If the order of $x$ is at most 3, i.e., 1 or 3, then $x$ is either equal to 1, $\alpha$, or $\alpha^2$, since $p \equiv 2 \bmod 3$. Thus, the trace of $x$ is equal to 3, $3\alpha$, or $3\alpha^2$. For the first possibility, $x = 1$, a trace value of 3 leads to two simultaneous polynomial equations $(a^2 - 1)^3 - (a^6 - a^3 + 1) = 0$ and $a^3(a^3 - 3a + 1) - (a^6 - a^3 + 1) = 0$; since these polynomials are relatively prime, $x$ cannot be equal to 1. For the other two possibilities, $x = \alpha$ or $x = \alpha^2$, the corresponding trace values lead to $a = 0$ or $a = \pm 1$, respectively, which are excluded by assumption.

It follows from Corollary 4.4 with $a = 2$ and $a = 1/2$ that $(-27\alpha - 24\alpha^2)/19$ and $(27\alpha + 3\alpha^2)/19$, respectively, are trace values of elements of $\mathrm{GF}(p^6)^*$ of order dividing $p^2 - p + 1$ and $> 3$. This leads to the following algorithm to find $Tr(g)$.

**Algorithm 4.5 (Computation of $Tr(g)$)**
1. Let $c = (27\alpha + 3\alpha^2)/19 \in \mathrm{GF}(p^2)$ and compute $c_{(p^2-p+1)/q}$ (cf. Section 2).
2. If $c_{(p^2-p+1)/q} \neq 3$, then let $Tr(g) = c_{(p^2-p+1)/q}$ and return success.
3. Otherwise, if $c_{(p^2-p+1)/q} = 3$, then replace $c$ by $(-27\alpha - 24\alpha^2)/19 \in \mathrm{GF}(p^2)$ and recompute $c_{(p^2-p+1)/q}$.
4. If $c_{(p^2-p+1)/q} \neq 3$, then let $Tr(g) = c_{(p^2-p+1)/q}$ and return success.
5. Otherwise, if $c_{(p^2-p+1)/q} = 3$, then return failure.

The probability of failure of Algorithm 4.5 may be expected to be $q^{-2}$, i.e., negligibly small. If this is a matter of concern, Algorithm 4.5 can trivially be extended and include more 'hard-wired' choices for $c$ (corresponding to $a \neq 0, \pm 1, 2, 1/2$). In the very unlikely event that Algorithm 4.5 fails, which so far has not happened in our test implementation, a different $q$ and $p$ can be selected. On average one may expect that Algorithm 4.5 finds the trace of a generator of the XTR group in about $8 \log_2((p^2 - p + 1)/q)$ plus a small constant number of multiplications in $\mathrm{GF}(p)$. This is almost twice as fast as the method based on Algorithm 3.5 (cf. Theorem 3.7), but Algorithm 4.5 applies only to the case $p \not\equiv 8 \bmod 9$.

## 5    Key Size Reduction

In this section we show that $Tr(g^{k+1})$ and $Tr(g^{k-1})$ can be derived from $Tr(g)$ and $Tr(g^k)$, assuming the (unknown) private key $k$ is properly chosen. Throughout this section let $c = Tr(g)$ and $c_n = Tr(g^n)$ for $n \in \mathbf{Z}$. We first show that $c_{k-1}$ (or $c_{k+1}$) follows directly from $c$, $c_k$ and $c_{k+1}$ (or $c_{k-1}$) using surprisingly simple formulas.

**Theorem 5.1**
1. If $k \neq p, 1 - p \bmod q$ then $c^p c_{k-1} - c c_k \neq 0$ and

$$c_{k+1} = \frac{c_k^p(c^2 - 3c^p) - c_{k-1}^p(c^{2p} - 3c) - c_{k-1}^2 c + c_k^2(c^p - c^2) + c_k c_{k-1} c^{p+1}}{c^p c_{k-1} - c c_k}.$$

2. *If $k \neq -p, p-1 \bmod q$ then $cc_{k+1} - c^p c_k \neq 0$ and*

$$c_{k-1} = \frac{c_k^p(c^{2p} - 3c) - c_{k+1}^p(c^2 - 3c^p) - c_{k+1}^2 c^p + c_k^2(c - c^{2p}) + c_k c_{k+1} c^{p+1}}{cc_{k+1} - c^p c_k}.$$

**Proof.** From Corollary $2.3.5.ii$ in [4] it follows that $c^p c_{k-1} - cc_k = Tr(g^{k-2}) - Tr(g^{k+1})$. Thus $c^p c_{k-1} - cc_k$ can only be zero if $Tr(g^{k-2}) = Tr(g^{k+1})$, which implies that $g^{k-2}$ and $g^{k+1}$ are conjugates. Thus, either $k-2 \equiv k+1 \bmod (p^2 - p+1)$, or $k-2 \equiv p^2(k+1) \bmod (p^2-p+1)$, $k-2 \equiv p^4(k+1) \bmod (p^2-p+1)$. The first equation has no solution, the second one leads to $k \equiv p \bmod (p^2 - p+1)$, and the third one to $k \equiv 1 - p \bmod (p^2 - p+1)$. Since $k \neq p, 1-p \bmod q$ and $q$ divides $p^2 - p + 1$ we find that $c^p c_{k-1} - cc_k$ is non-zero.

The polynomial $F(c, X)$ is the characteristic polynomial of the matrix $A = \begin{pmatrix} 0 & 0 & 1 \\ 1 & 0 & -c^p \\ 0 & 1 & c \end{pmatrix}$ (cf. Definition 2.4.1 in [4]). That is, the roots $g$, $g^{p-1}$, and $g^{-p}$ of $F(c, X)$ are the eigenvalues of $A$. Thus $g^k$, $g^{k(p-1)}$, and $g^{-kp}$ are the eigenvalues of the matrix $A^k$, so that the polynomial $F(c_k, X)$ with roots $g^k$, $g^{k(p-1)}$, and $g^{-kp}$ is the characteristic polynomial of $A^k$. From Lemma 2.4.6 in [4] we have that

$$A^k = \begin{pmatrix} Tr(g^2)^p & c^p & 3 \\ c^p & 3 & c \\ 3 & c & Tr(g^2) \end{pmatrix}^{-1} \begin{pmatrix} Tr(g^{k-2}) & c_{k-1} & c_k \\ c_{k-1} & c_k & c_{k+1} \\ c_k & c_{k+1} & Tr(g^{k+2}) \end{pmatrix}.$$

Computing the characteristic polynomial of $A^k$ using this expression, combined with the fact that $Tr(g^{k-2}) = c_{k+1} - cc_k + c^p c_{k-1}$, $Tr(g^{k+2}) = cc_{k+1} - c^p c_k + c_{k-1}$ and $Tr(g^2) = c^2 - 2c^p$ (cf. Corollary $2.3.5.ii$ and $i$ in [4]), one obtains a polynomial $\lambda^3 - c_k\lambda^2 + f_1\lambda + f_0$ with

$$Df_1 = (c^{2p} - 3c)c_{k+1}^2 + (3c^p c_k - 9c_{k-1} + 2c^2 c_k + c^{p+1}c_{k-1} - c^{2p+1}c_k)c_{k+1}$$

$$-3c^p c_{k-1}^2 + 9c_k^2 + c^{3p}c_k^2 + c^3 c_k^2 + c^2 c_{k-1}^2 + 3cc_k c_{k-1} - c^{p+2}c_k c_{k-1} + 2c^{2p}c_k c_{k-1} - 7c^{p+1}c_k^2.$$

Here $D = c^{2p+2} + 18c^{p+1} - 4(c^{3p} + c^3) - 27 \in GF(p)$ as in Lemma 2.4.4 of [4] and $D \neq 0$ (cf. Lemma 2.4.5 in [4]). Since also $f_1 = c_k^p$ we find that

$$c_{k+1}^2 = (c^{2p} - 3c)^{-1}((-3c^p c_k + 9c_{k-1} - 2c^2 c_k - c^{p+1}c_{k-1} + c^{2p+1}c_k)c_{k+1} - Dc_k^p$$

$$+3c^p c_{k-1}^2 - 9c_k^2 - c^{3p}c_k^2 - c^3 c_k^2 - c^2 c_{k-1}^2 - 3cc_k c_{k-1} + c^{p+2}c_k c_{k-1} - 2c^{2p}c_k c_{k-1} + 7c^{p+1}c_k^2).$$

Note that $c^{2p} - 3c = c^p Tr(g^{-1}) - cTr(1)$, which is non-zero based on the same argument why $c^p c_{k-1} - cc_k$ is non-zero.

Repeating the same argument for the matrix $A^{k-1}$ and its characteristic polynomial $F(c_{k-1}, X)$ (and using Corollary $2.3.5.ii$ of [4] to express $Tr(g^{k-3})$ in terms of $c$, $c_k$, $c_{k+1}$, and $c_{k-1}$) we obtain another expression for $c_{k+1}^2$:

$$c_{k+1}^2 = (c^2 - 3c^p)^{-1}(2c^3 c_k - 3cc_{k-1} - c^{p+2}c_{k-1} + 9c_k + 4c^{2p}c_{k-1} - 7c^{p+1}c_k)c_{k+1}$$

$$- Dc_{k-1}^p - c^{2p}c_k^2 - c^4 c_k^2 + 4c^{p+1}c_{k-1}^2 + 6c^p c_k c_{k-1} - 6cc_k^2 + 4c^{p+2}c_k^2$$

$$- c^3 c_{k-1}^2 + c^2 c_k c_{k-1} - 4c^{2p+1}c_k c_{k-1} - 9c_{k-1}^2 + c^{p+3}c_k c_{k-1}).$$

Here $c^2 - 3c^p$ is non-zero because its conjugate $c^{2p} - 3c$ over $GF(p)$ is non-zero. Subtraction of the two expressions for $c_{k+1}^2$ followed by multiplication by $c^{2p} - 3c$ and $c^2 - 3c^p$ and division by $D$, leads to the formula for $c_{k+1}$.

For a proof of the second formula, we apply the first one replacing $c_k$, $c_{k+1}$ and $c_{k-1}$ by $d_{-k} = Tr(g^{-k})$, $d_{-k+1} = Tr(g^{-k+1})$, and $d_{-k-1} = Tr(g^{-k-1})$, respectively. The proof then follows by observing that $c_{k-1}^p = Tr(g^{-k+1}) = d_{-k+1}$, $c_k^p = Tr(g^{-k}) = d_{-k}$ and $c_{k+1}^p = Tr(g^{-k-1}) = d_{-k-1}$ (since $c_n^p = c_{-n}$, cf. Section 2) and by taking the conjugate over $GF(p)$.

Because $p^{\text{th}}$ powering is free in $GF(p^2)$, computation of the formulas in Theorem 5.1 takes only a small constant number of operations in $GF(p)$, where the following algorithm can be used for the division.

**Algorithm 5.2 (Inversion in $GF(p^2)$)** Let $x = x_1\alpha + x_2\alpha^2 \in GF(p^2)$. Compute $t = (x_1 x_2 + (x_1 - x_2)^2)^{-1} \in GF(p)$, then $1/x = t(x_2\alpha + x_1\alpha^2) \in GF(p^2)$.

Theorem 5.1 shows that including both $c_{k-1}$ and $c_{k+1}$ in the XTR public key is never necessary, and that $c_{k+1}$ (or $c_{k-1}$) suffices (assuming of course that $c$ and $c_k$ are part of the public key). Actually, even $c_{k+1}$ (or $c_{k-1}$) does in principle not have to be included, because the recipient can determine it by finding the roots of $F(c, X)$ and $F(c_k, X)$, leading to 3 possible representations $c_{k+1}$ ($c_{k-1}$). Thus, two bits in the public key would suffice to indicate which of the three representations is the correct one, but this would come at the cost of a considerable computation for the recipient of the key.

We now show that if $p \not\equiv 8 \bmod 9$ then the results from Sections 3 and 4 can be used to formulate a fast method to compute $c_{k+1}$ given $c$ and $c_k$ (where, of course, $k$ is unknown) that does not require any additional bits in the public key. The method to compute $c_{k-1}$ given $c$ and $c_k$ is very similar and follows easily from the method for $c_{k+1}$. Roughly speaking the method works as suggested above, namely by computing explicit representations of $g$ and $g^k$ in $GF(p^6) = GF(p)[X]/(X^6 + X^3 + 1)$ (cf. Section 4) based on their representations $c$ and $c_k$, respectively, so that the value of $c_{k+1}$ follows as the trace over $GF(p^2)$ of $g * g^k \in GF(p^6)$.

More precisely, the owner of the private key $k$ computes $c_k = Tr(g^k)$ given $c = Tr(g)$ and $k$. The same $c_k$ is obtained for $kp^2 \bmod q$ and $kp^4 \bmod q$ since $g^k$, $g^{kp^2}$, and $g^{kp^4}$ are conjugates over $GF(p^2)$ and thus have the same trace over $GF(p^2)$, namely $c_k$. As a side result of the computation of $c_k$, the owner of the private key obtains $c_{k+1} = Tr(g^{k+1})$ (cf. Section 2). However, the value $c_{k+1}$ thus obtained is in general not the same as the value that would be obtained for $kp^2 \bmod q$ or $kp^4 \bmod q$, because $Tr(g^{k+1})$, $Tr(g^{kp^2+1})$, and $Tr(g^{kp^4+1})$ are not the same unless $k = 0 \bmod q$, despite the fact that $Tr(g^k)$, $Tr(g^{kp^2})$, and

$Tr(g^{kp^4})$ are the same. This is because $g^{k+1}$, $g^{kp^2+1}$, and $g^{kp^4+1}$ are not conjugates over $\mathrm{GF}(p^2)$ unless $k = 0 \bmod q$, despite the fact that $g^k$, $g^{kp^2}$, and $g^{kp^4}$ are conjugates over $\mathrm{GF}(p^2)$. It follows that for any pair $(c, c_k)$ there are three possible different values for $c_{k+1}$: one that corresponds to the proper secret value $k$, and two that correspond to the 'wrong' values $kp^2 \bmod q$ and $kp^4 \bmod q$.

Any method to recover $c_{k+1}$ from $(c, c_k)$ will have to resolve this ambiguity. To do this without requiring additional bits in the public key we do the following. The owner of the private key computes not only $Tr(g^{k+1})$, but $Tr(g^{kp^2+1})$ and $Tr(g^{kp^4+1})$ as well. Next he selects the secret key $k$ as $k$, $kp^2 \bmod q$, or $kp^4 \bmod q$ depending on which of the three values $Tr(g^{k+1})$, $Tr(g^{kp^2+1})$, $Tr(g^{kp^4+1})$ is the 'smallest' (or 'largest')[1]. It follows that $c_{k+1}$ is the 'smallest' possibility given the pair $(c, c_k)$. Obviously this way of changing an initially selected private key value $k$ does not have a negative impact on the security.

How this method enables the recipient of the pair $(c, c_k)$ to compute the proper $c_{k+1}$ without knowing $k$ is described in Algorithm 5.6 below. We first describe how the owner of the private key computes $Tr(g^{k+1})$, $Tr(g^{kp^2+1})$, and $Tr(g^{kp^4+1})$. A conceptually straightforward method would be for the owner of the private key to compute $c_m$ three times, once for $m = k$ itself, once for $m = kp^2 \bmod q$, and once for $m = kp^4 \bmod q$, and to pick the $k$ corresponding to the smallest $c_{m+1}$ (the three $c_m$'s are the same, as noted above). A more complicated but faster method is as follows. Suppose that $(c_{k-1}, c_k, c_{k+1})$ and $(c_{-p-1}, c_{-p}, c_{-p+1})$ have been computed, at the cost of $16 \log_2(q)$ multiplications in $\mathrm{GF}(p)$ (cf. Section 2). The values $c_{k\pm2}$ can then easily be obtained and $c_2 = c^2 - 2c^p$ (cf. [4]). To compute $Tr(g^{kp^2+1})$ we observe that $Tr(g^{kp^2+1}) = Tr(g^{kp^2-p^3}) = Tr(g^{(k-p)p^2}) = Tr(g^{k-p})$. We then use Lemmas 2.4.2 and 2.4.5 from [4] and find that

$$\begin{pmatrix} Tr(g^{k-p-1}) \\ Tr(g^{k-p}) \\ Tr(g^{k-p+1}) \end{pmatrix}^T = \begin{pmatrix} c_{-p-1} \\ c_{-p} \\ c_{-p+1} \end{pmatrix}^T \begin{pmatrix} c_2^p & c^p & 3 \\ c^p & 3 & c \\ 3 & c & c_2 \end{pmatrix}^{-1} \begin{pmatrix} c_{k-2} & c_{k-1} & c_k \\ c_{k-1} & c_k & c_{k+1} \\ c_k & c_{k+1} & c_{k+2} \end{pmatrix},$$

so that $Tr(g^{kp^2+1})$ follows after a small constant number of multiplications in $\mathrm{GF}(p)$. A similar matrix identity involving $(c_{p-1}, c_p, c_{p+1})$ (obtained using $c_{-n} = c_n^p$, cf. Section 2) is used to compute $Tr(g^{kp^2-1}) = Tr(g^{k+p})$. Given $(Tr(g^{kp^2-1}), Tr(g^{kp^2}), Tr(g^{kp^2+1}))$ (with $Tr(g^{kp^2}) = c_k$) and $(c_{-p-1}, c_{-p}, c_{-p+1})$, the same method is then used to compute $Tr(g^{kp^4+1})$.

The corresponding method to compute the 'smallest' $c_{k+1}$ given just $(c, c_k)$ but without knowing the secret $k$ relies on Algorithm 3.1, Scipione del Ferro's method. We need two auxiliary algorithms, the correctness of which follows by inspection (cf. [Lemma 2.1.1] in [4]).

---

[1] For $x \in \mathrm{GF}(p)$ let $\pi_0(x) \in \{0, 1, \ldots, p-1\}$ be the image of $x$ under the 'natural' bijection between $\mathrm{GF}(p)$ and $\{0, 1, \ldots, p-1\}$. For $x = x_1\alpha + x_2\alpha^2 \in \mathrm{GF}(p^2)$, using the representation of elements of $\mathrm{GF}(p^2)$ from [4], let $\pi(x) = \pi_0(x_1) + p * \pi_0(x_2)$. We use the ordering on $\mathrm{GF}(p^2)$ induced by $\pi$.

**Algorithm 5.3 (Exponentiation in $\mathrm{GF}(p^2)$)** Let $x \in \mathrm{GF}(p^2)$ and let $e$ be an integer. To compute $x^e \in \mathrm{GF}(p^2)$ do the following.

1. Compute $e_0, e_1 \in \{0, 1, \ldots, p-1\}$ such that $e_0 + e_1 p = e \bmod (p^2 - 1)$ and let $e_i = \sum_j e_{ij} 2^j$, with $e_{ij} \in \{0, 1\}$ for $i = 0, 1$ and $j \geq 0$, be the binary representations of $e_0$ and $e_1$.
2. Let $n$ be the largest index such that $e_{in} \neq 0$ for $i = 0$ or $1$.
3. Compute $x' = x * x^p \in \mathrm{GF}(p)$.
4. Let $y = 1$ in $\mathrm{GF}(p^2)$. For $j = n, n-1, \ldots, 0$ in succession do the following:
   - if $e_{0j} = 1$ and $e_{1j} = 1$, then replace $y$ by $y * x'$;
   - if $e_{0j} = 1$ and $e_{1j} = 0$, then replace $y$ by $y * x$;
   - if $e_{0j} = 0$ and $e_{1j} = 1$, then replace $y$ by $y * x^p$;
   - if $j > 0$, then replace $y$ by $y^2$.
5. Return $y = x^e \in \mathrm{GF}(p^2)$.

**Lemma 5.4** *The expected cost of Algorithm 5.3 is $4\log_2(p)$ multiplications in* $\mathrm{GF}(p)$.

**Algorithm 5.5 (Cube root in $\mathrm{GF}(p^2)$ if $p \not\equiv 8 \bmod 9$)** To compute a cube root in $\mathrm{GF}(p^6)$ of $r \in \mathrm{GF}(p^2)$ perform the following steps.

1. Use Algorithm 5.3 to compute $t = r^{(8p^2-5)/9} \in \mathrm{GF}(p^2)$ if $p \equiv 2 \bmod 9$ or $t = r^{(p^2+2)/9} \in \mathrm{GF}(p^2)$ if $p \equiv 5 \bmod 9$.
2. Compute $s = t^3 \in \mathrm{GF}(p^2)$ and determine $j = 0, 1$ or $2$ such that $\alpha^j s = r$.
3. Return a cube root $\zeta^j t \in \mathrm{GF}(p^6)$ of $r$ (the result is in $\mathrm{GF}(p^2)$ if $j = 0$).

**Algorithm 5.6 (Key recovery)** To compute the 'smallest' $c_{k+1}$ corresponding to $(c, c_k)$, perform the following steps.

1. Use Algorithm 3.1 to compute a root $g \in \mathrm{GF}(p^6) = \mathrm{GF}(p)[X]/(X^6 + X^3 + 1)$ of the polynomial $F(c, X)$, using Algorithm 5.5 to compute a cube root in Step 3. Note that Algorithm 5.2 can be used for the division by $u$ in Step 3, since $u$ is a $\mathrm{GF}(p^2)$-multiple of a power of $\zeta$.
2. Use Algorithm 3.1 to compute the three roots $y_1, y_2, y_3 \in \mathrm{GF}(p^6)$ of $F(c_k, X)$, with $w = \alpha$ in Step 4.
3. For $i = 1, 2, 3$ compute the trace $t_i$ over $\mathrm{GF}(p^2)$ of $g y_i \in \mathrm{GF}(p^6)$ (cf. Lemma 4.1).
4. Let $c_{k+1}$ be the 'smallest' of $t_1$, $t_2$, and $t_3$.

**Theorem 5.7** *Algorithm 5.6 can be expected to require $10.6\log_2(p)$ multiplications in* $\mathrm{GF}(p)$.

**Proof.** The square-root computation in Step 2 of Algorithm 3.1 can be expected to require $1.3\log_2(p)$ multiplications in $\mathrm{GF}(p)$ (cf. Proof of Theorem 3.7). The application of Algorithm 5.5 in Step 3 of Algorithm 3.1 requires a call to Algorithm 5.3, at an expected cost of $4\log_2(p)$ multiplications in $\mathrm{GF}(p)$ (cf. Lemma

5.4). Thus, a single call to Algorithm 3.1 can be expected to require $5.3\log_2(p)$ multiplications in $GF(p)$, from which the proof follows.

We conclude that $Tr(g^{k-1})$ and $Tr(g^{k+1})$ do not have to be included in the XTR public key data $(p, q, Tr(g), Tr(g^k))$ for digital signature or authentication applications, if

1. the owner of the private key has selected its private exponent $k$ in the proper fashion as explained above, and if
2. the recipient of the public key is willing and able to perform Algorithm 5.6 to compute $Tr(g^{k+1})$ followed by an application of Theorem 5.1 to compute $Tr(g^{k-1})$.

To summarize, there are three options for XTR public keys used for digital signatures or authentication, namely to include one, two, or all three of the values $Tr(g^{k-1})$, $Tr(g^k)$, $Tr(g^{k+1})$. In some applications, e.g. issuance of a certificate by a Certificate Authority, it may be required that the relative correctness of these components can be verified by a third party. A method to do this will be published at a later date (cf. [5]).

**Acknowledgment.** The method from Section 4 is based on a more general argument from H.W. Lenstra, Jr. We gratefully acknowledge his assistance.

# References

1. A.E. Brouwer, R. Pellikaan, E.R. Verheul, *Doing more with fewer bits*, Proceedings Asiacrypt99, LNCS 1716, Springer-Verlag 1999, 321-332.
2. H. Cohen, A. Miyaji, T. Ono, *Efficient elliptic curve exponentiation using mixed coordinates*, Proceedings Asiacrypt'98, LNCS 1514, Springer-Verlag 1998, 51-65.
3. A.K. Lenstra, *Using cyclotomic polynomials to construct efficient discrete logarithm cryptosystems over finite fields*, Proceedings ACISP97, LNCS 1270, Springer-Verlag 1997, 127-138.
4. A.K. Lenstra, E.R. Verheul, *The XTR public key system*, Proceedings Crypto 2000, LNCS 1880, Springer-Verlag 2000, 1-19; available from www.ecstr.com.
5. A.K. Lenstra, E.R. Verheul, *Fast irreducibility testing for XTR*, in preparation.
6. W.K. Nicholson, *Introduction to abstract algebra*, PWS-Kent Publishing Company, Boston, 1993.

# Security of Cryptosystems Based on Class Groups of Imaginary Quadratic Orders

Safuat Hamdy* and Bodo Möller

TU Darmstadt, Fachbereich Informatik
{hamdy,moeller}@cdc.informatik.tu-darmstadt.de

**Abstract.** In this work we investigate the difficulty of the discrete logarithm problem in class groups of imaginary quadratic orders. In particular, we discuss several strategies to compute discrete logarithms in those class groups. Based on heuristic reasoning, we give advice for selecting the cryptographic parameter, i.e. the discriminant, such that cryptosystems based on class groups of imaginary quadratic orders would offer a similar security as commonly used cryptosystems.

## 1  Introduction

Cryptosystems based on class groups of imaginary quadratic orders (IQC) have been first proposed by Buchmann and Williams [3,4] in 1988 and 1990. Since then, there was no clear advice on how to select the cryptographic parameter, i.e. the discriminant of the quadratic order. The goal of this work is to close this gap. In particular, we demonstrate how large $\Delta$ must be selected such that computing logarithms in $Cl(\Delta)$ is as hard as factoring an integer $n$ of given size. We consider several strategies for computing discrete logarithms in class groups, such as reductions to other computational problems, index-calculus algorithms, Pollard's $\lambda$ algorithm, and the Pohlig-Hellman algorithm in connection with an algorithm similar to the $(p-1)$-factoring method. We obtain the result that, in order to get the same security with IQC as with RSA with 1024 bit moduli, the discriminant should have at least 687 bits.

The security of IQC is based on the apparent difficulty of computing discrete logarithms in class groups of imaginary quadratic orders (Cl-DLP). The Cl-DLP can be extended to class groups of orders of number fields with arbitrarily high degree, and in furthermore, there is a generalization of the discrete logarithm problem [2]. However, in this work we shall focus only on imaginary quadratic fields, and whenever the term "class groups" appears in the sequel, we actually mean class groups of imaginary quadratic orders.

It is well known that solving the Cl-DLP is at least as hard as solving the integer factorization problem (IFP); we shall describe the reduction later in this work. However, it is still unknown whether the Cl-DLP is really harder than the IFP. The Cl-DLP can be solved with a subexponential index-calculus algorithm

---

* Supported by the Deutsche Forschungsgemeinschaft (DFG)

T. Okamoto (Ed.): ASIACRYPT 2000, LNCS 1976, pp. 234–247, 2000.

due to Hafner and McCurley [11]. This algorithm was improved by Düllmann [9]. Recently, in [30] it has been rigorously proven under the Generalized Riemann Hypothesis that for solving the Cl-DLP using index-calculus algorithms one can expect a running time proportional to $L_{|\Delta|}\left[\frac{1}{2}, \frac{3}{4}\sqrt{2} + o(1)\right]$ where $\Delta$ is the discriminant of the imaginary quadratic order. Moreover, Jacobson [16] has applied the ideas of the MPQS to class group computations. In fact, the machinery behind his algorithm is the same as that of the original MPQS, and although this algorithm has not been analyzed, empirical data suggest a running time proportional to $L_{|\Delta|}\left[\frac{1}{2}, 1 + o(1)\right]$.

The best known algorithm to solve the IFP is the GNFS with asymptotic expected running time proportional to $L_n\left[\frac{1}{3}, \sqrt[3]{\frac{64}{9}}\right]$ where $n$ is the number to be factored; the best known algorithm to solve the GF-DLP (DLP in multiplicative groups of Galois fields) is a variant of the GNFS with a similar asymptotic expected running time where $n$ is the order of the group. Thus, currently the IFP or the GF-DLP can be solved asymptotically faster than the Cl-DLP. This means that the Cl-DLP is apparently harder than the IFP or the GF-DLP.

Hence class groups form another potential alternative to finite fields for DL-based cryptographic protocols. Unfortunately, popular signature protocols such as DSA can't be used with class groups in a direct way, because DSA requires the knowledge of the group order. Computing the order of an arbitrary class group appears to be as hard as computing discrete logarithms in class groups because no efficient algorithm is known that computes the class number. In [22] a variant of the Schnorr signature scheme that doesn't require knowledge of the group order has been proposed.

Computing roots without knowing the class number also appears to be intractable. This makes the Guillou-Quisquater signature protocol [10] suitable for class groups, since in this protocol even the signer does not need to know the class number. Moreover, in [1] a variant of DSA was presented that is based on the intractability of computing roots in finite abelian groups.

This paper is organized as follows: In Section 2 we recall the background we need, and in Section 3 we give advice for selecting the security parameters.

## 2   Class Groups

Recall that we consider class groups of imaginary quadratic fields only. We shall only state some necessary facts without proofs; for details we refer to [12]. Let $\Delta$ be a negative integer such that $\Delta \equiv 0, 1 \pmod 4$. Then $\Delta$ is the discriminant of a unique order of $\mathbb{Q}(\sqrt{\Delta})$, namely $\mathcal{O}_\Delta = \mathbb{Z} + \mathbb{Z}(\Delta + \sqrt{\Delta})/2$. $\mathcal{O}_\Delta$ is maximal if and only if $\Delta$ is fundamental, i.e. if $\Delta$ is square free in case $\Delta \equiv 1 \pmod 4$ or if $\Delta/4$ is square free in case $\Delta \equiv 0 \pmod 4$.

Let $\mathcal{O}_\Delta$ be any (not necessarily maximal) order. The class group of $\mathcal{O}_\Delta$ is denoted by $Cl(\Delta)$, its elements are equivalence classes of invertible ideals of $\mathcal{O}_\Delta$. The group order of $Cl(\Delta)$ is the class number $h(\Delta)$. Later in this work we shall need the odd parts of class groups. We denote the odd part of a class group $Cl(\Delta)$ by $Cl_{\mathsf{odd}}(\Delta)$ and its cardinality by $h_{\mathsf{odd}}(\Delta)$.

Any integral ideal of $\mathcal{O}_\Delta$ can be expressed as $\mathbb{Z}a + \mathbb{Z}(b + \sqrt{\Delta})/2$ such that $a, b \in \mathbb{Z}$, $a > 0$ and $4a \mid (b^2 - \Delta)$, that is, such that there exists a positive integer $c$ such that $\Delta = b^2 - 4ac$. Thus we represent ideals as pairs $(a, b)$ of integers.

Observe that if $b = 0$ or $b = a$, then $\Delta = -4ac$ or $\Delta = a(a - 4c)$, respectively, and if $a = c$, then $\Delta = (b - 2a)(b + 2a)$. Ideals of any of these forms are called *ambiguous*, and their classes have order two in $Cl(\Delta)$.

An ideal is said to be *reduced* if $\gcd(a, b, c) = 1$, $-a < b \leq a \leq c$, and $b \geq 0$ if $a = c$. Each equivalence class of $\mathcal{O}_\Delta$ contains exactly one reduced ideal. Thus the elements of $Cl(\Delta)$ can be represented by the reduced ideals of $\mathcal{O}_\Delta$, and checking equality of two ideal classes means comparing the representatives. The neutral element of $Cl(\Delta)$ is represented by $(1, \Delta \bmod 2)$. The group operation of $Cl(\Delta)$ is ideal multiplication followed by reduction (e.g. see [16] or [6, Chap. 5]). It can be shown that a group operation requires $O(\log^2 |\Delta|)$ bit operations. The inverse of the ideal class represented by $(a, b)$ under this operation is the ideal class represented by $(a, -b)$. If an ideal $(a, b)$ is reduced, then $a < \sqrt{|\Delta|/3}$, therefore $a, |b| = O(\sqrt{|\Delta|})$.

# 3    Selecting the Class Group

In this section we shall see that the discriminant is the cryptographic parameter. We shall discuss how to select a discriminant such that, based on heuristic grounds, computing discrete logarithms or the order of arbitrary elements in the corresponding class group is intractable. In particular,

- $\Delta$ must be chosen so that there is no efficient reduction of the CL-DLP to simpler problems,
- $|\Delta|$ must be large enough to preclude attacks with index-calculus algorithms,
- $h(\Delta)$ must be large enough to preclude attacks with $\rho$ or $\lambda$ algorithms,
- $h(\Delta)$ must not be smooth in order to preclude the computation of $h(\Delta)$ by an algorithm similar to the $(p-1)$-factoring algorithm with subsequent application of the Pohlig-Hellman algorithm.

It is tempting to ask whether the discriminant can be chosen such that its class number has properties selected a priori. However, we do not have much control over the class number; there is not even a probabilistic efficient algorithm known that outputs a fundamental discriminant whose class number has certain interesting properties, e.g. contains a large prime factor.

We shall show in the following subsections that if $\Delta$ is chosen appropriately, then the above conditions hold with high probability. In particular, in Sect. 3.1 we show that selecting $\Delta = -p$ or $\Delta = -8pq$ where $p, q$ are primes precludes reductions to the GF-DLP and keeps the two-part of $Cl(\Delta)$ small. In Sect. 3.2 we show how large $\Delta$ must be to preclude index-calculus attacks. In Sect. 3.3 we show how large the class group must be to preclude attacks with the aid of Pollard's $\lambda$-method; based on the Brauer-Siegel theorem we deduce the required size of the discriminant. In Sect. 3.4 we describe the relevance of the Pohlig-Hellman algorithm for class groups and discuss a possible application on class groups of smooth order in conjunction with an algorithm similar to the $(p-1)$-factoring method. Let a smoothness bound $B$ be given; in Sect. 3.5, based on heuristic assumptions we show how $\Delta$ must be chosen so that the class number is $B$-smooth only with negligible probability.

It turns out that asymptotically the selection of the discriminant size depends only on the index-calculus methods. Moreover, since the best known algorithm to compute class numbers of fundamental discriminants are again index-calculus methods, it is infeasible to compute the class number of fundamental discriminants if these are large. Therefore, the Pohlig-Hellman algorithm plays no role for class groups of maximal orders, unless the class number is smooth, because then an algorithm similar to the $(p-1)$-factoring algorithm can be applied to compute the class number.

## 3.1   Class Group Computation by Reduction to Other Problems

Let $\Delta$ be a negative fundamental discriminant and let $f$ be a positive integer. Then, if $\Delta \neq -3, -4$,

$$h(\Delta f^2) = h(\Delta) f \prod_{p|f} \left( 1 - \left( \frac{\Delta}{p} \right) \frac{1}{p} \right) , \tag{1}$$

where $(\Delta/p)$ denotes the Kronecker symbol. For instance, $h(-8) = 1$ and $h(-8p^2) = p - (-8/p)$. Since in general it is intractable to compute class numbers of large *fundamental* discriminants (see below), this could be a nice way to avoid such computations altogether and yet know the class number.

However, the Cl-DLP in $Cl(-8p^2)$ can be reduced in polynomial time to the GF-DLP in $\mathbb{F}_p$ [14]. Currently no similar efficient reductions for maximal orders are known. Therefore we shall use only class groups of maximal orders, and in the sequel $\Delta$ will always be fundamental and thus $\mathcal{O}_\Delta$ will be maximal.

**Selection of a Fundamental Discriminant**  In order to check whether an arbitrary discriminant $\Delta$ is fundamental, it must be checked whether $\Delta$ (if $\Delta \equiv 1 \pmod 4$) or $\Delta/4$ (if $\Delta \equiv 0 \pmod 4$) is square free. This can be achieved by factoring the discriminant, but this is infeasible if the discriminant under consideration is large. A better method is to construct $D$ from distinct prime factors, and set $\Delta = -D$ if $D \equiv 3 \pmod 4$ and $\Delta = -4D$ otherwise.

Some of the simplest cases are

1. $\Delta = -p$ where $p \equiv 3 \pmod 4$ is prime; and
2. $\Delta = -8pq$ where $p$ and $q$ are primes such that $p \equiv 1 \pmod 8$, $p + q \equiv 8 \pmod{16}$, and $(p/q) = -1$, where $(p/q)$ denotes the Legendre symbol.

Discriminants selected like this have the additional advantage that the two-part of the class number is known to be small: In case 1, $h(\Delta)$ is odd; in case 2, the even part of $h(\Delta)$ is exactly 8 (see [17, Proposition B$_9'$]).

Observe that $\Delta = -8pq$ is attractive by a complexity theoretic argument, because if $\Delta$ is composite, then $Cl(\Delta)$ has non-trivial ambiguous elements, whose components lead immediately to a factorization of $\Delta$; these ambiguous elements can be obtained by computing discrete logarithms in $Cl(\Delta)$ [25], therefore IFP $\leq$ Cl-DLP. This means that if $\Delta$ is chosen like this and $p$ and $q$ are not disclosed, then solving the Cl-DLP for $\Delta$ is at least as hard as breaking IFP-based cryptosystems such as RSA with modulus $pq$.

## 3.2 Class Group Computations by Index-Calculus Techniques

Let $L_x[e, c]$ be defined as usual, that is

$$L_x[e, c] \stackrel{\text{def}}{=} \exp\left(c(\log x)^e (\log \log x)^{1-e}\right) \tag{2}$$

for real positive $x$, real positive $c$, and $0 \leq e \leq 1$. In practice, instead of the term $L_x[e, c]$ we often see $L_x[e, c + o(1)]$, but in the sequel we shall ignore the $o(1)$ term.

We want to compare the expected computational work for solving the IFP and the Cl-DLP. In the following, we assume the expected running time for factoring an integer $n$ by the GNFS to be proportional to $L_n\left[\frac{1}{3}, \sqrt[3]{\frac{64}{9}}\right]$. For the Cl-DLP, index-calculus algorithms with an expected running time proportional to $L_{|\Delta|}\left[\frac{1}{2}, \frac{3}{4}\sqrt{2}\right]$ have been presented in [30]. However, Jacobson [16] showed that one can use a variant of the MPQS for DL-computations in $Cl(\Delta)$. The MPQS factoring algorithm has a conjectured expected running time proportional to $L_n\left[\frac{1}{2}, 1\right]$, while the MPQS DL-computation algorithm hasn't been analyzed, yet (not even heuristically). Empirical data suggests an expected running time of $L_{|\Delta|}\left[\frac{1}{2}, 1\right]$, so we shall base our arguments on this running time. In terms of security and efficiency, this will yield slightly larger keys: If we underestimate the running time of the Cl-MPQS, we overestimate the size of the security relevant parameters. This conservative approach is quite common practice.

The usual approach to estimate running times of an algorithm for large input parameters is to start from the empirical running time for smaller input parameters. If $x_1$ and $x_2$ are inputs for an algorithm with expected running time $L_x[e, c]$ and $t_1$ is the running time of the algorithm when executed with $x_1$, then the running time $t_2$ of the algorithm with input $x_2$ can be estimated by the equation

$$\frac{L_{x_1}[e, c]}{L_{x_2}[e, c]} = \frac{t_1}{t_2} \tag{3}$$

(cf. [21] or [18]). However, this holds only if the sizes of $x_1$ and $x_2$ do not differ too much; otherwise it can't be ignored that $o(1) \to 0$. Thus, if $x_2$ is much larger than $x_1$, then $t_2$ will be a significant overestimate. (For more precise estimates taking into account the $o(1)$ term, see [13]. We stick to (3) since the estimates presented here differ only slightly from those given in [13].)

Table 1 shows some extrapolated running times for the GNFS. They are based on data points of the factorization of RSA-155 (155 decimal digits, 512 bits) with

**Table 1.** Estimated expected computational work of the GNFS for larger inputs

| magnitude of $n$ | expected no. of MIPS-years to factor $n$ |
|---|---|
| $2^{512}$ | $8.00 \times 10^3$ |
| $2^{768}$ | $4.91 \times 10^7$ |
| $2^{1024}$ | $5.99 \times 10^{10}$ |
| $2^{1280}$ | $2.68 \times 10^{13}$ |
| $2^{1536}$ | $5.97 \times 10^{15}$ |
| $2^{1792}$ | $7.91 \times 10^{17}$ |
| $2^{2048}$ | $6.98 \times 10^{19}$ |
| $2^{2560}$ | $2.16 \times 10^{23}$ |
| $2^{3072}$ | $2.64 \times 10^{26}$ |
| $2^{3584}$ | $1.63 \times 10^{29}$ |
| $2^{4096}$ | $5.87 \times 10^{31}$ |

the GNFS [28]. In particular, it was estimated that about 8000 MIPS-years were spent.

To estimate the expected running time of the MPQS for DL-computations in class groups for large groups, we made extensive experiments where we computed discrete logarithms in 20 class groups of different negative discriminants for each magnitude tabulated below. The computations were carried out on a Sparc with Ultra-170 processor using Jacobson's MPQS implementation, which is part of the C++ library LiDIA [19]. The results are summarized in Table 2.

Table 2 supports the conjectured running time of $L_{|\Delta|}\left[\frac{1}{2}, 1\right]$ for the MPQS. Note also that the standard deviation is almost always about half the running time. This shows that the running times are pretty spread, which in turn confirms our suspicions of taking just a single sample.

SUN Microsystems does not publish MIPS ratings for its machines, and in fact, the unit MIPS-year is actually not appropriate [27]. However, it is widely used, so for simplicity we assume 100 MIPS, which is a value of reasonable order of magnitude for the machine that we used. By Table 2 let us assume that $L_{|\Delta|}\left[\frac{1}{2}, 1\right]/t_\Delta = 1.8 \times 10^7$ sec$^{-1}$. Then we obtain the extrapolations in Table 3.

When we align the parameters of the IFP and of the Cl-DLP in such a way that the expected running time for solving the Cl-DLP roughly equals the expected running time for solving the IFP for $n$ of some particular magnitudes, we arrive at Table 4.

### 3.3   Class Group Computations by Pollard's $\lambda$ Method

We now consider Pollard's $\lambda$ method for computing discrete logarithms, orders of group elements and hence roots of group elements. From [29] it is known that the unparallelized version of this algorithm takes $\sqrt{\pi|G|/2}$ group operations (ignoring lower order terms) for cyclic groups $G$. Moreover, $r$-fold parallelization speeds up the $\lambda$-method by factor $r$.

**Table 2.** Empirical computational work of the Cl-MPQS for relatively small inputs

| magnitude of $|\Delta|$ | mean running time (sec) $\overline{t_\Delta}$ | standard deviation | $L_{|\Delta|}\left[\frac{1}{2},1\right]/\overline{t_\Delta}$ (sec$^{-1}$) |
|---|---|---|---|
| $2^{140}$ | $8.59 \times 10^1$ | $3.58 \times 10^1$ | $1.65 \times 10^7$ |
| $2^{142}$ | $1.29 \times 10^2$ | $8.66 \times 10^1$ | $1.31 \times 10^7$ |
| $2^{144}$ | $1.36 \times 10^2$ | $5.32 \times 10^1$ | $1.50 \times 10^7$ |
| $2^{146}$ | $1.32 \times 10^2$ | $3.87 \times 10^1$ | $1.85 \times 10^7$ |
| $2^{148}$ | $1.98 \times 10^2$ | $6.98 \times 10^1$ | $1.47 \times 10^7$ |
| $2^{150}$ | $2.20 \times 10^2$ | $1.38 \times 10^2$ | $1.59 \times 10^7$ |
| $2^{152}$ | $2.63 \times 10^2$ | $1.44 \times 10^2$ | $1.59 \times 10^7$ |
| $2^{154}$ | $3.26 \times 10^2$ | $1.82 \times 10^2$ | $1.53 \times 10^7$ |
| $2^{156}$ | $3.52 \times 10^2$ | $1.64 \times 10^2$ | $1.69 \times 10^7$ |
| $2^{158}$ | $4.90 \times 10^2$ | $3.28 \times 10^2$ | $1.44 \times 10^7$ |
| $2^{160}$ | $4.41 \times 10^2$ | $1.98 \times 10^2$ | $1.90 \times 10^7$ |
| $2^{162}$ | $7.67 \times 10^2$ | $4.21 \times 10^2$ | $1.30 \times 10^7$ |
| $2^{164}$ | $6.84 \times 10^2$ | $2.20 \times 10^2$ | $1.73 \times 10^7$ |
| $2^{166}$ | $8.79 \times 10^2$ | $3.22 \times 10^2$ | $1.60 \times 10^7$ |
| $2^{168}$ | $1.07 \times 10^3$ | $4.12 \times 10^2$ | $1.56 \times 10^7$ |
| $2^{170}$ | $1.49 \times 10^3$ | $8.25 \times 10^2$ | $1.33 \times 10^7$ |
| $2^{172}$ | $1.74 \times 10^3$ | $8.99 \times 10^2$ | $1.34 \times 10^7$ |
| $2^{174}$ | $1.54 \times 10^3$ | $9.83 \times 10^2$ | $1.79 \times 10^7$ |
| $2^{176}$ | $1.61 \times 10^3$ | $8.45 \times 10^2$ | $2.03 \times 10^7$ |
| $2^{178}$ | $2.77 \times 10^3$ | $1.37 \times 10^3$ | $1.39 \times 10^7$ |
| $2^{180}$ | $2.73 \times 10^3$ | $1.39 \times 10^3$ | $1.67 \times 10^7$ |
| $2^{184}$ | $3.37 \times 10^3$ | $1.82 \times 10^3$ | $1.87 \times 10^7$ |
| $2^{188}$ | $4.07 \times 10^3$ | $1.95 \times 10^3$ | $2.14 \times 10^7$ |
| $2^{192}$ | $5.96 \times 10^3$ | $2.86 \times 10^3$ | $2.02 \times 10^7$ |
| $2^{196}$ | $9.23 \times 10^3$ | $3.80 \times 10^3$ | $1.79 \times 10^7$ |
| $2^{200}$ | $1.30 \times 10^4$ | $5.13 \times 10^3$ | $1.74 \times 10^7$ |
| $2^{210}$ | $2.63 \times 10^4$ | $8.49 \times 10^3$ | $1.87 \times 10^7$ |
| $2^{220}$ | $6.28 \times 10^4$ | $3.78 \times 10^4$ | $1.68 \times 10^7$ |

By the heuristics of Cohen and Lenstra [7,8], the probability that $Cl_{odd}(\Delta)$ is cyclic is $0.9775\ldots$. Moreover, it can be deduced from the heuristics that if $Cl_{odd}(\Delta)$ is not cyclic, then with high probability $Cl_{odd}(\Delta)$ has a cyclic subgroup $G_{cyc}$ such that $|G_{cyc}|$ is of nearly the same order of magnitude as $h_{odd}(\Delta)$, and therefore, by our selection of $\Delta$, the even part is 1 or 8 and thus $|G_{cyc}|$ and $h(\Delta)$ have nearly the same order of magnitude.

In order to provide a lower bound for $\Delta$ we need an (asymptotic) lower bound for $h(\Delta)$ that depends on $\Delta$ only. The best *proven* explicit lower bound is $h(\Delta) > 1/55 \ln|\Delta| \prod_{p|\Delta}\left(1 - \frac{2\sqrt{p}}{p+1}\right)$ [6, Sect. 5.10.1], which is too weak for our purposes. By the Brauer-Siegel Theorem we know that $\ln h(\Delta) \sim \ln \sqrt{|\Delta|}$

**Table 3.** Estimated expected computational work of the Cl-MPQS for larger inputs

| magnitude of $\|\Delta\|$ | expected no. of MIPS-years for solving the Cl-DLP in $Cl(\Delta)$ |
|---|---|
| $2^{256}$ | 2.58 |
| $2^{348}$ | $9.75 \times 10^3$ |
| $2^{512}$ | $1.18 \times 10^7$ |
| $2^{640}$ | $6.74 \times 10^9$ |
| $2^{768}$ | $2.24 \times 10^{12}$ |
| $2^{896}$ | $4.94 \times 10^{14}$ |
| $2^{1024}$ | $7.79 \times 10^{16}$ |
| $2^{1280}$ | $8.90 \times 10^{20}$ |
| $2^{1536}$ | $4.56 \times 10^{24}$ |
| $2^{1792}$ | $1.26 \times 10^{28}$ |
| $2^{2048}$ | $2.13 \times 10^{31}$ |
| $2^{2560}$ | $1.92 \times 10^{37}$ |
| $2^{3072}$ | $5.30 \times 10^{42}$ |
| $2^{3584}$ | $5.88 \times 10^{47}$ |
| $2^{4096}$ | $3.15 \times 10^{52}$ |

**Table 4.** Estimated expected computational work of the GNFS for factoring integers and the Cl-MPQS for computing discrete logarithms in class groups aligned

| magnitude of $n$ | magnitude of $\|\Delta\|$ | expected no. of MIPS-years |
|---|---|---|
| $2^{768}$ | $2^{540}$ | $4.99 \times 10^7$ |
| $2^{1024}$ | $2^{687}$ | $6.01 \times 10^{10}$ |
| $2^{1536}$ | $2^{958}$ | $5.95 \times 10^{15}$ |
| $2^{2048}$ | $2^{1208}$ | $7.05 \times 10^{19}$ |
| $2^{3072}$ | $2^{1665}$ | $2.65 \times 10^{26}$ |
| $2^{4096}$ | $2^{2084}$ | $5.87 \times 10^{31}$ |

as $\Delta \to -\infty$, that is, $\sqrt{|\Delta|}^{1-\epsilon} \le h(\Delta) \le \sqrt{|\Delta|}^{1+\epsilon}$ for any positive real $\epsilon$ and sufficiently large $\Delta$, but no explicit constants are known to make this statement effective. However, if one assumes the Extended Riemann Hypothesis, then it is possible to show [20] that

$$h(\Delta) > c_1 \frac{(1 + o(1))\sqrt{|\Delta|}}{\log\log|\Delta|} \tag{4}$$

for $\Delta \neq -3, -4$ where $c_1 = \pi/(12e^\gamma) \approx 0.147$. Moreover, it is possible to show that $h(\Delta)$ is *on average* $c_2\sqrt{|\Delta|}$ where $c_2 = 0.461559\ldots$ [6, Sect. 5.10.1]. This result has been proven for averages taken over class numbers of fundamental dis-

**Table 5.** Estimated expected computational work of the $\lambda$-method

| magnitude of | | expected no. of Group operations | expected no. of MIPS-years |
|---|---|---|---|
| $h(\Delta)$ | $|\Delta|$ | $\sqrt{\pi h(\Delta)/2}$ | |
| $2^{108}$ | $2^{218}$ | $2^{54}$ | $4.56 \times 10^7$ |
| $2^{129}$ | $2^{260}$ | $2^{64}$ | $6.60 \times 10^{10}$ |
| $2^{162}$ | $2^{326}$ | $2^{81}$ | $6.12 \times 10^{15}$ |
| $2^{189}$ | $2^{380}$ | $2^{94}$ | $7.09 \times 10^{19}$ |
| $2^{233}$ | $2^{468}$ | $2^{116}$ | $2.97 \times 10^{26}$ |
| $2^{268}$ | $2^{538}$ | $2^{134}$ | $5.51 \times 10^{31}$ |

criminants. In this work we make the assumption that this result is not affected by the restriction to the special discriminants given in section 3.1.

*Example* The time to perform a single group operation in $Cl(\Delta)$ depends on $\Delta$, yet let us assume a fixed time of 1 ms on a machine with a computing power of 100 MIPS. Then the computational work of a single MIPS-year is equivalent to about $2^{28.23}$ group operations. Based on this assumption and on the assumed average for the class number of a prime discriminant, in Table 5 we present some samples for (prime) discriminants, their average class number, and the expected computing amount for computing discrete logarithms by the $\lambda$-method.

## 3.4    Class Group Computations and the Pohlig-Hellman Algorithm

The Pohlig-Hellman algorithm utilizes the prime factorization of the group order in order to simplify DL computations. However, the best known algorithm for computing the class number is a variant of MPQS for DL computations in class groups and has heuristically the same expected asymptotic running time as the original MPQS. Thus, if $|\Delta|$ is large, it is infeasible to compute $h(\Delta)$ or even odd multiples or factors (in particular the smooth part) of $h(\Delta)$. Moreover, there is no efficient method known that checks whether a particular odd prime divides $h(\Delta)$. Consequently, the Pohlig-Hellman algorithm is not applicable to class groups in general. There are also cryptographic protocols (e.g. the Guillou-Quisquater signature protocol) that depend explicitly on the fact that the group order is unknown.

We now consider the special case when $h(\Delta)$ is smooth. If the class number is smooth, then it is possible to compute the order of an arbitrary element by a method similar to the $(p-1)$-factoring algorithm. That is, given $\gamma \in Cl(\Delta)$, set $\alpha_0 = \gamma$ and successively compute $\alpha_i = \alpha_{i-1}^{p_i^{e(p_i,B)}}$ for all $p_i \leq B$, where $p_i$ is the $i$th prime, $B$ is a smoothness bound, and $e(p_i, B)$ depends only on $p_i$ and $B$. For instance, if $e(p_i, B) = \log_{p_i} B$ for each $p_i$, then the algorithm will cover each possible prime power below the smoothness bound. A similar method is used in the factoring algorithm of Schnorr and Lenstra [25].

If $h(\Delta)$ is $B$-smooth, then this computation may yield $1_{Cl(\Delta)}$. If this happens, then there is an $i$ such that $\alpha_i = 1_{Cl(\Delta)}$ but $\alpha_{i-1} \neq 1_{Cl(\Delta)}$, and we immediately know that $p_i$ is the largest prime factor of $\mathrm{ord}_{Cl(\Delta)} \gamma$. If we set $\gamma' = \gamma^{p_i^{e(p_i)}}$ where $e(p_i)$ is the smallest positive integer such that $\alpha_{i-1}^{p_i^{e(p_i)}} = 1_{Cl(\Delta)}$ and repeat the complete procedure with $\gamma'$, then we obtain the second largest prime factor, and eventually we get the complete prime factorization of $\mathrm{ord}_{Cl(\Delta)} \gamma$. Then we are able to compute roots as well as discrete logarithms in $\langle \gamma \rangle$ by applying the Pohlig-Hellman algorithm.

Assume that the $(p-1)$-like method above succeeds for an element $\gamma$ and a bound $B$, and let $q$ denote the largest prime factor of $\mathrm{ord}_{Cl(\Delta)} \gamma$. It is obvious that if we use a fast exponentiation method, then we have to perform at least $\sum_{p<q} e(p, B) \log_2 p$ group operations to find $q$. In order to find a smoothness bound, we must consider the easiest case, i.e. $e(p_i, B) = 1$ for all $p_i$. Now $\sum_{p<q} \log_2 p = \theta(q)/\ln 2$, where $\theta$ is the Chebyshev $\theta$-function. In [24] it has been shown that $0.998684\,x < \theta(x) < 1.001102\,x$ for all $x \geq 1319007$ (under assumption of the Riemann hypothesis, it is even possible to show that $|\theta(x) - x| = 1/(8\pi)\sqrt{x}\ln^2 x$ for $x \geq 599$, cf. [26]). Therefore, to find $q$ we have to perform about $q/\ln 2$ group operations. (Note that we get the same result even in the case $e(p_i, q) = \log_{p_i} q$, because $\sum_{p<q} \log_2 p_i \log_{p_i} q \approx \pi(q) \log_2 q \approx q/\ln 2$ as $q \to \infty$, where $\pi(q)$ denotes the number of primes up to $q$.) In Sect. 3.5, we will use this result to determine lower bounds for the required size of $\Delta$.

*Example* We use part of the example from the previous section, namely, that $2^{64}$ group operations require about $6\times 10^{10}$ MIPS-years (similar to the computational work to factor a 1024 bit integer with the aid of the GNFS). If we assume that this amount of work is infeasible, then it is safe to select a 64 bit smoothness bound. At the end of the next section we will see that a smaller smoothness bound is sufficient.

## 3.5   The Smoothness Probability of Class Numbers

The estimates in this section are based on the heuristics of Cohen and Lenstra [7,8], although our derivation is not rigorous at all. A more rigorous derivation should be done as in [8]; this is work in progress, and we shall present the results in a future work. In this work we compare class numbers and ordinary integers with respect to smoothness, and we argue that under reasonable assumptions the probability to get a smooth class number of a random fundamental discriminant is not much larger than the probability that a random integer is smooth.

Consider the set of all negative fundamental discriminants $\Delta$ such that $|\Delta| \leq N$ for some bound $N$. Based on the heuristics of Cohen and Lenstra we assume that, given an odd prime $p$ much smaller than $N$ and a positive integer $i$, the proportion of such discriminants satisfying $p^i \mid h(\Delta)$ (or the "probability" that $p^i \mid h(\Delta)$) is at most $1/p^i + 1/p^{i+1} = (1+1/p)/p^i$. The conjectures of Cohen and Lenstra [8] predict that for $N \to \infty$, the probability that $p \mid h(\Delta)$ converges to

$$1 - \prod_{j \geq 1} \left(1 - \frac{1}{p^j}\right) = \frac{1}{p} + \frac{1}{p^2} - \frac{1}{p^5} - \frac{1}{p^7} + \frac{1}{p^{12}} + \frac{1}{p^{15}} - \cdots . \tag{5}$$

Our assumption for $i \geq 2$ is accordance with computational experiments [5].

We cannot use similar heuristics for primes that are not small compared to $N$. However, we know by the Brauer-Siegel theorem that $\ln h(\Delta) \sim \ln \sqrt{|\Delta|}$ for $\Delta \to -\infty$, thus class numbers are usually not small themselves.

Which power of 2 divides $h(\Delta)$ depends on the factorization of $\Delta$. As discussed in section 3.1, we will restrict to special discriminants in order to control the two-part of $h(\Delta)$. In extension to the heuristics of Cohen and Lenstra, we assume that such restrictions do not affect the probabilities discussed above.

For $x$ uniformly chosen from a sufficiently large interval of integers, the probability that $p^i \mid x$ is only about $1/p^i$. Comparing this with the above estimates for class numbers, we obtain

$$\frac{\Pr\left(p^i \mid h(\Delta)\right)}{\Pr\left(p^i \mid x\right)} \leq 1 + \frac{1}{p} \tag{6}$$

for small odd primes, which suggests that it must be expected to occur more frequently for negative fundamental discriminants to have smooth class numbers than for uniformly chosen integers to be smooth. We will now argue, however, that this increase in smoothness does not imply that a significant proportion of class numbers will be smooth.

Let $k$ be any odd smooth integer. We write $k$ as $\prod_{p|k} p^{e_p(k)}$. If $k$ is not so large that $k \mid h(\Delta)$ is actually impossible, then $k$ will have only a few different prime factors. Thus, it is conceivable that the probabilities discussed above will be reasonably close to being statistically independent over the different $p$ dividing $k$. Under this presumption, we obtain

$$\frac{\Pr\left(k \mid h(\Delta)\right)}{\Pr\left(k \mid x\right)} = \frac{\prod_{p|k} \Pr\left(p^{e_p(k)} \mid h(\Delta)\right)}{\prod_{p|k} \Pr\left(p^{e_p(k)} \mid x\right)} \leq \prod_{p|k} \left(1 + \frac{1}{p}\right) \stackrel{\text{def}}{=} F_k . \tag{7}$$

We now want to estimate the maximum value that this product can take for $k$ not exceeding the order of $\sqrt{|\Delta|}$ (as suggested by the Brauer-Siegel Theorem). In order to reach the maximum, $k$ obviously must be of the form $k = \prod_{p < t} p$, i.e. the product of the smallest primes up to some bound. We have $\prod_{p<t} p \approx e^t$ as $t$ tends to infinity (e.g. see [23, Chap. 12]), i.e. $t \approx \ln k \approx \ln \sqrt{|\Delta|}$; and thus we estimate the maximum for $F_k$ as

$$\prod_{p<t} \left(1 + \frac{1}{p}\right) \approx \prod_{p < \ln \sqrt{|\Delta|}} \left(1 + \frac{1}{p}\right) \approx \ln \ln \sqrt{|\Delta|} , \tag{8}$$

where the latter approximation can be seen as follows: $(1+1/p) = (1-1/p^2)/(1-1/p)$, and $\prod_{p<t}(1 - 1/p) = e^{-\gamma}/\ln t + O(1/\ln^2 t)$ (Mertens' theorem, cf. [23, Chap. 12]), while $\prod_p (1 - 1/p^2) = 1/\zeta(2) = 6/\pi^2$, thus $\prod_{p<t}(1 + 1/p) \approx 6e^\gamma/\pi^2 \ln t \approx 1.08 \ln t$ as $t$ tends to infinity.

Now if we choose $|\Delta|$ so large that random integers of the expected order of $h(\Delta)$ are smooth only with probability close to 0, then the modest maximum size of $F_k$ indicates that the tendency of the class number towards having small factors does not mean it will be smooth with non-negligible probability.

Specifically, let $B = M^{1/u}$; then the probability that a random positive integer less than $M$ is $B$-smooth is approximately $\rho(u)$, where $\rho$ is Dickmann's $\rho$-function [15]. We arrive at an estimated probability of at most $\rho(u) \ln \ln M$ for the class number being $B$-smooth by requiring $M \approx \frac{h_{\text{odd}}(\Delta)}{h(\Delta)} c \sqrt{|\Delta|}$ where $\frac{h_{\text{odd}}(\Delta)}{h(\Delta)}$ is either 1 or $\frac{1}{8}$ depending on how $\Delta$ is chosen (Sect. 3.1) and where $c = 0.461559\ldots$ [6, Sect. 5.10.1]. I.e.,

$$|\Delta| \approx 2^2 B^{2u} \tag{9}$$

if $h(\Delta)$ is odd and

$$|\Delta| \approx 2^8 B^{2u} \tag{10}$$

if the even part of $h(\Delta)$ is 8. Note that if $|\Delta| < 2^{4600}$, then $\ln \ln \sqrt{|\Delta|} < 2^3$ so that $8\rho(u)$ is an upper bound for the probability estimate.

Assume that an attacker applies the algorithm from the preceding section to class groups of random discriminants of a certain length (chosen as described in Sect. 3.1). Further assume that he will spend at most $W_{\text{max}}$ computational work for a single class group until he gives up, and that $B$ is the smoothness bound for which he can succeed with this amount of work. Then he can expect one case of success for an investment of computational work $W = W_{\text{max}}/\Pr(h(\Delta) \text{ is } B\text{-smooth})$. We will determine lower bounds for the size of $\Delta$ based on this attack scenario.

Recall that 1 MIPS-year is approximately equivalent to about $2^{29}$ group operations. Let $W = 2^{64}$ group operations which is comparable to the expected computational work to factor a composite 1024 bit integer by the GNFS; then $W$ is currently infeasible (see the example in Sect. 3.4). Let $W_{\text{max}} = 2^{42}$ group operations (corresponding to a smoothness bound of approximately $2^{42}/\ln 2$, see Sect. 3.4), which is comparable to the expected work to factor a 512 bit integer by the GNFS. Then a smoothness probability of up to $2^{-22}$ is acceptable, thus we need $u$ such that $\rho(u) \approx 2^{-22}/8$, and this is satisfied by $u = 8$. Since $B \approx 2^{41.5}$, the discriminant should have at least 666 bits for case 1 of Sect. 3.1 and at least 672 bits for case 2 of Sect. 3.1 according to (9) and (10).

If $W_{\text{max}}$ is larger or if a smaller smoothness probability is demanded, then the order of magnitude of the discriminant will increase accordingly. For instance, if we choose $\Pr(h(\Delta) \text{ is } B\text{-smooth}) = 2^{-30}$ with $W_{\text{max}}$ (and hence $B$) as before, then $u = 9.6$, and thus the discriminant should have at least 799 (case 1) or 805 (case 2) bits.

## 4   Conclusion

Based on the investigation of several strategies to solve the CL-DLP and based on heuristic reasoning, we have shown how to select the discriminant such that the security of cryptosystems based on class groups offer a comparable security as commonly used cryptosystems (such as RSA). In particular, we have shown that the size of the discriminant asymptotically depends only on index-calculus algorithms (see Table 4). Thus, since index-calculus algorithms for solving the Cl-DLP are asymptotically much slower than index-calculus algorithms to solve the IFP (such as the GNFS), the discriminant can be selected smaller than an RSA modulus.

In a future work we shall demonstrate the impact of this result on the efficiency and performance of IQC. As a further research project we would also like to replace the heuristic reasoning of Sect. 3.5 by a more rigorous reasoning.

## Acknowledgements

We would like to thank Johannes Buchmann and Ulrich Vollmer for valuable discussions and comments, as well as the anonymous ASIACRYPT 2000 referee for his comments and suggestions.

## References

1. BIEHL, I., BUCHMANN, J., HAMDY, S., AND MEYER, A. A signature scheme based on the intractability of extracting roots. Tech. Rep. TI-1/00, Technische Universität Darmstadt, Fachbereich Informatik, 2000.
   http://www.informatik.tu-darmstadt.de/TI/Veroeffentlichung/TR/.
2. BUCHMANN, J., AND PAULUS, S. A one way function based on ideal arithmetic in number fields. In *Advances in Cryptology — CRYPTO '97* (1997), B. S. Kaliski, Ed., vol. 1294 of *Lecture Notes in Computer Science*, Springer–Verlag, pp. 385–394.
3. BUCHMANN, J., AND WILLIAMS, H. C. A key-exchange system based on imaginary quadratic fields. *Journal of Cryptology 1*, 3 (1988), 107–118.
4. BUCHMANN, J., AND WILLIAMS, H. C. Quadratic fields and cryptography. In *Number Theory and Cryptography*, J. H. Loxton, Ed., vol. 154 of *London Mathematical Society Lecture Note Series*. Cambridge University Press, 1990, pp. 9–25.
5. BUELL, D. A. The expectation of success using a Monte Carlo factoring method — some statistics on quadratic class numbers. *Mathematics of Computation 43*, 167 (1984), 313–327.
6. COHEN, H. *A Course in Computational Algebraic Number Theory*, vol. 138 of *Graduate Texts in Mathematics*. Springer–Verlag, 1995.
7. COHEN, H., AND LENSTRA, JR., H. W. Heuristics on class groups. In *Number Theory, New York 1982*, vol. 1052 of *Lecture Notes in Mathematics*. Springer–Verlag, 1984, pp. 26–36.
8. COHEN, H., AND LENSTRA, JR., H. W. Heuristics on class groups of number fields. In *Number Theory, Noordwijkerhout 1983*, vol. 1068 of *Lecture Notes in Mathematics*. Springer-Verlag, 1984, pp. 33–62.
9. DÜLLMANN, S. *Ein Algorithmus zur Bestimmung der Klassengruppe positiv definiter binärer quadratischer Formen*. PhD thesis, Universität des Saarlandes, Saarbrücken, Germany, 1991. German.
10. GUILLOU, L. C., AND QUISQUATER, J.-J. A practical zero-knowledge protocol fitted to security microprocessors minimizing both transmission and memory. In *Advances in Cryptology — EUROCRYPT '88* (1988), C. G. Günther, Ed., vol. 330 of *Lecture Notes in Computer Science*, Springer–Verlag, pp. 123–128.

11. HAFNER, J. L., AND MCCURLEY, K. S. A rigorous subexponential algorithm for computation of class groups. *Journal of the American Mathematical Society 2* (1989), 837–850.

12. HUA, L. K. *Introduction to Number Theory*. Springer–Verlag, 1982.

13. HÜHNLEIN, D. Quadratic orders for NESSIE — overview and parameter sizes of three public key families. Tech. Rep. TI-3/00, Technische Universität Darmstadt, Fachbereich Informatik, 2000.
    http://www.informatik.tu-darmstadt.de/TI/Veroeffentlichung/TR/.

14. HÜHNLEIN, D., AND TAKAGI, T. Reducing logarithms in totally non-maximal imaginary quadratic orders to logarithms in finite fields. In *Advances in Cryptology — ASIACRYPT '99* (1999), K. Y. Lam, E. Okamato, and C. Xing, Eds., vol. 1716 of *Lecture Notes in Computer Science*, Springer–Verlag, pp. 219–231.

15. HUNTER, S., AND SORENSON, J. Approximating the number of integers free of large prime factors. *Mathematics of Computation 66*, 220 (1997), 1729–1741.

16. JACOBSON, JR., M. J. *Subexponential Class Group Computation in Quadratic Orders*. PhD thesis, Technische Universität Darmstadt, Fachbereich Informatik, Darmstadt, Germany, 1999.

17. KAPLAN, P. Sur le 2-groupe des classes d'idéaux des corps quadratiques. *Journal für die reine und angewandte Mathematik 283/284* (1976), 313–363. French.

18. LENSTRA, A. K., AND VERHEUL, E. R. Selecting cryptographic keysizes. In *Practice and Theory in Public Key Cryptography, PKCS 2000* (2000), H. Imai and Y. Zheng, Eds., vol. 1751 of *Lecture Notes in Computer Science*, Springer–Verlag, pp. 446–465. Full version available from http://www.cryptosavvy.com/.

19. LiDIA — a C++ library for computational number theory.
    http://www.informatik.tu-darmstadt.de/TI/LiDIA/. The LiDIA Group.

20. LITTLEWOOD, J. E. On the class number of the corpus $P(\sqrt{-k})$. *Proceedings of the London Mathematical Society, 2nd series 27* (1928), 358–372.

21. ODLYZKO, A. M. The future of integer factorization. *CryptoBytes 1*, 2 (1995).
    http://www.rsa.com/rsalabs/pubs/cryptobytes/.

22. POUPARD, G., AND STERN, J. Security analysis of a practical "on the fly" authentication and siganture generation. In *Advances in Cryptology – EUROCRYPT '98* (1998), K. Nyberg, Ed., vol. 1403 of *Lecture Notes in Computer Science*, Springer–Verlag, pp. 422–436.

23. ROSE, H. E. *A Course in Number Theory*, 2 ed. Oxford University Press, 1994.

24. ROSSER, J. B., AND SCHOENFELD, L. Sharper bounds for the Chebyshev functions $\theta(x)$ and $\psi(x)$. *Mathematics of Computation 29*, 129 (1975), 243–269.

25. SCHNORR, C. P., AND LENSTRA, JR., H. W. A Monte Carlo factoring algorithm with linear storage. *Mathematics of Computation 43*, 167 (1984), 289–311.

26. SCHOENFELD, L. Sharper bounds for the Chebyshev functions $\theta(x)$ and $\psi(x)$, ii. *Mathematics of Computation 30*, 134 (1976), 337–360.

27. SILVERMAN, R. D. Exposing the mythical MIPS year. *IEEE Computer 32*, 8 (1999), 22–26.

28. TE RIELE, H. J. J. Factorization of a 512-bits RSA key using the number field sieve. Announcment on the Number Theory List (NMBRTHRY@listserv.nodak.edu), August 1999.

29. VAN OORSCHOT, P. C., AND WIENER, M. J. Parallel collusion search with cryptanalytic applications. *Journal of Cryptology 12*, 1 (1999), 1–28.

30. VOLLMER, U. Asymptotically fast discrete logarithms in quadratic number fields. In *Algorithmic Number Theory, ANTS IV* (2000), W. Bosma, Ed., vol. 1838 of *Lecture Notes in Computer Science*, Springer–Verlag, pp. 581–594.

# Weil Descent of Elliptic Curves over Finite Fields of Characteristic Three

Seigo Arita

NEC, Kawasaki Kanagawa, Japan
arita@ccm.cl.nec.co.jp

**Abstract.** The paper shows that some of elliptic curves over finite fields of characteristic three of composite degree are attacked by a more effective algorithm than Pollard's $\rho$ method. For such an elliptic curve $E$, we construct a $C_{ab}$ curve $D$ on its Weil restriction in order to reduce the discrete logarithm problem on $E$ to that on $D$. And we show that the genus of $D$ is small enough so that $D$ is attacked by a modified form of Gaudry's variant for a suitable $E$. We also see such a weak elliptic curve is easily constructed.

## 1 Introduction

An elliptic curve cryptosystem(ECC) is a discrete-logarithm-based public key cryptosystem using the Jacobian group of an elliptic curve[9,12]. In ECC, we must be careful to choose an elliptic curve. Many classes of week elliptic curves have been found since ECC was presented [11,4,19,15,18,16,14].

Recently, Gaudry, Hess and Smart[7] found new week elliptic curves. They show that some of elliptic curves over finite fields of characteristic two of composite degree are attacked by a more effective algorithm than Pollard's $\rho$ method. They construct a hyperelliptic curve $H$ on the Weil restriction of such an elliptic curve $E$, and show that the discrete logarithm problem(DLP) on $E$ is reduced to that on $H$. Moreover they observe that for some such $E$, the genus of the corresponding $H$ becomes small enough for the DLP on $H$ to be attacked by Gaudry's variant [6].

This paper treats elliptic curves over finite fields of characteristic *three* of composite degree, and shows some of such elliptic curves are also attacked by a more effective algorithm than Pollard's $\rho$ method.

We construct a $C_{ab}$ curve [13,3] $D$ on the Weil restriction of an elliptic curve $E$ over a finite field of characteristic three of composite degree, and reduce the discrete logarithm problem(DLP) on $E$ to that on $D$. Moreover, we clarify the condition for an elliptic curve $E$ to correspond to a $C_{ab}$ curve $D$ of small genus, as well as the method to construct such $E$. Since Gaudry's variant is also effective for $C_{ab}$ curves with a slight modification [2], this means that some of elliptic curves of characteristic three of composite degree are also attacked by a more effective algorithm than Pollard's $\rho$ method, and that we can construct such weak elliptic curves effectively.

T. Okamoto (Ed.): ASIACRYPT 2000, LNCS 1976, pp. 248–258, 2000.

## 2    Computation of Weil Descent

We treat Weil descent of an elliptic curve $E_a$

$$Y^2 + Y = X^3 + aYX \tag{1}$$

defined over a finite field $\mathbb{F}_{q^n}$ of characteristic three. Here, for $q = 3^d$, we assume

$$\gcd(d, n) = 1. \tag{2}$$

Note $E_a$ is not supersingular for nonzero $a$ (Theorem 4.1. on [17]).

Let $\Omega = \{\omega, \omega^3, \cdots, \omega^{3^{n-1}}\}$ be a normal basis for $\mathbb{F}_{3^n} \mid \mathbb{F}_3$. By the condition (2), $\Omega$ is a basis also for $\mathbb{F}_{q^n}$ over $\mathbb{F}_q$. Substituting $Y = y_0\omega + y_1\omega^3 + \cdots + y_{n-1}\omega^{3^{n-1}}$, $X = x_0\omega + x_1\omega^3 + \cdots + x_{n-1}\omega^{3^{n-1}}$ for the defining equation (1) of $E_a$, and comparing coefficients of $\omega^i$, we get $n$ equations among $2n$ variables $\{y_0, \ldots, y_{n-1}, x_0, \ldots, x_{n-1}\}$. An abelian variety $A_a = \prod_{\mathbb{F}_{q^n} \mid \mathbb{F}_q} E_a$ defined by these $n$ equations is called Weil restriction of $E_a$ [5]. Moreover, taking an intersection of $A_a$ and $(n-1)$ hyperplanes $y_0 = y_i (i = 1, \ldots, n-1)$, we get an algebraic curve $C_a$. $C_a$ is an algebraic curve defined by $n$ equations in $(n+1)$-dimensional affine space.

For an element $a \in \mathbb{F}_{q^n}$, let $A(a) \in M_n(\mathbb{F}_q)$ be a regular representation of $a$ with respect to $\Omega$ :

$$a \cdot [\omega, \omega^3, \cdots, \omega^{3^{n-1}}] = [\omega, \omega^3, \cdots, \omega^{3^{n-1}}] \cdot A(a).$$

Using $A := A(a)$, the defining equations for $C_a$ are given by

$$C_a : \begin{cases} x_{n-1}^3 - c_1y(A_{11}x_0 + A_{12}x_1 + \cdots + A_{1n}x_{n-1}) = -c_1y^2 + y \\ x_0^3 - c_1y(A_{21}x_0 + A_{22}x_1 + \cdots + A_{2n}x_{n-1}) = -c_1y^2 + y \\ \cdots \\ x_{n-2}^3 - c_1y(A_{n1}x_0 + A_{n2}x_1 + \cdots + A_{nn}x_{n-1}) = -c_1y^2 + y \end{cases} \tag{3}$$

Here, we put $y = y_i (i = 0, \ldots, n-1)$, and let the minimal polynomial of $\omega$ be $T^n + c_1T^{n-1} + \cdots c_n$.

Putting

$$\boldsymbol{x} = \begin{pmatrix} x_0 \\ x_1 \\ \vdots \\ x_{n-1} \end{pmatrix}, \quad \boldsymbol{e} = \begin{pmatrix} 1 \\ 1 \\ \vdots \\ 1 \end{pmatrix}, \quad P = \begin{pmatrix} 0 & 0 & \cdots & 0 & 1 \\ 1 & 0 & \cdots & 0 & 0 \\ \vdots & \vdots & \ddots & \vdots & \vdots \\ 0 & 0 & \cdots & 1 & 0 \end{pmatrix}$$

($P$ is a matrix for a cyclic permutation), Equations (3) become

$$P\boldsymbol{x}^3 - c_1y A\boldsymbol{x} = (-c_1y^2 + y)\boldsymbol{e}. \tag{4}$$

Here, $\boldsymbol{x}^3$ denotes an vector gotten by cubing every components of $\boldsymbol{x}$.

Regular representations $A(a)$ $(a \in \boldsymbol{F}_{q^n})$ are diagonalized simultaneously using a matrix $T$ with the eigenvectors for the Frobenius automorphism $x \mapsto x^q$ as columns:

$$T^{-1}A(a)T = D(a^{(0)}, a^{(1)}, \cdots, a^{(n-1)}), \tag{5}$$

where $D(a, b, \ldots, z)$ denotes a diagonal matrix with $a, b, \ldots, z$ as diagonal elements, and $a^{(0)}, a^{(1)}, \cdots, a^{(n-1)}$ $(a^{(i)} := a^{q^i})$ is a whole of elements conjugate to $a$ in $\mathbb{F}_{q^n}$ over $\mathbb{F}_q$.

Putting

$$\boldsymbol{x} = T\boldsymbol{w}, \tag{6}$$

equation (4) becomes

$$T^{-1}PT^{(3)}\boldsymbol{w}^3 - c_1 y D(a^{(0)}, a^{(1)}, \cdots, a^{(n-1)})\boldsymbol{w} = (-c_1 y^2 + y)T^{-1}\boldsymbol{e}, \tag{7}$$

where $T^{(3)}$ denotes a matrix gotten by cubing every elements of $T$.

**Lemma 1.** $T^{-1}PT^{(3)}$ *is a diagonal matrix over* $\mathbb{F}_{q^n}$.

*Proof.* For any element $a \in \mathbb{F}_{q^n}$, by the definition of $A$,

$$a \cdot [\omega, \omega^3, \cdots, \omega^{3^{n-1}}] = [\omega, \omega^3, \cdots, \omega^{3^{n-1}}] \cdot A(a).$$

Cubing two sides,

$$a^3 \cdot [\omega^3, \omega^9, \cdots, \omega] = [\omega^3, \omega^9, \cdots, \omega] \cdot A(a)^{(3)}.$$

The left-hand side is equal to $a^3 \cdot [\omega, \omega^3, \cdots, \omega^{3^{n-1}}]P = [\omega, \omega^3, \cdots, \omega^{3^{n-1}}]A(a)^3 P$, and the right-hand side is $[\omega, \omega^3, \cdots, \omega^{3^{n-1}}]PA(a)^{(3)}$. So, we get

$$A(a)^3 = PA(a)^{(3)}P^{-1}.$$

Therefore we have

$$T^{-1}A(a)^3T = T^{-1}PA(a)^{(3)}P^{-1}T = T^{-1}PT^{(3)} \cdot T^{(3)^{-1}}A(a)^{(3)}T^{(3)} \cdot T^{(3)^{-1}}P^{-1}T.$$

Thus, for any $a \in \boldsymbol{F}_{q^n}$,

$$T^{-1}A(a)^3T \cdot T^{-1}PT^{(3)} = T^{-1}PT^{(3)} \cdot T^{(3)^{-1}}A(a)^{(3)}T^{(3)}.$$

However, $T^{-1}A(a)^3T = T^{(3)^{-1}}A(a)^{(3)}T^{(3)} = D(a^{(1)^3}, \cdots, a^{(n-1)^3})$. So, $T^{-1}PT^{(3)}$ must be a diagonal matrix. $\square$

In equation (7), putting

$$D(b_0, \cdots, b_{n-1}) = T^{-1}PT^{(3)} \quad (b_i \in \mathbb{F}_{q^n}) \tag{8}$$
$$\boldsymbol{d} = T^{-1}\boldsymbol{e}, \tag{9}$$

we get defining equations of $C_a$ over $\mathbb{F}_{q^n}$ :

$$w_i^3 - b_i^{-1}c_1 a^{(i)}yw_i = b_i^{-1}d_i(-c_1 y^2 + y) \quad (i = 0, 1, \ldots, n-1). \tag{10}$$

We note that $b_i, c_1, d_i$ are determined only by $n$ and $d$, independent from $a \in \mathbb{F}_{q^n}$.

## Example: d=5,n=4

Let $d = 5$, $n = 4$. Let $\kappa$ be a root of the irreducible polynomial $T^5 + T^4 + T^3 + T^2 - T + 1$ over $\mathbb{F}_3$. $\kappa$ is a primitive element of $\mathbb{F}_q$. Let $\omega$ be a root of the irreducible polynomial $T^4 - T^3 + T^2 + T - 1$ over $\mathbb{F}_3$ (i.e. $c_1 = -1$). $\Omega = \{\omega, \omega^3, \omega^{3^2}, \omega^{3^3}\}$ is a normal basis of $\mathbb{F}_{3^n}$ over $\mathbb{F}_3$. Since $d$ and $n$ are prime to each other, $\Omega$ is a basis also for $\mathbb{F}_{q^n}$ over $\mathbb{F}_q$.

For

$$a = \kappa^{216}\omega^3 + \kappa^{95}\omega^2 + \kappa^{95}\omega, \tag{11}$$

defining equations of $C_a$ over $\mathbb{F}_{q^n}$ are given by

$$\begin{cases} w_0^3 + (\kappa^{86}\omega^3 + \kappa^{168}\omega^2 + \kappa^{200}\omega + \kappa^{62})yw_0 = (\kappa^{162}\omega^3 + \kappa^{239}\omega^2 + \omega + \kappa^{19})(y^2 + y) \\ w_1^3 + (\kappa^{181}\omega^3 + \kappa^{207}\omega^2 + \kappa^{168}w + \kappa^{182})yw_1 = (\kappa^{142}\omega^3 + \kappa^{41}\omega^2 + \kappa^{239}w + \kappa^{238})(y^2 + y) \\ w_2^3 + (\kappa^{79}\omega^3 + \kappa^{60}\omega^2 + \kappa^{207}\omega + \kappa^{85})yw_2 = (\kappa^{121}\omega^3 + \kappa^{21}\omega^2 + \kappa^{41}w + \kappa^{201})(y^2 + y) \\ w_3^3 + (\kappa^{47}\omega^3 + \kappa^{200}\omega^2 + \kappa^{60}w + \kappa^8)yw_3 = (\kappa^{118}\omega^3 + \omega^2 + \kappa^{21}w + \kappa^{220})(y^2 + y) \end{cases}$$

## 3   A Component $D_a$ of the Curve $C_a$

We show that the curve $C_a$ has a component $D_a$ with small genus for a suitable $a \in \mathbb{F}_{q^n}$. We use notations in section 2.

**Lemma 2.** *For an element $h$ in a function field of $C_a$ over $\mathbb{F}_{q^n}$, let $h^q$ denote the image of $h$ by the Frobenius automorphism with respect to $q$ (i.e. the generator of the Galois group $\mathrm{Gal}(\mathbb{F}_{q^n}(y, x_0, \ldots, x_{n-1}) \mid \mathbb{F}_q(y, x_0, \ldots, x_{n-1})) \simeq \mathrm{Gal}(\mathbb{F}_{q^n} \mid \mathbb{F}_q)$ ). We have*

$$\begin{aligned} &w_0^q = w_1, w_1^q = w_2, \cdots, w_{n-1}^q = w_0 \\ &a^{(0)q} = a^{(1)}, a^{(1)q} = a^{(2)}, \cdots, a^{(n-1)q} = a^{(0)} \\ &b_0^q = b_1, b_1^q = b_2, \cdots, b_{n-1}^q = b_0 \\ &d_0^q = d_1, d_1^q = d_2, \cdots, d_{n-1}^q = d_0 \end{aligned}$$

*Proof.* As $a^{(i)} = a^{q^i}$, claims for $a_i$ are obvious. In equation (5), the $i$-th column of the matrix $T$ is gotten by taking $q$-th power of every elements of the $(i-1)$-th column of $T$. So, the $i$-th row of the matrix $T^{-1}$ is gotten by taking $q$-th power of every elements of the $(i-1)$-th row of $T^{-1}$. From this, we obtain claims for $w_i$ and $d_i$. Claims for $b_i$ are also gotten from equation (8) □

Putting

$$\alpha_i = -b_i^{-1}c_1a^{(i)}, \quad \beta_i = b_i^{-1}d_i, \quad f = -c_1y^2 + y \quad (i = 0, 1, \ldots, n-1), \tag{12}$$

defining equations (10) become

$$w_i^3 + \alpha_i yw_i = \beta_i f \quad (i = 0, 1, \ldots, n-1). \tag{13}$$

By Lemma 2, we have

$$\begin{aligned} &\alpha_0^q = \alpha_1, \alpha_1^q = \alpha_2, \ldots, \alpha_{n-1}^q = \alpha_0, \\ &\beta_0^q = \beta_1, \beta_1^q = \beta_2, \ldots, \beta_{n-1}^q = \beta_0. \end{aligned} \tag{14}$$

For defining equations (13), put $F_0 = \mathbb{F}_{q^n}(y, w_0)$, $F_1 = \mathbb{F}_{q^n}(y, w_0, w_1)$, $\cdots$, $F = F_{n-1} = \mathbb{F}_{q^n}(y, w_0, w_1, \cdots, w_{n-1})$. $F$ is a function field of $C_a$ over $\mathbb{F}_{q^n}$. Put

$$I_i = \{\gamma \in \mathbb{F}_{q^n} \mid \gamma f = \delta^3 + \alpha_i y \delta \ (\exists \delta \in F_{i-1})\} \quad (i = 1, \ldots, n - 1). \tag{15}$$

$I_i$ is a vector space over $\mathbb{F}_3$.

**Proposition 1.** *For* $i = 1, \ldots, n-1$, *put* $J_i = \langle \alpha_0^{\frac{3}{2}(q^i - 1)} \beta_0, \ \ldots, \ \alpha_{i-1}^{\frac{3}{2}(q-1)} \beta_{i-1} \rangle_{\mathbb{F}_3}$. *Then we have* $I_i \supseteq J_i$ $(i = 1, \ldots, n - 1)$. *Here, for $i$ and $j$ with $j < i$,* $\alpha_i^{\frac{3}{2}(q^{i-j} - 1)} \beta_j \in I_i$ *corresponds to* $\delta = \alpha_i^{\frac{1}{2}(q^{i-j} - 1)} w_j$ *( see equation (15)).*

*Proof.* Let $i > j$. For $\gamma = \left(\frac{\alpha_i}{\alpha_j}\right)^{\frac{1}{2}} = \alpha_j^{\frac{1}{2}(q^{i-j} - 1)}$, we have

$$
\begin{aligned}
(\gamma w_j)^3 + \alpha_i y (\gamma w_j) &= \gamma^3 (w_j^3 + \tfrac{\alpha_i}{\gamma^2} y w_j) \\
&= \alpha_j^{\frac{3}{2}(q^{i-j} - 1)} (w_j^3 + \alpha_j y w_j) \\
&= \alpha_j^{\frac{3}{2}(q^{i-j} - 1)} \beta_j f.
\end{aligned}
$$

So, $\alpha_j^{\frac{3}{2}(q^{i-j} - 1)} \beta_j \in I_i$. $\square$

**Theorem 1.** *If $\beta_i \in J_i$ holds for some $i$, then $C_a$ has a component*

$$
D_a : \begin{cases}
w_0^3 + \alpha_0 y w_0 = \beta_0(-c_1 y^2 + y) \\
\quad \cdots \\
w_{i-1}^3 + \alpha_{i-1} y w_{i-1} = \beta_{i-1}(-c_1 y^2 + y) \\
\qquad w_i = \delta_i \\
\quad \cdots \\
\qquad w_{n-1} = \delta_{n-1}
\end{cases}
$$

$(\exists \delta_i, \ldots, \delta_{n-1} \in F_{i-1})$.

*Proof.* Suppose $\beta_i \in J_i$ holds for some $i$. For $j$ with $j \geq i$, we have $\beta_j = \beta_i^{q^{j-i}} \in J_i^{q^{j-i}} \subset J_j$ by (14). So, by Proposition 1, $\beta_j \in I_j$ $(\forall j \geq i)$. Then, by the definition of $I_j$, this means that the equation $w_j^3 + \alpha_j y w_j = \beta_j f$ $(j \geq i)$ for $w_j$ has a root $w_j = \delta_j$ already in $F_{i-1}$. $\square$

From Theorem 1, we see that $C_a$ has a component $D_a$ of the smaller genus if we choose $a \in \mathbb{F}_{q^n}$ such that $\beta_i \in J_i$ holds for the smaller $i$.

**Proposition 2.** *Suppose $n$ is a multiple of 4. Let $\omega \in \mathbb{F}_{q^n}$ be a root of the irreducible polynomial $T^4 - T^3 + T^2 + T - 1$ over $\mathbb{F}_3$, and $\gamma$ be any $(q - 1)/2$-th root of unity in $\mathbb{F}_q$, and $\delta$ be a root of $\delta^{\frac{3}{2}(q-1)} = \omega - \omega^3 - \omega^9$ in $\mathbb{F}_{q^n}$ (the root exists since the order of the right-hand side is a divisor of $2(q^n - 1)/(q - 1)$). Then for $a = -b_0 c_1^{-1} \beta_0^{\frac{2}{3}} \gamma \delta$, we have $\beta_2 \in J_2$.*

*Proof.* By equation (12), we have $\alpha_0 = -b_0^{-1}c_1 a$. We will find $\alpha_0$ such that

$$\beta_2 = \alpha_0^{\frac{3}{2}(q^2-1)}\beta_0 + \alpha_1^{\frac{3}{2}(q-1)}\beta_1. \tag{16}$$

By (14), we see $\beta_2 = \beta_0^{q^2}, \beta_1 = \beta_0^{q}, \alpha_1 = \alpha_0^{q}$. So, equation (16) becomes

$$\beta_0^{q^2} = \alpha_0^{\frac{3}{2}(q^2-1)}\beta_0 + \alpha_0^{\frac{3}{2}(q^2-q)}\beta_0^{q}.$$

Putting $\epsilon = \beta_0^{-\frac{2}{3}}$, $\delta = \epsilon\alpha_0$, this becomes

$$\delta^{\frac{3}{2}(q^2-1)} + \delta^{\frac{3}{2}(q^2-q)} = 1.$$

Moreover, putting $z = \delta^{\frac{3}{2}(q-1)}$, this is

$$z^q + z^{q+1} = 1. \tag{17}$$

By condition (2), the extension $\mathbb{F}_{q^n} \mid \mathbb{F}_q$ and the extension $\mathbb{F}_{3^n} \mid \mathbb{F}_3$ has the isomorphic Galois group. So, Frobenius automorphism $x \mapsto x^q$ in $\mathbb{F}_{q^n}$ becomes $x \mapsto x^3$ when restricted to $\mathbb{F}_{3^n}$. Therefore, equation (17) becomes $z^4 + z^3 = 1$ over $\mathbb{F}_{3^n}$. This has a root in $\mathbb{F}_{3^n}$ when $n$ is a multiple of 4. For example, with $\omega$ as above, we can take $z = \omega - \omega^3 - \omega^9$ □

## Example: d=5,n=4

Let $d = 5$, $n = 4$. We constructed $a$ in equation (11) using Proposition 2. In fact, for $a$ in equation (11), $C_a$ has a component

$$D_a : \begin{cases} w_0^3 + (\kappa^{86}\omega^3 + \kappa^{168}\omega^2 + \kappa^{200}\omega + \kappa^{62})yw_0 = (\kappa^{162}\omega^3 + \kappa^{239}\omega^2 + \omega + \kappa^{19})(y^2 + y) \\ w_1^3 + (\kappa^{181}\omega^3 + \kappa^{207}\omega^2 + \kappa^{168}\omega + \kappa^{182})yw_1 = (\kappa^{142}\omega^3 + \kappa^{41}\omega^2 + \kappa^{239}\omega + \kappa^{238})(y^2 + y) \\ w_2 = (\kappa^{198}\omega^3 + \kappa^{50}\omega^2 + \kappa^{186}\omega + \kappa^{223})w_0 + (\kappa^{128}\omega^3 + \kappa^{163}\omega^2 + \kappa^{135}\omega + \kappa^{223})w_1 \\ w_3 = (\kappa^{168}\omega^3 + \kappa^{184}\omega^2 + \kappa^{95}\omega + \kappa^{179})w_0 + (\kappa^{184}\omega^3 + \kappa^{198}\omega^2 + \kappa^{171}\omega + \kappa^{199})w_1 \end{cases}. \tag{18}$$

## 4   $C_{ab}$ Model of the Component $D_a$

In this section, we assume that the curve $C_a$ has the following form of component $D_a$ (see Proposition 2):

$$D_a : \begin{cases} w_0^3 + \alpha_0 yw_0 = \beta_0(-c_1 y^2 + y) \\ w_1^3 + \alpha_1 yw_1 = \beta_1(-c_1 y^2 + y) \\ \quad\quad w_2 = \gamma_2 \\ \quad\quad \cdots \\ \quad w_{n-1} = \gamma_{n-1} \end{cases}, \tag{19}$$

where, $\gamma_2, \ldots, \gamma_{n-1} \in F_1 = \mathbb{F}_{q^n}(y, w_0, w_1)$. $D_a$ has a unique point $P_\infty$ at infinity as a space curve in the space of $y, w_0, w_1$. In this section, we construct a nonsingular model of the component $D_a$ by a $C_{ab}$ curve [13,3] over $\mathbb{F}_q$, and determines its genus. In the below, we call a model by a $C_{ab}$ curve just as $C_{ab}$ model.

Because $D_a$ has a singular point (at the origin), we need some tasks to construct its nonsingular $C_{ab}$ model. Theoretically, by computing the integral closure $\tilde{R}$ of the coordinate ring $R$ of $D_a$ using the algorithm of Jong [8] and by determining functions in $\tilde{R}$ with small pole numbers at $P_\infty$, we can construct a nonsingular $C_{ab}$ model of $D_a$ using those functions [10]. However, we do the task more directly and easily as seen in Algorithm 1.

Let $v_{P_\infty}(h)$ denote an order of a function $h$ on $D_a$ at the point $P_\infty$. Since $P_\infty$ is totally ramified over $\mathbb{F}_{q^n}(y, w_0)$, we see $v_{P_\infty}(y) = -9$, $v_{P_\infty}(w_0) = -6$, $v_{P_\infty}(w_1) = -6$. Comparing the values of $w_0$ and $w_1$ at $P_\infty$, we get $v_{P_\infty}(\beta_1^{\frac{1}{3}} w_0 - \beta_0^{\frac{1}{3}} w_1) = -m$, $m < 6$.

By Lemma[Determination of defining equations](p1410) in [13], we can construct a singular $C_{m,6,9}$ model of $D_a$ over $\mathbb{F}_{q^n}$ using three functions $\beta_1^{\frac{1}{3}} w_0 - \beta_0^{\frac{1}{3}} w_1$, $w_0$, and $y$. In order to get a singular $C_{m,6,9}$ model $\overline{R}$ of $D_a$ over $\mathbb{F}_q$, we can use three functions

$$s := \mathrm{Tr}(\beta_1^{\frac{1}{3}} w_0 - \beta_0^{\frac{1}{3}} w_1), \ t := \mathrm{Tr}(w_0), \ w := y, \tag{20}$$

where, Tr is a trace of an extension

$$\mathbb{F}_{q^n}(y, w_0, \ldots, w_{n-1}) = \mathbb{F}_{q^n}(y, x_0, \ldots, x_{n-1}) \mid \mathbb{F}_q(y, x_0, \ldots, x_{n-1}).$$

Note $Tr(w_0) = w_0 + w_1 + \cdots + w_{n-1}$ by Lemma 2.

We normalize the singular $C_{ab}$ model $\overline{R}$ as follows:

**Algorithm 1 (Normalization of a singular $C_{ab}$ model)**
*Input: $R = \mathbb{F}_q[x_1, \ldots, x_n]/I$: $C_{a_1,\ldots,a_n}$ model*
*Output: its normalization $R$*

$J \leftarrow$ the radical of the ideal of singular points in $R$
**WHILE** $J \neq (1)$ **DO**
    $y \in \mathrm{Hom}_R(J, J) \setminus R$
    $n \leftarrow n + 1$
    $x_n \leftarrow y$
    $a_n \leftarrow -v_{P_\infty}(y)$
    $R \leftarrow \mathbb{F}_q[x_1, \ldots, x_n]/I$; $C_{a_1,\ldots,a_n}$ model constructed by $x_1, \ldots, x_n$
    $J \leftarrow$ the radical of the ideal of singular points in $R$

For the method for computation of $\mathrm{Hom}_R(J, J)$ ($\subset \tilde{R}$), see [20] Section 2.2.

**Example: d=5,n=4**

Let $d = 5$, $n = 4$. For $a = \kappa^{216} \omega^3 + \kappa^{95} \omega^2 + \kappa^{95} \omega$, the component $D_a$ was given by equation (18). In this case, functions $s, t, w$ in (20) are calculated as

$$\begin{cases} s = (\kappa^6 \omega^3 + \kappa^{49} \omega^2 + \kappa^{100} \omega + \kappa^{71}) w_0 + (\kappa^{190} \omega^3 + \kappa^5 \omega^2 + \kappa^{89} \omega + \kappa^{192}) w_1 \\ t = (\kappa^{151} \omega^3 + \kappa^{200} \omega^2 + \kappa^{195} \omega + \kappa^{66}) w_0 + (\kappa^{53} \omega^3 + \kappa^{113} \omega^2 + \kappa^{221} \omega + \kappa^{35}) w_1 \ . \\ w = y \end{cases}$$

First, assuming $m = 5$, we construct a $C_{5,6,9}$ model of $D_a$ using functions $s, t, w$ (If $m < 5$ in fact, then we would fail in constructing the $C_{5,6,9}$ model and we would know it):

$$\begin{cases} \kappa^{88} sw + \kappa^{60} s^3 + tw = 0 \\ w + \kappa^{176} sw + \kappa^{64} s^3 + \kappa^{22} t^3 + w^2 = 0 \\ \kappa^{159} s^3 + \kappa^{131} s^4 + \kappa^{42} s^3 t + \kappa^{88} st^3 + \kappa^{159} s^3 w + t^4 = 0 \end{cases}.$$

This model has a single singular point at the origin, and the radical $J$ of its ideal is $(w, t, s)$. Calculating $\mathrm{Hom}(J, J)$, we get $x := (w^2 + w)/s \in \tilde{R} \setminus R$. Since $v_{P_\infty}(x) = -13$, now we can construct a $C_{5,6,9,13}$ model of $D_a$ using $s, t, w$, and $x$:

$$\begin{cases} \kappa^{88} sw + \kappa^{60} s^3 + tw = 0 \\ \kappa^{154} sw + \kappa^{42} s^3 + \kappa^{220} sx + t^3 = 0 \\ w - sx + w^2 = 0 \\ \kappa^{60} s^2 + \kappa^{88} sx + \kappa^{60} s^2 w + tx = 0 \\ \kappa^{55} w + \kappa^{176} sx + \kappa s^2 w + \kappa^{137} s^4 + \kappa^{170} s^3 t + \kappa^{203} s^2 t^2 + wx = 0 \\ \kappa^{110} w + \kappa^{137} s^3 + \kappa^{170} s^2 t + \kappa^{203} st^2 + \kappa^{231} sx + \kappa^{56} s^2 w + \kappa^{192} s^4 + \kappa^{225} s^3 t \\ \quad + \kappa^{16} s^2 t^2 + \kappa s^2 x + \kappa^{230} s^5 + \kappa^{142} s^4 t + x^2 = 0 \end{cases}.$$

This model also has a single singular point at the origin, and the radical $J$ of its ideal is $(s, t, w, x)$. Calculating $\mathrm{Hom}(J, J)$, we get $u := (\kappa^{13} stw + \kappa^{13} st)/x, v := (\kappa^{170} stw + \kappa^{203} t^2 w + \kappa^{170} st + \kappa^{203} t^2)/x \in \tilde{R} \setminus R$. Since $v_{P_\infty}(u) = -7, v_{P_\infty}(v) = -8$, now we can construct a $C_{5,6,7,8,9}$ model of $D_a$ using $s, t, u, v, w$:

$$\begin{cases} w^2 + s^2 v + \kappa^{198} s^2 t + \kappa^{64} s^3 + \kappa^{176} sw + w = 0 \\ vw + \kappa^8 s^2 u + \kappa^{170} s^2 t = 0 \\ uw + \kappa^{134} s^2 t = 0 \\ v^2 + \kappa^{142} s^2 t + \kappa^{230} s^3 + \kappa^{137} sw + \kappa sv + \kappa^{110} su + \kappa^{166} st + \kappa^{230} s^2 + \kappa^{129} u + \kappa^{49} t + \kappa^{16} s = 0 \\ uv + \kappa^{194} s^3 + \kappa^{222} sw + \kappa^8 sv + \kappa^{95} su + \kappa^{189} st + \kappa^{13} t = 0 \\ tw + \kappa^{60} s^3 + \kappa^{88} sw = 0 \\ u^2 + \kappa^{65} sw + \kappa^{93} sv + \kappa^{129} su + \kappa^{190} st + \kappa^{37} s^2 + \kappa^{65} s = 0 \\ tv + \kappa^{181} sw + \kappa^{88} sv + \kappa^{124} su + \kappa^{64} st + \kappa^{153} s^2 + \kappa^{181} s = 0 \\ tu + \kappa^{173} sv + \kappa^{209} su = 0 \\ t^2 + \kappa^{47} su + \kappa^{88} st = 0 \end{cases}$$

(21)

This is a nonsingular $C_{ab}$ model.

Thus, for $a = \kappa^{216} w^3 + \kappa^{95} w^2 + \kappa^{95} w$, we succeeded in constructing a nonsingular $C_{ab}$ model (21) of $D_a$. Since the gap sequence at $P_\infty$ of (21) is $(1,2,3,4)$, we know its genus is four.

# 5  The Reduction

We constructed the $C_{ab}$ curve $D_a$ of genus 4 over $\mathbb{F}_q$ on the Weil restriction $A_a = \prod_{\mathbb{F}_{q^n} | \mathbb{F}_q} E_a$ for the value of $a$ given by Proposition 2. Tracing the route, we can construct the morphism $\Phi$ from $D_a$ to $A_a$ over $\mathbb{F}_q$ easily. From the definition of Weil restriction, the morphism $\Phi$ is also the morphism from $D_a$ to $E_a$ over $\mathbb{F}_{q^n}$. So, $\Phi$ induces the morphism $\Phi^*$ between jacobians over $\mathbb{F}_{q^n}$;

$$\Phi^* : E_a(\mathbb{F}_{q^n}) \to J_{D_a}(\mathbb{F}_{q^n}).$$

By taking a composition with the norm map, we get the morphism $\Psi$ from $E_a(\mathbb{F}_{q^n})$ to $J_{D_a}(\mathbb{F}_q)$;

$$\Psi = \mathrm{Norm}_{\mathbb{F}_{q^n}|\mathbb{F}_q} \circ \Phi^* : E_a(\mathbb{F}_{q^n}) \to J_{D_a}(\mathbb{F}_q),$$

which reduces DLP on $E_a(\mathbb{F}_{q^n})$ to DLP on $J_{D_a}(\mathbb{F}_q)$. Since the genus of $D_a$ is 4, Gaudry's variant against $J_{D_a}(\mathbb{F}_q)$ is more effective than Pollard's $\rho$ method against $E_a(\mathbb{F}_{q^n})$ [7,2].

## Example: d=5,n=4

Let $d = 5$, $n = 4$. For $a = \kappa^{216}\omega^3 + \kappa^{95}\omega^2 + \kappa^{95}\omega$, we constructed a nonsingular $C_{ab}$ model (21) of $D_a$. The morphism $\Phi$ from (the $C_{ab}$ model of) $D_a$ to $A_a$ is given by

$$\Phi: \begin{array}{ccc} D_a & \longrightarrow & A_a, \\ (s,t,u,v,w) & \mapsto & (\kappa^{55}s + \kappa^{209}t, \kappa^{223}s + \kappa^{209}t, \kappa^{193}s + \kappa^{209}t, \kappa^{55}s + \kappa^{209}t, w, w, w, w). \end{array}$$

As the morphism from $D_a$ to $E_a$, $\Phi$ can be written as

$$\Phi: \begin{array}{ccc} D_a & \longrightarrow & E_a, \\ (s,t,u,v,w) & \mapsto & ((\kappa^{81}\omega^3 + \kappa^{202}\omega^2 + \kappa^{193})s + \kappa^{209}t, w). \end{array}$$

For example, take a point $P_1 = (\kappa^4\omega^3 + \kappa^{225}\omega^2 + \kappa^{42}\omega + \kappa^{187}, \kappa^{187}\omega^3 + \kappa^{94}\omega^2 + \kappa^{197}\omega + \kappa^{239})$ of the prime order 78427 on $E_a$. Then $P_1$ is pulled back to $J_{D_a}(\mathbb{F}_{q^n})$ by $\Phi$ (In the below, an element in the jacobian of $D_a$ is expressed by a Gröebner basis w.r.t. $C_{5,6,7,8,9}$ order of the corresponding ideal ([1])) ;

$$\begin{aligned}
&\Phi^*(P_1) \\
&= (\quad u^2 + (\kappa^{231}\omega^3 + \kappa^{107}\omega^2 + \kappa^{70}\omega + \kappa^2)u + (\kappa^{194}\omega^3 + \kappa^{204}\omega^2 + \kappa^{12}\omega + \kappa^{229})s \\
&\qquad + \kappa^{205}\omega^3 + \kappa^{43}\omega^2 + \kappa^{203}\omega + \kappa^{118}, \\
&\quad su + (\kappa^4\omega^3 + \kappa^{66}\omega^2 + \kappa^{229}\omega + \kappa^{34})u + (\kappa^{201}\omega^3 + \kappa^{228}\omega^2 + \kappa^{236}\omega + \kappa^{221})s \\
&\qquad + \kappa^7\omega^3 + \kappa^{87}\omega^2 + \kappa^{78}\omega + \kappa^{55}, \\
&\quad s^2 + (\kappa^{62}\omega^3 + \kappa^{190}\omega^2 + \kappa^{33}\omega + \kappa^{64})u + (\kappa^{125}\omega^3 + \kappa^{187}\omega^2 + \kappa^{108}\omega + \kappa^{155})s \\
&\qquad + \kappa^{70}\omega^3 + \kappa^{40}\omega^2 + \kappa^{163}\omega + \kappa^{191}, \\
&\quad w + \kappa^{66}\omega^3 + \kappa^{215}\omega^2 + \kappa^{76}\omega + \kappa^{118}, \\
&\quad v + (\kappa^{183}\omega^3 + \kappa^{62}\omega^2 + \kappa^{183})u + (\kappa^{208}\omega^3 + \kappa^{72}\omega^2 + \kappa^{69}\omega + \kappa^{88})s + \kappa^{168}\omega^3 \\
&\qquad + \kappa^{86}\omega^2 + \kappa^{202}\omega + \kappa^{36}, \\
&\quad t + (\kappa^{114}\omega^3 + \kappa^{235}\omega^2 + \kappa^{226})s + \kappa^{158}\omega^3 + \kappa^{137}\omega^2 + \kappa^{196}\omega + \kappa^{99} \quad ).
\end{aligned}$$

By taking its norm to $\mathbb{F}_q$-coefficients, we get the element $j_1$ in $J_{D_a}(\mathbb{F}_q)$ corresponding to $P_1$;

$$
\begin{aligned}
j_1 =\ & \Psi(P_1) \\
=\ & \mathrm{Norm}_{\mathbb{F}_{q^n}|\mathbb{F}_q}(\Phi^*(P_1)) \\
=\ & (\ u^2 + \kappa^{230}u + \kappa^7 t + \kappa^{45}s + \kappa^{11}, \\
& tu + \kappa^{106}u + \kappa^{203}t + \kappa^{194}s + \kappa^{227}, \\
& su + \kappa^{50}u + \kappa^{98}t + \kappa^8 s + \kappa^{154}, \\
& t^2 + \kappa^{119}u + \kappa^{95}t + \kappa^{90}s + \kappa^{100}, \\
& st + \kappa^{111}u + \kappa^{13}t + \kappa^{38}s + \kappa^{70}, \\
& s^2 + \kappa^{13}u + \kappa^{76}t + \kappa^6 s + \kappa^{132}, \\
& w + \kappa^{125}u + \kappa^{193}t + \kappa^{192}s + \kappa^{188}, \\
& v + \kappa^{131}u + \kappa^{135}t + \kappa^{30}s + \kappa^{56}\ ).
\end{aligned}
$$

Similarly, for the point $P_2 = 45821 \cdot P_1 = (\kappa^{188}\omega^3 + \kappa^{141}\omega^2 + \kappa^{10}\omega + \kappa^{238}, \kappa^{34}\omega^3 + \kappa^{186}\omega^2 + \kappa^{234}\omega + \kappa^{82})$, we have

$$
\begin{aligned}
j_2 =\ & \Psi(P_2) \\
=\ & (\ u^2 + \kappa^{118}u + \kappa^{150}t + \kappa^{127}s + \kappa^{130}, \\
& tu + \kappa^{208}u + \kappa^{31}t + \kappa^{145}s + \kappa^{118}, \\
& su + \kappa^{192}u + \kappa^{42}t + \kappa^{27}s + \kappa^{134}, \\
& t^2 + \kappa^{217}u + \kappa^{17}t + \kappa^{136}s + \kappa^{12}, \\
& st + \kappa^{231}u + \kappa^{168}t + \kappa^{144}s + \kappa^6, \\
& s^2 + \kappa^{229}u + \kappa^{70}t + \kappa^{132}s + \kappa^{26}, \\
& w + \kappa^{234}u + \kappa^{185}t + \kappa^{157}s + \kappa^{106}, \\
& v + \kappa^{215}u + \kappa^{119}t + \kappa^{142}s + \kappa^{37}\ ).
\end{aligned}
$$

We verified that $j_2$ is actually equal to $45821 \cdot j_1$, using the addition algorithm in the jacobian of $C_{ab}$ curve [1].

## 6   The Cryptographic Implications

We saw an example of an elliptic curve $E_a$ over a finite field of characteristics 3, DLP on which is reduced to DLP on $C_{ab}$ curve $D_a$ of genus 4, and is attacked by Gaudry's variant effectively than by Pollard's $\rho$ method. The values of $a$ giving such week elliptic curves $E_a$ are obtained by Proposition 2. Proportion of such values of $a$ is small. So, a randomly generated $E_a$ is safe.

However, consider the following scenario. First we construct such a weak elliptic curve $E_a$ by Proposition 2. Then, we apply some isogeny against $E_a$ to get a new elliptic curve $E'$. In the almost case, $E'$ itself cannot be attacked by Weil descent technique. However, since we know the isogeny, we can reduce DLP on $E'$ to DLP on $E_a$, and so we can solve DLP on $E'$ more effectively than the others without the knowledge of the isogeny.

It seems difficult to check whether the given elliptic curve is obtained as the image of some isogeny of such a week $E_a$, or not.

# References

1. S. Arita, "Algorithms for computations in Jacobian group of $C_{ab}$ curve and their application to discrete-log-based public key cryptosystems," Conference on The Mathematics of Public Key Cryptography, Toronto, 1999.
2. S. Arita, "Gaudry's variant against $C_{ab}$ curve," LNCS 1751, Proceedings of PKC 2000, pp. 58-67, Melbourne, 2000.
3. S. Arita, "Construction of Secure $C_{ab}$ Curves Using Modular Curves," ANTS VI, pp.113–126, Leiden, 2000.
4. G.Frey and H.-G.Rück, "A remark concerning m-divisibility and the discrete logarithm in the divisor class group of curves", Mathematics of Computation, 62 (1994), 865-874.
5. S. Galbraith and N. Smart, "A Cryptographic Application of Weil Descent," HP Labs Tech. Report, HPL-1999-70 .
6. P.Gaudry, "A variant of the Adleman-DeMarris-Huang algorithm and its application to small genera," Conference on The Mathematics of Public Key Cryptography, Toronto, 1999.
7. P. Gaudry, F. Hess and N. Smart, "Constructive and destructive facets of Weil descent on elliptic curves," HP Labs Tech. Report, HPL-2000-10.
8. T. de Jong, "An algorithm for computing integral closure," J. Symbolic Comp., vol. 26, no.3, pp. 36-47, 1998.
9. N. Koblitz, "Elliptic curve cryptosystems," Mathematics of Computation, 48(1987), pp. 203-209.
10. R. Matsumoto, "Constructing Algebraic Geometry Codes on the Normalization of a Singular $C_{ab}$ Curve," Trans. of IEICE, vol. E82-A, no. 9, 1981-1985, Sep. 1999.
11. A.Menezes, T.Okamoto, and S.Vanstone, "Reducing elliptic curve logarithms to logarithms in a finite field", Proceedings of the 23rd Annual ACM Symposium on Theory of Computing, 80-89, 1991.
12. V.S. Miller, "Use of elliptic curves in cryptography," CRYPTO '85(LNCS 218), pp. 417-426, 1986.
13. S. Miura, "Linear Codes on Affine Algebraic Curves", Trans. of IEICE, vol. J81-A, no. 10, 1398-1421, Oct. 1998.
14. H.-G.Rück, "On the discrete logarithm in the divisor class group of curves," Math. Comp.,68(226),pp.805-806,1999.
15. T.Satoh, K.Araki, "Fermat Quotients and the Polynomial Time Discrete Log Algorithm for Anomalous Elliptic Curves", COMMENTARII MATHEMATICI UNIVERSITATIS SANCTI PAULI, vol. 47, No. 1, 81-92, 1998.
16. I.A.Semaev, "Evaluation of discrete logarithms in a group of p-torsion points of an elliptic curves in characteristic p," Math. Comp. 67, pp. 353-356, 1998.
17. J.H.Silverman, "The Arithmetic of Elliptic Curves", Springer-Verlag, 1986.
18. P.N.Smart, "The discrete logarithm problem on elliptic curves of trace one," J. Cryptology 12, 193-196 (1999).
19. S. Uchiyama, T. Saitoh "A Note on the Discrete Logarithm Problem on Elliptic Curves of Trace Two," Proc. of the 1998 Engineering Sciences Society Conference of IEICE, pp. 231-232.
20. W. V. Vasconcelos, "Computational Methods in Commutative Algebra and Algebraic Geometry", Springer, 1998.

# Construction of Hyperelliptic Curves with CM and Its Application to Cryptosystems

Jinhui Chao[1], Kazuto Matsuo[2], Hiroto Kawashiro[3], and Shigeo Tsujii[3]

[1] Dept. of Electrical, Electronic, and Communication Engineering,
Faculty of Science and Engineering, Chuo University,
1-13-27 Kasuga, Bunkyo-ku, Tokyo, 112-8851 Japan
jchao@elect.chuo-u.ac.jp
[2] Toyo Communication Equipment Co., Ltd.,
1-1, Koyato 2, Samukawa-machi, Koza-gun,
Kanagawa-pref., 253-0192 Japan
matuo@toyocom.co.jp
[3] Dept. of Information and System Engineering,
Faculty of Science and Engineering, Chuo University,
1-13-27 Kasuga, Bunkyo-ku, Tokyo, 112-8851 Japan
tsujii@ise.chuo-u.ac.jp

**Abstract.** Construction of secure hyperelliptic curves is of most important yet most difficult problem in design of cryptosystems based on the discrete logarithm problems on hyperelliptic curves. Presently the only accessible approach is to use CM curves. However, to find models of the CM curves is nontrivial. The popular approach uses theta functions to derive a projective embedding of the Jacobian varieties, which needs to calculate the theta functions to very high precision. As we show in this paper, it costs computation time of an exponential function in the discriminant of the CM field. This paper presents new algorithms to find explicit models of hyperelliptic curves with CM. Algorithms for CM test of Jacobian varieties of algebraic curves and to lift from small finite fields both the models and the invariants of CM curves are presented. We also show that the proposed algorithm for invariants lifting has complexity of a polynomial time in the discriminant of the CM field.

## 1 Introduction

Hyperelliptic curves and more general Jacobian varieties over finite fields have been used to build cryptosystems in recent years e.g. [20]. The cryptosystems based on these curves are recently under intensive investigation on their integrity. The generic square-root attacks works for arbitrary Abelian groups but cost exponential time in general. Various "reduction" attack initiated by the MOV attacks[23] intended to transform the discrete logarithm problem on the Jacobian varieties to some simpler and easier problems, e.g. the discrete logarithm problems on the multiplicative or the additive group of the ground field. Such attacks are effective to certain curves with special properties[30]. Another generic attack, the "smooth divisor attack" [2],[24] solves the discrete logarithm

T. Okamoto (Ed.): ASIACRYPT 2000, LNCS 1976, pp. 259–273, 2000.
© Springer-Verlag Berlin Heidelberg 2000

problem on curves of large genera in subexponential time. In particular, an attack on hyperelliptic curves of genus six is reported recently [14][11]. In spite of these researches, the discrete logarithm problems on generic curves with small genera and almost-prime orders still seemed to be at least as intractable as on the elliptic curves.

Besides, cryptosystems based on Abelian varieties of genus $g > 1$ will have also shorter word-length for the same key size than the systems on elliptic curves, which means advantages in processing, transmission and implementation. Moreover, since there is much richer isogenous classes of such curves than elliptic curves, more secure and flexible application of the cryptosystems can be expected. However, construction of secure hyperelliptic curves seems far more nontrivial than elliptic curves.

The order-counting algorithms or Schoof's algorithm for elliptic curves is extended to hyperelliptic curves. e.g. [1] shown an algorithm to calculate orders of Jacobians for curves of genus 2 in random polynomial time. [29] presented a deterministic polynomial time order-counting algorithm of $O(\log p)^\Delta$. However, it is observed that the $\Delta > \exp(\exp(2g+1))$ where $g$ is the genus of the curve [17]. [17] also extended Schoof's algorithm to plane curves over algebraic number fields with arbitrary singularity, with cost of random polynomial time $O((\log p)^\delta)$, where the $\delta = (2g + 1)^{O(1)}$. The present record of these kind of algorithms is by [3] which gave a deterministic algorithm improving [29] and cost $O((\log q)^{O(g^6)})$. All these general order counting algorithms are still too costly to be used in practical calculation and seems difficult to implement. Besides, they have to repeat the whole order counting calculations many times until an almost prime Jacobian is found. In [20], the order of a Jacobian variety at small finite fields is counted then lift the curve by the Weil conjecture This method is very fast although the number of secure curves can be found seems limited. Besides, these curves are also subjected to attacks using large automorphism groups [11].

Another approach which has been pursued in recent years is to use the simple factors of the Jacobian varieties of a special kind of curves called modular curves [13][38][39] using analytical embedding by theta functions [25][26]. Besides the computation cost for high precision expansion of these modular functions, since their method to count the order by the Eichler-Shimura formula is of exponential time, it seems that Jacobian varieties can be built over finite fields with characteristic no more than ten digits. Considering that presently used curves are with genera less than or equal to four, one can only count Jacobian varieties with order of forty digits, still quite insufficient for cryptographic applications. A recent report shown a straightforward implementation of Schoof-like algorithm on hyperelliptic curves using Cantor's analogue of the division polynomials of elliptic curves, but it can only count the Jacobian varieties with orders less than, again, forty digits[15].

A hopeful direction is to use CM curves, or the algebraic curves defined over algebraic number fields whose Jacobian varieties with complex multiplications. In fact, fast algorithms which design secure Jacobian varieties over finite field using CM curves have been shown in [7][8]. These algorithms have complexity as a polynomial in the characteristic of the finite ground fields[8]. The CM curves have nice properties to make order-counting easy and enough randomness for security as well. Furthermore, they have a potential advantage in implementation of cryptosystems. In particular, once one has an arbitrary CM curve over an algebraic number field, he will be able to design different secure curves or their Jacobian varieties over finite fields based on the same CM curve very quickly by changing the definition finite fields. It is then convenient in practice since when one wishes to update the system periodically by changing the curves he needs not to transmit a new curve over an insecure channel.

Recently, [11] presents an attack on curves with large automorphism groups and applied it to a genus six curve with CM fields as cyclotomic fields. In fact, these kind of fields contain the roots of unity of order $2g + 1$ where $g$ is the genus. However, generic CM curves has only trivial automorphism groups so these curves are among very special CM curves. These CM fields were used simply because the order calculation of the Jacobian varieties could be easy by using the Jacobi sums in the cyclotomic fields[20][27][4]. Since the polynomial time algorithm to calculate the order of Jacobian varieties for general CM field is available already [7][8], one can readily avoid such non-generic curves.

Thus, the remain problem is how to find CM curves as fast and as many as possible. More concretely, to find CM curves with small genera and large discriminants of their CM fields. Until now, the main approach to build CM curves is to use theta function theory to build a projective embedding of the Jacobian varieties[25][26]. [33] built two CM hyperelliptic curves of genus two. This approach is then improved by [38][39][36]. In fact, the nineteen CM hyperelliptic curves defined over $Q$ are built recently in [36]. This approach however needs exponentially high precision computation in the theta series expansion in order to cope with potential approximate errors. As we shown in this paper, this algorithm costs exponential time in the discriminant of the CM field.

In this paper, we present new algorithms to find explicit models of hyperelliptic curves with CM. We avoid the numerical difficulty of the analytical embedding by using only algebraic manipulations on small finite fields. Both models and invariants of the CM curves are lifted with CRT from these finite fields. The algorithm 2 which lifts models of CM curves has no restriction on genera and shapes of definition equations of curves. In the algorithm 3, 4 which lifts invariants of CM curves, we restrict ourselves to a subfamily so that one can always obtain the model of the curve from their invariants. Besides, these algorithms can also be used to other curves of genera larger than two if an explicit definition of their moduli invariants is available. Being probabilistic algorithms, the CM tests proved to be very simple and sharp. The lifting algorithms are of deterministic, and a complexity analysis shows that the invariant lifting algorithm has complexity of a polynomial in the discriminant of the CM field.

This paper is organized as follows. In the chapter two, we give some notations and definitions. In the chapter three, we show algorithms for CM test of Jacobian varieties of algebraic curves. In the chapter four, an algorithm to lift from finite fields the models of algebraic curves with CM is presented. In the chapter five, an algorithm is presented to lift the invariants of CM hyperelliptic curves. In the chapter six, we show a complexity analysis for both the analytical embedding algorithm and the invariant lifting algorithm. Finally, we show an examples to compare the proposed algorithm with the analytical embedding algorithm and also application to design a secure hyperelliptic cryptosystems. In the appendix we show an algorithm to find the model of a curve from its moduli invariants using a subfamily of curves.

## 2   Preliminary

A hyperelliptic curve over a field $F$ of genus $g$ is defined by

$$C : Y^2 + Y h(X) = f(X)$$

with the point at infinity, where $\deg h \leq g$, $\deg f = 2g + 1$. For char$F \neq 2$, one can use the definition as

$$C : Y^2 = f(X).$$

A $F$-rational point $P \in C(F)$ is defined by both $P = (x, y)$ such that $x, y \in F$ and $y^2 + yh(x) = f(x)$ or the point at infinity. A (Weil) divisor $D$ on $C$ is defined as a finite formal sum of form $\sum_i m_i P_i$, $m_i \in \mathbf{Z}$, $P_i \in C(\bar{F})$ . The degree of $D$ is defined as $\deg(D) = \sum_i m_i$. In particular, the divisors with degree zero form a subgroup $\mathcal{D}^0(C)$ of the divisor group whose elements are algebraically equivalent to zero. The function field of $C$ is consisted of $\{p/q\}, p, q \in \bar{F}[u, v], q \neq 0$ mod $v^2 + vh(u) - f(u)$. The divisor of a function $p/q$ on $C$ is defined as $\sum_i m_i P_i - \sum_j n_j Q_j$, here $P_i, Q_j \in C(\bar{F})$ are zeros and poles of the function and $m_i, n_j$ are the multiplicity of the zeros and the poles. It can be shown that all the divisors of functions over $C$ have degree zero and will be called as principal divisors, or linearly equivalent to zero. Obviously the principal divisors form a subgroup $\mathcal{D}^l(C)$ of $\mathcal{D}^0(C)$. The Jacobian variety of $C$ is then defined as follows.

$$\mathcal{J} = \mathcal{D}^0(C)/\mathcal{D}^l(C)$$

For $F = \mathbf{F}_q$, a $\mathbf{F}_q$-rational divisor is defined as a divisor which is fixed under the Galois action on $\mathbf{F}_q$ and the group of $\mathbf{F}_q$-rational points $\mathcal{J}(\mathbf{F}_q)$ is generated by $\mathbf{F}_q$-rational divisors.

It is known that a Jacobian variety is an Abelian variety or a complete and nonsingular variety with the commutative addition law as an algebraic group. As proved by A. Weil for curves of genus $g$, the orders of their Jacobian varieties over finite fields fall in the following range.

$$(q^{1/2} - 1)^{2g} \leq \#\mathcal{J}(\mathbf{F}_q) \leq (q^{1/2} + 1)^{2g}$$

The Jacobian varieties of hyperelliptic curves can then be used to defined discrete logarithm problem as to find $m \in \mathbf{Z}$ given two divisors $D_1, D_2 \in \mathcal{J}(\mathbf{F}_q)$ such that $D_1 = mD_2$.

We now introduce an important property of the endomorphism rings of Abelian varieties. Let $F$ be an algebraic number field, $A/F$ a $g$-dimensional Abelian variety, $\mathrm{End}_F A$ its endomorphism ring. It is known that for a simple Abelian variety $A$, $\mathrm{End}_F A$ is a division algebra of finite rank over $\boldsymbol{Q}$ with an involution $x \mapsto x'$ such that if $x \neq 0$, $\mathrm{Tr}_{F/\boldsymbol{Q}}(xx') > 0$. Define $K = \mathrm{End}^\circ A :=$ $\mathrm{End}_F A \otimes_{\boldsymbol{Z}} \boldsymbol{Q}$. When $K$ is isomorphic to a totally imaginary quadratic extension of a totally real extension of $\boldsymbol{Q}$ of degree $2g$, $A$ is called with complex multiplications or CM. $K$ is called the CM field of $A$. It is known that ordinary Abelian varieties over finite fields are all CM, and any CM Abelian variety is isogenous to an Abelian variety over finite fields. Further details of notations are referred to e.g. [21], [32].

# 3   CM Tests of Jacobian Varieties

In this section, we show an efficient algorithm to test whether the Jacobian variety of an algebraic curve has CM, which proves very useful in later chapters. This probabilistic algorithm is based on certain interesting relation between the reduction of an Abelian variety over an algebraic number field to a finite field modulo a prime ideal in the integral ring of the number field, lying over a prime number, and the decomposition of the principal ideal generated by the prime number in the integral ring of the definition field [32][21].

**Definition 1.** *A pseudo-CM algebraic curve is defined as one whose Jacobian variety passed one of the following CM tests.*

Consider a curve $C/F$, $F$ an algebraic number field, we denote the residue field of a prime $p$ of $F$ as $\boldsymbol{F}_q$, $Z(X)$ the characteristic polynomial of the Frobenius endomorphism on $\boldsymbol{F}_q$-rational divisors of the Jacobian variety $\mathcal{J}/\boldsymbol{F}_q$ of $C/\boldsymbol{F}_q$. To simplify treatment involved with the reflex CM field and reflex CM type, we will hereafter assume that all the CM fields are abelian and the $Z(X)$'s are irreducible.

Below, we will use the algorithms in e.g. [8][28] to calculate CM field and CM type of a Jacobian variety with CM.

**Algorithm 1**
**Procedure 1    (Ordinary reduction test)**

**Input** A random curve $C/F$ of genus $g$, $N \in \boldsymbol{N}$;
**Output** If $C/F$ is a pseudo-CM curve, and when it is, the CM field $K$;
**Step 1** Find the CM field and the discriminant $d_1$ of $Z_1(X)$ of $C/\boldsymbol{F}_{q_1}$ with ordinary reduction for a small prime $p_1$;
**Step 2** Choose small primes $p_i, i = 2, \cdots, N$ such that $Z_1(X)$ mod $p_i$ splits. For $\mathcal{J}/\boldsymbol{F}_{q_i}$ find the discriminant $d_i$ of $Z_i(X)$, if the square-free part of $d_1$ equals not that of $d_i$ for some $i$, output that $C/F$ has no CM;
**Step 3** Output $C$ as a pseudo-CM curve whose CM field $K$ has a minimal polynomial $Z_1(X)$;

**Procedure 2     (Supersingular/bad reduction tests)**

**Input** A random curve $C/F$ of genus $g$, $N \in \mathbf{N}$;

**Output** If $C/F$ is a pseudo-CM curve, its CM field $K$;

**Step 1** Find the CM field and the characteristic polynomial $Z_1(X)$ of $\mathcal{J}/\mathbf{F}_{q_1}$ with ordinary reduction for a small prime $p_1$;

**Step 2** Choose small primes $p_i, i = 2, \cdots, N$ such that $Z_1(X) \bmod p_i$ is irreducible or $p_i | d_1$, if neither $\mathcal{J}/\mathbf{F}_{q_i}$ is supersingular nor $C/\mathbf{F}_{q_i}$ is singular for any $i$, output that $C/F$ has no CM;

**Step 3** Output $C$ as a pseudo-CM curve whose CM field has a minimal polynomial $Z_1(X)$;

# 4   Lifting Models of Curves with CM

It is known that every CM Abelian variety $A$ has a projective model over $\bar{\mathbf{Q}}$. The definition field $F$ of equations is contained in the definition field of the model $(A, \iota, \mathcal{C})$, where $\iota : K \hookrightarrow \mathrm{End}_F^\circ A$ an embedding and $\mathcal{C}$ a polarization, which is however not easy to find. One may then use the latter for $F$ instead and also denote it as $F$. Furthermore, one may choose that the definition fields of the models are coincide with the so-called fields of moduli under certain conditions[40], which can be built from the class field of $K$. In particular, when $A$ is simple which is the case we are dealing with, and $A$ is principal e.g. $\mathrm{End}(A)$ equals the maximal order of $\mathcal{O}_K$, the definition field of the model can derived under minor conditions from the Hilbert class field, which we denote as $K_{ab}$. [32][40].

In this section, we show how to lift from small finite fields the models of curves with CM defined e.g. over the class field of $K$.

## Algorithm 2

**Input** : A model of equations of a curve family $\{C/\bar{\mathbf{Q}}\}$ with genus $g$;

**Output** : Curves in the family with CM over $K_{ab}$ and their CM fields $K$;

**Step 1** For small prime $p_1$, choose models of all non-isomorphic curves $C_1/\mathbf{F}_{q_1}$ among the family over $\mathbf{F}_{q_1}$, e.g. in the case of hyperelliptic curves, one may use

$$y^2 \equiv x^{2g+1} + a_1 x^{2g} +, \cdots, + a_{2g} \bmod p_1$$

such that $\mathcal{J}_1/\mathbf{F}_{q_1}$ are ordinary, their CM fields $K$ are abelian and $Z(X)$ irreducible. Calculate $K$, its discriminant $d_K$ and the class number $h$ of $\mathcal{O}_K$;

**Step 2** For each of the pairs $(C_1/\mathbf{F}_{q_1}, K)$, choose small prime $p_i$ such that $p_i | d_K$, then find the curves $C_i/\mathbf{F}_{q_i}$ such that either $C_i/\mathbf{F}_{q_i}$ is singular or $\mathcal{J}_i/\mathbf{F}_{q_i}$ is supersingular. The conjugates of the coefficients in each definition equation are collected to compose the reduction of the minimal polynomial of the same coefficients in the curve $C/K_{ab}$ with CM field $K$, modulo the prime ideal over $p_i$;

**Step 3** For each of the pairs $(C_1/\mathbf{F}_{q_1}, K)$, choose small prime $p_i, i = 2, \dots M$ such that $p_i$ is inert in $K$, then find the curves $C_i/\mathbf{F}_{q_i}$ such that either $C_i/\mathbf{F}_{q_i}$ is singular or $\mathcal{J}_i/\mathbf{F}_{q_i}$ is supersingular. The conjugate of coefficients in the equation are collected to compose the reduction of the minimal polynomial of the same coefficients in the curve $C/K_{ab}$ with CM field $K$;

**Step 4** For each of the pair $(C_1/\boldsymbol{F}_{q_1}, K)$, choose small prime $p_i, i = 2, \ldots M$ such that $p_i$ splits completely in $K$, then find the curves $C_i/\boldsymbol{F}_{q_i}$ such that $End^\circ_{\boldsymbol{F}_{q_i}} \mathcal{J}_i \cong K$. Again, the conjugate coefficients are collected to compose the minimal polynomial of the same coefficient in the curve $C/\boldsymbol{Q}$ with CM field $K$;

**Step 5** Choose one candidate curve $C_i$ for each $i$ and apply the CRT to each coefficients to recover the equation $C/K_{ab}$;

**Step 6** If the $C/K_{ab}$ passed the CM tests, then output it as a pseudo-CM curves, if not goto step 2-4 to try the other combinations or add one more prime;

*Remark 1.* The family of curves is not limited to hyperelliptic curves or genus two curves. A reasonable choice for such a family is the de Jong-Noot family [10], which is known to contain infinite number of CM curves.

*Remark 2.* For fast implementation, one-parameter family would be desirable. More efficient approach is to use to select candidate curves by determination of the isomorphic type of the endomorphism ring of $J/\boldsymbol{F}_q$, using the algorithm such as generalization of the Kohel's algorithm for ordinary reduction in [22].

*Remark 3.* It is also possible to use certain convenient properties in the super-singular reduction to raise the lift efficiency. Especially, choose $p_i$ carefully the reduction of the Jacobian will be isomorphic to product of supersingular elliptic curves (supersingular Abelian varieties) then calculation over elliptic curves can be made use of [19].

## 5   Lifting of Invariants of Hyperelliptic Curves with CM Jacobians

It can be observed in the lifting of the models of curves that it is desirable if one can lift the invariants instead of the models in order to reduce the number of candidates. This is possible if an explicit definition of moduli invariants is known, such as in the genus two case.

**Algorithm 3**

**Input** A model of curve family $C/\bar{\boldsymbol{Q}}$ of which their invariants $\mathcal{I} = (I_1, \ldots, I_m)$ in their moduli space is explicitly defined;

**Output** : Invariants $\mathcal{I}$ of curves in the family with CM over $K_{ab}$ and their CM fields $K$;

**Step 1** For small prime $p_1$, choose among the family of all non-isomorphic curves $C_1/\boldsymbol{F}_{q_1}$ such that $\mathcal{J}_{1s}/\boldsymbol{F}_{q_1}$ are ordinary reductions, their CM fields $K$ are abelian and $Z(X)$ irreducible. Calculate $K$ and the discriminants $d_K$, the class number $h$ of $\mathcal{O}_K$ and their invariants $\mathcal{I}_1/\boldsymbol{F}_{q_1}$;

**Step 2** For each of the pairs $(C_1/\boldsymbol{F}_{q_1}, K)$, choose small prime $p_i$ such that $p_i | d_K$, then find the curves among the family $C_i/\boldsymbol{F}_{q_i}$ such that either $C_i/\boldsymbol{F}_{q_i}$ is singular or $\mathcal{J}_i/\boldsymbol{F}_{q_i}$ is supersingular. Calculate all conjugates of their invariants $\mathcal{I}_i/\boldsymbol{F}_{q_i}$. These invariants are collected to compose the reduction of the minimal polynomial of the same invariant $\mathcal{I}$ of the curve $C/K_{ab}$ with CM field $K$, modulo the prime ideal over $p_i$;

**Step 3** For each of the pairs $(C_1/\boldsymbol{F}_{q_1}, K)$, choose small prime $p_i, i = 2, \ldots M$ such that $p_i$ is inert in $K$, then find the curves among the family $C_i/\boldsymbol{F}_{q_i}$ such that either $C_i/\boldsymbol{F}_{q_i}$ is singular or $\mathcal{J}_i/\boldsymbol{F}_{q_i}$ is supersingular. Calculate all conjugates of their invariant $\mathcal{I}_i/\boldsymbol{F}_{q_i}$. Then compose the reduction of the minimal polynomial the same invariants $\mathcal{I}$ of the curve $C/K_{ab}$ with CM field $K$;

**Step 4** For each of the pair $(C_1/\boldsymbol{F}_{q_1}, K)$, choose small prime $p_i, i = 2, \ldots M$ such that $p_i$ splits completely in $K$, then find the curves among the 1-parameter family $C_i/\boldsymbol{F}_{q_i}$ such that $End^{\circ}_{\boldsymbol{F}_{q_i}} \mathcal{J}_i \cong K$. Calculate all conjugates of their invariants $\mathcal{I}_i/\boldsymbol{F}_{q_i}$. Then compose the reduction of the minimal polynomial of the same invariants $\mathcal{I}$ of the curve $C/K_{ab}$ with CM field $K$;

**Step 5** Choose one candidate minimal polynomial of $\mathcal{I}_i$ for each $i$ and use the CRT to lift each coefficient of the minimal polynomial to $K_{ab}$;

**Step 7** Test if the model $C/K_{ab}$ with the invariant $\mathcal{I}$ passed the CM tests, then output it as a pseudo-CM curve, if not goto Steps 2-4 to try the other combinations or add one more prime;

*Remark 4.* Again efficient identification of the isomorphism type of the endomorphism ring of $J/\boldsymbol{F}_q$ could substantially accelerate the calculation. The only algorithm available presently is in [22] which generalizes Kohel's algorithm for determination of the isomorphic type of endomorphism ring of ordinary elliptic curves over finite fields, which uses the Cantor's analogue of division polynomials for elliptic curves.

*Remark 5.* One can lift either the integral (relative ) or the absolute invariants. The absolute invariants is known as algebraic numbers but may not be algebraic integers. To lift such numbers, one may use the algorithm in [37] which needs the CRT of double size of the maximum between the numerator or the denominator.

*Remark 6.* It is known that usually to find the equation of a curve from its invariants is very difficult. In projective embedding using theta functions, this problem is solved by using Mestre's trick, which however does not apply here. We show an algorithm to overcome this problem using one-parameter family. It is shown as Algorithm 4 in Appendix. To find equations of curves one may apply its Steps 1-2 before the Step 1 of the algorithm 3 and its Step2 3-4 will be used after the Step 5 in the algorithm 3. The example to be shown bellow used a new approach to find curve equation from its invariant based on polynomial resultant, which will be reported in the near future.

*Remark 7.* Further approaches to reduce the number of candidates so as to accelerate the whole calculation are discussed in [22][16].

## 6    Complexity Analysis

We give analysis of both the analytical embedding of CM Jacobian varieties using the theta functions and the proposed algorithm for invariant lifting.

Below, we follow the notations and the algorithms in e.g. [36] and assume $g = 2$.

**Theorem 1.** *The analytical embedding using the theta functions costs exponential time of the discriminant of the CM field $K$: $O(|d_K|^{9/8}2^{3/2}\sqrt{|d_K|})$.*

*Proof.* For simplicity we assume the CM type $(K, \{\phi_i\})$ is self-dual, the endomorphism ring is the maximal order of $\mathcal{O}_K$. The discriminant of $K$ is denoted as $d_K$. Assume the principal polarization of the embedding is given by the Riemann form

$$E(z, w) = \sum_i \phi_i(\xi)(\bar{z}_i w_i - z_i \bar{w}_i), \quad \xi \in K, K = K^+(\xi), \ \phi_i(\xi^2) < 0.$$

The theta functions can be estimated by Minkowsky's lemma, using minimal the sum of abstract values to approximate the minimal type trace. In particular, for $\delta, \epsilon \in \mathbf{R}^2, \Omega = \mathrm{diag}[\phi_i(\xi)]$

$$\left|\theta\begin{bmatrix}\delta\\\epsilon\end{bmatrix}(\Omega)\right| = \left|\sum_{m\in\mathbf{Z}^2} e^{\pi i(m+\delta)^t\Omega(m+\delta)} \times e^{2\pi i(m+\delta)^t\epsilon}\right|$$

$$= O\left(\sum_{m\in\mathbf{Z}} e^{-\pi\ \mathrm{Im}T_\Phi(\xi)m^2}\right)$$

$$= O(e^{-(96\pi^3)^{1/4}|d_K|^{1/8}}).$$

Then in the Rosenhain normal form

$$y^2 = x(x-1)(x-\lambda_1)(x-\lambda_2)(x-\lambda_3)$$

the roots $\lambda_i$ can be estimated from the theta constants or the values of theta functions on particular choices of $\delta, \epsilon$: $\lambda_i = O(e^{\ 4(96\pi^3)^{1/4}|d_K|^{1/8}})$. Thus, the integral Igusa Invariants which are defined by the Rosenhain normal form can be estimated as $I_i = O(e^{120(96\pi^3)^{1/4}|d_K|^{1/8}})$. Since the absolute invariants are homogeneous ratios of the integral invariants, the calculations in the embedding by the theta functions will be dominated by those for the integral invariants, we will use the estimate of integral invariants in analysis of the whole algorithm.

Next, assume that the Igusa invariant is defined over a ray class field, for simplicity a Hilbert class field. To calculate an algebraic integer with a minimal polynomial of degree $h$ will generally cause precision of

$$\mathrm{Prec}(d_K) = O(120(96\pi^3)^{1/4}h\binom{h}{\lfloor\frac{h}{2}\rfloor}|d_K|^{1/8})$$

due to the error accumulation mainly in the middle term of the minimal polynomial (see also [5]). Using the Sterling formula and take a upper bound of the class number $h$ as $\sqrt{|d_K|}$,

$$\mathrm{Prec}(d_K) = O(2^{\sqrt{|d_K|}} |d_K|^{\frac{3}{8}}).$$

Take the number of the terms in the theta series expansion as

$$\sqrt{\frac{\mathrm{Prec}(d_K)}{|d_K|^{1/8}}} = 2^{1/2\sqrt{|d_K|}}|d_K|^{1/4}$$

the complexity of the whole embedding is of $O(|d_K|^{9/8}2^{3/2\sqrt{|d_K|}})$.     □

**Theorem 2.** *The invariant lifting algorithm find the model of a CM curve in cost of polynomial time of the discriminant of the CM field $K$: $O(|d_K|^{135/8})$.*

*Proof.* Consider the lifting of the Igusa invariants over the Hilbert class field of the CM field $K$. The largest coefficient, which is of the middle degree term, but its order can also be estimated as the highest degree coefficient. Since $I_i = O(e^{120(96\pi^3)^{1/4}|d_K|^{1/8}})$, the order of the largest coefficient is about $O(I_i^h)$. By the Chinese remainder theorem and the theorem of prime number, one knows that in order to lift such integers, it is enough to repeat calculations of its shadows or reductions on $L = O(|d_K|^{5/8})$ finite fields $\boldsymbol{F}_q$. The sizes of these finite fields are also of the same order $q = O(L)$. (Here lifting of a rational numbers requires CRT in twice size of denominators and numerators, but the order remains the same [37].)

Determination of isomorphic types of the endomorphism ring over $\boldsymbol{F}_q$ using the generalized Kohel algorithm in [22] required $O(q^{20})$ computations. If this algorithm is applied to all $q^6$ curves over each finite fields, the calculations will be $L^{26} = O(|d_K|^{65/4})$. The overall cost is then $L^{27} = O(|d_K|^{135/8})$.

Lifting of the minimal polynomials of the absolute Igusa invariants for $h$ coefficients from the $L$ residues over finite fields of size $L$ requires $hL^3$ computations . The whole calculation is $L = O(|d_K|^{19/8})$. Thus, the whole complexity is dominated by $L^{27} = O(|d_K|^{135/8})$.     □

## 7   Example

We show an example of construction of a secure hyperelliptic curve using the CM field $K = \boldsymbol{Q}(\alpha)$, where $\alpha = \sqrt{-61 + 6\sqrt{61}}$. One can shown that $Gal(K/\boldsymbol{Q}) \cong \boldsymbol{Z}/4\boldsymbol{Z}$, its class number $h = 1$ and the minimal polynomial of $\alpha$ is $Z(X) = X^4 + 4X^2 + 2$.

Firstly, we construct by ordinary lifting the absolute Igusa invariants of a curve of which the endomorphism ring is isomorphic to the maximal order $O_K$ of $K$. We chose some small primes $l$ such that $Z$ splits completely modulo $l$ and we can compute the absolute Igusa invariants of curves of which the endomorphism ring over $\mathbf{F}_l$ is isomorphic to $\mathcal{O}_K$. Then they were lifted to $\mathbf{Q}$ by the CRT and Wang's algorithm in [37]. The following table shows the process in which the set of the primes $l$, used in residue collecting and lifting steps of the invariants by CRT, is enlarged one by one. The places marked by "−" denotes when no rational number is output by Wang's algorithm, which means one has to use more primes and the residues because Wang's algorithm requires the product of all the primes used in CRT greater than the square of the maximum between the numerator and the denominator.

| $l$ | $(i_1, i_2, i_3)$ |
|---|---|
| 13 | $(-, \frac{1}{2}, 1)$ |
| 47 | $(8, \frac{-4}{5}, -)$ |
| 73 | $(\frac{-11}{75}, \frac{31}{114}, \frac{-139}{95})$ |
| 83 | $(\frac{-782}{1277}, \frac{32}{32}, \frac{-960}{977})$ |
| 103 | $(\frac{5212}{957}, \frac{909}{9098}, \frac{12548}{2655})$ |
| 131 | $(\frac{45725}{81556}, -, -)$ |
| 137 | $(-, \frac{179573}{119374}, -)$ |
| 179 | $(-, -, \frac{-11412461}{1383790})$ |
| 199 | $(\frac{-77531094}{109697555}, -, \frac{24918900}{5235127})$ |
| 239 | $(\frac{2186985284}{3193941857}, -, \frac{-1640487534}{417031347})$ |
| 241 | $(\frac{23480370079}{7651764248}, -, \frac{-3798052317}{1459572997})$ |
| 257 | $(\frac{24829901496}{677918816561}, -, -)$ |
| 269 | $(-, \frac{-73355585483}{60532937290}, \frac{13309544621557}{7113235908976})$ |
| 317 | $(-, \frac{-1310651902222379}{527755546859371}, -)$ |
| 347 | $(\frac{-2134258375735}{380855313817042}, \frac{4795562804619412}{416652904406961}, \frac{437038183335891}{1181681515448155})$ |
| 367 | $(\frac{-9544691354733067}{1335850947268042676}, \frac{5771153292886870746}{495445778110039265}, -)$ |
| 379 | $(\frac{4815475531661448664}{556460730772033016}, \frac{793418475254355983}{2138841242118446}, \frac{-1724244246804899840}{957123916241973161})$ |
| 439 | $(\frac{6401196155012958989}{382383046061239901463}, \frac{5088610825173732293315}{375740404519468288884}, -)$ |
| 443 | $(-, \frac{-61811314844862517342}{95297355758955401148675}, \frac{10729788160492482025270}{847878261510243193250})$ |
| 449 | $(-, \frac{-35103170735291450086810}{7392748713701866063586}, \frac{-123886307667533380144430}{17187104504207396501870})$ |
| 461 | $(\frac{-29927487013532718098126716}{979914900416525246051843316}, \frac{1638494059002328300160}{7515286520993930649211}, \frac{-363114988047137371773170}{1981252142710456062498})$ |
| 503 | $(\frac{6148917421576784421994809}{\,}, -, -)$ |
| 569 | $(-, -, -)$ |
| 571 | $(-, -, \frac{-5963566557743154949796078203}{1536547059300225516350695163})$ |
| 607 | $(-, \frac{-4040478469371641363774098636368}{\,}, \frac{-63585703806382049374801395712}{663049476357702087656258025625})$ |
| 619 | $(-, \frac{-288633656196550173439682001}{18400583279728414407612609559}, -)$ |
| 683 | $(-, \frac{-82310585301850115990999586471322}{6883986318383374838069483160436853}, \cdot)$ |
| 691 | $(-, \cdot, \cdot)$ |
| 727 | $(\frac{9002218257568433062920010702880679}{7319341342731625939019371833767634839}, \cdot, \cdot)$ |
| 733 | $(-, \cdot, \cdot)$ |
| 757 | $(-, \cdot, \cdot)$ |
| 809 | $(\frac{28793780960554991224537212069644628066355163}{14177662602226964645619958031257350556163}, \cdot, \cdot)$ |
| 827 | $(\frac{-144498005853290597609676324874208359622}{\,}, \cdot, \cdot)$ |
| 863 | $(\frac{1075013978414957123575367744211073508857}{459886998583272734726451530263685776195060605}, \cdot, \cdot)$ |
| 911 | $(\frac{1007454955893534349749734524587766180992187}{64362932426160170239524338594955645950915199}, \cdot, \cdot)$ |
| 937 | $(\frac{3164988533830060995355023583058091910369064}{1161522064576509496050413807041682945649672192}, \cdot, \cdot)$ |
| 977 | $(\frac{-1161522064576509496050413807041682945649672192}{6402508672323911307856854498377929685875}, \cdot, \cdot)$ |

Secondly, we construct a secure Jacobian defined over a finite field by the fast algorithm of [8]. Specifically, using prime ideal factorization of the Frobenius endomorphism, we found a principal prime ideal of $K$

$$(\omega) = \begin{pmatrix} 1 & \frac{\alpha+1}{2} & \frac{\alpha^2+7}{12} & \frac{\alpha^3+5\alpha^2+7\alpha+35}{120} \end{pmatrix} \begin{pmatrix} -438577 \\ -3748 \\ 284050 \\ 124962 \end{pmatrix} \mathcal{O}_K$$

such that $N_{K/\mathbf{Q}}(\omega) = p$, where $p = 523126243402421378838738 7$. Then we obtained a secure Jacobian with order

$$\#\mathcal{J}(\mathbf{F}_p) = 2^4 \times p_{max}$$

where $p_{max}$ is a $160 bits$ prime number

$$p_{max} = 1710381665854894312958517262601197350921820022483.$$

Finally, we construct a secure curve over $\mathbf{F}_p$, of which Jacobian has above order, from invariants calculated above:

$$i_1 = \frac{-1161522664578509496050138070416829456496721 92}{640250862723239113078565449877929687 5}$$
$$i_2 = \frac{-40404764693716413637740986368}{2210164921192340292187 5}$$
$$i_3 = \frac{-63585703806382049374801395712}{6630494767357702087656 25}.$$

The equation of curves is restricted here in form of

$$Y^2 = X^5 + X^3 + a_2 X^2 + a_1 X + a_0.$$

Notice that this restriction does not exclude any possible isomorphism classes of the curves.

By an algorithm mentioned before using polynomial resultant computation, we obtained coefficients of a curve with given invariants as

$$a_0 = 4179295909743236969433 68$$
$$a_1 = 2257561965032447596454 492$$
$$a_2 = 2418466578595705463946 119$$

over $\mathbf{F}_p$. The twisted curve of the above curve has equation as

$$C/\mathbf{F}_p : Y^2 = X^5 + c^2 X^3 + c^3 a_2 X^2 + c^4 a_1 X + c^5 a_0,$$

where $c = 2$. It has the same secure order constructed above.

**Acknowledgment**: The authors wish to thank Prof. Fumiyuki Momose for helpful discussions, Prof. Gerhard Frey for interesting comments on [33] and Dr. Michael Müller for sending us a copy of Dr. Spallek's thesis.

# References

1. L.M.Adleman, M.D.A.Huang, "Primality Testing and Abelian Varieties Over Finite Fields," Springer-Verlag , (1992.)
2. L.M.Adleman , J.DeMarrais, M.D.Huang, "A Subexponential Algorithms for Discrete Logarithms over the Rational Subgroup of the Jacobians of Large Genus Hyperelliptic Curves over Finite Fields," Proc. of ANTS95, Springer, (1995)

3. L.M.Adleman, M.D.Huang, "Counting rational points on curves and Abelian varieties over finite fields" Henri Cohen (Ed) "Algorithmic number theory" Lecture Notes in Computer Science, 1122, Second International Symposium, ANTS-II, Proceedings, p.1-16. 1996

4. S.Arita, "Public key cryptosystems with $C_{ab}$ curve (II)" IEICE, Symposium on Cryptography and Information Security, SCIS'98, 7.1-B, 1998-1.

5. A.O.L.Atkin, F.Morain, "Elliptic Curves and Primality Proving" , Research Report 1256, INRIA, (1990).

6. D.Cantor, "Computing in the jacobian of hyperelliptic curve," Math. Comp., vol.48, p.95-101, (1987)

7. J.Chao, N.Matsuda, S.Tsujii, "Efficient construction of secure hyperelliptic discrete logarithm problems" Springer-Verlag Lecture Notes on Computer Science, Vol.1334, pp.292-301, "Information and Communication Security" Y. Han, T. Okamoto, S. Qing (Eds.) Proceedings of First International Conference ICICS'97, Beijing, China, Nov. 1997.

8. J.Chao, K.Matsuo, S.Tsujii "Fast construction of secure discrete logarithm problems over Jacobian varieties," Information Security for Global Information Infrastructures: IFIP TC 11 16th Annual Working Conference on Information Security, S.Qing and J.Eloff(Eds.), Kluwer, July 2000.

9. H. Cohen "A course in computational algebraic number theory," Springer, GTM-138, 1995.

10. J.de Jong, R.Noot, "Jacobians with complex multiplication," Arithmetic Algebraic Geometry, Birkhäuser PM89, pp.177-192, 1991.

11. Duursma, Gaudry, Morain, "Speeding up the discrete log computation on curves with automorphism", Proceeding. Asiacrypt-99, 1999.

12. G. Frey, H.G. Rück, "A remark concerning m-divisibility and the discrete logarithm in the divisor class group of curves," Math. Comp., 62, 865-874, 1994.

13. G.Frey, M.Müller, "Arithmetic of modular curves and applications," Preprint, 1998.

14. P.Gaudry "A variant of the Adelman-DeMarrais-Huang algorithm and its application to small genera," Preliminary version, June 1999.

15. P.Gaudry, R.Harley, "Counting points on hyperelliptic curves over finite fields," Preprint, 2000.

16. T.Haga, K.Matsuo, J.Chao, S.Tsujii, "Construction of CM hyperelliptic curves using ordinary lifting," Proc. of SCIS'2000, IEICE Japan, 2000.

17. M.D.Huang, D.Ierardi, "Counting Rational Point on Curves over Finite Fields," Proc. 32nd IEEE Symp. on the Foundations of Computers Science, 1993.

18. J.Igusa, "Arithmetic variety of for genus two," Ann. of Math. , vol.72, No.3, p.612-649, (1960)

19. K. Kamio, H. Kawashiro, J. Chao, S. Tsujii, "A fast algorithm of model lifting for CM hyperelliptic curves," Proc. SCIS'99, IEICE, Japan, 1999.

20. N.Koblitz, "Hyperelliptic cryptosystems," J. of Cryptology, vol.1, p.139-150, (1989)

21. S.Lang, "Complex multiplication," Springer-Verlag, (1983)

22. K.Matsuo, J.Chao and S.Tsujii, "On lifting of CM hyperelliptic curves," Proc. of SCIS'99, W3-1.4, IEICE Japan (1999).

23. A.Menezes, S.Vanstone, T.Okamoto, "Reducing Elliptic Curve Logarithms to Logarithms in a Finite Fields," Proc. of STOC, p.80-89, (1991).

24. V. Müller, A. Stein, C. Thiel, "Computing discrete logarithms in real quadratic congruence function fields of large genus," Preprint, Nov. 13, (1997)

25. D.Mumford, "Tata Lectures on Theta I," Birkhäuser, Boston , (1983).

26. D.Mumford, "Tata Lectures on Theta II," Birkhäuser, Boston , (1984).
27. K.Nagao, "Construction of the Jacobians of Curves $Y^2 = X^5 + k$ /$\boldsymbol{F}_p$ with Prime Order," Manuscript, 1998.
28. O. Nakamura, N. Matsuda, J. Chao, S. Tsujii, "On cryptosystems based on abeian varieties with CM," IEICE, Symposium on Cryptography and Information Security, SCIS'97, 12-E, 1997-1. IEICE, Tech. rep. ISEC-96-81, 1997-3.
29. J.Pila, "Frobenius maps of abelian varieties and finding roots of unity in finite fields," Math. Comp., vol.55 , p. 745-763, (1990)
30. H.G. Rück, "on the discrete logarithm problem in the divisor class group of curves," Preprint, 1997.
31. J.P.Serre, J.Tate, "Good reduction of abelian varieties," Ann. of Math. (2), 88 (1968), page 492-517.
32. G.Shimura : "Abelian Varieties with Complex Multiplication and Modular Functions", Princeton Univ. Press, 1998.
33. A-M.Spallek, "Kurven vom Geschlecht 2 und ihre Anwendung in Public-Key-Kryptosystemen," Dissertation, preprint, No. 18, 1994.
34. J.Tate, "Endomorphisms of Abelian varieties over finite fields," Invent. Math. 2, p.134-144, (1966)
35. E.J.Volcheck, "Computing in the Jacobian of a plane algebraic curve," Proc. of ANT-1, p.221-233, LNCS-877, (1994).
36. P.V.Wamelen, "Examples of genus two CM curves defined over the rationals," Math. Comp., 68(225), pp. 308-320, 1999.
37. P.S.Wang, "A $p$-adic algorithm for univariate partial fractions," Proc. of ACM SYMSAC'81, ACM, 212-217, 1981.
38. X. Wang, "2-dimensional simple factors of $J_0(N)$," Manuscripta Mathematica, 87:179-197, 1995.
39. H.J. Weber, "Hyperellptic simple factors of $J_0(N)$ with dimension at least 3," Experimental Math. vol. 6, No.4, 273-287, 1997.
40. H. Yoshida, "Hecke characters and models of abelian varieties with complex multiplication," J. Fac. Sci. Univ. of Tokyo, Sec. IA, 28 , 633-649(1982).

## Appendix: Subfamilies Whose Models Can Be Determined from Their Invariants

Since it is generally difficult to find an explicit model of a curve with given invariants, we will use the following algorithm to find the models of curves from a point in their moduli space.

### Algorithm 4

**Input** : A model of a family $\mathcal{C}/K$ with $r$ parameter $\alpha_1, \cdots, \alpha_r$, $\{f(x, y, \alpha_1, \cdots, \alpha_r) = 0\}$, and a point $\mathcal{I} = (I_1, \ldots, I_m)$ in their moduli space;

**Output** : Definition field $F$ and a model of $C/F : h(x, y, \alpha) = 0$ corresponding to $\mathcal{I}$, where $\alpha \in F$;

**Step 1** Choose $r - 1$ constraints $c_i(\alpha_1, \cdots, \alpha_r) = 0, i = 1, \ldots, r - 1$ in the parameter space to obtain a subfamily $h(x, y, \alpha)$ with 1-parameter $\alpha$;

**Step 2** Reduce the definition equations of the invariants $I_i := g_i(\alpha_1, \cdots, \alpha_r)$ to $J_i(\alpha) = 0, i = 1, \ldots, m$;

**Step 3** Calculate $\gcd(J_1, \ldots, J_m) =: J(x) \in K[I_1, \ldots, I_m][x]$;

**Step 4** If $J(x) \neq const$, output $h(x, y, \alpha)$ as the model over definition field $F := K(\alpha)$ with a minimal polynomial as $J(x)$;

We can then apply this algorithm to obtain models of hyperelliptic curves of genus two from their Igusa invariants.

# Provable Security for the Skipjack-like Structure against Differential Cryptanalysis and Linear Cryptanalysis

Jaechul Sung[1], Sangjin Lee[1], Jongin Lim[1], Seokhie Hong[1], and Sangjoon Park[2]

[1] Center for Information and Security Technologies(CIST),
Korea University, Anam Dong, Sungbuk Gu,
Seoul, KOREA
{sjames, sangjin, jilim, hsh}@cist.korea.ac.kr
[2] National Security Research Institute(NSRI), Taejon, KOREA
sjpark@etri.re.kr

**Abstract.** In this paper we introduce a structure iterated by the rule A of Skipjack and show that this structure is provably resistant against differential or linear attacks. It is the main result of this paper that the upper bound of r-round ($r \geq 15$) differential(or linear hull) probabilities are bounded by $p^4$ if the maximum differential (or linear hull) probability of a round function is $p$, and an impossible differential of this structure does not exist if $r \geq 16$. Application of this structure which can be seen as a generalized Feistel structure in a way to block cipher designs brings out the provable security against differential and linear attacks with some upper bounds of probabilities. We also propose an interesting conjecture.

## 1 Introduction

The most powerful known attacks on block ciphers are Differential Cryptanalysis(DC) [2,3] and Linear Cryptanalysis(LC) [10,11]. Since such cryptanalyses have been proposed, designers of block ciphers have tried to give the provable security against DC and LC. Kanda et al [7] classified four measures to evaluate the security of a cipher against DC and LC as follows;

1. Precise measure : The maximum average of differential and linear hull probabilities.
2. Theoretical measure : The upper bounds of the maximum average of differential and linear hull probabilities.
3. Heuristic measure : The maximum average of differential characteristic and linear approximation probabilities.
4. Practical measure : The upper bounds of the maximum average of differential characteristic and linear approximation probabilities.

Among the above four measures, the first two are the measures of the theoretical point of view and the last two are the measures of the practical point of

T. Okamoto (Ed.): ASIACRYPT 2000, LNCS 1976, pp. 274–288, 2000.

view. If the number of rounds increases, it is computationally infeasible to compute exactly with the point of the precise and heuristic measures. Therefore the theoretical and practical measures are important measures to show the security of a cipher against DC and LC. However the practical measure does not give a sufficient condition for the security of a cipher against DC and LC. It is only a necessary condition, so the theoretical measure is the only left one to give the provable security against DC and LC.

K.Nyberg and R.Knudsen showed that the r-round differential(or linear hull) probabilities in the Feistel structure are bounded by $2p^2$ if the maximal differential(or linear hull) probability of round function is $p$ and $r \geq 4$ [16]. Furthermore, the probability can be reduced to $p^2$ if the round function is bijective and $r \geq 3$ [1]. So the construction of a round function with a small maximal probability of differentials(linear hull) is a very important factor to give the provable structure against DC(LC). M.Matsui gave an example of such a construction using the iterative nested Feistel structures [12,13].

In this paper we will prove the security of an iterated cipher which follows the rule A of Skipjack structure against DC and LC. The r-round($r \geq 15$) differential probabilities are bounded by $p^4$ if the maximal differential probability of the round function is $p$. Since the proof of linear hull probabilities in LC is almost same as that of differential probabilities [12,16,17], we will just prove the upper bound of differential probabilities of the structure. Furthermore we will show that there does not exists an impossible differential if $r \geq 16$ in the generalized Feistel structure and Skipjack-like structure. Also we give some conjectures in the generalized Feistel and Skipjack-like structures.

## 2   Preliminaries

Differential cryptanalysis uses the non-uniformity of the output differences given input differences and linear cryptanalysis relies on the correlations of input/ouput bits and key bits. Block ciphers are usually constructed iteratively with the same round function. So in order to avoid DC and LC it needs to use the round functions which have the good properties against such attacks with sufficient rounds.

In this section we consider a round function $F\colon GF(2)^n \to GF(2)^n$. We assume that round keys are independent and uniformly random. Furthermore, input data are also independent and uniformly random.

**Definition 1.** *[12] For any given $\Delta X, \Delta Y, \Gamma X, \Gamma Y \in GF(2)^n$, the differential and linear hull probabilities of a round function $F$ are defined as;*

$$DP^F(\Delta X \to \Delta Y) = \frac{\#\{x \in GF(2)^n \mid F(X) \oplus F(X \oplus \Delta X) = \Delta Y\}}{2^n}$$

$$LP^F(\Gamma X \to \Gamma Y) = \left( \frac{\#\{x \in GF(2)^n \mid \Gamma X \bullet X = \Gamma Y \bullet F(Y)\}}{2^{n-1}} - 1 \right)^2$$

*where $\Gamma x \bullet \Gamma y$ denotes the parity of bitwise exclusive-or of $\Gamma x$ and $\Gamma y$.*

In the above definitions the probabilities mean the average probabilities for all the possible keys. To give the provable security against DC and LC with the theoretical measure we need the following definitions.

**Definition 2.** *The maximal differential and linear hull probability of $F$ are defined by*

$$DP_{max}^{F} = \max_{\Delta x \neq 0, \Delta y} DP^{F}(\Delta x \to \Delta y)$$

*and*

$$LP_{max}^{F} = \max_{\Gamma x, \Gamma y \neq 0} LP^{F}(\Gamma x \to \Gamma y)$$

*respectively.*

On the point of view of the provable security $DP_{max}^{F}, LP_{max}^{F}$ are the very important factors. With above two definitions, we can easily get the following two theorems.

**Theorem 1.** *[12] (i) For any function $F$,*

$$\sum_{\Delta Y} DP^{F}(\Delta X \to \Delta Y) = 1, \quad \sum_{\Gamma X} LP^{F}(\Gamma X \to \Gamma Y) = 1.$$

*(ii) For any bijective function $F$,*

$$\sum_{\Delta X} DP^{F}(\Delta X \to \Delta Y) = 1, \quad \sum_{\Gamma Y} LP^{F}(\Gamma X \to \Gamma Y) = 1.$$

If $F_1$ and $F_2$ which are functions from $GF(2)^n$ to $GF(2)^n$ are used as consecutive round functions and relatively independent, we can calculate differential and linear hull probabilities with the following theorem.

**Theorem 2.** *[12] For any $\Delta X, \Delta Z, \Gamma X, \Gamma Z \in GF(2)^n$,*

$$DP^{F_1,F_2}(\Delta X \to \Delta Z) = \sum_{\Delta Y} DP^{F_1}(\Delta X \to \Delta Y) \cdot DP^{F_2}(\Delta Y \to \Delta Z)$$

*and*

$$LP^{F_1,F_2}(\Gamma X \to \Gamma Z) = \sum_{\Gamma Y} LP^{F_1}(\Gamma X \to \Gamma Y) \cdot LP^{F_2}(\Gamma Y \to \Gamma Z).$$

Since the method of calculating linear hull probabilities can be calculated with the reverse order of the method of calculating differential probabilities [12,16,17], we will only consider the differential probabilities in this paper.

## 3    Provable Security for Block Cipher Structures against DC and LC

Structures of block ciphers can be roughly classified by the Feistel structure and the SPN structure. Since there has been much progress in the structures of bijective functions with good properties, the interest in the SPN structure has been increased. There are Square [5], Riindael [6], and Crypton [8] which are constructed by considering the branch number [18] in the SPN structure from the practical point of view. However the Feistel structure has been used more widely since it has no limit of round function. In this section we consider the security of the Feistel structure and its modifying structure against DC and LC. We assume that the round keys of round function $F$ are mutually independent and uniformly distributed and the maximal differential probability of the round function $F$, $DP^F_{max}$, is $p$.

K.Nyberg and R.Knudsen showed that the r-round differential(or linear hull) probabilities in the Feistel structure are bounded by $2p^2$ if the maximal differential(or linear hull) probability of round function is $p$ and $r \geq 4$ in Feistel structure. Furthermore, the probability can be reduced to $p^2$ if the round function is bijective and $r \geq 3$. So the smaller probability $p$ is, the better security level against DC and LC we can give. For example, consider the Feistel structure block cipher which has bijective round function $F : GF(2)^{32} \rightarrow GF(2)^{32}$ with more than or equal to 3 round. If the maximal differential probability is close to $2^{-32}$, then the upper bound of differential of the cipher is close to $2^{-64}$. So we can give the almost perfect security against DC. M.Matsui gave the example of such a construction using the iterated nested Feistel structures [12,13].

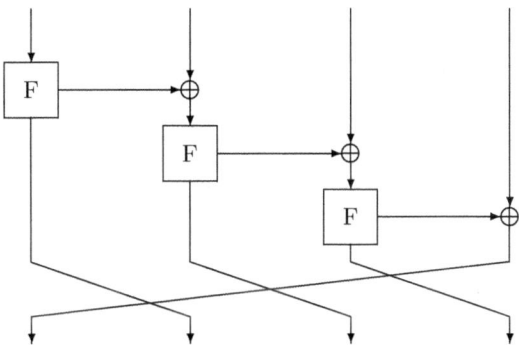

**Fig. 1.** Skipjack-like structure

Since AES(Advanced Encryption Standard) have been proposed, the 128-bit block ciphers are usually adopted. If we construct 128-bit block ciphers with the Feistel structure, we need to design 64-bit round function. However, to construct 64-bit round function are usually more difficult than to design 32-bit round function and it is also a hard problem to give the provable security against DC and LC. So the generalized Feistel structure which divides input blocks by 4 was proposed and used in MARS, RC6, TWOFISH, and etc. We also have Skipjack [19] which is the 64-bit block cipher with the generalized Feistel structure dividing input blocks by 4 and it has 32 rounds where half of them are ruled by A type and the others by B type. Fig. 1 describe the structure of iterated ciphers using the rule A of Skipjack. Since the output block of a round function $F$ has effects on the next block and its own block, Skipjack-like structure is different to the generalized Feistel structure and data randomization is faster than the generalized Feistel structure. However the Skipjack-like structure needs a bijective round function. In the next section we will prove the upper bound of differential probabilities of the Skipjack-like structure as in the Feistel structure case.

## 4    The Main Result - Provable Security against DC and LC in the Skipjack-like Structure

In this section we prove the upper bound of differential probabilities in the iterated Skipjack-like structure from the theoretical point of view. We assume that a round function $F$ is bijective and the maximal differential of $F$ is $p$.

Now we consider the 15-round Skipjack-like iterated block cipher. In Fig. 2 the $\alpha_i$'s mean the input block differences, $\beta_i$'s mean the output block differences and $\delta_i$'s are variables which mean i-th round output differences. Set an input difference to $\alpha = (\alpha_1, \alpha_2, \alpha_3, \alpha_4)$ and an output difference $\beta = (\beta_1, \beta_2, \beta_3, \beta_4)$. By the assumption that a round function is bijective we just consider $\alpha \neq 0$ and $\beta \neq 0$. Therefore for the given nonzero input difference $\alpha$ the probability that an output difference is $\beta$ is calculated as following. We denote the 15-round differential probability as $DP(\alpha \rightarrow \beta)$ and $DP^F(\Delta X \rightarrow \Delta Y)$ as $DP(\Delta X \rightarrow \Delta Y)$.

$$
\begin{aligned}
DP(\alpha \rightarrow \beta) \;=\; \sum_{\delta_i, 1 \leq i \leq 11} & DP(\alpha_1 \rightarrow \delta_1) \cdot DP(\alpha_2 \oplus \delta_1 \rightarrow \delta_2) \cdot DP(\alpha_3 \oplus \delta_2 \rightarrow \delta_3) \\
& \cdot DP(\alpha_4 \oplus \delta_3 \rightarrow \delta_4) \cdot DP(\delta_1 \oplus \delta_4 \rightarrow \delta_5) \cdot DP(\delta_2 \oplus \delta_5 \rightarrow \delta_6) \\
& \cdot DP(\delta_3 \oplus \delta_6 \rightarrow \delta_7) \cdot DP(\delta_4 \oplus \delta_7 \rightarrow \delta_8) \cdot DP(\delta_5 \oplus \delta_8 \rightarrow \delta_9) \quad (1) \\
& \cdot DP(\delta_6 \oplus \delta_9 \rightarrow \delta_{10}) \cdot DP(\delta_7 \oplus \delta_{10} \rightarrow \delta_{11}) \cdot DP(\delta_8 \oplus \delta_{11} \rightarrow \beta_3 \oplus \beta_4) \\
& \cdot DP(\delta_9 \oplus \beta_3 \oplus \beta_4 \rightarrow \beta_1) \cdot DP(\delta_{10} \oplus \beta_1 \rightarrow \beta_2) \cdot DP(\delta_{11} \oplus \beta_2 \rightarrow \beta_3)
\end{aligned}
$$

Using the equation(1) we prove the following main theorem.

**Theorem 3.** *If a round function of the Skipjack-like structure is bijective and $r \geq 15$, then r-round differential probabilities are bounded by $p^4$ where $p$ is the maximal average differential probability of a round function.*

*Proof.* We prove the case r=15. If r is greater than 15, we can easily prove by the case r=15 and the Theorem 1,2. We will prove the theorem case by case and the cases are classified by 8, i.e. , the $\beta_i$'s ($1 \leq i \leq 3$) are zero or not.

Case 1( $\beta_1 = 0, \beta_2 = 0, \beta_3 = 0$ )
Since we do not consider the case $\beta \neq 0$, $\beta_4$ is nonzero. By the case assumption, we have $\delta_7 = \delta_{10} = \delta_{11} = 0$ and $\delta_3 = \delta_6 = \delta_9 = \beta_4 \neq 0$. Therefore $\delta_3, \delta_6, \delta_7, \delta_9, \delta_{10}, \delta_{11}$ is fixed and variable $t = \{\delta_1, \delta_2, \delta_4, \delta_5, \delta_8\}$ will be only summed over in equation (1). So we have the following;

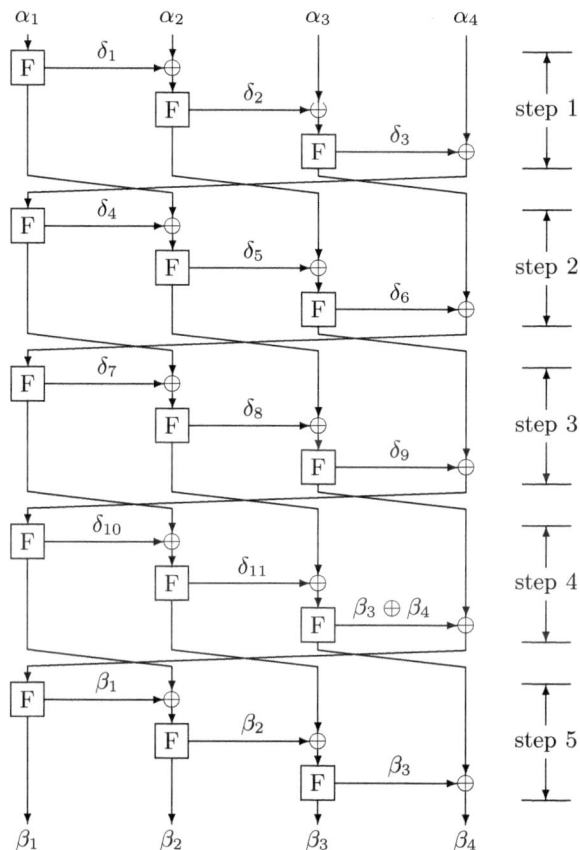

**Fig. 2.** Notations of 15-round differential

**Table 1.** Notations of Proof

| Relations | | | |
|---|---|---|---|
| Variable $t$ | | | |
| step 1 | $DP(\alpha_1 \rightarrow \delta_1)$ | $DP(\alpha_2 \oplus \delta_1 \rightarrow \delta_2)$ | $DP(\alpha_3 \oplus \delta_2 \rightarrow \delta_3)$ |
| step 2 | $DP(\alpha_4 \oplus \delta_3 \rightarrow \delta_4)$ | $DP(\delta_1 \oplus \delta_4 \rightarrow \delta_5)$ | $DP(\delta_2 \oplus \delta_5 \rightarrow \delta_6)$ |
| step 3 | $DP(\delta_3 \oplus \delta_6 \rightarrow \delta_7)$ | $DP(\delta_4 \oplus \delta_7 \rightarrow \delta_8)$ | $DP(\delta_5 \oplus \delta_8 \rightarrow \delta_9)$ |
| step 4 | $DP(\delta_6 \oplus \delta_9 \rightarrow \delta_{10})$ | $DP(\delta_7 \oplus \delta_{10} \rightarrow \delta_{11})$ | $DP(\delta_8 \oplus \delta_{11} \rightarrow \beta_3 \oplus \beta_4)$ |
| step 5 | $DP(\delta_9 \oplus \beta_3 \oplus \beta_4 \rightarrow \beta_1)$ | $DP(\delta_{10} \oplus \beta_1 \rightarrow \beta_2)$ | $DP(\delta_{11} \oplus \beta_2 \rightarrow \beta_3)$ |

$$DP(\alpha \to \beta) = \sum_t DP(\alpha_1 \to \delta_1) \cdot DP(\alpha_2 \oplus \delta_1 \to \delta_2) \cdot DP(\alpha_3 \oplus \delta_2 \to \delta_3)$$
$$\cdot DP(\alpha_4 \oplus \beta_3 \to \delta_4) \cdot DP(\delta_1 \oplus \delta_4 \to \delta_5) \cdot DP(\delta_2 \oplus \delta_5 \to \beta_4)$$
$$\cdot DP(\delta_4 \to \beta_4) \cdot DP(\delta_5 \oplus \delta_8 \to \beta_4) \cdot DP(\delta_8 \to \beta_4)$$

Among the above equation $DP(\alpha_3 \oplus \delta_2 \to \delta_3), DP(\delta_2 \oplus \delta_5 \to \beta_4), DP(\delta_5 \oplus \delta_8 \to \beta_4)$ and $DP(\delta_8 \to \beta_4)$ are bounded by $p$ since the output differences are nonzero and F is bijective. So we have

$$DP(\alpha \to \beta) \le p^4 \cdot \sum_t DP(\alpha_1 \to \delta_1) \cdot DP(\alpha_2 \oplus \delta_1 \to \delta_2) \cdot DP(\alpha_4 \oplus \beta_3 \to \delta_4)$$
$$\cdot DP(\delta_1 \oplus \delta_4 \to \delta_5) \cdot DP(\delta_4 \to \beta_4) \le p^4.$$

From now on we will use the table such as Table 1. In the Table 1 relations mean the relations of variable $\alpha_i$'s,$\beta_i$'s and $\delta_i$'s. Therefore the variables of the relations in the table relations are fixed and variables($= t$) are only summed over in the equation (1). Using the notations of Table 1 we can represent the proof of Case 1 by the following table.

**Table 2.** Proof of Case 1 : $\beta_1 = 0, \beta_2 = 0, \beta_3 = 0$

| Relations | $\delta_7 = \delta_{10} = \delta_{11} = 0,\ \delta_3 = \delta_6 = \delta_9 = \beta_4 \ne 0$ | | |
|---|---|---|---|
| Variable $t$ | $\delta_1, \delta_2, \delta_4, \delta_5, \delta_8$ | | |
| step 1 | sum over $\delta_1$ | sum over $\delta_2$ | $\le p$ |
| step 2 | sum over $\delta_4$ | sum over $\delta_5$ | $\le p$ |
| step 3 | 1 | sum over $\delta_8$ | $\le p$ |
| step 4 | 1 | 1 | $\le p$ |
| step 5 | 1 | 1 | 1 |

Case 2( $\beta_1 = 0, \beta_2 = 0, \beta_3 \ne 0$ )
We divide Case 2 by 2 cases whether $\beta_3 \oplus \beta_4$ is zero or not. In the Case 2-1 $DP(\alpha \to \beta)$ is bounded by $p^5$ and in the Case 2-2 $DP(\alpha \to \beta)$ is bounded by $p^4$.

Proofs of other cases can be proved in the similar way. More details are in the Appendix. All the cases $DP(\alpha \to \beta)$ is bounded by $p^4$.

Since the Skipjack-like structure can be regarded as one of the generalizations of the Feistel structure in a way, provable security against LC is also obtained as in [12,16,17].

**Theorem 4.** If a round function of the Skipjack-like structure is bijective and $r \ge 15$, then $r$-round linear hull probabilities are bounded by $q^4$ where $q$ is the maximal average linear hull probability of a round function.

**Table 3.** Proof of Case 2-1 : $\beta_1 = 0, \beta_2 = 0, \beta_3 \neq 0, \beta_3 \oplus \beta_4 = 0$

| Relations | $\delta_6 = \delta_9 = \delta_{10} = 0,\ \delta_2 = \delta_5 = \delta_7 = \delta_{11}$ | | |
|---|---|---|---|
| Variable $t$ | $\delta_1, \delta_3 \neq 0, \delta_4, \delta_{11} \neq 0$ | | |
| step 1 | sum over $\delta_1$ | $\leq p$ | sum over $\delta_3$ |
| step 2 | sum over $\delta_4$ | $\leq p$ | 1 |
| step 3 | $\leq p$ | $\leq p$ | 1 |
| step 4 | 1 | $\leq p$ | 1 |
| step 5 | 1 | 1 | sum over $\delta_{11}$ |

**Table 4.** Proof of Case 2-2 : $\beta_1 = 0, \beta_2 = 0, \beta_3 \neq 0, \beta_3 \oplus \beta_4 \neq 0$

| Relations | $\delta_{10} = 0,\ \delta_6 = \delta_9 = \beta_3 \oplus \beta_4$ | | |
|---|---|---|---|
| Variable $t$ | $\delta_1, \delta_2, \delta_3, \delta_4, \delta_5, \delta_7 \neq 0, \delta_8, \delta_{11} \neq 0$ | | |
| step 1 | sum over $\delta_1$ | sum over $\delta_2$ | sum over $\delta_3$ |
| step 2 | sum over $\delta_4$ | sum over $\delta_5$ | $\leq p$ |
| step 3 | $\leq p$ | sum over $\delta_7$ | sum over $\delta_8$ |
| step 4 | 1 | $\leq p$ | $\leq p$ |
| step 5 | 1 | 1 | sum over $\delta_{11}$ |

Now, let's consider one of the generalization of the Feistel structure as Fig. 3.

Assume that the round function is bijective. In the case $m = 2$ (the Feistel structure), if $r \geq 3 = 1 \cdot 3$, then r-round differential probabilities are bounded by $p^2$. In the case $m = 3$, S.Sung [21] proved that r-round differential probabilities are bounded by $p^3$ if $r \geq 8 = 2 \cdot 4$. Also in the case $m = 4$, r-round differential probabilities are bounded by $p^4$ if $r \geq 15 = 3 \cdot 5$. So we can conjecture the following.

*Conjecture 1.* In the generalized Feistel structure and Skipjack-like structure, r-round differential probabilities are bounded by $p^m$ if $r \geq (m - 1)(m + 1)$.

## 5  Impossible Truncated Differential of the Generalized Feistel Structure and Skipjack-like Structure

In this section we consider an impossible truncated differential of $(i)$ the generalized Skipjack-like structure whose one-round transformation is $F_k(x_1, x_2, \cdots, x_m) = (f_k(x_1) \oplus x_2, x_3, , \cdots, x_m, f_k(x_1))$, and $(ii)$ the generalized(CAST256-like) Feistel structure whose one-round transformation is $F_k(x_1, x_2, \cdots, x_m) = (f_k(x_1) \oplus x_2, x_3, \cdots, x_m, x_1)$, where $f_k : \{0, 1\}^n \to \{0, 1\}^n$ is a keyed-round function.

Now we can consider the relation of an impossible truncated differential and a number of round in $(i), (ii)$. We assume that round functions are bijective, random, and pairwise independent. Let $\Delta X = (\Delta X_1, \cdots, \Delta X_m)$ and $\Delta Y =$

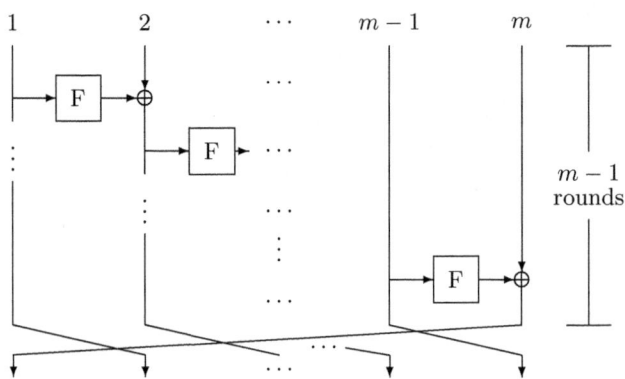

**Fig. 3.** Generalized Feistel Structure

$(\Delta Y_1, \cdots, \Delta Y_m)$ be an input and output difference respectively. Then we have the following results.

**Proposition 1.** *If* $r = m^2 - 1$*, there exist an impossible truncated differential whose form is* $(0, 0, \cdots, 0, \Delta\alpha) \nrightarrow (\Delta\beta, 0, \cdots, 0)$ *in* (*i*) *and* (*ii*)*, where* $\Delta\alpha$ *and* $\Delta\beta$ *are nonzero.*

**Note** : In the case of $m = 3$ , we can find the 8-round impossible truncated differential whose form is $(0, 0, \Delta\alpha) \nrightarrow (\Delta\beta, 0, 0)$ in (*ii*), where $\Delta\alpha$ and $\Delta\beta$ are nonzero (similarly it holds in (*i*)). Consider the following figure. Since we assumed that round functions are bijective and $\Delta\alpha$ is nonzero, $\Delta t$ is nonzero. But the four round output differential is zero. This is the contradiction.

With Proposition 1 and the notion of pseudorandomness in Luby and Rackoff [9], we can conjecture that impossible differentials and the pseudorandomness are closely related. However, the number of queries in the impossible differential attack model are more than that in the distinguishing attack model [14]. Also we can conjecture the followings.

*Conjecture 2. If* $r \geq m^2$*, there does not exist an impossible truncated differential in* (*i*) *and* (*ii*)*.*

Conjecture 2 can be proved by a computer programming if m is small enough, say less than 32. A similar method can be seen in [20]. However, since we could not find a general rule of proof, we just do conjecture it in the case that m is large. So we need further works.

We can find the impossible differential whose form is $(0, 0, 0, \Delta\alpha) \nrightarrow (\Delta\beta, 0, 0, 0)$ in the Skipjack-like structure($m = 4$) if $r = 15$. Skipjack is the 64-bit block cipher with 80-bit key and 32-round($A^8 B^8 A^8 B^8$) using rules A and B iteratively. There has been the impossible differential attack [4] which use the weakness of this cipher to apply the rule B only after 8-round of rule A. These attacks only

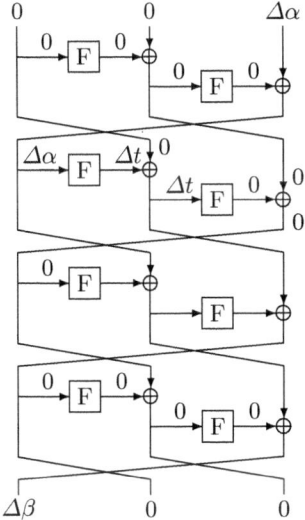

**Fig. 4.** 8-round impossible truncated differential in the case of $m = 3$

use the structural weakness. However, if Skipjack algorithm use $A^{16}B^{16}$ or $A^{32}$ then the impossible differential attack can not be applied any more by Conjecture 2 in case $m = 4$.

## 6   Conclusion

In this paper we give the provable security for the Skipjack-like cipher against DC from the theoretical point of view. If the maximal differential of a round function of a Skipjack-like cipher is $p$ and $r \geq 15$, then r-round differential probabilities are bounded by $p^4$. Also we suggest the conjecture that r-round differential probabilities are bounded by $p^m$ if $r \geq (m-1)(m+1)$ and there does not exists an impossible differential if $r \geq m^2$ in the generalized Feistel structure and Skipjack-like structure.

It seems a hard problem to give the provable security against DC and LC in the block cipher. Until now, there have been no 128-bit block cipher with the provable security against DC and LC from the theoretical point of view. So we believe our result to be very helpful to design provably secure block ciphers against DC and LC.

## References

1. K. Aoki and K. Ohta, *Stict evaluation for the maximum average of differential probability and the maximem average of linear probability*, IEICE Transcations fundamentals of Elections, Communications and Computer Sciences, No.1, pp 2-8, 1997.

2. E. Biham and A. Shamir, *Differential cryptanalysis of DES-like cryptosystems*, Advances in Cryptology - CRYPTO'90, LNCS 537, Springer-Verlag, 1991, pp. 2–21.
3. E. Biham and A. Shamir, *Differential cryptanalysis of the full 16-round DES*, Advances in Cryptology - CRYPTO'92, LNCS 740, Springer-Verlag, 1992, pp. 487–496.
4. E. Biham, A. Biryukov, and A. Shamir, *Cryptanalysis of skipjack reduced to 31 rounds using impossible differentials*, Advances in Cryptology - EUROCRYPT'99, LNCS 1592, Springer-Verlag, 1992, pp. 12–23.
5. J. Daemen, Lars R.Knudsen, and Vincent Rijmen. *The block cipher SQUARE*, Fast Software Encryption Workshop 97, 1997, pp 137–151.
6. J. Daemen and V. Rijndael, *The Rijndael block cipher*, AES proposal, 1998.
7. M. Kanda, Y. Takashima, T. Matsumoto, K. Aoki, and K. Ohta, *A strategy for constructing fast functions with practical security against differential and linear cryptanalysis*, Selected Areas in Cryptography, LNCS 1556, 1999, pp 264–279.
8. C.H. Lim, *CRYPTON: A new 128-bit block cipher*, AES proposal, 1998.
9. M. Luby and C. Rackoff, *How to construct pseudorandom permutations from pseudorandom functions*, SIAM J.Comput., vol. 17, pp.373-386, 1988.
10. M. Matsui, *Linear cryptanalysis method for DES cipher*, Advances in Cryptology - EUROCRYPT'93, LNCS 765, Springer-Verlag, 1994, pp. 386–397.
11. M. Matsui, *The first experimental cryptanalysis of the Data Encryption Standard*, Advances in Cryptology - CRYPTO'94, LNCS 839, Springer-Verlag, 1994, pp. 1–11.
12. M. Matsui, *New structure of block ciphers with provable security against differential and linear cryptanalysis*, Fast Software Encryption Workshop 96, 1996, pp. 205–218.
13. M. Matsui, *New Block Encryption Algorithm MISTY*, Fast Software Encryption Workshop 97, 1997.
14. Shiho Moriai and Serge Vaudenay, *Comparision of Randomness Provided by Several Schemes for Block Ciphers*, Presented at Third AES Workshop, April 2000.
15. M. Naor and O. Reingold, *On the construction of pseudorandom permutations : Luby-Rackoff Revisited*, J.Cryptology, pp.29-66, 1999.
16. K. Nyberg and Lars R. Knudsen, *Provable security against differential cryptanalysis*, Advances in Cryptology - CRYPTO'92, LNCS 740, Springer-Verlag, 1992, pp. 566–574.
17. K. Nyberg, *Linear approximation of block ciphers*, Presented at rump session, Eurocrypt'94, May 1994.
18. Vincent Rijmen, J. Daemen, Bart Preneel, Antoon Bosselaers, and Erik De Win, *The cipher SHARK*, Fast Software Encryption Workshop 96, 1996, pp. 99–112.
19. *Skipjack and KEA Algorithm Specifications, version 2.0*, Technical report, Available at the National Institute of Standard and Technology web page, http://crsc.nist.gov/encryption/skipjack-kea.htm, May 1998.
20. M. Sugita, K Kobara, K. Uehara, S. Kubota, and H. Imai, *Relations among Differential, Truncated Differential, Impossible Differential Cryptanalyses against Word-Oriented Block Ciphers like Rijndael, E2*, Presented at Third AES Workshop, April 2000.
21. Suhak Sung, Private Communications, 1999.

# Appendix: Proof of Theorem 3

**Table 5.** Proof of Case 3-1 : $\beta_1 = 0, \beta_2 \neq 0, \beta_3 = 0, \beta_4 = 0$

| Relations | $\delta_9 = 0,\ \delta_5 = \delta_8 = \delta_{11} = \beta_2 \neq 0$ | | |
|-----------|-------------------|-------------------|-------------------|
| Variable $t$ | $\delta_1, \delta_2, \delta_3, \delta_4, \delta_6, \delta_7, \delta_{10}$ | | |
| step 1 | sum over $\delta_1$ | sum over $\delta_2$ | sum over $\delta_3$ |
| step 2 | sum over $\delta_4$ | $\leq p$ | sum over $\delta_6$ |
| step 3 | sum over $\delta_7$ | $\leq p$ | 1 |
| step 4 | sum over $\delta_{10}$ | $\leq p$ | 1 |
| step 5 | 1 | $\leq p$ | 1 |

**Table 6.** Proof of Case 3-2 : $\beta_1 = 0, \beta_2 \neq 0, \beta_3 = 0, \beta_4 \neq 0$

| Relations | $\delta_9 = \beta_4 \neq 0,\ \delta_{11} = \beta_2 \neq 0$ | | |
|-----------|-------------------|-------------------|-------------------|
| Variable $t$ | $\delta_1, \delta_2, \delta_3, \delta_4, \delta_5, \delta_6, \delta_7, \delta_8, \delta_{10}$ | | |
| step 1 | sum over $\delta_1$ | sum over $\delta_2$ | sum over $\delta_3$ |
| step 2 | sum over $\delta_4$ | sum over $\delta_5$ | sum over $\delta_6$ |
| step 3 | sum over $\delta_7$ | sum over $\delta_8$ | $\leq p$ |
| step 4 | $\leq p$ | sum over $\delta_{10}$ | $\leq p$ |
| step 5 | 1 | $\leq p$ | 1 |

**Table 7.** Proof of Case 4-1 : $\beta_1 \neq 0, \beta_2 = 0, \beta_3 = 0, \beta_4 = 0$

| Relations | $\delta_8 = \delta_{11} = 0,\ \delta_4 = \delta_7 = \delta_{10} = \beta_1 \neq 0$ | | |
|-----------|-------------------|-------------------|-------------------|
| Variable $t$ | $\delta_1, \delta_2, \delta_3, \delta_5, \delta_6, \delta_9$ | | |
| step 1 | sum over $\delta_1$ | sum over $\delta_2$ | sum over $\delta_3$ |
| step 2 | $\leq p$ | sum over $\delta_5$ | sum over $\delta_6$ |
| step 3 | $\leq p$ | 1 | sum over $\delta_9$ |
| step 4 | $\leq p$ | 1 | 1 |
| step 5 | $\leq p$ | 1 | 1 |

**Table 8.** Proof of Case 4-2 : $\beta_1 \neq 0, \beta_2 = 0, \beta_3 = 0, \beta_4 \neq 0$

| Relations | $\delta_{11} = 0,\ \delta_7 = \delta_{10} = \beta_1 \neq 0$ | | |
|---|---|---|---|
| Variable $t$ | $\delta_1, \delta_2, \delta_3, \delta_4, \delta_5, \delta_6, \delta_8, \delta_9,$ | | |
| step 1 | sum over $\delta_1$ | sum over $\delta_2$ | sum over $\delta_3$ |
| step 2 | sum over $\delta_4$ | sum over $\delta_5$ | sum over $\delta_6$ |
| step 3 | $\leq p$ | sum over $\delta_8$ | sum over $\delta_9$ |
| step 4 | $\leq p$ | 1 | $\leq p$ |
| step 5 | $\leq p$ | 1 | 1 |

**Table 9.** Proof of Case 5-1 : $\beta_1 \neq 0, \beta_2 \neq 0, \beta_3 = 0, \beta_4 = 0$

| Relations | $\delta_8 = \delta_{11} = \beta_2 \neq 0$ | | |
|---|---|---|---|
| Variable $t$ | $\delta_1, \delta_2, \delta_3, \delta_4, \delta_5, \delta_6, \delta_7, \delta_9 \neq 0, \delta_{10}$ | | |
| step 1 | sum over $\delta_1$ | sum over $\delta_2$ | sum over $\delta_3$ |
| step 2 | sum over $\delta_4$ | sum over $\delta_5$ | sum over $\delta_6$ |
| step 3 | sum over $\delta_7$ | $\leq p$ | $\leq p$ |
| step 4 | sum over $\delta_{10}$ | $\leq p$ | 1 |
| step 5 | sum over $\delta_9$ | $\leq p$ | 1 |

**Table 10.** Proof of Case 5-2 : $\beta_1 \neq 0, \beta_2 \neq 0, \beta_3 = 0, \beta_4 \neq 0$

| Relations | $\delta_{11} = \beta_2 \neq 0$ | | |
|---|---|---|---|
| Variable $t$ | $\delta_1, \delta_2, \delta_3, \delta_4, \delta_5, \delta_6, \delta_7, \delta_8, \delta_9, \delta_{10}$ | | |
| step 1 | sum over $\delta_1$ | sum over $\delta_2$ | sum over $\delta_3$ |
| step 2 | sum over $\delta_4$ | sum over $\delta_5$ | sum over $\delta_6$ |
| step 3 | sum over $\delta_7$ | sum over $\delta_8$ | sum over $\delta_9$ |
| step 4 | sum over $\delta_{10}$ | $\leq p$ | $\leq p$ |
| step 5 | $\leq p$ | $\leq p$ | 1 |

**Table 11.** Proof of Case 6 : $\beta_1 \neq 0, \beta_2 = 0, \beta_3 \neq 0$

| Relations | $\delta_{10} = \beta_1 \neq 0$ | | |
|---|---|---|---|
| Variable $t$ | $\delta_1, \delta_2, \delta_3, \delta_4, \delta_5, \delta_6, \delta_7, \delta_8, \delta_9, \delta_{11}$ | | |
| step 1 | sum over $\delta_1$ | sum over $\delta_2$ | sum over $\delta_3$ |
| step 2 | sum over $\delta_4$ | sum over $\delta_5$ | sum over $\delta_6$ |
| step 3 | sum over $\delta_7$ | sum over $\delta_8$ | sum over $\delta_9$ |
| step 4 | $\leq p$ | $\leq p$ | sum over $\delta_{11}$ |
| step 5 | $\leq p$ | 1 | $\leq p$ |

**Table 12.** Proof of Case 7-1 : $\beta_1 = 0, \beta_2 \neq 0, \beta_3 \neq 0, \beta_3 \oplus \beta_4 = 0$

| Relations | $\delta_9 = 0,\ \delta_5 = \delta_8 = \delta_{11}$ | | |
|-----------|-----------------------|-----------------------|-----------------------|
| Variable $t$ | $\delta_1, \delta_2, \delta_3, \delta_4, \delta_6 \neq 0, \delta_7, \delta_{10}, \delta_{11} \neq 0$ | | |
| step 1 | sum over $\delta_1$ | sum over $\delta_2$ | sum over $\delta_3$ |
| step 2 | sum over $\delta_4$ | $\leq p$ | $\leq p$ |
| step 3 | sum over $\delta_6$ | sum over $\delta_7$ | 1 |
| step 4 | $\leq p$ | sum over $\delta_{10}$ | 1 |
| step 5 | 1 | $\leq p$ | sum over $\delta_{11}$ |

**Table 13.** Proof of Case 7-2 : $\beta_1 = 0, \beta_2 \neq 0, \beta_3 \neq 0, \beta_3 \oplus \beta_4 \neq 0$

| Relations | $\delta_9 = \beta_3 \oplus \beta_4 \neq 0$ | | |
|-----------|-----------------------|-----------------------|-----------------------|
| Variable $t$ | $\delta_1, \delta_2, \delta_3, \delta_4, \delta_5, \delta_6, \delta_7, \delta_8, \delta_{10}, \delta_{11}$ | | |
| step 1 | sum over $\delta_1$ | sum over $\delta_2$ | sum over $\delta_3$ |
| step 2 | sum over $\delta_4$ | sum over $\delta_5$ | sum over $\delta_6$ |
| step 3 | sum over $\delta_7$ | sum over $\delta_8$ | $\leq p$ |
| step 4 | sum over $\delta_{10}$ | sum over $\delta_{11}$ | $\leq p$ |
| step 5 | 1 | $\leq p$ | $\leq p$ |

**Table 14.** Proof of Case 8-1 : $\beta_1 \neq 0, \beta_2 \neq 0, \beta_3 \neq 0, \alpha_1 \neq 0$

| Relations | | | |
|-----------|-----------------------|-----------------------|-----------------------|
| Variable $t$ | $\delta_1 \neq 0, \delta_2, \delta_3, \delta_4, \delta_5, \delta_6, \delta_7, \delta_8, \delta_9, \delta_{10}, \delta_{11}$ | | |
| step 1 | $\leq p$ | sum over $\delta_1$ | sum over $\delta_2$ |
| step 2 | sum over $\delta_3$ | sum over $\delta_4$ | sum over $\delta_5$ |
| step 3 | sum over $\delta_6$ | sum over $\delta_7$ | sum over $\delta_8$ |
| step 4 | sum over $\delta_9$ | sum over $\delta_{10}$ | sum over $\delta_{11}$ |
| step 5 | $\leq p$ | $\leq p$ | $\leq p$ |

**Table 15.** Proof of Case 8-2 : $\beta_1 \neq 0, \beta_2 \neq 0, \beta_3 \neq 0, \alpha_1 = 0, \alpha_2 \neq 0$

| Relations | $\delta_1 = 0$ | | |
|-----------|-----------------------|-----------------------|-----------------------|
| Variable $t$ | $\delta_2 \neq 0, \delta_3, \delta_4, \delta_5, \delta_6, \delta_7, \delta_8, \delta_9, \delta_{10}, \delta_{11}$ | | |
| step 1 | 1 | $\leq p$ | sum over $\delta_2$ |
| step 2 | sum over $\delta_3$ | sum over $\delta_4$ | sum over $\delta_5$ |
| step 3 | sum over $\delta_6$ | sum over $\delta_7$ | sum over $\delta_8$ |
| step 4 | sum over $\delta_9$ | sum over $\delta_{10}$ | sum over $\delta_{11}$ |
| step 5 | $\leq p$ | $\leq p$ | $\leq p$ |

**Table 16.** Proof of Case 8-3 : $\beta_1 \neq 0, \beta_2 \neq 0, \beta_3 \neq 0, \alpha_1 = 0, \alpha_2 = 0, \alpha_3 \neq 0$

| Relations | $\delta_1 = \delta_2 = 0$ | | |
|---|---|---|---|
| Variable $t$ | $\delta_3 \neq 0, \delta_4, \delta_5, \delta_6, \delta_7, \delta_8, \delta_9, \delta_{10}, \delta_{11}$ | | |
| step 1 | 1 | 1 | $\leq p$ |
| step 2 | sum over $\delta_3$ | sum over $\delta_4$ | sum over $\delta_5$ |
| step 3 | sum over $\delta_6$ | sum over $\delta_7$ | sum over $\delta_8$ |
| step 4 | sum over $\delta_9$ | sum over $\delta_{10}$ | sum over $\delta_{11}$ |
| step 5 | $\leq p$ | $\leq p$ | $\leq p$ |

**Table 17.** Proof of Case 8-4 : $\beta_1 \neq 0, \beta_2 \neq 0, \beta_3 \neq 0, \alpha_1 = 0, \alpha_2 = 0, \alpha_3 = 0, \alpha_4 \neq 0$

| Relations | $\delta_1 = \delta_2 = \delta_3 = 0$ | | |
|---|---|---|---|
| Variable $t$ | $\delta_4 \neq 0, \delta_5, \delta_6, \delta_7, \delta_8, \delta_9, \delta_{10}, \delta_{11}$ | | |
| step 1 | 1 | 1 | 1 |
| step 2 | $\leq p$ | sum over $\delta_4$ | sum over $\delta_5$ |
| step 3 | sum over $\delta_6$ | sum over $\delta_7$ | sum over $\delta_8$ |
| step 4 | sum over $\delta_9$ | sum over $\delta_{10}$ | sum over $\delta_{11}$ |
| step 5 | $\leq p$ | $\leq p$ | $\leq p$ |

# On the Pseudorandomness of Top-Level Schemes of Block Ciphers

Shiho Moriai[1] and Serge Vaudenay[2*]

[1] NTT Laboratories
1-1 Hikarinooka, Yokosuka, 239-0847 Japan
shiho@isl.ntt.co.jp
[2] Swiss Federal Institute of Technology (EPFL)
1015 Lausanne, Switzerland
serge.vaudenay@epfl.ch

**Abstract.** Block ciphers are usually based on one top-level scheme into which we plug "round functions". To analyze security, it is important to study the intrinsic security provided by the top-level scheme from the viewpoint of randomness: given a block cipher in which we replaced the lower-level schemes by idealized oracles, we measure the security (in terms of best advantage for a distinguisher) depending on the number of rounds and the number of chosen plaintexts. We then extrapolate a sufficient number of secure rounds given the regular bounds provided by decorrelation theory.

This approach allows the comparison of several generalizations of the Feistel schemes and others. In particular, we compare the randomness provided by the schemes used by the AES candidates.

In addition we provide a general paradigm for analyzing the security provided by the interaction between the different levels of the block cipher structure.

## 1 Introduction

From the attacker's viewpoint, the block cipher used by a given user can be considered as an instance of a random permutation over a message block space: since he only knows how the secret key has been chosen, he only has probabilistic information (in a Shannon sense) on the key and the permutation. In this setting, security can be formalized by pseudorandomness: if there is no way to distinguish the block cipher from an ideal random permutation, then we cannot attack it. Pseudorandomness more precisely means that no oracle circuit with polynomially many oracle gates can distinguish between the encryption function and a truly random permutation.

A block cipher usually made from a top-level oracle circuit that we call "scheme" (for instance the circuit of the Feistel scheme [4]) into which we plug lower-level circuits that we call "primitives" like round functions, S-boxes, and so

---

[*] Part of this work was done while the author was visiting NTT Laboratories.

T. Okamoto (Ed.): ASIACRYPT 2000, LNCS 1976, pp. 289–302, 2000.

on. An attack may succeed if it "bypasses" some of the primitives by using some intrinsic weaknesses of the scheme. For instance, differential cryptanalysis [1] can investigate the differentials in which some S-boxes play no role at all. This idea motivated this paper: we consider ideal models of the block ciphers by replacing the primitives by truly random functions and study the pseudorandomness provided by the scheme.

In this paper we investigate the randomness of several of the schemes used in many block ciphers. The target schemes are the Feistel scheme, variants of the Feistel scheme (the CAST256-like Feistel scheme, the MARS-like Feistel scheme, and the RC6-like Feistel scheme), and the SQUARE-like scheme used in SQUARE, Rijndael and Crypton.

The pseudorandomness of some general schemes were discussed in previous papers e.g. [9,17]. In this paper we show how we can reach these kind of results and extensions in an easier and more systematic way by using the decorrelation theory introduced in [13,14,15].

In order to compare the schemes we study the threshold number of rounds needed to achieve randomness, a theoretically sufficient number of secure rounds against attacks that are limited to two chosen plaintexts or ciphertexts (which plays a crucial role in the security against differential and linear cryptanalysis), and the sufficient number of secure rounds, in practice, when we use a practical decorrelation module (as in DFC [5]) for primitives instead of an ideal primitive.

## 2    Decorrelation Theory and Randomness of Iterated Ciphers

### 2.1    Definitions and Basic Properties

The goal of decorrelation theory is to provide some kind of formal proof of security on block ciphers. This section describes the essential definitions and lemmas in decorrelation theory to prove the randomness of iterated ciphers.

**Definition 1 (d-wise distribution matrix).** *Given a random function $F^1$ from a set $\mathcal{M}_1$ to a set $\mathcal{M}_2$ and an integer $d$, we define the "d-wise distribution matrix" of $F$ as the following $\mathcal{M}_1^d \times \mathcal{M}_2^d$-matrix.*

$$[F]^d_{(x_1,\ldots,x_d),(y_1,\ldots,y_d)} = \Pr[F(x_1) = y_1, \ldots, F(x_d) = y_d],$$

*where $x_i \in \mathcal{M}_1$ and $y_i \in \mathcal{M}_2$ for $i = 1, \ldots, d$*

**Definition 2 (d-wise decorrelation bias).** *Given a random function $F$ from a set $\mathcal{M}_1$ to a set $\mathcal{M}_2$, a canonical idealized version $F^*$ of $F$, an integer $d$, and a*

---

[1] Throughout this paper, "a random function $F$" means a random variable $F$ which takes values in a set of functions, following regular probability theory. The same holds for "a random permutation $C$".

*distance $D$ over the matrix space $\mathbf{R}^{\mathcal{M}_1^d \times \mathcal{M}_2^d}$, we define the "d-wise decorrelation bias of $F$" as being the distance*

$$\text{Dec}_D^d(F) = D([F]^d, [F^*]^d).$$

*In cases where the canonical idealized version $F^*$ is not explicit, we will use the notation DecF in order to make implicit that $F^*$ is a uniformly distributed random function, and DecP in order to make implicit that $F^*$ is a uniformly distributed random permutation.*

For instance, when talking about a block cipher as a random permutation $C$, the canonical idealized version $C^*$ is a random permutation with uniform distribution. This canonical idealized version should be clear from the context.

Given two random functions $F$ and $G$ from $\mathcal{M}_1$ to $\mathcal{M}_2$ we call "a distinguisher between $F$ and $G$" any oracle Turing machine $\mathcal{A}^O$ that can send $\mathcal{M}_1$-element queries to the oracle $O$ and receive $\mathcal{M}_2$-element responses, and which finally outputs 0 or 1. In particular, the Turing machine can be probabilistic. In the following, the number of queries to the oracle will be limited to $d$. The distributions of $F$ and $G$ induce a distribution of $\mathcal{A}^F$ and $\mathcal{A}^G$, thus we can compute the probability that these probabilistic Turing machines output 1. We call the function

$$\text{Adv}_{\mathcal{A}}(F, G) = \Pr[\mathcal{A}^F = 1] - \Pr[\mathcal{A}^G = 1].$$

the advantage $\mathcal{A}^O$ achieves in distinguishing $F$ from $G$.

We consider the classes $\text{Cl}_{na}^d$ (resp. $\text{Cl}_a^d$) of non adaptive (resp. adaptive) distinguishers limited to $d$ queries. Similarly, when $F$ and $G$ are permutations, we also consider the extension $\text{Cl}_s^d$ of distinguishers that are limited to $d$ queries but who can query either the function $F/G$ or its inverse $F^{-1}/G^{-1}$. For any class of distinguishers Cl we will denote

$$\text{BestAdv}_{\text{Cl}}(F, G) = \max_{\mathcal{A} \in \text{Cl}} \text{Adv}_{\mathcal{A}}(F, G).$$

**Lemma 1 (Equivalence between best advantage and decorrelation distance [13,15]).** *For any random functions $F$ and $G$ and any integer $d$, we have*

$$|||[F]^d - [G]^d|||_\infty = 2 \cdot \text{BestAdv}_{\text{Cl}_{na}^d}(F, G)$$

$$||[F]^d - [G]^d||_a = 2 \cdot \text{BestAdv}_{\text{Cl}_a^d}(F, G)$$

$$||[F]^d - [G]^d||_s = 2 \cdot \text{BestAdv}_{\text{Cl}_s^d}(F, G)$$

*where $||.||_a$ and $||.||_s$ are special matrix norms defined in [15] and $|||.|||_\infty$ is the regular infinity associated matrix norm (the maximum of row sums).*

**Lemma 2 (Multiplicativity).** *For any $f$ and $g$, we denote by $f \circ g$ their composition. For any independent random functions $F_1, \ldots, F_r$, any integer $d$ and any matrix norm we have*

$$\mathrm{DecF}^d(F_1 \circ \cdots \circ F_r) \leq \mathrm{DecF}^d(F_1) \cdots \mathrm{DecF}^d(F_r).$$

*For any independent random permutations $C_1, \ldots, C_r$ we have*

$$\mathrm{DecP}^d(C_1 \circ \cdots \circ C_r) \leq \mathrm{DecP}^d(C_1) \cdots \mathrm{DecP}^d(C_r).$$

Some known functions have quite small decorrelation biases called decorrelation modules. An example of decorrelation module is the NUT-IV decorrelation module.

**Lemma 3 (NUT-IV decorrelation module with $d = 2$ [15]).** *For an injection $r$ from $\{0,1\}^m$ to $\mathrm{GF}(q)$ and a surjection $\pi$ from $\mathrm{GF}(q)$ to $\{0,1\}^m$, it has been shown that the random function $F$, defined on $\{0,1\}^m$ by*

$$F(x) = \pi(r(K_0) + r(K_1)x)$$

*for $(K_0, K_1)$ uniformly distributed in $\{0,1\}^{2m}$, provides quite good decorrelation. Namely,*

$$\mathrm{DecF}^2_{||.||_a}(F) \leq 2(q^2.2^{-2m} - 1).$$

For better implementation efficiency, we will only consider prime integers $q$ in this paper. The reader can refer to Noilhan [11] for implementation issues. For instance, DFC uses $q = 2^{64} + 13$ for which we obtain $\mathrm{DecF}^2_{||.||_a}(F) \leq 2^{-58.3}$ (see [7]).

## 2.2   Basic Tools

The randomness of a cipher constructed using random primitives such as decorrelation modules can be proven using decorrelation theory. In order to deduce an upper bound on the decorrelation bias of the cipher from an upper bound on the decorrelation bias of these primitives, we use the following lemma.

**Lemma 4 (Reduction to the randomness of ideal constructions [15]).** *Let $d$ be an integer, $F_1, \ldots, F_r, C_1, \ldots, C_s$ be $r + s$ independent random function oracles which are idealized by $F_1^*, \ldots, F_r^*, C_1^*, \ldots, C_s^*$ respectively, where the $C_j$ and $C_j^*$ are permutations. We let $\Omega^{F_1, \ldots, F_r, C_1, \ldots, C_s}$ be an oracle that can access the previous oracles and from each query $x$ define an output $G(x)$. We assume that $\Omega$ is such that the number of queries to $F_i$ is limited to some integer $a_i$, and the number of queries to $C_j$ or $C_j^{-1}$ is limited to $b_j$ in total for any $i = 1, \ldots, r$ and $j = 1, \ldots, s$. We let $G^*$ be the function defined by $\Omega^{F_1^*, \ldots, F_r^*, C_1^*, \ldots, C_s^*}$. We have*

$$\mathrm{Dec}^d_{||.||_a}(G) \leq \sum_{i=1}^{r} \mathrm{Dec}^{a_i d}_{||.||_a}(F_i) + \sum_{j=1}^{s} \mathrm{Dec}^{b_j d}_{||.||_s}(C_j) + \mathrm{Dec}^d_{||.||_a}(G^*)$$

*In addition, if the $\Omega$ construction defines a permutation $G$, assuming that computing $G^{-1}$ leads to the same $a_i$, $b_j$ and $c_k$ limits, we have*

$$\mathrm{Dec}^d_{||.||_s}(G) \leq \sum_{i=1}^{r} \mathrm{Dec}^{a_i d}_{||.||_a}(F_i) + \sum_{j=1}^{s} \mathrm{Dec}^{b_j d}_{||.||_s}(C_j) + \mathrm{Dec}^d_{||.||_s}(G^*).$$

**Lemma 5 ([16]).** *Let $d$ be an integer. Let $F$ be a random function from a set $\mathcal{M}_1$ to a set $\mathcal{M}_2$. We let $\mathcal{X}$ be the subset of $\mathcal{M}_1^d$ of all $(x_1, \ldots, x_d)$ with pairwise different entries. We let $F^*$ be a uniformly distributed random function from $\mathcal{M}_1$ to $\mathcal{M}_2$. We know that for all $x \in \mathcal{X}$ and $y \in \mathcal{M}_2^d$ the value $[F^*]^d_{x,y}$ is the constant $p_0 = (\#\mathcal{M}_2)^{-d}$. We assume there exists a subset $\mathcal{Y} \subseteq \mathcal{M}_2^d$ and two positive real values $\epsilon_1$ and $\epsilon_2$ such that*

- *$(\#\mathcal{Y})p_0 \geq 1 - \epsilon_1$*
- *$\forall x \in \mathcal{X} \quad \forall y \in \mathcal{Y} \quad [F]^d_{x,y} \geq p_0(1 - \epsilon_2)$.*

*This yields $\mathrm{DecF}^d_{||.||_a}(F) \leq 2\epsilon_1 + 2\epsilon_2$.*

This lemma intuitively means that if $[F]^d_{x,y}$ is close to $[F^*]^d_{x,y}$ for all $x$ and almost all $y$, then the decorrelation bias of $F$ is small. We have a twin lemma for the $||.||_s$ norm. Here, since we can query $y$ as well, the approximation must hold for all $x$ and $y$.

**Lemma 6 ([16]).** *Let $d$ be an integer. Let $C$ be a random permutation on a set $\mathcal{M}$. We let $\mathcal{X}$ be the subset of $\mathcal{M}^d$ of all $(x_1, \ldots, x_d)$ with pairwise different entries. We let $F^*$ be a uniformly distributed random function on $\mathcal{M}$. We let $C^*$ be a uniformly distributed random permutation on $\mathcal{M}$. We have*

- *if $[C]^d_{x,y} \geq [C^*]^d_{x,y}(1 - \epsilon)$ for all $x$ and $y$ in $\mathcal{X}$*
  *then $\mathrm{DecP}^d_{||.||_s}(F) \leq 2\epsilon$*
- *if $[C]^d_{x,y} \geq [F^*]^d_{x,y}(1 - \epsilon)$ for all $x$ and $y$ in $\mathcal{X}$*
  *then $\mathrm{DecP}^d_{||.||_s}(F) \leq 2\epsilon + 2d^2(\#\mathcal{M})^{-1}$.*

## 2.3   Examples

First this section studies how many rounds are required for Luby-Rackoff's randomness assuming round functions to be random ones. This is related to the "lack of randomness" provided by the upper-level design. The required numbers of rounds for the Feistel scheme and some generalized Feistel schemes are shown in [17, Section 3.2].

Hereafter we use the following notations. $I_n$ denotes the set of all $n$-bit strings, $\{0,1\}^n$. $H_n$ denotes the set of all $I_n \mapsto I_n$ functions and $P_n$ denotes the set of all such permutations. By $x \in_U X$ we mean that $x$ is drawn randomly and uniformly from a finite set $X$.

**Lemma 7 (Luby-Rackoff 1986 [9]).** *Let $(F_1^*, F_2^*, F_3^*, F_4^*) \in_U \left(H_{\frac{m}{2}}\right)^4$ be four independent random functions. We have*

$$\mathrm{DecF}^d_{||\cdot||_a}(\Psi(F_1^*, F_2^*, F_3^*)) \le 2d^2 \cdot 2^{-\frac{m}{2}}$$

$$\mathrm{DecP}^d_{||\cdot||_a}(\Psi(F_1^*, F_2^*, F_3^*)) \le 2d^2 \cdot 2^{-\frac{m}{2}}$$

$$\mathrm{DecP}^d_{||\cdot||_s}(\Psi(F_1^*, F_2^*, F_3^*, F_4^*)) \le 2d^2 \cdot 2^{-\frac{m}{2}}$$

*Here $\Psi(F_1, \ldots, F_r)$ is the notation introduced by Luby and Rackoff in order to denote a Feistel scheme where the i-th round function is $F_i$.[2]*

This lemma is tight in the sense that 2 rounds are not enough for pseudorandomness and 3 rounds are not enough for super-pseudorandomness. Indeed, we can make a simple distinguisher against a 2-round Feistel scheme with $d = 2$ queries with an advantage equal to $1 - 2^{-\frac{m}{2}}$ by querying random $(a, b)$ and $(a, c)$ plaintexts and checking that the right half difference is equal to $b \oplus c$. The same holds for super-pseudorandomness with 3 rounds (see Patarin [12]): we can query for the encryption of $(a, b)$ and $(a, b \oplus \delta)$, obtain $(x, y)$ and $(x', y')$ respectively, query for the decryption of $(x, y \oplus \delta)$ and $(x', y' \oplus \delta)$, and check that the obtained left halves are equal.

This lemma can be formally proven by using Lemma 5 and 6. From Lemma 2 and 4 this is generalized for a permutation on $\{0, 1\}^m$ consisting of $r$ rounds of Feistel transformations:

$$\mathrm{DecP}^d_{||\cdot||_a}(\Psi(F_1, \ldots, F_r)) \le \left(2d^2 \cdot 2^{-\frac{m}{2}} + 3 \max_i \mathrm{DecF}^d_{||\cdot||_a}(F_i)\right)^{\lfloor \frac{r}{3} \rfloor}$$

$$\mathrm{DecP}^d_{||\cdot||_s}(\Psi(F_1, \ldots, F_r)) \le \left(2d^2 \cdot 2^{-\frac{m}{2}} + 4 \max_i \mathrm{DecF}^d_{||\cdot||_a}(F_i)\right)^{\lfloor \frac{r}{4} \rfloor}$$

for any independent functions $F_1, \ldots, F_r \in H_{\frac{m}{2}}$. This leads to the following conclusions about the regular Feistel scheme with $m = 128$.

- The *threshold number of rounds* for achieving the security result is 3 for pseudorandomness and 4 for super-pseudorandomness, when $d \ll 2^{32}$.
- The *theoretical sufficient number of secure rounds* for achieving the decorrelation bias of $2^{-m}$ is $\frac{\alpha m}{\frac{m}{2} - 1 - 2 \log_2 d}$ with $\alpha = 3$ for pseudorandomness and $\alpha = 4$ for super-pseudorandomness, when $d \ll 2^{32}$. This leads to 9 and 12 rounds, respectively, for $d = 2$.
- When using the NUT-IV decorrelation module with $d = 2$, $m = 128$ and $q = 2^{64} + 13$ in each round (as for instance DFC), these numbers of rounds provide decorrelation biases less than $2^{-m}$ for the corresponding norms.

Here we used an arbitrary threshold of $2^{-m}$ for the decorrelation bias used in order to compare different schemes. Since $2^{-m}$ yields a level of security given by exhaustive search on $m$ bits, we believe it is a relevant objective criterion for comparing schemes. We also focused on $d = 2$ which leads to security against differential and linear cryptanalysis.

---

[2] In order to be consistent with further schemes, the first round here maps the left half through $F_1$ and add to the right half.

**Fig. 1.** CAST256-like Feistel Scheme

# 3    Several Cases

## 3.1    CAST256-like Feistel Scheme

CAST-256 is an AES candidate based on a generalized Feistel scheme called "Type-1 transformation" by Zheng-Matsumoto-Imai [17] and denoted by $\Psi_1$. Formally, we define $\Psi_1 \in H_m$ as $\Psi_1()(x) = x$ and

$$\Psi_1(f_1, \ldots, f_r)(x_1, \ldots, x_k) =$$
$$\Psi_1(f_2, \ldots, f_r)(f_1(x_1) + x_2, x_3, x_4, \ldots, x_k, x_1)$$

for any primitive set $f_1, \ldots, f_r \in H_{\frac{m}{k}}$. Here $k$ is the number of branches and $r$ is the number of rounds.

**Lemma 8 (Zheng-Matsumoto-Imai 1989 [17]).** *For independent and uniformly distributed random functions $F_1^*, \ldots, F_{2k-1}^* \in_U H_{\frac{m}{k}}$ and an integer $d$, we have*

$$\mathrm{DecP}_{||\cdot||_a}^d (\Psi_1(F_1^*, \ldots, F_{2k-1}^*)) \le 2(k-1)d^2 \cdot 2^{-\frac{m}{k}}$$

It can easily be shown that the number of rounds of $2k-1$ for pseudorandomness is actually minimal. For instance, if we take $2k-2$ rounds and $d = 2$, we can submit two chosen plaintexts for which only the input of the rightmost branch has changed. The input difference in this branch will always be equal to the output difference in the second branch, which leads to a distinguisher of advantage $1 - 2^{-k}$.

We however notice that a number of rounds of $k^2 - k$ is not enough for super-pseudorandomness. With $k(k-1)$ rounds, we can decrypt $(y_1, y_2, \ldots, y_k)$ and $(y_1', y_2, \ldots, y_k)$, obtain $(x_1, \ldots, x_k)$ and $(x_1', \ldots, x_k')$ respectively, and check that $x_1 \oplus x_1' = y_1 \oplus y_1'$. This actually shows that the inverse of the $\Psi_1$ scheme is not pseudorandom unless the number of rounds is very large. Actually, the CAST256 cipher is a construction like

$$\left(\Psi_1(f_r, \ldots, f_{\frac{r}{2}+1})\right)^{-1} \circ \Psi_1(f_1, \ldots, f_{\frac{r}{2}}).$$

We can show that the above attack generalizes to this scheme for $r \le 4k - 6$, that $r = 4k - 4$ is enough for pseudorandomness, and that $r = 4k - 2$ is enough for super-pseudorandomness.

*Proof (sketch).* We use Lemma 5 for evaluating $\text{DecF}^d_{||.||_a}$.

For $\text{DecP}_{||.||_a}$ we let $\mathcal{Y}$ be the set of all $y = (y_1, \ldots, y_d)$ where $y_i = (y_i^1, \ldots, y_i^k)$ such that $y_i^j \neq y_{i'}^j$ for $j > 1$ and $i < i'$. We get $\epsilon_1 = (k-1)\frac{d(d-1)}{2}2^{-\frac{m}{k}}$. We then consider the event in which the first entry after the $(k-1)$th round takes pairwise different values for $x_1, \ldots, x_d$. Upper bounding the probability when this event occurs we get $\epsilon_2 = (k-1)\frac{d(d-1)}{2}2^{-\frac{m}{k}}$. Thus $\text{DecF}^d_{||.||_a}(F) \leq 2(k-1)d(d-1)2^{-\frac{m}{k}}$.

Here, $\epsilon_2$ is evaluated as the number of unexpected equalities between two outputs from a single circuit of depth $k-1$ with $k$ inputs and internal $F_j^*$ and additions times the probability it occurs, which is at most the depth $k-1$ times $2^{-\frac{m}{k}}$.

Now to get DecP from DecF, from $\text{DecF}^d_{||.||_a}(C^*) \leq d(d-1)2^{-m}$ and the triangular inequality we have

$$\text{DecP}^d_{||.||_a}(F) \leq \text{DecF}^d_{||.||_a}(F) + \text{DecP}^d_{||.||_a}(F^*) \leq \text{DecF}^d_{||.||_a}(F) + d^2 2^{-m}.$$

We then notice that the obtained upper bound for $\text{DecF}^d_{||.||_a}$ can be written $\text{DecF}^d_{||.||_a}(F) \leq Ad(d-1)2^{-\frac{m}{k}}$ for some $A \geq 2$. For $d \leq A2^{m-\frac{m}{k}}$ we thus obtain $\text{DecP}^d_{||.||_a}(F) \leq Ad^2 2^{-\frac{m}{k}}$. For larger $d$, this bound is greater than $A^3 2^{m(2-\frac{3}{k})}$ which is greater than 8 since $m \geq k \geq 2$. Since $\text{DecP}^d_{||.||_a}(F)$ is always less than 2, the bound is thus still valid.    □

Thus the required number of rounds for the CAST256-like scheme is proven to be $2k-1$, where $k$ is the number of branches. That is, the required numbers of rounds for the Feistel scheme and the CAST256-like scheme are 3 and 7, respectively.

This leads to the following conclusions about the CAST256-like scheme with $k = 4$ branches and $m = 128$.

- The *threshold number of rounds* is 7 for pseudorandomness when $d \ll 2^{16}$. For super-pseudorandomness, this threshold is larger than 13.
- For $d = 2$, the *theoretical sufficient number of secure rounds* is 35 for pseudorandomness.
- For the NUT-IV decorrelation module with $d = 2$, $m = 128$ and $q = 2^{32}+15$, the sufficient number of rounds is 42 pseudorandomness.

## 3.2    MARS-like Feistel Scheme

Similarly, we define the MARS-like generalized Feistel scheme denoted by $\Psi_1' \in H_m$ as $\Psi_1'()(x) = x$ and

$$\Psi_1'(f_1, \ldots, f_r)(x_1, \ldots, x_k) = $$
$$\Psi_1'(f_2, \ldots, f_r)(f_1^2(x_1) + x_2, f_1^3(x_1) + x_3, \ldots, f_1^k(x_1) + x_k, x_1)$$

where $f_i = (f_i^2, \ldots, f_i^k)$, $f_i^2, \ldots, f_i^k \in H_{\frac{m}{k}}$.

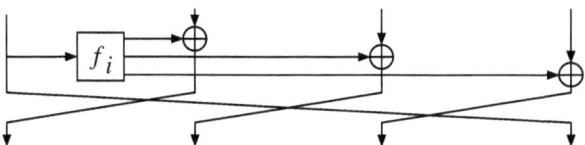

**Fig. 2.** MARS-like Feistel Scheme

**Lemma 9.** *For independent uniformly distributed random functions $F_i^* \in_U H_{\frac{m}{k}}$ for $i = 1, \ldots, 2k$ and $j = 2, \ldots, k$ and an integer $d$, we have*

$$\mathrm{DecP}^d_{||\cdot||_a}(\Psi'_1(F_1^*, \ldots, F_{k+1}^*)) \leq 2d^2 \cdot 2^{-\frac{m}{k}}$$

$$\mathrm{DecP}^d_{||\cdot||_s}(\Psi'_1(F_1^*, \ldots, F_{2k}^*)) \leq 2d^2 \cdot 2^{-\frac{m}{k}}$$

It can easily be shown that the number of rounds of $k+1$ for pseudorandomness is actually minimal since a difference in the last input branch only remains unchanged after $k$ rounds. Similarly, for $2k - 1$ rounds, we can merge the first $k - 1$ branches and consider that we have a regular 3-round Feistel scheme, and we can apply the same attack for proving it is not super-pseudorandom.

*Proof (sketch).* Using Lemma 5 we let $\mathcal{Y}$ be the set of all $(y_1, \ldots, y_d)$ such that $y_i^k \neq y_j^k$ for $i \neq j$. We get $\epsilon_1 = \frac{d(d-1)}{2}2^{-\frac{m}{k}}$. We focus on the event that the first output after $k - 1$ rounds leads to no collision. We get $\epsilon_2 = \frac{d(d-1)}{2}2^{-\frac{m}{k}}$.

For $\mathrm{DecP}^d_{||\cdot||_s}$ we use the same event. $\square$

This leads to the following conclusions about the MARS-like scheme with $k = 4$ branches and $m = 128$.

- The *threshold number of rounds* is 5 for pseudorandomness and 8 for super-pseudorandomness, when $d \ll 2^{16}$.
- For $d = 2$, the *theoretical sufficient number of secure rounds* is 25 for pseudorandomness and 40 for super-pseudorandomness.
- For the NUT-IV decorrelation module with $d = 2$, $m = 128$ and $q = 2^{32}+15$, the sufficient number of rounds is as for the ideal case.

### 3.3   RC6-like Feistel Scheme

The RC6 block cipher is designed to be secure by mixing operations that are efficiently implemented on most modern processors.

One controversial additional operation is the data dependent rotation. Such a scheme cannot provide pseudorandomness nor super-pseudorandomness.[3] Indeed, the attack in Gilbert et al. [6] exhibits an efficient polynomial time distinguisher.

---

[3] As was mentioned by Joux during the third Advanced Encryption Standard workshop, although Iwata and Kurosawa had claimed the opposite two days before at the FSE00 workshop [8].

**Fig. 3.** RC6'-like Feistel Scheme

However, we can consider RC6', a transformation of RC6 WITHOUT the data dependent rotations. The structure of RC6' can be regarded as a generalized Feistel scheme, which is similar to "Type-2 transformation" named by Zheng-Matsumoto-Imai [17] assuming that primitives are independent random functions. Formally, as the RC6'-like Feistel scheme $\Psi_2 \in H_m$ is defined for $k$ even and $r$ a multiple of $\frac{k}{2}$, by $\Psi_2()(x) = x$ and

$$\Psi_2(f_1, \ldots, f_r)(x_1, \ldots, x_k) =$$
$$\Psi_2(f_{\frac{k}{2}+1}, \ldots, f_r)(x_2, f_2(x_4) + x_3, \ldots, x_{k-2}, f_{\frac{k}{2}}(x_k) + x_{k-1}, x_k, f_1(x_2) + x_1),$$

where $f_1, \ldots, f_r \in H_{\frac{m}{k}}$. We consider this as $r$ rounds which are processed in bunch of $\frac{k}{2}$ parallel rounds.

**Lemma 10.** *For independent uniformly distributed random functions $F_1^*, \ldots,$ $F_{k^2}^* \in_U H_{\frac{m}{k}}$ and an integer $d$, we have*

$$\mathrm{DecP}_{||\cdot||_a}^d(\Psi_2(F_1^*, \ldots, F_{\frac{k}{2}(k+1)}^*)) \leq \frac{k^2}{2}d^2 \cdot 2^{-\frac{m}{k}}$$

$$\mathrm{DecP}_{||\cdot||_s}^d(\Psi_2(F_1^*, \ldots, F_{k^2}^*)) \leq \frac{k^2}{2}d^2 \cdot 2^{-\frac{m}{k}}$$

It can easily be shown that the number of rounds of $\frac{k}{2}(k+1)$ for pseudorandomness is actually minimal. Tightness of the $k^2$ bound for super-pseudorandomness is still open. (We already know that it is tight for $k = 2$.)

*Proof (sketch).* Similarly, we use Lemma 9 for evaluating $\mathrm{DecP}_{||\cdot||_a}^d$. For $\Psi_2$ we let $\mathcal{Y}$ be the set of all $y$ such that $y_i^j \neq y_{i'}^j$ for odd $j$ and $i < i'$. We get $\epsilon_1 = \frac{k}{2} \times \frac{d(d-1)}{2}2^{-\frac{m}{k}}$. We consider the event in which all even entries after the $(k-1)$th bunch of rounds takes pairwise different values for $x_1, \ldots, x_d$. We get $\epsilon_2 = \frac{k}{2}(k-1) \times \frac{d(d-1)}{2}2^{-\frac{m}{k}}$. Thus $\mathrm{DecF}_{||\cdot||_a}^d(F) \leq \frac{k^2}{2}d(d-1)2^{-\frac{m}{k}}$. For $\mathrm{DecP}_{||\cdot||_s}^d$, we add $k-1$ more bunch of rounds and study the probability that we get $\mathcal{Y}$ if we invert them on $y_1, \ldots, y_d$. The result comes from Lemma 6.     □

This leads to the following conclusions about the RC6'-like scheme with $k = 4$ branches and $m = 128$.

- The *threshold number of rounds* is 5 for pseudorandomness and between 5 and 8 for super-pseudorandomness, when $d \ll 2^{16}$.

- For $d = 2$, the *theoretical sufficient number of secure rounds* is 25 for pseudorandomness and between 25 and 40 for super-pseudorandomness.
- For the NUT-IV decorrelation module with $d = 2$, $m = 128$ and $q = 2^{32} + 15$, the sufficient number of rounds as the ideal case.

### 3.4   SQUARE-like Scheme

In this paper we discuss only the Rijndael scheme. The pseudorandomness of other SQUARE-like schemes will be described in the full paper. Let us formalize the Rijndael scheme on $k^2$ values by

$$\Sigma(f_1, \ldots, f_r)(x_1, \ldots, x_{k^2}) =$$
$$\Sigma(f_2, \ldots, f_r)(\text{MixCol}(\text{ShiftRow}(f_1^1(x_1), \ldots, f_1^{k^2}(x_{k^2}))))$$

where $f_i = (f_i^1, \ldots, f_i^{k^2})$, $f_i^1, \ldots, f_i^{k^2} \in H_{\frac{m}{k^2}}$, the ShiftRow transformation is a fixed linear transformation on the rows of a $k \times k$ matrix which consists in mixing them, and the MixCol transformation is a fixed linear transformation on the columns [3].

**Lemma 11.** *For independent uniformly distributed random functions $F_1^*, \ldots, F_5^*$ and an integer $d$, we have*

$$\text{DecP}_{||\cdot||_a}^d(\Sigma(F_1^*, \ldots, F_3^*)) \leq 2k^2 d^2 \cdot 2^{-\frac{m}{k^2}}$$
$$\text{DecP}_{||\cdot||_s}^d(\Sigma(F_1^*, \ldots, F_5^*)) \leq 2k^2 d^2 \cdot 2^{-\frac{m}{k^2}}$$

Thus achieving decorrelation to the order $d \geq \frac{1}{k\sqrt{2}} 2^{\frac{m}{2k^2}}$ does not seem possible with this design. (For $m = 128$ and $k = 4$, this is $d = 2\sqrt{2}$.)

It can easily be shown that the number of rounds of 3 for pseudorandomness is actually minimal. The tightness of the 5 bound depends on the instance of the cipher.

*Proof (sketch).* We use Lemma 9 for evaluating $\text{DecP}_{||\cdot||_a}^d$. We let $\mathcal{Y}$ be the set of all $y = (y_1, \ldots, y_d)$ that take different values on all positions before the last MixCol and ShiftRow transformations. We have $\epsilon_1 = k^2 \frac{d(d-1)}{2} 2^{-\frac{m}{k^2}}$. We consider the event that after two rounds we obtain different values on all positions. Provided that the MixCol transformation has good diffusion properties we obtain $\epsilon_2 = k^2 \frac{d(d-1)}{2} 2^{-\frac{m}{k^2}}$. □

This leads to the following conclusions about the Rijndael scheme with $k^2 = 4^2$ branches and $m = 128$.

- The *threshold number of rounds* is 3 for pseudorandomness and between 3 and 5 for super-pseudorandomness, when $d < 3$.
- For $d = 2$, the *theoretical sufficient number of secure rounds* is 384 for pseudorandomness and between 384 and 640 for super-pseudorandomness.
- For the NUT-IV decorrelation module with $d = 2$, $m = 128$ and $q = 2^8 + 1$, the bounds of decorrelation theory cannot guaranty any low decorrelation bias for any number of rounds.

**Table 1.** Randomness of several schemes (when $d = 2, k = 4, m = 128$)

| Scheme | Feistel | CAST256-like | MARS-like | RC6'-like | Rijndael |
|---|---|---|---|---|---|
| Threshold number of rounds for p.r. | 3 | 7 | 5 | 5 | 3 |
| sufficient number of rounds for p.r. (ideal) | 9 | 35 | 25 | 25 | 384 |
| sufficient number of rounds for p.r. (NUT-IV) | 9 | 42 | 25 | 25 | $\infty$ |
| Threshold number of rounds for s.p.r. | 4 | $\geq 13$ | 8 | 5–8 | 3–5 |
| sufficient number of rounds for s.p.r. (ideal) | 12 | | 40 | 25–40 | 384–640 |
| sufficient number of rounds for s.p.r. (NUT-IV) | 12 | | 40 | 25–40 | $\infty$ |
| Example | Twofish, DFC | CAST-256 | MARS | | Rijndael |

Note: "p.r." and "s.p.r." mean pseudorandomness and super-pseudorandomness, respectively.

## 4    Conclusion

We studied the randomness provided by several schemes used in block ciphers. We focused on the schemes for AES candidates in particular (see Table 1). The randomness so discovered is a good measure for evaluating the security from a randomness viewpoint but the readers should take care to note that it doesn't show the actual security of a cipher based on one of the schemes. To study the intrinsic security provided by the general schemes, we decomposed the ciphers into a general scheme and internal primitives, ignoring the components that we considered do not affect its randomness. We also assumed that internal primitives are ideal random ones.

The results in Table 1 show that the regular Feistel scheme is the best in that it requires the fewest number of rounds for pseudorandomness and super-pseudorandomness. However, when comparing the randomness of several schemes we should take account of the computational cost of random primitives. For example, for the Feistel scheme we assume the random functions on $\{0,1\}^{64}$, and for the CAST256-like[4], MARS-like, and RC6-like schemes, we assume the random functions on $\{0,1\}^{32}$, whose computational cost is much cheaper than the former. Under the same assumption of the computational cost of random functions on $\{0,1\}^{32}$, the MARS-like scheme is the best. Table 1 separates the schemes according to the size of the internal random functions.

Our results show that the schemes that use random primitives with smaller input/output sizes are less secure, which is not surprising because the randomness bias is larger in these cases. We should interpret these conclusions with great care. Indeed, our results do not mean that Rijndael (or Serpent[5]) is not

---

[4] Table 1 considers the $\Psi_1$ structure only and not the $\Psi_1^{-1} \circ \Psi_1$ scheme on which CAST256 is based. This latter scheme increases the threshold number of rounds for p.r. to 12.

[5] A preliminary study suggested that the Serpent scheme requires too many rounds for randomness, because the size of primitives is too small (4 bits).

secure, or less secure than regular Feistel schemes. Rather they mean that the latter can benefit from stronger security arguments: we can prove that an efficient attack against — say Twofish — must use an unexpected property of the round function, whereas an attack against Serpent may hold for any set of (random) S-boxes.

# References

1. E. Biham, A. Shamir. *Differential Cryptanalysis of the Data Encryption Standard*, Springer-Verlag, 1993.
2. L. Carter, M. Wegman. Universal Classes of Hash Functions. *Journal of Computer and System Sciences*, vol.18, pp.143–154, 1979.
3. J. Daemen, V. Rijmen. AES Proposal: Rijndael.
   URL: `http://www.esat.kuleuven.ac.be/~rijmen/rijndael/`
4. H. Feistel. Cryptography and Computer Privacy. *Scientific American*, vol. 228, pp. 15–23, 1973.
5. H. Gilbert, M. Girault, P. Hoogvorst, F. Noilhan, T. Pornin, G. Poupard, J. Stern, S. Vaudenay. Decorrelated Fast Cipher: an AES Candidate. (Extended Abstract.) In *Proceedings from the First Advanced Encryption Standard Candidate Conference*, National Institute of Standards and Technology (NIST), August 1998.
6. H. Gilbert, H. Handschuh, A. Joux, S. Vaudenay. A Statistical Attack on RC6. To appear in the proceedings of FSE00.
7. L. Granboulan, P. Nguyen, F. Noilhan, S. Vaudenay. DFCv2. To appear in the proceedings of SAC00.
8. T. Iwata, K. Kurosawa. On the Pseudorandomness of AES Finalists — RC6, Serpent, MARS and Twofish. To appear in the proceedings of FSE00.
9. M. Luby, C. Rackoff. How to Construct Pseudorandom Permutations from Pseudorandom Functions. *SIAM Journal on Computing*, vol. 17, pp. 373–386, 1988.
10. M. Matsui. The First Experimental Cryptanalysis of the Data Encryption Standard. In *Advances in Cryptology CRYPTO'94*, Santa Barbara, California, U.S.A., Lectures Notes in Computer Science 839, pp. 1–11, Springer-Verlag, 1994.
11. F. Noilhan. Software Optimization of Decorrelation Module. In *Selected Areas in Cryptography*, Kingston, Ontario, Canada, Lectures Notes in Computer Science 1758, pp. 175–183, Springer-Verlag, 2000.
12. J. Patarin. *Etude des Générateurs de Permutations Basés sur le Schéma du D.E.S.*, Thèse de Doctorat de l'Université de Paris 6, 1991.
13. S. Vaudenay. Provable Security for Block Ciphers by Decorrelation. In *STACS 98*, Paris, France, Lectures Notes in Computer Science 1373, pp. 249–275, Springer-Verlag, 1998.
14. S. Vaudenay. On the Lai-Massey Scheme. *Advances in Cryptology — ASIACRYPT'99*, Singapore, Lecture Notes in Computer Science 1716, pp.8–19, Springer-Verlag, 1999.
15. S. Vaudenay. Adaptive-Attack Norm for Decorrelation and Super-Pseudorandomness. Technical report LIENS-99-2, Ecole Normale Supérieure, 1999. In *Selected Areas in Cryptography*, Kingston, Ontario, Canada, Lectures Notes in Computer Science 1758, pp. 49–61, Springer-Verlag, 2000.
16. S. Vaudenay. On Provable Security for Conventional Cryptography. Invited talk. To appear in the proceedings of ICISC' 99, LNCS, Springer-Verlag.

17. Y. Zheng, T. Matsumoto, H. Imai. On the Construction of Block Ciphers Provably Secure and Not Relying on Any Unproved Hypotheses (Extended Abstract). *Advances in Cryptology — CRYPTO'89*, Santa Barbara, California, U.S.A., Lecture Notes in Computer Science 435, pp.461–480, Springer-Verlag, 1990.

# Exploiting Multiples of the Connection Polynomial in Word-Oriented Stream Ciphers

Philip Hawkes[1] and Gregory G. Rose[1]

Qualcomm Australia, Suite 410 Birkenhead Point, Drummoyne NSW 2047 Australia,
{phawkes, ggr}@qualcomm.com

**Abstract.** This paper describes some attacks on word-oriented stream ciphers that use a linear feedback shift register (LFSR) and a non-linear filter. These attacks rely on exploiting linear relationships corresponding to multiples of the connection polynomial that define the LFSR.
**Keywords**: stream ciphers, cryptanalysis, SOBER, t-class, SSC-II.

## 1 Introduction

This paper presents new attacks on word-oriented stream ciphers constructed from a *linear feedback shift register* (LFSR) and a *non-linear filter* (NLF). These ciphers are constructed from operations on blocks of bits called *words*, where the length of a word is denoted by $w$. In particular this paper analyses what we call SOBER-like ciphers (based on the SOBER family of ciphers [8,12,13,14]) and SSC-like ciphers (as used in SSC [15], and SSC-II [16]).

The LFSR of a SOBER-like cipher produces a stream $\{s_t\}$ of $w$-bit words using operations over the Galois field of order $2^w$, which is denoted $GF(2^w)$. The words $s_t$ are called *L-words* and the stream is called the *L-stream*. The L-words $(s_0, \ldots, s_{r-1})$ are initialised from the secret key (some ciphers also initialise using a resynchronisation value). The remaining words are produced by iterating a linear recurrence $s_{t+r} = \sum_{i=0}^{r-1} \alpha_i s_{t+i}$, where $\alpha_i \in GF(2^w)$ are constant, and multiplication and addition are performed over $GF(2^w)$. Addition over $GF(2^w)$ is equivalent to bit-wise exclusive-OR (XOR). The LFSR is represented by the *connection polynomial*: $p(x) = x^r + \sum_{i=0}^{r-1} \alpha_i x^i$, where, once more, multiplication and addition are performed over $GF(2^w)$. The set of exponents of $p(x)$ with nonzero coefficients is called the *LFSR tapset*, denoted $T$. The LFSR of an SSC-like cipher differs in that it uses bit rotations rather than field multiplications and is based on a bit-wise LFSR (more details are given in Sect. 2). The vector $\sigma_t = (s_t, \ldots, s_{t+r-1})$ in either cipher is known as the *state* of the LFSR at time $t$.

The L-stream is fed through an NLF to produce the *N-stream* $\{v_t = F(\sigma_t)\}$. The words $v_t$ are called *N-words*. SOBER-like ciphers use an LFSR with a large state $\sigma_t$, and the NLF relies on a small, fixed subset of the words in $\sigma_t$. That is, we can write $v_t = F(s_{t+\gamma_1}, \ldots, s_{t+\gamma_a})$, where $\Gamma = \{\gamma_1, \ldots, \gamma_a\} \subset \{0, \ldots, r-1\}$, is the *NLF tapset*. SSC-like ciphers, on the other hand, use an LFSR with a small state, and the NLF relies on the entire state.

T. Okamoto (Ed.): ASIACRYPT 2000, LNCS 1976, pp. 303–316, 2000.

SOBER-like ciphers use an LFSR, an NLF and a form of decimation called stuttering (described in Sect. 3). The resulting stream, denoted $\{z_n\}$, is the *key stream*. The stuttering chooses which N-words will be output to the key stream. The stuttering is intended to, and appears to, defeat attacks requiring large amounts of output, such as correlation attacks [4,10]. However, the stuttering merely adds an almost constant factor to the complexity of the attacks described below.

In the analysis of stream ciphers based on bit-wise LFSRs, cryptanalysts found that attacks could be improved by exploiting linear relationships in the L-stream other than that expressed by the linear recurrence (see for example [4,6,10]). Such linear relationships correspond to multiples of the connection polynomial: the polynomial $r(x) = p(x) \cdot q(x) = \sum_{i=0}^{a} \epsilon_i x^i$, corresponds to a linear relationship of the form $\sum_{i=0}^{a} \epsilon_i s_{t+i} = 0$. For the remainder of the paper, a *multiple* refers to either a multiple of the connection polynomial or the linear relationship corresponding to that multiple. The main purpose of this paper is to provide examples of word-oriented stream ciphers for which the multiples can lead to low complexity attacks.

The first example is a component of the word-oriented stream cipher SSC-II [16]. SSC-II consists of two *half-ciphers* producing streams that are XORed to form the output. One of these half-ciphers is based on a 4-word LFSR (each word consists of 32 bits), with an NLF and no stuttering. The LFSR is based on a simple 127-bit, bit-wise linear recurrence that appears difficult to exploit due to the word-oriented structure of the NLF. However, a power of the bit-wise connection polynomial results in a linear relationship between corresponding bits of $s_t$, $s_{t+63}$ and $s_{t+127}$. This paper describes how this relationship can be exploited in an attack of complexity $c(2^{41.7})$ against the LFSR-half cipher, where $c(N)$ indicates that the complexity is expected to be a small multiple of $N$. The authors would like to emphasise that this attack on the half-cipher does not defeat the entire SSC-II cipher.

The attack on the SSC-II half-cipher is due to the bit-wise connection polynomial of the LFSR having extremely low weight (that is, a low number of terms). If the LFSR was based on a higher-weight connection polynomial, but there was some low-weight, low-degree multiple $r(x)$, then a similar attack could be applied using this multiple. The linear recursion over $GF(2^w)$ in a SOBER-like cipher can be shown to be equivalent to implementing $w$ parallel bit-wise LFSRs of length $wr$ over $GF(2)$, see [9]. The constants $\alpha_i$ are chosen so that the bit-wise LFSR has many terms (high weight). This property defeats attacks similar to the above attack, as well as defeating other attacks designed for stream ciphers employing bit-wise LFSRs. The most successful attacks against SOBER-like ciphers have been what we call *guess-and-determine* (GD) attacks [1,2,3,7,8,12,13]. These GD attacks are based on exploiting two relationships: the linear relationship between L-words described by the LFSR; and the relationship between L-words and the key stream defined by the NLF. However, previous attacks have not exploited any further linear relationships.

The latest edition SOBER ciphers, the t-class [8], contains three ciphers: t8, t16 and t32. The cipher t16 is currently being assessed for use in "third generation" mobile communication systems, while t32 is being implemented for encryption in mail transfer sessions between e-mail servers. Thus far, our research into the t-class has not found any GD attacks exploiting further linear relationships that can decrease the complexity below that of previously known GD attacks. However, we have observed that multiples can lead to low-complexity GD attacks on other SOBER-like ciphers. This is demonstrated by a dummy SOBER-like cipher, TIPSY, for which the best GD attacks exploiting only the LFSR and NLF have complexity $c(2^{150})$. Our search method found a GD attack exploiting further linear relationships for which the complexity is reduced to $c(2^{117})$.

The paper is arranged as follows. Section 1.1 introduces some definitions. Section 2 describes the analysis of the LFSR half-cipher in SSC-II. Section 3 introduces GD attacks and the cipher TIPSY is analysed. Section 4 describes our method for finding GD attacks. Conclusions and areas for further research are discussed in Sect. 5.

## 1.1   Definitions

For any $t \geq 0$, we define a *candidate L-word* $u_t$ to be a guess for the value of the L-word $s_t$, and define a *candidate state* $\mu_t = (u_t, \ldots, u_{t+r-1})$ to be a guess for the value of $\sigma_t$. We consider that an LFSR-based stream cipher is broken once the initial state of the LFSR has been determined. One method by which a stream cipher can be attacked is to search through every candidate $\mu_t$ until the value of $\sigma_t$ is found (this process is commonly known as *guessing*). A candidate state $\mu_t$ is *tested* (to see if it is correct) by constructing a key stream using this value $\mu_t$, and comparing the resulting key stream with the observed key stream. If the two streams match then the candidate is correct. In general, the large size of the register and the corresponding large number of possible candidate states make any such attack prohibitive.

## 2   Analysis of SSC-II

SSC-II [16] was proposed by Zhan, Carroll and Chan, and is based on $w = 32$-bit operations and $w$-bit words. The cipher consists of two *half-ciphers*: each half cipher produces a stream of 32-bit words and these streams are XORed to form the output. One half-cipher uses a lagged Fibonacci Generator which is based on addition modulo $2^{32}$ and is not considered here. The other half-cipher is based on a four-word LFSR. This LFSR produces an L-stream of 32-bit L-words $\{s_t\}$ by iterating the linear recurrence: $s_{t+4} = s_{t+2} \oplus (s_{t+1} << 31) \oplus (s_t >> 1)$, where $a << b$ ($a >> b$) denotes left (right) shifting of $a$ by $b$ bits. The bit-shifts are not cyclic: the remaining values are filled with zero bits. We denote the corresponding bit-stream by $\{b_i\}$ where $b_{32t+j} = s_t[j]$, the $j$-th bit of $s_t$, $0 \leq j \leq 31, t \geq 0$. The bit stream $\{b_i\}$ can be produced by a bit-wise LFSR with linear recurrence $b_{i+127} = b_{i+63} + b_i$ (mod 2). The LFSR in SSC-II calculates

32 bits of the L-stream simultaneously. SSC [15] employs an LFSR based on a similar principle.

The LFSR half-cipher has an NLF containing: addition modulo $2^{32}$, denoted by $\boxplus$; 32-bit XOR; swapping the higher and lower order halves of the 32-bit word, denoted by $SWAP$; and including the carry resulting from adding words, where $\rightarrow$ denotes outputting this carry. Let $s_t^*$ denote the word $s_t$ with the least significant bit (LSB) set to one. The N-word $v_t$ is determined from the state $\sigma_t = (s_t, \ldots, s_{t+3})$ as follows:

$$A = s_t^* \boxplus s_{t+3} \rightarrow c_1, \qquad\qquad B = SWAP(A),$$
$$C = B \boxplus (s_{t+2} \oplus (c_1 \cdot s_t^*)) \rightarrow c_2, \ v_t = c_2 \boxplus (s_{t+1} \oplus s_{t+2}) \boxplus C,$$

where $c_1 \cdot s_t^* = 0$ if $c_1 = 0$ and $c_1 \cdot s_t^* = s_t^*$ if $c_1 = 1$.

*Note 1.* Let $\hat{p}(x) = x^{127} + x^{63} + 1$ denote the connection polynomial for the bit stream $\{b_i\}$. Due to cancellation of terms, $\hat{p}^2(x) = x^{127 \cdot 2} + x^{63 \cdot 2} + 1$, $\hat{p}^4(x) = x^{127 \cdot 4} + x^{63 \cdot 4} + 1$ and so forth. Thus, $\hat{p}^{32}(x) = x^{127 \cdot 32} + x^{63 \cdot 32} + 1$, indicating that $b_{i+127 \cdot 32} = b_{i+63 \cdot 32} + b_i$. This implies that $s_{t+127}[m] = s_{t+63}[m] + s_t[m]$, for each $m$, $0 \le m \le 31$, and thus $s_{t+127} = s_{t+63} \oplus s_t$.

This linear relationship is likely to lend the LFSR half-cipher to a fast correlation attack. The authors are currently analysing SSC-II to assess the complexity of such an attack. The following attack illustrates an alternative method of exploiting this linear relationship. The 32-bit words are first divided into two 16-bit half-words: for example, $s_{t+i} = sH_{t+i}\|sL_{t+i}$ and $v_{t+j} = vH_{t+j}\|vL_{t+j}$. Note that the half-word N-words $vH_t$ and $vL_t$ are functions of the half-words $sH_{t+i}$ and $sL_{t+i}$, $0 \le i \le 3$, using addition modulo $2^{16}$ (denoted by $\boxplus$), 16-bit XOR and carries $d_i$ from the addition of the lower half-words:

$$AL = sL_t^* \boxplus sL_{t+3} \rightarrow d_1 , \qquad\qquad AH = sH_t \boxplus sH_{t+3} \boxplus d_1 \rightarrow c_1 ,$$
$$CL = AH \boxplus (sL_{t+2} \oplus (c_1 \cdot sL_t^*)) \rightarrow d_2 ,$$
$$CH = AL \boxplus (sH_{t+2} \oplus (c_1 \cdot sH_t)) \boxplus d_2 \rightarrow c_2 ,$$
$$vL_t = c_2 \boxplus (sL_{t+1} \oplus sL_{t+2}) \boxplus CL \rightarrow d_3 ,$$
$$vH_t = (sH_{t+1} \oplus sH_{t+2}) \boxplus CH \boxplus d_3 .$$

(The $SWAP$ step is integrated into the evaluation of $CL$ and $CH$). If the values of $c_1 \in \{0, 1\}$, $(c_2 \boxplus d_1) \in \{0, 1, 2\}$ and $(d_2 \boxplus d_3) \in \{0, 1, 2\}$ are known, then the NLF half-word outputs can be written as:

$$vL_t = sH_t \boxplus sH_{t+3} \boxplus (sL_{t+2} \oplus (c_1 \cdot sL_t^*)) \boxplus (sL_{t+1} \oplus sL_{t+2}) \boxplus (c_2 \boxplus d_1) ,$$
$$vH_t = sL_t^* \boxplus sL_{t+3} \boxplus (sH_{t+2} \oplus (c_1 \cdot sH_t)) \boxplus (sH_{t+1} \oplus sH_{t+2}) \boxplus (d_2 \boxplus d_3) .$$

For fixed values of $c_1$, $(c_2 \boxplus d_1)$ and $(d_2 \boxplus d_3)$, the expression for the LSB of $vL_t$ provides a linear relationship between the LSBs of $sL_t^*$, $sH_t$, $sL_{t+1}$ and $sH_{t+3}$. Similarly, the expression for the LSB of $vH_t$ provides a linear relationship between the LSBs of $sL_t^*$, $sH_t$, $sH_{t+1}$ and $sL_{t+3}$. The LSB of $sL_t^*$ is one, so this can be ignored.

Consider the sets $X = \{0, 1, 2, 3, 63, 64, 65, 66, 126, 189\}$,

$Y = \{0, 63, 126, 127, 189, 190, 253, 254, 317, 381\}$, and

$Z = \{0, 1, 2, 3, 63, 64, 65, 66, 126, 127, 128, 129, 130, 189, 190, 191, 192, 193,$
$\qquad 253, 254, 255, 256, 257, 317, 318, 319, 320, 381, 382, 383, 384\}$ .

The values of L-words $s_{t+j}$, $j \in Z$, can be derived from values of the L-words $s_{t+i}$, $i \in X$, by applying the equation in Note 1. For example, $s_{t+127} = s_{t+63} \oplus s_t$, and $s_{t+191} = s_{t+127} \oplus s_{t+64}$. Thus, each L-word $s_{t+j}$, $j \in Z$, can be expressed as $s_{t+j} = \bigoplus_{i \in X} \beta_{j,i} s_{t+i}$, where $\beta_{j,i} \in \{0,1\}$ for $i \in X$. Furthermore, these equations relate bits of the L-words $s_{t+j}$, $j \in Z$, to corresponding bits of the L-words $s_{t+i}$, $i \in X$: $s_{t+j}[m] = \bigoplus_{i \in X} \beta_{j,i} s_{t+i}[m]$, for each $m$, $0 \le m \le 31$. Note that the values of the 10 N-words $v_{t+j}$, $j \in Y$ rely on the set L-words $s_{t+j}$, $j \in Z$. Each bit of these L-words $s_{t+j}$, $j \in Z$ is, in turn, a linear function of the corresponding bits in 10 L-words $s_{t+i}$, $i \in X$. Candidates $u_{t+i}$, $i \in X$, for the L-words $s_{t+i}$, $i \in X$, are determined as follows.

¿From the expressions for the 20 half-word outputs $vL_{t+j}$ and $vH_{t+j}$, $j \in Y$, we get 20 linear equations in the LSBs of $uH_{t+j}$, $uL_{t+j+1}$, $uH_{t+j+1}$, $uL_{t+j+3}$ and $uH_{t+j+3}$, $j \in Y$. The attacker guesses the values of of $c_1$, $(c_2 \boxplus d_1)$ and $(d_2 \boxplus d_3)$ in the expression for each N-word $v_{t+j}$, $j \in Y$. For each of the 10 N-words there are 2 possible values for $c_1$, and 3 possible values each for $(c_2 \boxplus d_1)$ and $(d_2 \boxplus d_3)$. Therefore, the total number of guesses is $(2 \cdot 3^2)^{10} = 2^{41.7}$. These values are subtracted from the expressions for the 20 half-word outputs $vL_{t+j}$ and $vH_{t+j}$, $j \in Y$, to get 20 linear equations in the LSBs of $uH_{t+j}$, $uL_{t+j+1}$, $uH_{t+j+1}$, $uL_{t+j+3}$ and $uH_{t+j+3}$, $j \in Y$. As noted above, each of these LSBs is, in turn, a linear equation in the LSBs of $uL_{t+i}$ and $uH_{t+i}$, $i \in X$. Thus the attacker obtains 20 linear equations in the LSBs of $uL_{t+i}$ and $uH_{t+i}$, $i \in X$ (these LSBs represent a total of 20 bits). These equations are solved to obtain the LSBs of $uL_{t+i}$ and $uH_{t+i}$, $i \in X$. ¿From these LSBs, the attacker determines $uL_{t+j}$ and $uH_{t+j}$, $j \in Z$, which enables the attacker to determine the carries up to the second LSBs of $vL_{t+j}$ and $vH_{t+j}$, $j \in Y$. After subtracting these carries, the attacker now has 20 linear equations in the second LSBs of $uH_{t+j}$, $uL_{t+j+1}$, $uH_{t+j+1}$, $uL_{t+j+3}$ and $uH_{t+j+3}$, $j \in Y$. Once again, each of these bits is a linear equation in the second LSBs of $uL_{t+i}$ and $uH_{t+i}$, $i \in X$. The attacker obtains the system of 20 linear equations in the second LSBs of $uL_{t+i}$ and $uH_{t+i}$, $i \in X$ (20 in total), and solves this system to obtain these values. This process is repeated to obtain all of the bits in $uL_{t+i}$ and $uH_{t+i}$, $i \in X$. These candidates ($uL_{t+j}$ and $uH_{t+j}$, $j \in X$) combine to form several full states, any of which may be tested (by producing some of the N-stream and comparing it with the observed key stream).

As mentioned above, the total number of guesses is $2^{41.7}$, so the process complexity of the attack is $c(2^{41.7})$. The data complexity of the attack is small: the attacker requires $v_{t+j}$, $j \in Y$, for a single $t$, which will require observing 382 consecutive key-stream words. This attack is feasible for one primary reason: the bit-wise connection polynomial has a small number of terms. The attack would also have been feasible if there was a low-weight, low-degree multiples of the bit-wise connection polynomial. However, the attack cannot be applied if the weight of the multiple is sufficiently high, or the degree is sufficiently large, for the following reasons. A high-weight multiple of the bit-wise connection polynomial would require more equations in the N-words before system of bit-wise linear equations was solvable. Consequently, more values of $c_1, (c_2 \boxplus d_1), (d_2 \boxplus d_3)$ would be guessed, increasing the complexity and rendering the attack infeasible. On the other hand, if the degree of the multiple exceeds the maximum number of key-stream words produced from a single initial state, then this relationship cannot be exploited, regardless of weight.

## 3   Guess-and-Determine Attacks

The LFSRs of SOBER-like ciphers correspond to bit-wise connection polynomials with extremely large numbers of terms. For example, the LFSR of t16 has a corresponding bit-wise connection polynomial with approximately 136 terms. This property helps SOBER-like ciphers resist the kind of attack described in the previous section. The most successful attacks [1,2,3,7,14,13,12] against SOBER-like ciphers have been GD attacks (there is no common name for these attacks). The following example describes a dummy SOBER-like cipher which is used to demonstrate how GD attacks are performed, and how GD attacks can, in some cases, be improved by exploiting multiples.

*Example 1.* TIPSY is a SOBER-like cipher designed for $w = 16$-bit processors, so the words are 16-bits long and all operations are 16-bit operations. TIPSY uses the LFSR tapset $T = \{0, 1, 4, 13\}$ and the NLF tapset $\Gamma = \{0, 5, 10, 11\}$. The linear recursion is of the form $s_{t+13} = s_{t+4} + s_{t+1} + \alpha s_t$, where $\alpha = \text{0xEDED}$, and addition and multiplication are performed over $GF(2^{16})$. The corresponding connection polynomial is $p(x) = x^{13} + x^4 + x + \alpha$. The NLF is of the form: $v_t = F(s_t, s_{t+5}, s_{t+10}, s_{t+11}) = f(s_t \boxplus s_{t+11}) \boxplus s_{t+5} \boxplus s_{t+10}$, where $\boxplus$ denotes addition modulo $2^{16}$ and $f$ is a fixed, nonlinear, one-to-one 16-bit S-box. TIPSY decimates the N-stream to form the key stream using the same stuttering as t16 (the stuttering is described in Sect. 3.1).

As mentioned in Sect. 1.1, a stream cipher can be broken by guessing the value of any state $\sigma_t$, but the large size of the register and the corresponding large number of possible candidate states make any such attack prohibitive. GD attacks guess only a small set of candidate L-words, rather than an entire state. These attacks then use some observed N-stream words, and the relationships resulting from the LFSR and the NLF, to determine an entire state from this smaller set of L-words.

*Example 2.* In attacking TIPSY, if $u_t$, $u_{t+1}$ and $u_{t+13}$ are guessed, then $u_{t+4}$ can be determined as $u_{t+4} = u_{t+13} + u_{t+1} + \alpha u_t$. Alternatively, if $u_{t+5}, u_{t+10}$ and $u_{t+11}$ are guessed then $u_t$ can be determined from $v_t$; if $\boxminus$ denotes subtraction modulo $2^{16}$, then $u_t = f^{-1}(v_t \boxminus (u_{t+5} \boxplus u_{t+10})) \boxminus u_{t+11}$.

These two processes of determining L-words are called *D-exploiting* the LFSR and NLF respectively (the 'D' is for 'determine'). Note that, for TIPSY, D-exploiting the LFSR and NLF is computationally equivalent to $c(1)$ encryption. The same applies to the t-class ciphers. D-exploiting the NLF is not a new concept: inversion attacks [6] and the generalised inversion attacks [5] are based on a similar approach.

Given a suitable portion of the N-stream,[1] previous GD attacks were based on **guessing** candidates for a small set of L-words, D-exploiting the LFSR and NLF to **determine** a full candidate state, and then **testing** this candidate state. These analyses of SOBER-like ciphers examined only those GD attacks that exploit the relationships explicitly defined by the LFSR and NLF. This paper extends the range of GD attacks by D-exploiting further multiples. There are simply too many multiples to begin searching for all attacks exploiting all possible multiples. Consequently, a method has been developed for reducing the amount of work by considering multiples that are more likely to lead to improved attacks: the rationale behind the authors' approach is described in Sect. 4. Using this method, the authors conducted a search for attacks exploiting polynomials of degree $2r$ (twice the degree of $p(x)$) or less and with 10 or less terms. This method cannot be guaranteed to find the best attack, as there may be some other high-weight or high-degree polynomial which can be exploited in a low complexity attack. However, the existence of such an attack becomes more unlikely as the weight and degree of the polynomials increases.

When applied to the t-class ciphers, the analysis described in Sect. 4 revealed that the additional linear relationships did not provide an attack of lower complexity than was already known. However, the analysis of TIPSY did find improvements by exploiting further multiples. The lowest complexity GD-attack D-exploiting only the LFSR and NLF of TIPSY has complexity $c(2^{128})$, given a suitable portion of the N-stream. Using the method described in Sect. 4, the authors found the following attack of complexity $c(2^{96})$, given a suitable portion of N-stream, a significant improvement.

*Example 3.* Table 1 describes an GD attack on TIPSY that D-exploits the LFSR, the NLF and the following multiples:

$$p^2(x) = x^{26} + x^8 + x^2 + \alpha^2 \ ,$$
$$r_1(x) = (x^9 + x^6 + x^3 + 1) \cdot p(x)$$
$$= x^{22} + x^{19} + x^{16} + \alpha x^9 + \alpha x^6 + \alpha x^3 + x + \alpha \ ,$$
$$r_2(x) = (x^{12} + \alpha x^{11} + \alpha^2 x^{10} + x^6 + x^3 + \alpha x^2 + \alpha^2 x + 1) \cdot p(x)$$
$$= x^{25} + \alpha x^{24} + \alpha^2 x^{23} + x^{19} + (\alpha^3 + 1)x^{10} + \alpha^2 x^5 + (\alpha^3 + 1)x + \alpha \ .$$

To perform the attack, a portion of the N-stream must be observed, including $v_{t+i}$, $i \in \{4, 7, 11, 12, 17, 18, 22, 23\}$ for some value of $t$. Let $\phi_t$ denote the six-

---

[1] The problem of obtaining a suitable portion of N-stream from the key stream is addressed in Sect. 3.1.

word candidate vector $\phi_t = (u_{t+12}, u_{t+14}, u_{t+15}, u_{t+17}, u_{t+22}, u_{t+27})$. For a given value of $\phi_t$, Steps 2 to 18 in Table 1 determine candidates for the 17 L-words:

$$s_{t+i}, \ i \in \{4, 5, 6, 7, 8, 9, 18, 21, 23, 25, 28, 29, 30, 32, 33, 34, 41\} \ .$$

For example, in Step 2, the value of $u_{t+23}$ is determined from the values of $v_{t+12}$, $u_{t+12}$ $u_{t+17}$, and $u_{t+22}$ by D-exploiting the NLF:

$$u_{t+23} = f^{-1}(v_{t+12} \boxminus (u_{t+17} \boxplus u_{t+22})) \boxminus u_{t+12}.$$

**Table 1.** A GD attack on TIPSY exploiting the LFSR, the NLF, $p^2(x)$, $r_1(x)$ and $r_2(x)$, given $v_{t+i}, i \in \{4, 7, 11, 12, 17, 18, 22, 23\}$. "Action" indicates the following actions: C, perform an NLF check; G, guess values; L, D-exploit the LFSR; N, D-exploit the NLF; $r_1$, D-exploit the multiple $r_1(x)$; $r_2$, D-exploit the multiple $r_2(x)$; S, D-exploit the square of the connection polynomial ($p^2(x)$); T, test the given candidate state. In the next two columns a candidate L-word $u_{t+i}$ is indicated using the value of $i$. "Used" indicates those values used to determine or check the value indicated in the "Det." column.

| Step | Act. | Values Used | Value Det. | Step | Act. | Values Used | Value Det. |
|------|------|-------------|-----------|------|------|-------------|-----------|
| 1 | G | | 12,14,15, 17,22,27 | 13 | L | 4,5,17 | 8 |
| | | | | 14 | L | 8,9,12 | 21 |
| 2 | N | $v_{t+12}$, 12, 17, 22 | 23 | 15 | L | 17,18,21 | 30 |
| 3 | N | $v_{t+17}$, 17, 22, 27 | 28 | 16 | S | 4,12,30 | 6 |
| 4 | S | 15,17,23 | 41 | 17 | $r_1$ | 6,7,9,12,15,22,28 | 25 |
| 5 | L | 14,15,27 | 18 | 18 | L | 21,22,25 | 34 |
| 6 | N | $v_{t+7}$, 12, 17, 18 | 7 | 19 | C | 23,28,33,34 | $v_{t+23}$ |
| 7 | N | $v_{t+18}$, 18, 23, 28 | 29 | 20 | G | | 11 |
| 8 | L | 28,29,41 | 32 | 21 | N | $v_{t+11}$, 11, 21, 22 | 16 |
| 9 | N | $v_{t+22}$, 22, 27, 32 | 33 | 22 | L | 12,16,25 | 13 |
| 10 | S | 7,15,33 | 9 | 23 | S | 8,16,34 | 10 |
| 11 | N | $v_{t+4}$, 9, 14, 15 | 4 | 24 | T | $\mu_{t+4}$ | |
| 12 | $r_2$ | 4,9,14,23,27,28,29 | 5 | | | | |

Note that the L-words $s_{t+i}, i \in \{23, 28, 33, 34\}$, are the inputs to the NLF producing $v_{t+23}$, and candidates for all these inputs are known after Step 18 is performed. However, $v_{t+23}$ has not been used to determine any of these values when exploiting the NLF, so these candidates are independent of the value of $v_{t+23}$. Clearly, if the candidates in $\phi_t$ are correct, then $F(u_{t+23}, u_{t+28}, u_{t+33}, u_{t+34}) = v_{t+23}$. If $F(u_{t+23}, u_{t+28}, u_{t+33}, u_{t+34}) \neq v_{t+23}$, then at least one of the candidates in $\phi_t$ is incorrect, and there is no use in completing any further steps. This information can be used to eliminate incorrect values of $\phi_t$ using a process called an

*NLF check.* If $F(u_{t+23}, u_{t+23}, u_{t+33}, u_{t+34}) = v_{t+23}$, then the vector $\phi_t$, is said to *pass* the NLF check, otherwise it *fails*. If $\phi_t$ fails the NLF check in Step 19, then the attack returns to Step 1 and tries another guess for $\phi_t$, otherwise the attack proceeds to Step 20.

At Step 20, a candidate $u_{t+11}$ for $s_{t+11}$ is guessed. Steps 21, 22 and 23 determine candidates $u_{t+16}$, $u_{t+13}$ and $u_{t+10}$. Thus after Step 23, a candidate state $\mu_{t+4} = (u_{t+4}, \ldots, u_{t+16})$ for the state $\sigma_{t+4}$ has been determined. This candidate state $\mu_{t+4}$ is then tested in Step 24. If $\mu_{t+4}$ is incorrect, then the attack returns to Step 20 and guesses another value for $u_{t+11}$, unless all values for $u_{t+11}$ have been tested for a given value of $\phi_t$, in which case the attack returns to Step 1 and guesses another value for $\phi_t$.

There are $2^{6w} = 2^{96}$ possible values for $\phi_t$, so performing Steps 1 to 19 is computationally equivalent to $c(2^{96})$ encryptions. As the NLF is balanced, only one in $2^{w} = 2^{16}$ values of $\phi_t$ will pass the NLF check. Thus, only $2^{80}$ values of $\phi_t$ will proceed to Step 20. There are $2^{16}$ values for $u_{t+11}$, so Steps 20 to 24 are performed $2^{80} \cdot 2^{16} = 2^{96}$ times: equivalent to $c(2^{96})$ encryptions. Therefore, the total complexity of the attack is equivalent to only $c(2^{96}) + c(2^{96}) = c(2^{96})$ encryptions.

*Note 2.* This attack clearly exploits the property that TIPSY has two pairs of NLF taps which are 5 words apart, contravening criteria suggested by Golic [6] and Löhlien [11].

## 3.1 Accounting for the Stuttering

The stuttering decimates the N-stream $\{v_t\}$ as follows. The first output of the NLF $(v_1)$ is the first *stutter control word* (SCW). Each SCW is partitioned into eight pairs of bits (each pair is called a *dibit*). Beginning with the least significant dibit, the stuttering reads the value of the dibit and performs one of four actions according to the value of the dibit. The actions corresponding to the dibits are shown in Table 2. When all the dibits have been read, the LFSR is cycled, and the output of the NLF becomes the next SCW. The resulting stream, denoted $\{z_n\}$, is the key stream.

The stuttering decimates the N-stream in a random manner, so that consecutive key-stream words may or may not be consecutive N-stream words. This results in some uncertainty in relating the position of N-words to position of key-stream words. Furthermore, this uncertainty increases with the distance (in words) between key-stream words. This helps defeat attacks which require large amounts of key stream, such as correlation attacks. However, the stuttering does not add much resistance against GD attacks.

*Example 4.* Consider the attack in Example 3. This attack requires the attacker to know the values of $v_{t+i}$, $i \in \{4, 7, 11, 12, 17, 18, 22, 23\}$. To perform this attack, the attacker must assume that at a certain point in the key stream, one or more SCWs have a particular value or values which allow the appropriate N-words to

**Table 2.** The actions of the stuttering corresponding to the four possible values of the dibits.

| | |
|---|---|
| 00: | Cycle the LFSR, but do not output anything. |
| 01: | Cycle the LFSR, output the NLF output XOREd with 0x6996, then cycle the LFSR again (without producing another output). |
| 10: | Cycle the LFSR once (without producing any output), then cycle the LFSR again and output the NLF output. |
| 11: | Cycle the LFSR and output the NLF output XORed with the bit-wise complement of 0x9669. |

be obtained from the key stream. Given a suitably large amount of key stream, an attacker can assume that for some values of $t$, $v_{t+3} = (01, 10, ab, 01, 10, cd, 10, 01)$ where $ab, cd \in \{01, 10\}$, and $v_{t+3}$ is an SCW. The key stream output by this SCWs will be:

$$z_n = v_{t+4} \oplus 0x6996, \qquad z_{n+1} = v_{t+7},$$
$$z_{n+2} = v_{t+8} \oplus 0x6996 \text{ OR } z_{n+2} = v_{t+9},$$
$$z_{n+3} = v_{t+11}, \qquad z_{n+4} = v_{t+12} \oplus 0x6996,$$
$$z_{n+5} = v_{t+14} \oplus 0x6996 \text{ OR } z_{n+5} = v_{t+15},$$
$$z_{n+6} = v_{t+17}, \qquad z_{n+7} = v_{t+18} \oplus 0x6996,$$

The next SCW will $v_{t+20}$. The attacker can assume that for some value of $t$, not only is $v_{t+3}$ of the above form, but $v_{t+20}$ is also of the form $v_{t+20} = (\ldots, 01, 10)$, If this is the case, then the next key-stream words are $z_{n+8} = v_{t+22}$ and $z_{n+9} = v_{t+23} \oplus 0x6996$.

Thus, assuming that the values of the SCWs are correct, the attacker is able to determine the N-words from the key stream, and perform the attack in Example 3. There are two obstacles. First, the attacker does not know when the SCWs have these values, and second, the attacker does not even know where in the key stream the SCWs occur. As a result, the attacker proceeds through the key stream assuming that each sequence of 10 key-stream words was derived from the N-stream using the SCWs in Example 4, and performs the steps in Example 3 until the correct state is found. Let $N$ denote the data complexity, equal to the number of times that the process in Example 3 is repeated. The expected value of $N$ is the inverse of the probability that a random portion of key stream was obtained from the N-stream using the SCWs in Example 4. This probability is determined as follows. Firstly, consider the probability that the first key-stream word is the first word output by an SCW. There are an average of 6 key-stream words output for every SCW, so this is $1/6$. Secondly, ignoring the requirement that $v_{t+4}$ be an SCW, the values of $v_{t+4}$ and $v_{t+20}$ are of the correct form (in this example) with probability $2^{-18}$. The combined probability is $\frac{1}{6} \cdot 2^{-18} \approx 2^{-20.6}$. Consequently, $N = 2^{20.6}$ is the expected data complexity and the expected process complexity of the attack is $c(2^{20.6} \cdot 2^{96}) = c(2^{116.6})$. The GD attack on TIPSY exploiting only the LFSR and NLF (of process complexity $c(2^{128})$, given the N-stream) would correspond to an attack of process complexity $c(2^{150})$, when considering the stuttering.

## 4   Searching for GD Attacks

This section provides a brief description of the authors' method of searching for GD attacks. In this section, the *tapset* of any polynomial $r(x) = \sum_{i=0}^{r+k} \epsilon_i x^i$, is defined to be $T[r(x)] = \{i : \epsilon_i \neq 0\}$, and the number of non-zero coefficients of $r(x)$ (equal to $|T[r(x)]|$) is called the *weight* of $r(x)$.[2] A GD attack is defined by a set of steps where the LFSR, the NLF and other multiples are D-exploited to determine a candidate state from a small set of candidate L-words. It is the tapsets (of the LFSR, NLF and multiples) that determine which candidate L-words can be determined from a given set of candidate L-words. Thus, the existence of a GD attack is determined by the tapsets of the LFSR, NLF and multiples, and not other details of the relationship such as the coefficients. In the case of a bit-wise LFSR, finding the tapsets for the multiples is simple because the tapsets of the the factors $p(x)$ and $q(x)$ define the polynomials and hence define the tapset of the product $r(x) = p(x) \cdot q(x)$. However, in a word-oriented LFSR, there can be many factors $q(x)$ with the same tapsets (but different coefficients) for which the products $p(x) \cdot q(x)$ have different tapsets. This adds significant complication to the search for GD attacks. In addition to this complication, there is a very large set of multiples (and their tapsets). Consequently, the task of searching for the optimal GD attack (the GD attack of lowest complexity) is still an open problem.

The search for GD attacks can be approached from two directions. One approach is to have a growing set of multiples to exploit, where the search program constantly tests for all multiples that can be D-exploited given the set of L-words that are currently known. This approach has not yet been implemented, although the authors are in the process of developing such a program.

The second approach divides the search into two parts: a *polynomial search*, that determines a set of multiples $B$ to exploit; and a *B-attack search*, that examines the GD attacks exploiting the NLF and the polynomials in $B$. The set $B$ is called an *GD basis* and is always assumed to contain $p(x)$.

### 4.1   The $B$-Attack Search

The $B$-attack search finds a GD attack which minimises the complexity of the GD attacks exploiting the NLF and the polynomials in $B$. The $B$-attack search chooses a subset of L-words to guess, and finds the position of all L-words that could be determined by exploiting the NLF and the polynomials in $B$. If these

---

[2] Note that D-exploiting $r(x)$ is computationally equivalent to at most $c(|T[r(x)]|)$ encryptions.

L-words do not comprise a full state, then an additional L-word is guessed, and the process repeated. This continues until all L-words in an entire state are determined. Alternatively, if guessing an additional word will result in an attack with complexity larger than that of the best known attack, then the $B$-attack search tries another subset of L-words. To ensure that the $B$-attack search does not proceed indefinitely, the authors bounded the distance between the first L-word guessed and any determined L-words to a maximum of four register lengths.

## 4.2    The Polynomial Search

The speed of the $B$-attack search decreases as the size of $B$ increases, so the aim of the polynomial search is to find a small set of multiples that are likely to find the best attack. Intuition suggests that a multiple $r(x)$ is more likely to be D-exploited if the corresponding linear relationship is between a small number of L-words. That is, $r(x)$ is more likely to be exploited if it has low weight. Consequently, the first criterion used for selecting multiples for the set $B$ is that they have low weight. Now, suppose that the polynomial search is considering adding a multiple $r(x)$ to $B$. Suppose that whenever $r(x)$ is D-exploited, some combination of multiples can be D-exploited to determine the same L-word. Such multiples are *redundant* and should not be added to $B$. Hence, the polynomial search looks for a set of low-weight, non-redundant multiples of $p(x)$. The polynomial search takes a polynomial $p(x)$, and two bounds $D$ and $W$ on the degree and weight of the polynomials to be added to the GD basis. The polynomial search looks through the multiples of degree $\leq D$ and with weight $\leq W$: any non-redundant multiples are added to the GD basis. The polynomial search fixes a tapset $T'$ and considers the tapsets of $r(x) = p(x) \cdot q(x)$ when $T[q(x)] = T'$. Note that for a given $T'$, all these multiples $r(x)$ will share some similar characteristics. There will be some coefficients of $r(x)$ which will be certain to be zero (in the *zero* positions), there will be some coefficients which will be certain to be nonzero (in the *nonzero* positions), and the remaining coefficients could be either zero or nonzero, depending on the cancellation of terms in the expansion of $p(x) \cdot q(x)$, (the *zero-or-nonzero* positions). ¿From these sets of coefficients we can determine a superset of the possible tapsets for multiples $p(x) \cdot q(x)$ with $T[q(x)] = T'$, by considering all possible combinations of the nonzero positions and the zero-or-nonzero positions. The polynomial search only considers those tapsets with weight less than the bound $W$. For each resulting tapset, the polynomial search conducts tests for redundancy, and then confirm that the tapset corresponds to a multiple $p(x) \cdot q(x)$ with $T[q(x)] = T'$. This requires less processing than determining if the tapset corresponds to a multiple and then conducting the tests for redundancy.

The greatest restriction on the authors' polynomial search is the weight of the tested multiples. Our fastest algorithm employed fixed arrays containing the subsets of $b$ elements from a set of $a$ elements. This method worked best for us. As $a$ and $b$ increases, the necessary storage requirements increase significantly, placing constraints on $a$ and $b$. The authors restricted the polynomial search to finding multiples of degree less than $2r$ (twice the degree of $p(x)$) and weight 10 or less. The tests for redundancy then reduced this set of multiples. Given these restrictions, the polynomial search and $B$-attack search require less than a day of processing each.

### 4.3  Results

The polynomial search on the LFSR of TIPSY found 123 multiples within the above constraints (maximum degree $D = 26 = 2r$ and maximum weight $W = 10$). Using this basis, the $B$-attack search found an attack on TIPSY of complexity $c(2^{96})$ (ignoring stuttering): this is the attack described in Example 3. Given the improved attack on TIPSY, the authors considered that t-class might also weaker than first claimed. A polynomial search on the LFSR of t16 was conducted to find the GD basis $B$ within the aforementioned constraints (maximum degree $D = 34 = 2r$ and maximum weight $W = 10$). This search revealed a GD basis of 63 multiples. The $B$-attack search using this basis found only GD-attacks of complexity $c(2^{160})$ (ignoring stuttering). Such attacks offer no improvement over previous GD attacks (such attacks are simple variants of the attacks in [2,7], discussed in [8]). A similar analysis of t8 and t32 revealed that the additional linear relationships did not provide an attack of lower complexity than was already known.

## 5  Conclusion

This paper provides two examples of how multiples can be exploited in attacks against various word-oriented ciphers. In the first example, powers of the bit-wise connection polynomial reveal a weakness in SSC-II. This supports the well-known criteria that stream ciphers (even word-oriented stream ciphers) should avoid using connection polynomials for which there exists low-degree, low-weight multiples. In the second example, multiples of the connection polynomial over $GF(2^w)$ are used in a low complexity GD attack against a dummy SOBER-like cipher, TIPSY. However, the t-class ciphers appear to resist attacks exploiting multiples. The authors continue to examine how multiples can be exploited against SOBER-like ciphers, and consider how SOBER-like ciphers resist such attacks. It is hoped that this will lead to a method of determining the best possible GD attack on a given SOBER-like cipher.

## References

1. S. Blackburn, S. Murphy, F. Piper, and P. Wild. A SOBERing remark. Technical report, Information Security Group, Royal Holloway University of London, Egham, Surrey TW20 0EX, U.K., 1998.
2. D. Bleichenbacher, W. Meier, and S Patel. Analysis of the SOBER stream cipher. Technical Report TR45.AHAG.08.30.12, TR45 Ad Hoc Authentication Group, 1999.

3. D. Bleichenbacher and S Patel. SOBER cryptanalysis. *Fast Software Encryption, FSE'99 Lecture Notes in Computer Science, vol. 1636, L. Knudsen ed., Springer-Verlag*, pages 305–316, 1999.

4. V. Chepyzhov and B. Smeets. On a fast correlation attack on certain stream ciphers. *Advances in Cryptology, EUROCRYPT'91, Lecture Notes in Computer Science, vol. 547, D. W. Davies ed., Springer-Verlag*, pages 176–185, 1991.

5. J. Golić, A. Clark, and E. Dawson. Inversion attack and branching. *Information Security and Privacy, Fourth Australasian Conference, ACISP'99, Lecture Notes in Computer Science, vol. 1587, J. Pieprzyk, R Safavi-Naini, J. Seberry eds., Springer-Verlag*, pages 88–102, 1999.

6. J. Dj. Golić. On the security of nonlinear filter generators. *Fast Software Encryption, Lecture Notes in Computer Science, vol. 1039, D. Gollmann ed., Springer*, pages 173–188, 1995.

7. P. Hawkes. An attack on SOBER-II. Technical report, QUALCOMM Australia, Suite 410, Birkenhead Point, Drummoyne NSW 2137, Australia, 1999.

8. P. Hawkes and G. Rose. The t-class of SOBER stream ciphers. Technical report, QUALCOMM Australia, Suite 410, Birkenhead Point, Drummoyne NSW 2137, Australia, 1999. See http://www.home.aone.net.au/qualcomm.

9. T. Herlestam. On functions of Linear Shift Register Sequences. *Advances in Cryptology, EUROCRYPT'85, Lecture Notes in Computer Science, vol. 219, F. Pichler ed., Springer-Verlag*, 1986.

10. T. Johansson and F Jönsson. Improved fast correlation attacks on stream ciphers via convolutional codes. *Advances in Cryptology, EUROCRYPT'99, Lecture Notes in Computer Science, vol. 1592, J. Stern ed., Springer-Verlag*, pages 347–362, 1999.

11. B. Löhlein. Analysis and modifications of the conditional correlation attack. 1999. Accepted at 3rd IEEE/ITG Conference on Source and Channel Coding, 17-19 Jan. 2000, Munich.

12. G. Rose. S32: A fast stream cipher based on linear feedback over $GF(2^{32})$. Technical report, QUALCOMM Australia, Suite 410, Birkenhead Point, Drummoyne NSW 2137, Australia, 1998.

13. G. Rose. SOBER: A stream cipher based on linear feedback over $GF(2^8)$. Technical report, QUALCOMM Australia, Suite 410, Birkenhead Point, Drummoyne NSW 2137, Australia, 1998. See http://www.home.aone.net.au/qualcomm.

14. G. Rose. A stream cipher based on linear feedback over $GF(2^8)$. *Information Security and Privacy, Third Australasian Conference, ACISP'98, Lecture Notes in Computer Science, vol. 1438, C. Boyd, E. Dawson eds., Springer-Verlag*, pages 135–146, 1998.

15. M. Zhang, C. Carroll, and A. Chan. SSC. Technical Report TR45.AHAG.99.02.09.15, TR45 Ad Hoc Authentication Group, 1999.

16. M. Zhang, C. Carroll, and A. Chan. The software-oriented stream cipher SSC-II. In *Proceedings of Fast Software Encryption Workshop 2000*, pages 39–56, 2000.

# Encode-Then-Encipher Encryption: How to Exploit Nonces or Redundancy in Plaintexts for Efficient Cryptography

Mihir Bellare[1] and Phillip Rogaway[2]

[1] Dept. of Computer Science & Engineering, University of California at San Diego,
9500 Gilman Drive, La Jolla, CA 92093, USA
mihir@cs.ucsd.edu, www-cse.ucsd.edu/users/mihir
[2] Dept. of Computer Science, Engineering II Building, One Shields Avenue,
University of California at Davis, Davis, CA 95616, USA,   and
Dept. of Computer Science, Faculty of Science, Chiang Mai University, Thailand
rogaway@cs.ucdavis.edu, www.cs.ucdavis.edu/~rogaway

**Abstract.** We investigate the following approach to symmetric encryption: first *encode* the message via some keyless transform, and then *encipher* the encoded message, meaning apply a permutation $F_K$ based on a shared key $K$. We provide conditions on the encoding functions and the cipher which ensure that the resulting encryption scheme meets strong privacy (eg. semantic security) and/or authenticity goals. The encoding can either be implemented in a simple way (eg. prepend a counter and append a checksum) or viewed as modeling existing redundancy or entropy already present in the messages, whereby encode-then-encipher encryption provides a way to exploit structured message spaces to achieve compact ciphertexts.

## 1   Introduction

ENCIPHERING VS. ENCRYPTING. Many popular books on cryptography describe "encryption" as applying a key-indexed permutation $F_K$ to the plaintext $M$, thereby obtaining the ciphertext $C = F_K(M)$. Yet, if the goal of encryption is privacy (as it is usually assumed to be), then our community has long since recognized that, being deterministic, this realization of encryption cannot possibly achieve the strong security guarantees that one would hope for, namely, semantic security under chosen-plaintext attack and beyond [9,7,13]. (For example, if the same message is encrypted twice an adversary will be able to detect this.)

From this point forward, a family of permutations $F = \{F_K\}$ will be called a *cipher*. Applying one of these functions, $F_K$, is *enciphering* (not encrypting). Applying $F_K^{-1}$ is *deciphering* (not decrypting). In this paper, "good" for an enciphering method means approximating (in the usual ways [11]) a family of random permutations. On the other hand, "good" for an encryption scheme means achieving privacy properties at least as strong as semantic security. As indicated above, good enciphering never, by itself, makes for good encryption.

T. Okamoto (Ed.): ASIACRYPT 2000, LNCS 1976, pp. 317–330, 2000.

Despite the last statement, there seems to be a widespread belief that enciphering a message is, somehow, almost as good as encrypting it. When messages are somehow "structured," or the message space has "enough entropy," maybe enciphering does the job. Is there some scientific basis for such a belief?

In this paper we investigate the circumstances under which good enciphering really *does* make for good encryption. This leads us to introduce *encoding schemes* as a way to conceptualize what is happening when you encipher structured messages. Let us describe what are encoding schemes, and how they relate to enciphering.

ENCODE-THEN-ENCIPHER ENCRYPTION. Start with a good cipher that operates on messages of any length at all. (In other words, $F_K$, for a random $K$, "looks like" a random length-preserving permutation.) Now to encrypt $M$, first "encode" it into some string $M^*$. The encoding might be extremely simple—like prepending a counter, or appending some 0-bits, or maybe doing both. The encoding might even be the identity function. All that is demanded of an encoding method is that it does not "lose" information: you can "decode" $M^*$ to recover $M$, and you can recognize when a string is and is not the encoding of any message. Now to encrypt message $M$ under key $K$, encipher the encoded message $M^*$ using $F_K$, yielding ciphertext $C = F_K(M^*)$. To decrypt a ciphertext $C$ decipher it to find $M^* = F_K^{-1}(C)$, and then decode $M^*$ to get either a message $M$ or an indication that $M^*$ is not the encoding of any message. We call this style of encryption "encode-then-encipher encryption." This is not a popular way to encrypt, though it is certainly a very natural paradigm.

OUR RESULTS. In this paper we investigate how properties of the encoding scheme and the enciphering scheme can give rise to security properties of the resulting encryption scheme.

Suppose first that the encoding scheme adds in a *nonce*—usually a counter or a random value. The nonce can be added into the message in any way at all. All one needs is that the "collision probability"—the chance that two encoded messages come out the same"—be small. We prove in Theorem 1 that enciphering such encodings provides *semantic security*.

Next we look at encoding schemes which result in encoded messages which have enough *redundancy*. This means that "most" strings $M^*$ will be considered "bad." We prove in Theorem 2 that the resulting encryption scheme will now achieve *message authenticity*. It is as though the sender had sent a MAC along with his transmission. Interestingly, this theorem requires that the cipher be a *strong* pseudorandom permutation [11]. We show in Theorem 3 that an ordinary pseudorandom permutation won't do.

The actual results are quantitative. They show how much privacy and authenticity is guaranteed as a function of (easily-calculated) numbers associated to the encoding scheme, and as a function of the (quantified) security of the underlying cipher.

JUSTIFYING SOME OLD INTUITION. At some level it would seem to be folklore that enciphering strings which employ nonces or redundancy makes for good encryption. In the security literature one sees many statements to the effect that *we assume that messages to be encrypted employ adequate redundancy*, or *we avoid replay attacks by including a nonce in the messages we encrypt*. Our results help formalize what such authors may have had in mind, since the statements above become meaningful and true when "encryption" means "enciphering" and when the roles of nonces and redundancy are formally defined.

IS THE ENCODING STEP "REAL"? In some applications of encode-then-encipher encryption we imagine that the encoding step will be an ostensible part of encrypting: the piece of software which encrypts $M$ will encode it first, and then encipher the encoded message. For example, the encryption engine might take in a message $M$, prepend a counter, append a checksum, and encipher the resulting string. But encode-then-encipher encryption is actually more interesting when the encoding and decoding operations do *not* occur within the customary boundary of the encryption engine. For example, the encryption software may be presented with an already-formatted IP packet $M^*$. Its payload is the message $M$ one should get on decoding $M^*$, but the encryption software itself knows nothing about where is the payload or how to extract it. Still, the encoding and decoding processes really did occur, albeit within a different piece of code. Finally, the encoding step may exist purely as a conceptualization. For example, if messages are supposed to be English-language sentences then the encoding step can be regarded as the the identity function on the space of proper English-language sentences, while the decoding function takes a string $M^*$ and returns $M = M^*$ if it is English, or else an indication that this is not an English sentence. Probably this decoding operation can only performed by a human! Nonetheless, even in this case the language of encodings makes sense.

In general, the encoding of messages should be seen as a *model* for how the messages that we are enciphering might arise. This model is a more useful and general approach than trying to equip an unknown message space with a distribution. For example, a distribution on messages can not handle ideas like inserting a counter into the message, and it is quite artificial to try to equip English-language utterances with some distribution. The encoding/decoding model lets us discuss, in a natural and simple way, all the relevant properties about how messages might look.

WHY ENCODE-THEN-ENCIPHER? Encode-then-encipher encryption can be used to provide short ciphertexts with a high degree of independence on message-formatting conventions. As such, it can be used to provide a convenient migration path for legacy protocols. Let us explain.

In various application, particularly in networking, a "packet format" will have been defined, where this packet format includes redundancy and/or nonces, but has no fields for cryptographic purposes (eg., fields for an IV or MAC). Now suppose a need arises to add in privacy or authenticity features. At the same time, there will often be a real-world constraint not to grow or re-define the packet format.

Using encode-then-encipher you probably do not have to. If packets are known to repeat rarely or not at all (eg., packets always contain a sequence number) then semantic security is automatically guaranteed just by applying a good cipher. And if packet formats already include redundancy (which they typically do if for no other reason than to simplify parsing) then there may be no need to add in a separate MAC; once again, good enciphering (this time, with a *strong* pseudorandom permutation) is enough. And because it is irrelevant how and where the nonce and redundancy appeared in the packet, privacy and authenticity will be retained, with no protocols changes at all, if packet formats should subsequently change in some details.

The result is that encode-then-encipher encryption would leave packet sizes alone (our ciphers are understood to be length-preserving), and they would leave packets looking identical (after deciphering) to the way they looked before. This allows for modular software changes with minimal code disruption. The code which enciphers as a way to encrypt doesn't know (or care) where is the sequence number, say, where other fields are, or what values these fields can take. Such indifference makes for robust and simple software, and thus an easier migration path for adding in security features.

CONSTRUCTING VARIABLE-INPUT-LENGTH CIPHERS. To encrypt messages using the encode-then-encipher approach you need to encipher strings which may be long or short, and whose lengths may vary from one enciphering to the next. The cipher should look like a random length-preserving permutation $\pi : \mathcal{M}^* \to \mathcal{M}^*$. This may sound just like a block cipher, but it is actually quite different, because the domain includes strings of different lengths. One construction is given in [5], and others are possible, building on work like [11] and [12].

A NOTION OF AUTHENTICITY FOR ENCRYPTION SCHEMES. We note a final contribution of this paper, which is the notion of authenticity defined in Section 2. The usual way that message authenticity has been defined (eg., [2]) assumes that each message $M$ is accompanied by a tag (the message authentication code) $\tau$. The adversary wants to produce a hitherto unseen message $M'$ and a valid tag $\tau'$ for it. But this setting does not apply to us, where the messages being authenticated are never made visible. In the new setting the adversary's goal is to get the receiver to accept as authentic a string $C$—with a possibly unknown "meaning" $M$—where the adversary has not already witnessed $C$. This necessitates a new notion (or measure) of security for a symmetric encryption scheme.

While several definitions of privacy for symmetric encryption schemes are given in [1], here we are suggesting a notion of authenticity for an encryption scheme. Namely, consider a symmetric encryption scheme in which the decryption algorithm is allowed to reject ciphertexts to indicate that they are unauthentic. We take the setting of [1] in which the adversary gets to see (via an oracle) ciphertexts of messages of her choice encrypted under a key $K$. We then say that the adversary wins if she can produce a valid ciphertext (meaning one which the decryption function under $K$ does not reject) which was never an output of the encryption oracle.

Early (submitted) versions of this paper date to December 1998. Since then, definitions of authenticity for symmetric encryption schemes have appeared elsewhere [10]. We refer the reader to [4] for a comprehensive treatment of different notions of authenticity for symmetric encryption schemes and their relations to the notions of privacy.

## 2   Definitions

We provide definitions for PRFs, PRPs and SPRPs over arbitrary message spaces, and definitions of privacy and authenticity for symmetric encryption schemes.

HISTORY AND COMPARISONS. The basic definition of a PRF (pseudorandom function), as given by [8], sets the domain, range and keyspace to be the set of strings of length equal to the security parameter, and then defines security asymptotically. We adopt concrete versions of these definitions, as per [2], in order to model block-cipher based construction, and also to allow for a domain (which we call the message space) containing strings of different lengths. Our notion of a PRP (pseudorandom permutation) follows [2,3] and differs from that of [11] in that we measure distinguishability versus a random permutation rather than a random function, which is important when concrete security is considered. The notion of an SPRP (strong pseudorandom permutation) is that of [11] concretized in the style of [2] and extended with regard to domains. The definition of privacy for symmetric encryption schemes is from [1].

NOTATION AND CONVENTIONS. A *message space* $\mathcal{M}$ is a subset of $\{0,1\}^*$ for which $x \in \mathcal{M}$ implies that $x' \in \mathcal{M}$ for all $x'$ of the same length of $x$, and for which there exists an efficient (say linear time) algorithm to decide membership. A *ciphertext space* $\mathcal{C}$ is a subset of $\{0,1\}^*$. A *key space* $\mathcal{K}$ is a set together with a probability measure on that set. Writing $K \leftarrow \mathcal{K}$ means to choose $K$ at random according to this probability measure. The notation $|X|$ denotes the length of $X$ if $X$ is a string and the number of elements in $X$ if $X$ is a set.

CIPHERS. Let $\mathcal{K}$, $\mathcal{M}$ and $\mathcal{C}$ be a key space, message space, and ciphertext space. A *family of functions* is a map $F\colon \mathcal{K} \times \mathcal{M} \to \mathcal{C}$. If $K \in \mathcal{K}$ then we let $F_K(\cdot) = F_K(\cdot)$ and call this an *instance* of $F$. We let $f \leftarrow F$ denote the operation of picking a function from $F$ at random. (This is shorthand for $K \leftarrow \mathcal{K}; f \leftarrow F_K$.) We assume that $|F_K(M)| = \ell(|M|)$ depends only on $|M|$ and call $\ell$ the *length function* of the family. A *cipher* is a family of functions $F\colon \mathcal{K} \times \mathcal{M} \to \mathcal{C}$ in which each $F_K \colon \mathcal{M} \to \mathcal{C}$ is one-to-one and onto. In this case, $F_K^{-1}$ denotes the inverse of $F_K(\cdot)$. A cipher is *length-preserving* if $F_K(M) = |M|$ for all $K \in \mathcal{K}$ and $M \in \mathcal{M}$. For simplicity, all ciphers in this paper are assumed to be length-preserving. A *block-cipher* is a cipher with domain and range $\{0,1\}^n$. The number $n$ is called the *block length*.

We let $\mathbf{Rand}(\mathcal{M}, \ell)$ denote the family of all functions $f\colon \mathcal{M} \to \{0,1\}^*$ that satisfy $|f(M)| = \ell(|M|)$ for all $M \in \mathcal{M}$. A random function $f$ from $\mathbf{Rand}(\mathcal{M}, \ell)$ is determined as follows: for each $M \in \mathcal{M}$, $f(M)$ is a random string of length $\ell(|M|)$. Also let $\mathbf{Perm}(\mathcal{M})$ denote the cipher consisting of all length-preserving, one-to-one and onto functions on $\mathcal{M}$. A random function $\pi$ from $\mathbf{Perm}(\mathcal{M})$ is determined as follows: for each number $i$ such that $\mathcal{M}$ contains strings of length $i$, let $\pi_i$ be a random permutation on $\{0,1\}^i$. Then define $\pi(M) = \pi_i(M)$, where $i = |M|$.

PRFs, PRPs AND SPRPs. A *distinguisher* is a (possibly probabilistic) algorithm $A$ which has access to an oracle. If $F \colon \mathcal{K} \times \mathcal{M} \to \mathcal{C}$ is a function family with length function $\ell$ we let

$$\mathbf{Adv}_F^{\mathrm{prf}}(A) \;=\; \Pr[K \leftarrow \mathcal{K} : A^{F_K(\cdot)} = 1] - \Pr[f \leftarrow \mathbf{Rand}(\mathcal{M}, \ell) : A^{f(\cdot)} = 1]$$

denote the advantage of $A$ in distinguishing $F$ from a random function. We let

$$\mathbf{Adv}_F^{\mathrm{prp}}(A) \;=\; \Pr[K \leftarrow \mathcal{K} : A^{F_K(\cdot)} = 1] - \Pr[\pi \leftarrow \mathbf{Perm}(\mathcal{M}) : A^{\pi(\cdot)} = 1]$$

denote the advantage of $A$ in distinguishing $F$ from a random permutation. Define

$$\mathbf{Adv}_F^{\mathrm{prf}}(t, q, \mu) = \max_A \{\mathbf{Adv}_F^{\mathrm{prf}}(A)\}$$
$$\mathbf{Adv}_F^{\mathrm{prp}}(t, q, \mu) = \max_A \{\mathbf{Adv}_F^{\mathrm{prp}}(A)\}$$

where the maximum is taken over all adversaries having time-complexity at most $t$ and asking at most $q$ oracle queries, these queries totaling at most $\mu$ bits. (The time-complexity, here and hereafter, refers to the execution time of the experiment underlying the definition of the advantage, plus the size of the description of the adversary.)

To define SPRPs we give the distinguisher not only an oracle for the function, but also one for its inverse. Let $F \colon \mathcal{K} \times \mathcal{M} \to \mathcal{C}$ be a PRP with length function $\ell$. Then we let

$$\mathbf{Adv}_F^{\mathrm{sprp}}(A) \;=\;$$
$$\Pr[K \leftarrow \mathcal{K} : A^{F_K(\cdot), F_K^{-1}(\cdot)} = 1] - \Pr[\pi \leftarrow \mathbf{Perm}(\mathcal{M}) : A^{\pi(\cdot), \pi^{-1}(\cdot)} = 1]$$

denote the advantage of $A$ in distinguishing $F$ from a random permutation. Define

$$\mathbf{Adv}_F^{\mathrm{sprp}}(t, q, \mu) \;=\; \max_A \{\mathbf{Adv}_F^{\mathrm{sprp}}(A)\}$$

where the maximum is taken over all adversaries having time-complexity at most $t$ and asking at most $q$ oracle queries, these queries totaling at most $\mu$ bits.

Throughout, if the distinguisher inquires as to the value of oracle $f$ at a point $M \notin \mathcal{M}$ then the oracle responds with the distinguished point $\perp$. Since we assume that there is a (simple) algorithm to decide membership in $\mathcal{M}$ there is in fact no point for the adversary to make such inquiries.

ENCAPSULATION SCHEMES. Fix a key space $\mathcal{K}$, a message space $\mathcal{M}$, and a ciphertext space $\mathcal{C}$. An *encapsulation scheme* $\mathcal{SE} = (\mathcal{K}, \mathcal{E}, \mathcal{D})$ is a triple of algorithms. The probabilistic key-generation algorithm $\mathcal{K}$ produces a key $K \in \mathcal{K}$; we write $K \leftarrow \mathcal{K}$. The encryption algorithm $\mathcal{E}$ can be either probabilistic or stateful. It takes a key $K \in \mathcal{K}$ and a message $M \in \mathcal{M}$ and returns ciphertext $C = \mathcal{E}_K(M, r) \in \mathcal{C} \cup \{\perp\}$. If probabilistic, $r \in \{0,1\}^*$ is its coins tosses, which are taken anew upon each invocation. If stateful, $r$ is the internal state, which the encryption algorithm updates upon each invocation and which is securely maintained across invocations. (The state is typically a counter, which is incremented by some message-dependent amount.) The value $\perp$ is returned if $M \notin \mathcal{M}$ or (if this is a stateful encryption scheme) the state $r$ indicates that the message $M$ can not be sent (when, for example, too many messages have already been sent). Algorithm $\mathcal{D}$ takes $K \in \mathcal{K}$ and $C \in \{0,1\}^*$ and computes $M = \mathcal{D}_K(C)$ where $M$ is either a string in $\mathcal{M}$ or the distinguished symbol $\perp$. A return value of $\perp$ is used to indicate that $C$ is regarded as unauthentic. We call $C$ *valid* if $\mathcal{D}_K(C) \in \mathcal{M}$ and we call $C$ *invalid* if $\mathcal{D}_K(C) = \perp$. We also permit applying $\mathcal{E}_K$ to $(\perp, r)$, which results in a return value of $\perp$. Likewise, applying $\mathcal{D}_K$ to $\perp$ is permitted and this gives a return value of $\perp$. We require that if $C = \mathcal{E}_K(M, r)$ and $C \neq \perp$ then $\mathcal{D}_K(C) = M$.

When we think of the goal of $\mathcal{SE}$ as privacy, or a combination of privacy and message authenticity, we typically call it an *encryption scheme*. When we think of the goal of $\mathcal{SE}$ as authenticating messages then we call it an *authentication scheme*. But we emphasize that there is no syntactic distinction between an encryption scheme and an authentication scheme under this formalization: they are both encapsulation schemes.

PRIVACY. Several formulations for the privacy of a symmetric encryption scheme under chosen-plaintext attack were provided in [1] and compared in terms of concrete security. We will use one of these notions, namely "real-or-random" security. The idea is that an adversary cannot distinguish the encryption of text from the encryption of an equal-length string of garbage. For the formalization, let $\mathcal{SE} = (\mathcal{K}, \mathcal{E}, \mathcal{D})$ be an encryption scheme and let $A$ be an adversary with an encryption oracle. If the encryption scheme is probabilistic then fresh random choices are made for each query. If the encryption scheme is stateful then the state is properly initialized and then adjusted with each query. Define

$$\mathbf{Adv}_{\mathcal{SE}}^{\mathrm{priv}}(A) \;=\; \Pr\left[K \leftarrow \mathcal{K}: A^{\mathcal{E}_K(\cdot)} = 1\right] - \Pr\left[K \leftarrow \mathcal{K}: A^{\mathcal{E}_K(\$^{|\cdot|})} = 1\right] .$$

In the first game, the oracle, given a message, returns an encryption of it under key $K$; in the second game the oracle, given a message, ignores it except to record its length $n$, and then returns an encryption of a random message of length $n$. The advantage of $A$ is a measure of the adversary's ability to tell these two worlds apart. We let

$$\mathbf{Adv}_{\mathcal{SE}}^{\mathrm{priv}}(t, q, \mu) \;=\; \max_A \{\mathbf{Adv}_{\varPi}^{\mathrm{priv}}(A)\}$$

where the maximum is over all adversaries which have time-complexity at most $t$ and ask at most $q$ oracle queries, where these queries total at most $\mu$ bits.

AUTHENTICITY. Consider parties sharing a key $K$ and sending messages using an encapsulation scheme $\mathcal{SE} = (\mathcal{K}, \mathcal{E}, \mathcal{D})$. We are interested in authenticity: the receiver wants to be confident that a received ciphertext (and underlying message) really did originate with the sender. To formalize this an adversary will be given a way to generate authenticated messages of her choice: $M_1 \mapsto C_1, M_2 \mapsto C_2, \ldots, M_q \mapsto C_q$. She will "win" if she computes a new string $C$ (that is, $C \notin \{C_1, \ldots, C_q\}$) which would be deemed authentic by the receiver.

Authenticity in the context of an encapsulation scheme is a more general concept than that of a message authentication code (MAC). A MAC makes explicit a particular mechanism, namely the attachment of a tag to the transmission. (The tag, computed using the key, is created by the sender and checked by the receiver.) An encapsulation scheme may use a MAC, or may not, and consideration of authenticity for such a scheme cannot make assumptions about the presence of any type of mechanism. But there is a deeper difference between a MAC and a general authentication scheme. In formalizing the security of a MAC the adversary makes a number of queries to a MAC-generation oracle, with each query mapping the message $M_i$ to its tag $t_i$. After that the adversary has to come up with a new message $M$ and a tag $t$ such that the receiver will deem $(M, t)$ authentic. In particular, the adversary must "know" the message $M$ that is being forged, insofar as the adversary outputs it along with $t$. In contrast, an adversary attacking an authentication scheme in the general sense we are defining wins *even if she does not know what is the message $M$ which is being forged.* All that is required is that there *is* such a message underlying $C$—that is, the receiver will recover *something* in the message space $\mathcal{M}$ (and not an indication that $C$ is bogus).

Formally, let $\mathcal{SE} = (\mathcal{K}, \mathcal{E}, \mathcal{D})$ be an authentication scheme and let $A$ be an adversary who is given oracle access to $\mathcal{E}$. After interacting with that oracle the adversary outputs a string $C$. We say $C$ is *new* if $C$ was not the response to any earlier oracle query asked by $A$. Adversary $A$ is said to be *successful* if $C$ is new and valid, and we measure the probability of this:

$$\mathbf{Adv}_{\mathcal{SE}}^{\mathrm{auth}}(A) = \Pr[K \leftarrow \mathcal{K} \, ; \, C \leftarrow A^{\mathcal{E}_K(\cdot)} \; : \; C \text{ is new and } \mathcal{D}_K(C) \neq \perp ] \, .$$

The quality of $\mathcal{SE}$ in authenticating messages is measured by the function

$$\mathbf{Adv}_{\mathcal{SE}}^{\mathrm{auth}}(t, q, \mu) = \max_A \{ \mathbf{Adv}_{\Pi}^{\mathrm{auth}}(A) \}$$

where the maximum is over all adversaries who have time-complexity at most $t$ and make at most $q - 1$ oracle calls, these totaling a most $\mu - |C|$ bits, where $C$ is the length of $A$'s output. For simplicity, we assume that an adversary $A$ attacking the authenticity of $\mathcal{SE}$ will only output a string which is new.

The above notion is called "integrity of ciphertexts" in [4] who provide a comprehensive picture of how it relates to other notions of privacy and authenticity for encapsulation schemes. In particular they show that integrity of ciphertexts plus privacy against chosen-plaintext attack imply privacy under chosen-ciphertext attack.

## 3    Encoding Schemes

SYNTAX. Fix message spaces $\mathcal{M}, \mathcal{M}^*$. An *encoding scheme* ("on $\mathcal{M}$", or "from $\mathcal{M}$ to $\mathcal{M}^*$") is a pair of algorithms Encode $= (Encode, Decode)$ as we now describe.

Algorithm *Encode* can be either probabilistic or stateful, while *Decode* is neither. First assume that *Encode* is probabilistic (not stateful). Then each time *Encode* is called on an input $M \in \mathcal{M}$ the algorithm flips some coins, $r$, and returns a string $M^* = Encode(M, r) \in \mathcal{M}^*$. We assume that for any string $M \in \mathcal{M}$ and any coins $r$, we have that $|Encode(M, r)| = \ell(|M|)$ for some function $\ell$, the "length function" of the encoding scheme.

Algorithm $Decode$ takes as input $M^* \in \{0,1\}^*$. It returns either a binary string $M \in \mathcal{M}$ or the distinguished symbol $\perp$. If $Decode(M^*)$ is a binary string we say that $M^*$ is *valid*, while we say that $M^*$ is *invalid* if $Decode(M^*) = \perp$. We demand that for any $M \in \mathcal{M}$ and any $r$, we have that $Decode(Encode(M, r)) = M$.

We allow that $Encode$ and $Decode$ be presented with any string at all, even ones outside of $\mathcal{M}$ and $\mathcal{M}^*$. If you try to encode a string $M \notin \mathcal{M}$ then the result is the distinguished value $\perp$. If you try to decode a string $M^* \notin \mathcal{M}^*$ then the result is the distinguished value $\perp$. We further establish the convention that you can encode or decode $\perp$, which once again returns $\perp$.

For simplicity in theorem statements we assume that $Encode$ and $Decode$ are efficiently computable, say in linear time.

RARE-COLLISION ENCODINGS. Let Encode $= (Encode, Decode)$ be an encoding scheme and let $\ell(n)$ be its length function. Let $\epsilon \colon \mathsf{N} \to \mathbf{R}$ be a function. We say that Encode is $\epsilon$-**colliding** if for and any number $q$ and any (even computationally unbounded) adversary $A$ who asks $q$ queries, the probability that some two of these queries receive the same valid response is at most $\epsilon(q)$.

$$\Pr[(M_1^*, \ldots, M_q^*) \leftarrow \text{Responses } A^{Encode(\cdot)} : \exists\, i < j \text{ s.t. } M_i^* \neq \perp,$$
$$M_j^* \neq \perp, \text{ and } M_i^* = M_j^*] \leq \epsilon(q) .$$

We shall say that $\langle M_1^*, \ldots M_q^* \rangle$ "collide" if some pair of these strings are the same and are different from $\perp$. The reader may prefer to think of $M_1 = M_2 = \cdots = M_q$ since typically this would be the adversary's best strategy when trying to produce a collision (as $M \neq M'$ implies that their encodings, if valid, have to be different).

*Example 1.* Encoding scheme Prepend-128-Random-Bits works as follows. The message space is $\mathcal{M} = \{0,1\}^*$. Function $Encode$ takes an input $M$ and outputs $r \parallel M$, where $r$ is a sequence of 128 random bits. Function $Decode$ takes an input $M^*$ and behaves as follows. If $M^*$ is at least 128 bits, then $Decode$ outputs all but the first 128 bits of $M^*$. If $M^*$ is less than 128 bits then $Decode(M^*)$ outputs $\perp$. Then Prepend-128-Random-Bits is $C(q, 2^{128})$-colliding, where $C(q, m)$ denotes the probability of at least one collision in the experiment of throwing $q$ balls, independently and at random, into $m$ bins. ∎

COLLISION-FREE ENCODINGS. For algorithm $Encode$ to be stateful means that it maintains state across invocations. The initial value of that state is some fixed constant, $r_0$. Typically there will be a limit, $N$, on the number of times that $Encode$ may be used. After that number of invocations $Encode$ will return $\perp$ even when the inquiry is in $\mathcal{M}$. We require that for all messages $M$ and all internal states $r$, if $Encode(M, r)$ returns a binary string $M^*$ then $Decode(M^*) = M$. We emphasize that decoding is stateless.

Stateful encoding schemes are of interest because with them we can make an encoding scheme **collision free**, meaning 0-colliding, in the language above. Note that getting two $\perp$ values does *not* count as a collision. Here is an example.

*Example 2.* Encoding scheme Prepend-64-Bit-Counter works as follows. The message space is $\mathcal{M} = \{0,1\}^*$. A counter *ctr* is initialized to 0. The $i$-th message is encoded as follows. If $i \geq 2^{64}$ then the encoding is $\perp$. Otherwise the encoding is $M^* = \langle i \rangle \parallel M$, where $\langle i \rangle$ the number $i$ written as a 64-bit binary string. Function *Decode* takes an input $M^*$ and behaves as follows. If $|M^*| < 64$ then *Decode* returns $\perp$. Otherwise it returns $M^*$ after having expunged the first 64-bits. Clearly Prepend-64-Bit-Counter is collision free: the counter guarantees that no two encodings can collide. ∎

SPARSE ENCODINGS. Let Encode $= (Encode, Decode)$ be an encoding scheme and let $\delta$ be a real number. We say that encoding scheme Encode is $\delta$-**dense** if for all $n \in \mathsf{N}$,

$$\Pr[\, M^* \leftarrow \{0,1\}^n : Decode(M^*) \in \{0,1\}^*\,] \leq \delta \,.$$

That is, for every message length, at most a $\delta$-fraction of all strings of that length are valid (they decode to strings in $\mathcal{M}$). The rest are invalid encodings (they decode to $\perp$).

*Example 3.* The encoding scheme Prepend-32-Zeros works as follows. Let $\mathcal{M} = \{0,1\}^*$. Define $Encode(M) = 0^{32} \parallel M$. Define $Decode(M^*)$ to be $M^*$ after stripping away its first 32 bits, assuming that $M^*$ has at least 32 bits, and set $Decode(M^*) = \perp$ otherwise. Then Prepend-32-Zeros is $2^{-32}$-dense: a string is valid (it starts with 32 zeros) with probability at most $2^{-32}$. Indeed the probability that a random string $M^*$ is valid is exactly $2^{-32}$ if the length of $M^*$ is at least 32 bits, while the probability is 0 if the length of $M^*$ is less than 32 bits. ∎

*Example 4.* Let the message space $\mathcal{M}$ be odd-parity-adjusted ASCII strings of length at least 50 bytes. This means that a message $M \in \mathcal{M}$ is a sequence of bytes $M = b_1 \parallel \cdots \parallel b_n$, for $n \geq 50$, where each $b_i$ is a byte having its low 7 bits arbitrary and its high bit whatever is necessary so that the number of 1-bits in $b_i$ will be odd. Encoding scheme Odd-Parity is defined as follows. Function *Encode* is the identity function. Function *Decode* checks that the bit length of its input is divisible by 8, that the input is at least 50 bytes, and that each byte has odd parity. If these conditions are satisfied then *Decode* returns its input. Otherwise it returns $\perp$. Then Odd-Parity is $2^{-50}$-dense: a random string is valid with probability at most $2^{-50}$. Indeed the probability that a random $n$-byte string is valid is $2^{-n}$ if $n \geq 50$, and 0 if $n < 50$ or if the input is not a byte string at all. ∎

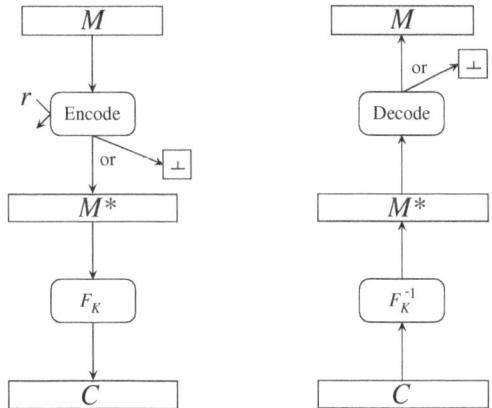

**Fig. 1.** *Scheme* $F \circ$ Encode: *encrypting (left-hand side) and decrypting (right-hand side) using the encode-then-encipher paradigm. The plaintext is* $M$, *the ciphertext is* $C$, *the cipher is* $F = \{F_K\}$, *and the encoding scheme is* Encode $=$ *(Encode, Decode).*

## 4   Enciphering Encoded Messages

Let Encode $= $ *(Encode, Decode)* be an encoding scheme from $\mathcal{M}$ to $\mathcal{M}^*$ and let $F = \{F_K : \mathcal{M}^* \to \mathcal{M}^*\}$ be a cipher with key space $\mathcal{K}$. Then we define the following encapsulation scheme $F \circ$ Encode $= (\mathcal{K}, \mathcal{E}, \mathcal{D})$:

(1)  $\mathcal{K}$ chooses a random key $K \leftarrow \mathcal{K}$ and outputs it.

(2)  $\mathcal{E}_K(M)$ sets $M^* \leftarrow Encode(M)$, returns $\perp$ if $M^* = \perp$, and otherwise computes $C \leftarrow F_K(M^*)$ and returns that. Algorithm $\mathcal{E}$ is stateful if and only if *Encode* is. If *Encode* is stateful then the initial state for $\mathcal{E}$ is the initial state mandated by *Encode*, and $\mathcal{E}$ maintains the state needed by the encoding scheme.

(3)  $\mathcal{D}_K(C)$ returns $\perp$ if $C \notin \mathcal{M}^*$, and otherwise computes $M^* \leftarrow F_K^{-1}(C)$, sets $M \leftarrow Decode(M^*)$, and returns $M$.

For a pictorial representation, see Figure 3.

PRIVACY FROM RARE/COLLISION-FREE ENCODINGS. We show that encryption scheme $F \circ$ Encode is private if encoding scheme Encode has rare or no collisions and $F$ is a secure cipher, in the sense of being a good PRP. The following theorem makes this formal and quantitative.

**Theorem 1.** *Let* Encode $= $ *(Encode, Decode) be an encoding scheme from* $\mathcal{M}$ *to* $\mathcal{M}^*$ *and let* $F = \{F_K : \mathcal{M}^* \to \mathcal{M}^*\}$ *be a cipher with key space* $\mathcal{K}$. *Suppose that* Encode *is* $\epsilon$-*colliding. Then* $F \circ$ Encode $= (\mathcal{K}, \mathcal{E}, \mathcal{D})$ *has security*

$$\mathbf{Adv}^{\mathrm{priv}}_{F \circ \mathsf{Encode}}(t, q, \mu) \leq \mathbf{Adv}^{\mathrm{prf}}_{F}(t', q, \mu) + \epsilon(q)$$

*where* $t' = t + O(\mu)$.

*Proof.* Let $B$ be an adversary attacking the privacy of $F \circ \mathsf{Encode}$. Let $t$ be its running time, $q$ the number of queries it makes, and $\mu$ the length of all its queries put together, plus the length of $B$'s output. Our goal is to upper bound $\mathbf{Adv}^{\mathrm{priv}}_{F \circ \mathsf{Encode}}(B)$. To this end we introduce a couple of more algorithms and some associated probabilities.

Algorithm $D$ is a distinguisher for $F$. It is given an oracle for a permutation $f \in \mathbf{Perm}(\mathcal{M}^*)$. It runs $B$. When $B$ makes an oracle query $M$, distinguisher $D$ computes $M^* \leftarrow \mathsf{Encode}(M)$ and $C \leftarrow f(M^*)$. It returns $C$ to $B$ as the answer to the query. When $B$ terminates, $D$ outputs whatever $B$ outputs.

Algorithm $A$ is a collision finding adversary for $\mathsf{Encode}$. It is given oracle $\mathsf{Encode}$. It picks a permutation $f$ from $\mathbf{Perm}(\mathcal{M}^*)$ at random. (Or simulates such a permutation. The difference is technically immaterial since the running time of $A$ is not restricted.) It then runs $B$. When $B$ makes an oracle query $M$, algorithm $A$ computes $M^* \leftarrow \mathsf{Encode}(M)$ and $C \leftarrow f(M^*)$. It returns $C$ to $B$ as the answer to the query. When $B$ terminates, so does $A$.

We now define the following probabilities:

$$p_1 = \Pr[K \leftarrow \mathcal{K} : B^{\mathcal{E}_K(\cdot)} = 1]$$
$$p_2 = \Pr[K \leftarrow \mathcal{K} : B^{\mathcal{E}_K(\$^{|\cdot|})} = 1]$$
$$p_3 = \Pr[K \leftarrow \mathcal{K} : D^{F_K(\cdot)} = 1]$$
$$p_4 = \Pr[\pi \leftarrow \mathbf{Perm}(\mathcal{M}^*) : D^{\pi(\cdot)} = 1]$$
$$p_5 = \Pr[(M_1^*, \ldots, M_q^*) \leftarrow \mathsf{Responses}\ A^{\mathsf{Encode}(\cdot)} : \exists\, i < j \text{ s.t. } M_i^* = M_j^* \neq \bot].$$

Note that $\mathbf{Adv}^{\mathrm{priv}}_{F \circ \mathsf{Encode}}(B) = p_1 - p_2$. To upper bound it we use the following claims. First, $p_1 = p_3$. Second, $p_2 \geq p_4 - p_5$. The proofs of these claims are omitted here for lack of space but can be found in the full version of this paper [6]. Given these claims we have

$$\mathbf{Adv}^{\mathrm{priv}}_{F \circ \mathsf{Encode}}(B) = p_1 - p_2 \leq p_3 - (p_4 - p_5) = (p_3 - p_4) + p_5 \leq \mathbf{Adv}^{\mathrm{prp}}_F(D) + \epsilon(q).$$

This concludes the proof of Theorem 1. $\blacksquare$

AUTHENTICITY FROM SPARSE ENCODINGS. We show that $F \circ \mathsf{Encode}$ is an authenticated encryption scheme if encoding $\mathsf{Encode}$ adds adequate redundancy and $F$ is a strong PRP. The following theorem makes this formal and quantitative. We remark that this result requires that the PRP be strong, which the previous result did not, and we subsequently show this extra requirement is necessary.

**Theorem 2.** *Let $\mathsf{Encode} = (Encode, Decode)$ be an encoding scheme from $\mathcal{M}$ to $\mathcal{M}^*$ and let $F = \{F_K : \mathcal{M}^* \to \mathcal{M}^*\}$ be a cipher with key space $\mathcal{K}$. Suppose that $\mathsf{Encode}$ is $\delta$-dense and that $q \leq \frac{1}{2\delta}$. Then $F \circ \mathsf{Encode} = (\mathcal{K}, \mathcal{E}, \mathcal{D})$ has security*

$$\mathbf{Adv}^{\mathrm{auth}}_{F \circ \mathsf{Encode}}(t, q, \mu) \leq \mathbf{Adv}^{\mathrm{sprp}}_F(t', q, 2\mu) + 2\delta$$

*where $t' = t + O(\mu)$.*

*Proof.* Let $B$ be an adversary attacking the authenticity of $F \circ \mathsf{Encode}$. Let $t$ be its running time, $q - 1$ the number of queries it makes, and $\mu$ the total length of all its queries put together, and its final output. Our goal is to upper bound $\mathbf{Adv}^{\mathrm{auth}}_{F \circ \mathsf{Encode}}(B)$. To this end we introduce an algorithm $D$ and some probabilities.

Algorithm $D$ is a distinguisher for $F$. It is given two oracles: $f$ and $f^{-1}$, where $f \in \mathbf{Perm}(\mathcal{M}^*)$ is a permutation. It runs $B$. When $B$ makes an oracle query $M$, distinguisher $D$ computes $M^* \leftarrow Encode(M)$ and $C \leftarrow f(M^*)$. It returns $C$ to $B$ as the answer to the query. When $B$ terminates, it outputs a ciphertext $\mathbf{C}$, which is supposed to its forgery. Algorithm $D$ outputs 0 if $\mathbf{C} \notin \mathcal{M}^*$. Otherwise $D$ computes $\mathbf{M}^* \leftarrow f^{-1}(\mathbf{C})$ (this is the one and only time it uses its $f^{-1}$ oracle). Algorithm $D$ then computes $\mathbf{M} \leftarrow Decode(\mathbf{M}^*)$. If $\mathbf{M} = \perp$ then $D$ outputs 0, else $D$ outputs 1.

We now define the following probabilities:

$$p_1 = \Pr[K \leftarrow \mathcal{K} ; \mathbf{C} \leftarrow B^{\mathcal{E}_K(\cdot)} : \mathbf{C} \text{ is new and } \mathcal{D}_K(\mathbf{C}) \neq \perp]$$
$$p_2 = \Pr[K \leftarrow \mathcal{K} : D^{F_K(\cdot), F_K^{-1}(\cdot)} = 1]$$
$$p_3 = \Pr[\pi \leftarrow \mathbf{Perm}(\mathcal{M}^*) : D^{\pi(\cdot), \pi^{-1}(\cdot)} = 1]$$

Note that $\mathbf{Adv}^{\mathrm{auth}}_{F \circ \mathsf{Encode}}(B) = p_1$. To upper bound it we use the following claims. First, $p_1 = p_2$. Second, $p_3 \leq 2\delta$. The proofs of these claims are omitted here for lack of space but can be found in the full version of this paper [6]. Given these claims we have

$$\mathbf{Adv}^{\mathrm{auth}}_{F \circ \mathsf{Encode}}(B) = p_1 = p_2 = (p_2 - p_3) + p_3 \leq \mathbf{Adv}^{\mathrm{prp}}_F(D) + 2\delta .$$

This concludes the proof of Theorem 2. ∎

We now discuss the necessity of the extra requirement on the PRP above, namely that it be strong. The following indicates that without this requirement, the authenticity does not hold. Using the bounds found in the proof, the informal theorem statement below is easily adapted to give a more precise (but less understandable) quantitative assertion. A proof of the following can be found in [6].

**Theorem 3.** *If there exists a secure PRP then there exists a secure PRP $F$ (that is not a strong-PRP) and a $\delta$-dense encoding scheme* Encode *for which the scheme $F \circ$ Encode does not achieve authenticity.* ∎

## Acknowledgments

Mihir Bellare was supported in part by NSF CAREER AWARD CCR-9624439 and a Packard Foundation Fellowship in Science and Engineering. Phillip Rogaway was supported in part under NSF CAREER Award CCR-962540, and under MICRO grants 97-150 and 98-129, funded by RSA Data Security, Inc..

# References

1. M. BELLARE, A. DESAI, E. JOKIPII, AND P. ROGAWAY, "A concrete security treatment of symmetric encryption." *Proceedings of the 38th Symposium on Foundations of Computer Science*, IEEE, 1997.

2. M. BELLARE, J. KILIAN AND P. ROGAWAY, "On the security of cipher block chaining." *Advances in Cryptology – Crypto '94*, Lecture Notes in Computer Science Vol. 839, Y. Desmedt ed., Springer-Verlag, 1994.

3. M. BELLARE, T. KROVETZ AND P. ROGAWAY, "Luby-Rackoff backwards: Increasing security by making block ciphers non-invertible." *Advances in Cryptology – Eurocrypt '98*, Lecture Notes in Computer Science Vol. 1403, K. Nyberg ed., Springer-Verlag, 1998.

4. M. BELLARE AND C. NAMPREMPRE, "Authenticated encryption: Relations among notions and analysis of the generic composition paradigm." *Advances in Cryptology – Asiacrypt '00*, Lecture Notes in Computer Science, T. Okamoto, ed., Springer-Verlag, 2000.

5. M. BELLARE AND P. ROGAWAY, "On the construction of variable-input-length ciphers." *Fast Software Encryption '99*, Lecture Notes in Computer Science Vol. 1636, L. Knudsen ed., Springer-Verlag, 1999.

6. M. BELLARE AND P. ROGAWAY, "Encode-then-encipher encryption: How to exploit nonces or redundancy in plaintexts for efficient cryptography." Full version of this paper, available via `http://www-cseucsd.edu/users/mihir`.

7. D. DOLEV, C. DWORK AND M. NAOR. "Non-malleable cryptography," *Proceedings of the 23rd Annual Symposium on the Theory of Computing*, ACM, 1991. To appear in *SIAM J. on Computing*.

8. O. GOLDREICH, S. GOLDWASSER AND S. MICALI, "How to construct random functions." *Journal of the ACM*, Vol. 33, No. 4, 210–217, (1986).

9. S. GOLDWASSER AND S. MICALI, "Probabilistic encryption." *Journal of Computer and System Sciences* **28**, 270-299, April 1984.

10. J. KATZ AND M. YUNG, "Unforgeable encryption and adaptively secure modes of operation." *Fast Software Encryption '00*, Lecture Notes in Computer Science, B. Schneier, ed., Springer-Verlag, 2000.

11. M. LUBY AND C. RACKOFF, "How to construct pseudorandom permutations from pseudorandom functions." *SIAM J. Computing*, Vol. 17, No. 2, April 1988.

12. M. NAOR AND O. REINGOLD, "On the construction of pseudo-random permutations: Luby-Rackoff revisited." *J. of Cryptology*, vol. 12, 1999, pp. 29–66.

13. C. RACKOFF AND D. SIMON, "Non-interactive zero-knowledge proof of knowledge and chosen ciphertext attack." *Advances in Cryptology – Crypto '91*, Lecture Notes in Computer Science Vol. 576, J. Feigenbaum ed., Springer-Verlag, 1991.

14. R. RIVEST, "All-or-nothing encryption and the package transform." *Fast Software Encryption '97*, Lecture Notes in Computer Science Vol. 1267, E. Biham ed., Springer-Verlag, 1997.

15. C. SHANNON, "Communication theory of secrecy systems." Bell Systems Technical Journal, 28(4), 656–715 (1949).

# Verifiable Encryption, Group Encryption, and Their Applications to Separable Group Signatures and Signature Sharing Schemes
## (Extended Abstract)

Jan Camenisch[1] and Ivan Damgård[2*]

[1] IBM Research
Zurich Research Laboratory
CH–8803 Rüschlikon
jca@zurich.ibm.com
[2] Department of Computer Science, BRICS
University of Aarhus
DK-8000 Aarhus C, Denmark
ivan@daimi.au.dk

**Abstract.** We generalize and improve the security and efficiency of the verifiable encryption scheme of Asokan et al., such that it can rely on more general assumptions, and can be proven secure without assuming random oracles. We extend our basic protocol to a new primitive called verifiable group encryption. We show how our protocols can be applied to construct group signatures, identity escrow, and signature sharing schemes from a wide range of signature, identification, and encryption schemes already in use. In particular, we achieve perfect separability for all these applications, i.e., all participants can choose their signature and encryption schemes and the keys thereof independent of each other, even without having these applications in mind.

## 1   Introduction

A *verifiable encryption scheme* is in its basic form a two-party protocol between a prover $P$ and a verifier $V$. Their common inputs are a public encryption key $E$, a public value $x$, and a binary relation $\mathcal{R}$. As a result of the protocol, $V$ either rejects or obtains the encryption of some value $w$ under $E$ such that $(x, w) \in \mathcal{R}$ holds. For instance, $\mathcal{R}$ could be defined such that $(x, w) \in \mathcal{R}$ if and only if $w$ is a signature on message $x$ w.r.t. to some fixed public key. In other words, $P$ claims to have given $V$ the encryption of a valid signature on $x$.

The protocol should ensure that $V$ accepts an encryption of an invalid $w$ with only negligible probability. Moreover, $V$ should learn nothing except the encryption of $w$ and the fact that $w$ is valid w.r.t. $x$. In particular, if the encryption scheme is semantically secure, the protocol should be zero-knowledge.

---

* BRICS: Basic Research in Computer Science, Center of the Danish National Research Foundation

T. Okamoto (Ed.): ASIACRYPT 2000, LNCS 1976, pp. 331–345, 2000.

The encryption key $E$ can belong to $P$, but typically belongs to a third party in which case the third party should not need to take part in the protocol, in other words, $P$ does not need to know the secret key corresponding to $E$.

Verifiable encryption schemes are employed in many cryptographic protocols (although the term "verifiable encryption" is not always used). Examples are digital payment systems with revocable anonymity (e.g., [7,21]), verifiable signature sharing (e.g., [22]), (publicly) verifiable secret sharing (e.g., [32]), escrow schemes [30,34], or fair exchange of signatures [1,2,4]. However, only the schemes presented in [2,26,33] do not apply ad-hoc constructions using a specific encryption scheme that suits the particular application; in fact, our protocols can be seen as a generalization of the protocols employed in [2,26,33].

The concept of verifiable encryption was introduced in [32] in the context of publicly verifiable secret sharing schemes, and in a more general form in [2], for the purpose of fair exchange of signatures. Micali [27] also proposes the use of provable encryption of data for third parties to solve several variants of the fair exchange problem.

In this paper, we first show how to modify and generalize the verifiable encryption scheme from [2] to achieve the following:

- The relation $\mathcal{R}$ can be any relation possessing a three-move proof of knowledge which is an Arthur-Merlin game, i.e., as the second message, the verifier sends a random challenge. This proof should be honest-verifier zero-knowledge, and a cheating prover should be unable to answer more than one challenge correctly.
- It can be based on any public-key encryption scheme.
- The verifier needs to store only $O(\log k)$ encryptions of the underlying encryption scheme.
- In its interactive form, our scheme can be proved secure without relying on random oracles.

In comparison, the scheme from [2] only works for relations $\mathcal{R}$ containing pairs of form $(x, f^{-1}(x))$, where $f$ is a one-way group homomorphism and the verifier is required store $k$ encryptions of the underlying encryption scheme. Finally, the scheme from [2] is only secure in the random oracle model, even in its interactive form — in other words, it relies on random oracles for more than just standard removal of interaction à la Fiat-Shamir[1].

Our results are especially suited for a situation where a public-key infrastructure already exists, i.e., users already have (certified) public keys for encryption, signature, or identification schemes. However, we assume that these keys are not necessarily generated with other and more advanced primitives in mind, such as group signatures, identity escrow, fair contract signing, or blind signatures with revocable anonymity. We believe this is a very realistic scenario.

---

[1] We are referring here to the proceedings version of [2]. Having been informed of our results, the authors of [2] later modified their protocols such that they do not rely on random oracles. These protocols appear in the journal version of their paper [3].

We provide examples of how our verifiable encryption schemes can be applied in this scenario to build such advanced primitives, where their security can be proved based only on security of the existing infrastructure: No special assumptions are needed on the existing encryption scheme. The signature scheme only needs to satisfy that one can prove knowledge of a signature on a given message by a so-called $\Sigma$-protocol. All standard signature schemes (RSA, El-Gamal, DSS) satisfy this. Any identification scheme that is a three-move honest-verifier proof of knowledge is also suitable.

All our solutions for these applications possess *perfect separability*, i.e., all participants can choose (or renew) their keys independently of each other or even use different kinds of encryption and signature schemes which is a requirement met for the first time for these applications. The term separability originates from [26] and was diversified in [8]. There exist weaker forms of separability, i.e., *partial* separability allows only a subset of the participants to choose their keys independently whereas *weak* separability requires that all participants use common system parameters, e.g., they must all use DSS and choose their keys from the same algebraic group (cf. [8]). Clearly, both these forms of separability are not sufficient if one accepts no special purpose key-setup procedure.

Furthermore, we introduce a new primitive called *verifiable group encryption* involving $n > 1$ third parties (called *proxies*) where only certain subsets of them can jointly decrypt the secret. This new primitive is an extension of our verifiable encryption scheme: we are given $n$ public encryption keys $E_1, \ldots, E_n$ and the prover and the verifier agree on any monotone access structure over $\{1, \ldots, n\}$. Then, if $V$ accepts, he is convinced that he has obtained an encryption of a valid secret which a subset of the proxies can decrypt if and only if that subset is contained in the access-structure. For example, one can decide that for some $t < n$, any subset of at least $t$ players can decrypt, whereas less than $t$ players cannot. This notion of group encryption should not be confused with the notion of threshold encryption [17]. In the latter case, a number of parties publish a *single* public key and the access structure is determined by these parties during the setup of the system. In contrast, a group encryption scheme allows to choose a (possibly different) access structure each time when (verifiably) encrypting a message. Another distinguishing feature is that a group encryption enables the proxies to choose their encryption keys independently of each other and allows them even to use different encryption schemes (perfect separability).

We also show how to get verifiable signature sharing from verifiable group encryption, yielding more general solutions for this problem than what were previously known and thereby solve an open problem raised in [10].

We believe that our verifiable encryption primitives facilitate the design of provably secure protocols for many applications such that their setup remains minimal and perfectly separable, i.e., protocols that are tailored for a public key infrastructure as described above. Previously, one had to resort to general zero-knowledge techniques for such solutions and thus accept a prohibitive loss of efficiency. Also, much more efficient schemes can sometimes be obtained by ad-hoc constructions, but this requires relying on particular properties of the encryption and/or signature scheme involved, and sometimes means that no proofs of security can be given, and always means that separability cannot be provided. For a comparison to some concrete previous schemes of this type, please refer to Section 5.2.

## 2 Preliminaries

### 2.1 $\Sigma$-Protocols

A $\Sigma$-protocol [13,15] for a boolean relation $\mathcal{R} \subseteq \{0,1\}^* \times \{0,1\}^*$ is a three-move honest-verifier zero-knowledge proof of knowledge for $\mathcal{R}$. That is, a string $x$ is the common input to a prover $P$ and a verifier $V$, and $P$ demonstrates knowledge of a $w$ such that $(x, w) \in \mathcal{R}$. We call $w$ a *witness* for $x$, and the set of $x$'s that have witnesses is called $L_{\mathcal{R}}$. Of course, carrying out a $\Sigma$-protocol for some relation $R$ makes sense only if it is a hard relation for (at least) one of the two involved parties. A $\Sigma$-protocol can be defined by three (probabilistic) procedures $\sigma_t$, $\sigma_s$, and $\sigma_v$ as follows. On input $x$ and $w$, the prover uses $\sigma_t$ to compute a so-called *commitment* $t$ and some side-information $r$. The prover sends the verifier $t$ as the first message. Then, the verifier sends back a random bit string $c$, called a *challenge*. The prover uses $x$, $w$, $r$, and $c$ as input to $\sigma_s$ to compute the *response* $s$ which he sends the verifier. Finally, $\sigma_v$ is a predicate taking $x$, $t$, $c$, and $s$ as input that $V$ uses to check whether $s$ is a valid response, i.e., $V$ accepts if $\sigma_v(x, t, c, s) = 1$ holds. A triple $(t, c, s)$ such that $\sigma_v(x, t, c, s) = 1$ is called an *accepting* triple for $x$.

We require that if $P$ and $V$ follow the protocol, $V$ always accepts, whereas a cheating prover can answer at most one challenge correctly per commitment. More precisely, there is some polynomial-time procedure $\rho$ that, given two accepting triples $(t, c_1, s_1)$ and $(t, c_2, s_2)$ with $c_1 \neq c_2$, computes $w$ such that $(x, w) \in \mathcal{R}$.

We also require that a $\Sigma$-protocol is honest verifier perfect zero-knowledge in the particular sense that there is a simulator which, given input $x$ and challenge $c$, computes a $t$ and an $s$ such that $(t, c, s)$ is accepting w.r.t. $x$, and has a distribution equal to (or computationally indistinguishable from) that of real conversations with the honest verifier where $c$ occurs as challenge.

Cramer et al. [15] show that different $\Sigma$-protocols can be composed to obtain $\Sigma$-protocols for statements such as "I know a witness to $x_1 \in L_{\mathcal{R}_1}$ or a witness to $x_2 \in L_{\mathcal{R}_2}$" while retaining efficiency. Note that the zero-knowledge property in particular implies that $V$ does not learn whether $P$ knows a witness to $x_1$ or to $x_2$. More generally, we have the following lemma.

**Lemma 1 (Composition of $\Sigma$-protocols [15]).** *Given $\Sigma$-protocols for relations $\mathcal{R}_1, \ldots, \mathcal{R}_n$, then one can construct $\Sigma$-protocols for the relations*

- $\mathcal{R}_{\binom{n}{1}} = \{((x_1, \ldots, x_n), w) \mid \exists i : (x_i, w) \in \mathcal{R}_i\}$ *and*
- $\mathcal{R}_\Gamma = \{((x_1, \ldots, x_n), (w_1, \ldots, w_n)) \mid \exists S \in \Gamma : \forall i \in S : (x_i, w_i) \in \mathcal{R}_i\}$,

*where $\Gamma$ is a monotone access structure over $\{1, \ldots, n\}$.*

Probably, the best-known special case of relations $\mathcal{R}$ with $\Sigma$-protocols are public-key identification schemes such as the ones by Feige, Fiat, and Shamir [19], by Guillou and Quisquater [25], or by Schnorr [31]. Furthermore, many proofs of knowledge of or about discrete logarithms found in literature are $\Sigma$-protocols. A protocol that is of interest in the context of this paper is a proof of knowledge of a pre-image under a group homomorphism. It turns out that this protocol is very useful in practice because demonstrating (in zero-knowledge) that one knows a signature on given message reduces to demonstrating that one knows a pre-image under a group homomorphism. This is true for any of the standard signature schemes in use today (RSA, DSA, etc.) (see [2,22]).

## 2.2   Probabilistic Encryption Schemes

A triple $(G, E, D)$ of probabilistic polynomial-time algorithms is a polynomially secure public key encryption system (see for instance [24,28]) if we have the following:

1. For every output $(E, D) \in G(1^k)$ and all messages $m \in \{0,1\}^k$ we have $D(E(m)) = m$.

2. For all probabilistic algorithms $T$ and $M$, all polynomials $p(\cdot)$, and all sufficiently large $k$ we have

$$Pr[T(1^k, E, m_0, m_1, \alpha) = m : (E, D) := G(1^k); (m_0, m_1) := M(E, 1^k);$$

$$m \in_R \{m_0, m_1\}; \alpha := E(m)] < \frac{1}{2} + \frac{1}{p(k)} .$$

For convenience, $E$ denotes the public key as well as the actual encryption algorithm, and $D$ the secret key as well as the decryption algorithm. Furthermore, we will write sometimes $E(r, m)$ rather than $E(m)$, where $r$ contains all coinflips to be made for encryption; hence $E$ will in these cases denote a "deterministic" algorithm.

## 3   Verifiable Encryption

### 3.1   Definition of Verifiable Encryption

We give a definition of a secure verifiable encryption scheme for a relation $\mathcal{R}$ following [2].

**Definition 1 (Secure Verifiable Encryption).** *Let $\mathcal{R}$ be a binary relation and let $L_{\mathcal{R}} = \{x | \exists w : (x, w) \in \mathcal{R}\}$. A secure verifiable encryption scheme for a relation $\mathcal{R}$ consists of a two-party protocol $(P, V)$ and a recovery algorithm $R$. Let $V_P(E, x, k)$ denote the output of $V$ when interacting with $P$ on input $(E, x, k)$, where $k$ is a security parameter. The following properties must hold:*

Completeness: *If $P$ and $V$ are honest then $V_P(E, x, k) \neq \perp$ for all $(E, D) \in G(1^k)$ and for all $x \in L_{\mathcal{R}}$.*

Validity: *For all polynomial time $\tilde{P}$, all positive polynomials $p(\cdot)$, all sufficiently large $k$, and all $(E, D) = G(1^k)$ we have*

$$Pr\big[(x, R(D, \alpha)) \notin \mathcal{R} \text{ and } \alpha \neq \bot \; : \; \alpha := V_{\tilde{P}}(E, x, k)\big] < \frac{1}{p(k)} \; .$$

Computational Zero-Knowledge: *For every $\tilde{V}$ there exists a expected polynomial-time simulator $S_{\tilde{V}}$ with black-box access to $\tilde{V}$ s.t. for all distinguishers $A$, all positive polynomials $p$, all $x \in L_{\mathcal{R}}$, and all sufficiently large $k$, we have*

$$Pr\big[A(E, x, \alpha_i) = i \; : \; (E, D) := G(1^k); \alpha_0 := S_{\tilde{V}}(E, x, k);$$
$$\alpha_1 := \tilde{V}_P(E, x, k); i \in_R \{0, 1\}\big] < \frac{1}{2} + \frac{1}{p(k)} \; .$$

The completeness property should need no discussion. Validity ensures that it almost never happens that an honest verifier accepts simultaneously with recovery failing to compute a witness for $x \in L_{\mathcal{R}}$. A consequence of this is that a verifiable encryption scheme is necessarily also a proof of knowledge of such a witness. Finally, the zero-knowledge condition assures the prover that no verifier can learn anything beyond the fact that $x \in L_{\mathcal{R}}$ and that a witness for $x$ can be recovered from the output.

We note that computational zero-knowledge is the best we can achieve due to the requirement that a witness is (and should be) recoverable from the conversation. Also note that the encryption scheme must be semantically secure (cf. Section 2.2) in order for the zero-knowledge property to be satisfied.

Our zero-knowledge definition allows the simulator to rewind the verifier. In some applications it may be desirable to require that simulation can be done without rewinding, since this implies that the protocol is *concurrent* zero-knowledge [18], i.e., even an arbitrary interleaving of different instances of the protocol is simulatable. We note that one variant of our protocol in fact has this stronger property.

The verifiable encryption scheme for group homomorphisms described in [2] can be seen as a special case of our definition with respect to the relations $\mathcal{R}$ that are considered. In particular, the relation defined by $\{(x, w) | x = f(w)\}$, where $f$ is a group homomorphism, is a subclass of the relations having a $\Sigma$-protocol.

## 3.2   A Verifiable Encryption Scheme

We first present a very simple but not very efficient scheme, and then move on with improvements.

Let $(G, E, D)$ be a semantically secure crypto system. Let $(E, D) := G(1^k)$ be the public and secret key of a third party. Also, we are given a relation with a $\Sigma$-protocol defined by procedures $\sigma_t$, $\sigma_s$, and $\sigma_v$. Assume for simplicity that the verifier can choose only between 0 and 1 as challenges. The idea is now simply that given $x$, the prover will start a conversation in the $\Sigma$-protocol. Using his knowledge of a witness $w$ he can compute answers to both $c = 0$ and $c = 1$, and

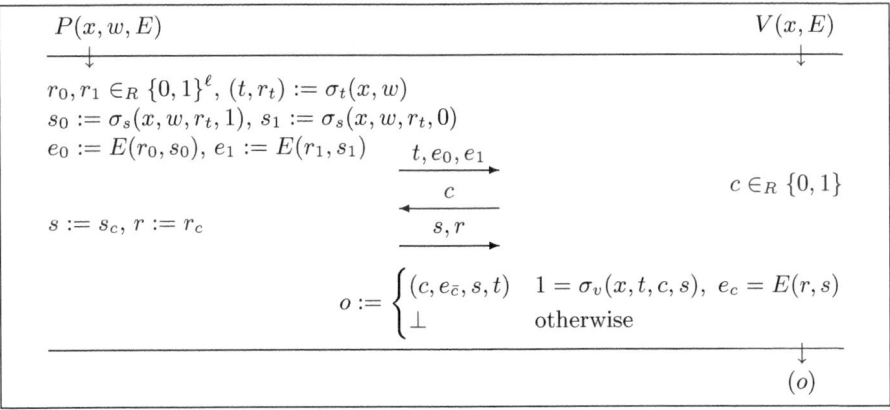

**Fig. 1.** One round of the basic verifiable encryption scheme. Recovery takes place by decrypting $e_{\bar{c}}$ and using the soundness property of the $\Sigma$-protocol to compute $w$. The parameter $\ell$ is defined by the encryption algorithm $E$.

he supplies encryptions under $E$ of both of these. The verifier can now ask $P$ to open one of these encryptions to check if it contains a valid answer. If this is true for both encryptions, decrypting the other one allows to recover $w$ (due to the properties of the $\Sigma$-protocol), whereas if at least one of them contains garbage, the prover will be caught with probability $1/2$. Concretely, we have the following. (The proof can be found in the full version of this paper [5].)

**Theorem 1.** *Let $\mathcal{R}$ be a relation that has a $\Sigma$-protocol. The protocol depicted in Figure 1 is a secure verifiable encryption scheme for $\mathcal{R}$ when sequentially repeated $k$ times.*

The main drawback of our basic scheme is that the verifier must store an encryption and a conversation (accepting triple) in the $\Sigma$-protocol for *each* repetition and that it needs to be repeated $\Theta(k)$ times sequentially. In Figure 2 an improved scheme is depicted, that allows to store much less encryptions and triples. The idea is that in the basic step, the prover will supply a valid triple where the challenge is 0, but where the prover's response is encrypted. The verifier can then ask the prover either to open the encryption or to supply a valid answer to challenge 1. The point is that the verifier only needs to remember the unopened encryptions and the related triples.

Furthermore, the protocol is made constant-round by relying on a commitment scheme, i.e., a function $Com$ that takes as input the string $\alpha$ to commit to and an additional input string $\beta$. Then a player can commit by sending $T := Com(\alpha, \beta)$, where $\beta$ is randomly chosen (we write just $Com(\alpha)$ in the following). The commitment scheme must possess the following three properties:

*Hiding:* The distributions of commitments to different $\alpha$'s must be computationally indistinguishable.

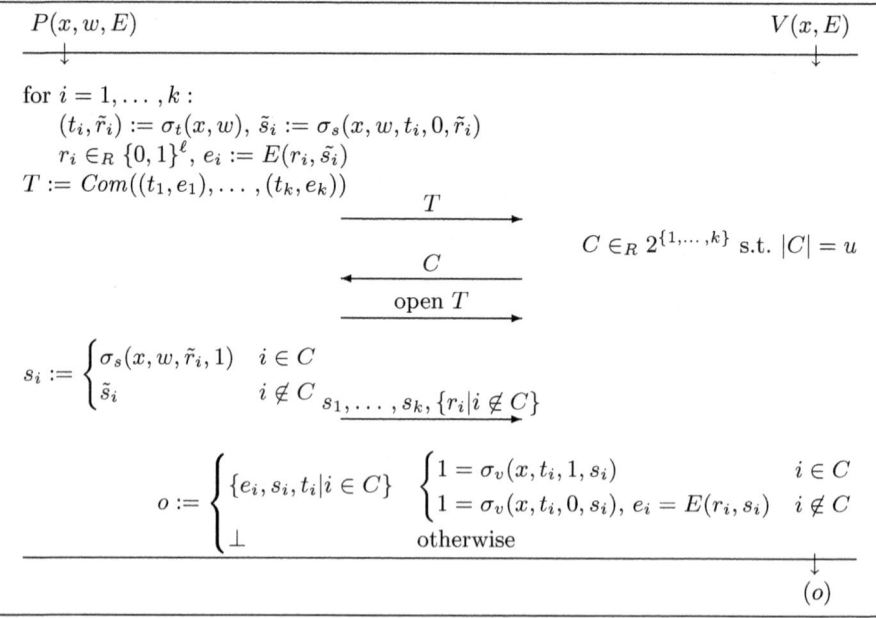

**Fig. 2.** An improved verifiable encryption scheme using a commitment scheme. Reconstruction is straightforward by decrypting the $e_i$'s in the output. The integers $k$, $u$, and $\ell$ are security parameters and $|C|$ denotes the cardinality of the set $C$.

*Binding:* It should be computationally hard to open a commitment in two different ways, i.e., to find $\alpha \neq \alpha'$ and $\beta, \beta'$ such that $Com(\alpha, \beta) = Com(\alpha', \beta')$.

*Trapdoor:* There is a piece of trapdoor information the knowledge of which allow to open a commitment in an arbitrary way.

There are numerous efficient constructions known of such schemes (see for instance [14]), but in fact our assumptions in this scenario are already sufficient to ensure their existence: $\mathcal{R}$ must be a hard relation, in order for our scenario to make sense and it is known that a $\Sigma$-protocol for a hard relation implies the existence of a secure commitment scheme with these three properties [16].

We require a once-and-for-all set-up phase for the commitment schemes where the verifier generates the $Com$-function together with the trapdoor information, sends the function to the prover, and proves knowledge of the trapdoor. For simplicity, we do not include this set-up phase explicitly in the protocol description; nevertheless, this set-up phase needs to be considered in the proof of security.

**Theorem 2.** *Let $\mathcal{R}$ be a relation possessing a $\Sigma$-protocol. The protocol depicted in Figure 2, when using a secure commitment scheme Com, is a secure verifiable encryption scheme for $\mathcal{R}$ for any $u$ such that $\log k < u < k/2$.*

The proof can be found in the full version of this paper [5].

There is another variant of our protocol which requires 4 messages, but has the advantage of being secure against an unbounded prover. Here, $V$ commits to his choice of $C$ in advance, $P$ sends $(t_1, e_1), ..., (t_k, e_k)$, $V$ opens the commitment, and $P$ responds to the $C$ revealed as above. This can be proven zero-knowledge using techniques from [23].

In practice one is often interested in a particular error probability $\epsilon$. There will then be many values of $k$ and $u$ with $1/\binom{k}{u} \leq \epsilon$, and one should then of course choose whatever particular $k$ and $u$ fit the application best.

It is easy to make a non-interactive variant of this construction using the Fiat-Shamir heuristic: the prover computes $(t_1, e_1), \ldots, (t_k, e_k)$, determines the challenge $C$ from $h((t_1, e_1), \ldots, (t_k, e_k))$, where $h$ is some suitable (hash) function, and finally appends valid responses $(s_1, r_1), \ldots, (s_k, r_k)$. All this can be verified as before by $V$. It is straightforward to show that this is secure in the random oracle model, replacing calls to $h$ by calls to the oracle.

# 4   Verifiable Group Encryption

In our basic primitive, the prover and the verifier have to trust the third party to behave as expected. That is, the prover must trust him/her to recover the encrypted witness when appropriate only, whereas the verifier relies on the third party to decrypt the witness when required. To achieve higher security against fraudulent third parties, we extend our basic primitive to verifiable *group* encryption, which on its own is a useful concept in many cases. Here, the witness gets encrypted for $n$ third parties, called *proxies*, such that only designated subsets of them can jointly recover the witness. Although it superficially looks more complicated, it turns out to be trivial to implement using our basic verifiable encryption scheme, as we shall see.

Informally, verifiable group encryption takes place in a similar model as ordinary verifiable encryption, that is, $P$ and $V$ interact on common input $x$, where $P$ knows $w$ such that $(x, w) \in \mathcal{R}$. As before, it is instructive to think of $w$ as being a signature on $x$ w.r.t. some fixed public key. Now, however, $n$ public encryption keys $E_1, \ldots, E_n$ are involved, and a monotone access structure $\Gamma$ over $\{1, \ldots, n\}$ is agreed upon by $P$ and $V$. Then an honest $V$ obtains from $P$ encryptions $E_1(w_1), \ldots, E_n(w_n)$ such that a valid $w$ can be reconstructed from a subset $A$ of the $w_i$'s, if $A \in \Gamma$, whereas a set $A \notin \Gamma$ gives no information. Finally, if honest proxies forming a set $A \in \Gamma$ decide to reconstruct $w$, they can do so successfully, even if dishonest proxies also participate.

This notion of "group encryption" should not be confused with the notion of threshold encryption [17]. In the latter case, a number of parties publishes a *single* public key and the access structure is determined by these parties during the setup of the system. In contrast, a group encryption scheme allows to choose a (possibly different) access structure each time when (verifiably) encrypting a message. Another distinguishing feature is that a group encryption enables the proxies to choose their encryption keys independent of each other and allows them even to use different encryption schemes (perfect separability). We finally note that one could also use any threshold encryption scheme on its own to achieve higher security against fraudulent third parties for the basic verifiable encryption scheme. However, such a solution will not offer perfect separability and not fit the framework of an existing public key infrastructure we are targeting.

### 4.1   Realization

A verifiable group encryption scheme can be realized using a secret sharing scheme for the chosen access structure $\Gamma$. We just need to observe that our verifiable encryption scheme from before works based on *any* public key encryption scheme. In particular, we will execute it using the following encryption scheme: given input $s$, use the given secret sharing scheme with $\Gamma$ to get shares $s_1, \ldots, s_n$ of $s$, and then let the output ciphertext be $E_1(s_1), \ldots, E_n(s_n)$. Clearly, one can compute $s$ from a correctly formed ciphertext, if one can decrypt a subset of the $s_i$'s corresponding to a set in $\Gamma$.

From the construction it is clear that, due to the properties of secret sharing schemes, the proxies can reconstruct the witness when given $E_1(s_1), \ldots, E_n(s_n)$ and $\Gamma$ by decrypting, pooling the shares, reconstructing $s$, and then using the properties of the basic verifiable encryption protocol. This is possible even with the participation of malicious proxies as long as the honest proxies form a set $A \in \Gamma$, however, they might not be able to do this efficiently in the presence of malicious proxies who provide incorrect values for the $s_i$'s. If the encryption scheme used allows proxy $i$ to prove that $s_i$ is indeed the value encrypted in $E_i(s_i)$ directly, the problem is trivial to solve. If this is not the case, we can modify the encryption scheme and encrypt the random choices used for encryption of the shares as well for the respective proxies, i.e., our encryption scheme would output $\big(E_1(r_1, s_1), E_1(\tilde{r}_1, r_1)\big), \ldots, \big(E_n(r_n, s_n), E_n(\tilde{r}_n, r_n)\big)$. Now, proxy $i$ can prove correct decryption by providing $r_i$ and $s_i$.

In case the scheme is made non-interactive using the Fiat-Shamir heuristic [20], the access structure $\Gamma$ should also be included in the hash-function.

### 4.2   Verifiable Group Encryption vs. Verifiable Signature Sharing

The concept of *verifiable signature sharing* (V$\Sigma$S) [10,22] involves a *signature receiver* who distributes shares of a signature on a public message to a set of proxies such that all proxies can verify that this has been done correctly, and a qualified set of proxies can always reconstruct the signature, even in the presence of malicious proxies. Trivially, a verifiable group encryption scheme can be used to implement V$\Sigma$S: we simply execute verifiable group encryption of a signature on the given message, such that the signature receiver plays the role of the prover, and the proxies together play the role of the verifier. If the interactive variant is used, this is done by having proxies generate challenges for the prover by collective coin-flipping. With the non-interactive variant, no special precautions are needed, and the scheme becomes even publicly verifiable. The possibility to use any encryption scheme for the individual proxies solves an open problem raised in [10]. Moreover, a (publicly) verifiable secret sharing scheme, e.g., see [32], can be obtained along the same lines.

## 5    Application to Group Signatures and Identity Escrow

A *group signature scheme* [6,8,9,11,12,29] allows a member of a group of users to sign a message on the group's behalf. The scheme protects the privacy of signers in that the verifier should not be able to find the identity of the signer. However, to handle special cases where the scheme is misused by some user, there is a *revocation manager* who can indeed find the identity of the signer from the signature. In some schemes, there is also a *membership manager* who takes care of the key set-up and enrollment of users in the group. No collusion of users (even including the membership manager) should be able to forge signatures such that it is not possible for the revocation manager to reveal the identity of the signer. Furthermore, no collusion of users (even including both managers) should be able to forge signatures such that upon revocation they seem to originate from another user. Moreover, we want to minimize the involvement of parties in the protocols. This means that managers should not be involved in creating a signature, and ideally also that the revocation manager is not involved in establishing the group, and is in fact completely inactive until revocation of anonymity is needed.

The interactive equivalent of a group signature scheme is a group identification scheme with revocable anonymity (also called identity escrow scheme [26]). Here, the goal is for a group member to anonymously identify himself as a member — rather than being able to sign a message. Except for this, the security properties of group signatures carry over directly.

For both kind of schemes, it is of course desirable that they possess perfect separability, i.e., that they can be implemented based on keys that users and managers already have established, even if those keys were not intended to be used in these schemes.

In the following, we show how to use a verifiable encryption scheme to design a separable group identification scheme with revocable anonymity. This scheme can be proved secure, assuming only security of the encryption, signature, and identification schemes involved. We then modify this to a group signature scheme using the Fiat-Shamir heuristic [20]. This scheme is secure assuming in addition that the heuristic is valid for the protocol and hash function involved. It can be proved secure with no additional assumptions in the random oracle model. For a formal model of group signatures and identity escrow we refer to [9,26].

### 5.1    Realizations

We describe the basic idea for the case where the users' public keys are of some signature schemes and then extend the scheme such that the users' public keys can also be of some interactive identification schemes.

We assume we are given public keys $P_1, \ldots, P_n$ of secure signature schemes for $n$ players, and that for each signature scheme employed, there is a $\Sigma$-protocol for the relation $\{(x, w)|\ w$ *is a valid signature on message* $x\}$. As mentioned earlier, this is true of any signature scheme for which reduction functions to a group-homomorphism exists [2]. To prove our scheme's security formally, we need that the signature schemes are secure against chosen message attacks. Finally, we assume that a revocation manager has been selected, who has published a public key $E$ of a semantically secure encryption scheme. By Lemma 1 and Theorem 2 we get a group identification scheme with revocable anonymity as follows:

The group's public key consists just of the membership manager's signature (certificate) on the tuple $(P_1, \dots, P_n)$. To authenticate himself as a group member, the prover computes the signature on message $x$ (randomly chosen by the verifier) with respect to his/her public key. Then the prover and the verifier carry out the verifiable encryption protocol for the relation $\{(x, w)|w$ *is a valid signature on* $x$ *with respect to one of the public keys* $P_1, \dots, P_n\}$, where encryption public key used is $E$, the one of the revocation manager. We can do this because, by Lemma 1, an efficient $\Sigma$-protocol for this relation can be derived from the $\Sigma$-protocols we assumed exist for each single signature scheme.

This derived $\Sigma$-protocol proves that $w$ is a valid signature w.r.t. one of the public keys, but yields no information about which one is involved. This and the zero-knowledge property of the verifiable encryption scheme implies anonymity for the prover. Furthermore, if some coalition of users could impersonate another user, then they could also forge this user's signatures. Finally, due to the validity of the verifiable encryption, the anonymity can be revoked just by decryption. With respect to the revocation manager's ability to prove correct decryption of the witness, the remarks of the previous section applies: either the underlying encryption scheme allows this or, if not, the verifiable encryption scheme is modified as described there.

Clearly, applying the Fiat-Shamir heuristic [20] yields a group *signature* scheme from this construction: the message $x$ which was chosen by the verifier before, will now be the message to be signed. And we hash $x$ and the prover's first message in the verifiable encryption protocol to get a challenge. This can be proved secure in the random oracle model. In summary, we have argued:

**Theorem 3.** *Given any set of secure signature schemes with an associated $\Sigma$-protocol each and any secure public key encryption scheme, a secure separable group identification scheme with revocable anonymity and a separable group signature scheme secure in the random oracle model can be constructed. The complexities of the schemes are linear in the group's size, the security parameter, and in the complexity of the signature and encryption schemes.*

The full version of this paper [5] discusses how to cope with the situation were we are not given a number of signature public keys, but instead public keys of interactive identification protocols. The pitfall here is that we cannot use these schemes directly as revocation would reveal a user's secret. The full version also provides a method to exploit the $\Sigma$-protocol of any identification scheme such that one can nevertheless build a group identification/signature scheme from it.

## 5.2   Extensions and Related Work

Extensions to generalized group identification schemes and schemes with higher protection against fraudulent anonymity revocation are discussed in the full version of this paper [5].

Comparing our group signature and identity escrow schemes with previous proposals, we find that ours are the only ones that provide perfect separability. The identity escrow scheme in [26] provides only partial separability, i.e., separability w.r.t. the revocation manager; the group signature scheme in [8] provides only weak separability, i.e., group members can only use RSA or discrete logarithm based keys of a given size and the membership manager can only use the RSA signature scheme together with a particular algebraic padding function.

The schemes proposed in [8,9,26] have the property that the group's public key does not depend on the size of the group. These scheme are an order of magnitude more efficient than ours. However, our schemes have some advantages over them. Namely, in order to exclude group members, the membership manager has to interact with all remaining group members in these schemes whereas in our solution the membership manager just publishes a new group public key (our scheme does not require the group members to participate in the setup at all). Furthermore, only the kind of schemes as ours allows to make it publicly known who the group members are (note signatures are still anonymous) while this is not possible for the other type of scheme. Finally, generalized group signature schemes can only be realized this way. Therefore, we believe that schemes where the group's public key reflects the group's size and structure are indeed the only solution for certain applications.

It remains an open problem to find efficient schemes providing perfect separability and where the group's public key does not depend on the group's size.

## Acknowledgments

The authors are grateful to Christian Cachin and Anna Lysyanskaya for their comments on earlier versions of this paper.

## References

1. G. Ateniese Efficient Verifiable Encryption (and Fair Exchange) of Digital Signatures, In *6th ACM CCS*, pp. 138–146, 1999.
2. N. Asokan, V. Shoup, and M. Waidner. Optimistic fair exchange of digital signatures. In *EUROCRYPT '98*, vol. 1403 of *LNCS*, pp. 591–606, 1998.
3. N. Asokan, V. Shoup, and M. Waidner. Optimistic fair exchange of digital signatures. *IEEE Journal on Selected Areas in Communications*, 18(4):591–610, Apr. 2000.
4. F. Bao. An Efficient Verifiable Encryption Scheme for the Encryption of Discrete Logarithms, In *CARDIS '98* vol. 1820 of *LNCS*, 2000.
5. J. Camenisch and I. Damgård. Verifiable encryption and applications to group signatures and signature sharing. Technical Report RS-98-32, BRICS, Department of Computer Science, University of Aarhus, Dec. 1998.

6. J. Camenisch. Efficient and generalized group signatures. In *EUROCRYPT '97*, vol. 1233 of *LNCS*, pp. 465–479, 1997.

7. J. Camenisch, U. Maurer, and M. Stadler. Digital payment systems with passive anonymity-revoking trustees. In *Computer Security — ESORICS 96*, vol. 1146 of *LNCS*, pp. 33–43. Springer Verlag, 1996.

8. J. Camenisch and M. Michels. Separability and Efficiency for Generic Group Signature Schemes In M. Wiener, *CRYPTO '99*, vol. 1666 of *LNCS*, 1998.

9. J. Camenisch and M. Stadler. Efficient group signature schemes for large groups. In *CRYPTO '97*, vol. 1296 of *LNCS*, pp. 410–424, 1997.

10. D. Catalano and R. Gennaro. New efficient and secure protocols for verifiable signature sharing and other applications. In *CRYPTO '98*, vol. 1642 of *LNCS*, pp. 105–120, Berlin, 1998. Springer Verlag.

11. D. Chaum and E. van Heyst. Group signatures. In *EUROCRYPT '91*, vol. 547 of *LNCS*, pp. 257–265. Springer-Verlag, 1991.

12. L. Chen and T. P. Pedersen. New group signature schemes. In *EUROCRYPT '94*, vol. 950 of *LNCS*, pp. 171–181, 1995.

13. R. Cramer. *Modular Design of Secure yet Practical Cryptographic Protocol*. PhD thesis, University of Amsterdam, 1997.

14. R. Cramer and I. Damgård. Zero-knowledge proof for finite field arithmetic, or: Can zero-knowledge be for free? In *CRYPTO '98*, vol. 1642 of *LNCS*, 1998.

15. R. Cramer, I. Damgård, and B. Schoenmakers. Proofs of partial knowledge and simplified design of witness hiding protocols. In *CRYPTO '94*, vol. 839 of *LNCS*, pp. 174–187. Springer Verlag, 1994.

16. I. B. Damgård. On the existence of bit commitment schemes and zero-knowledge proofs. In *CRYPTO '89*, vol. 435 of *LNCS*, pp. 17–27, 1990.

17. Y. Desmedt and Y. Frankel. Threshold cryptography. In *CRYPTO '89*, vol. 435 of *LNCS*, pp. 307–315. Springer-Verlag, 1990.

18. C. Dwork, M. Naor, and A. Sahai. Concurrent zero knowledge. In *Proc. 30th Annual ACM Symposium on Theory of Computing (STOC)*, 1998.

19. U. Feige, A. Fiat, and A. Shamir. Zero-knowledge proofs of identity. *Journal of Cryptology*, 1:77–94, 1988.

20. A. Fiat and A. Shamir. How to prove yourself: Practical solution to identification and signature problems. In *CRYPTO '86*, vol. 263 of *LNCS*, pp. 186–194, 1987.

21. Y. Frankel, Y. Tsiounis, and M. Yung. "Indirect discourse proofs:" Achieving efficient fair off-line e-cash. In *ASIACRYPT '96*, vol. 1163 of *LNCS*, 1996.

22. M. Franklin and M. Reiter. Verifiable signature sharing. In *EUROCRYPT '95*, vol. 921 of *LNCS*, pp. 50–63. Springer Verlag, 1995.

23. O. Goldreich and A. Kahan. How to construct constant-round zero-knowledge proof systems for NP. *Journal of Cryptology*, 9(3):167–190, 1996.

24. S. Goldwasser and S. Micali. Probabilistic encryption. *Journal of Computer and System Sciences*, 28(2):270–299, Apr. 1984.

25. L. C. Guillou and J.-J. Quisquater. A practical zero-knowledge protocol fitted to security microprocessor minimizing both transmission and memory. In *EURO-CRYPT '88*, vol. 330 of *LNCS*, pp. 123–128, 1988.

26. J. Kilian and E. Petrank. Identity escrow. In *CRYPTO '98*, vol. 1642 of *LNCS*, pp. 169–185, Berlin, 1998. Springer Verlag.

27. S. Micali. Efficient certificate revocation and certified e-mail with transparent post offices. Presentation at the 1997 RSA Security Conference.

28. S. Micali, C. Rackoff, and B. Sloan. The notion of security for probabilistic cryptosystems. *SIAM Journal on Computing*, 17(2):412–426, April 1988.

29. H. Petersen. How to convert any digital signature scheme into a group signature scheme. In *Security Protocols Workshop*, Paris, 1997.
30. G. Poupard and J. Stern, Fair Encryption of RSA Keys. In *EUROCRYPT 2000*, LNCS, pp. 173–190. Springer Verlag, 2000.
31. C. P. Schnorr. Efficient signature generation for smart cards. *Journal of Cryptology*, 4(3):239–252, 1991.
32. M. Stadler. Publicly verifiable secret sharing. In *EUROCRYPT '96*, vol. 1070 of *LNCS*, pp. 191–199. Springer Verlag, 1996.
33. M. Stadler, J.-M. Piveteau, and J. Camenisch. Fair blind signatures. In *EURO-CRYPT '95*, vol. 921 of *LNCS*, pp. 209–219, 1995.
34. A. Young and M. Yung. Auto-Recoverable Auto-Certifiable Cryptosystems. In *EUROCRYPT '98*, vol. 1403 of *LNCS*, pp. 17–31, 1998.

# Addition of ElGamal Plaintexts

Markus Jakobsson[1] and Ari Juels[2]

[1] Information Sciences Research Center
Bell Labs
Murray Hill, New Jersey 07974
www.bell-labs.com/user/markusj/
[2] RSA Laboratories
RSA Security Inc.
Bedford, MA 01730, USA
ajuels@rsasecurity.com

**Abstract.** We introduce an efficient method for performing computation on encrypted data, allowing addition of ElGamal encrypted plaintexts. We demonstrate a solution that is robust and leaks no information to a minority of colluding cheaters. Our focus is on a three-player solution, but we also consider generalization to a larger number of players. The amount of work is exponential in the number of players, but reasonable for small sets.

**Keywords:** addition, directed decryption, ElGamal, fast-track, repetition robustness

## 1 Introduction

For a number of public-key encryption schemes, it possible to compose plaintexts multiplicatively through simple manipulation of their corresponding ciphertexts, without knowledge of an associated decryption key. For ElGamal encryption, this is achieved by component-wise multiplication of individual ciphertexts. If the ciphertexts have been encrypted under the same public key, the result is an ElGamal encryption of the product of the plaintexts. It is known that there can not exist a *non-interactive* method for computing the *sum* of plaintexts in this setting given only the ciphertexts. If so, then it would be possible to break the Decision Diffie-Hellman Assumption.[1] Furthermore, even allowing interaction, and working in a setting where the decryption key is held distributively, it is not known how to compute the sum of plaintexts without invocation of costly general secure multiparty computation methods. In this paper, we propose the first efficient solution for performing the operation of distributed ElGamal plaintext

---

[1] The ability to perform non-interactive plaintext addition implies the ability to build a comparitor that determines whether a ciphertext corresponds to a particular plaintext. This would contradict the semantic security of the cipher, which is known to hold if the DDH assumption holds. The comparitor would add an encryption of the additive inverse of the assumed plaintext, and determine whether the result of the addition is an encryption of the value zero, which is the only recognizable ciphertext.

T. Okamoto (Ed.): ASIACRYPT 2000, LNCS 1976, pp. 346–358, 2000.
© Springer-Verlag Berlin Heidelberg 2000

addition. However, we note that our solution is only efficient for small number of participants, as its asymptotic costs are exponential in the number of players. We hope that improvements to our techniques will be able to remove this restriction.

Our goal is therefore to solve the following problem: Two ElGamal ciphertexts, both encrypted using a public key $y$, are given to a set of $n$ players, among whom the corresponding secret key $x$ is shared using a standard $(k, n)$-threshold secret sharing scheme. The players wish to generate an encryption of the sum of the encrypted plaintexts. We want the computation to be done in a robust manner, i.e., such that cheating will be detected and traced with an overwhelming probability. At the same time, it must be performed without revealing any information to a minority of dishonest and colluding players. In the main body of the paper, we focus on a $(2, 3)$-threshold solution, but this solution generalizes rather straightforwardly to more players. (However, as mentioned, the solution is only practical for small sets.)

We want our multiplicative group to be closed under addition. Working in the field $F[2^t]$ gives us a close approximation of this. In order not to leak the residuosity of results, we additionally require the size of the multiplicative group to be prime. We review the related security aspects of this modification.

Our method is based on iterated additive resharing of secrets and on a handful of scheduling tricks. Since an important goal is for the protocol to be computationally efficient, we avoid traditional zero-knowledge based robustness methods, and employ instead the so-called *repetition robustness* method [9]. We expand on the use of this method by offering a new type of setting in which it is beneficial. We also employ the ideas of *fast-track computation*, optimizing the scheme for the case where no players attempt to corrupt the computation. (See [7] for the introduction of the term and, e.g., [5] for an early introduction to the idea.)

Apart from addressing a long open problem, we introduce methods for efficient computation that may be of independent interest. For example, our result might serve as a possible alternative basis for general secure multiparty computation. It is interesting to note that this would give us a solution where one pays for *addition*, while *multiplications* are (almost) for free. This is in contrast to the usual case for general secure multiparty computation, where addition is free while multiplication is relatively costly. It appears that for some types of computations, our approach would lower the cost of the function evaluation.

Our building block may also find applications in more specific multi-party settings, such as certification schemes, payment schemes, and election schemes. This may be both to allow new functionality, and to lower computational costs by allowing for alternative ways of arriving at a given result. It is interesting to note that it sometimes may lower computational costs to perform computation on encrypted data, compared to traditional secret-sharing methods in which each party needs to prove that he performed the correct computation.

We therefore see our work as an interesting step in a direction that may provide new types of solutions for multi-party computation. We do not, however, view the work presented here to be conclusive in any manner. Instead, we want to bring the attention to several remaining open problems. Most notable among these are the questions of whether there exists an efficient and secure two-player solution, and whether an efficient solution can be developed for large sets of players, ours being efficient only for moderate sized sets.

**Outline.** We begin with a review of related work (Section 2). We then introduce our building blocks and design methods (Section 3), and present a description of a non-robust solution (Section 4). In Section 5, we show how to incorporate robustness into that solution. We prove that our solution satisfies our stated requirements in Appendix A.

## 2    Related Work

We introduce the notion of one-splitting. This is superficially related to standard secret-sharing methods [17], and in particular to zero-sharing (see, e.g., [11]). The latter involves superimposition of several polynomials, each encoding the value zero, in order to create an unknown polynomial that likewise encodes a zero. While the technical differences are substantial, the principles are related. Our methods involve composition of multiple ciphertexts, resulting in a set of new ciphertexts such that the sum of their plaintexts equals the value one.

We use the notion of repetition robustness, introduced in [9] and later also employed in [10], to make our protocol robust without excessive use of zero-knowledge proofs. If zero-knowledge proofs were to be used instead, it would appear necessary to invoke computationally costly cut-and-choose methods. (While there is nothing inherent in the setting that requires a cut-and-choose approach, we are not aware of any other kind of zero-knowledge method that can be used to perform the required proof.)

The version of repetition robustness outlined in [9,10] relies on the use of permutation, and potentially requires a large number of repetitions in order to render the success probability of an attacker sufficiently small. Our flavor of the principle, however, does not involve permutation, and only requires two runs of the protocol to ensure a negligible probability of success for an attacker. This is because our protocol by its very nature destroys the homomorphic properties exploitable by an attacker in [9,10].

Our solution is based on the principles of fast-track computation [7,5]. The guiding aim here is to streamline the efficiency of the protocol for the case in which no player attempts to corrupt the output, in the hope that this is the most common case. This translates to inexpensive methods for detection of errors, followed by a conditional execution of potentially expensive methods for determining which player misbehaved.

We use standard methods for proving correctness of exponentiation. These are related to verification of undeniable signatures [2,1] and to discrete log based signatures. We refer to [8] for methods relating to the latter approach.

All computation takes place in $F[2^t]$, where $q = 2^t - 1$ is prime. This is done to avoid leaks of information relating to the Jacobi symbol of the result of the computation, and to achieve "approximate closure" of the multiplicative group under addition. It should be noted that performing the computation in this structure allows for more efficient attacks than for the standard parameter choices, as shown by Coppersmith [3]. Coppersmith's result improves the asymptotic running time of the special number field sieve. Implementations [12] suggest that if we perform all arithmetic in a field $F[2^{2203} - 1]$ (corresponding to the smallest Mersenne prime of the approximate size we want), instead of the standard choice of 1024 bit moduli, the computational hardness of computing discrete logs would be maintained. We note that the speed of multiplication in this structure is about a quarter of the speed compared to $F[p]$ for $|p| = 1024$ if software is used, and about half if special-purpose hardware is employed.

# 3    Building Blocks and Design Methods

We employ the standard cryptographic model in which players are modeled by polynomial-time Turing Machines. We make the common assumption that a minority of players may be dishonest and may collude. A further requirement is the existence of an authenticated broadcast channel among players. We first review standard building blocks presented in the literature, and then introduce some building blocks peculiar to the protocols in this paper.

## 3.1    Standard Building Blocks

**Group structure.** The ElGamal cryptosystem, which we will use, may operate over any of a number of choices of group in which the discrete log problem is infeasible (See [13], for a discussion and list of some proposed group choices.) We let $g$ denote a generator of $F^*[2^t]$, where $p = 2^t - 1$ is prime. We note that the additive and multiplicative groups overlap on all elements but 0.

**Secret sharing.** We assume that the private decryption key $x$ is shared among players using $(k, n)$-threshold secret sharing [17], and denote by $x_i$ the secret key share of $x$ held by player $i$. Here, $x \equiv \sum_{i \in S} x_i \lambda_{Si}$, where $S$ is a set of $k$ players, and $\lambda_{Si}$ is the Lagrange coefficient for the set $S$ and player $i$. It is possible to generate and distribute $x$ using any of a number of well studied protocols. See, e.g., [6] for a brief overview and some caveats.

**ElGamal encryption.** In the ElGamal cryptosystem, the private key is an integer $x$ selected uniformly at random from $Z_p$. The corresponding public key is $y = g^x$. In order to encrypt a value $m \in F^*[2^t]$ under public key $y$, we pick a random element $\alpha \in_u Z_p$, and compute the ciphertext as $(a, b) = (y^\alpha m, g^\alpha)$. (We note that we allow a more "liberal" choice of message encodings than for standard ElGamal encryption, since every element but zero is in the multiplicative group.)

**Standard and directed decryption.** The plaintext message can be computed from a ciphertext $(a, b)$ by computing $a/b^x$, where $x$ is the secret decryption key. Note that this trivially allows distribution, where each player, holding an additive share $x_i$ of $x$, computes and publishes $B_i = b^{x_i}$, from which $B = b^x$ can easily be constructed. We obtain a *directed decryption* for a player $i$ by having all players but $i$ publish their shares of $B$, after which player $i$ locally computes $B_i$, then $B$, and finally the plaintext $m = a/B$.

**Multiplication and division of ciphertexts.** Let $E(m_1) = (a_1, b_1)$ and $E(m_2) = (a_2, b_2)$ be two ciphertexts, corresponding to plaintexts $m_1$ and $m_2$. We say that $E(m_3) = E(m_1)E(m_2) = (a_1a_2, b_1b_2)$ is the *product* of $E(m_1)$ and $E(m_2)$, since its plaintext $m_3 = m_1m_2$. Similarly, we say that the *quotient* of two ciphertexts is $E(m_3) = E(m_1)/E(m_2) = (a_1a_2^{-1}, b_1b_2^{-1})$, where the resulting ciphertext corresponds to the plaintext $m_3 = m_1m_2^{-1}$.

**Distributed blinding.** For $k$ players to blind a ciphertext $E(m)$, each player $i$ selects a random number $r_i \in F^*[2^t]$ and computes and publishes $E(r_i)$. Then, the players compute $E(\overline{m}) = E(m)\prod_{i=1}^{k} E(r_i)$, for which we have $\overline{m} = m \prod_{i=1}^{k} r_i$. The players unblind $E(\overline{m})$ by computing $E(\overline{m})/(\prod_{i=1}^{k} E(r_i))$, where $E(r_i)$ are the individual blinding factors applied in the blinding step.

**Plaintext equality test.** This is a distributed protocol for determining whether the plaintexts corresponding to two ElGamal ciphertexts are equal. Given two ciphertexts, $E_1$ and $E_2$, we compute $(a, b) = E_1/E_2$. Using, e.g., the techniques described in [8], we then determine in a distributed fashion whether $log_y a = log_g b$. If this equality holds, then the two plaintexts are determined to be equal.

## 3.2 Special Building Blocks

We now introduce some building blocks peculiar to our protocols in this paper.

**One-splitting.** Two parties can compute a one-splitting, i.e., a set of ciphertexts for which the plaintext sum is congruent to 1, using the following approach.

1. The first player selects a random value $w_1 \in_u F[2^t] - \{0, 1\}$, and computes $\overline{w}_1 \equiv 1 - w_1$. He encrypts these two plaintexts, resulting in the ciphertexts $E_1$ and $\overline{E}_1$. We call this portion of the one-splitting the "root".
2. The second player selects two random numbers, $w_{2i} \in_u F[2^t] - \{0, 1\}$, where $1 \leq i \leq 2$, and computes $\overline{w}_{2i} \equiv 1 - w_{2j}$. He encrypts these four plaintexts, giving him $E_{21}$, $\overline{E}_{21}$, $E_{22}$ and $\overline{E}_{22}$. We call this portion of the one-splitting the "leaves".
3. Both parties commit to their ciphertexts, and then decommit and compute the new ciphertext quadruple $(\mathcal{E}_1, \mathcal{E}_2, \mathcal{E}_3, \mathcal{E}_4 = (E_1 E_{21}, E_1 \overline{E}_{21}, \overline{E}_1 E_{22}, \overline{E}_1 \overline{E}_{22})$. These constitute a one-splitting: It is easy to see that $\sum_{i=1}^{4} \mathcal{E}_i \equiv 1$.

**Remark:** The one-splitting protocol generalizes straightforwardly to any number of players, but incurs costs exponential in the number of players. It is therefore only suited to small numbers of players. In this paper, we consider a setting in which only two (out of three) players engage in the one-splitting.

**Blinded round-robin addition.** Let $(A, B, C)$ correspond to three participating players, and let $(M_A, M_B, M_C) \in (F[2^t])^3$ correspond to their respective private inputs. In this protocol, a player "sends" a message $m$ to another player by publishing an encryption $E(m)$, which all players then directively decrypt for the target player. The effect of this procedure is to establish the ciphertext $E(m)$ as a commitment to the transmitted message. This commitment can later be used to trace cheaters if necessary. The blinded round-robin addition protocol is now as follows.

1. $A$ selects $\Delta \in_u F[2^t]$, and then sends $S_1 = M_A + \Delta$ to $B$ and $\Delta$ to $C$.
2. $B$ computes $S_2 = S_1 + M_B$ and sends it to $C$.
3. $C$ computes $S_3 = S_2 + M_C - \Delta$, and publishes an ElGamal encryption $E(S_3)$ under public key $y$.

**Remark 1:** We note that the above protocol will fail to hide the result if $S_3 = 0$. This only happens with a negligible probability for independent and uniformly distributed inputs. For inputs of "dangerous" distributions, we need to split each input value into two portions before performing the addition. This will be described later in the section.

**Remark 2:** The addition protocol can be extended to $k > 3$ players by having each of the $k$ players compute a $(k, k)$-threshold sharing of her value. Each player then distributes the pieces of her sharing among all $k$ players. Then, in a round-robin addition, the final player obtains the sum of all shares, for which she outputs the corresponding ciphertext.

**Repetition robustness.** While standard zero-knowledge based methods can be employed to achieve robustness, the cost for doing so would be substantial. We show how to use a recently introduced method, so-called *repetition robustness* [9], to obtain robustness at low cost. This method works by performing portions of the computation twice, using different random strings for each invocation, and comparing the resulting outputs. We repeat a portion relating to the one-splitting once, and a portion relating to the addition of partial results once, giving us a robust result with a cost less than three times that of the non-robust version of the protocol.

**Scheduling tricks.** In the primitives we develop, different relations are learned by the different players, and it becomes of vital importance to schedule carefully what player performs which tasks. This is to prevent any player from ending up with a fully determined set of equations and thereby learning information about the plaintexts. The resulting scheduling techniques are remotely related to standard blinding methods. The intuition behind our scheduling methods is as follows: In the different building blocks we have presented, it can be seen that the different parties learn different amounts or relations. For example, the third player in the blinded round-robin addition protocol we presented learns the product of the blinding factor and the message, whereas the other players do not learn this piece of information. In order to ascertain that no player learns any function of the secret information, it is important to schedule the execution of the different building blocks in a manner that does not allow any one participant (or more generally, any set controlled by the adversary) to collect enough relations that he can compute any non-trivial function of secret information. (We will solidify this in the appendix by showing how each party can produce a simulation that is coherent with his view and any set of secret inputs.)

**Avoiding zero.** ElGamal encryption has the property that the zero plaintext cannot be encrypted, but must be avoided or otherwise encoded, as its corresponding ciphertext is distinguishable from ciphertexts of other plaintexts. Depending on the use of our proposed scheme, and depending on its input, a related problem may be that the whereas no inputs are encryptions of a zero, an intermediary value or an output may still be. In order to avoid this problem, one can represent every item of the computation as a pair of ciphertexts, such that their plaintexts, if added, correspond to the value to be manipulated. We note that it is easy, given our methods, to produce such a "pair representation" of each already encrypted input; this is done plainly by selecting a random ciphertext from the correct distribution, and subtracting this from the initial ciphertext. The result, along with the random ciphertext, is a pair whose plaintext sum corresponds to the plaintext of the original ciphertext. This can be done to all values, after which the desired computation is performed on the ciphertext pairs instead of on the original ciphertexts. (Note that addition can simply be done element-wise, whereas multiplication becomes more laborious.) In our basic solution, we do not consider these issues.

## 4     A Non-robust Solution (Protocol $\mathcal{P}_1$)

Using the building blocks introduced in the previous section, we now present a preliminary solution for addition of plaintexts. This solution is correct and complete and implements privacy, but is not robust. Our solution involves three players, two that are active (i.e., are involved in choosing random values) and one that is passive (i.e., only involved in adding values given to him.)

Let $E(m_1)$ and $E(m_2)$ be the input ciphertexts, and $E(m_3)$ the output ciphertext. We assume that the players share the secret decryption key $x$. Since our protocol is only secure against dishonest minorities, we will use a threshold scheme that reflects the same trust setting. In the three-player setting we consider here, that means that a $(2,3)$-threshold scheme is employed. Call the following protocol $\mathcal{P}_1$:

1. The two active parties compute a blinding factor $E(r)$ using the methods of *distributed blinding*.
2. The two active parties compute two independent one-splittings. Call these $(\mathcal{E}_{11}, \mathcal{E}_{12}, \mathcal{E}_{13}, \mathcal{E}_{14})$ and $(\mathcal{E}_{21}, \mathcal{E}_{22}, \mathcal{E}_{23}, \mathcal{E}_{24})$. We let the first player set the root of the first one-splitting, and the leaves of the second.

3. In this step we perform the first robustness check; we will elaborate on this in the next section.

4. Let $\mu_{j\kappa} = \mathcal{E}_{j\kappa}E(r)E(m_j)$ for $1 \leq j \leq 2$. The two parties use the methods of *directed decryption* to decrypt the resulting ciphertexts $\mu_{j\kappa}$, giving the first active player the plaintexts of the ciphertexts with $\kappa = 1$, and the second active player the plaintexts of those with $\kappa = 3$. The passive player gets the remaining plaintexts, i.e., those with $\kappa = 2$ and $\kappa = 4$.

5. Each player computes the sum of the above plaintexts. All parties then add these using the blinded round-robin addition protocol. The scheduling order here is $(2, 3, 1)$, i.e., the second active player begins and the first active player finishes, sandwiching the passive player. We denote the result of this step by $E(M)$.

6. The unblinded result $E(m_3) = E(M)/E(r)$ is computed and output.

## 5    Robustness

The above protocol has three weaknesses with respect to robustness. First, it is possible for a cheater to publish ciphertext pairs for which it is not the case that the respective plaintexts add up to one; second, it is possible for a cheater to cause incorrect decryption; and third, it is possible for a cheater to publish a value which is not the sum of the plaintexts she received. We note that it is not possible to corrupt the computation in other places, as the blinding factor $E(r)$ applied in the first step is cancelled in the last, and both of these computations are performed "in public".

We address avoidance of the first and the third attack in the following two subsections, starting with how to guarantee correct one-splittings, followed by a method for guaranteeing correct addition of plaintexts. The second attack is easily avoided by use of proofs of correct exponentiation (see e.g., [8]).

### 5.1    Attaining Robustness I (Protocol $\mathcal{P}_2$)

Let us consider how to guarantee that a one-splitting is correctly performed. Let $(\mathcal{E}_{j1}, \mathcal{E}_{j2}, \mathcal{E}_{j3}, \mathcal{E}_{j4})$ be the previously described one-splittings. The players run the following protocol to verify once for each such one-splitting $1 \leq j \leq 2$:

3a. The two active parties compute a blinding factor $E(\rho_j)$.

3b. Let $\beta_{j\kappa} = \mathcal{E}_{j\kappa}E(\rho_j)$, for $1 \leq \kappa \leq 4$. Using directed decryption, the parties decrypt these ciphertexts, giving (as before) the first player the plaintext with $\kappa = 1$, the second that with $\kappa = 3$, and the third player the remaining two plaintexts.

3c. Using the blinded round-robin addition method, they compute the sum of the plaintexts they have been given. Here, we use the scheduling order $(1, 3, 2)$. This corresponds to a change in the order of the active players with respect to the main protocol. The resulting ciphertext is $E(B_j)$.

3d. The players determine if $E(B_j)$ and $E(\rho_j)$ correspond to the same plaintexts, using the *plaintext equality test* building block. They accept iff the plaintexts are equal.

As suggested by the enumeration of the above steps, this protocol is meant to be inserted in place of step 3 in protocol $\mathcal{P}_1$. We call the resulting protocol $\mathcal{P}_2$. Protocol $\mathcal{P}_2$ implements privacy, as stated in the following lemma, whose proof is sketched in Appendix A.

**Lemma 1:** Protocol $\mathcal{P}_2$ implements privacy. More precisely, we can construct a simulator $\Sigma$ such that an adversary $\mathcal{A}$ controlling a minority of the players cannot distinguish the view of a real protocol run from the view generated by $\Sigma$, assuming that the adversary only corrupts a minority of participants, and that the DDH problem over $F^*[2^t]$ is hard.    □

If the parties accept in the above protocol, then the one-splitting must be correct with overwhelming probability; otherwise, somebody must have cheated. In other words, the protocol $\mathcal{P}_2$ for proving valid one-splitting has following the property, relating to the robustness of the final scheme.

**Lemma 2:** If the parties accept in the protocol described by steps 3a-3d, then with overwhelming probability, for each quadruple $(\mathcal{E}_{j1}, \mathcal{E}_{j2}, \mathcal{E}_{j3}, \mathcal{E}_{j4})$, $1 \leq j \leq 2$, the sum of the corresponding plaintexts is congruent to 1.    □

Note that the above protocol only *detects* cheating, but does not determine who cheated. In order to reveal the identity of a cheater, all the players publish their protocol-specific inputs, after which the computation of each player is verified by each other player, and the cheater pinpointed. This procedure is, of course, only performed in case that the above protocol results in a reject.

## 5.2    Attaining Robustness II (Protocol $\mathcal{P}_3$)

The protocol we called $\mathcal{P}_1$ is not robust, as it allows a cheating player to use an arbitrary value as input to the blinded round-robin addition step without being detected. Again, we can use the principles of repetition robustness to avoid this problem. More precisely, we can run the previously described protocols twice, using the same inputs but different random strings. The output can be shown (and will be shown) to be correct with an overwhelming probability if the two resulting ciphertexts correspond to the same plaintext values.

In fact, we do not have to run the partially robust protocol $\mathcal{P}_2$ twice. We can, instead, execute the protocol $\mathcal{P}_3$, which is as follows.

1. Run one instance of $\mathcal{P}_1$ and one instance of $\mathcal{P}_2$ on the same inputs, but using independent random strings.
2. Let $E(m_3)$ be the output of the above invocation of $\mathcal{P}_1$ and $E(m_4)$ be the output of the above invocation of $\mathcal{P}_2$. These are compared using the *plaintext equality test* building block. If the equality holds, then the players output $E(m_3)$. Otherwise they must perform a protocol $\mathcal{P}_4$ for identifying cheaters.

We note that the protocol $\mathcal{P}_4$, which we do not elaborate on, can use general multi-party computation, and therefore be computationally expensive. However, since it is only employed in what are presumably rare cases of cheating, this is not a concern. (In other words, we take a fast-track or "optimistic" approach to robustness.) It is easy to see that protocol $\mathcal{P}_3$ is correct and complete. Furthermore, as will be proven in Appendix A, it also implements privacy, and is robust. Thus we have the following two theorems.

**Theorem 1:** Assuming $\mathcal{P}_4$ is private (which will follow from its zero-knowledge properties), we have that protocol $\mathcal{P}_3$ also is private. More precisely, we can construct a simulator $\Sigma$ such that an adversary $\mathcal{A}$ controlling exactly one of the players cannot distinguish the view of a real protocol run from the view generated by $\Sigma$. □

**Theorem 2:** Protocol $\mathcal{P}_3$ is robust. That is, if $E(m_1)$ and $E(m_2)$ are the input ciphertexts, then the output of $\mathcal{P}_3$ will be $E(m)$, where $m \equiv m_1 + m_2$. This is under the assumption that the adversary only corrupts a minority of participants, and that the DDH problem over $F^*[2^t]$ is hard. □

## Acknowledgements

Many thanks to Daniel Bleichenbacher, Kevin McCurley and Moti Yung for helpful discussions.

# References

1. D. Chaum, "Zero-Knowledge Undeniable Signatures," in *Advances in Cryptology – EUROCRYPT '90*, I Damgård, ed., pp. 458–464, Springer-Verlag, 1990. LNCS No. 473.
2. D. Chaum and H. van Antwerpen, "Undeniable Signatures," in *Advances in Cryptology – CRYPTO '89*, G. Brassard, ed., pp. 212–216, Springer-Verlag, 1989. LNCS No. 435.
3. D. Coppersmith, "Fast evaluation of logarithms in fields of characteristic two," IEEE Transactions of Information Theory, 30 (1984), 587–594.
4. T. ElGamal, "A Public Key Cryptosystem and a Signature Scheme Based on Discrete Logarithms," *IEEE Transactions on Information Theory*, vol. 31, pp. 469-472, 1985.
5. M. Franklin and M. Yung, "Communication Complexity of Secure Computation", in *Proc. 24th Annual Symp. on the Theory of Computation (STOC)*, pp. 699-710, ACM Press, 1992.
6. R. Gennaro, S. Jarecki, H. Krawczyk, and T. Rabin, "Secure Distributed Key Generation for Discrete-Log Based Cryptosystems", in *Advances in Cryptology – EUROCRYPT '99*, J. Stern, ed., pp. 295–310, Springer-Verlag, 1999. LNCS No. 1592.
7. R.Gennaro, M.Rabin, and T.Rabin, "Simplified VSS and Fast-track Multiparty Computations with Applications to Threshold Cryptography", in *Proc. 1998 ACM Symposium on Principles of Distributed Computing (PODC)*, ACM Press, 1999.

8. M. Jakobsson and C.-P. Schnorr, "Efficient Oblivious Proofs of Correct Exponentiation," in *Communications and Multimedia Security (CMS) '99*, B. Preneel, ed., pp. 71–84, Kluwer Academic Publishers.

9. M. Jakobsson, "A Practical Mix," in *Advances in Cryptology – EUROCRYPT '98*, K. Nyberg, ed., pp. 448–461, Springer-Verlag, 1998. LNCS No. 1403.

10. M. Jakobsson, "Flash Mixing," in *Proc. 1999 ACM Symposium on Principles of Distributed Computing (PODC)*, pp. 83–89, ACM Press, 1999.

11. A. Herzberg, S. Jarecki, H. Krawczyk, M. Yung, "Proactive Secret Sharing or How to Cope With Perpetual Leakage," in *Advances in Cryptology – Crypto '95*, D. Coppersmith, ed., pp. 339–352, Springer-Verlag, 1995. LNCS No. 963.

12. K. McCurley, personal communication.

13. A.J. Menezes, P.C. van Oorschot, and S.A. Vanstone, *Handbook of Applied Cryptography*, CRC Press, 1996.

14. T.P. Pedersen, "Distributed Provers with Applications to Undeniable Signatures," in *Advances in Cryptology – EUROCRYPT '91*, D.W. Davies, ed., Springer-Verlag, pp. 221–242, 1991. LNCS No. 547.

15. T.P. Pedersen, "Non-Interactive and Information-Theoretic Secure Verifiable Secret Sharing," in *Advances in Cryptology – CRYPTO '91*, J. Feigenbaum, ed., pp. 129–140, Springer-Verlag, 1991. LNCS NO. 576.

16. C.-P. Schnorr, "Efficient Signature Generation for Smart Cards," *Journal of Cryptology*, vol. 4, pp. 161-174, 1991.

17. A. Shamir, "How to share a secret", *Communications of the ACM*, vol. 22, pp. 612-613, 1979.

# A   Proofs

**Proof of Lemma 1:** *(Sketch)*
Our approach is to show for each player that, given a view and a guess of an input pair $(\hat{m}_1, \hat{m}_2)$, these two are consistent with each other. If this is the case for each possible pair $(\hat{m}_1, \hat{m}_2)$, then each such pair is equally likely, given the view of the player, and thus, the protocol does not leak any information. For simplicity, we mark known, derived, and assumed quantities with a hat where it is not obvious from the context that they are known.

*Player 1:* The first active player knows $M$ and can therefore compute $\hat{r} = \hat{M}/(\hat{m}_1 + \hat{m}_2)$. He also knows $w_1^1$, $w_{21}^2$, $w_{22}^2$. He has that

$$
\begin{cases}
\hat{\mu}_{11} = \hat{w}_1^1 w_{21}^1 \hat{r}\hat{m}_1 \\
\hat{\mu}_{21} = w_1^2 \hat{w}_{21}^2 \hat{r}\hat{m}_2 \\
\hat{\beta}_{14} = \overline{\hat{w}}_1^1 \overline{w}_{22}^1 \rho_1 \\
\hat{\beta}_{24} = \overline{w}_1^2 \hat{\overline{w}}_{22}^2 \rho_2
\end{cases}
$$

For these four equations, there are five unknowns, $w_{21}^1$, $w_{22}^1$, $w_1^2$, $\rho_1$, and $\rho_2$. No matter what the first player's view is, it is consistent with any pair $(\hat{m}_1, \hat{m}_2)$.

*Player 2:* Similarly, the second active player knows $R_j$. Thus, he knows $\hat{\rho}_j = \hat{R}_j$. He also knows $w_{21}^1$, $w_{22}^1$, $w_1^2$. He has that

$$
\begin{cases}
\hat{\mu}_{13} = \overline{w}_1^1 \hat{w}_{22}^1 r \hat{m}_1 \\
\hat{\mu}_{23} = \hat{\overline{w}}_1^2 w_{22}^2 r \hat{m}_2 \\
\hat{\beta}_{12} = w_1^1 \hat{\overline{w}}_{21}^1 \hat{\rho} \\
\hat{\beta}_{22} = \hat{w}_1^2 \overline{w}_{21}^2 \hat{\rho}
\end{cases}
$$

For these equations, there are four unknowns, $w_1^1$, $w_{21}^2$, $w_{22}^2$, and $r$. Again, we get that this view is consistent with $(\hat{m}_1, \hat{m}_2)$.

*Player 3:* Finally, the passive player knows the following eight equations:

$$
\begin{cases}
\hat{\mu}_{12} = w_1^1 \overline{w}_{21}^1 r \hat{m}_1 \\
\hat{\mu}_{14} = \overline{w}_1^1 \overline{w}_{22}^1 r \hat{m}_1 \\
\hat{\mu}_{22} = w_1^2 \overline{w}_{21}^2 r \hat{m}_2 \\
\hat{\mu}_{24} = \overline{w}_1^2 \overline{w}_{22}^2 r \hat{m}_2 \\
\hat{\beta}_{11} = w_1^1 w_{21}^1 \rho_1 \\
\hat{\beta}_{13} = \overline{w}_1^1 w_{22}^1 \rho_1 \\
\hat{\beta}_{21} = w_1^2 w_{21}^2 \rho_2 \\
\hat{\beta}_{23} = \overline{w}_1^2 w_{22}^2 \rho_2
\end{cases}
$$

For the above eight equations, there are nine unknowns, corresponding to all of the values that went into making the two one-splittings (namely $w_1^1$, $w_1^2$, $w_{21}^1$, $w_{22}^1$, $w_{21}^2$, and $w_{22}^2$); and the blinding factors, $r$, $\rho_1$, and $\rho_2$. Therefore, the passive player's view is also consistent with any pair $(\hat{m}_1, \hat{m}_2)$, and thus, $\mathcal{P}_2$ is private. $\square$.

*Remark 1.* In this proof sketch and those that follow, we do not fully consider information that the players may derive from published ciphertexts. Under the DDH assumption, the semantic security of ElGamal assures that this information is negligible. We shall treat this issue formally in proofs provided in the full version of the paper.

**Corollary 1:** Each component of $\mathcal{P}_2$ is private, and in particular, $\mathcal{P}_1$ is.

**Claim 1:** Any composition of private protocols with independent random strings for the players is private.

**Proof of Lemma 2:** *(Sketch)*
Assume that there is a polynomial-time cheating algorithm $\mathcal{A}$ for generating the transcripts of steps 1-3 of $\mathcal{P}_2$, so that the plaintext sum is not congruent to 1, and the honest players accept in step 3d with a non-negligible probability. We will show how to use $\mathcal{A}$ to break the DDH assumption in $F^*[2^t]$. The input to the algorithm will be the ciphertexts that constitute the output from the honest players, namely those for the generation of $(\mathcal{E}_1, \mathcal{E}_2, \mathcal{E}_3, \mathcal{E}_4)$, and those for the generation of $E(\rho)$. For $E(B)/E(\rho)$ to correspond to the plaintext 1 requires

that the plaintexts in step 3c add up to $\rho$. Assume that the portions held by the honest players are known by the adversary. If $\mathcal{A}$ could produce a share so that the sum of the shares equals $\rho$ with a non-negligible probability, $\mathcal{A}$, together with simulations of the honest players (which takes a suspected value $\rho$ as input), could be used to determine if $E(\rho)$ is an encryption of $\rho$ with a non-negligible success probability. This would break the DDH assumption in $F^*[2^t]$, as it would show that the standard ElGamal encryption scheme is not semantically secure in this group. $\square$.

**Proof of Theorem 1:** *(Sketch)*
Consider first the case in which no cheater is detected: Since $\mathcal{P}_1$ and $\mathcal{P}_2$ both are private, so must be a composition of the two. Consider now the case in which a cheater is detected. If a cheater is detected in step 3 of $\mathcal{P}_2$, then the entire protocol run is halted after each player reveals his random strings. Since no player has computed any function of his secret inputs at this point, that cannot leak any information, and this event must be simulable. The privacy of $\mathcal{P}_3$ therefore follows from the our choice of a good secure function evaluation protocol for $\mathcal{P}_4$. If the latter is zero-knowledge, $\mathcal{P}_3$ will also be zero-knowledge.
$\square$

**Proof of Theorem 2:** *(Sketch)*
We know from Lemma 2 that $\mathcal{P}_2$ is robust, and that if the players accept in this sub-protocol, that the one-splittings with an overwhelming probability are correct. The robustness of step 4 of $\mathcal{P}_1$ follows from the soundness of the proof protocol for proving correct decryption. We will now show that if the result of the addition of plaintexts (step 5 of $\mathcal{P}_1$) is corrupted, then it will be detected by the honest players with an overwhelming probability. In order for the results of the two computations (of $\mathcal{P}_1$ resp. $\mathcal{P}_2$) to be equal, we have the following: the adversary must add a value $vr_1$ in the addition step of the first protocol, and a value $vr_2$ in the second, where $r_1$ is the blinding factor used in the first protocol, and $r_2$ that used in the second. Following the argument in the proof of Lemma 2, this would allow him to break the DDH assumption on $F^*[2^t]$, as only encryptions of these values are available to him. Therefore, the output will, with overwhelming probability, only be accepted when it is correct. $\square$

# Improved Methods to Perform Threshold RSA

Brian King

University of Wisconsin-Milwaukee, Milwaukee,WI 53201 USA
Motorola Labs, Schaumburg IL 60010, USA
brking@execpc.com

**Abstract.** A $t$ out of $n$ threshold scheme is such that shares are distributed to $n$ participants so that any set of $t$ participants can compute the secret, whereas any set of less than $t$ participants gain no information about the secret. In [4], Desmedt and Frankel introduced a threshold scheme that can be used with any finite Abelian group. Hence it can be used to provide threshold RSA. In this scheme, the size of the share is on the order $n$ times the size of the secret. Further, due to a complicated algebraic setting, and the large shares, this schemes requires a "large" amount of computations. Recent work have addressed how to reduce the resource requirements. Within this paper we provide improved methods and demonstrate the computational requirements of the Desmedt-Frankel scheme using our method is, in many cases, better than other existing threshold RSA signature schemes.

**Keywords:** threshold secret sharing, threshold cryptography, threshold RSA, cyclotomic polynomials

## 1 Introduction

RSA [18] is an important cryptographic scheme. The development of threshold RSA was problematic due to the fact that the modulus $\phi(N)$, [1] [2] as well any multiple, cannot be leaked to any of the shareholders. Threshold RSA has been examined in [10,5,6], then in [13,4,3], and most recently in [11,12,17,20]. The Desmedt-Frankel scheme [4] was the first secure threshold RSA sharing scheme. This is a zero-knowledge threshold scheme. Further this scheme is a group independent scheme. That is, the shareholder reconstruction of the secret key is independent of the group.[3] Group independent schemes provide a flexible threshold secret sharing. However, there is a disadvantage when using this scheme. The disadvantage of the scheme is the amount of resources it requires, in the sense of memory (share size) and processing (computational time). The memory requirement is caused by share expansion. That is, the share expansion is such

---

[1] Here $N = p_1 p_2$, the product of two distinct primes, and $\phi(N) = (p_1 - 1) \cdot (p_2 - 1)$.

[2] The true modulus is the Carmichael function $\lambda(N)$, which is a divisor of $\phi(N)$

[3] In most applications of threshold cryptography, the reconstructed value is a function of the secret key, not the secret key. For example a signature. In such cases the shareholder will obviously have to perform algebraic operations within the ring $\mathbf{Z}_N$.

T. Okamoto (Ed.): ASIACRYPT 2000, LNCS 1976, pp. 359–372, 2000.
© Springer-Verlag Berlin Heidelberg 2000

that shares will consist of $O(n)$ subshares drawn from the keyspace. The processing cost comes from the computing requirements. Computations will need to be performed on these large shares. Moreover, computations will need to be performed in what appears to be a complicated algebraic structure. These resource requirements (and interest in development of robust, proactive, and/or verifiable threshold RSA) have led to developing other schemes. However the algebraic structure of the Desmedt-Frankel scheme is used in some of these other schemes (for example [3,13]). Work has been initiated in relieving some of the computational requirements for the Desmedt-Frankel scheme. In [8], the authors established that within the Desmedt-Frankel scheme, the share size for each participant could be halved. In [15], we showed how to speed up the required computations when using the Desmedt-Frankel scheme to achieve a threshold RSA signature scheme. Most of the computation improvements were developed for the shareholder. Further, we provided a computation comparison of the Desmedt-Frankel signature scheme with the signature scheme recently developed by Shoup in [20]. Using this comparison we pointed out that in many cases the Desmedt-Frankel scheme performed better than the signature scheme developed by Shoup. Here we will provide further improvement in the performance of the Desmedt-Frankel scheme. Our analysis will show that the computational requirements on the shareholder, as described here, is always less than or equal to the computational requirement for the shareholder as described in [15]. Furthermore, we will show that the computational requirement for the distributor is significantly better than those computation requirements stated by both [4] and [15].

## 1.1   The Desmedt Frankel Scheme

$\mathcal{K}$ represents a finite abelian group. The secret $k$ is selected from $\mathcal{K}$. A prime $q$ is chosen such that $q \geq n + 1$. (We can assume, due to Bertrand's postulate [16], that $O(q) = O(n)$.) Let $u$ represent a root of the cyclotomic polynomial $m(x) = \sum_{j=0}^{q-1} x^j$. Many of the computations are performed in the ring $\mathbf{Z}[u] \cong \mathbf{Z}[x]/m(x)$. Notice that $\alpha_i = \sum_{j=0}^{i-1} u^j$ is a unit (an invertible element in $\mathbf{Z}[u]$) for each $i$ ($1 \leq i \leq q-1$) and that $\alpha_i - \alpha_j$ is a unit for all distinct $i, j$, with $1 \leq i, j \leq q-1$.

Consider the group $\mathcal{K}^{q-1}$ where $\mathcal{K}^{q-1} = \mathcal{K} \times \mathcal{K} \times \cdots \times \mathcal{K}$. If $\boldsymbol{x} \in \mathcal{K}^{q-1}$ then $\boldsymbol{x} = [x_0, x_1, \ldots, x_{q-2}]$. For all $\boldsymbol{x_1}, \boldsymbol{x_2} \in \mathcal{K}^{q-1}$

$$\boldsymbol{x_1} + \boldsymbol{x_2} = [x_{1,0} + x_{2,0}, x_{1,1} + x_{2,1}, \ldots, x_{1,q-2} + x_{2,q-2}].$$

Let $\boldsymbol{0} = [0, 0, \ldots, 0]$, where $0$ denotes the identity in $\mathcal{K}$. For all $b \in \mathbf{Z}$, $b\boldsymbol{k} = [bk_0, bk_1, \ldots, bk_{q-2}]$. where $bk_i$ represents the element in $\mathcal{K}$, formed by applying $k_i$ to itself $b$ times. For $u \in \mathbf{Z}[u]$, $u\boldsymbol{k} = [0, k_0, k_1, \ldots, k_{q-3}] + [-k_{q-2}, \ldots, -k_{q-2}] = [-k_{q-2}, -k_{q-2}+k_0, \ldots, -k_{q-2}+k_{q-3}]$. Then $u^{i+1}\boldsymbol{k} = u(u^i\boldsymbol{k})$. For all polynomials $f$ in $u$ with integer coefficients, $f(u) = b_0 + b_1 u + \cdots + b_k u^k$, $f(u)$ is defined by $f(u)\boldsymbol{k} = \sum_{i=0}^{k} b_i(u^i\boldsymbol{k})$. Then $\mathcal{K}^{q-1}$ is a module over $\mathbf{Z}[u]$ (for more information see [1]).

## 1.2   How Shares Are Computed

Given secret $k$, we represent the secret by $\boldsymbol{k} = [k, 0, \ldots, 0] \in \mathcal{K}^{q-1}$. There are two alternative to generate shares.

Each shareholder $P_i$ $(i = 1, \ldots, n)$ is given share $\boldsymbol{s}_i$ in the following manner: First, $\boldsymbol{s}_1, \boldsymbol{s}_2, \ldots, \boldsymbol{s}_{t-1}$ are chosen uniformly random from $\mathcal{K}^{q-1}$. For $i \geq t$, $\boldsymbol{s}_i$ is defined by:

$$\boldsymbol{s}_i = y_{i,C_i}^{-1} \cdot \left( \boldsymbol{k} - \sum_{\substack{j \neq i \\ j \in C_i}} y_{j,C_i} \cdot \boldsymbol{s}_j \right) \tag{1}$$

where $C_i = \{1, 2, \ldots, t - 1, i\}$ and for each $j \in C_i$

$$y_{j,C_i} = \frac{\displaystyle\prod_{\substack{h \in C_i \\ h \neq j}} (0 - \alpha_h)}{\displaystyle\prod_{\substack{h \in C_i \\ h \neq j}} (\alpha_j - \alpha_h)} \tag{2}$$

An alternate manner to compute the shares is to choose $\boldsymbol{c}_1, \boldsymbol{c}_2, \ldots, \boldsymbol{c}_{t-1}$ uniformly random from $\mathcal{K}^{q-1}$, such that share $\boldsymbol{s}_i$ is determined by

$$\begin{bmatrix} \boldsymbol{s}_1 \\ \boldsymbol{s}_2 \\ \vdots \\ \boldsymbol{s}_l \end{bmatrix} = \begin{bmatrix} 1 & \alpha_1 & \cdots & \alpha_1^{t-1} \\ 1 & \alpha_2 & \cdots & \alpha_2^{t-1} \\ \vdots & \vdots & & \vdots \\ 1 & \alpha_l & \cdots & \alpha_l^{t-1} \end{bmatrix} \begin{bmatrix} \boldsymbol{k} \\ \boldsymbol{c}_1 \\ \vdots \\ \boldsymbol{c}_{t-1} \end{bmatrix} \tag{3}$$

Therefore for each $i$, $\boldsymbol{s}_i = \boldsymbol{k} + \alpha_i \cdot \boldsymbol{c}_1 + \cdots + \alpha_i^{t-1} \boldsymbol{c}_{t-1}$. Hence $\boldsymbol{s}_i = g(\alpha_i)$ where $g(x) = \boldsymbol{k} + x \cdot \boldsymbol{c}_1 + \cdots + x^{t-1} \boldsymbol{c}_{t-1}$. (In this case it is possible to compute the shares using Horner's algorithm.)

## 1.3   How the Secret $k$ Is Computed

When a set $B$ of $t$ participants wish to compute $\boldsymbol{k} \in \mathcal{K}$, they can determine $\boldsymbol{k}$, of which the first component of $\boldsymbol{k}$, is the secret $k$. The participants determine $\boldsymbol{k}$ by

$$\boldsymbol{k} = \sum_{i \in B} y_{i,B} \cdot \boldsymbol{s}_i \tag{4}$$

where $y_{i,B}$ is defined by (2). We will use the $F_0$ to denote the function which maps any $(q-1)$ tuple to its first coordinate. Thus $k = F_0(\boldsymbol{k})$.

## 1.4   How Much Time Is Needed to Perform the Necessary Algebraic Operations

One of our concerns is the amount of time each shareholder uses to perform algebraic operations, another is the amount of time the distributor needs to perform algebraic operations. As described by equations (1) and (4), a required computation appears to be $y_{i,B}$. (Where in the case of distributor, they need to calculate such a $y_{i,B}$ "$t(n-t)$" times.)

There are two approaches to perform the needed calculation. The first approach is described as follows: to perform $y_{i,B} \cdot s_i$, perform

$$y_{i,B} \cdot s_i = \prod_{\substack{h \in C_i \\ h \neq j}} (0 - \alpha_h) \left( \prod_{\substack{h \in C_i \\ h \neq j}} \left( \frac{1}{\alpha_j - \alpha_h} \cdot s_i \right) \right).$$

The running time to compute $y_{i,B} \cdot s_i$ in this manner, as stated by [4] is:

**Theorem 1.** [4] The shareholder performs $O(tn^2)$ group operations, and $O(n)$ inverses. In addition to the time to choose randomly $t-1$ shares. The distributor performs $O(t^2 n^2 (n-t))$ group operations and $O(tn(n-t))$ inverse operations.

A different method is suggested if group operation is slower than integer multiplication. Instead of performing a series of group operations, a series of $\mathbf{Z}[u]$ operations are performed, until $y_{i,B}$ is formed, then one group operation is performed. In [4], the authors established the following:

**Theorem 2.** [4] Each shareholder performs $O(nt \log n)$ group operations, $O(n)$ inverses, and $O(t^3 n^2 (\log n)^2)$ elementary integer operations. In addition to the time needed to choose $t-1$ random shares, the distributor performs $O(t^2 n(n-t) \log_2 n)$ group operations, $O(tn(n-t))$ inverse operations, and $O(t^4 n^2 (n-t)(\log n)^2)$ elementary integer operations.

## 2 Algorithms Which Improved Performance in the Desmedt-Frankel Scheme

In [15], it was illustrated how to improve the performance of the Desmedt-Frankel scheme. Here, a series of results and algorithms were introduced in order to compute more efficiently. We discuss now those results that we will utilize. We will let $\Gamma_0$ denote all sets of $t$ participants. (In order to avoid confusion between an inverse of an $\alpha_x$ with an inverse of an $x \in \mathbf{Z}_q^*$, we will represent an inverse of $x \in \mathbf{Z}_q^*$ by $r_x$.)

**Theorem 3.** [15]
(1) For all $i, j$, with $i \neq j$, $\alpha_i - \alpha_j = u^j \alpha_{i-j}$.
(2) For all $x, a \in \mathbf{Z}_q^*$, $\dfrac{\alpha_x}{\alpha_a} = \dfrac{\alpha_{ka}}{\alpha_a} = 1 + u^a + \cdots + u^{a(k-1)}$, where $k = xr_a \bmod q$.
(3) If $f$ is the product of $r$ many cyclotomic polynomials of the form $\frac{\alpha_x}{\alpha_a}$, then all coefficients of $f$ are bounded by $-q^{r-1}$ and $q^{r-1}$.
(4) For all $i$, $\alpha_x^{-1} = \dfrac{\alpha_{kx}}{\alpha_x}$ where $k = r_x \bmod q$.
(5) For all $i$ and $B \in \Gamma_0$, $y_{i,B} = u^{-(t-1)i} \prod_{\substack{h \in B \\ h \neq i}} \dfrac{\alpha_h}{\alpha_{h-i}}$.

*(6) There exists an algorithm that will calculate the product of $r$ many cyclotomic polynomials of the form $\frac{\alpha_{ax}}{\alpha_a}$ with a running time of $O(rn \log_2 n)$.*
*(7) There exists an algorithm that will calculate $\alpha_x^{-1}$ with a running time of $O(n \log_2 n)$.*
*(8) There exist an algorithm that will calculate $y_{i,B}$ with a running time $O(t^2 n \log_2 n)$*

As discussed in [15], consider the Desmedt-Frankel scheme implementing a threshold RSA signature. $N$ is the product of two distinct primes, and $\phi(N)$ is the Euler totient function. Then $\mathcal{K}$ is $\mathbf{Z}_{\phi(N)}$. In threshold RSA no participant can be given any information concerning $\phi(N)$. Shares are actually $(q-1)$ dimensional vectors in the module $\mathbf{Z}_{\phi(N)}^{q-1}$. From a shareholder's view this will look like a $(q-1)$ dimensional integer vector. Thus computations can be performed using integer addition. The secret is $d$, where the RSA public key is $(e, N)$ and $ed = 1$ (mod $\phi(N)$). Therefore $\boldsymbol{k} = [d, 0, \ldots, 0] = \sum_{i \in B} y_{i,B} \boldsymbol{s}_i$. If a set of $t$ participants would like to sign a message $m$, then they will not send their subshares of $d$ (i.e. they will NOT send $y_{i,B} \cdot \boldsymbol{s}_i$), but rather they will send partial signatures. They could send $m^{y_{i,B} \cdot \boldsymbol{s}_i}$. If all $t$ participants sent $m^{y_{i,B} \cdot \boldsymbol{s}_i}$ then a combiner would get $m^d$ by

$$m^d = F_0(m^{[d,0,\ldots,0]}) = F_0(\prod_{i \in B} m^{y_{i,B} \cdot \boldsymbol{s}_i}).$$

However we must point out that this method wastes resources. That is, the combiner is actually computing $m^{[d,0,\ldots,0]} = [m^d, 1, 1, \ldots, 1]$, whereas the only element of interest is $m^d$. Consider the following method of computation. Recall that $F_0$ is a function which maps a $j$-tuple to its first coordinate. Now, suppose that $B$ is a set of participants, with $|B| = t$, and that for all $j \in B$, $\boldsymbol{z}_j \in \mathcal{K}^{q-1}$, $\boldsymbol{z}_j = [z_{j,0}, z_{j,1}, \ldots, z_{j,q-2}]$. Then $F_0(\prod_{j \in B} m^{\boldsymbol{z}_j}) = F_0(m^{\sum z_{j,0}}, m^{\sum z_{j,1}}, \ldots,$ $m^{\sum z_{j,q-2}}) = \prod_{j \in B} F_0(m^{\boldsymbol{z}_j}) = m^{F_0(\sum \boldsymbol{z}_j)} = m^{\sum z_{j,0}}$. So we see that to compute $m^d$, all that is needed is $F_0(y_{i,B} \cdot \boldsymbol{s}_i)$.

Using a result developed by Desmedt-Frankel in [10], [15] established the following.

**Theorem 4.** *[15] To compute a partial signature, a shareholder must compute an integer then exponentiate $m$ to this integer. The amount of time required for shareholder computations can be expressed as: The time to compute the integer is $O(nt(t \log_2 n + \log_2 n \log_2 \phi(N))$. The time to exponentiate is $O((t \log_2 n + \log_2 \phi(N))(\log_2 N)^2)$.*

**Theorem 5.** *[15] The distributor is required to determine $O((t-1)n)$ elements from $\mathbf{Z}_{\phi(N)}$. The distributor needs to compute $O((n-t+1)n)$ elements from $\mathbf{Z}_{\phi(N)}$. The amount of time required for the distributor computations is $O(nt^2(n-t)\log_2 n \log_2 \phi(N)))$.*

Both Theorem 4 and Theorem 5 incorporate a technique which calls for the computation of $y_{i,B}$ and then applying $y_{i,B} \cdot \boldsymbol{s}$ (a technique proposed by Theorem 2). We now provide a better alternative. This method will apply a technique very similar to the technique proposed by Theorem 1.

## 3  Applying a Cyclotomic Polynomial to a Share

Given a share $s = [s_0, s_1, \ldots, s_{q-2}]$, we will refer to $s_0$ as the *zeroth* term, $s_1$ as the first term, etc. Recall the application of $u$ to a share $s$ as: $us = [-s_{q-2}, s_0 - s_{q-2}, s_1 - s_{q-2}, \ldots, s_{q-3} - s_{q-2}]$. Then $u^2 s = [s_{q-2} - s_{q-3}, -s_{q-3}, s_0 - s_{q-3}, \ldots, s_{q-4} - s_{q-3}]$, $u^3 s = [s_{q-3} - s_{q-4}, s_{q-2} - s_{q-4}, \ldots, s_{q-5} - s_{q-4}]$, and so forth. In general

$$u^a s = [s_{q-a} - s_{q-a-1}, s_{q-a+1} - s_{q-a-1}, s_{q-a+2} - s_{q-a-1} \cdots, s_{q-a-2} - s_{q-a-1}] \quad (5)$$

Where all but one of the terms is a difference of two subshares, this term which is not the difference of two subshares is the $a - 1^{st}$ term and is the negative of the subshare $s_{q-a-1}$. Further observe that when considering all positive terms of the difference each one of the original subshares occurs once, except for the subshare $s_{q-a-1}$ (which is the negative term for all the differences). Extend the definitions of the subshares $s_i$ by defining $s_q = s_0$, and for any integer $i$, $s_i = s_{i \bmod q}$. Due to space, we have omitted many of the proofs to our results.

**Theorem 6.** *To compute $u^a s_i$ requires $O(n)$ additions and 1 inverse within the group $\mathcal{K}$.*

In an effort to develop symmetry we will define an artificial subshare $s_{q-1}$. We are only creating this artificial subshare to illustrate a property concerning cyclotomic polynomials applied to $s$. In order for our computations to remain correct there is only one choice for $s_{q-1}$, we define $s_{q-1} = 0$ for all participants.

Now consider the effect of our definition of $s_{q-1}$, we have $u^a s = [s_{q-a} - s_{q-a-1}, s_{q-a+1} - s_{q-a-1}, s_{q-a+2} - s_{q-a-1} \cdots, s_{q-a-2} - s_{q-a-1}]$, where the $j^{th}$ term of $u^a s$ is $s_{q-a+j} - s_{q-a-1}$. Then $\alpha_i s = [\sum_{j=0}^{i-1} s_{q-j} - cst_i, \sum_{j=0}^{i-1} s_{q-j+1} - cst_i, \sum_{j=0}^{i-1} s_{q-j+2} - cst_i, \ldots, \sum_{j=0}^{i-1} s_{q-j+q-2} - cst_i]$ where $cst_i = \sum_{j=0}^{i-1} s_{q-1-j}$. (Observe that in our definition of $cst_i$ we have included $i - 1$ additions, in reality there are only $i - 2$ additions for the case for $j = 0$ recall that $s_{q-1-j} = s_{q-1} = 0$.) If we represent $\alpha_i s = [a'_0, a'_1, \ldots, a'_{q-2}]$, then

$$a'_0 = s_{q-0} + s_{q-1} + \cdots + s_{q-(i-1)} - cst_i,$$

$$\vdots$$

$$a'_j = s_j + s_{j-1} + s_{j-2} + \cdots + s_{j-(i-1)} - cst_i.$$

Thus, for $1 \leq j \leq q - 2$,

$$a'_j = a'_{j-1} + s_{j-1} - s_{q-(i-j)}. \quad (6)$$

Now observe that $a'_0 = s_q - s_{q-i} = s_0 - s_{q-i}$.

**Theorem 7.** *To compute $\alpha_i s$ requires $O(n)$ additions and $O(n)$ inverses within the group $\mathcal{K}$.*

We now consider the cost of computing $\frac{\alpha_{ax}}{\alpha_a}s$. To this end, we generalize $us$ to include the artificial subshare $s_{q-1}$. So for all $s$ we extend $s$ by

$$s = [s_0, \ldots, s_{q-2}] = [s_0, \ldots, s_{q-2}, 0]_{ext} = [s_0, \ldots, s_{q-2}, s_{q-1}]_{ext} \qquad (7)$$

Then for $s = [s_0, \ldots, s_{q-2}, s_{q-1}]_{ext}$, and any $b \in \mathbf{Z}$, $bs = [bs_0, \ldots, bs_{q-2}, bs_{q-1}]_{ext}$. Further, extend $us = [s_{q-1} - s_{q-2}, s_0 - s_{q-2}, \ldots, s_{q-3} - s_{q-2}, s_{q-2} - s_{q-2}]_{ext}$. Notice that both $bs$ and $us$ are the same as the true $bs$ and $us$ with a 0 appended at the end.

**Theorem 8.** *For all $a$, with $1 \leq a \leq q - 1$, $u^a s = [s_{q-a} - s_{q-a-1}, s_{q-a+1} - s_{q-a-1}, s_{q-a+2} - s_{q-a-1} \ldots, s_{q-a-2} - s_{q-a-1}, s_{q-a-1} - s_{q-a-1}]_{ext}$. Thus the $j^{th}$ term of $u^a s$ is $s_{q-a+j} - s_{q-a-1}$.*

Observe that for each $x \in \{0, \ldots, q-1\}$ there exists an integer $i \in \{0, \ldots, q-1\}$ such that $x = ia \bmod q$. Of course if $x = 0$ then $i = 0$, otherwise $i = xr_a \bmod q$, where we have used $r_a$ to represent the inverse of $a$ in the field $\mathbf{Z}_q$. Next observe from the above theorem that if we represent $u^a s$ by $u^a s = [s'_0, s'_1, \ldots, s'_{q-1}]_{ext}$, then $s'_x = s_{q-a+x} - s_{q-a-1}$. Therefore $s'_{ia} = s_{q-a+ia} - s_{q-a-1}$. Now observe that $q - a = (qa - a) \bmod q = (q-1)a \bmod q$. Let $\epsilon_a \in \mathbf{Z}_q^*$ such that $\epsilon_a a = q - 1 \bmod q$. Then $(\epsilon_a - 1)a = q - a - 1 \bmod q$. Hence $s'_x = s'_{ia} = s_{(q-1)a+ia} - s_{(\epsilon_a-1)a} = s_{(q-1+i)a} - s_{(\epsilon_a-1)a} = s_{(i-1)a} - s_{(\epsilon_a-1)a}$.

**Corollary 9.** *For all integers $j$, with $1 \leq j \leq q-1$, $u^{aj}s = [s_0^{(j)}, s_1^{(j)}, \ldots, s_{q-1}^{(j)}]_{ext}$, where $s_{ia}^{(j)} = s_{(i-j)a} - s_{(\epsilon_a-j)a}$.*

**Theorem 10.** *For all $j$, with $1 \leq j \leq q - 1$, $(1 + u^a + u^{2a} + \cdots + u^{ja})s = [b_0, b_1, \ldots, b_{q-1}]_{ext}$, such that $b_{ia} = s_{ia} + \sum_{x=1}^{j} s_{(i-x)a} - \sum_{x=1}^{j} s_{(\epsilon_a-x)a}$. (Recall that $\epsilon_a = q - 1 \bmod q$.)*

We make the following observations. Consider the $b_i$ from Theorem 10. First $b_0 = s_0 + \sum_{x=1}^{j} s_{(q-x)a} - \sum_{x=1}^{j} s_{(\epsilon_a-x)a}$. Next $b_a = s_a + \sum_{x=1}^{j} s_{(1-x)a} - \sum_{x=1}^{j} s_{(\epsilon_a-x)a} = b_0 + s_a - s_{(-j)a}$. In general

$$b_{(i+1)a} = b_{ia} + s_{(i+1)a} - s_{(i-j)a}. \qquad (8)$$

Also observe that for all $i$,

$$0 \leq |b_{ia}| \leq q \cdot \max_j |s_j|. \qquad (9)$$

**Theorem 11.** *For each $a$ and $x$, with $1 \leq a, x \leq q - 1$ it requires $O(n)$ additions and $O(n)$ inverses to compute $\frac{\alpha_{ax}}{\alpha_a}s$. It also requires $O(n \log_2 n)$ elementary operations. (The elementary operations cost represents the time required to increment $ia$ for $i = 0, \ldots, q - 1$.)*

Regarding shareholder computations, we have not referred to the arithmetic operations as "arithmetic operations in the group $\mathcal{K}$". The shareholder may not be given enough information concerning the group $\mathcal{K}$ to perform arithmetic operations in $\mathcal{K}$. (For example in threshold RSA the group is $\mathbf{Z}_{\phi(N)}$. In this case $\phi(N)$ cannot be revealed to any shareholder, otherwise they can compute the secret key $d$.) The shareholder can perform the arithmetic operations as integer operations. The cost of an integer addition is equal to the logarithm of the maximum addend.

**Corollary 12.** *The cost for a shareholder to perform the required additions to compute $\frac{\alpha_{ax}}{\alpha_a}s$ is $O(n \cdot \log_2(\max_{1 \leq i \leq q-2} s_i))$. If $s$ is the original share dealt to the shareholder and we are implementing the scheme to perform threshold RSA then this cost is $O(n \log_2 \phi(N))$.*

Observe that if $s$ was the shareholder's original share then the subshare of $\frac{\alpha_{ax}}{\alpha_a}s$ is bounded by $-q\phi(N)$ to $q\phi(N)$ (as stated by equation (9)). Hence a bound on the size of such shares would be $\log_2(q\phi(N)) = \log_2 q + \log_2 \phi(N)$. If one successively applies two distinct cyclotomic polynomials of the form $\frac{\alpha_{ax}}{\alpha_a}$ to an original share $s$ then the bound on resulting shares would be $-q^2\phi(N)$ to $q^2\phi(N)$. Thus if one applies a product of $t-1$ cyclotomic polynomial to an original share, the bound is $-q^{t-1}\phi(N)$ to $q^{t-1}\phi(N)$.

We are now ready to describe an alternate way a shareholder may compute their partial result (signature). We will refer to this as Method $\mathcal{M}$. Method $\mathcal{M}$ requires less computational time in comparison to the method developed in [15], which we will refer to a Method $\mathcal{A}$.

**Theorem 13.** *Shareholder $P_i$ needs to perform $O(nt)$ integer additions to compute $y_{i,B}s_i$. In addition, the shareholder will need to perform $O(tn \log_2 n)$ elementary operations.*

*Proof.* In [15], it was established that $y_{i,B} = u^{-(t-1)i} \prod_{\substack{h \in B \\ h \neq i}} \frac{\alpha_h}{\alpha_{h-i}}$. Then $y_{i,B}s = u^{-(t-1)i} \prod_{\substack{h \in B \\ h \neq i}} \frac{\alpha_h}{\alpha_{h-i}}s$. For each $h \in B$, $h \neq i$, $\frac{\alpha_h}{\alpha_{h-i}} = \frac{\alpha_{x(h-i)}}{\alpha_{h-i}}$, where $x = r_{h-i}h \bmod q$. Consider

$$\prod_{\substack{h \in B \\ h \neq i}} \frac{\alpha_h}{\alpha_{h-i}}s = \frac{\alpha_{**}}{\alpha_*}\left(\cdots\left(\frac{\alpha_{**}}{\alpha_*}\left(\frac{\alpha_{**}}{\alpha_*}s\right)\right)\cdots\right). \tag{10}$$

This represents $t-1$ successive applications of a cyclotomic polynomial to a share. To give a definitive cost we will assume we are performing threshold RSA then the cost is

$$\begin{aligned}
\text{cost} &= q\log_2 \phi(N) + q(\log_2(q\phi(N)) + \cdots + q(\log_2(q^{t-2}\phi(N)) \\
&= O(q(t^2 \log_2 q + t \log_2 \phi(N)) \\
&= O(nt(t \log_2 n + \log_2 \phi(N)).
\end{aligned}$$

We then need to apply $u^{-(t-1)i}$ to equation (10), this cost is $n$ integer additions which costs $n \cdot (\log_2(q^{t-1}\phi(N))$ which is $O(n(t \log_2 n + \log_2 \phi(N))$. Overall this cost is $O(nt(\log_2 n + \log_2 \phi(N))$. The cost of integer operations can be characterized as follows: $(t-1)$ inverses in $\mathbf{Z}_q^*$ need to be computed, i.e. $r_{h-i}$, $(t-1)$ multiplications need to be computed, $t$ incrementations in $\mathbf{Z}_q^*$ by a $ia$ for $a = h-i$. This cost can be measured as $O(t(\log_2 q)^2 + t(\log_2 q)^2 + tq \log_2 q) = O(nt \log_2 n)$.

There is one more step that the shareholder must perform and that is to exponentiate the $zero^{th}$ term, i.e. $m^{F_0(y_{i,B}s_i)}$ (we are assuming a signature scheme). Since the size of the coefficient is bounded by $-q^{t-1}\phi(N)$ to $q^{t-1}\phi(N)$, this cost is $O((t \log n + log\phi(N))(\log N)^2)$.

Shareholder's computations

|  | Method $\mathcal{A}$ [15] | Method $\mathcal{M}$ |
|---|---|---|
| Time to compute a partial share | $nt(t \log n + \log n \log \phi(N))$ | $nt(t \log_2 n + \log_2 \phi(N))$ |
| Time to exponentiate | $(t \log n + \log \phi(N))(\log N)^2$ | $(t \log n + \log \phi(N))(\log N)^2$ |

Method $\mathcal{M}$ reflects the cost of $nt$ additions, similar to the approach discussed in Theorem 1. However our method is $1/n$ of the amount of time required by Theorem 1. Method $\mathcal{M}$ is always superior to Method $\mathcal{A}$. However it maybe that the most time consuming operation is the exponentiation which in both methods costs the same amount of time. Note that the $nt^2 \log_2 n$ of Method $\mathcal{A}$ (in "compute a partial share") represents the cost of computing $y_{i,B}$. The $nt^2 \log_2 n$ of Method $\mathcal{M}$ represents the cost of incrementing and other $\mathbf{Z}_q$ operations. The $nt \log n \log \phi(N)$ of Method $\mathcal{A}$ (in "compute a partial share") represents the cost of the group operations and $nt \log \phi(N)$ of Method $\mathcal{M}$ represents the costs of group operations. Thus we see that the improvement in time (when comparing our method to Method $\mathcal{A}$) is a factor of $\log_2 n$.

## 4    How a Distributor Computes the Shares

A distributor knows enough information concerning the group $\mathcal{K}$, to perform additions in the group $\mathcal{K}$.

**Theorem 14.** *The cost for a distributor to perform the required additions to compute $\frac{\alpha_{ax}}{\alpha_a}s$ is $O(n \cdot \log_2(|\mathcal{K}|))$ and $O(n \log_2 n)$ elementary operations. If $s$ is a share dealt to the shareholder and we are implementing the scheme to perform threshold RSA then the cost for the distributor to compute $\frac{\alpha_{ax}}{\alpha_a}s$ is $O(n \log_2 \phi(N))$.*

The following is a simple result which we use later.

**Corollary 15.** *If the distributor applies two cyclotomic polynomials to a share dealt to a shareholder then the cost of performing the additions is $O(n \log_2(\phi(N)))$.*

**Theorem 16.** *For a distributor to compute $y_{i,B}s$. The distributor needs to perform $O(nt)$ additions in the group $\mathcal{K}$. In terms of threshold RSA, the cost is $O(nt \log_2 \phi(N))$. In addition, the distributor will need to perform $O(tn \log_2 n)$ elementary operations.*

We are ready to describe how the distributor can compute all the shares $s_1, \ldots, s_n$. There are two alternatives. As we will establish, the second alternative is the preferred method.

### 4.1   Distributor–Method 1

Recall that the shares $s_1, \ldots, s_n$ can be determined by randomly selecting vectors $c_1, \ldots, c_{t-1}$ from $\mathcal{K}^{q-1}$ and apply the representation given by equation (3). Then for all $i$, $s_i = k + \alpha_i c_1 + \cdots + \alpha_i^{t-1} c_{t-1}$. Observe that we can apply Horner's algorithm, and we get

$$s_i = \alpha_i \left( \cdots \alpha_i \left( \alpha_i c_{t-1} + c_{t-2} \right) \cdots + c_1 \right) + k \tag{11}$$

**Theorem 17.** *To compute $s_i$ using equation (11) requires $t$ multiplications of an $\alpha_i$ to a vector and $t$ additions of vectors. The cost of Horner's algorithm to compute the shares for threshold RSA is $O(tn^2 \log_2 \phi(N))$.*

*Proof.* Using Theorem 7 the cost of computing $s_i$ is $nt$ additions in $\mathbf{Z}_{\phi(N)}$, which is $O(nt \log_2 \phi(N))$. The distributor need to compute all $n$ shares. Hence a total cost of $O(tn^2 \log_2 \phi(N))$.

### 4.2   Distributor–Method 2

Recall the following manner in which a distributor may distribute shares. The distributor selects $t - 1$ random shares $s_1, \ldots, s_{t-1}$. The remaining $n - t$ shares are computed using equation (1), therefore

$$s_i = y_{i,C_i}^{-1} k - \sum_{\substack{j \in C_i \\ j \neq i}} \frac{y_{j,C_i}}{y_{i,C_i}} s_j \tag{12}$$

where $t \leq i \leq n$ and $C_i = \{1, 2, \ldots, t - 1, i\}$.

Observe that to compute $s_i$ (for $i = t, \ldots, n$) we need to compute $\frac{y_{j,C_i}}{y_{i,C_i}}$, for $j = 1, \ldots, t - 1$, and $y_{i,C_i}^{-1}$. Equation (12) illustrates how a combination of the secret "$y_{i,C_i}^{-1} k$" is salted with randomness "$\sum_{\substack{j \in C_i \\ j \neq i}} \frac{y_{j,C_i}}{y_{i,C_i}} s_j$" (recall that in this distribution method, $s_1, \ldots, s_{t-1}$ are random vectors) to form the share $s_i$.

**Lemma 18.** *For each $i, j$ and $B$, $\frac{y_{j,B}}{y_{i,B}} = -y_{j-i,B'(j,i)}$, where $B'(j,i) = \{-i\} \cup \{h - i : h \in B, h \neq i\}$.*

Observe that to compute $\frac{y_{j,B}}{y_{i,B}} s$ reduces to computing a $y_{*,**} s$. Note that this ratio will need to be determined by the distributor $(t - 1)(n - t)$ times, where $B = C_i$.

**Lemma 19.** *For all $i$ and all $i \in B$, $y_{i,B}^{-1} = u^{(t-1)i} =, y_{-i,B''(i)}$, where $B''(i) = \{-i\} \cup \{h - i : h \in B, h \neq i\}$.*

Observe that to compute $y_{i,B}^{-1}s$ reduces to computing a $y_{*,**}s$.

**Lemma 20.** *For all $i \in \{1, \ldots, t-1\}$ and $t \leq j \leq n$*

$$\frac{y_{j+1,C_i}}{y_{i,C_i}} = -u^{j-(t-1)} \frac{\alpha_{t-1-j}\alpha_{i-j}}{\alpha_{j+1}\alpha_{i-1-j}} \frac{y_{j,C_i}}{y_{i,C_i}}$$

The table below, together with Lemma 20, illustrate how a distributor can compute successive ratios of $\frac{y_{*,*}}{y_*}$'s by starting in upper left hand corner and moving to the right, using the previous entry together with the product of two cyclotomic polynomials and a $u$ to a power.

|       | 1 | $\cdots$ | $\cdots$ | $t-1$ |
|-------|---|----------|----------|-------|
| $t$ | $\dfrac{y_{1,C_t}}{y_{t,C_t}}$ | $\dfrac{y_{2,C_{t+1}}}{y_{i,C_i}}$ | $\cdots$ | $\dfrac{y_{t-1,C_i}}{y_{i,C_i}}$ |
| $t+1$ | $\dfrac{y_{1,C_{t+1}}}{y_{t+1,C_{t+1}}}$ | $\dfrac{y_{2,C_{t+1}}}{y_{t+1,C_{t+1}}}$ | $\cdots$ | $\dfrac{y_{t-1,C_{t+1}}}{y_{t+1,C_{t+1}}}$ |
| $\vdots$ | $\vdots$ | $\vdots$ | $\cdots$ | $\vdots$ |
| $n$ | $\dfrac{y_{1,C_n}}{y_{n,C_n}}$ | $\dfrac{y_{2,C_n}}{y_{n,C_n}}$ | $\cdots$ | $\dfrac{y_{t-1,C_n}}{y_{n,C_n}}$ |

$$(13)$$

From Lemma 20 we see that from the product of two cyclotomic polynomials and a $u$ to a power applied to $\frac{y_{1,C_i}}{y_{i,C_i}}$ generates $\frac{y_{2,C_i}}{y_{i,C_i}}$. Similarly, the product of two cyclotomic polynomials and a $u$ to a power applied to $\frac{y_{2,C_i}}{y_{i,C_i}}$ generates $\frac{y_{3,C_i}}{y_{i,C_i}}$ and so on. For each $i, j$ with $1 \leq j \leq t - 1$ and $t \leq i \leq n$, define

$$\gamma_{ij} = u^{j-t-1} \frac{\alpha_{t-1-j}}{\alpha_{j+1}} \frac{\alpha_{i-j}}{\alpha_{i-1-j}}. \text{ Then}$$

$$\frac{y_{2,C_i}}{y_{i,C_i}} = \gamma_{i2} \frac{y_{1,C_i}}{y_{i,C_i}}$$

$$\frac{y_{3,C_i}}{y_{i,C_i}} = \gamma_{i3} \frac{y_{2,C_i}}{y_{i,C_i}} = \gamma_{i3}\gamma_{i2} \frac{y_{1,C_i}}{y_{i,C_i}}$$

$$\vdots$$

$$\frac{y_{j,C_i}}{y_{i,C_i}} = \gamma_{ij} \frac{y_{j-1,C_i}}{y_{i,C_i}} = \cdots = \gamma_{ij} \cdots \gamma_{i2} \frac{y_{1,C_i}}{y_{i,C_i}}$$

For the distributor to compute $s_i$ ($t \leq i \leq n$), they must compute two parts, one they must compute $y_{i,B}^{-1}k$. (By Lemma 19, this has cost $O(nt \log \phi(N) + nt \log_2 n)$.) Also, the distributor must compute $\sum_{j=1}^{t-1} \frac{y_{j,C_i}}{y_{i,C_i}} s_j$. The distributor can compute

$$\sum_{j=1}^{t-1} \frac{y_{j,C_i}}{y_{i,C_i}} s_j = \frac{y_{1,B}}{y_{i,B}} \left( \gamma_{i2} \left( \cdots \gamma_{it-2} \left( \gamma_{it-1}s_{t-1} + s_{t-2} \right) \cdots + s_2 \right) + s_1 \right). \quad (14)$$

Now note that the time to compute $\gamma_{ij}$ times a share is equivalent to applying two cyclotomic polynomials on a vector then apply $u$ to a power times a share. By Theorem 6,Theorem 14, and Corollary 15, the cost is $2n \log \phi(N) + 2n \log n = O(n(\log n + \log \phi(N))$. There are $t - 1$ of these multiplications by a $\gamma_{**}$ to compute. In addition, there are $t - 1$ vector additions to be performed. The total running time is $O(tn(\log n + \phi(N)) + tn \log \phi(N)) = O(tn(\log n + \phi(N)))$. Then we must perform a $y_{1,C_i}$ to this resulting vector. This cost is $O(tn(\log n + \phi(N)))$. Thus to compute this portion of $s_i$ has cost $O(tn(\log n + \phi(N)))$. To compute $y_{i,C_i}^{-1} k$ is $O(nt \log_2 \phi(N) + nt \log_2 n)$. To compute both portions has cost $O(tn(\log n + \phi(N)))$.

Now the distributor has to compute $s_t, \ldots, s_n$. Altogether, to compute all $n - t$ shares has a total cost of $O(tn(n - t)(\log n + \phi(N)))$.

Distributor's computations

|  | Method $\mathcal{A}$ [15] | Method 1 (Horner) | Method 2 |
|---|---|---|---|
| Randomness required | $(t-1)(q-1)$ | $(t-1)(q-1)$ | $(t-1)(q-1)$ |
| Time to compute remaining shares | $nt^2(n-t)\times$ $\log_2 n \log_2 \phi(N)$ | $tn^2\times$ $\log_2 \phi(N)$ | $tn(n-t)\times$ $(\log_2 \phi(N) + \log_2 n)$ |

When $n \log \phi(N) < (n-t)(\log \phi(N) + \log n)$, Method 2 performs better than Method 1. Although we omit the lengthy argument, the addition of $\log_2 n$ in Method 2 can be dropped, if one wants to store previously computed $ia$'s. This is achieved with a large memory cost.

## 5    Comparison with Other Schemes

Comparing Method $\mathcal{M}$ with Method $\mathcal{A}$,we have seen that in terms of Shareholder computations, Method $\mathcal{M}$ always performs better or the same than Method $\mathcal{A}$. Regarding Distributor computations Method $\mathcal{M}$ always performs better than Method $\mathcal{A}$.

Similar to what was done in [15], we compare the performance of Shoup signature scheme [20] with the Desmedt-Frankel scheme using our method. Regarding Shareholder computations, there are various cases to consider. It is clear that Method $\mathcal{M}$ is superior to Shoup's scheme (regarding Shareholder computations) when $(t \log_2 n + \log_2 \phi(N))(\log_2 N)^2 \geq nt(t \log_2 n + \log_2 \phi(N))$ Regarding Distributor computations, Method $\mathcal{M}$ is superior to Shoup's scheme whenever $(\log_2 \phi(N))^2 \geq (n - t)(\log_2 n + \log_2 \phi(N))$. Thus in many cases the Desmedt-Frankel scheme performs better than other threshold RSA signature scheme. This scheme still has disadvantages, in that it has large shares and it requires a large amount of randomness for the distributor to compute the shares. Shoup's scheme, like the schemes by Rabin [17] and Frankel, Gemmel, Mackenzie and Yung [12], may perform worse than the Desmedt-Frankel scheme, because they utilize a technique of exponentiating to a large exponent, an exponent greater than $n!$, rather than using expanded shares. Whereas in the Desmedt-Frankel scheme the largest the exponent can be is $q^t \phi(N)$. In reality $q^t \phi(N)$ is an upper bound and is dependent on the $i$ and $B$. In many cases, the exponent will be much smaller. We do note that threshold schemes like [12] have other benefits that the Desmedt-Frankel scheme does not possess.

| | Shoup's scheme | Desmedt-Frankel scheme Method $\mathcal{M}$ |
|---|---|---|
| Size of share | 1 | $n$ |
| Shareholder time required | $(n \log_2 n + \log_2 \phi(N)) \times$ $(\log_2 N)^2$ | $max\{nt(t \log_2 n + \log_2 \phi(N)),$ $[t \log_2 n + \log_2 \phi(N)](\log_2 N)^2\}$ |
| Distributor randomness required | $t - 1$ | $(t-1)O(n)$ |
| Distributor time required | $nt(\log_2 \phi(N))^2$ | $nt(n - t)(\log_2 n + \log_2 \phi(N))$ |
| Combiner time required | $(t + \log_2 a + \log_2 b) \times$ $(\log_2 N)^2$ | $t(\log_2 N)^2$ |

## 6   Conclusion

We have described algorithms which effectively speed up computations in the threshold RSA scheme developed by [4]. We have pointed out that there are many occasions when the Desmedt-Frankel scheme will perform better than "efficient" schemes. However, our work has not reduced share size nor the randomness required. It is ironic that with a large share, computations may be maybe smaller than schemes with small share size. In the end, in most threshold examples, it is response time, i.e. time required for the shareholder to respond to a signature request, that is most important. Further, we point out that although shares are large, the partial signatures sent to the Combiner are the same size as an RSA signature. If the distributor has enough random resources, it appears that the remaining computations is comparable (or better) than distributor work for other threshold schemes. Lastly the Combiner's amount of computations within the Desmedt-Frankel scheme, are less than or equal, to the Combiner's work in Shoup's scheme.

## References

1. W. Adkins and S. Weintrab. *Algebra, an approach via module theory*. Springer-Verlag, NY, 1992.
2. G. Blakley. "Safeguarding cryptographic keys." In *Proc. Nat. Computer Conf. AFPIPS Conf. Proc., 48* pp. 313-317, 1979.
3. A. De Santis, Y. Desmedt, Y. Frankel, and M. Yung. "How to share a function". In *Proceedings of the twenty-sixth annual ACM Symp. Theory of Computing (STOC)*, pp. 522-533, 1994.

4. Y. Desmedt and Y. Frankel. "Homomorphic zero-knowledge threshold schemes over any finite abelian group". In *Siam J. Disc. Math. vol 7, no. 4* pp. 667-679, SIAM, 1994.
5. Y. Desmedt and Y. Frankel. Threshold Cryptosystems *In Advances of Cryptology- Crypto '89*, pp. 307-315, 1989.
6. Y. Desmedt and Y. Frankel. Shared generation of authenticators and signatures. In *Advances of Cryptology- Crypto '91*, 1991.
7. Y. Desmedt and S. Jajodia. Redistributing secret shares to new access structures and its applications. Tech. Report ISSE-TR-97-01, George Mason University, July 1997 ftp://isse.gmu.edu/pub/techrep/97.01.jajodia.ps.gz
8. Y. Desmedt, B. King, W. Kishimoto, and K. Kurosawa, "A comment on the efficiency of secret sharing scheme over any finite abelian group", In *Information Security and Privacy*, ACISP'98 (Third Australasian Conference on Information Security and Privacy), LNCS 1438, 1998, 391-402.
9. Y.Frankel. A practical protocol for large group oriented networks. In *Advances of Cryptology- Eurocrypt '89*, Lecture Notes in Computer Science 434, Springer Verlag, 1990, pp 56-61.
10. Y.Frankel and Y. Desmedt. Parallel reliable threshold multisignature. *Tech. report TR-92-04-02*, Univ. of Wisconsin-Milwaukee, 1992.
11. Y. Frankel, P. Gemmel, P. Mackenzie, and M. Yung. Proactive RSA In *Advances of Cryptology- Crypto '97*, 1997 Lecture Notes in Computer Science 1294, Springer Verlag, 1997, pp 440-454.
12. Y. Frankel, P. Gemmel, P. Mackenzie, and M. Yung. Optimal-Resilience Proactive Public-key Cryptosystems In *Proc. 38th FOCS*, pp 384-393, IEEE, 1997
13. R. Gennaro, S. Jarecki, H. Krawczyk, and T. Rabin. Robust and efficient sharing of RSA functions. In *Advances of Cryptology- Crypto '96,* Lecture Notes in Computer Science 1109, Springer Verlag, 1996, pp 157-172.
14. T. Hungerford. *Algebra*. Springer-Verlag, NY, 1974.
15. B. King. Algorithms to speed up computations in threshold RSA, Australasian Conference on Information Security and Privacy 2000.
16. H.L. Keng. *Introduction to Number Theory*. Springer Verlag, NY 1982
17. T.Rabin. A Simplified Approach to threshold and proactive RSA. In *Advances of Cryptology- Crypto '98*, 1998
18. R. Rivest, A. Shamir, and L. Adelman, A method for obtaining digital signatures and public key cryptosystems, *Comm. ACM,* 21(1978), pp 294-299.
19. A. Shamir, How to share a secret, *Comm. ACM*, 22(1979), pp 612-613.
20. V. Shoup. "Practical Threshold Signatures" In *Advances of Cryptology- Eurocrypt 2000,* pp 207-220.

# Commital Deniable Proofs and Electronic Campaign Finance

Matt Franklin[1]* and Tomas Sander[2]

[1] Department of Computer Science
University of California, Davis
One Shields Avenue
Davis, CA 95616-8562, USA
franklin@cs.ucdavis.edu
[2] InterTrust Technologies
STAR Lab,
4750 Patrick Henry Drive
Santa Clara, CA 95054, USA
sander@intertrust.com

**Abstract.** In a recent Stanford Law Review article, Ayres and Bulow [1] propose a radical anonymity-based solution to disrupt the "market" for monetary influence in political campaigns. To realize their proposal, we propose new cryptographic protocols for commital deniable proofs and deniable payment schemes.

*"[T]here is little reason to doubt that sometimes large contributions will work actual corruption of our political system, and no reason to question the existence of a corresponding suspicion among voters."*
*– U.S. Supreme Court Justice David Souter, Nixon v. Shrink Missouri Government PAC, Jan 24, 2000.*

*"[Spiritually] lower than this is one who gives to the poor in a way that the giver does not know to whom he is giving and the poor person does not know who he took from. Lower than this is where the giver knows who he is giving to and the poor does not know who he is receiving from. Lower than this is where the poor knows who he is receiving from but the giver does not." – Maimonides, Laws of Gifts to the Poor 10:7-14, 12th Century.*

## 1   Introduction

The success of political candidates in U.S. elections depends critically on the amount of money they can spend on their campaign. Candidates may thus become vulnerable to influence buying by wealthy citizens, corporations, or Political Action Committees (i.e., groups that are able to raise and bundle significant

---

* Work on this paper was done while author was at Xerox PARC.

T. Okamoto (Ed.): ASIACRYPT 2000, LNCS 1976, pp. 373–387, 2000.

amounts). Influence buying can range from simply buying time with the candidate, to buying the opportunity to express opinions on particular political issues, to outright quid pro quo corruption where political positions are traded for donations. Candidates may also extort donations from potential donors, by threatening them with punitive treatment or indifference. The potential for political corruption has led to regular attempts to reform the system of campaign finance. Mainstream proposals include mandated disclosure of campaign donations (to expose suspicious correlations between the candidate's positions and the donors' interests) and limits on the amount of donations.

Ayres and Bulow [1] propose a more radical approach to disrupt the "market" for monetary influence. Any donor can contribute any amount to any candidate's campaign, but must not be able to prove to the candidate that he made a donation. Since a true influence buyer has no more credibility than a fake influence claimer, potential influence buyers have no incentive to actually make a contribution. Furthermore a candidate who tries to extort donations (or "launder" funds through phony donors) has no way to verify that the extorted party in fact followed his blackmailing. We refer the reader directly to [1] for a more detailed discussion of "mandated donor anonymity" and its consequences, constitutionality, and political feasibility.

To implement their proposal, Ayres and Bulow offer only a trusted third party design called the "Blind Trust." All donations are made through the Blind Trust, which has a policy of never revealing the identity of the donors. This complete reliance on a trusted third party is unsatisfying. Moreover, if donations to the blind trust are made by check or other traditional payment mechanisms, then external paper trails and bank records could later be used by the donor to prove to a candidate that a certain donation was made.

We view this as a cryptographic problem. In this paper, we will introduce the problem of deniable electronic payment mechanisms, and show how they can be applied to the mandated donor anonymity problem. Our proposal is practical, and improves on the original Blind Trust of Ayres and Bulow in several critical aspects.

Most importantly our protocol achieves verifiability: although donations are deniable it can still be publicly verified that no donations were withheld from the candidate. To achieve this we introduce the notion of commital deniable proofs and show that every statement can be proved in a commital deniable way. Commital deniable proofs allow a player to prove knowledge of certain decommitals that satisfy a predicate, without revealing which commitments he is using. Later, given witnesses to *any* decommitals that satisfy the predicate, he can later claim that these were the ones used to produce the proof. This fairly general construction seems to be applicable in many situations in which provability and deniability should be combined. Our protocol builds on ideas of Cramer and Damgard [12] for efficient zero-knowledge proofs, which in turn builds on ideas of Cramer, Damgard, and Schoenmakers [13]. In essence, we show that these earlier protocols have our new property of commital deniability.

We also elaborate on the differences of anonymity and deniability for payment schemes and show how the basic Chaumian ecash system [11] can be made deniable. Although the deniability properties of this variant of Chaumian ecash are not strong enough to give a solution to the campaign financing problem it still offers stronger privacy protecting guarantees than "solely" anonymous ecash as it allows payers to convincingly lie about how they spent their electronic money under coercion, which may be of independent interest.

### 1.1   Organization of the Paper

In Section 2 we describe related work. In Section 3 we state the requirements and assumptions for the donation protocol. In Section 4, we give definitions and protocols for commital deniable proofs. In Section 5, we show how to use commital deniable proofs in a deniable payment scheme. In Section 6, we elaborate on the similarities and differences of anonymity and deniability for payment schemes. Conclusions are given in Section 7.

## 2   Related Work

The previous discussion motivates the design of a payment mechanism that protects against the "adversarial" behavior of

1. A donor who wishes to prove to a candidate that he made donation. This influence buying donor is an example of a self-coercing adversary. If a protocol protects against this attack, we say that it is "receipt-free" for the donor.
2. A candidate who tries to extort contributions from a donor, or tries to extract information from the trust. If a protocol protects against this attack, we say that it is "incoercible" for the donor or trust.

Our deniable payment protocol will be receipt-free for the donor, and incoercible for the trust. There will be other requirements as well, e.g., public verifiability for the trust. This is all discussed in more detail in Section 3.

A scheme for "deniable encryption" was introduced by Canetti et al. [7]. Consider encryption to be a one-round, one-message protocol from the sender to the receiver. Consider deniability with respect to the message that was encrypted, i.e., a coercer who saw the ciphertext wants to know what plaintext was actually encrypted and sent. They use the term "sender-deniability" (or "receiver-" or "sender-and-receiver-") for what we are calling incoercibility. They give a solution to these problems which require that the size of the ciphertext is linear in $1/\delta$, where $\delta$ is the probability that the adversary can distinguish the real plaintext-plus-coinflips of the sender from the phony plaintext-plus-coinflips. Canetti et al. also consider a setting they call "flexibly deniable encryption". In this setting the sender chooses the fake message already before encryption. They give a good solution for this case. Intuitively, our notion of commital deniability can be said to lie somewhere between deniability and flexible-deniability. Canetti et al. do not consider receipt-freeness.

Building on the primitive of deniable encryption Canetti and Gennaro [9] show that any function can be securely evaluated in an incoercible way, i.e. the parties can lie about their inputs to the secure function evaluation under external coercion.

Even if one does not care about efficiency the problem with applying these incoercible protocols to the campaign donation problem is that they are not receipt-free, i.e. a self-coercing adversary can still prove what his input was. The well known reason (see [4,5]) is that the ability of a party to lie about its input relies on its ability to produce fake random coin flips that were used for the (probabilistic) encryption of the input. Thus a party who commits itself to the used randomness by choosing it as the output of a hash function lost its ability to lie and can thereby prove what its input was.

Nevertheless, incoercible protocols can be useful for the design of receipt-free protocols. We use the commital deniable proofs to ensure public verifiability in our protocol. The protocol is still receipt-free for the donors.

Benaloh and Tuinstra [4] initiated the study of receipt-freeness and inco-ercibility for secret ballot election schemes (see also [18,16,17,15]). To achieve incoercibility these protocols typically have a "voting booth" (or make other physical assumptions) that guarantee that the voter is isolated from the coercer for one phase of the voting protocol. In principle every receipt-free or incoercible voting scheme can be mapped into a deniable payment scheme. Each potential contributor registers as many times as he likes, paying one unit of cash for each vote. The contributor later casts as many of these votes as he likes to contribute to a particular candidate. The voting authority moves cash to candidate ac-counts according to the outcome of the election. Our efficient deniable payment scheme is based on different principles than the previous ones for receipt-free and incoercible voting which use homomorphic encryption, Mixnets, or blind signatures.

Canetti and Ostrovsky [10] consider multiparty computation where all par-ties may diverge from the protocol as long as they can do so undetectably. They distinguish the cases of globally-honest-looking and locally-honest-looking mis-behavior, i.e., whether any party's deviation is undetectable by all parties or by any one party. The problems they face are similar to the problems of defending against a self-coercing adversary.

## 3    Requirements and Assumptions

Before describing our solution, we discuss the requirements and assumptions, and point to some of the potential problems one runs into when designing a deniable payment mechanism. Let's first restate the main requirement for our protocol:

**Req. 1: Receipt-Freeness for the Donor:** A donor should not be able to prove to the  candidate that he made a donation.

If we want to avoid the cumbersome necessity of a physical donation booth, in which the donor drops his dollars into a donation box we have to allow for mechanisms for remote donations. This creates potential problems that have been recognized in remote voting systems: a coercer may look over the voter's shoulder while he casts his ballot. This applies even more so to a donation system: Candidate and donor can always get together while the donor writes the check or initiates a payment with electronic cash. The effect of this behavior in the donation setting is potentially very effective as a few donors may already account for a large total sum of donations (in contrast to remote voting in a large scale election where it is much less feasible "to watch over the shoulder" of a sufficient number of voters to influence the outcome). This includes attacks where a donor tries to prove a donation was made via covert timing channels in the financial system. For example a donor could announce to the candidate that a specific amount is about to be contributed.

A way to defeat this kind of attacks is to give a donor the ability to cancel his donation. To enable the donor to send cancellation messages for his to donation to the blind trust we make the physical assumption of the availability of an:

**Ass. 1.: Untappable Channel:** The donor has the ability to send one (unnoticed) message to the trust via an untappable channel in some time window (e.g., two weeks).

This assumption seems to be well implementable in real life systems. Note that the trust could set up various channels to receive cancellation messages, e.g., via (anonymized) email, the phone system, postal mail ... It seems realistic to assume that not all of these channels are under the control of parties colluding with a candidate.

Thus the first main ingredient to achieve receipt-freeness for the donor is to allow for (possible) overpayment via cancellation.

To ensure the overall correctness of the deniable donation process we require

**Req. 2: Verifiability:** It can be publicly verified that the trust paid out no less money to the candidate than if it would have followed the protocol honestly.

Note that accountability is an important feature to ensure the overall trustworthiness of a deniable payment mechanism. As individual payers must not have individualized receipts and as, e.g., in our campaign finance application large sums of money may be involved, the system would else become a lucrative target for insider attacks that may be hard to detect.

In our protocol each cancellation message consists of a secret that is sent by the donor to the trust. For public verification the trust will construct a proof of how many secrets it has learned in the cancellation phase. (The proof does not reveal which secrets the trust has learned). Once this proof has been publicly verified there is no longer a need that it can be derived from the trusts records which the actual secrets were that the trust has learnt. This should allow to secure the trust against later coercion attempts (e.g., by a curious and powerful politician *after* he has been elected). A simple, but reasonable solution to achieve this is that the trust "forgets" which secrets it has learnt during the cancellation phase, i.e. it erases all its corresponding records.

Note that however reliable erasure of records in the presence of multiple operators seems to be hard to ensure. (To realize the untappable channel in our system records will e.g. be received by parties operating phones, postal mail and email). The alternative solution we suggest is based on deniable proofs of knowledge. It is no longer vulnerable to the revelation of accidentally or intentionally kept copies of the revealed secrets. Accidental or coerced revelation of a record after the deniability phase of the protocol does no longer prove anything.

To ensure this deniability property of the protocol it is sufficient that the trust makes sure that random bits used for the construction of the proof were in fact randomly chosen (and not e.g. as images of a hash function).

**Req. 3: Incoercibility for the Trust Without Erasure:** There is a deniability phase in the protocol such that after its completion the honest trust cannot be forced to prove to anybody who the actual individual donors were or what the individually donated amounts were, even if the trust performs no erasures.

We make the following assumption about the trust:

**Ass. 2: No Pre-Coercion of Trust:** The trust is not coerced (or corrupted) up to the deniability phase.

This assumption implies in particular that the trust itself does not collude with the candidate for whom it is collecting the donations.

The following requirement is motivated by the fact that, e.g., foreign nationals are not allowed to make donations to U.S. campaigns. Furthermore it should be prevented that money from criminal organizations is funneled to a candidate, or that the deniable payment mechanism is abused for money laundering activities.

**Req. 4: Legitimacy of Funds:** The candidate should only receive donations from "legitimate sources".

What exactly "legitimate sources" are is beyond the (technical) scope of this paper and may depend on the particular election situation. We assume that the legitimacy of the origin of (non-anonymized) funds can be determined by traditional means.

## 4   Commital Deniable Proofs

In this section we introduce the notion of commital deniable proofs. We show that every predicate has a commital deniable proof, using techniques of Cramer et al. [13,12]. That is, we show that the protocol from [12] has our new property of commital deniability. The protocol requires constant rounds and message complexity proportional to the size of the formula for the predicate. This protocol is a useful building block for our deniable payment scheme.

The general intuition behind our notion is that there is a predicate and a set of unconditionally hiding commitments. A party should be able to prove knowledge of certain decommitals that satisfy the predicate, without revealing which commitments he is using. Later, given witnesses to *any* decommitals that satisfy the predicate, he can later claim that these were the ones used to produce the proof. More formally, there is a faking algorithm that takes as input the new (claimed) decommitals, the old (actually used) decommitals, the transcript of the proof, and the coin flips of the prover during the proof. The output of the faking algorithm is a new set of coin flips that is consistent with the old transcript together with the new decommitals.

Let $z_1, \ldots, z_n$ be boolean variables. A "boolean circuit" is a directed acyclic graph where every node has in-degree 0 or 2, and one node has out-degree 0. A node with in-degree 0 is called an input node, and is labeled with some $z_i$ or $\bar{z}_i$ (possibly repeated). A node with in-degree 2 is called a gate, and is labeled with OR or AND. The node with out-degree 0 is called the output node. Let $E, I, G$ denote the edges, input nodes, and gates of a circuit. A "boolean formula" is a boolean circuit where no node has out-degree greater than 1. Let $p$ be a sufficiently large prime. Let $g$ and $h$ have large prime order $q$ in $Z_p^*$, where the discrete log of $h$ to the base $g$ is unknown to all parties. Let $y_{i,j} = g^{b_{i,j}} h^{r_{i,j}} \bmod p$, where $b_{i,j} \in \{0, 1\}$ and $r_{i,j} \in_R Z_q$ (unconditionally hiding boolean commitments). We say that $b_{i,j}, r_{i,j}$ is a "decommital" of $y_{i,j}$.

In a commital deniable proof of knowledge for a language $L$ both prover and verifier are given a formula $\phi$ over $n$ boolean variables and commitments $\{y_{i,j} : 1 \le i \le k, 1 \le j \le n\}$. The prover knows decommitals $\{(b_{i^*,j}, r_{i^*,j} : 1 \le j \le n\}$ for some $i^*$ such that $\phi(b_{i^*,1}, \ldots, b_{i^*,n}) = 1$.

**Definition 1.** *A proof system is called commital deniable if the following conditions hold:*

**Completeness** *When executed with an honest prover $P$, an honest verifier $V$ always accepts at the end of the protocol.*

**Soundness** *There is a knowledge extractor such that if $V$ accepts the proof then the knowledge extractor can find in polynomial time w.v.h.p. decommitals of $y_{i,1}, \ldots, y_{i,n}$ for some $i$ that satisfy $\phi$.*

**Commital Deniability** *There is a faking algorithm $F$ that takes as an input the real decommitals $b_{i^*,1}, r_{i^*,1}, \ldots, \ldots, b_{i^*,n}, r_{i^*,n}$ and the new decommitals $b_{i',1}, r_{i',1}, \ldots, b_{i',n}, r_{i',n}$, where $i^* \ne i'$ or $i^* = i'$ are both possible, and where $\phi(b_{i^*,1}, \ldots, b_{i^*,n}) = \phi(b_{i',1}, \ldots, b_{i',n}) = 1$. The faking algorithm is also given the transcript $T$ of the proof protocol and the internal coin flips of the prover. The output of $F$ is a new sequence of internal coin flips that make the (real) transcript consistent with the new decommitals.*

**Theorem 1.** *Every formula has a commital deniable proof of knowledge.*

Prover and Verifier both know a boolean formula $\phi$, cryptographic parameters $p, q, g, h$, and boolean commitments $\{y_{i,j} : 1 \le i \le k, 1 \le j \le n\}$ for some $k \ge 2$.

The Prover secretly knows $\{b_{i^*,j}, r_{i^*,j} : 1 \leq j \leq n\}$ for some $i^*$. The Prover wants to demonstrate to the Verifier that $\phi(\{b_{i^*,j} : 1 \leq j \leq n\}) = 1$, without revealing any useful information about $i^*$ or the satisfying assignment.

1. Prover $\rightarrow$ Verifier: $\{u_{i,v} : 1 \leq i \leq k, v \in I\}$.
2. Verifier $\rightarrow$ Prover: $c \in_R Z_q$.
3. Prover $\rightarrow$ Verifier: $\{c_i : 1 \leq i \leq k\}, \{c_{i,e} : 1 \leq i \leq k, e \in E\}, \{\alpha_{i,v} : 1 \leq i \leq k, v \in I\}$.
4. Verifier accepts if and only if the following:
   (a) $c = \sum_{i=1}^{k} c_i \bmod q$.
   (b) $c_{i,e_1} + c_{i,e_2} = c_{i,e_3} \bmod q$ for every internal OR gate with incoming edges $e_1, e_2$ and outgoing edge $e_3$, for every $1 \leq i \leq k$.
   (c) $c_{i,e_1} = c_{i,e_2} = c_{i,e_3}$ for every internal AND gate with incoming edges $e_1, e_2$ and outgoing edge $e_3$, for every $1 \leq i \leq k$.
   (d) If the output node is an OR gate with incoming edges $e_1, e_2$, then $c_{i,e_1} + c_{i,e_2} = c_i \bmod q$, for every $1 \leq i \leq k$.
   (e) If the output node is an AND gate with incoming edges $e_1, e_2$ then $c_{i,e_1} = c_{i,e_2} = c_i$, for every $1 \leq i \leq k$.
   (f) For every input node $v \in I$ with label $z_i$ and outgoing edge $e$, $u_{i,v}^{c_{i,e}} y_{i,j} = gh^{\alpha_{i,v}} \bmod p$, for every $1 \leq i \leq k$.
   (g) For every input node $v \in I$ with label $\bar{z}_i$ and outgoing edge $e$, $u_{i,v}^{c_{i,e}} y_{i,j} = h^{\alpha_{i,v}} \bmod p$, for every $1 \leq i \leq k$.

**Claim 1**: An honest prover can execute this protocol so that an honest verifier always accepts.

**Claim 2**: This protocol is a witness indistinguishable proof of knowledge of a $\phi$-satisfying decommital of $\{y_{ij} : 1 \leq j \leq n\}$ for some $i$.

**Claim 3**: This protocol is commital-deniable.

**Claim 4**: The message complexity of the protocol is $O(\#Ik)$. (The last message from Prover to Verifier appears to have size $O(\#Ek)$, but it was written this way for simplicity. In fact, all of the $c_{i,e}$ can be derived from $\{c_{i,e} : 1 \leq i \leq k, e \in E_I\}$ where $E_I$ are the out-edges of input nodes.)

Claim 1-4 yield Theorem 2.

**Proof of Claim 1**:

1. Prover prepares the first message to Verifier as follows:
   (a) Choose $c_i \in_R Z_q$ for all $i \neq i^*$.
   (b) Choose $c_{i,e} \in Z_q$ for all $i \neq i^*$ and for all $e \in E$, subject to constraints 4b-e, but otherwise drawn from the uniform distribution.
   (c) If input node $v$ has outgoing edge $e$ and label $z_j$:
      i. Choose $\alpha_{i,v} \in_R Z_q$ for all $i \neq i^*$.
      ii. Compute $u_{i,v} = (gh^{\alpha_{i,v}}/y_{i,j})^{-c_{i,e}} \bmod p$ for all $i \neq i^*$.
      iii. If $b_{i^*,j} = 0$, then choose $c_{i^*,e}, \alpha_{i^*,v} \in_R Z_q$ and compute $u_{i^*,v} = (gh^{\alpha_{i^*,v}}/y_{i^*,j})^{-c_{i^*,e}} \bmod p$.

  iv. If $b_{i^*,j} = 1$, then choose $s_{i^*,v} \in_R Z_q$ and compute $u_{i^*,v} = h^{s_{i^*,v}} \bmod p$.

(d) If input node $v$ has outgoing edge $e$ and label $\bar{z}_j$:

  i. Choose $\alpha_{i,v} \in_R Z_q$ for all $i \neq i^*$.

  ii. Compute $u_{i,v} = (h^{\alpha_{i,v}}/y_{i,j})^{-c_{i,e}} \bmod p$ for all $i \neq i^*$.

  iii. If $b_{i^*,j} = 1$, then choose $c_{i^*,e}, \alpha_{i^*,v} \in_R Z_q$ and compute $u_{i^*,v} = (h^{\alpha_{i^*,v}}/y_{i^*,j})^{-c_{i^*,e}} \bmod p$.

  iv. If $b_{i^*,j} = 0$, then choose $s_{i^*,v} \in_R Z_q$ and compute $u_{i^*,v} = h^{s_{i^*,v}} \bmod p$.

2. Prover receives challenge $c$ from Verifier.
3. Prover prepares his response to Verifier as follows:

  (a) Compute $c_{i^*} = c - \sum_{i \neq i^*} c_i \bmod q$.

  (b) Choose $c_{i^*,e} \in Z_q$ for all $e \in E$ for which this is still unassigned, subject to constraints 4b-e, and otherwise drawn from the uniform distribution. Note that this must be possible given that $\{b_{i^*,j} : 1 \leq j \leq n\}$ is a satisfying assignment for $\phi$.

  (c) If input node $v$ has outgoing edge $e$ and label $z_j$, and $b_{i^*,j} = 1$, then compute $\alpha_{i^*,v} = r_{i^*,j} + s_{i^*,v} c_{i^*,e} \bmod q$.

  (d) If input node $v$ has outgoing edge $e$ and label $\bar{z}_j$, and $b_{i^*,j} = 0$, then compute $\alpha_{i^*,v} = r_{i^*,j} + s_{i^*,v} c_{i^*,e} \bmod q$.

**Proof of Claim 2**: The proof follows from Theorem 8 of Cramer et al. [13]. Here the underlying secret sharing scheme is the dual of the Benaloh-Leichter scheme [3].

**Proof of Claim 3**: Given knowledge of $\{b_{i',j}, r_{i',j} : 1 \leq j \leq n\}$ for some $i'$, Prover can compute $s_{i',v}$ for every satisfied input node $v$ with outgoing edge $e$: $s_{i',v} = c_{i',e}(r_{i',j} - \alpha_{i',v})$. (Here "satisfied" means that $v$ is labeled with $z_j$ and $b_{i',j} = 1$, or labeled with $\bar{z}_j$ and $b_{i',j} = 0$.) This equation allows also to compute the needed $\alpha_{i^*,v}$ for previously satisfied nodes $v$. Given these discrete logs, the Prover can fake the internal transcript for its computation as in the proof of Claim 1.

**Corollary 1**: Claims 1-4 remain true when some of the boolean variables have a fixed assignment that is known to both the Prover and Verifier. Prover and Verifier simply replace $\phi$ with a smaller formula that "hardwires" the assignment $\{b_j : j \in F\}$, and then proceed with the earlier protocol.

**Corollary 2**: The protocol can be modified for the case where $\phi$ is a boolean circuit but not a formula. This can be viewed as applying a standard transformation to the circuit to convert it to an equivalent formula, and then executing the original protocol on the resulting formula.

**Corollary 3**: When $\phi = (z_1 \vee \bar{z}_1)$, then the protocol is a commital deniable proof of knowledge of a decommital for one of $k$ committed bits. This can easily be modified into a commital deniable proof of knowledge for $\ell \leq k$ committed bits, by modifying Verifier's test (4a): All of the $(i, c_i)$ should lie on a degree

$k - \ell$ polynomial that passes through $(0, c)$. It is this version of the protocol that we will need in the next section for our deniable payment scheme.

**Corollary 4**: Consider the formula $\Phi(x_{1,1}, \ldots, x_{k,n}) = \phi(x_{1,1}, \ldots, x_{1,n}) \vee \ldots \vee \phi(x_{k,1}, \ldots, x_{k,n})$. Then our protocol can be viewed as a commital deniable proof of knowledge of certain decommitals of inputs to $\Phi$ that guarantee that it is satisfied. Our protocol can be modified to allow a prover to demonstrate this for any sufficient subset of decommitals of any formula (instead of just the $k$ partition subsets that we need for our applications).

# 5    Our Deniable Payment Protocol

In this section we will describe a protocol that allows a party to receive deniable payments. The protocol is practical. As in [1] we call this receiving party a Blind Trust and describe a five phase protocol how such a trust can be used to collect deniable donations for one candidate.

## 5.1    The Protocol

After an initial system setup phase, our protocol has five consecutive phases: (1) Pre-donation, (2) Cancellation, (3) Verification, (4) Deniability, and (5) Reimbursement.

**System Setup:** A trusted organization chooses a field $F_p$ such that $G_q$ is a cyclic subgroup of large prime order $q$ of the multiplicative group of $F_p$ and that $DLOG$ is hard in $G_q$. Furthermore generators $g, h$ of $G_q$ are chosen, s.t. $\log_g h$ is unknown. $p, q, g, h$ are made public.

1. **Pre-Donation:** Every party $D_i, 1 \leq i \leq l$ who would like to make a (deniable) donation of $d_i$ dollars to the candidate selects $d_i$ elements $r_{i,j} \in_R Z_q$ and $d_i$ elements $b_{i,j} \in_R Z_2$ and computes $y_{i,j} := g^{b_{i,j}} h^{r_{i,j}}$. He transfers via a non-anonymous, payment mechanism that has receipts (e.g., checks) the amount of $d_i$ dollars to the trust. He additionally sends the list of elements $y_{i,1} \ldots, y_{i,d_i}$ to the trust. The trust verifies the legitimacy of the origin of the received funds by traditional means and enters the fact that party $D_i$ made a pre-donation of amount $d_i$ and the elements $y_{i,1}, \ldots y_{i,d_i}$ into a public database. $D_i$ checks that his predonated amount $d_i$ was correctly entered into the database. (If this is not the case he can complain to a third a party using the receipt.) After the pre-donation phase is closed no further pre-donations are accepted.

2. **Cancellation:** A donor who wishes to "cancel" an amount of $c_i \leq d_i$ dollars of his pre-donation sends a message to the trust that contains the quadruples $(i, j, b_{i,j}, r_{i,j}), 1 \leq j \leq c_i$. The trust stores secretly all the quadruples it receives during this phase.

3. **Verification:** Assume the trust received $k$ quadruples of discrete logs of elements in the database D during the cancellation phase. In the verification

phase the trust proves with a commital deniable proof of knowledge that it knows decommitals for $k$ of the $d := d_1 + \ldots + d_l$ elements in the public database. The trust makes a payment of $d-k$ dollars to the candidate it collects the donations for. This uses the commital deniable proof of knowledge from the previous section, as modified in Corollary 3.

4. **Deniability:** All donors are required to reveal their secret values $b_{i,j}, r_{i,j}$ to the public.

5. **Reimbursement:** In the reimbursement phase each donor who made a cancellation can contact the trust and arrange for reimbursement, e.g., with electronic cash. Here it is important that a user undeniably identifies himself to the trust (e.g., in a personal contact and with a picture I.D.) to avoid impersonation attacks of a blackmailing candidate who tries to check if actually a donation was made by trying to get reimbursed.

**Theorem 2.** *Under the assumption of an untappable channel and that the trust is not pre-coerced, the protocol is receipt-free for the donor. It is incoercible for the trust without erasure after completion of the deniability phase. Every system participant can verify that the candidate did not receive less money than what a trust following the protocol honestly would have paid him. Furthermore the money the candidate obtains comes from legitimate sources.*

Before giving the proof sketch we make several remarks.

1. As the protocol is receipt-free for the donor, it defends against a blackmailing candidate as well as against an influence buying donor.

2. The receipt-freeness for the donor relies on the fact that he knows the representation of the elements he submitted during the pre-donation phase. Extra measures can be taken to assure this. E.g., if each donor holds a public/ secret key pair $(P_K, S_K)$ (of which one is sure the donors know the secret key), the protocol could require that the donor's additionally submit an encryption of the representations under their public key and a ZK proof that they encrypted the correct value. These data are additionally entered into the public database.

3. Using techniques from distributed cryptography [6,14] the trust can be distributed over several agencies that cooperate in the execution of the protocol.

4. As the donors and the values $g^b h^r$ they submitted are publicly known they can be forced to reveal their secret values in the revelation phase by external means.

5. Our cryptography based solution improves on the earlier physical implementation of [1] which does not offer verifiability. There it was suggested to achieve some form of auditability by having the trust keep all the records that could later (e.g., 10 years after the election) be publicly audited. Besides the delay, record keeping has further disadvantages. It would make the agency vulnerable to coercion, e.g., by the candidate after he won the presidential elections. Furthermore sensitive information about which donor cancelled would be revealed in the auditing process.

6. Our protocol improves on the physical implementation of deniable donations, where a donor steps into a donation booth and drops dollar bills into a donation box, as this implementation does not offer verifiability.

7. To minimize the information that can be derived about individual donations from the publicly known values $d_1, \ldots, d_l, k$, parties who have an interest in a well functioning deniable donation mechanism can deliberately predonate and cancel. (Note that if, e.g., $k$ were 0 it would be clear that each donor $D_i$ made an actual donation $d_i$.)

8. The complexity for construction and verification of the of the committal deniable proof depends linearly on the number of witnesses. Although this still seems to be feasible it requires significant computing power at the side of the trust and the parties that verify the proof. We think it would be interesting to improve the efficiency of the proofs (e.g., with probabilistic techniques).

9. Our protocol as it stands does not ensure that donors who cancelled get reimbursed. The trust may refuse to pay them. We sketch a variant of the protocol that prevents this: in the pre-donation phase donor $D_i$ sends a *pair* of commitments to values $(b_{i,j}, v_{i,j})$ for his $j$'th donated dollar to the trust which enters these pairs into the public database. In the cancellation phase the donor cancels his $j$'th dollar by sending as before $(b_{i,j}, r_{i,j})$ to the trust. Assuming the trust receives $k$ decommitals it pays $d - k$ dollars to the candidate. The reimbursement phase follows in this variant directly the cancellation phase. When $D_i$ obtains his $j$'th dollar back from the trust he reveals $(v_{i,j}, s_{i,j})$ in return to the trust ($s_{i,j}$ is the randomness used when committing to $v_{i,j}$). In the verifiability phase the trust proves with a commital deniable proof that it knows $k$ out of the $d$ pairs that were entered into the public database. Then follows the deniability phase where all donors are required to reveal their secret values $b_{i,j}, r_{i,j}, v_{i,j}, s_{i,j}$.

10. Our protocol does not protect against third party attempts (e.g., by a competing candidate) to force a donor to cancel his donation.

## Proof of Theorem 2 (Sketch)

Under the assumption that the donor has an untappable channel he could have sent an unnoticed cancellation notice to the trust. In particular he can not prove to the candidate that he did not send a cancellation message to the trust. A donor could be coerced to cancel a donation, but not to make a donation that cannot be canceled later. This gives receipt-freeness for the donor.

As the trust makes a sound proof of knowledge of $k$ out of $d$ representations during the verification phase public, and as the total amount of predonated funds $d$ is publicly known, the trust can not pay less money out to the candidate than $d$ minus the number of distinct secrets it received during the cancellation phase which proves public verifiability.

As we assumed the trust is not pre-coerced. Thus in particular the "random" bits used to produce the proof of knowledge in the verification phase were in fact chosen honestly chosen at random. After the deniability phase is completed the trust can "open" his proof of knowledge as coming from any $k$ elementary subset of the secret values by the previously proved properties of a commital deniable proof. This shows that the trust is later unable to prove any more information about who the actual donors (resp. the donated amounts were) than what can already be derived from the publicly known values $d_1, \ldots, d_l$ and $k$. As this holds even in the presence of accidentally or intentionally kept cancellation messages our protocol does not require erasure of cancellation messages.

As the trust accepts only, non-anonymous pre-donations the legitimacy of these funds and consequently also of the funds that get paid out to the candidate can be determined. This concludes the proof.

## 6    Deniability of Chaumian E-Cash

### 6.1    An Incoercible Payment System

In this section we study the deniability properties of anonymous electronic cash. We show how the basic anonymous ecash system can be made incoercible. Deniability can be seen as a much stronger privacy enhancing property than pure anonymity as it additionally preserves the privacy of payments under external coercion.

We briefly review the protocol for Chaumian ecash [11]. The bank has generated an RSA modulus $N$ and a public/secret key pair $(e, d)$. An electronic coin consists of a pair $(x, h(x)^d)$, where $h$ is a fixed hash function. During withdrawal the user A obtains a blind signature on $h(x)$ from the bank: A picks a random serial number $x$ and computes $h(x)$. A picks a random "blinding factor" $r$ and computes $m = r^e h(x)$. A sends $m$ to the bank. The bank computes the RSA signature $m^d$ on $m$ and sends $m^d$ back to A. A computes $r^{-1} m^d$ and has obtained an RSA signature on $h(x)$. During payment the user sends the coin $(x, h(x)^d)$ to the merchant who passes the coin on to the bank. The bank verifies the validity of the signature and that the coin has not been spent before. If both conditions are met the bank credits the merchant's account with the corresponding value and enters the serial number $x$ into its database of spent coins.

A user could be coerced by the bank, or by the government to reveal how he spent the coin obtained during a particular withdrawal session. Thus in order to make this protocol incoercible a user has to be able to "open" the message $m$ in a way that leads to a different coin than the one he actually withdrew. As the system is unforgeable, lying can not result in the presentation of a coin that has not in fact been obtained from the bank before. However we observe that a user can open the message $m = r^e h(x)$ he sent during withdrawal to come from any other coin $(y, h(y)^d)$ he is aware of, as the following simple algorithm shows:

FAKING-ALGORITHM:

**Input:** $r, h(x)^d, h(y)^d$
**Output:** an element $s$ , s.t. $s^e h(y) = r^e h(x)$
**Algorithm:**

  1. Compute: $s := \frac{r h(x)^d}{h(y)^d}$.
  2. Output $s$.

Thus the only modification needed to make the Chaumian ecash system incoercible is to require that the bank makes a list of all spent coins $(x, h(x)^d)$ public. Under coercion a user could then choose any coin of this list to open his withdrawal transcript.[1]

This incoercibility protects the privacy of payments even in the presence of a later coercion attempt. Another interpretation of this observation is that the classical version of Chaumian ecash does not allow users to prove or disprove how they spent their coins under external coercion, say if they are under investigation by the police.

### 6.2   Self-Coercion

The protocol is quite ineffective against a self-coercing user. A self-coercing user can deviate from the protocol by choosing his blinding factor $r$ not as a random value, but as the image of a cryptographically strong hash function H at a randomly chosen $t$. To prove now to somebody else that in fact a payment with a coin with serial number x was initiated by him, he presents his withdrawal record, the coin and the preimage t of the blinding factor r. He can no longer make his withdrawal record look like that of another coin, as this would require him to find a preimage of $rh(x)^d/h(y)^d$ under H. (Note that even the knowledge of the secret key d of the bank does not seem to help to create withdrawal records that can be opened as two different coins as the serial numbers are also images of hash functions.)

## 7   Conclusions

In conclusion, it is possible to use cryptographic methods to implement the radical campaign proposal of Ayres and Bulow. The building blocks of commital deniable proofs and deniable payment schemes are interesting in their own right, and may well find other applications. It would be interesting to find efficient deniable versions for other cryptographic applications, e.g., for anonymous remailing where the traffic from the client to the mix is observed by a potential future coercer.

## References

1. I. Ayres and J. Bulow, "The donation booth: Mandating donor anonymity to disrupt the market for political influence", Stanford Law Review 50(3), 1998.
2. D. Beaver, "Plug and play encryption", Crypto '97, pp. 75–89.

---

[1] We remark that extra measures can and should be taken to make this protocol more fault tolerant. One obvious attack could be that a bank could enter fake records into the list of spent coins and thereby catch a user who lies that he made that payment. To make lying of users more convincing it could be additionally required that the banks publishes in the database the recipient of the coin.

3. J. Benaloh and J. Leichter, "Generalized secret sharing and monotone functions", Crypto '88, pp. 27–35.

4. J. Benaloh and D. Tuinstra,"Receipt-free secret-ballot elections", ACM Symposium on the Theory of Computing, 1994, pp. 544–552.

5. J. Benaloh and D. Tuinstra,"Incoercible communication", Clarkson University Department of Mathematics and Computer Science Technical Report number TR-MCS-94-1. Feb. 94.

6. J. Benaloh and M. Yung, "Distributing the power of a government to enhance the privacy of voters", ACM Symposium on Principles of Distributed Computing, 1985.

7. R. Canetti, C. Dwork, M. Naor, and R. Ostrovsky, "Deniable encryption", Crypto '97, pp. 90–104.

8. R. Canetti, U. Feige, O. Goldreich, and M. Naor, "Adaptively secure computation", ACM Symposium on the Theory of Computing, 1996.

9. R. Canetti and R. Gennaro, "Incoercible multiparty computation", IEEE Foundations of Computer Science, 1996.

10. R. Canetti and R. Ostrovsky,"Secure computation with honest-looking parties: What if nobody is truly honest?", ACM Symposium on the Theory of Computing, 1999.

11. D. Chaum, "Blind signatures for untraceable payments", Crypto'82, pp. 199–203.

12. R. Cramer and I. Damgard, "Linear zero-knowledge – A note on efficient zero-knowledge proofs and arguments", ACM Symposium on the Theory of Computing, 1997, 436–445.

13. R. Cramer, I. Damgard, B. Schoenmakers, "Proofs of partial knowledge and simplified design of witness hiding protocols", Crypto '94, pp. 174–187.

14. O. Goldreich, S. Micali, and A. Wigderson, "How to play any mental game, or: A completeness theorem for protocols with honest majority", ACM Symposium on Theory of Computing, 1987, 218–229.

15. M. Hirt and K. Sako, "Efficient receipt-free voting based on homomorphic encryption", Eurocrypt 2000, pp.539–556.

16. V. Niemi and A. Renvall, "How to prevent buying of votes in computer elections", Asiacrypt'94, pp. 141–148.

17. T. Okamoto, "Receipt-free electronic voting schemes for large scale elections", Security Protocol Workshop '97, LNCS, pp. 25 –35.

18. K. Sako and J. Kilian, "Receipt-free mix-type voting schemes", Eurocrypt '95, pp. 393–403.

# Provably Secure Metering Scheme

Wakaha Ogata and Kaoru Kurosawa

Tokyo Institute of Technology,
2-12-1 O-okayama, Meguro-ku, Tokyo 152-8552, Japan
{wakaha, kurosawa}@ss.titech.ac.jp

**Abstract.** Naor and Pinkas introduced metering schemes at Eurocrypt '98 in order to decide on advertisement fees for web servers. In the schemes, any server should be able to construct a *proof* to be sent to an audit agency if and only if it has been visited by at least a certain number, say $k$, of clients. This paper first shows an attack for their schemes such that only two malicious clients can prevent a server from computing a correct proof. We next present provably secure metering schemes. Finally, an efficient robust secret sharing scheme is derived from our metering scheme.

## 1 Introduction

In the Internet, the amount of money paid to a web server from an advertisement company for hosting an ads should depend on the number of clients which have visited the server. A metering scheme is a protocol which measures this number.

We assume an audit agency as well as servers and clients. Any server should be able to construct a *proof* to be sent to the audit agency if and only if it has been visited by at least a certain number, say $k$, of clients during a certain time frame. It should be secure against fraud attempts by servers which inflate the number of their clients and against clients that attempt to disrupt the metering process.

A naive metering scheme could be implemented by using digital signature schemes; each client gives a digital signature to a server which confirms his visit when the clients visits the server. A server can present a list of the digital signatures as a *proof*. This system is, however, not efficient: both the size of the *proof* and the time to verify it are of the same order as $k$. Naor and Pinkas showed much more efficient metering schemes at Eurocrypt'98 [2].

This paper first shows an attack for their schemes such that only two malicious clients can prevent a server from computing a correct proof. We next present provably secure metering schemes, an unconditionally secure one and a computationally secure variant for multiple use under the computational Diffie-Hellman assumption.

We finally derive an efficient robust secret sharing scheme from our unconditionally secure metering scheme. In our robust secret sharing schemes, the size of shares is much smaller than those of the previous ones while the cheating probability is slightly larger.

T. Okamoto (Ed.): ASIACRYPT 2000, LNCS 1976, pp. 388–398, 2000.

# 2  Model and Goal

In the model of metering schemes, there exist *clients* (denoted by $C_i$), *servers* (denoted by $S_j$) and an *audit agency* (denoted by $A$). Each $S_j$ should be able to construct a *proof* to be sent to $A$ if and only if $k$ or more clients visit $S_j$ during a certain time frame.

Some clients and servers are malicious while the audit agency is honest. Assume that there exists an adversary which corrupts some clients and servers. Then our goal is to design metering schemes which satisfy the following two requirements.

**Security for servers** Suppose that $k$ or more clients visit a server $S_j$ during a time frame $t$. Then $S_j$ should be able to compute a *proof* with overwhelming probability even if the adversary corrupts all the clients and all the other servers than $S_j$.

**Security for the audit agency** Suppose that less than $k$ clients visit a server $S_j$ during a time frame $t$. Then the adversary should not be able to compute a *proof* with nonnegligible probability even if the adversary corrupts $k - 1$ clients and some number of servers.

It will be shown that the metering schemes of Naor and Pinkas do not satisfy the security for servers for an adversary who corrupts only two clients. On the other hand, the proposed metering schemes satisfy both of the above two requirements.

# 3  Metering Schemes of Naor and Pinkas

## 3.1  Basic Idea

The metering schemes of Naor and Pinkas are based on Shamir's secret sharing scheme. First, suppose that there exist a single server and a single time frame, and all the participants are honest. Then their basic scheme is described as follows. The audit agency chooses a random polynomial $f(x)$ of degree $k - 1$ over $GF(p)$. He gives each client $C_i$ a share $f(i)$. When a client visits a server, it gives it its share. When the server receives $k$ or more shares, he can compute $f(0)$ and it is the *proof* that he was visited by $k$ or more clients.

To accommodate many servers and many time frames, this scheme is generalized as follows. The audit agency chooses a random polynomial $P(x, y)$ over $GF(p)$ of degree $k - 1$ in $x$ and degree $d - 1$ in $y$. He gives $P(i, y)$ to each client $C_i$. When client $C_i$ visits server $S_j$ during time frame $t$, it gives $P(i, j \circ t)$ to $S_j$. When $S_j$ receives $k$ or more shares during time frame $t$, he can compute $P(0, j \circ t)$ and it is the *proof* that he was visited by $k$ or more clients.

This scheme is one-time use because the size of keys is proportional to the number of time frames for fixed $k$ and fixed number of servers.

### 3.2   Unconditionally Secure Scheme [2, Sec.3.3]

Naor and Pinkas then proposed the following unconditionally secure scheme to make the above scheme secure against malicious clients and servers. This is a one-time use scheme as mentioned above.

**Initialization:** The audit agency $\mathcal{A}$ chooses random polynomials $P(x, y), A(x, y)$ and $B(y)$ over $GF(p)$ such that
 - $P(x, y)$ is degree $k - 1$ in $x$ and degree $d - 1$ in $y$,
 - $A(x, y)$ is degree $c_k$ in $x$ and degree $c_d$ in $y$,
 - $B(y)$ is degree $c_d$ in $y$.

$\mathcal{A}$ computes

$$V(x, y) \stackrel{\triangle}{=} A(x, y)P(x, y) + B(y), \tag{1}$$

It then sends $(V(i, y), P(i, y))$ to $\mathcal{C}_i$ and $(A(x, j \circ 1), \ldots, A(x, j \circ T), B(j \circ 1), \ldots, B(j \circ T))$ to $\mathcal{S}_j$, where $\circ$ denotes concatenation of two strings.

**Interaction between client $\mathcal{C}_i$ and server $\mathcal{S}_j$:** To get a service from server $\mathcal{S}_j$ at time frame $t$, client $\mathcal{C}_i$ sends

$$(P(i, j \circ t), V(i, j \circ t))$$

to $\mathcal{S}_j$. The $\mathcal{S}_j$ checks if

$$V(i, j \circ t) = A(i, j \circ t)P(i, j \circ t) + B(j \circ t).$$

$\mathcal{S}_j$ offers a service to $\mathcal{C}_i$ if the above equality holds. Otherwise, $\mathcal{S}_j$ rejects.

**End of time frame:** If $\mathcal{S}_j$ has been visited by $k$ or more clients at time frame $t$, it can compute $P(0, j \circ t)$ from the received $P(i, j \circ t)$. The $P(0, j \circ t)$ is the *proof* that $\mathcal{S}_j$ has been visited by $k$ or more clients at time frame $t$. $\mathcal{A}$ who received $P(0, j \circ t)$ checks whether it is indeed $P(0, j \circ t)$.

### 3.3   Computationally Secure Scheme [2, Sec.3.5]

Naor and Pinkas further presented a computationally secure variant for multiple use under the computational Diffie-Hellman assumption.

Let $Z_p^*$ be the cyclic group modulo $p$, and let $g$ be a generator of a subgroup of $Z_p^*$ of order $q$, where $q$ is a prime.

**Initialization:** Similarly to the scheme of Sec.3.2, client $\mathcal{C}_i$ receives $P(i, y)$ and $V(i, y)$ and server $\mathcal{S}_j$ receives $A(x, j)$ and $B(j)$.

**Beginning of a time frame:** Each server receives a challenge $h = g^r$ from the audit agency, where $r$ is a random number.

**Interaction between client $\mathcal{C}_i$ and server $\mathcal{S}_j$:** To get a service from server $\mathcal{S}_j$, client $\mathcal{C}_i$ receives $h$ from $\mathcal{S}_j$ and sends

$$c_{i,j} \stackrel{\triangle}{=} (h^{P(i,j)}, h^{V(i,j)})$$

to server $\mathcal{S}_j$. $\mathcal{S}_j$ accepts $c_{i,j}$ if and only if

$$h^{V(i,j)} = (h^{P(i,j)})^{A(i,j)} h^{B(j)} \bmod p.$$

**End of time frame:** $\mathcal{S}_j$ can compute $h^{P(0,j)}$ if it has been visited by $k$ or more clients. The $h^{P(0,j)} (= g^{rP(0,j)})$ is the proof.

## 4    Attack for Naor and Pinkas Metering Schemes

In this section, we show an attack for both of the Naor and Pinkas metering schemes. In our attack, two clients, one who has special share and the other is arbitrary, can prevent a server from computing a proof. In other words, the security for servers is not satisfied. We describe our attack for the unconditionally secure scheme. It works for the computationally secure scheme similarly.

For some server $S_j$ and some time frame $t$, suppose that there exists two clients $C_{i_0}$ and $C_{i_1}$ such that

$$P(i_0, j \circ t) = 0$$
$$P(i_1, j \circ t) \neq 0.$$

Then, from Equation (1) they can compute $B(j \circ t)$ and $A(i_1, j \circ t)$ as follows.

$$V(i_0, j \circ t) = A(i_0, j \circ t)P(i_0, j \circ t) + B(j \circ t)$$
$$= B(j \circ t)$$
$$A(i_1, j \circ t) = (V(i_1, j \circ t) - B(j \circ t))/P(i_1, j \circ t)$$
$$= (V(i_1, j \circ t) - V(i_0, j \circ t))/P(i_1, j \circ t).$$

They next computes a random $(\tilde{P}, \tilde{V})$ such that

$$\tilde{P} \neq P(i_1, j \circ t)$$
$$\tilde{V} = A(i_1, j \circ t)\ \tilde{P} + B(j \circ t).$$

Finally, $C_{i_1}$ sends $(\tilde{P}, \tilde{V})$ to $S_j$ at time frame $t$ to get a service. Then $S_j$ accepts $(\tilde{P}, \tilde{V})$ because eq.(1) is satisfied.

However, at the end of time frame $t$. $S_j$ cannot compute the correct $P(0, j \circ t)$ even if it has been visited by $k$ clients because $\tilde{P} \neq P(i_1, j \circ t)$

## 5    Proposed Unconditionally Secure Metering Scheme

In this section, we present an unconditionally secure metering scheme which satisfies both the security for servers and the security for the audit agency. We assume that there are $T$ time frames. This scheme is one-time use and the size of keys is essentially proportional to $T$.

### 5.1    Proposed Scheme

**Initialization:** The audit agency $\mathcal{A}$ chooses a key polynomial $F(x, y, z)$ over $GF(p)$ with degree 1 in $x$, degree $d - 1$ in $y$ and degree $k - 1$ in $z$ randomly. He also chooses a random element $r_j \in Z_p \setminus \{0\}$ for each server $S_j$. Let

$$P_i(x, y) \stackrel{\triangle}{=} F(x, y, i)$$
$$A_j^t(z) \stackrel{\triangle}{=} F(r_j, j \circ t, z).$$

He then sends $c_i \stackrel{\triangle}{=} P_i(x, y)$ to client $C_i$ and $s_j \stackrel{\triangle}{=} (A_j^1(z), \ldots, A_j^T(z), r_j)$ to server $S_j$.

**Interaction between a client $C_i$ and a server $S_j$:** To get service from $S_j$ at time frame $t$, $C_i$ sends

$$c_{i,j}^t \stackrel{\triangle}{=} P_i(x, j \circ t)$$

to $S_j$. The $S_j$ checks if

$$A_j^t(i) = P_i(r_j, j \circ t).$$

$S_j$ offers a service to $C_i$ if the above equality holds. Otherwise, $S_j$ rejects.

**End of time frame:** Note that if $C_i$ visits $S_j$ during time frame $t$, then $S_j$ can compute

$$P_i(0, j \circ t) = F(0, j \circ t, i).$$

Therefore, if $S_j$ has been visited by $k$ or more clients during time frame $t$, then $S_j$ can compute $F(0, j \circ t, 0)$. The $F(0, j \circ t, 0)$ is the *proof* that $S_j$ has been visited by $k$ or more clients during time frame $t$.

The audit agency $A$ who received $F(0, j \circ t, 0)$ checks whether it is indeed $F(0, j \circ t, 0)$.

The scheme is illustrated in Fig. 1.

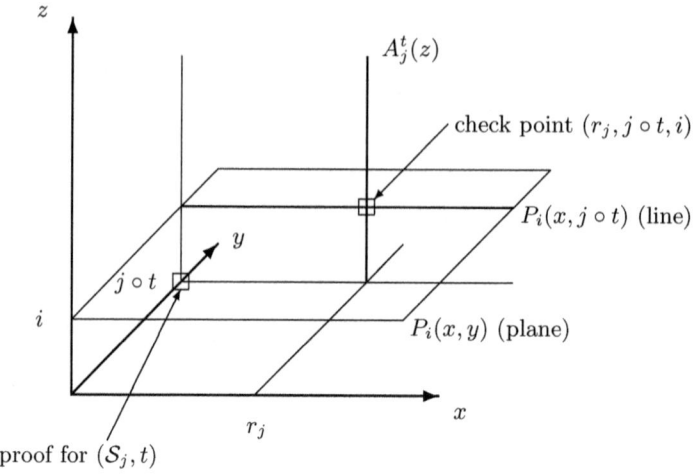

**Fig. 1.** Robust metering scheme

## 5.2   Security for Servers

In this subsection, we prove that the security for servers is satisfied for any infinitely powerful adversaries.

**Theorem 1.** *Suppose that $k$ or more clients visit server $\mathcal{S}_j$ during time frame $t$. Then $\mathcal{S}_j$ can compute the proof $F(0, j \circ t, 0)$ with probability more than $1 - 1/(p-1)$ for any adversary who corrupts all the clients and all the servers other than $\mathcal{S}_j$.*

*Proof.* Note that $\deg P_j(x, j \circ t) = 1$. Now at least one client $C_i$ must send $\tilde{P}(x)$ of degree 1 to $\mathcal{S}_j$ such that $\tilde{P}(x) \neq P_i(x, j \circ t)$ and $\tilde{P}(r_j) = P_i(r_j, j \circ t)$ to prevent $\mathcal{S}_j$ from computing the proof. For any fixed $\tilde{P}(x)$ such that $\tilde{P}(x) \neq P_i(x, j \circ t)$, we have $\tilde{P}(r_j) = P_i(r_j, j \circ t)$ with probability $1/(p-1)$ because $r_j$ is randomly chosen from $Z_p \setminus \{0\}$ and two lines intersect at one point. This holds for any adversary who corrupts all the clients and all the servers other than $\mathcal{S}_j$ because the adversary does not know $r_j$. $\qquad\square$

## 5.3   Security for the Audit Agency

In this subsection, we prove that the security for the audit agency is satisfied for any infinitely powerful adversaries.

**Theorem 2.** *Suppose that less than $k$ clients visit server $\mathcal{S}_j$ during time frame $t$. Then no adversary who corrupts $d/T$ servers and $k - 1$ clients can compute the proof $F(0, j \circ t, 0)$ for any $j$ and $t$ with probability more than $1/p$.*

*Proof.* Let $\alpha_s \overset{\triangle}{=} d/T$. Without loss of generality, we consider an adversary who corrupts $\alpha_s$ servers $\mathcal{S}_1, \ldots, \mathcal{S}_{\alpha_s}$ and $k - 1$ clients $\mathcal{C}_1, \ldots, \mathcal{C}_{k-1}$. The adversary tries to forge a false proof $\tilde{F}(0, \alpha_s \circ T, 0)$ for server $\mathcal{S}_{\alpha_s}$ and the last time frame $T$.

We assume that for any $j \leq \alpha_s$ and any $t \leq T$ such that $j \circ t \neq \alpha_s \circ T$, server $\mathcal{S}_j$ has been visited by $k$ or more clients during time frame $t$. Then all the information that the adversary has are (a) the initial secrets of the corrupted clients, (b) the initial secrets of the corrupted servers and (c) the information that the corrupted servers received from honest clients. (a) and (b) are

(a) $F(x, y, 1), \ldots, F(x, y, k - 1)$,
(b) $r_1, \ldots, r_{\alpha_s}$ and

$$\begin{cases} F(r_1, 1 \circ 1, z), & \ldots, & F(r_{\alpha_s - 1}, (\alpha_s - 1) \circ 1, z), & F(r_{\alpha_s}, \alpha_s \circ 1, z), \\ \quad\vdots & & \quad\vdots & \quad\vdots \\ F(r_1, 1 \circ T, z), & \ldots, & F(r_{\alpha_s - 1}, (\alpha_s - 1) \circ T, z), & F(r_{\alpha_s}, \alpha_s \circ T, z). \end{cases}$$

(c) is at most

$$\{F(x, j \circ t, i) \mid \text{ for all } i \geq k \text{ and for all } j \circ t \text{ such that } j \leq \alpha_s \text{ and } j \circ t \neq \alpha_s \circ T\}.$$

Suppose that the forged proof is $\beta$. For any value of $\beta$, we will show that there exists a key polynomial $\tilde{F}(x, y, z)$ which interpolates all the points of (a), (b), (c) and $\tilde{F}(0, \alpha_s \circ T, 0) = \beta$. This means that the probability that $\beta$ is the correct proof is $1/p$.

Note that $F(x, y, z)$ has degree 1 in $x$. Let $L_{\alpha_s \circ T}(x)$ be a line which interpolates $\tilde{F}(0, \alpha_s \circ T, 0) = \beta$ and $F(r_{\alpha_s}, \alpha_s \circ T, 0)$, where $F(r_{\alpha_s}, \alpha_s \circ T, 0)$ is obtained from (b). Next we can compute $F(0, j \circ t, 0)$ from (c) for all $j \circ t$ such that $j \leq \alpha_s$ and $j \circ t \neq \alpha_s \circ T$. Let $L_{j \circ t}(x)$ be a line which interpolates $F(0, j \circ t, 0)$ and $F(r_j, j \circ t, 0)$, where $F(r_j, j \circ t, 0)$ is obtained from (b).

Then we have $d$ lines

$$\{L_{j \circ t}(x) \mid j \leq \alpha_s \text{ and } t \leq T\}$$

because $\alpha_s T = d$. Next note that $F(x, y, z)$ has degree $d - 1$ in $y$. Let $B(x, y)$ be a polynomial of degree 1 in $x$ and degree $d - 1$ in $y$ such that $B(x, j \circ t) = L_{j \circ t}(x)$ for all $j \leq \alpha_s$ and $t \leq T$.

Finally, let $\tilde{F}(x, y, z)$ be a polynomial of degree 1 in $x$, degree $d - 1$ in $y$ and degree $k - 1$ in $z$ such that $\tilde{F}(x, y, 0) = B(x, y)$ and $\tilde{F}(x, y, i) = F(x, y, i)$ for $1 \leq i \leq k - 1$, where $F(x, y, i)$ is obtained from (a).

We have to show that $\tilde{F}(x, y, z) = F(x, y, z)$ for all the points of (a), (b) and (c). It is clear that the claim holds for (a). Next we prove the claim for (b). Fix $j \leq \alpha_s$ and $t \leq T$ arbitrarily. Then from our construction, it is easy to see that $\tilde{F}(r_j, j \circ t, i) = F(r_j, j \circ t, i)$ for $0 \leq i \leq k - 1$. Therefore,

$$\tilde{F}(r_j, j \circ t, z) = F(r_j, j \circ t, z). \tag{2}$$

Finally, we prove the claim for (c). Similarly to eq.(2), we can show that

$$\tilde{F}(0, j \circ t, z) = F(0, j \circ t, z). \tag{3}$$

From eq.(3), we have $\tilde{F}(0, j \circ t, i) = F(0, j \circ t, i)$. From eq.(2), we have $\tilde{F}(r_j, j \circ t, i) = F(r_j, j \circ t, i)$. Hence, it holds that $\tilde{F}(x, j \circ t, i) = F(x, j \circ t, i)$.    □

## 6    Proposed Computationally Secure Scheme

In this section, we present a computationally secure variant for multiple use under the computational Diffie-Hellman assumption.

### 6.1    Proposed Scheme

Let $Z_p^*$ be the cyclic group modulo $p$, and let $g$ be a generator of a subgroup of $Z_p^*$ of order $q$, where $q$ is a prime.

**Initialization:** Similarly to the scheme of Sec.5.1, the audit agency $\mathcal{A}$ chooses a key polynomial $F(x, y, z)$ over $GF(q)$ with degree 1 in $x$, degree $d - 1$ in $y$ and degree $k - 1$ in $z$ randomly. He also chooses a random element $r_j \in Z_q \setminus \{0\}$ for each server $\mathcal{S}_j$. Let

$$P_i(x, y) \stackrel{\triangle}{=} F(x, y, i), \quad A_j(z) \stackrel{\triangle}{=} F(r_j, j, z).$$

He then sends $P_i(x, y)$ to client $\mathcal{C}_i$ and $(A_j(z), r_j)$ to server $\mathcal{S}_j$.

**Beginning of a time frame :** At the beginning of time frame $t$, the audit agency $\mathcal{A}$ publishes a challenge $h_t = g^{u_t} \bmod p$, where $u_t$ is a random number.

**Interaction between a client and a server:** When client $\mathcal{C}_i$ gets a service from server $\mathcal{S}_j$ at time frame $t$, he computes

$$P_i(x, j) = a_{i,j} + b_{i,j}x,$$
$$d_{i,j}^t = h_t^{a_{i,j}} \bmod p$$
$$e_{i,j}^t = h_t^{b_{i,j}} \bmod p$$

He then sends $(d_{i,j}^t, e_{i,j}^t)$ to $\mathcal{S}_j$. The $\mathcal{S}_j$ accepts $(d_{i,j}^t, e_{i,j}^t)$ if and only if

$$h_t^{A_j(i)} = d_{i,j}(e_{i,j})^{r_j} \bmod p \ (= h_t^{P_i(r_j, j)} \bmod p).$$

**End of time frame:** Note that if $\mathcal{C}_i$ visits $\mathcal{S}_j$ during time frame $t$, then $\mathcal{S}_j$ obtains

$$d_{i,j}^t = h_t^{a_{i,j}} = h_t^{P_i(0,j)} = h_t^{F(0,j,i)}.$$

Therefore, if $\mathcal{S}_j$ has been visited by $k$ or more clients during time frame $t$, then $\mathcal{S}_j$ can compute $h_t^{F(0,j,0)}$ by using Lagrange formula. The $h_t^{F(0,j,0)}$ is the *proof* that $\mathcal{S}_j$ has been visited by $k$ or more clients during time frame $t$. The audit agency $\mathcal{A}$ who received $h_t^{F(0,j,0)}$ checks whether it is indeed $h_t^{F(0,j,0)}$.

## 6.2 Security for Servers

In our scheme, the security for servers is satisfied for any infinitely powerful adversaries.

**Theorem 3.** *Suppose that $k$ or more clients visit server $\mathcal{S}_j$ during time frame $t$. Then $\mathcal{S}_j$ can compute the proof $g^{u_t F(0,j,0)}$ with probability more than $1 - 1/(q-1)$ for any adversary who corrupts all the clients and all the servers other than $\mathcal{S}_j$.*

*Proof.* Similar to the proof of Theorem 1.                               □

## 6.3 Security for the Audit Agency

In this subsection, we consider probabilistic polynomial time adversaries who can corrupt $d$ servers and $k - 1$ clients.

**Theorem 4.** *Suppose that there exists an adversary $M_0$ who can compute the proof $h_t^{F(0,j,0)}$ for some $j$ and $t$ with nonnegligible probability. Then there exists a probabilistic polynomial time Turing machine $M_1$ which can solve the computational Diffie-Hellman problem.*

*Proof.* Without loss of generality, we assume that $M_0$ corrupts $d$ servers $\mathcal{S}_1,\dots,\mathcal{S}_d$ and $k-1$ clients $\mathcal{C}_1,\dots,\mathcal{C}_{k-1}$, and then compute the proof $h_T^{F(0,d,0)}$ for server $\mathcal{S}_d$ and time frame $T$ with nonnegligible probability.

Let the input to $M_1$ be $p, g, X(= g^\alpha \bmod p)$ and $Y(= g^\beta \bmod p)$. We will show that $M_1$ can generate a view of the adversary $M_0$ such that $\alpha = u_T$ and $\beta = F(0,d,0)$. Then by using $M_0$ as a subroutine, $M_1$ can obtain

$$h_T^{F(0,d,0)} = g^{u_T F(0,d,0)} = g^{\alpha\beta}$$

with nonnegligible probability. This means that $M_1$ can solve the computational Diffie-Hellman problem.

The view of the adversary $M_0$ consists of

(a) the initial secrets of the corrupted clients : $P_1(x,y),\dots,P_{k-1}(x,y)$,
(b) the initial secrets of the corrupted servers : $r_1,\dots,r_d$ and $A_1(z),\dots,A_d(z)$,
(c) the challenges : $h_1,\dots,h_T$,
(d) the information that the corrupted servers received from honest clients. This is at most

$$\{(d_{i,j}^t, e_{i,j}^t) \mid \text{for all } i \geq k \text{ and for all } (j,t) \text{ such that } j \leq d \text{ and } (j,t) \neq (d,T)\}.$$

Now $M_1$ generates (a),(b),(c) and (d) as follows.

(a) $M_1$ randomly chooses $P_1(x,y),\dots,P_{k-1}(x,y)$.
(b) $M_1$ randomly chooses $r_1,\dots,r_d$ and $A_1(0),\dots,A_d(0)$. These determine $A_1(z),\dots,A_d(z)$ because $\deg A_j(z) = k-1$ and

$$A_j(i) = P_i(r_j, j) \text{ for } 1 \leq i \leq k-1.$$

(c) $M_1$ randomly chooses $u_1,\dots,u_{T-1}$ and let $h_i = g^{u_i}$ for $1 \leq i \leq T-1$. Let $h_T = X$.

Finally, $M_1$ generates the elements of (d) as follows. $M_1$ has

$$Y = g^\beta = g^{F(0,d,0)}.$$

By slightly modifying the proof of Theorem 2, $M_1$ can compute $(a_{i,j}, b_{i,j})$ for $j < d$ and $(g^{a_{i,d}}, g^{b_{i,d}})$, where

$$a_{i,j} + b_{i,j}x = P_i(x,j) = F(x,j,i).$$

Therefore,

1. For $j = d$ and $t \leq T-1$, $M_1$ can compute $(d_{i,d}^t, e_{i,d}^t)$ from $(g^{a_{i,d}}, g^{b_{i,d}})$ and $u_t$.
2. For $j < d$ and $t = T$, $M_1$ can compute $(d_{i,j}^T, e_{i,j}^T)$ from $(a_{i,j}, b_{i,j})$ and $X = g^\alpha = g^{u_T}$.
3. For $j < d$ and $t \leq T-1$, $M_1$ can compute $(d_{i,j}^t, e_{i,j}^t)$ from $(a_{i,j}, b_{i,j})$ and $u_t$.

$\square$

# 7 New Robust Secret Sharing Scheme

## 7.1 Previous Schemes

In a $(k, n)$ threshold secret sharing scheme, a dealer distributes a secret $s$ to $n$ participants, $\mathcal{P}_1, \ldots, \mathcal{P}_n$ in such a way that $k$ or more participants can recover $s$ and $k - 1$ or less participants have no information on $s$. A piece of information held by $\mathcal{P}_i$ is called a share and it is denoted by $v_i$.

In the reconstruction phase, $\mathcal{P}_i$ opens $v_i$ if he is honest. However, he may lie about $v_i$ if he is a cheater. A secret sharing scheme is called robust if a cheater can be identified with overwhelming probability.

Let $S$ denote the set of secrets and $V_i$ denote the set of possible shares of $\mathcal{P}_i$. T.Rabin and Ben-Or [3] showed a robust $(k, n)$ threshold secret sharing scheme (RB scheme) such that

$$\log_2 |V_i| = (3n - 2) \log_2 |S|.$$

Carpentieri [1] showed a robust $(k, n)$ threshold secret sharing scheme such that

$$\log_2 |V_i| = (2n + k - 1) \log_2 |S|.$$

In these schemes, $n - 1$ cheaters cannot cheat an honest participant with probability more than $1/|S|$.

## 7.2 Proposed Scheme

In this section, we derive a new robust $(k, n)$ threshold secret sharing scheme from our metering scheme such that $|V_i|$ is much smaller than those of the previous schemes with slightly less cheater detection capability.

In our scheme,

$$\log_2 |V_i| = (2k + 1) \log_2 |S|.$$

and $k - 1$ cheaters cannot cheat an honest participant with probability more than $(k - 1)/(|S| - 1)$. Note that it is assumed that there are at most $k - 1$ cheaters instead of $n - 1$ cheaters. This is, however, not a problem because any $k$ participants (cheaters) can recover the secret.

Let $p$ be a prime and let $S \stackrel{\triangle}{=} GF(p)$.

**Distribution phase:** For a secret $s \in GF(p)$, the dealer chooses a bivariate random polynomial over $GF(p)$ such that

$$F(x, y) = \sum_{l=0}^{k-1} \sum_{m=0}^{k-1} a_{lm} x^l y^m$$

with $F(0, 0) = a_{00} = s$. He also chooses $n$ random elements $r_i \in Z_p \setminus \{0\}$. Let

$$B_i(x) \stackrel{\triangle}{=} F(x, i)$$
$$A_i(y) \stackrel{\triangle}{=} F(r_i, y),$$

The dealer then gives $v_i = (B_i(x), r_i, A_i(y))$ to the participant $\mathcal{P}_i$.

**Reconstruction phase:** Each $\mathcal{P}_i$ opens $B_i(x)$. Each other $\mathcal{P}_j$ accepts $B_i(x)$ if and only if

$$B_i(r_j) = A_j(i).$$

Note that $k$ or more correct $B_i(x)(= F(x, i))$ uniquely determine $s = F(0, 0)$.

**Theorem 5.** $k - 1$ *cheaters cannot cheat an honest participant with probability more than* $(k - 1)/(p - 1)$.

*Proof.* Suppose $k - 1$ participants $\mathcal{P}_1, \ldots, \mathcal{P}_{k-1}$ conspire and try to cheat $\mathcal{P}_k$. Suppose $\mathcal{P}_1$ opens $\tilde{B}(x)$ such that $\tilde{B}(x) \neq B_1(x)$. $\mathcal{P}_1$ succeeds if $\tilde{B}(r_j) = A_j(1)$. Since $\tilde{B}(x)$ can be written as $\tilde{B}(x) = B_1(x) + \Delta B(x)$ for some polynomial $\Delta B(x) \neq 0$ with degree $k - 1$ or less, this cheating probability is computed as follows.

$$
\begin{aligned}
\Pr[\tilde{B}(r_j) = A_j(1)] &= \Pr[B_1(r_j) + \Delta B(r_j) = A_j(1)] \\
&= \Pr[\Delta B(r_j) = 0] \\
&= |\{r \in Z_p \setminus \{0\} \mid \Delta B(r) = 0\}| \,/\, |\{r \in Z_p \setminus \{0\}\}| \\
&\leq (k - 1)/(p - 1)
\end{aligned}
$$

$\square$

# References

1. Carpentieri, M.: A perfect threshold secret sharing scheme to identify cheaters. Designs, Codes and Cryptography, **5** (1995) 183–187
2. Naor, M., Pinkas, B.: Secure and Efficient Metering. Proc. of Eurocrypt'98, Lecture Notes in Computer Science, **1403**, Springer Verlag (1998) 576–589
3. Rabin, T., Ben-Or, M.: Verifiable secret sharing and multiparty protocols with honest majority. Proc. 21st ACM Symposium on Theory of Computing (1989) 73–85
4. Shamir, A.: How to share a secret. Comm. ACM, **22** (1979) 612–613

# CRYPTREC Project
# - Cryptographic Evaluation Project for the Japanese Electronic Government -

Hideki Imai[1] and Atsuhiro Yamagishi[2]

[1] Institute of Industrial Science, The University of Tokyo,
Roppongi, Minato-ku, Tokyo 106-8558, Japan
`imai@iis.u-tokyo.ac.jp`
[2] Information-Technology Promotion Agency,Japan
Bunkyo Green Court Center Office
2-28-8 Honkomagome, Bunkyo-ku, Tokyo 113-6591, Japan
`a-yamagi@ipa.go.jp`

**Abstract.** We will describe the outline of the cryptographic technology evaluation project in Japan and those present conditions. The purpose of this project is that the cyptographic technology which the Japanese Government uses is evaluated and listed. Selected cryptographic technology will be used in the information security system which the Japanese Government will use in the future.

**Keywords.** Cryptographic technology, Symmetric ciphers, Asymmmetric ciphers,Evaluation

## 1 Background

Creating the common security basis is one of the most important tasks for the Japanese electronic government of which the infrastructure and primary systems will be constructed by FY 2003. Cryptographic techniques are particularly important and indispensable components of the electronic government because these not only provide information confidentiality and prevent information falsification, but also assure electronic authentication. Because of this importance, it has been pointed out domestically that the Japanese national government should adopt a cryptography usage policy in order to ensure that cryptography is integrated properly into the electronic government. Internationally, on the other hand, the ISO/IEC JTC1 has begun efforts aimed at standardizing cryptographic algorithms. CRYPTREC Project is an essential part of the MITI Action Plan for a Secure E- Government, which was announced by the Ministry of International Trade and Industry (MITI) in April 2000. MITI has entrusted the IPA with the implementation of this project.

## 2 Purposes and Project Implementation

The purpose of this project is to publish a technical report by the end of March 2001. It is to include a list of characteristics on cryptographic techniques that

T. Okamoto (Ed.): ASIACRYPT 2000, LNCS 1976, pp. 399–400, 2000.

will be proposed through a call for submission applicable to the Japanese electronic government. In order to make such a list, the action plan of the project contains investigation and evaluation of the proposed cryptographic techniques in terms of security, implementation and other characteristics from the objective viewpoints of various specialists. Four governmental offices (the Management and Coordination Agency, the Ministry of International Trade and Industry, the Ministry of Posts and Telecommunications and the Defense Agency, Japan) jointly organized CRYPTREC (CRYPTREC: CRYPTography Research & Evaluation Committee) to carry out this project. The committee, which is composed of prominent cryptography specialists in Japan, will evaluate the submitted cryptographic techniques.@ Cryptographic Techniques that will be evaluated are the four types of techniques considered indispensable in the electronic government:

**(1)** asymmetric cryptographic schemes
**(2)** symmetric ciphers
**(3)** hash functions
**(4)** pseudorandom number generators.

The evaluation will be conducted in two phases: the screening test phase and the detailed evaluation phase. The latter will be carried out on the proposals that have passed the screening tests. The evaluation guidelines are to be established by the committee.

Reports, including the evaluation results, will be compiled by the committee following due and proper consideration on fairness and transparency, and will be announced on web pages hosted by the IPA.

## 3    Status of Subscriptions

CRYPTREC received 48 proposals for Call for Cryptographic Techniques. The following table shows the number of submissions in each category. At present, CRYPTREC is evaluating these candidates from the viewpoints of both security and efficiency of implementation.

**Table 1.** Subscriptions

| Total Number of All Subscriptions | | 48 |
|---|---|---|
| Asymmetric Cryptographic Schemes | | 24 |
| | Confidentiality | 7 |
| | Authentication | 1 |
| | Signature | 10 |
| | Key-sharing | 6 |
| Symmetric Ciphers | | 19 |
| | Stream ciphers | 6 |
| | 64-bit block ciphers | 4 |
| | 128-bit block ciphers | 9 |
| Hash Functions | | 0 |
| Pseudo-random Number Generators | | 5 |

The detail of the project will be found at the following.
URL:http://www.ipa.go.jp/security/enc/CRYPTREC/index-e.html

# Anonymous Fingerprinting with Direct Non-repudiation

Birgit Pfitzmann and Ahmad-Reza Sadeghi

Universität des Saarlandes, Fachbereich Informatik,
D-66123 Saarbrücken, Germany
{pfitzmann, sadeghi}@cs.uni-sb.de

**Abstract.** Fingerprinting schemes support copyright protection by enabling the merchant of a data item to identify the original buyer of a redistributed copy. In asymmetric schemes, the merchant can also convince an arbiter of this fact. Anonymous fingerprinting schemes allow buyers to purchase digital items anonymously; however, identification is possible if they redistribute the data item.

Recently, a concrete and reasonably efficient construction based on digital coins was proposed. A disadvantage is that the accused buyer has to participate in any trial protocol to deny charges. Trials with direct non-repudiation, i.e., the merchant alone holds enough evidence to convince an arbiter, are more useful in real life. This is similar to the difference between "normal" and "undeniable" signatures.

In this paper, we present an equally efficient anonymous fingerprinting scheme with direct non-repudiation. The main technique we use, delayed verifiable encryption, is related to coin tracing in escrowed cash systems. However, there are technical differences, mainly to provide an unforgeable link to license conditions.

**Key words:** Fingerprinting, Digital Coin, Anonymity, Restrictiveness

## 1 Introduction

Protection of intellectual property in digital form has been a subject of research for many years and led to the development of various techniques. Fingerprinting schemes are an important class of these techniques. They are cryptographic methods applied to deter people from redistributing a data item by enabling the original merchant to trace a copy back to its original buyer. Dishonest buyers who redistribute the data item illegally are called traitors. The identifying information, called fingerprint, is embedded into copies of the original data item. The underlying watermarking techniques should guarantee that the embedded fingerprints are imperceptible and resistant to data manipulation as long as a traitor only uses one copy.

The first enhancement is collusion tolerance [BMP86, BS95, CKLS96], i.e., resistance even if traitors compare up to a certain number of different copies. A second addition is asymmetry [PS96a, PW97a, BM97]; here the merchant finds

T. Okamoto (Ed.): ASIACRYPT 2000, LNCS 1976, pp. 401–414, 2000.

an actual proof of the treachery in a redistributed copy, i.e., some data (similar to a signature "I redistributed") that only the identified buyer could have computed. The third addition is anonymity where buyers can stay anonymous in purchasing a fingerprinted data item. Only if they redistribute the data item, the identity is revealed. We mean anonymity in the strong sense of the original definition in [PW97b], i.e., any coalition of merchants, central parties and other buyers should not be able to distinguish purchases of the remaining buyers. A weak form can easily be achieved by using any asymmetric fingerprinting scheme under a certified pseudonym instead of a real identity. In the context of fingerprinting a distinction can be made whether one fingerprints the actual data item or a key for decrypting it. The latter, introduced in [CFN94], is typically called traitor tracing. Here we deal with anonymous asymmetric data fingerprinting with collusion tolerance.[1]

Anonymous fingerprinting was introduced in [PW97b], but only a construction using general theorems like "every NP-language has a zero-knowledge proof system" was presented there. In [PS99], an explicit construction based on digital coins was shown. It is fairly efficient in the sense that all operations are efficient computations with modular multiplications and exponentiations; however, at least in the collusion-tolerant case, the code needed for embedding is so long that the overall system cannot be called practical.

A remaining problem with the coin-based construction is that it does not offer direct non-repudiation, i.e., in the case of a dispute, the accused buyer has to participate in the trial to deny the charges if possible. Direct non-repudiation, where the merchant alone has enough information to convince any arbiter, is more useful in real life. This is obviously true when the buyer is not reachable. But it holds even if the accused buyer has to be found in any case for reasons outside the cryptographic system, e.g., for punishment, or simply because real-life trials require the accused person to be notified. The buyer could rightly or wrongly claim to have lost the information needed for the trial or the password to it, or it could happen that a dissolved company did not leave such information to its legal successors. The difference is similar to that between normal digital signatures (direct non-repudiation) and undeniable signatures [CA90] (signer needed in trial).

In this paper we remedy this problem. Our new construction is coin-based again and equally efficient as the previous one. The new part is based on methods from coin tracing, concretely [FTY96], in particular a technique we call delayed verifiable encryption. However, on the one hand the similarity is only at the technical level: recall that we do not require a trusted third party.[2] On the other hand, we need a closer binding between this encryption and the coin than in coin tracing to provide an unforgeable link to the license conditions.

---

[1] Omitting the collusion tolerance automatically makes the schemes significantly more efficient.

[2] Otherwise we could use the simple solution (weak form) mentioned above.

## 2    Overview of the Model

In this section, we briefly review the model of anonymous fingerprinting proposed in [PW97b]. It involves merchants $\mathcal{M}$, buyers $\mathcal{B}$, registration centers $\mathcal{RC}$ and arbiters $\mathcal{A}$. We assume that buyers can already digitally sign under their "real" identity $ID_\mathcal{B}$, i.e., that corresponding public keys $pk_\mathcal{B}$ have been distributed. Before the buyers can purchase fingerprinted data items, they must register with a registration center $\mathcal{RC}$. Registration centers will enjoy the minimum possible trust, i.e., the most a dishonest $\mathcal{RC}$ can do is to refuse a registration.[3] An arbiter $\mathcal{A}$ represents an arbitrary honest party who should be convinced by a proof.

The four main protocols of an anonymous fingerprinting scheme are registration, fingerprinting, identification, and trial. Besides, there are three protocols for registration center key distribution, where $\mathcal{RC}$ distributes specific parameters, data initialization, which a merchant carries out before the first sale of a specific data item, and enforced identification for the case where a merchant claims towards an arbiter that $\mathcal{RC}$ refuses to cooperate in identification.

The main security requirements on an anonymous fingerprinting scheme are the following (for more details see [PW97b] and [PS00] the section on security):

1. An honest merchant must be able to identify a traitor and win in the corresponding trial for every illegally redistributed copy of the data item he finds, unless the collusion is larger than the tolerated limit. The identified traitor may be $\mathcal{RC}$, in particular if it wrongly refuses identification. Moreover, even if there are more traitors, the merchant may want to be protected from damaging his reputation by making accusations and losing the trial. Hence it is required that if identification succeeds at all, he should also win the trial.

2. No honest buyer $\mathcal{B}$ or honest $\mathcal{RC}$ should be found guilty by an honest arbiter, not even if there are more traitors than the limit used in the security of the merchant. In particular, as some redistributions may be legal, a proof of redistribution must be unambiguously linked to a value *text* used during fingerprinting and typically designating the terms and conditions.

3. Purchases of honest buyers should not be linkable even by a collusion of all merchants, $\mathcal{RC}$, and other buyers.

## 3    General Ideas of Coin-Based Fingerprinting

In this section we recall the coin-based fingerprinting from [PS99]. The basic idea for using digital cash systems with double-spender identification to construct an anonymous fingerprinting scheme is as follows: Registration corresponds to withdrawing a coin. (The "coins" only serve as a cryptographic primitive and have no

---

[3] One may ask why $\mathcal{RC}$ is then needed, e.g., whether the merchants could not play this untrusted role themselves. However, buyers will only be anonymous among all people registered at the same registration center, and corresponding groups per merchant could be too small for meaningful anonymity.

monetary value.) During fingerprinting, the coin is given to the merchant, and in principle a first payment with this coin is made.[4] So far, the untraceability of the cash system should guarantee that the views of the registration center and the merchant are unlinkable. Then a second payment with the same coin is started. Now, instead of giving the buyer's response to the merchant, it is embedded in the data item. This embedding must be both secret and verifiable. After a redistribution, the merchant can extract the second response from the data item and carry out double-spender identification.

Apart from the efficient secret and verifiable embedding of the second payment response in the data, the main problem is the unambiguous link to a text describing the terms and conditions of the purchase that we required. Recall that in cash systems, double-spender identification has no such properties: the merchant simply obtains one fixed value $i$, called identity proof, independent of which coins were doublespent and how often. The first idea was to sign the text with a secret key whose corresponding public key $pk_{text}$ is included in the coin. However, the registration center, as the signer of the coins, can forge coins even in such a way that they can be linked to a certain withdrawal (where the buyer may have signed the withdrawal data). Hence the real problem is how to show that the particular coin with $pk_{text}$ is in fact one that the accused buyer has withdrawn. The solution idea in [PS99] was as follows: The buyer is able to repudiate an accusation with a wrong coin by presenting a different coin and the blinding elements that link it to the specific withdrawal from which this coin is supposed to come. For the case of Brands' payment system [Bra94], this was shown to be secure under a slightly stronger restrictiveness assumption than what would be needed for the pure payment system. Instead, we now want to give the merchant a direct proof that does not involve the buyer.

## 4    Ideas for Achieving Direct Non-repudiation

In this section we give an informal overview of the new construction with direct non-repudiation, i.e., where the merchant can convince an arbiter without participation of the accused buyer. As described in Section 3, we want to fix the actual terms and conditions $text$ by signing them with respect to a key $pk_{text}$ contained in the coin, and it remains to link this key unforgeably to a particular buyer after a redistribution.

The basic idea is to encrypt this coin key $pk_{text}$ during the registration, and such that the identity proof $i$ is the secret key needed for decryption. The buyer must sign this encryption $enc$ under his real identity so that he is bound to it. Hence, once the merchant learns $i$ due to a redistribution, it is possible to decrypt $enc$ and verify which coin key $pk_{text}$ the buyer planned to use. Note that the buyer is not needed in this step; this is essential for the direct non-repudiation.

---

[4] Actually the protocol is simpler, more like "zero-spendable" coins where the coin as such can be shown but any response to a challenge leads to identification. For intuitiveness, we nevertheless still call this response "second payment" in the informal part.

Each $i$ is only used for one coin so that the link between the particular coin and the corresponding encryption $enc$ will be clear. The next step is to force the buyer to encrypt the same $pk_{text}$ in $enc$ as he uses in the coin—clearly, if he can encrypt another value, his real coin will later not be attributed to him. Hence we need a kind of verifiable encryption. However, at this point there is nothing to verify the encryption against—$pk_{text}$ is deep inside the perfectly blinded coin. Here the ideas from coin tracing are applied, in particular from [FTY96] for Brands' cash scheme, where a similar problem exists with an encryption $enc^*$ for a trusted third party. The solution is to provide an additional specific encoding $M$ of $pk_{text}$ whose content is invariant under blinding. During registration (withdrawal), the buyer proves in zero-knowledge that $enc$ and $M$ have the same content. The registration center then blindly signs $M$ and the buyer transforms it to $M'$. Later, in fingerprinting (a payment), the merchant sees the real $pk_{text}$ used in the coin in clear. The buyer then opens the blinded encoding $M'$, which has the same content as $M$, and the merchant verifies that this content is really $pk_{text}$. Overall, this implies that also $enc$ contained the correct $pk_{text}$.

Apart from using the identity proof $i$ as a key instead of a trusted third party's key, we need another modification to this idea: In [FTY96], the coin and $M$ are blindly signed in two different signatures. If we did this, traitors could successfully attack the scheme by combining wrong pairs of coins and $M$'s. Hence we need a combined blind signature on the pair, where the pair can be uniquely decomposed both in the blinded and the unblinded form. Thus, while the coins and the encodings $M$ in [FTY96] are constructed using the same pair of generators in a discrete-logarithm setting, we use four generators and construct coins and $M$ using different pairs. The blind signature is made on the product. (More generators in conjunction with Brands' system have been used several times in the past, e.g., in [Bra93, BGK95, FTY98].) Restrictiveness of the blind signature scheme, together with proofs of knowledge that the values are formed over the correct generators, guarantees that a buyer cannot decompose the product in two non-corresponding ways at both sides. Here is also where the specific restrictiveness assumption comes in: The security of $\mathcal{RC}$ relies on the correct decomposition, and $\mathcal{RC}$ cannot trust the merchants to verify zero-knowledge proofs in fingerprinting correctly. Hence one aspect of the decomposition, (the fact that the buyer knows the discrete logarithm of $pk_{text}$ over the correct generator), is only substantiated by a Schnorr signature towards $\mathcal{RC}$. In our setting, even in the random oracle model we cannot easily define and prove this Schnorr signature to be a non-interactive proof of knowledge for lack of an initial common input and hence we have to accommodate for this immediately in the restrictiveness assumption, see Section 6.1. We believe that certain statements in papers on related coin systems must be formalized in the same way.

## 5   Construction

We now present the new construction step by step. There are no surprises given the informal description in the previous section. However, as there are no modular definitions for most components we use, and as we modify some of them internally, a concrete description of the overall system seems to be the easiest way to make everything precise and to get security proofs.

For simplicity, we assume that there is only one registration center. Once and for all, a group $G_q$ from a family of groups of prime order and generators $g$, $g_1, g_2, g_3, g_4 \in_R G_q \setminus \{1\}$ are selected. For concreteness, assume that $G_q$ is the unique subgroup of order $q$ of the multiplicative group $\mathbb{Z}_p^*$, where $p$ is another prime with $q|(p-1)$. Even $\mathcal{RC}$, who will typically make this choice, should not be able to compute discrete logarithms in $G_q$, and the generators must be truly random.[5] Hash functions $hash$ and $hash'$ for the underlying protocols (Brands and Schnorr signatures) must also be fixed. Finally, $\mathcal{RC}$ generates a secret signing key $x \in_R \mathbb{Z}_q^*$ and publishes the public key $h \equiv g^x \bmod p$.

## 5.1   Registration

An overview of the registration protocol is given in Figures 1 and 2. In the following, we relate the figures to the informal description and explain the correctness proof.

**1. Opening a one-time account.** $\mathcal{B}$ chooses the "identity proof" $i \in_R \mathbb{Z}_q^*$ randomly and secretly and computes $h_1 \equiv g_1^i$ (with $h_1 g_2 \neq 1$), the "account number" from Brands' system, and $h_3 \equiv g_3^i$, which we introduced specially as a public key for ElGamal encryption.

**2. Coin key and encryption.** The value $k$, also selected secretly and randomly by $\mathcal{B}$, serves as the secret coin key and $pk_{text} \equiv g_4^k \bmod p$ as the corresponding public key. $\mathcal{B}$ encrypts this public coin key into a ciphertext $enc$ using $h_3$ as the public key of ElGamal encryption. She computes a signature $sig_{coin} \leftarrow sig_{pk_\mathcal{B}}(h_1, h_3, enc)$ under her normal identity and sends it to $\mathcal{RC}$, who verifies it. This signature later shows that $\mathcal{B}$ is responsible for this "account" identified by the keys $h_1$ and $h_3$ and for the public key encrypted in $enc$.

**3. Encoding for delayed verifiable encryption.** The additional encoding of $pk_{text}$ is the pair $(M_1, M_2) = (g_3^j, pk_{text}^j)$ whose content is invariant under the following blinding operation. $\mathcal{RC}$ will verify that $M_1 \neq 1$. The content is uniquely defined because $M_1 \neq 1$ uniquely defines $j \neq 0$, and then $M_2$ and $j$ uniquely define $pk_{text}$.

**4. Correctness proofs.** Now $\mathcal{B}$ sends the public values to $\mathcal{RC}$ and gives certain correctness proofs. Intuitively, this is in particular that $h_1$ and $h_3$ contain the same identity proof $i$, and that the content of the encryption (which is uniquely defined given $h_3$) equals the content of the pair $(M_1, M_2)$ as defined above. Formally, $\mathcal{B}$ has to give a zero-knowledge proof of knowledge of the values $i, j, k, y$

---

[5] The randomness of the generators can be verified if $\mathcal{RC}$ proceeds as follows: Select a non-secret string $r$ of a certain length uniformly and randomly, e.g., by using an old random number table. Using $r$, generate primes $q$ and $p$ and elements $e_i \in \mathbb{Z}_p^*$ deterministically. Compute the generators as $g_i \equiv e_i^{(p-1)/q}$. If a $g_i$ is not a generator, repeat its choice.

such that the public values, i.e., $h_1, h_3, enc, M_1, M_2$ fulfill the prescribed equations.

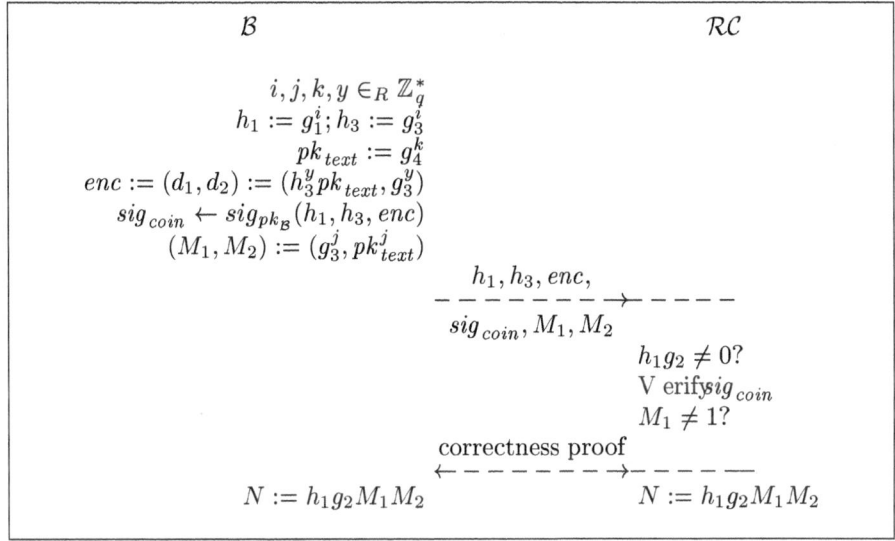

**Fig. 1.** The registration protocol before the blind signature

This can be done by using a simple protocol from [CEG88] for $i$ and the specific "indirect discourse proof" from [FTY96] for the remaining parameters. However, there is also a general efficient technique for proving low-degree polynomial relations in exponents [Cam98], Section 3.5, which comprises this and many similar situations. The protocol from [CEG88] for showing that $h_1$ and $h_3$ are correct is shown in [PS00]. Exactly the same type of proof is not possible for the other values because one equation is $M_2 = g_4^{jk}$, where neither $g_4^j$ nor $g_4^k$ can be public. Here is where the techniques for polynomials come in (e.g., Camenisch uses blinded versions of the required intermediate values, e.g., $g_5^r g_4^k$ to get back to the linear situation.).

**5. Withdrawal.** Now $\mathcal{RC}$ gives a blind signature on the combination of a coin and the encoding $(M_1, M_2)$. Let $m \equiv g_1^i g_2 = h_1 g_2$ be the value typically signed in Brands' scheme, $M \equiv M_1 M_2$, and $N \equiv mM$. This $N$ is the common input to the blind signing protocol (essentially from [CP93]). In [Bra94], an additional value is included in the hashing; we use $pk_{text}$ in that place. The resulting protocol is shown in Figure 2. As a result, $\mathcal{B}$ obtains the "coin" $coin' = (N', pk_{text}, \tau')$, where $N' \equiv (mM)^s$ and $\tau' = (z', a', b', r')$ is called the signature on $(N', pk_{text})$.[6] We denote the blinded versions of $m$ and $M$ by $m' \equiv m^s \equiv g_1^{is} g_2^s$ and $M' \equiv M^s \equiv g_3^{s'} pk_{text}^{s'}$, where $s' = sj$.

---

[6] In the sense of Section 4 this is not only the coin, but also still contains the blinded specific of $pk_{text}$. However, in the following, it is simpler to call this unit a coin.

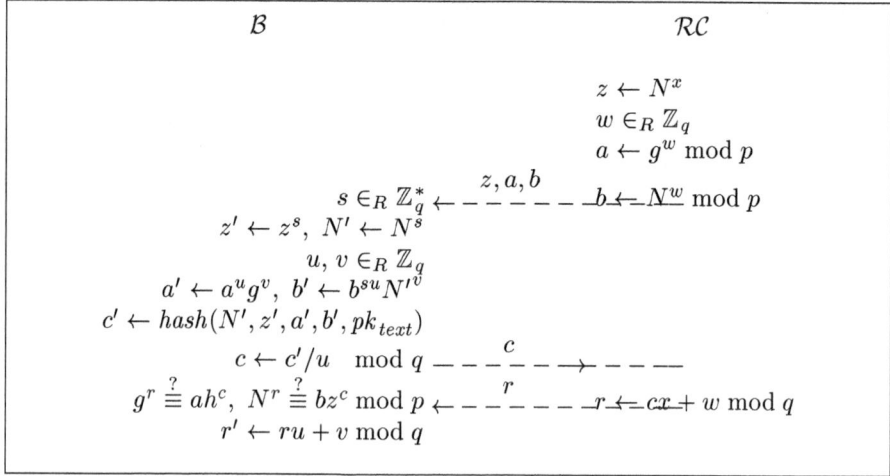

**Fig. 2.** The blind signature part of the registration protocol

## 5.2   Fingerprinting

The main common input in fingerprinting is the value *text* typically used to refer to the license conditions. We assume that each *text* is fresh for both buyer and merchant in this protocol, i.e., neither of them uses a value *text* twice. This can be achieved by a number of standard techniques.

**1. Text signing and coin verification.** $\mathcal{B}$ selects an unused coin $coin' = (N', pk_{text}, \tau')$. He uses the corresponding secret key $k$ to make a Schnorr signature $sig_{text}$ on $text$ (where we include $pk_{text}$ in the hashing) and sends $(coin', m', M', s', sig_{text})$ to $\mathcal{M}$. Now $\mathcal{M}$ first verifies the blind signature: He computes $c' \equiv hash(N', z', a', b', pk_{text}) \bmod q$ and tests whether $g^{r'} \equiv a'h^{c'}$ and $N'^{r'} \equiv b'z'^{c'} \bmod p$ hold. We say that a coin is valid if and only if it passes these tests. He then verifies $sig_{text}$ using $pk_{text}$ from $coin'$.

**2. Verification of decomposition.** $\mathcal{M}$ first verifies that $N' \equiv m'M'$, $N' \neq 1$ and $m' \neq 1$. Then $\mathcal{B}$ proves to $\mathcal{M}$ in zero-knowledge that he knows a representation of $m'$ with respect to $(g_1, g_2)$ and of $pk_{text}$ with respect to $g_4$ [CEG88].

**3. Delayed part of verifiable encryption.** $\mathcal{M}$ verifies whether $M' \equiv g_3^{s'} pk_{text}^{s'}$ holds. (Details why this verification is sufficient can be seen in the proof of the security of the registration center, see [PS00].)

**4. Embedding.** $\mathcal{B}$ takes the representation $(is, s)$ of $m' \equiv g_1^{is} g_2^{s}$ as the value *emb* to be embedded secretly and verifiably in the data item. This is the identical task as in [PS99] and thus from here on we can reuse the old protocol.

   For the overall security considerations later, note that in this protocol, additional commitments on $(is, s)$ are made. These are information-theoretically hiding discrete-logarithm commitments using generators chosen by the merchant and quadratic-residue commitments with respect to a number $n$ chosen by the buyer specially for this embedding. The rest are zero-knowledge protocols. Finally, the buyer decrypts quadratic-residue commitments provided by the merchant with respect to the buyer's $n$.

## 5.3 Identification

**1. Merchant-side retrievals.** $\mathcal{M}$ extracts a value $emb = (r_1, r_2)$ from the redistributed data item using the same extraction algorithm (consisting of a watermarking part and a decoding part) as in [PS99]. This pair should be $(is, s)$ with $s \neq 0$; thus he sets $s = r_2$ and $i = r_1/r_2$. He computes $m' \equiv g_1^{is} g_2^s \bmod p$ and uses it to retrieve $coin'$, $M'$, $text$ and $sig_{text}$ from the corresponding purchase record of the given data item. If any of these steps do not succeed, he gives up. (The collusion tolerance of the underlying code may be exceeded.) Otherwise he sends to $\mathcal{RC}$ the triple $proof_0 = (i, text, sig_{text})$.

**2. Registration center retrieval.** On input $proof_0$, the registration center searches in its registration database for a buyer who has registered the one-time account number $h_1 \equiv g_1^i$ and retrieves the values $(pk_{\mathcal{B}}, enc, sig_{coin})$, where $pk_{\mathcal{B}}$ corresponds to a real identity $ID_{\mathcal{B}}$. $\mathcal{RC}$ refuses identification if it is clear from $text$ that the redistribution was legal. Otherwise $\mathcal{RC}$ decrypts $enc$ using $i$ to obtain $pk_{text}$ and verifies that $sig_{text}$ is a valid signature on $text$ for this public key $pk_{text}$ with respect to the generator $g_4$. If positive, $\mathcal{RC}$ sends the retrieved values to $\mathcal{M}$.

**3. Merchant verification.** If $\mathcal{M}$ gets an answer $(pk_{\mathcal{B}}, enc, sig_{coin})$ from $\mathcal{RC}$, he first verifies that $sig_{coin}$ is a valid signature with respect to $pk_{\mathcal{B}}$ on the triple $(h_1 \equiv g_1^i, h_3 \equiv g_3^i, enc)$. He also verifies that $enc$ correctly decrypts to the value $pk_{text}$ contained in $coin'$ with respect to the secret key $i$ and the generator $g_3$. If one of these tests fails or $\mathcal{M}$ receives no answer, he starts enforced identification.

## 5.4 Enforced Identification

If $\mathcal{M}$ has to enforce the cooperation of $\mathcal{RC}$, he sends $proof_1 = (coin', s', i, s, text, sig_{text})$ to an arbiter $\mathcal{A}$. $\mathcal{A}$ verifies the validity of $coin'$ and calls its components $(N', pk_{text}, \tau')$ as usual. Then she verifies that $N' \equiv m'M'$ for $m' \equiv g_1^{is} g_2^s \bmod p$ and $M' \equiv g_3^{s'} pk_{text}^{s'}$. Finally, she verifies that $sig_{text}$ is a valid signature on $text$ for the public key $pk_{text}$ with respect to the generator $g_4$.[7]

If any of these tests fails, $\mathcal{A}$ rejects $\mathcal{M}$'s claim. Otherwise she sends $proof_0 = (i, text, sig_{text})$ to $\mathcal{RC}$ and requires values $(pk_{\mathcal{B}}, enc, sig_{coin})$. Then $\mathcal{A}$ verifies them as $\mathcal{M}$ does in Step 3 of identification.

---

[7] This is necessary for the security of $\mathcal{RC}$ by guaranteeing that the division of $N'$ into $m'$ and $M'$ is correct, even if $\mathcal{RC}$ is supposed to identify all redistributors independent of $text$.

## 5.5   Trial

Now $\mathcal{M}$ tries to convince an arbiter $\mathcal{A}$ that $\mathcal{B}$ redistributed the data item bought under the conditions described in *text*. The values $pk_{\mathcal{B}}$ and *text* are common inputs. Note that in the following no participation of $\mathcal{B}$ is required in the trial.

**1. Proof string.** $\mathcal{M}$ sends to $\mathcal{A}$ the proof string

$$proof = (coin', s', i, s, sig_{text}, enc, sig_{coin}).$$

**2. Verification of $i$.** $\mathcal{A}$ computes $h_1 \equiv g_1^i$ and $h_3 \equiv g_3^i \bmod p$ and verifies that $sig_{coin}$ is a valid signature on $(h_1, h_3, enc)$ with respect to $pk_{\mathcal{B}}$. If yes, it means that $i$, the discrete logarithm of an account number $h_1$ for which $\mathcal{B}$ was responsible, has been recovered by $\mathcal{M}$ and thus, as we will see, $\mathcal{B}$ has redistributed some data item. It remains to verify the link to the terms and conditions described by *text*.

**3. Verification of *text*.** $\mathcal{A}$ verifies the validity of $coin'$ and calls its components $(N', pk_{text}, \tau')$. She then verifies that $N' \equiv m'M'$ for $m' \equiv g_1^{is}g_2^s \bmod p$ and $M' = g_3^{s'}pk_{text}^{s'}$.[8] She also verifies the signature $sig_{text}$ on the disputed text with respect to $pk_{text}$ and the generator $g_4$. These verifications imply that if the accused buyer owned this coin, he must have spent it in the disputed purchase on *text*. Finally, $\mathcal{A}$ verifies that this coin belongs to $\mathcal{B}$: She tests whether $enc$ correctly decrypts to $pk_{text}$ if one uses $i$ as the secret key. If all verifications are passed, $\mathcal{A}$ finds $\mathcal{B}$ guilty of redistribution, otherwise $\mathcal{M}$ should be declared as the cheating party.

# 6   Security

Due to space restrictions we leave out the proofs and only highlight the security aspects which in our belief are of more theoretical importance. Hence we omit the security for buyers and merchants and sketch security for $\mathcal{RC}$ and buyer's anonymity. For analysis of all security issues we refer the interested reader to [PS00].

## 6.1   Security for the Registration Center

The security requirement is that if the registration center is honest, an honest arbiter will never output that $\mathcal{RC}$ is guilty.

For this, we need the restrictiveness of the underlying blind signature scheme for showing that the value $m'$ used in fingerprinting "contains" the same value $i$ as the original $m$, and also that the delayed verification of $pk_{text}$ works. In [Bra94], Brands only works with two generators $g_1, g_2$, while we use four. However, in the underlying report [Bra93] the same assumptions are made and heuristically explained for any number of generators $g_1, \dots, g_n$, and coin systems with more than two generators have also been presented in [BGK95, FTY98]. The exact assumption we need is the following:

---

[8] The latter verification is not essential, but otherwise $\mathcal{M}$ must include $M'$ in *proof*.

**Assumption 1** *(Restrictiveness with Schnorr signature). Let $A$ be a probabilistic polynomial-time adversary that can interact with a Brands signer as in Figure 2 several times for messages $N$ of its choice. $A$ also has to output representations of all these messages, i.e., quadruples $(i_1, \ldots, i_4)$ such that*

$$N = g_1^{i_1} \cdots g_4^{i_4}.$$

*At the end, $A$ has to output a message (coin) $N'$ with a valid signature and a representation of $N'$, except that it need not show $i'_4$, but only values $(h'_4, i''_4, msg, \sigma)$ such that $N' = g_1^{i'_1} \cdots h_4'^{i''_4}$ and $\sigma$ is a valid Schnorr signature on $msg$ for the public key $h'_4$ and the generator (with $h'_4$ included in the hashing). We then define $i'_4$ as $i''_4 \log_{g_4}(h'_4)$.*

*Then the probability that $A$ fulfills all the conditions and that the vector $(i'_1, \ldots, i'_4)$ is not a scalar multiple of one of the vectors $(i_1, \ldots, i_4)$ is negligible. (The probability is taken over the random choices of the signer and $A$.)*

**Discussion of the assumption.** In a simpler restrictiveness assumption, the adversary has to output complete representations of both the blinded and unblinded values, i.e., also $i'_4$. In our case, he only outputs a factor $i''_4$ of $i'_4$ and, instead of the other factor $k := i'_4/i''_4$, a Schnorr signature with respect to the corresponding public key $h'_4 = g_4^k$. The intuitive idea why this should be secure is that a Schnorr signature should be a non-interactive proof of the knowledge of the secret key. Such arguments are mentioned, e.g., in [Bra94] (Corollary 9) and [FTY96, FTY98]. However, really trying to prove our assumption from the simpler one, even in the random oracle model, leads to problems. First, the given situation does not fall under the most obvious way to define Schnorr signatures to be non-interactive zero-knowledge proofs in the random oracle model: One would take $g, h$ as common inputs and an extractor, allowed to simulate the random oracle (in a way indistinguishable for the adversary) would have to extract the secret $x$ with $h = g^x$. Under this definition, it is easy to prove that Schnorr signatures are proofs of knowledge. However, in our situation and many others where a non-interactive proof is needed, $h$ is not a common input, but chosen by the adversary in the same step as the signature serving as proof. Hence as to definitions, it is not clear what $x$ the extractor should extract—simply producing pairs $(x, h)$ with $h = g^x$ is trivial. The definition must therefore be made with respect to a scenario, i.e., in a joint probability space together with other variables. We can, e.g., define that the extractor must output pairs $(x, h)$ where $h$ has the same joint distribution with the other variables as the values $h$ output by the adversary.

Now, if the scenario is non-interactive, one can still prove the desired theorem by using the forking lemma from [PS96b] if one includes $h$ into the hashing in the Schnorr signature. However, in our scenario the adversary interacts with the bank as blind signer, in addition to the random oracle. This gives the same problems with exponential rewinding as in [PS96c] and [SG98], Section 2.4. It may be interesting to investigate how to modify either the proof techniques or the scheme so that some proof of this type goes through, but for the moment we had to make the stronger assumption.

In our scheme, $\mathcal{RC}$ could only be found guilty in enforced identification, because in a trial an honest arbiter $\mathcal{A}$ only finds either $\mathcal{B}$ or $\mathcal{M}$ guilty. Under Assumption 1 we can prove, as shown in [PS00] in detail, the security for an honest $\mathcal{RC}$ with an honest arbiter.

## 6.2   Anonymity

We assume that $\mathcal{RC}$ and $\mathcal{M}$ collude and both may deviate from their protocols, hence we call them $\mathcal{RC}^*$ and $\mathcal{M}^*$. We want to show that they learn nothing about the purchase behaviour of honest buyers, except for facts that can simply be derived from the knowledge of who registered and for what number of purchases, and at what time protocols are executed. This should even hold for the remaining purchases of a buyer if $\mathcal{RC}^*$ and $\mathcal{M}^*$ obtain some data items this buyer bought.

In our construction, the only information common to all registrations of a buyer is her global key pair $(sk_{\mathcal{B}}, pk_{\mathcal{B}})$ (recall that we use each $i$ only once). She only uses it to generate the signature $sig_{coin}$, and uses neither the keys nor this signature in fingerprinting. Thus other fingerprintings and possible redistributions of a buyer are statistically independent of one registration and the corresponding fingerprinting. Hence we focus on the question whether $view_{reg}$ and $view_{fing}$ from such a pair of corresponding protocols are linkable. For this, we let an adversary carry out two registrations and then the two corresponding fingerprintings in random order. The adversary is considered successful if it can guess with probability significantly better than $1/2$ which views correspond to each other.

More precisely, first the global parameters are generated (the group and generators in our construction), given a security parameter $l$. Then the two buyers generate their global keys. Next, the registration protocol $reg$ is run where $\mathcal{RC}^*$ inputs the buyer's public key and the buyer $\mathcal{B}$ her secret key. The outputs are $\mathcal{RC}^*$'s view and $\mathcal{B}$'s view $view_{\mathcal{B}}$. For $\mathcal{RC}^*$'s view we write $(traf_{reg}, aux_i)$, where $traf_{reg}$ ("traffic" in slight abuse of the term) denotes the messages from $\mathcal{B}$ to $\mathcal{RC}^*$, while the variables $aux_i$ model the adversary's entire memory between protocol executions. Now a bit $b$ is uniformly chosen; it denotes on which registration the first execution of fingerprinting is based, assuming that the registrations succeeded from the buyers' point of view. The notation for the fingerprinting protocol $fing$ is similar to that for $reg$. Finally, the adversary algorithm $A_{Link}$ outputs a guess $b^*$ for $b$ based on the adversary's memory, which may of course contain the traffic. The values sent by $\mathcal{B}$ are (for simplicity we included $pk_{\mathcal{B}}$ in $traf_{reg}$):

$$traf_{reg,0} = (pk_{\mathcal{B},0}, h_{1,0}, h_{3,0}, M_{1,0}, M_{2,0}, enc_0, sig_{coin,0}, c_0, traf_{reg,0}^{ZKP}),$$

$$traf_{fing,b} = (coin_b', m_b', M_b', s_b', sig_{text,b}, traf_{embed,b}, traf_{fing,b}^{ZKP}),$$

and similarly for $traf_{reg,1}$ and $traf_{fing,\bar{b}}$. Here $c_0$ is the only value sent in the withdrawal subprotocol, $coin_b' = (N_b', pk_{text,b}, \tau_b')$ the coin, $traf_{embed,b}$ the traffic from Step 4 of fingerprinting and $traf_{reg,0}^{ZKP}$, $traf_{fing,b}^{ZKP}$ that from all zero-knowledge protocols in registration and fingerprinting. The texts to be signed may be chosen adaptively by $\mathcal{M}^*$ in *fing*.

We can prove, as shown in detail in [PS00], that given a successful adversary as defined above, there are also successful adversaries in successive scenarios where the "buyer" sends fewer and fewer values. This finally leads to a contradiction. The anonymity of our scheme is based on the following assumption and the random oracle model for the hash function used in the blind signature protocol:

**Assumption 2** *(Strong Decisional Diffie-Hellman Assumption). No probabilistic polynomial-time algorithm $A_{SDDH}$, on inputs of the form*

$$(g, g^x, g^y, g^{y^{-1}}, u)$$

*where $u$ is either $g^{xy}$ or a random group element, can distinguish the two cases with probability significantly better than $1/2$.*

## Acknowledgment

We thank Mihir Bellare, Victor Shoup, Michael Steiner and Michael Waidner for interesting discussions.

## References

[BGK95]   Ernest Brickell, Peter Gemmell, David Kravitz: Trustee-based Tracing Extensions to Anonymous Cash and the Making of Anonymous Change; SODA 1995, ACM Press, New York 1995, 457-466.

[BM97]    Ingrid Biehl, Bernd Meyer: Protocols for Collusion-Secure Asymmetric Fingerprinting; STACS 97, LNCS 1200, Springer-Verlag, Berlin 1997, 399-412.

[BMP86]   G. R. Blakley, Catherine Meadows, George B. Purdy: Fingerprinting Long Forgiving Messages; Crypto'85, LNCS 218, Springer-Verlag, Berlin 1986, 180-189.

[Bra93]   Stefan Brands: An Efficient Off-line Electronic Cash System Based On The Representation Problem; Centrum voor Wiskunde en Informatica, Report CS-R9323, March 1993.

[Bra94]   Stefan Brands: Untraceable Offline Cash in Wallet with Observers; Crypto'93, LNCS 773, Springer-Verlag, Berlin 1994, 302-318.

[BS95]    Dan Boneh, James Shaw: Collusion-Secure Fingerprinting for Digital Data; Crypto'95, LNCS 963, Springer-Verlag, Berlin 1995, 452-465.

[CA90]    David Chaum, Hans van Antwerpen: Undeniable Signatures; Crypto'89, LNCS 435, Springer-Verlag, Berlin 1990, 212-216.

[Cam98]   Jan Camenisch: Group Signature Schemes and Payment Systems Based on the Discrete Logarithm Problem; Hartung-Gorre Verlag, Konstanz 1998.

[CEG88]   David Chaum, Jan-Hendrik Evertse, Jeroen van de Graaf: An improved protocol for demonstrating possession of discrete logarithms and some generalizations; Eurocrypt'87, LNCS 304, Springer-Verlag, Berlin 1988, 127-141.

[CFN94]    Benny Chor, Amos Fiat, Moni Naor: Tracing traitors; Crypto'94, LNCS 839, Springer-Verlag, Berlin 1994, 257-270.

[CKLS96]   Ingemar Cox, Joe Kilian, Tom Leighton, Talal Shamoon: A Secure, Robust Watermark for Multimedia; Information Hiding, LNCS 1174, Springer-Verlag, Berlin 1996, 185-206.

[CP93]     David Chaum, Torben Pryds Pedersen: Wallet Databases with Observers; Crypto'92, LNCS 740, Springer-Verlag, Berlin 1993, 89-105.

[FTY96]    Yair Frankel, Yiannis Tsiounis, Moti Yung: "Indirect Discourse Proofs": Achieving Efficient Fair Off-Line E-cash; Asiacrypt'96, LNCS 1163, Springer-Verlag, Berlin 1997, 287-300.

[FTY98]    Yair Frankel, Yiannis Tsiounis, Moti Yung: Fair Off-Line e-Cash Made Easy; Asiacrypt'98, LNCS 1514, Springer-Verlag, Berlin 1998, 257-270.

[PS96a]    Birgit Pfitzmann, Matthias Schunter: Asymmetric Fingerprinting; Eurocrypt'96, LNCS 1070, Springer-Verlag, Berlin 1996, 84-95.

[PS96b]    David Pointcheval, Jacques Stern: Security proofs for signature schemes; Eurocrypt'96, LNCS 1070, Springer-Verlag, Berlin 1996, 387-398.

[PS96c]    David Pointcheval, Jacques Stern: Provably Secure Blind Signature Schemes; Asiacrypt'96, LNCS 1163, Springer-Verlag, Berlin 1996, 252-265.

[PS99]     Birgit Pfitzmann, Ahmad-Reza Sadeghi: Coin-Based Anonymous Fingerprinting; Eurocrypt'99, LNCS 1592, Springer-Verlag, Berlin 1996, 150-164.

[PS00]     Birgit Pfitzmann, Ahmad-Reza Sadeghi: Anonymous Fingerprinting with Direct Non-Repudiation; Cryptology ePrint Archive, Report 2000/032, http://eprint.iacr.org/, 2000.

[PW97a]    Birgit Pfitzmann, Michael Waidner: Asymmetric Fingerprinting for Larger Collusions; 4th ACM CCS, ACM Press, New York 1997, 151-160.

[PW97b]    Birgit Pfitzmann, Michael Waidner: Anonymous Fingerprinting; Eurocrypt'97, LNCS 1233, Springer-Verlag, Berlin 1997, 88-102.

[Sch91]    Claus-Peter Schnorr: Efficient Signature Generation by Smart Cards; Journal of Cryptology 4/3 (1991) 161-174.

[SG98]     Victor Shoup, Rosario Gennaro: Securing threshold cryptosystems against chosen ciphertext attack; Eurocrypt'98, LNCS 1403, Springer-Verlag, Berlin 1998, 1-16.

# Efficient Anonymous Fingerprinting with Group Signatures

## (Extended Abstract)

Jan Camenisch

IBM Research
Zurich Research Laboratory
CH–8803 Rüschlikon
jca@zurich.ibm.com

**Abstract.** Fingerprinting schemes enable a merchant to identify the buyer of an illegally distributed digital good by providing each buyer with a slightly different version. Asymmetric fingerprinting schemes further prevent the merchant from framing a buyer by making the fingerprinted version known to the buyer only. In addition, an anonymous fingerprinting scheme allows the buyer to purchase goods without revealing her identity to the merchant. However, as soon as the merchant finds a sold version that has been (illegally) distributed, he is able to retrieve a buyer's identity and take her to court.

This paper proposes a new and more efficient anonymous fingerprinting scheme that uses group signature schemes as a building block. A byproduct of independent interest is an asymmetric fingerprinting scheme that allows so-called two-party trials, which is unmet so far.

## 1 Introduction

Today's computer networks allow the trading of digital goods in an easy and cheap way. However, they also facilitate the illegal distribution of (copyrighted) data. Fingerprinting schemes are a method for supporting copyright protection. The idea is that a merchant sells every customer a slightly different "copy" of the good. For instance, in the case of an image, the merchant could darken or lighten some pixels. Of course, the fingerprint must be such that a buyer cannot easily detect and remove it. When the merchant later finds an illegally distributed copy, he can recognize the copy by its fingerprints and then hold its buyer responsible. A number of authors (cf. [8]) have studied methods to achieve this for various kinds of digital goods. Research is ongoing in this area.

Whereas fingerprinting as such is a technique that was already used in the previous century, security against colluding buyers was achieved only recently [2,3]. Such schemes tolerate a collusion of buyers up to a certain size, i.e., a collusion cannot produce a copy such that the merchant cannot trace it back to at least one of the colluders. The first such schemes that were proposed are *symmetric*, meaning that the merchant knows which copy a buyer gets [2,3].

T. Okamoto (Ed.): ASIACRYPT 2000, LNCS 1976, pp. 415–428, 2000.

Thus a malicious merchant could spread himself the version sold to some buyer and then accuse that buyer of having done so.

This problem is overcome by *asymmetric* schemes [1,12,14]. Here, the buyer sends the merchant a commitment to a secret she chose. Then the two carry out a protocol at the end of which the buyer possesses the desired digital good, fingerprinted with the chosen secret, whereas the merchant does not learn anything. Hence, whenever the merchant is able to present a sufficiently large fraction of the secret contained in a buyer's commitment, he must have found the copy a buyer bought (and distributed) and the buyer is therefore considered guilty.

In both symmetric and asymmetric schemes, the merchant needs to know a buyers' identity to be able to take her to court if she distributes the purchased copy. To protect buyers' privacy and match with anonymous digital payment systems, Pfitzmann and Waidner introduce *anonymous* asymmetric fingerprinting [13]. Here, a buyer must no longer identify herself for purchasing and remains anonymous as long as she keeps the purchased good secret, i.e., does not distribute it. More precisely, the merchant can learn a buyer's identity only if he obtains her purchased copy. This kind of scheme involves a further party, called registration center, at which all buyers are required to register prior to any purchase. Pfitzmann and Waidner also provide a general modular construction consisting of two building blocks. One handles the registration of buyers and the generation of the to-be-embedded information and the other building block is a method to embed committed information into the to-be-sold data. More precisely, the latter uses an error and erasure-correcting code together with an asymmetric fingerprinting scheme to guarantee that at least for one of the colluders *all* her committed secret bits can be extracted from a copy found.

The first building block uses general zero-knowledge proof techniques and renders the resulting scheme rather inefficient and hence it is merely considered a "proof of existence" [13]. The second building block can be realized efficiently in term of computations [11]. However, the use of the error and erasure-correcting code prohibitively enlarges the number of bits that need to be embedded.

Recently, Pfitzmann and Sadeghi [11] presented an efficient replacement for the first part of this construction. It is derived from the anonymous e-cash scheme by Brands [4]. More precisely, it uses its property that coins are anonymous when spent once but reveal a user's identity when spent twice. However, the resulting scheme has the drawback that a buyer must register once for each purchase and that the merchant has to contact the registration center to retrieve the identity of a malicious buyer.

This paper presents an anonymous fingerprinting scheme that overcomes these drawbacks using group signature schemes as its main building block. A *group signature scheme* (e.g., [7,9]) allows a member of a group of users to sign a message on the group's behalf. The scheme protects the privacy of signers in that the verifier has no means to determine which member originated a signature or whether two signatures stem from the same signer. However, to handle special cases of misuse by some user, there is a designated *revocation manager* who can indeed find the identity of a signature's originator.

The idea underlying our fingerprinting scheme is to have the buyer issuing a group signature on a message describing the deal. Opposed to an ordinary group signature scheme there is no (fixed) revocation manager. Instead, the buyer chooses a secret and public key pair for the revocation manager; this public key is then used for issuing the group signature, whereas the secret key gets embedded into the sold good. Thus, finding an illegally distributed copy puts the merchant in the position of the revocation manager for that particular group signature and he can retrieve the identity of the culprit. Due to the properties of group signature schemes, each buyer must register only once (registering basically amounts to join the group) and the merchant can retrieve a culprit's identity directly. One version of our scheme can even do without a registration center.

We also improve on the second building block: we exhibit a method for circumventing the use of error and erasure correction assuming a trusted third party (TTP). This TTP, however, needs only to be involved in the case that a malicious buyer is taken to court. This method can also be used to get a two-party trial for the asymmetric fingerprinting schemes [1,14]. We refer to Section 4 for an explanation of two- and three-party trials. Combining both our new building blocks gives an anonymous fingerprinting scheme that can tolerate larger collusions than previous ones and requires less administration from buyers and merchants.

## 2    Model of Anonymous Fingerprinting

Let $P_0 \in \{0,1\}^*$ denote some digital good (bit-string) that is fingerprintable, i.e., some of its bits can be changed such that (1) the result remains "close" to $P_0$ but (2) without knowing which particular bits where changed, altering "a good portion" of these bits is impossible without rendering the good useless. We refer to [3] for a formal definition of this "marking assumption". Finally, let $\mathcal{P}$ denote the set of all "close copies" of $P_0$ and $\ell$ be a security parameter (from now on we implicitly assume that $\ell$ is an input to all algorithms and protocols).

**Definition 1.** *An anonymous fingerprinting scheme involves a merchant, a buyer, and a registration center. Let c denote the maximal size of a collusion of buyers against which the scheme is secure. An anonymous fingerprinting scheme consists of the following five procedures.*

FKG-RC: *A probabilistic key setup algorithm for the registration center. Its output are the center's secret key $x_C$ and its public key $y_C$, which is published authentically.*

FReg: *A probabilistic two-party protocol (FReg-RC, FReg-B) between the registration center and the buyer. Their common input are the buyer's identity $ID_B$ and the center's public key $y_C$. The center's secret input is its secret key $x_C$. The buyer's output consists of some secret $x_B$ and related information $y_B$. The center obtains and stores $y_B$ and $ID_B$.*

FPri: *A two-party protocol* (FPri-M, FPri-B) *between the merchant and the buyer. Their common input consists of $y_C$. The merchant's secret input is $P_0$ and a transaction number $j$ and his output is a transaction record $t_j$. The buyer's secret input is $x_B$ and $y_B$ and her output consists of a copy $P_B \in \mathcal{P}$.*

FRec: *A two-party protocol between the merchant and the registration center. The merchant's input is a copy $\tilde{P} \in \mathcal{P}$, $P_0$, and all transaction records $t_i$. The center's input consists of its secret key $x_C$ and its list of $y_B$'s and $ID_B$'s. The merchant's output is a/the fraudulent buyer's identity together with a proof $p$ that this buyer indeed bought a copy of $P_0$, or $\perp$ in case of failure (e.g., if more than $c$ buyers colluded to produce $\tilde{P}$).*

FVer: *A verification algorithm, that takes as input the identity $ID_B$ of an accused buyer, the public key $y_C$ of the registration center, and a proof $p$ and outputs 1 iff the proof is valid.*

*We require that the following conditions hold.*

Correctness: *All protocols should terminate successfully whenever its players are honest (no matter how other players behaved in other protocols).*

Anonymity and unlinkability: *Without obtaining a particular $P_B$, the merchant (even when colluding with the registration center) cannot identify a buyer. Furthermore, the merchant must not be able to tell whether two purchases were made by the same buyer. In other words, all data stored by the merchant and registration center and the merchant's view of a run of FPri must be (computationally) independent of the buyer's secret input $ID_B$, $x_B$, and $y_B$.*

Protection of innocent buyers: *No coalition of buyers, the merchant, and the registration center should be able to generate a proof $\tilde{p}$ such that FVer($ID_B, y_C, \tilde{p}$) = 1, if buyer $ID_B$ was not present in the coalition.*

Revocability and collusion resistance: *There exist no polynomial-time algorithms FCol, FPri-B\*, and FReg-B\* such that for any $ID_1, \ldots, ID_c$ we have FRec($P_0$, FCol($\tilde{\mathcal{P}}, \mathcal{U}$)) $\notin \{ID_1, \ldots, ID_c\}$ with non-negligible probability, where*
$$\mathcal{U} = \{\mathsf{FReg\text{-}B}^*_{\mathsf{FReg\text{-}RC}(i,y_C,x_C)}(y_C) \mid i \in \{ID_1, \ldots, ID_c\}\} \quad and$$
$$\tilde{\mathcal{P}} = \{\mathsf{FPri\text{-}B}^*_{\mathsf{FPri\text{-}M}(P_0,y_C)}(y_C, \mathcal{U}, i) \mid i = 1, \ldots, c\}.$$

Some fingerprinting schemes allow the merchant to recover the identity of a fraudulent user without the help of the registration center, i.e., FRec is not a protocol but an algorithm.

Realizations of the procedures FPri and FRec typically involve a pair of sub-protocols, one to embed some secret, committed to by the buyer, into the digital good and one to recover the embedded data again. Let Com be a commitment scheme, i.e., a (deterministic) function that takes as input the string $x$ to commit to and an additional (randomizing) input string $\alpha$. A buyer can commit to some $x$ by $C = \mathsf{Com}(x, \alpha)$, where $\alpha$ is randomly chosen. We require that the distributions of commitments to different $x$'s are (computationally) indistinguishable. A commitment $C$ can be opened by revealing $x$ and $\alpha$. We require that it is (computationally) infeasible to open a commitment in two ways, i.e., to find pairs $(x, x')$ and $(\alpha, \alpha')$ such that $\mathsf{Com}(x, \alpha) = \mathsf{Com}(x', \alpha')$. If the value of the parameter $\alpha$ is not essential, we drop it for notational convenience.

**Definition 2.** *Let $P_0$ be a fingerprintable good known only to the merchant and let $y = \mathsf{Com}(x)$ be the buyer's commitment to some secret $x \in \{0, 1\}^\ell$. An embedding method for $P_0$, $x$, and $\mathsf{Com}$ consists of the following two procedures:*

Emb: *A two-party protocol $(\mathsf{Emb\text{-}M}, \mathsf{Emb\text{-}B})$ between the merchant and the buyer. The merchant's secret input is $P_0$, the buyer's secret input is $x$ and their common input is $y := \mathsf{Com}(x)$. The buyer's output is $P_B \in \mathcal{P}$.*

Rec: *An algorithm that takes as input and $P_0$ and a fingerprinted copy $\tilde{P}$ of it. The algorithm's output is the data $x$ embedded into $\tilde{P}$.*

*We require that the following properties are fulfilled.*

Correctness: $\forall x, P_0 : x = \mathsf{Rec}(P_0, \mathsf{Emb\text{-}B}_{\mathsf{Emb\text{-}M}(P_0, y)}(x, y))$, *where $y = \mathsf{Com}(x)$.*

Recovery and Collusion-Resistance: *There are no polynomial-time algorithms $\mathsf{Col}$ and $\mathsf{Emb\text{-}B}^*$ such that there is a set $\mathcal{U}$ of at most $c$ bit-strings of length $\ell$ for which $\mathsf{Rec}(P_0, \mathsf{Col}(\tilde{\mathcal{P}}, \mathcal{U})) \notin \mathcal{U}$ with non-negligible probability, where $\tilde{\mathcal{P}} = \{\mathsf{Emb\text{-}B}^*_{\mathsf{Emb\text{-}M}(P_0, y)}(x, y) \mid y = \mathsf{Com}(x), \ x \in \mathcal{U}\}$.*

Zero-Knowledgeness: *For all $\mathsf{Emb\text{-}M}^*$ there exists a simulator such that for all $x \in \{0, 1\}^\ell$ the output of the simulator and the view of $\mathsf{Emb\text{-}M}^*$ are (perfect/statistically/computationally) indistinguishable.*

# 3 Group Signature Schemes

**Definition 3.** *A group signature scheme consists of the following procedures:*

GKG-M: *A key setup algorithm for the membership manager $M$ that outputs her secret key $x_M$ and public key $y_M$.*

GKG-R: *A key setup algorithm for the revocation manager $R$ that outputs her secret key $x_R$ and public key $y_R$.*

GReg: *A probabilistic interactive protocol $(\mathsf{GReg\text{-}M}, \mathsf{GReg\text{-}U})$ between the membership manager and a group member $U$. Their common input is the group member's identity $ID_U$ and $y_M$. If both parties accept, the group member's output is her secret key $x_U$ and their common output is $U$'s membership key $y_U$.*

GSig: *A probabilistic algorithm that on input of $x_U$, $y_M$, $y_R$, and a message $m$ outputs a group signature $s$ on $m$.*

GVer: *An algorithm that on input of the group public key $Y$, an alleged signature $s$, and a message $m$ outputs 1 if and only if the signature is valid.*

GTrace: *A algorithm which on input of the revocation manager's secret key $x_R$, the group's public key $Y$, a message $m$, and a signature $s$ on $m$ outputs the identity $ID_U$ of the originator of the signature and a proof $V$ that $ID_U$ is indeed the originator.*

*The following security requirements must hold:*

Correctness of signature generation: *All signatures on any messages generated by any honest group member using* GSig *will be accepted by the verification algorithm.*

Anonymity and unlinkability of signatures: *Given two signature-message pairs, it is only feasible for the revocation manager to determine which group member(s) generated any of the signatures or whether the signatures have been generated by the same group member.*

Unforgeability of signatures: *It is feasible to sign messages only to group members (i.e., users that have run the registration protocol with the membership manager) or to the membership manager herself.*[1]

Unforgeability of tracing: *The revocation manager cannot falsely accuse a group member of having originated a given signature.*

No framing: *No coalition of group members, the revocation manager, and the group manager can produce a signature that will be associated with a group member not part of the coalition.*

Unavoidable traceability: *No coalition of group members and the revocation manager (but excluding the membership manager) can generate a valid signature that, when its anonymity is revoked, cannot be associated with a group member.*

To use the group signature scheme for our construction in Section 5, we require that the key setup algorithm GKG-R for the revocation manager can be run after the algorithms GKG-M and GReg. That is, we require that the revocation manager can change her keys after the scheme has been set up and without requiring group members to reselect their key material. This property is provided by many group-signature schemes (e.g., [5,7,10]).

## 4   Previous Fingerprinting Schemes

All current anonymous and asymmetric fingerprinting schemes [1,12,13,14] are based on the symmetric scheme of Boneh and Shaw [3]. This section presents this scheme briefly, giving only those details that are needed to describe our results.

### 4.1   Symmetric Fingerprinting

A symmetric fingerprinting scheme consists of a set of binary codewords (or marking patterns) $W = w_1, \ldots, w_n$ that can be embedded into the digital good [3]. Each time a copy is sold a different word is embedded and thereby

---

[1] The membership manager can always invent a fake identity and register it as a group member. It is understood that if a signature turns out to originate from a fake identity, the membership manager is considered guilty.

assigned to the copy's buyer. Let $\overline{W} \subseteq W$ denote all assigned codewords. If a redistributed copy is found later it must contain some word $\widetilde{w}$ that is a combination of words from $\overline{W}$ due to the marking assumption. A scheme is called $c$-secure if there exists an algorithm $A$ such that if a coalition $C$ of at most $c$ buyers generates a copy that contains a word $\widetilde{w}$, then $A(\widetilde{w}) \in \overline{W}$. It is said to have $\epsilon$-error if the probability that $A(\widetilde{w})$ outputs a codeword that was not assigned to any buyer in $C$ (but might have been assigned to an honest buyer) is at most $\epsilon$. Boneh and Shaw [3] show that $\epsilon = 0$ is not possible. They provide a binary code $\Gamma_0$ that is $c$-secure, has $n = c$ codewords, and whose length $l$ is polynomial in $n$ (i.e., $O(n^3 \log(n/\epsilon))$). Because the number of codewords equals the maximal number of colluding users tolerated, this code has the property that, no matter what a collusion does, the merchant will be able to extract a codeword assigned to one of the colluders with high probability (i.e., greater than $1 - \epsilon$, with $\epsilon = 2n2^{-l/(2n^2(n-1))}$).

Based on this code, Boneh and Shaw construct a random code $\Gamma_1$ over $\Gamma_0$, i.e., each codeword in $\Gamma_1$ consists of the concatenation of, say, $L$ randomly chosen codewords from $\Gamma_0$. Extraction of an embedded word from an illegally distributed copy will now in general no longer yield a codeword assigned to one of the colluders but only a word whose components (codewords from $\Gamma_0$) stem from codewords assigned to the colluding users. Because at least $L/c$ components of the extracted word must stem from a codeword assigned to one of the colluders, the extracted word must match that colluder's codeword in at least $L/c$ positions, provided that the (malicious) buyers do not know any of the codewords. Therefore, a member of the collusion can be found by comparing all assigned codewords with the extracted word (provided that the number of codewords in $\Gamma_1$ is not too large, cf. [3]). The resulting code $\Gamma_1$ has length $c^{O(1)} \log(n)$, where $n$ is the number of codewords (or, equivalently, the number of possible buyers).

*Remark.* As the amount of bits that can be embedded in a particular good is usually fixed, the length of a codewords translate into a maximum size of collusions that can be tolerated and how many buyers the good can be sold to.

## 4.2 Asymmetric Fingerprinting

In a nutshell, the idea behind an asymmetric scheme is as follows. First the buyer commits to some secret. Then merchant and buyer engage in a secure two-party protocol (henceforth called APri), at the end of which the buyer has obtained a copy of the good with her secret and some serial number (chosen by the merchant) embedded, whereas the merchant obtains the buyer's signature on a text describing their deal and on a commitment to the buyer's secret. Later, when the merchant finds an illegally distributed copy, he should be able to extract one of the colluding buyers' secret and the serial number from that copy. Being able to produce a buyer's secret will presumably convince a judge of her guilt. This approach is proposed by Pfitzmann and Schunter [12] for use with the code $\Gamma_0$, in which case the protocol APri is reasonably efficient. However, when used with $\Gamma_1$, the protocol APri is rendered prohibitively inefficient with this approach.

Pfitzmann and Waidner [14] solve this problem as follows (Biehl and Meyer [1] independently proposed a similar solution): During protocol APri, merchant and buyer construct on the fly a code similar to $\Gamma_1$, that is, they together choose $L$ random codewords $w_1, \ldots, w_L$ from $\Gamma_0$ such that the first half of each $w_i$ consists of bits chosen by the merchant (but not known to the buyer) and the second half of consists of bits chosen by the buyer (of which the merchant gets to know the only commitments $C_i$). At the end of the protocol, the buyer obtains a copy of the digital good with the codewords $w_1, \ldots, w_L$ embedded in it, whereas the merchant gets commitments $C_1, \ldots, C_L$ of the parts the buyer chose. When finding an illegally distributed copy, the merchant extracts the embedded word, and then can use a similar decoding strategy as the one described earlier for $\Gamma_1$ (restricted to the parts of the codewords known to him). Thus he will by able to identify one of the colluding buyers and also learn about $L/c$ of the values committed to by $C_1, \ldots, C_L$ of the identified colluder and hence prove her guilt.

All these schemes have the property that the judge will not be able to tell on her or his own whether the commitments indeed contain the values that the merchant presents (this is a property of every secure commitment scheme). Therefore, the accused buyer must take part in the trial (which seems a natural requirement) and will be found guilty only if she is not able to prove that most of her commitments do not contain the value presented by the merchant. This is called a three-party trial [14].

Subsequently, Pfitzmann and Waidner [13] improve on this and exhibit a new asymmetric fingerprinting scheme that allows two-party trials. This scheme has the property that the merchant can extract all secret bits of one of the colluding buyers. This is achieved by using an error and erasure-correcting code (EECC) on top of the scheme described in the previous paragraph [14]. In addition, the buyer now also signs the result of some one-way function applied to her secret bits, and thus the judge will be able to verify whether the merchant indeed presents a malicious buyer's secret bits by testing whether these bits are the function's pre-image of the value the buyer signed. Hence a trial could be held without the accused buyer. The price for this improvement is that the use of the EEEC increases significantly the number of bits that need to be embedded because the code must be able to handle a large number of erasures. To give a rough idea of this increase, in this scheme the underlying code $\Gamma_0$ must have $n = O(1)c^2\ell$ codewords, whereas it is $n = O(1)c$ in the one described previously [14], where $c$ is the size of the tolerated collusion and $\ell$ is the bit length of the buyer's (whole) secret to be embedded. (Recall that the bit length of codewords from $\Gamma_0$ is polynomial in $n$.) However, the purchase protocol for this new scheme can be realized quite efficiently [11]. Finally, we note that this asymmetric fingerprinting scheme in fact realizes the two procedures Emb and Rec of Definition 2, thus allowing the construction of an anonymous fingerprinting scheme as we will see.

### 4.3   Anonymous Fingerprinting

Anonymous fingerprinting takes asymmetric fingerprinting one step further in that the merchant no longer gets to know an honest buyer's identity. Of course, if the merchant finds an illegally distributed copy, he must nevertheless be able to retrieve the identity of a malicious buyer.

Building on the asymmetric fingerprinting scheme proposed in the same paper, Pfitzmann and Waidner [13] construct an anonymous fingerprinting scheme as follows. They introduce an additional party, a registration center, at which a buyer has to register beforehand under her real identity. To do so, the buyer chooses a pseudonym and a public/secret key pair of any signature scheme and receives a certificate for the public key and pseudonym. When purchasing some digital good from the merchant, the buyer commits to the certificate, the public key, the pseudonym, and a signature under this public key on a text describing their deal and sends these commitments to the merchant. She then proves to the merchant (in zero knowledge) that what is contained in the commitment is sound. Upon this, the two parties use the embedding protocol Emb as realized by the asymmetric scheme [13] described last in the previous section. As a result, the buyer obtains a copy with all the committed information embedded into it, whereas the merchant learns nothing. Apart from the commitment, the merchant obtains no information about the buyer during the transaction, but is assured that if he later finds an illegally distributed copy he will obtain all identifying information. However, the only known realization of this approach requires general zero-knowledge proof techniques, which are rather inefficient and thus the resulting anonymous fingerprinting scheme is considered an existence result [13].

Pfitzmann and Sadeghi [11] replace this general construction by an explicit and efficient one derived from the digital payment system by Brands [4]. Coins in that payment system are anonymous, but contain some identifying information that can be extracted as soon as a user spends a coin more than once (and only then). This information will then allow the bank to obtain the double-spender's identity. Pfitzmann and Sadeghi exploit this property as follows. The registration center plays the role of the bank and issues anonymous coins to registering buyers. Then, when purchasing some digital good, the buyer presents such a coin to the merchant. If the coin is valid, the merchant will be convinced that it contains information that will allow the registration center to retrieve the buyer's identity. Finally, they use the asymmetric fingerprinting scheme [13] such that the identifying information contained in the coin will be embedded in the sold copy. Owing to the algebraic properties of the payment system, the resulting protocol is quite efficient. However, two disadvantages remain: (1) a buyer must register with the center before each purchase and (2) the merchant must contact the center to learn the identity of a malicious buyer.

In the following section we provide another replacement for the general construction by Pfitzmann and Waidner [13] that uses a group signature scheme, is efficient, and overcomes these two restrictions. That is, our construction allows the merchant to directly identify the buyer of an illegally distributed copy and the buyer needs to register only once (or even not at all, depending on the kind of group signature scheme used, as we shall see).

# 5  Anonymous Fingerprinting Using Group Signatures

In this section we show how the asymmetric fingerprinting scheme [13] and any suitable group signature scheme can be combined to achieve an anonymous fingerprinting scheme. Suitable means that the key setup of the revocation manager can run after the registration of group members. This is true for many known group signature schemes (e.g., [5,7,10]).

The idea underlying our construction is that the registration center in the anonymous fingerprinting scheme plays the role of the *membership* manager in the group signature scheme. Every user that registers at the center then becomes a member of the group in the group signature scheme, i.e., the group consists of all registered buyers. When a user wants to buy some digital good $P_0$, he first runs the group signature scheme's key-generation protocol for the *revocation* manager and gets a key pair, say $y_R$ and $x_R$. Then the buyer signs a document describing the deal using the group signature scheme, where $y_R$ is used as the revocation manager's public key. Note that a different revocation manager's key pair is used for every instance of the purchase protocol. Finally, the merchant and the buyer carry out the asymmetric fingerprinting protocol with respect to $x_R$, i.e., such that $x_R$ is embedded into $P_0$. Whenever the merchant obtains on an (illegally distributed) copy $\tilde{P}$, he can extract[2] $x_R$—the secret key corresponding the $y_R$. This puts him into the position of the revocation manager for the instance of the group signature scheme that used $y_R$, and hence he can revoke the buyer's anonymity and identify her.

More formally, our anonymous fingerprinting scheme is as follows. Let (GKG-M, GKG-R, GReg, GSig, GVer, GTrace) be a suitable group signature scheme and (Emb, Rec) be an embedding protocol and a recovery algorithm for a commitment scheme Com as provided by the asymmetric fingerprinting scheme [13]. Let $P_0$ denote the digital good for sale.

FKG-RC: The registration center runs GKG-MM to get the key pair $(y_M, x_M)$ and publishes $y_M$.

FReg: The center and the buyer run (GReg-M, GReg-U). The buyer gets $y_U$ and $x_U$. The center gets and stores $ID_U$ and $y_U$.

FPrint: Let $m$ be the text that describes the deal. The buyer first runs GKG-R to obtain a key pair $(y_R, x_R)$, signs $m$ by computing $\sigma := \mathsf{GSig}(y_U, (y_R, y_M), m)$, and sends the merchant $\sigma$, $y_R$, and $y = \mathsf{Com}(x_R)$. The buyer proves to the merchant that $y$ indeed commits to the secret key corresponding to $y_R$. The merchant verifies $\sigma$ using GVer and, if it was successful, the two parties engage in the protocol Emb, where the merchant's input is $P_0$ and $y$, the buyer's input is $x_R$ and $y$, and the buyer's output is a copy $P_B$ of $P_0$.

FRec: Let $\tilde{P}$ be a copy of $P_0$ produced by at most $c$ dishonest buyers. Running Rec on $\tilde{P}$, the merchant obtains some $x_R$. This allows him to compute $y_R$ and find the group signature $\sigma$ in his database. Running $\mathsf{GTrace}(x_R, (y_R, y_M), m, \sigma)$,

---

[2] Here, we assume that fewer than $c$ buyers colluded to generate the distributed copy.

the merchant learns the identity of one of the buyers in the collusion that produced $\tilde{P}$.

**Theorem 1.** *Given a secure group signature scheme where the key setup of the revocation manager can be run after the registration of group members and a secure and collusion-resistant embedding method, the above construction is a secure anonymous asymmetric fingerprinting scheme.*

*Proof. Correctness:* By inspection.

*Anonymity and unlinkability:* All information the merchant obtains during a purchase is a group signature on a message that describes the deal. Owing to the properties of the embedding scheme, the merchant gets no information about the secret key $x_R$ corresponding to $y_R$. Because group signatures are anonymous and unlinkable for everybody but the one knowing $x_R$, purchases are anonymous and unlinkable.

*Protection of innocent buyers:* In order to frame an innocent buyer a coalition would either have to produce a group signature with respect to some public key $y'_R$ they can choose, or they would have to come up with a fingerprinted copy containing the secret key for some $y_R$ that the buyer used in a purchase. The first attack is prevented by the "no-framing" property of the group signature scheme. The second attack is infeasible due to zero-knowledge property of the embedding protocol.

*Revocability and collusion resistance:* Given the collusion resistance and the correctness of the embedding scheme, the merchant can recover at least the secret key for one of the $y_R$'s that was used by a member of the collusion if it contains fewer than $c$ buyers. Knowing the secret key of some $y_R$ places the merchant in the position of the revocation manager in the group signature scheme and hence he can revoke the anonymity of the buyer/group member.

## 5.1   Discussion and Comparison with Previous Solutions

It is easy to see that buyers in our anonymous fingerprinting scheme need to register only once and can then buy many goods without these transactions being linkable. Whether the merchant is able to retrieve the identity of a malicious buyer on his own depends on the group signature scheme chosen. We discuss this briefly as well as other properties the fingerprinting scheme will have as a function of the type of group signature scheme that is applied.

Most newer group signature schemes (including [6,7]) can be used for our construction. These schemes have the property that the group's public key and the length of signature are independent of the group's size. A signature in those schemes typically contains a randomized encryption of identifying information under the revocation manager's public key. If a group signature scheme is used that allows the revocation manager to trace a signature without any interaction with the membership manager, it follows that the merchant need not interact with the registration center to identify a malicious buyer. This is possible for instance with the recent group signature scheme by Camenisch and Michels [6]. There, a group member chooses her own RSA modulus that upon signing is encrypted by the revocation manager's public key. Thus, when the membership manager (aka registration center) enforces that the most significant bits are set to the identity of the group member (aka buyer), a direct identification is possible. The efficiency of the resulting fingerprinting scheme scheme is governed by the embedding protocol Emb. Because this is the same for the scheme by Pfitzmann and Sadeghi [11], the two schemes have about the same efficiency. Thus, the main advantage of our scheme is that is overcomes the latter's drawback that a buyer must register prior to each purchase and that the merchant needs to contact the registration center to identify a malicious buyer.

If we apply the group signature scheme described in [5] and assume a public key infrastructure, we do not even need a registration center. This group signature scheme works for any semantically secure public key encryption scheme for which the revocation manager can be used and the buyer can have any public key signature or identification scheme that fulfills certain properties [5]. This includes for instance the RSA, DSS, or Schnorr signature schemes. The group's public key in this scheme consists of a list of users' public keys and certificates on them. Thus, using this scheme, a buyer can simply present the merchant with any list of public keys and certificates among which she would like to hide and chooses some public and secret key of an encryption scheme. Then, using the group signature scheme, the buyer signs the purchase contract and engages with the merchant in protocol FPrint. The resulting fingerprinting scheme will not need a registration center at all and the merchant is able to identify a malicious buyer on his own. However, the merchant needs to store the list of all the public keys and certificates the buyer presents as well as the group signature, which is about the same size as this list. As long as the number of public keys presented by the buyer is not too large (i.e., much smaller than the number of bits of the sold good), the scheme's efficiency is governed by the embedding protocol Emb.

## 6   Replacing Error and Erasure Correction by TTPs

As described in Section 4, the asymmetric fingerprinting scheme [13] underlying our (and all other known) anonymous fingerprinting scheme uses an error and erasure-correcting code to guarantee the full recovery of one of the colluders' committed secret from a found copy. As mentioned earlier, this error and erasure correction significantly increases the number of bits that ultimately need to be embedded.

To be able to base our anonymous fingerprinting scheme on the more efficient asymmetric fingerprinting schemes [1,14] and thereby circumvent error and erasure correction, we extend the model by a trusted third party (TTP). This TTP will be responsible for identifying malicious buyers. Of course, the TTP must not be involved in normal operations but only in the case that it comes to a trial. Moreover, the trust to be put in the TTP shall be minimal, i.e., buyers need to trust the TTP only that it does not reveal identities at will and the merchant needs to trust only that the TTP cooperates for identifying malicious players. Other than that, the TTP is not trusted, e.g., a coalition of the TTP and the merchant must not be able to frame an honest buyer. Serving as a TTP could for instance the judge who has to take part in the trial anyway. To reduce the risk of fraudulent behavior, the TTP could be distributed (techniques for this are standard).

The general idea for our scheme with a TTP is to use the group signature scheme in the way originally conceived: the TTP plays the role of the revocation manager and the role of the membership manager is assumed by the registration center. Then, we use one of the asymmetric fingerprinting schemes [1,14] but without the buyer identifying herself and with her signature scheme being replaced by the group signature scheme. Now, if the merchant finds an illegally distributed copy and extracts the embedded information, the trial can take place as in the asymmetric scheme with the difference that the TTP must first identify the accused buyer via the revocation mechanism of the group signature scheme. Thus, we have an anonymous fingerprinting scheme.

Owing to the three-party-trial nature of efficient asymmetric fingerprinting schemes [1,14], the merchant cannot provide evidence to the TTP (aka revocation manager) that the buyer he wishes to identify is indeed malicious. A dishonest merchant could take advantage of this and learn the identity of an honest buyer simply by accusing her. This can be prevented by doubling the length of the parts of the $L$ codewords from $\Gamma_0$ that the buyer chooses and then requiring the anonymous buyer to verifiably encrypt (see, e.g., [5]) the first half of each of her (secret) parts under the TTP's public key. Then, the merchant stores these encryptions as part of his transcript. The rest of the scheme remains unchanged. Later, when finding an illegally distributed copy, extracting the embedded information and thereby linking the copy to a purchase transcript, the merchant sends the transcript together with the first half of the extracted buyer parts of codewords to the TTP. Receiving this, the TTP decrypts the verifiable encryptions and compares the result with the corresponding parts that the merchant claims to have extracted from the copy. If most of these match (the merchant must be allowed a certain error rate, see [1,14]), the TTP reveals the identity of the buyer (who can then be taken to court); otherwise the TTP refuses. After finding out the identity of one of the colluders, the merchant can take her to court as before.

It is easy to see that, as long as the TTP is honest, the merchant is guaranteed to learn the identity of a malicious buyer, whereas an honest buyer's anonymity is protected. Finally, the probability that a collusion of the TTP and the merchant can frame a buyer is the same as for the merchant in the underlying asymmetric scheme. With respect to efficiency, the number of bits that are embedded is at most a factor of 2 greater for the original asymmetric scheme.

We briefly describe how a TTP could also be used to achieve an asymmetric fingerprinting scheme with a two-party trial. The drawback of the asymmetric fingerprinting schemes [1,14] with a three-party trial is that a (malicious) merchant can accuse any buyer of misconduct, causing the buyer the inconvenience of going to court to prove her innocence. This is because in these schemes, it is not possible for the judge to check whether the evidence provided by the merchant is real. Thus the judge must always start a trial.

This can be overcome by using a TTP (which could be the judge himself) in the same way as described earlier in this section, neglecting the group signature scheme entirely. This results in an asymmetric fingerprinting scheme where the judge could use the TTP to check the evidence before opening a trial. Hence, a merchant can no longer accuse an honest buyer, as long as the TTP remains honest. Moreover, if the buyer trusts the TTP, then she can also discard her purchase transcript.

# References

1. I. Biehl and B. Meyer. Protocols for collusion-secure asymmetric fingerprinting. In *Proc. 14th STACS*, vol. 1200 of *LNCS*, pp. 213–222. Springer Verlag, 1997.
2. G. R. Blakley, C. Meadows, and G. B. Prudy. Fingerprinting long forgiving messages. In *Advances in Cryptology — CRYPTO '85*, vol. 218 of *LNCS*, pp. 180–189. Springer-Verlag, 1986.
3. D. Boneh and J. Shaw. Collusion-secure fingerprinting for digital data. In *Advances in Cryptology — CRYPTO '95*, vol. 963 of *LNCS*, pp. 452–465, 1995.
4. S. Brands. Electronic cash systems based on the representation problem in groups of prime order. In *Preproceedings of Advances in Cryptology — CRYPTO '93*, pp. 26.1–26.15, 1993.
5. J. Camenisch and I. Damgård. Verifiable encryption and applications to group signatures and signature sharing. Technical Report RS-98-32, BRICS, Department of Computer Science, University of Aarhus, Dec. 1998.
6. J. Camenisch and M. Michels. Separability and efficiency for generic group signature schemes. In *Advances in Cryptology — CRYPTO '99*, vol. 1296 of *LNCS*, pp. 413–430. Springer Verlag, 1999.
7. J. Camenisch and M. Stadler. Efficient group signature schemes for large groups. In *Advances in Cryptology — CRYPTO '97*, vol. 1296 of *LNCS*, pp. 410–424, 1997.
8. G. Caronni. Assuring ownership rights for digital images. In *Verlässliche IT-Systeme – VIS '95*, pp. 251–263. Vieweg Verlag, 1995.
9. D. Chaum and E. van Heyst. Group signatures. In *Advances in Cryptology — EUROCRYPT '91*, vol. 547 of *LNCS*, pp. 257–265. Springer-Verlag, 1991.
10. L. Chen and T. P. Pedersen. New group signature schemes. In *Advances in Cryptology — EUROCRYPT '94*, vol. 950 of *LNCS*, pp. 171–181, 1995.
11. B. Pfitzmann and A.-R. Sadeghi. Coin-based anonymous fingerprinting. In *Advances in Cryptology — EUROCRYPT '99*, vol. 1592 of *LNCS*, pp. 150–164. Springer Verlag, 1999.
12. B. Pfitzmann and M. Schunter. Asymmetric fingerprinting. In *Advances in Cryptology — EUROCRYPT '96*, vol. 1070 of *LNCS*, pp. 84–95. Springer Verlag, 1996.
13. B. Pfitzmann and M. Waidner. Anonymous fingerprinting. In *Advances in Cryptology — EUROCRYPT '97*, vol. 1233 of *LNCS*, pp. 88–102. Springer Verlag, 1997.
14. B. Pfitzmann and M. Waidner. Asymmetric fingerprinting for larger collusions. In *4th ACM CCS*, pp. 57–66. ACM press, Apr. 1997.

# Increasing the Power of the Dealer in Non-interactive Zero-Knowledge Proof Systems

Danny Gutfreund and Michael Ben-Or

Computer Science Department, Hebrew University, Jerusalem, Israel

**Abstract.** We introduce weaker models for non-interactive zero knowledge, in which the dealer is not restricted to deal a truly random string and may also have access to the input to the protocol (i.e. the statement to prove). We show in these models a non-interactive statistical zero-knowledge proof for every language that has (interactive) statistical zero-knowledge proof, and a computational zero-knowledge proof for every language in $NP$. We also show how to change the latter proof system to fit the model of non-interactive computational zero-knowledge with preprocessing to improve existing results in term of the number of bit commitments that are required for the protocol to work.

## 1 Introduction

When zero-knowledge proofs were first introduced by Goldwasser, Micali and Rackoff [10] it seemed that interaction played a crucial role in those proof systems. Indeed zero-knowledge was shown to exist only for languages in $BPP$ in the most straightforward non-interactive model ([12]). Blum, Feldman and Micali [1] showed however that if we change the model slightly then zero-knowledge can be achieved for languages not known to be trivial. In their model they assumed that both prover and verifier are dealt with a truly random string called the reference string. The proof consists of one message sent from the prover to the verifier and then the verifier decides whether to accept or reject according to this message, the input and the reference string.

Non-interactive zero-knowledge proofs are not only communication efficient, they also have several applications not offered by interactive zero-knowledge proofs. They have been used in applications like digital signature schemes secure against adaptive chosen message attack ([2]), public key cryptosystems secure against chosen cipher text attack ([5], [18]), and memoryless key distributions ([2]).

Two notions of non-interactive zero-knowledge proofs have been studied: statistical zero-knowledge where the distribution over the real protocol is statistically close to the distribution induced by the simulator, and computational zero-knowledge where these two distributions are computationally indistinguishable.

**Statistical zero-knowledge.** The study of non-interactive statistical zero-knowledge has been recently initiated by [6]. They showed a complete promise problem for the class of languages that have non-interactive statistical zero-knowledge proofs (denoted $NISZK$). They were followed by [14] who studied

T. Okamoto (Ed.): ASIACRYPT 2000, LNCS 1976, pp. 429–442, 2000.

the relationships between the class of languages that have interactive statistical zero-knowledge proofs (denoted $SZK$) and $NISZK$. They showed conditions under which these two classes are equal. In particular that if $NISZK$ is closed under complementation then $NISZK = SZK$.

**Computational zero-knowledge.** Blum et. al ([1]) showed that every language in $NP$ has a non-interactive computational zero-knowledge proof based on a number theoretic assumption. Since then various researchers improved this result by both relaxing the assumptions needed and making the proofs more efficient in term of the number of committed bits ([3], [4], [8], [17]). The most recent results are based on the assumption that one-way permutations exists. In [7] a weaker model was introduced called non-interactive zero-knowledge with preprocessing. They showed that every language in $NP$ has a proof system in this model based on the assumption that one-way functions exists.

**Our work.** In this paper we investigate relaxed versions of the non-interactive zero-knowledge with random reference string model. Specifically, we consider models in which the dealer (we refer to the *dealer* as the entity that provides the reference string to the protocol) is not restricted to deal a string of independent unbiased coin flips to the prover and the verifier (a private coins dealer rather than a public coins one that can only publish his coins flips). Two models are considered, in the first, the reference string is a sample from a distribution that can be sampled efficiently. In the second model, this distribution can also depend on the input to the protocol.

For statistical zero-knowledge we show that the class of languages that have non-interactive statistical zero-knowledge proof system with a (polynomial-time) dealer that has access to the input equals to the class $SZK$. This result not only gives a new characterization of the class $SZK$ but it also shows that if the dealer is given sufficient abilities (i.e. access to the input and the ability to compute) then every language in $SZK$ has a communication efficient statistical zero-knowledge proof (with a reference string). In the traditional model of statistical zero-knowledge, the known generic protocols for $SZK$ require polynomial number of communication rounds ([13]).

For computational zero-knowledge we show for every language in $NP$ a non-interactive zero-knowledge proof system with a dealer that can make polynomial-time computations but does not have access to the input. The proof has perfect completeness, perfect soundness, and it is based on the weakest cryptographic assumption of the existence of one-way functions. We then show how this proof can be changed to fit the non-interactive zero-knowledge with preprocessing model, to improve (in term of the number of bit commitments) a protocol by [7]. We also overview some known applications of non-interactive zero-knowledge and check what additional assumptions or changes should be done (if at all) in order to replace the random reference string model with the relaxed models in these applications. In particular we argue that the digital signature scheme of [2] which is secure against adaptive chosen message attack can be done with our model.

## 2   Definitions

Let us first recall the definition of non-interactive zero-knowledge with a random reference string ([1]).

**Definition 1.** A non-interactive computational (resp. statistical) zero-knowledge proof system with a random reference string for a language $L$ is defined by a computationally unbounded TM $P$ (the prover), a probabilistic polynomial-time TM $V$ (the verifier), a probabilistic polynomial-time TM $S$ (the simulator) and a polynomial $q$. On an input $x$ both $P$ and $V$ have access to a shared random reference string $\sigma$, where $\sigma \in_R \{0,1\}^{q(|x|)}$. The proof consists of one message sent from $P$ to $V$, and then $V$ based on $x$, $\sigma$ and this message either accepts or rejects. The following should hold:

1. (completeness) if $x \in L$ then $Pr(V(x, \sigma, P(x, \sigma)) = accept) > 2/3$
2. (soundness) if $x \notin L$ then for every prover's strategy $P^*$, $Pr(V(x, \sigma, P^*(x, \sigma)) = accept) < 1/3$
3. (zero-knowledge) if $x \in L$ then the following two distributions are computationally indistinguishable (resp. have statistical difference bounded by a negligible function):
   (a) $(\sigma, P(x, \sigma))$
   (b) $S(x)$

We define now relaxed versions of the shared random reference string model. The first relaxation we introduce is that the shared reference string need not be truly random, we only require that it can be sampled in polynomial-time.

**Definition 2.** A non-interactive computational (resp. statistical) zero-knowledge proof system with a protocol-dependent reference string for a language $L$ is defined by $P$, $V$, $S$, and $q$ as above, and the reference string $\sigma$ is $f(r)$, where $f$ is a polynomial-time computable function, and $r \in_R \{0,1\}^{q(|x|)}$.

The second relaxation is that the shared reference string can not only be non-uniformly distributed, but it can also depend on the input.

**Definition 3.** A non-interactive computational (resp. statistical) zero-knowledge proof system with an input-dependent reference string for a language $L$ is defined by $P$, $V$, $S$, and $q$ as above, and the reference string $\sigma$ is $f(x, r)$, where $f$ is a polynomial-time computable function, $x$ is the input to the protocol, and $r \in_R \{0,1\}^{q(|x|)}$.

Note that non-interactive zero-knowledge under all the definitions is closed under parallel repetitions, therefore the error bound can be brought down to be exponentially small in the length of the input.

We denote by $NICZK$ (resp. $NISZK$), $Protocol - Dependent\ NICZK$ (resp. $Protocol - Dependent\ NISZK$), and $Input - Dependent\ NICZK$ (resp. $Input - Dependent\ NISZK$) the class of languages possessing a non-interactive computational (resp. statistical) zero-knowledge proof system with a shared random, protocol-dependent and input-dependent string respectively. $SZK$ is the class of all the languages that have statistical-zero knowledge proof system as defined by [10].

# 3    Statistical Zero-Knowledge

In this section we show that if we relax the model of non-interactive proof systems then every language in $SZK$ has a proof system in this relaxed model:

**Theorem 4.** $SZK = Input - Dependent\ NISZK.$

## 3.1    Motivation

It is an interesting question to understand how much more can be proven in non-interactive statistical zero-knowledge as we gradually increase the power of the dealer. By referring to the power of the dealer we mean the type of computations he can do, and does he have access to the input (the statement to prove). Clearly, if the dealer is computationally unbounded and has access to the input then everything computable can be proven non-interactively with perfect zero-knowledge, by using the dealer as an unbounded trusted prover that tells the verifier whether the statement is correct or not. What happens if we do not give the dealer so much power? We can divide the languages into classes according to the power of the dealer in the non-interactive statistical-zero knowledge proof systems for them. We have the following hierarchy of classes, each one containing the class above it:

- No dealer: this class equals to $BPP$ ([12]).
- The dealer can just toss coins (a public coins dealer): this is the class $NISZK$, languages not known to be in $BPP$ were shown to be in this class ([6], [14]).
- The dealer can toss coins and make polynomial-time computations (a private coins polynomial-time dealer): this is the class $protocol - dependent\ NISZK$.
- A private coins polynomial-time dealer with access to the input: this is the class $input - dependent\ NISZK$ which equals to $SZK$ (Theorem 4).
- A private coins unbounded dealer with access to the input: everything computable is in this class.

By showing the exact location of $SZK$ in this hierarchy we get a strong connection between the question of how much interaction is needed for statistical zero-knowledge and how much power the dealer must have in non-interactive statistical zero-knowledge. Better understanding of this hierarchy can shed light on the $SZK$ vs. $NISZK$ question ([14]).

## 3.2    Separating Distributions

An important notion in the proof of theorem 4 will be the statistical difference between two distributions. Let us first define the statistical difference and some notations concerning it.

**Definition 5.** Let $X$ and $Y$ be two distributions (or random variables) over a discrete space $D$. The statistical difference between $X$ and $Y$, denoted as $\|X - Y\|$ is:
$$\|X - Y\| = MAX_{S \subseteq D}|Pr(X \in S) - Pr(Y \in S)|$$

Through out this paper we consider distributions with "succinct" description, i.e. distributions produced by circuits (with multiple output gates) when feeding them a uniformly chosen input. We write $C$ when we refer both to the circuit itself and to the distribution it induces. We will use the notation $x \leftarrow C$ to denote that $x$ is a sample taken from the distribution $C$, i.e. the output of the circuit $C$ when feeding it a uniformly chosen input.

In our proof of theorem 4 we will use the langauge STATISTICAL-DIFFERENCE shown to be complete for the class $SZK$ ([21]), and a "separated" version of this language.

**Definition 6.** STATISTICAL-DIFFERENCE $(SD)$ is the following promise problem:
$SD_Y = \{(D_0, D_1) : ||D_0 - D_1|| > 2/3\}$
$SD_N = \{(D_0, D_1) : ||D_0 - D_1|| < 1/3\}$
Where $(D_0, D_1)$ is a pair of distributions with "succinct" description.

**Definition 7.** "Separated" STATISTICAL-DIFFERENCE $(SD')$ is the following promise problem:
$SD'_Y = \{(D_0, D_1) : \text{for } x \leftarrow D_0, \ Pr(x \in Range(D_1)) < f(n)\}$
$SD'_N = \{(D_0, D_1) : ||D_0 - D_1|| < f(n)\}$
Where $n = |D_0, D_1|$ and $f$ is a negligible function.
In other words, a pair of circuits is in $SD'_Y$ if the probability that a sample taken from the first circuit is in the range of the second is negligible, and a pair of circuits is in $SD'_N$ if the distributions induced by the circuits have a negligible statistical difference. Note that $SD'_Y \subseteq SD_Y$ and $SD'_N \subseteq SD_N$.

Our main tool will be the following lemma:

**Lemma 8.** $SD$ reduces to $\overline{SD'}$.
*Specifically, given a pair of circuits $(D_0, D_1)$ there is a polynomial-time computable function that maps them to a new pair of circuits $(C_0, C_1)$ s.t:*
$||D_0 - D_1|| > 2/3 \longrightarrow ||C_0 - C_1|| < f(n)$
$||D_0 - D_1|| < 1/3 \longrightarrow \text{for } x \leftarrow C_0, \ Pr(x \in Range(C_1)) < f(n)$
*Where $f$ is a negligible function.*

Sahai and Vadhan showed a reduction from $SD$ to $\overline{SD}$ ([22]). The same reduction accomplishes Lemma 8, although it is implicit in their proof. We do not give the proof here and refer the reader to [22].

### 3.3    Proof of the Main Theorem

**Claim 9.** $SZK \subseteq Input - Dependent\ NISZK$

*Proof.* $SD$ is a complete promise problem for the class $SZK$ ([21]) and $SD$ reduces to $\overline{SD'}$ (Lemma 8), therefore it is enough to show an input-dependent non-interactive statistical zero-knowledge proof system for the language $\overline{SD'}$.

Let $l$ be the number of input gates of $D_0$ and $D_1$ (w.l.o.g they have the same number of input gates).

**The proof system**:

Common input: $(D_0, D_1)$

Shared reference string: $\sigma = D_0(r)$, where $r \in_R \{0,1\}^l$

The protocol:

1. $P$ sends $r' \in \{0,1\}^l$.
2. $V$ accepts if and only if $D_1(r') = \sigma$

**Completeness**: Recall that $(D_0, D_1) \in \overline{SD'}_Y$ means that $||D_0 - D_1|| < f(n)$ ($f$ is a negligible function). Let $k$ be the number of output gates of $D_0$ and $D_1$ (again w.l.o.g this number is the same for $D_0$ and $D_1$). Then by the definition of statistical difference, for $x \leftarrow D_0$, $y \leftarrow D_1$, and for every $S \subseteq \{0,1\}^k$:

$|Pr(x \in S) - Pr(y \in S)| < f(n)$

Let $T = Range(D_0) \setminus Range(D_1)$, $x \leftarrow D_0$ and $y \leftarrow D_1$. Then the following holds:

$f(n) > |Pr(x \in T) - Pr(y \in T)| = Pr(x \in T)$

In other words the probability that a sample taken from $D_0$ is not in the range of $D_1$ is at most $f(n)$. Therefore the probability that $\sigma$ is in the range of $D_1$ is at least $1 - f(n)$ and with this probability (over $r$) the (computationally unbounded) prover will be able to find $r'$ s.t $D_1(r') = \sigma$.

**Soundness**: $(D_0, D_1) \in \overline{SD'}_N$ means that the probability that a sample taken from $D_0$ is in the range of $D_1$ is at most $f(n)$, so the probability that $\sigma$ is not in the range of $D_1$ is at least $1 - f(n)$. In this case there is no $r'$ s.t $D_1(r') = \sigma$ and the prover will fail.

**Simulation**:

$S$: Choose $r \in_R \{0,1\}^l$, output $(D_1(r), r)$.

If $(D_0, D_1) \in \overline{SD'}_Y$ then the distributions $D_0, D_1$ has statistical difference at most $f(n)$ therefore $\sigma$ and the first message of the simulator has statistical difference $f(n)$ at the most. The second message both in the protocol and the simulator is determined by the first message (the shared reference string) to be a random input (to $C_1$) in the preimage of the first message. So for a given first message, the second message has the same distribution in the protocol and the simulator. $\square$

**Claim 10.** $SZK \supseteq Input - Dependent\ NISZK$

*Proof.* : We will show a reduction from any language in $Input - Dependent$ $NISZK$ to the language $\overline{SD}$. As $SD$ is complete for $SZK$, and $SZK$ is closed under complementation [19] this will suffice to prove the claim. Let $(P, V)$ be an input-dependent $NISZK$ protocol for a language $L$ with exponentially small error bound. Let $S$ be the simulator for the protocol, $f$ the polynomial-time computable function that produces the reference string and $q$ the polynomial (in the input length) that defines the length of the input to $f$. Define $\mu$ to be the negligible function bounding the statistical difference between the outputs of $(P, V)$ and $S$. For an input $x$, define the following pair of distributions:

$D_0$: choose $r \in_R \{0,1\}^{q(|x|)}$, output $f(x,r)$.
$D_1$: run the simulator $S$ on $x$ to obtain $(\sigma, p)$, if $V(x, \sigma, p) = 'accept'$ output $\sigma$, otherwise output $0^{q(|x|)}$ (here $0^{q(|x|)}$ is canonic for a string outside the range of $f$).

First we show that if $x \in L$ then $||D_0 - D_1|| < 1/3$. Let $n = |x|$, the statistical difference between the first message of the simulator and the real reference string is bounded by $\mu(n)$. Also, $Pr((P,V)(x) = 'accept') > 1 - 2^{-n}$, therefore with probability $1 - 2^{-n} - \mu(n)$ $D_1$ will output the first message of the simulator and thus $||D_0 - D_1|| < 2\mu(n) + 2^{-n}$.
Next we show that if $x \notin L$ then $||D_0 - D_1|| > 2/3$. Define:
$T = \{\sigma : \exists r \in \{0,1\}^{q(n)} \text{ and } \exists p \text{ s.t } f(x,r) = \sigma \text{ and } V(x,\sigma,p) = 'accept'\}$
That is, $T$ is the set of all reference strings for which there exist a proof that will convince $V$. Since the soundness error is bounded by $2^{-n}$, $Pr_{r \in \{0,1\}^{q(n)}}(f(x,r) \in T) < 2^{-n}$. Since $D_1$ only outputs $\sigma \in T$ or $0^{q(n)}$, and $D_0$ outputs a real reference string, $||D_0 - D_1|| > 1 - 2^{-n}$. $\square$

## 4    Computational Zero-Knowledge

### 4.1    A Protocol-Dependent $NICZK$ Proof for $NP$

In this section we show a protocol-dependent non-interactive $CZK$ proof system for every language in $NP$ with perfect completeness and perfect soundness. It is based on the assumption that one-way functions exists. The proof system is for the $NP$-complete language $3 - COL$ of all the 3-colorable graphs.
In our protocol we will make use of *characters*. *Characters* were used by [17] in their non-interactive zero-knowledge proof for $NP$ in the random reference string model. A *character* is an object that can have one of four possible values: 1, 2, 3 or WC (wild card). The value of a *character* is unknown to the verifier unless the prover reveals it for him. It can be revealed according to the following rules: if the value is 1, 2, or 3 then it can only be revealed to this value. If the value is WC then it can either be revealed to 1, 2 or 3 (what ever the prover chooses).
*Characters* can be implemented in the following way: a *character* will be the commitment on a triplet of bits that can have one of the following values: 001 (the *character* 1), 010 (the *character* 2), 100 (the *character* 3), or 111 (WC). The security of the bit-commitment scheme ensures that the value of a *character* is unknown to the polynomial-time verifier, unless the (computationally unbounded) prover reveals it for him. To reveal the value of a *character*, the prover opens one bit from the triplet and the verifier checks that the value of this bit is 1. The location of the revealed bit determines the value of the *character*. Clearly, $\{1,2,3\}$ *characters* can only be revealed to their real value and WC can be revealed to any value.
Next we define a test to check whether two nodes of an edge in the graph are colored in different colors. The test will be conducted in the following way: with each node we associate a triplet of *characters*. Two of them are WC and one

is a non-WC *character*, thus we have two non-WC *characters* associated with the edge (one in each node). We make sure that these two *characters* have different values (that is, the dealer prepares them in this manner). We define the color of a node to be the position of the non-WC *character* within the triplet of *characters* associated with it. We say that two such triplets of *characters* are consistent if they are representing the same color. By reordering the *characters* in each triplet, the prover can determine the colors of the two nodes. In order to prove that the two nodes were given different colors, the prover will reveal the two triplets of *characters*, and the verifier will accept if and only if the two triplets are the same permutation of $\{1, 2, 3\}$. We call this test the inconsistency test. We prove now some properties of this test.

**Claim 11 (completeness).** *If the two nodes were given different colors then the prover will always pass the test.*

*Proof.* The fact that the two nodes have two different colors means that the two non-WC *characters* are in different positions in the triplets associated with the nodes. That is, if we align the two triplets, against each non-WC *character* in one triplet there will be a WC *character* in the other, and in one position there will be a WC against a WC. The prover will reveal each WC *character* which is aligned against a non-WC *character* to the value of this *character*. The two WC *characters* which are aligned against each other will be revealed to the value which is not used in the other positions in the triplets (recall that the two non-WC *characters* have different values, therefore this value is determined). So in each position the same value will be revealed in the two triplets and each value in $\{1, 2, 3\}$ will be revealed exactly once. □

**Claim 12 (soundness).** *If the two nodes were given the same color then the prover will always fail the test.*

*Proof.* The fact that the two nodes have the same color means that if we align the two triplets, the two non-WC *characters* will be aligned against each other. Since they have two different values and the prover can not reveal them to any other value, the verifier will always see different values in this position in the two triplets and will reject the test. □

**Claim 13 ("zero-knowledge").** *If the values of the non-WC characters are chosen uniformly (with the restriction that they have different values) then if the nodes were given different colors, the triplets will be revealed to a random permutation of $\{1, 2, 3\}$ (with probability 1/6 for each permutation).*

*Proof.* Assume w.l.o.g that the first node receives the color 1 (the non-WC *character* is in the first position), the second receives the color 2, and we first choose the value of the first *character* with probability 1/3 for each possible value and then we choose the value of the second *character* with probability

1/2 for each remaining possible value. Clearly each pair can be chosen with probability $1/6$. After the values of the non-WC *characters* are chosen, the permutation is determined. This is because each WC *character* which is aligned against a non-WC *character* must be revealed to its value, and the value of the WC *characters* which are aligned against each other is determined to be the value which was not chosen for the two non-WC *characters*. Furthermore, each choice of values for the non-WC *characters* defines a different permutation. Therefore each permutation of $\{1, 2, 3\}$ can be revealed with probability $1/6$. □

We can now present a protocol-dependent non-interactive computational zero-knowledge proof system for the language $3 - COL$:

**The proof system:**
Input: A graph $G = (V, E)$ (denote $n = |V|$).
Reference string: An independently and randomly chosen inconsistency tests for each one of the edges of the complete graph $K_n$[1], where for each node all the triplets of *characters* associated with it are consistent. In other words, with each node we associate a $(n-1) \times 3$ matrix of *characters*, where two columns contains only WC *characters* and one contains only non-WC *characters*. For each matrix the position of the non-WC column is chosen randomly and independently.
The proof:

1. $P$'s proof is divided into two stages:
   (a) For each node in $V$, $P$ swaps two columns of the matrix associated with it to create a new matrix.
   (b) For each edge in $E$, $P$ performs the inconsistency test associated with it (with the new matrices from stage 1(a)).
2. $V$ accepts if and only if $P$ passes all the inconsistency tests.

Proof of correctness:
**Completeness**: Let $\gamma$ be a 3-coloring of $G$. For each node there is a $(n - 1) \times 3$ matrix of *characters* associated with it. With each such matrix we associate a color according to the position of the non-WC column. In stage 1(a), for each node, if the matrix does not represent the color of the node according to $\gamma$, $P$ swaps the non-WC column with a WC one so that the new matrix will represent the right color. If the matrix does represent the right color, $P$ swaps the two WC columns. Since $\gamma$ is a legal 3-coloring of $G$, after stage 1(a) all the matrices of adjacent nodes are inconsistent and $P$ will always pass all the inconsistency tests (claim 11).
**Soundness**: The fact that there is no legal 3-coloring of $G$ means that it is impossible to bring the matrices to a state where for every two adjacent nodes the matrices associated with them are inconsistent (otherwise the state of the matrices will define a valid 3-coloring). Therefore for at least one edge in $E$, $P$ will always fail the inconsistency test (claim 12).

---

[1] Note that in the protocol-dependent model the dealer does not have access to the input, therefore he must prepare an inconsistency test for every possible edge.

**Simulation:**

1. (simulation of the reference string) Associate with each node a $(n-1) \times 3$ matrix of WC *characters*. This is done by committing on $9n(n-1)$ bits of value 1.
2. (simulation of stage 1(a)) For each node in $V$, choose randomly two columns in the matrix associated with it and swap them.
3. (simulation of stage 1(b)) For each $e \in E$, choose a random permutation of $\{1, 2, 3\}$ and reveal the two triplets of *characters* associated with $e$ (i.e. the simulation of the inconsistency test for $e$) to be this permutation.

The security property of the bit-commitment scheme ensures that the real reference string and the simulated one are computationally indistinguishable.

Let $\gamma$ be a 3-coloring of $G$. Let $c \in \{1, 2, 3\}$ be the color of $v \in V$ according to $\gamma$. Let $A_v$ be the matrix associated with $v$ in the real proof. The position of the non-WC column in $A_v$ is chosen randomly. If $A_v$ represents the color $c$, and this happens with probability $1/3$, then the two WC columns (the columns that do not represent $c$) will be swapped in stage 1(a). Otherwise $A_v$ represents one of the other two colors, with probability $1/3$ for each one, and then the column that represents $c$ will be swapped with the non-WC column. So in stage 1(a) the two columns that will be swapped in $A_v$ are random and since the position of the non-WC column was chosen independently for all the matrices this is also the case for all the matrices together.

After all the matrices were brought in stage 1(a) to a state where every two adjacent nodes have inconsistent matrices, in every inconsistency test in stage 1(b) a random permutation of $\{1, 2, 3\}$ will be revealed (claim 13). Since the values for the inconsistency tests were chosen independently, all the permutations that will be revealed in stage 1(b) will be random and independent. $\square$

**Remark:** The proof system presented above requires that the reference string will contain $O(n^2)$ bit commitments. The reason for this is that the structure of the graph is unknown in advance and the dealer must prepare an inconsistency test for each possible edge (i.e. for every edge in the complete graph with $n$ nodes). However, if we use a more "structured" $NP$-complete problem such as the coloring problem on a wrapped de Bruijn graph ([20]), where only $O(n\log n)$ local tests are needed we can improve the number of bit-commitments to $O(n\log n)$. A proof system for this problem will be presented in the full version of this paper.

## 4.2  Non-interactive Proofs with Preprocessing

The notion of non-interactive zero-knowledge proofs with preprocessing was first introduced by [7]. In their model, the proof system is divided into two stages: first, before there is a statement to prove, the prover and the verifier execute an interactive protocol which ends with both of them agreeing on a common reference string. Then, when there is a statement to prove, the prover sends a single message to the verifier, and the latter decides whether to accept or reject according to the input, the prover's message, and the reference string. The original proof system required that the reference string will contain $O(n^3)$ bit commitments to prove that a 3-CNF formula of length $n$ is satisfiable, and it was based on the assumption that one-way functions exist. [16] gave a non-interactive zero-knowledge proof system with preprocessing that requires less bit commitments and also has the property that multiple statements can be proven based on a single reference string. However, this came at the expense of a stronger cryptographic assumption, namely oblivious transfer.

The protocol-dependent model seems to be close to the non-interactive with preprocessing model. In both cases the prover's message is based on a non-random reference string. In the protocol-dependent model, a trusted (polynomial-time) dealer provides the reference string and in the preprocessing model the prover and the verifier agree on it in advance. Indeed, the proof system we introduced in the previous section can be easily changed to work in the preprocessing model. To see that, notice that the following language is in $NP$:

$L = \{(s, 1^n) : s \in \{0, 1\}^{poly(n)}, s$ is a valid reference string for the protocol-dependent proof system for $3 - COL$ (of a graph with $n$ nodes)$\}$

The proof system (in the preprocessing model) will be:

**Stage 1 (interactive):** The prover chooses a reference string in the same way the dealer chose it in the protocol-dependent model. Then the prover proves interactively and in zero-knowledge that this string is in $L$. Since $L \in NP$, such a proof exists under the assumption that one-way functions exists ([11]).

**Stage 2 (non-interactive):** Continue as in the protocol-dependent model.

This proof system improves [7] as it is based on the same assumption and it requires that the reference string will contain only $O(nlogn)$ bit commitments to prove a statement of size $n$.

## 4.3 Efficient Provers and Applications

Note that the protocol in section 4.1 requires that the prover will have computational power sufficient to reverse the bit commitments of the dealer, i.e. to compute the inverse of a one way function. For the protocol to be applicable we would like it to work for an efficient prover, that is, a polynomial-time prover with an auxiliary input containing a witness for the $NP$ statement. For this we will have to change the cryptographic assumption, and instead of the bit commitment to be based on any one way function we would like it to be based on a family of unapproximable trapdoor predicates ([9]).

The proof system requires now an additional preliminary step: the prover sends to the dealer a predicate for which he knows the trapdoor to, and the dealer uses it for his bit commitments. Now the prover can reverse the bit commitments, and the protocol continues as before.

Non-interactive zero-knowledge proofs were shown to be useful in many applications. For applications where the prover is also the dealer (e.g. digital signature schemes secure against adaptive chosen message attack [2] and memoryless key distribution [2]) the protocol-dependent model will still work. This is because the prover will commit on the bits and therefore will be able to open them at a later stage. It is in the prover's best interest that the reference string will be correctly prepared. For public key cryptosystems secure against chosen ciphertext attack ([5], [18]) this is not the situation. The key-generator can not generate the public key (that includes the reference string) without knowing which predicate to use for the bit-commitments (the one that the prover knows the trapdoor for). Therefore the prover must notify it in advance which predicate to use. This is a major drawback because for each prover we will need a different public key.

## 5    Concluding Remarks

We showed that if we assume that the (polynomial-time) dealer has access to the input to the protocol as well as private coins then every language that has a statistical zero-knowledge proof system also has a non-interactive statistical zero-knowledge proof system. It would be very interesting to understand whether these assumptions are required for $SZK$ to be done non-interactively.

In the Computational zero-knowledge setting, we showed an efficient (in term of the number of bit-commitments) $protocol - dependent\ NICZK$ protocol for every language in $NP$. The protocol is based on the assumption of the existence of one-way functions (for unbounded provers), or on the assumption of the existence of a family of unapproximable trapdoor predicates (for efficient provers). We also showed how this model can replace the traditional model in some applications such as secure digital signatures. For secure public-key cryptosystems the use of our model raises problems, namely that a public-key is generated for a particular prover (sender) and can not be used by anyone. If we could avoid this problem, we would get a very interesting result, that the use of non-interactive zero-knowledge proofs in order to get a public-key cryptosystem that is secure against chosen cipher text attack does not impose a stronger cryptographic assumption than the public-key encryption itself. This is due to [9] who showed that the existence of a (semantically secure) public-key cryptosystem is equivalent to the existence of a family of unapproximable trapdoor predicates. Current use of non-interactive zero-knowledge in the random reference string model is based on the stronger assumption that there is a family of trapdoor permutations.

## 6    Acknowledgements

We would like to thank Salil Vadhan and Oded Goldreich for valuable discussions about zero-knowledge. We would also want to thank them as well as Yonatan Bilu and anonymous referees for helpful comments on the paper. The first author also wishes to express his gratitude to Avi Wigderson who directed him to this topic of research.

This research was supported in part by the Leibniz Center, a US-Israel Binational research grant, and an EU Information Technologies grant (IST-FP5).

# References

1. Manuel Blum, Paul Feldman, Silvio Micali: Non-interactive zero-knowledge and its applications (extended abstract). In Proceedings of the 20th ACM Symposium on the Theory of Computing, 103-112, Chicago, Illinois, 2-4 May 1988.
2. Mihir Bellare, Shafi Goldwasser: New paradigms for digital signatures and message authentication based on non-interactive zero-knowledge proofs. In Advances in Cryptology CRYPTO '89, LNCS volume 435, pages 194-211. Springer-Verlag, 1990.
3. Manual Blum, Alfredo De Santis, Silvio Micali, Guiseppe Persiano: Non-interactive zero-knowledge. SIAM Journal on Computing, 20(6):1084-1118, December 1991.
4. Ivan Damgard: Non-interactive circuit-based proofs and non-interactive perfect zero knowledge with preprocessing. In Advances in Cryptology EUROCRYPT '92. LNCS volume 658. Springer Verlag. 1992.
5. Danny Dolev, Cynthia Dwork, Moni Naor: Non-malleable cryptography (extended abstract). In Proceedings of the 23rd Annual ACM Symposium on the Theory of Computing, pages 542-552, New Orleans, Louisiana, 6-8 May, 1991.
6. Alfredo De Santis, Giovanni Di Crescenzo, Giuseppe Persiano, Moti Yung. Image density is complete for non-interactive SZK. In Automata, Languages and Programming, 25th International coloquium , LNCS, Springer-Verlag. 1998.
7. Alfredo De Santis, Silvio Micali, Giuseppe Persiano: Non-interactive zero-knowledge proof-systems with preprocessing. In Advances in Cryptology, CRYPTO 88, LNCS volume 403, Springer Verlag.
8. Uriel Feige, Dror Lapidot, Adi Shamir: Multiple non-interactive zero-knowledge proofs based on a single random string. In Proceedings of the 31st Annual Symposium on Foundations of Computer Science, pages 308-317, 1990.
9. Shafi Goldwasser, Silvio Micali: Probabilistic encryption. Journal of Computer and System Sciences, Vol. 28, pages: 270-299, April 1984.
10. Shafi Goldwasser, Silvio Micali, Charles Rackoff: The knowledge complexity of interactive proof systems. SIAM journal of computing, 18(1): 186-208, 1989.
11. Oded Goldreich, Silvio Micali, Avi Wigderson: Proofs that yield nothing but their validity or all languages in NP have zero-knowledge proof systems. Journal of the association for computing machinery, 38(1):691-729, 1991.
12. Oded Goldreich, Yair Oren: Definitions and properties of zero-knowledge proof systems. Journal of chryptology, 7(1):1-32, Winter 1994.
13. Oded Goldreich, Amit Sahai, Salil Vadhan: Honest-verifier statistical zero-knowledge equals general statistical zero-knowledge. In Proceedings of the 30th Annual ACM Symposium on the Theory of Computing, pages 399-408,1998.
14. Oded Goldreich, Amit Sahai, Salil Vadhan: Can statistical zero-knowledge be made non-interactive? or on the relationship of SZK and NISZK. In Michael Wiener, editor, Advances in Cryptology CRYPTO '99, LNCS, Springer-Verlag, 1999.
15. Russell Impaliazzo, Leonid A. Levin, Michael Luby. Pseudo-random generation from one-way functions (extended abstract). In Proceedings of the 21st Annual ACM Symposium on the Theory of Computing, pages 12-24, 1989.
16. Joe Kilian, Silvio Micali, Rafail Ostrovsky: Minimum resource zero-knowledge proofs. In Proceedings of the 30th annual Symposium on Foundations of Computer Science, pages 474-479, 1989.

17. Joe Kilian, Erez Petrank: An efficient non-interactive zero-knowledge proof system for NP with general assumptions. Journal of Cryptology, 11(1):1-27, Winter 1998.
18. Moni Naor, Moti Yung: Public-key cryptosystems provably secure against chosen ciphertext attacks. In Proceedings of the 22nd Annual ACM Symposium on the Theory of Computing, pages 427-437, Baltimore, Maryland, 14-16 May 1990.
19. Tatsuaki Okamoto: On relationships between statistical zero-knowledge proofs. In Proceedings of the 28th Annual Symposium on the Theory of Computing, 1996.
20. Dan Spielman: Computationally efficient error-correcting codes and holographic proofs. PhD thesis, 1995.
21. Amit Sahai and Salil Vadhan: A complete promise problem for statistical zero-knowledge. In Proceedings of the 38th Annual Symposium on Foundations of Computer Science, pages 448-457, 1997.
22. Amit Sahai, Salil Vadhan: Manipulating statistical difference. In Panos Pardalos, Sanguthevar Rajaseekaran, and Jose Rolim, editors, Proceedings of the DIMACS Workshop on Randomization Methods in Algorithm Design, Princeton, NJ, 1998.

# Zero-Knowledge and Code Obfuscation

Satoshi Hada

Tokyo Research Laboratory, IBM Research,
1623-14, Shimotsuruma, Yamato, Kanagawa 242-8502, Japan.
satoshih@jp.ibm.com

**Abstract.** In this paper, we investigate the gap between auxiliary-input zero-knowledge (AIZK) and blackbox-simulation zero-knowledge (BSZK). It is an interesting open problem whether or not there exists a protocol which achieves AIZK, but not BSZK. We show that the existence of such a protocol is closely related to the existence of secure code obfuscators. A code obfuscator is used to convert a code into an equivalent one that is difficult to reverse-engineer. This paper provides security definitions of code obfuscation. By their definitions, it is easy to see that the existence of the gap implies the existence of a cheating verifier such that it is impossible to obfuscate any code of it. Intuitively, this means that it is possible to reverse-engineer any code of such a cheating verifier. Furthermore, we consider the actual behavior of such a cheating verifier. In order to do so, we focus on two special cases in which the gap exists: (1) there exists a *constant round public-coin* AIZK interactive argument for a language outside of $\mathcal{BPP}$. (2) there exists a *3-round secret-coin* AIZK interactive argument for a language outside of $\mathcal{BPP}$. In the former case, we show that it is impossible to securely obfuscate a code of a cheating verifier behaving as a pseudorandom function. A similar result is shown also in the latter case. Our results imply that any construction of constant round public-coin or 3-round secret-coin AIZK arguments for non-trivial languages essentially requires a computational assumption with a reverse-engineering property.

**Keywords:** Zero-knowledge, code obfuscation, reverse-engineering, interactive proof, interactive argument.

## 1    Introduction

In this paper, we investigate the gap between two definitions of zero-knowledge (ZK): auxiliary-input zero-knowledge and blackbox-simulation zero-knowledge. We will show that the gap is closely related to code obfuscation.

### 1.1    Zero-Knowledge

ZK is one of the most important notions in modern cryptography. The original definition of ZK, which we call GMR-ZK, is given in [GMR85] as follows: For

T. Okamoto (Ed.): ASIACRYPT 2000, LNCS 1976, pp. 443–457, 2000.

any probabilistic polynomial-time (PPT) cheating verifier, there exists a PPT machine (called the *simulator*) which simulates the interaction, i.e., produces a probability distribution which is computationally indistinguishable from the distribution of the real interaction between the prover and the cheating verifier. This definition is not suitable for cryptographic applications since it is not closed under sequential composition [GoKr96]. In cryptographic applications, the cheating verifier may have some additional a-priori information. However, GMR-ZK did not deal with this stronger verifier. In order to overcome this problem, Goldreich and Oren introduced an alternative definition called auxiliary-input zero-knowledge (AIZK) [GoOr94]. AIZK is a stronger formulation in which the simulation requirement is extended to deal with stronger (non-uniform) verifiers, namely, verifiers with some additional a-priori information. They showed that AIZK is closed under sequential composition.

The above two definitions only require that, for each cheating verifier, there exists a simulator. That is, in both definitions, the simulator is allowed to examine the internal state of the cheating verifier. On the other hand, blackbox-simulation zero-knowledge (BSZK) requires that the existence of a single universal simulator which uses any non-uniform cheating verifier as a blackbox to simulate the interaction. That is, the simulator is only allowed to simply observe the input/output behavior of the cheating verifier. BSZK is most restrictive among three definitions. Nevertheless, almost all known ZK protocols are BSZK [1]. It is an interesting open problem whether there exists a protocol which achieves AIZK, but not BSZK.

## 1.2    Code Obfuscation

Given a code, how can we make it hard to reverse-engineer it ? This is one of major open problems concerning computer practice. Code obfuscation is the most viable method for preventing reverse-engineering. There are many heuristic and ad-hoc obfuscation techniques for particular programming languages such as C, C++, Java and so on [CTL97]. However, to the best of our knowledge, no theoretical treatment have been provided so far. In this paper, we provide the definitions of secure code obfuscators and show that the existence of secure obfuscators for some code is closely related to the gap between AIZK and BSZK.

Take pseudorandom function ensembles (PRFEs) for example [GGM86]. PRFEs are function ensembles that can not be distinguished from truly random functions by any efficient procedure (any adversary) which can get the value of the function at arguments of its choice, provided its seed is chosen randomly. However, the pseudorandomness is guaranteed only when the randomly chosen seed is unknown to adversaries. This means that if a code of a PRFE is given to an adversary and the seed is embedded into the code, it may no longer satisfy the pseudorandomness. This is because some information about the seed may be extracted from the given code. A code obfuscator can be used to solve this problem. It converts the given code into another code that is functionally

---

[1] The one exception appeared in [HT98]. See Section 1.3.

identical to the original code so that the seed remain unknown to the adversary who is allowed to analyze the obfuscated code.

We sketch a definition of secure code obfuscators from the above point of view. We consider code obfuscation for a function ensemble $\mathcal{F} = \{F_n\}_{n \in \mathbb{N}}$ such that $F_n = \{f_s\}_{s \in \{0,1\}^{l_s(n)}}$. $s$ is the seed that we want to remain unknown to an adversary. A code obfuscator $C$ for $\mathcal{F}$, given a code $\pi(f_s)$, produces another code denoted by $\Pi(f_s)$ that is functionally identical to $\pi(f_s)$. We want to guarantee that the adversary should not gain any information about $s$ given $\Pi(f_s)$ when $s$ is chosen randomly. That is, we say that $C$ is secure against an adversary $A$ if whatever can be gained by $A$ having access to the code $\Pi(f_s)$, can also be gained by a PPT machine having only blackbox-access to $f_s$. Roughly speaking, this guarantees that $A$ can not reverse-engineer the code $\Pi(f_s)$ produced by $C$. This is formalized based on the simulation paradigm [GM84][GMR85] [Go99, Section 1.2.3]. In Section 3, we discuss defining secure code obfuscators in more detail.

## 1.3   Motivation and Results

As described in Section 1.1, it is an interesting open problem whether it holds that $Cl(BSZK) \subset Cl(AIZK)$, where $Cl(def)$ denotes the class of all interactive arguments satisfying the requirements of ZK definition $def$. Hada and Tanaka have constructed a 3-round *secret-coin* AIZK argument for an $\mathcal{NP}$-complete language, that is, they have shown that it holds that $Cl(BSZK) \subset Cl(AIZK)$ unless $\mathcal{NP} \subseteq \mathcal{BPP}$[HT98]. However, their construction requires a non-standard computational assumption with a strong reverse-engineering property. Roughly, it requires that given any code of any cheating verifier, one can efficiently extract the secret-coin used by the cheating verifier while one having only blackbox-access to the code can not do it. This paper addresses a question of whether the reverse implication holds, i.e., whether some reverse-engineering property is essential for the gap between AIZK and BSZK. If it is true then $Cl(BSZK) \subset Cl(AIZK)$ implies some negative result for code obfuscation. The purpose of this paper is to give such negative results.

We discuss the gap between AIZK and BSZK in more detail. We start by reviewing the definition of universal simulation zero-knowledge (USZK) introduced by Oren [Or87]. Oren showed that it is equivalent to AIZK, i.e., $Cl(USZK) = Cl(AIZK)$. In the definition of BSZK, given any cheating verifier, the simulator is required to output a simulated conversation simply by observing input/output behavior of the cheating verifier. On the other hand, the definition of USZK allows the simulator to take as input the code of the cheating verifier and analyze it. Therefore, we can say that if it holds that $Cl(BSZK) \subset Cl(USZK)$ (equivalent to $Cl(BSZK) \subset Cl(AIZK)$), then there exists a cheating verifier for which the simulation is possible by analyzing the code of it, but impossible by simply observing its input/output behavior. This will imply that it is impossible to obfuscate the code of such a cheating verifier. Indeed, by our security definition of code obfuscators, if it holds that $Cl(BSZK) \subset Cl(USZK)$ then there exists a cheating verifier (which can be viewed as a function ensemble) for which secure code obfuscation is impossible (Theorem 3). However, this does not say anything about the actual behavior of such a cheating verifier.

We focus on the following two statements in order to consider the actual behavior of such a cheating verifier:

$\mathsf{Gap}_{pc}$: There exists a constant round public-coin AIZK argument for a language outside of $\mathcal{BPP}$.

$\mathsf{Gap}_{sc}$: There exists a 3-round secret-coin AIZK argument for a language outside of $\mathcal{BPP}$.

If either statement is true then it holds that $Cl(BSZK) \subset Cl(AIZK)$. Note that the analogous statements regarding BSZK are false [GoKr96] [2]. For each statement, we prove impossibility of secure code obfuscation for a specific cheating verifier (a specific function ensemble):

1. If $\mathsf{Gap}_{pc}$ is true, then there exists a cheating verifier behaving as a PRFE for which secure code obfuscators do not exist. In other words, there exists no secure code obfuscator for PRFEs (Theorem 4).

2. If $\mathsf{Gap}_{sc}$ is true, there exists a cheating verifier for which secure code obfuscators do not exist. The cheating verifier is equivalent to the prescribed (honest) verifier except that it computes its message as a pseudorandom function of the first message sent by the prover (Theorem 5).

We don't know whether these non-existence results are reasonable or not. However, our result is a good reason to believe that any construction of constant round public-coin or 3-round secret-coin AIZK arguments for non-trivial languages essentially requires a computational assumption with a reverse-engineering property. Indeed, the 3-round AIZK argument constructed in [HT98] required a computational assumption with a reverse-engineering property.

## 1.4    Related Works

Goldreich and Ostrovsky gave a definition of *software-protecting compilers* and an efficient construction of it [GoOs96]. It is quite different from the code obfuscator considered in this paper. In their setting, a code is encrypted and can be executed by a CPU having the corresponding decryption key and adversaries try to reconstruct the code from the encrypted one. The adversary is allowed to execute the program on the random-access machine (RAM) on arbitrary inputs of its choice and modify the data between the CPU and the memory. On the other hand, our code obfuscator never encrypts a given code.

In [DNRS99], Dwork et. al. showed the relationships among 3-round public-coin ZK arguments, selective decommitment and Fiat-Shamir methodology. They pointed out that the problem they studied is closely related to code obfuscation. However, they gave no results.

---

[2] The discussion in [GoKr96] are for ZK interactive proofs. However, their results extend to ZK interactive arguments. See Remarks 6.4 and 6.5 in that paper.

## 2    Preliminaries

We say that a function $\nu(\cdot) : \mathsf{N} \to \mathsf{R}$ is negligible in $n$ if for every polynomial $poly(\cdot)$ and all sufficiently large $n$'s, it holds that $\nu(n) < 1/poly(n)$. We often omit the expression "in $n$" when the definition of $n$ will be clear by the context.

If $S$ is any probability distribution then $x \leftarrow S$ denotes the operation of selecting an element randomly according to $S$. If $S$ is a set then we use the same notation to denote the operation of picking an element $x$ randomly from $S$. If $A$ is a probabilistic machine then $A(x_1, x_2, \cdots, x_k)$ denotes the output distribution of $A$ on inputs $(x_1, x_2, \cdots, x_k)$. Also, $\{x_1 \leftarrow S_1; x_2 \leftarrow S_2; \cdots x_k \leftarrow S_k : A(x_1, x_2, \cdots, x_k)\}$ denotes the output distribution of $A$ on inputs $(x_1, x_2, \cdots, x_k)$ when the processes $x_1 \leftarrow S_1, x_2 \leftarrow S_2, \cdots, x_k \leftarrow S_k$ are performed in order. Let $\Pr[x \leftarrow S_1; x_2 \leftarrow S_2; \cdots; x_k \leftarrow S_k : E]$ denote the probability of the event $E$ after the processes $x_1 \leftarrow S_1, x_2 \leftarrow S_2, \cdots, x_k \leftarrow S_k$ are performed in order.

We begin with reviewing the definitions of (non-uniform) computational indistinguishability and PRFEs.

**Definition 1 (computational indistinguishability).** *We define two types of computational indistinguishability.*

1. *Two distribution ensembles indexed by* $\mathsf{N}$ *and* $\{0,1\}^*$, $X = \{X_{n,w}\}_{n \in \mathsf{N}, w \in \{0,1\}^*}$ *and* $Y = \{Y_{n,w}\}_{n \in \mathsf{N}, w \in \{0,1\}^*}$ *are* computationally indistinguishable *if for every PPT machine* $D$ *(the distinguisher), every polynomial* $poly(\cdot)$, *all sufficiently large* $n$'s *and every string* $z \in \{0,1\}^*$,

$$\left| \Pr \begin{bmatrix} x \leftarrow X_{n,w}; \\ b \leftarrow D(1^n, w, x, z) \end{bmatrix} : b = 1 \right] - \Pr \begin{bmatrix} y \leftarrow Y_{n,w}; \\ b \leftarrow D(1^n, w, y, z) \end{bmatrix} : b = 1 \right| < \frac{1}{poly(n)}.$$

2. *Two distribution ensembles indexed by a string set* $S$ *and* $\{0,1\}^*$, $X = \{X_{s,w}\}_{s \in S, w \in \{0,1\}^*}$ *and* $Y = \{Y_{s,w}\}_{s \in S, w \in \{0,1\}^*}$ *are* computationally indistinguishable *if for every PPT machine* $D$ *(the distinguisher), every polynomial* $poly(\cdot)$ *and all sufficiently long* $s$'s *and every string* $z \in \{0,1\}^*$,

$$\left| \Pr \begin{bmatrix} x \leftarrow X_{s,w}; \\ b \leftarrow D(s, w, x, z) \end{bmatrix} : b = 1 \right] - \Pr \begin{bmatrix} y \leftarrow Y_{s,w}; \\ b \leftarrow D(s, w, y, z) \end{bmatrix} : b = 1 \right| < \frac{1}{poly(|s|)}.$$

**Definition 2 (function ensembles).** *An* $(l_{in}, l_{out}, l_s)$-*function ensemble is a sequence* $\mathcal{F} = \{F_n\}_{n \in \mathsf{N}}$ *of function family* $F_n = \{f_s : \{0,1\}^{l_{in}(n)} \to \{0,1\}^{l_{out}(n)}\}_{s \in \{0,1\}^{l_s(n)}}$, *such that there exists a polynomial-time machine* $Eval_{\mathcal{F}}$ *(called the evaluator) so that for all* $s \in \{0,1\}^{l_s(n)}$ *and* $x \in \{0,1\}^{l_{in}(n)}$, $Eval_{\mathcal{F}}(s, x) = f_s(x)$. *In the sequel, we call* $s$ *the* seed *of the function* $f_s$. *Also, we say that* $\mathcal{F}$ *is non-uniformly computable if the evaluator* $Eval_{\mathcal{F}}$ *is a non-uniform polynomial-time machine [Go99, Section A.2.3].*

**Definition 3 (pseudorandom function ensembles (PRFEs)).** *Let* $\mathcal{U}_{l_{in}, l_{out}} = \{U_{l_{in}(n), l_{out}(n)} : \{0,1\}^{l_{in}(n)} \to \{0,1\}^{l_{out}(n)}\}_{n \in \mathsf{N}}$ *be a uniform function* ensemble, i.e., $U_{l_{in}(n),l_{out}(n)}$ is uniformly distributed over the set of $\{0,1\}^{l_{in}(n)} \rightarrow \{0,1\}^{l_{out}(n)}$ functions. An $(l_{in}, l_{out}, l_s)$-function ensemble $\mathcal{F}$ is called pseudo-random if for every PPT machine $M$, the following two distribution ensembles are computationally indistinguishable: $\{s \leftarrow \{0,1\}^{l_s(n)} : M^{f_s}(1^n)\}_{n \in \mathbb{N}}$ and $\{u \leftarrow U_{l_{in}(n),l_{out}(n)} : M^u(1^n)\}_{n \in \mathbb{N}}$.

## 2.1   Interactive Arguments

We consider two probabilistic interactive machines called the prover and the verifier. The verifier is always a PPT machine. Initially both machines have access to a common input tape which includes $x$ of length $n$. The prover and the verifier send messages to one another through two communication tapes. After exchanging a polynomial number of messages, the verifier stops in an accept state or in a reject state. Each machine only sees its own tapes, namely, the common input tape, the random tape, the auxiliary-input tape and the communication tapes. We denote by $A(x, y, m)$ the next message of $A$ when $x$ is the common input, $y$ the auxiliary-input and $m$ the messages exchanged so far. When we want to make explicit the random coins $R$ used, we denote it by $A(x, y, m; R)$.

Let $\langle P_{x,w}, V_{x,y} \rangle$ denote the distribution of the decision (over $\{\mathsf{Acc}, \mathsf{Rej}\}$) of the verifier $V$ having an auxiliary-input $y$ when interacting on a common input $x$ with the prover $P$ having an auxiliary-input $w$, where the probability is taken over the random tapes of both machines. When auxiliary inputs $w$ or $y$ are empty, we omit them from $\langle P_{x,w}, V_{x,y} \rangle$ (e.g. $\langle P_{x,w}, V_x \rangle$ and $\langle P_x, V_{x,y} \rangle$).

There are two kinds of interactive protocols. One is "interactive proof" and the other is "interactive argument". The former requires that even a computationally unrestricted prover should be unable to make the verifier accept $x \notin L$, except with negligible (in $n$) probability [GMR85]. On the other hand, the latter requires that any cheating prover restricted to PPT should be unable to make the verifier accept $x \notin L$, except with negligible (in $n$) probability [BrCr86]. In this paper, we deal with interactive arguments.

**Definition 4 (interactive arguments [Go99, page 62]).** *Let $P, V$ be two PPT interactive machines. The verifier $V$ does not take any auxiliary-input. We say that $(P, V)$ is an* interactive argument *for $L$ if the following two conditions hold: (1)* Efficient Completeness: *For every polynomial $poly(\cdot)$, all sufficiently long $x \in L$, there exists an auxiliary-input $w$ such that $\Pr[b \leftarrow \langle P_{x,w}, V_x \rangle : b = \mathsf{Acc}] > 1 - 1/poly(|x|)$. (2)* Computational Soundness: *For every PPT machine $\hat{P}$ (the polynomial-time bounded cheating prover), every polynomial $poly(\cdot)$, all sufficiently long $x \notin L$ and every auxiliary-input $w$, $\Pr[b \leftarrow \langle P_{x,w}, V_x \rangle : b = \mathsf{Rej}] > 1 - 1/poly(|x|)$.*

## 2.2   Zero-Knowledge

We recall the three definitions of AZIK, USZK and BSZK. A *view* of the verifier is a distribution ensemble which consists of the common input, the verifier's auxiliary input, the verifier's random coins and the sequence of messages sent by the prover and the verifier during the interaction. Let $\mathsf{View}(P_x, V_{x,y}) = [x, y, m; R]$ denote $V$'s view after interacting with $P$, where $x$ is the common input, $y$ the auxiliary input to $V$, $R$ the random coins of $V$ and $m$ the sequence of messages sent by $P$ and $V$. When the auxiliary input $y$ to $V$ is empty, we write $\mathsf{View}(P_x, V_x)$. When the random coins $R$ used by $V$ is fixed, we write $\mathsf{View}(P_x, V_{x,y}^R)$ or $\mathsf{View}(P_x, V_x^R)$. For simplicity, when we talk about ZK, we omit the auxiliary-input to the prover $P$.

**Definition 5 (AIZK [GoOr94]).** *Let $P, V$ be two probabilistic interactive machines. We say that $(P, V)$ is* auxiliary-input *zero-knowledge for $L$ if for every PPT machine $\hat{V}$ (the cheating verifier), there exists a PPT machine $S_{\hat{V}}$ (the simulator) such that the following two distribution ensembles are computationally indistinguishable: $\{S_{\hat{V}}(x, y)\}_{x \in L, y \in \{0,1\}^*}$ and $\{\mathsf{View}(P_x, \hat{V}_{x,y})\}_{x \in L, y \in \{0,1\}^*}$.*

Next, we recall the definition of USZK. For every polynomial $Q(\cdot)$, we denote by $V_{P,Q}$ the set of probabilistic *non-uniform* polynomial-time machines whose running time when interacting with the prover $P$ is bounded by $Q$. It is important to note that $V \in V_{P,Q}$ is allowed to access an infinite advice sequence $a_1, a_2, \cdots$ such that $|a_n| < Q(n)$ [3]. $V$ is not allowed to take any auxiliary-input, but instead allowed to use the advice string $a_{|x|}$ when $x$ is a common input (So the auxiliary-input in $V$'s view is always empty). Note that the encoding of a non-uniform polynomial-time machine $V$ is an infinite sequence $EN_1(V), EN_2(V), \cdots$. Denote by $EN_n(V)$ an encoding of a machine $V$ running on a common input $x$ of length $n$, where $a_n$ is incorporated into the encoding [4]. We denote by $EN(V)$ this sequence.

USZK allows the simulator to take as input the encoding of a cheating verifier.

**Definition 6 (USZK [Or87]).** *Let $P, V$ be two probabilistic interactive machines. We say that $(P, V)$ is* universal simulation *zero-knowledge for $L$ if there exists a PPT machine $US$ (the universal simulator) such that for every polynomial $Q(\cdot)$ and every $\hat{V} \in V_{P,Q}$, the following two distribution ensembles are computationally indistinguishable: $\{US(x, EN_{|x|}(\hat{V}))\}_{x \in L, EN_{|x|}(\hat{V}) \in EN(\hat{V})}$ and $\{\mathsf{View}(P_x, \hat{V}_x)\}_{x \in L, EN_{|x|}(\hat{V}) \in EN(\hat{V})}$, where $EN$ is an arbitrary encoding.*

Finally, we recall the definition of BSZK, where the simulator is only allowed to use the cheating verifier as a blackbox.

**Definition 7 (BSZK [GoOr94]).** *Let $P, V$ be two probabilistic interactive machines. We say that $(P, V)$ is* blackbox-simulation *zero-knowledge for $L$ if there exists a PPT machine $BS$ (the blackbox-simulator) such that for every polynomial $Q(\cdot)$ and every $\hat{V} \in V_{P,Q}$, the following two distribution ensembles are computationally indistinguishable even when the distinguishers are allowed blackbox access to $\hat{V}$: $\{BS^{\hat{V}}(x)\}_{x \in L}$ and $\{\mathsf{View}(P_x, \hat{V}_x)\}_{x \in L}$.*

---

[3]  Refer to [Go99, Section A.2.3] for more detail of non-uniform polynomial-time machines.

[4]  Refer to [HU79, Section 8.3] for an example of the encoding of machines.

The following theorem says that USZK is equivalent to AIZK [Or87].

**Theorem 1 (Oren [Or87]).** $Cl(AIZK) = Cl(USZK)$.

# 3    Defining Secure Code Obfuscator

In this section, we provide security definitions of code obfuscators. We deal with the code obfuscators for function ensembles [5]. For simplicity, we identify a function $f_s$ with its evaluator, i.e., a (non-uniform) polynomial-time machine which evaluates it. That is, the encoding of a function means the encoding of the machine which evaluates it. If the function is non-uniformly computable, the encoding must be done depending on the input length.

**Definition 8 (code obfuscator).** *Let $\mathcal{F}$ be a function ensemble. Let $\pi(\cdot)$ be any encoding. A* code obfuscator *$C$ for $\mathcal{F}$ is a PPT machine, which takes as input a code $\pi(f_s)$ of $f_s$ and outputs another code $\Pi(f_s)$ which is also a code of $f_s$.*

As sketched in Section 1.2, we try to define the security of code obfuscators based on the simulation paradigm. We require that whatever can be gained by an adversary having access to the code $\Pi(f_s)$ produced by a code obfuscator, can also be gained by a PPT machine having only blackbox-access to the function $f_s$.

We first give two unsatisfactory definitions.

**Definition 9.** *A code obfuscator $C$ for $\mathcal{F}$ is* semantically secure *if for every encoding $\pi$ and every PPT machine $A$ (the adversary), there exists a PPT machine $M$ (the simulator) such that the following two distribution ensembles are computationally indistinguishable: $\{s \leftarrow \{0,1\}^{l_s(n)}; \Pi(f_s) \leftarrow C(\pi(f_s)) : A(1^n, \Pi(f_s), z)\}_{n \in \mathsf{N}, z \in \{0,1\}^*}$ and $\{s \leftarrow \{0,1\}^{l_s(n)} : M^{f_s}(1^n, z)\}_{n \in \mathsf{N}, z \in \{0,1\}^*}$.*

Consider a simulator which chooses a seed $s'$ randomly, produces a obfuscated code $\Pi(f_{s'})$ and outputs $A(1^n, \Pi(f_{s'}))$. Clearly, this simulator can *perfectly* simulate the output of any adversary. Therefore, this definition does not make sense. By adding the obfuscated cod $\Pi(f_s)$ to two distributions, we can prevent a simulator from taking such a strategy.

**Definition 10.** *A code obfuscator $C$ for $\mathcal{F}$ is* semantically secure *if for every encoding $\pi$ and every PPT machine $A$ (the adversary), there exists a PPT machine $M$ (the simulator) such that the following two distribution ensembles are computationally indistinguishable: $\{s \leftarrow \{0,1\}^{l_s(n)}; \Pi(f_s) \leftarrow C(\pi(f_s)) : (\Pi(f_s), A(1^n, \Pi(f_s), z))\}_{n \in \mathsf{N}, z \in \{0,1\}^*}$ and $\{s \leftarrow \{0,1\}^{l_s(n)} : (\Pi(f_s), M^{f_s}(1^n, z))\}_{n \in \mathsf{N}, z \in \{0,1\}^*}$.*

---

[5] Although we can define the security for functions rather than function ensembles, we don't deal with them in this paper.

Again, this definition does not make sense. Consider an adversary who outputs the given obfuscated code as it is. For such an adversary, we can't expect the existence of a simulator which outputs some code of $f_s$ by accessing $f_s$ in blackbox fashion. Such a simulator exists only for a function ensemble that can be learned efficiently and such a function ensemble is not interesting here. For example, consider a function ensemble $\mathcal{F}_\oplus = \{F_n\}_{n \in \mathsf{N}}$ such that $F_n = \{f_s(x) = x \oplus s : \{0,1\}^n \to \{0,1\}^n\}_{s \in \{0,1\}^n}$. We can easily compute $s$ by accessing $f_s$ in blackbox fashion. Therefore, there is a trivial secure code obfuscator $C$ for $\mathcal{F}_\oplus$, that is, $C$ outputs a given code $\pi(f_s)$ as it is and the simulator $M$ computes the seed $s$ and outputs $\pi(f_s)$.

Due to the above failure, we restrict our attention to a particular adversary (rather than every adversary) and define the security only against it. We give two definitions: one is based on the simulation paradigm like the above and the other is based on indistinguishability of obfuscations.

**Definition 11 (semantic security against an adversary).** *A code obfuscator $C$ for $\mathcal{F}$ is semantically secure against an adversary $A$ (a PPT machine) if for every encoding $\pi$, there exists a PPT machine $M$ (the simulator) such that the following two distribution ensembles are computationally indistinguishable:*
$\{s \leftarrow \{0,1\}^{l_s(n)}; \Pi(f_s) \leftarrow C(\pi(f_s)) : (\Pi(f_s), A(1^n, \Pi(f_s), z))\}_{n \in \mathsf{N}, z \in \{0,1\}^*}$ *and*
$\{s \leftarrow \{0,1\}^{l_s(n)} : (\Pi(f_s), M^{f_s}(1^n, z))\}_{n \in \mathsf{N}, z \in \{0,1\}^*}$.

Another definition is based on indistinguishability of obfuscated codes of a given pair of functions $(f_s, f_{s'})$.

**Definition 12 (indistinguishable security against an adversary).** *A code obfuscator $C$ for $\mathcal{F}$ is indistinguishably secure against an adversary $A$ (a PPT machine) if for every encoding $\pi$, the following two distribution ensembles are computationally indistinguishable:* $\{s \leftarrow \{0,1\}^{l_s(n)}; \Pi(f_s) \leftarrow C(\pi(f_s)) : (\Pi(f_s), A(1^n, \Pi(f_s), z))\}_{n \in \mathsf{N}, z \in \{0,1\}^*}$ *and* $\{s \leftarrow \{0,1\}^{l_s(n)}; s' \leftarrow \{0,1\}^{l_s(n)}; \Pi(f_{s'}) \leftarrow C(\pi(f_{s'})) : (\Pi(f_s), A(1^n, \Pi(f_{s'}), z))\}_{n \in \mathsf{N}, z \in \{0,1\}^*}$.

Consider an adversary who outputs the size of a given code. Because there is a code obfuscator such that the size of an obfuscated code is uniquely determined from the size of the seed. Therefore, there exists a trivial secure code obfuscator which is indistinguishably secure against such an adversary.

It is easy to see that indistinguishable security implies semantic security. However, we don't know whether the reverse implication holds.

**Theorem 2.** *If a code obfuscator $C$ for a function ensemble $\mathcal{F}$ is indistinguishably secure against an adversary $A$ then it is semantically secure against $A$.*

We don't know whether there is a "non-trivial" function ensemble for which there is a code obfuscator semantically secure against a "non-trivial" adversary. As shown in the next section, this problem is related to the gap between AIZK and BSZK.

# 4    Zero-Knowledge and Code Obfuscation

In this section, we present our results. First of all, we show that the gap between AIZK and BSZK implies the existence of a cheating verifier for which no code obfuscators is semantically secure against an adversary.

Let $(P, V)$ be an interactive argument in $Cl(USZK)$ but not in $Cl(BSZK)$. We consider a cheating verifier $\hat{V} \in V_{P,Q}$. Denote by $l_r(\cdot)$, $l_m(\cdot)$ and $l_s(\cdot)$ the polynomials bounding the number of rounds, the length of a message and the number of random coins used by $\hat{V}$, respectively. We can identify $\hat{V}$ with a *non-uniformly computable* function ensemble $\hat{\mathcal{V}} = \{\hat{V}_n\}_{n \in \mathbb{N}}$, $\hat{V}_n = \{\hat{v}_s : \{0,1\}^{n+l_r(n)l_m(n)} \rightarrow \{0,1\}^{l_m(n)}\}_{s \in l_s(n)}$ such that $\hat{v}_s(x, m) = V(x, -, m; s)$.

Since $(P, V)$ achieves USZK, for every code obfuscator $C$ for $\hat{V}$, every encoding $\pi$ and every cheating verifier $\hat{V} \in V_{P,Q}$, there exists a universal simulator $US$ such that for every polynomial $Q$, the distribution ensemble $\{s \leftarrow \{0,1\}^{l_s(|x|)}; \Pi_{|x|}(\hat{v}_s) \leftarrow C(\pi_{|x|}(\hat{v}_s)) : (\Pi_{|x|}(\hat{v}_s), US(x, \Pi_{|x|}(\hat{v}_s)))\}_{x \in L}$ is computationally indistinguishable from a view of $\hat{V}$ interacting with $P$, that is, $\{s \leftarrow \{0,1\}^{l_s(|x|)}; \Pi_{|x|}(\hat{v}_s) \leftarrow C(\pi_{|x|}(\hat{v}_s)) : (\Pi_{|x|}(\hat{v}_s), \mathsf{View}(P_x, \hat{V}_x^s)) \}_{x \in L}$. We consider the universal simulator $US$ as an adversary.

On the other hand, since $(P, V)$ does not achieve BSZK, there exists a polynomial $Q$ and a non-uniform cheating verifier $\hat{V} \in V_{P,Q}$ such that for every blackbox-simulator $BS$, the distribution ensemble $\{s \leftarrow \{0,1\}^{l_s(|x|)}; \Pi_{|x|}(\hat{v}_s) \leftarrow C(\pi_{|x|}(\hat{v}_s)) : (\Pi_{|x|}(\hat{v}_s), BS^{\hat{v}_s}(x))\}_{x \in L}$ is **NOT** computationally indistinguishable from $\{s \leftarrow \{0,1\}^{l_s(|x|)}; \Pi_{|x|}(\hat{v}_s) \leftarrow C(\pi_{|x|}(\hat{v}_s)) : (\Pi_{|x|}(\hat{v}_s), \mathsf{View}(P_x, \hat{V}_x^s))\}_{x \in L}$.

From the above, the theorem follows:

**Theorem 3.** *Assume that it holds that $Cl(BSZK) \subset Cl(AIZK)$. Then no code obfuscator for $\hat{V}$ is semantically secure against $US$.*

Now we know that $Cl(BSZK) \subset Cl(AIZK)$ implies the existence of a function ensemble for which no code obfuscator is semantically secure against an adversary. However, we don't know the actual behavior of such a function ensemble. In the rest of this section, we focus on two cases $\mathsf{Gap}_{pc}$ and $\mathsf{Gap}_{sc}$ to prove non-existence of secure code obfuscators for specific function ensembles.

## 4.1    The Case of $\mathsf{Gap}_{pc}$

**Theorem 4.** *Assume that $\mathsf{Gap}_{pc}$ is true. Let $\mathcal{F}$ be any PRFE (both input and output length functions are specified in the proof). Then, no code obfuscator for $\mathcal{F}$ is semantically secure against an adversary (The behavior of the adversary is specified in the proof).*

*Proof.* Let $(P_0, V_0)$ be a constant round public-coin AIZK argument for a language $L$ outside of $\mathcal{BPP}$. We use the following notations for $(P_0, V_0)$. Denote by $x$ the common input and by $n$ the length of $x$. For simplicity of the exposition we make some assumptions on the form of the protocol without loss of generality. We assume both the first and last messages are sent by $P_0$. By adding dummy message any protocol can be converted into one of this form. Note that in such a protocol, the number of rounds is always an odd number $2m + 1$, where $m$ is a constant. The messages sent by $P_0$ are denoted by $\alpha_1, \alpha_2, \cdots, \alpha_m$ and $\gamma$ ($\alpha_i$ is $i$th message and $\gamma$ is $m + 1$th message). The messages sent by $V_0$ are denoted by $\beta_1, \beta_2, \cdots, \beta_m$ ($\beta_i$ is $i$th message). We assume that for every $i$, $\alpha_i$ and $\beta_i$ have length $l_\alpha(n)$ and $l_\beta(n)$, respectively. We also assume that for every $i$, $V_0$ chooses $\beta_i$ randomly in $\{0,1\}^{l_\beta(n)}$. The predicate computed by $V_0$ in order to decide whether to accept or reject is denoted by $\rho(x, \alpha_1, \beta_1, \cdots, \alpha_m, \beta_m, \gamma)$. That is, $V_0$ accepts $x$ if and only if $\rho(x, \alpha_1, \beta_1, \cdots, \alpha_m, \beta_m, \gamma) = \mathsf{Acc}$. This predicate may be a randomized function.

Let $\mathcal{F}$ be an $(l_\alpha, l_\beta, l_s)$-PRFE. We transform $(P_0, V_0)$ into another interactive argument $(P_1, V_1)$ using $\mathcal{F}$. $P_1$ is same as $P_0$. For every $i$ ($1 \le i \le m$), $V_1$ computes $\beta_i = f_s(\alpha_i)$ instead of choosing it randomly, where $s$ is randomly chosen from $\{0,1\}^{l_s(n)}$ at the beginning of the protocol only at once. From the pseudorandomness of $\mathcal{F}$, it follows that $(P_1, V_1)$ is also a $(2m + 1)$-round public-coin AIZK interactive argument for $L$.

Let $C$ be a code obfuscator $C$ for $\mathcal{F}$. We further transform $(P_1, V_1)$ into a 2-round protocol $(P_2, V_2)$ using $C$. The idea behind this transformation is that $V_2$ sends to the prover $P_2$ a code required for the computation of $V_1$ and makes $P_2$ compute the messages of $V_1$. Let $\pi$ be any encoding.

**Protocol:** $(P_2, V_2)$, where $x$ is a common input of length $n$ and $w$ is an auxiliary input to $P_2$.

**R1:** $V_2$ randomly chooses $s$ from $\{0,1\}^{l_s(n)}$ to get $\pi(f_s)$. Then $V_2$ use the code obfuscator $C$ to produce a code $\Pi(f_s)$ and send it to $P_2$.

**R2:** Using the code $\Pi(f_s)$, $P_2$ computes $\alpha_1 \leftarrow P_0(x, w, -)$, $\beta_1 = f_s(\alpha_1)$, $\alpha_2 \leftarrow P_0(x, w, \alpha_1\beta_1)$, $\beta_2 = f_s(\alpha_2)$, $\alpha_3 \leftarrow P_0(x, w, \alpha_1\beta_1\alpha_2\beta_2)$, $\cdots$, $\beta_m = f_s(\alpha_m)$, $\gamma \leftarrow P_0(x, w, \alpha_1\beta_1 \cdots \alpha_m\beta_m)$. Then $P_2$ sends $(\alpha_1, \cdots, \alpha_m, \gamma)$ to $V_2$.

**Decision:** $V_2$ outputs $\rho(x, \alpha_1, f_s(\alpha_1), \cdots, \alpha_m, f_s(\alpha_m), \gamma)$.

*Claim.* $(P_2, V_2)$ achieves AIZK. $(P_2, V_2)$ satisfies the efficient completeness, but doesn't satisfy the computational soundness.

*Proof.* Firstly, we show that $(P_2, V_2)$ achieves AIZK. We have a universal simulator $US$ guaranteed by the USZK or AIZK property of $(P_1, V_1)$. For every cheating verifier $\hat{V}_2 \in V_{P_2, Q}$, we can use $US$ to simulate the conversation between $P_2$ and $\hat{V}_2$. The simulation is as follows: (i) Simulate $\hat{V}_2$ to get a code $\Pi(f_s)$. (ii) Produce the code of $V_1$ using $f_s$ from $\Pi(f_s)$. We denote it by $\Pi(V_1^{f_s})$. (iii) Output $US(x, \Pi(V_1^{f_s}))$. It is easy to see that this output distribution is computationally indistinguishable from the real interaction between $P_2$ and $\hat{V}_2$.

The efficient completeness is clearly satisfied. From the triviality result regarding AIZK [GoOr94], it follows that if $(P_2, V_2)$ satisfies the computational soundness, then $L \in \mathcal{BPP}$. This contradicts our assumption. Therefore, the computational soundness is not satisfied. □

Now we return to the proof of Theorem 4. We construct an adversary $A$ for which any simulator fails to satisfy the requirement in Definition 11. Recall that $(P_2, V_2)$ does not satisfy the computational soundness (Claim 4.1). Let $\hat{P}_2$ be a cheating prover who can violate the computational soundness. Let $x$ be a string of length $n$ such that $x \notin L$. Let $w$ be an auxiliary-input to $\hat{P}_2$. Given a code of a function $f_s$, $A$ tries to output an accepting conversation. Given an input $(1^n, \Pi(f_s), (x, w))$, $A$ simply outputs $\hat{P}_2(x, w, \Pi(f_s))$, i.e., the messages $(\alpha_1, \alpha_2, \cdots, \gamma)$. Since $\hat{P}_2$ violates the computational soundness, it follows that for every code obfuscator $C$ for $\mathcal{F}$,

$$\Pr \left[ \begin{array}{l} s \leftarrow \{0,1\}^{l_s(n)}; \Pi(f_s) \leftarrow C(\pi(f_s)); \\ (\alpha_1, \alpha_2, \cdots, \alpha_m, \gamma) \leftarrow A(1^n, \Pi(f_s), (x, w)); \; : b = \mathsf{Acc} \\ b \leftarrow \rho(x, \alpha_1, f_s(\alpha_1), \cdots, \alpha_m, f_s(\alpha_m), \gamma) \end{array} \right]$$

is **NOT** negligible in $n$.

On the other hand, from the argument in [GoKr96, Proof of Lemma 6.4], it follows that for every PPT machine $M$ (simulator), every auxiliary-input $w$ and every code obfuscator $C$ for $\mathcal{F}$,

$$\Pr \left[ \begin{array}{l} u \leftarrow U_{l_\alpha(n), l_\beta(n)}; (\alpha_1, \alpha_2, \cdots, \alpha_m, \gamma) \leftarrow M^u(1^n, (x, w)); \\ b \leftarrow \rho(x, \alpha_1, f_s(\alpha_1), \cdots, \alpha_m, f_s(\alpha_m), \gamma) \end{array} \; : b = \mathsf{Acc} \right]$$

is negligible in $n$. By the pseudorandomness of $\mathcal{F}$, we can replace the uniform function $u$ by the function $f_s$. It follows that for every PPT machine $M$, every auxiliary-input $w$ and every code obfuscator $C$,

$$\Pr \left[ \begin{array}{l} s \leftarrow \{0,1\}^{l_s(n)}; (\alpha_1, \alpha_2, \cdots, \alpha_m, \gamma) \leftarrow M^{f_s}(1^n, (x, w)); \\ b \leftarrow \rho(x, \alpha_1, f_s(\alpha_1), \cdots, \alpha_m, f_s(\alpha_m), \gamma) \end{array} \; : b = \mathsf{Acc} \right]$$

is negligible in $n$. Since any simulator fails to simulate $A$, no code obfuscator $C$ for $\mathcal{F}$ can be semantically secure against $A$. $\qquad\square$

The theorem does not extend to the case of non-constant rounds since we use the argument in [GoKr96, Proof of Lemma 6.4].

## 4.2   The Case of $\mathsf{Gap}_{sc}$

In this section, we consider the second case. The argument here is essentially equivalent to the one in the previous section.

Let $(P_0, V_0)$ be a 3-round secret-coin AIZK argument for a language $L$ outside of $\mathcal{BPP}$. We denote by $\alpha$ and $\gamma$ the messages sent by $P_0$. We denote by $\beta$ the message sent by $V_0$. Denote by $R_{sc}$ the secret-coin used by $V_0$ to compute $\beta \leftarrow V_0(x, -, \alpha)$. The length functions of $\alpha, \beta, R_{sc}$ are denoted by $l_\alpha(\cdot), l_\beta(\cdot)$ and $l_{R_{sc}}(\cdot)$, respectively. The predicate computed by $V_0$ in order to decide whether to accept or reject is denoted by $\rho(x, \alpha, \gamma, R_{sc})$.

Let $\mathcal{F}$ be an $(l_\alpha, l_{R_{sc}}, l_s)$-PRFE. We transform $(P_0, V_0)$ into another interactive argument $(P_1, V_1)$ using $\mathcal{F}$. $P_1$ is same as $P_0$. $V_1$ computes $\beta$ as follows: Chooses $s$ randomly from $\{0,1\}^{l_s(n)}$, computes $R_{sc} = f_s(\alpha_i)$ and $\beta = V_0(x, -, \alpha; R_{sc})$. We denote by $CV = \{CV_n\}_{n \in \mathbb{N}}$ a function ensemble such that $CV_n = \{CV_s : CV_s(x, \alpha) = V_0(x, -, \alpha; f_s(\alpha))\}_{s \in \{0,1\}^{l_s(n)}}$. From the pseudo-randomness of $\mathcal{F}$, it follows that $(P_1, V_1)$ is a 3-round secret-coin AIZK interactive argument for $L$.

Now we prove our second theorem.

**Theorem 5.** *Assume that $\mathsf{Gap}_{sc}$ is true. Let $CV$ be a function ensemble specified as the above. Then, no code obfuscator for $CV$ is semantically secure against an adversary (The behavior of the adversary is specified in the proof).*

*Proof.* Let $C$ be a code obfuscator $C$ for $CV$. We further transform $(P_1, V_1)$ into a 2-round protocol $(P_2, V_2)$ using $C$. Let $\pi$ be any encoding.

**Protocol:** $(P_2, V_2)$, where $x$ is a common input of length $n$ and $w$ is an auxiliary input to $P_2$.

**R1:** $V_2$ randomly chooses $s$ from $\{0,1\}^{l_s(n)}$ to get $\pi(CV_s)$. Then $V_2$ use the code obfuscator $C$ to produce a code $\Pi(CV_s)$ and send it to $P_2$.

**R2:** Using the code $\Pi(CV_s)$, $P_2$ computes $\alpha \leftarrow P_0(x, w, -)$, $\beta = CV_s(x, \alpha) = V_0(x, -, \alpha; f_s(\alpha))$ and $\gamma \leftarrow P_0(x, w, \alpha\beta)$. Then $P_2$ sends $(\alpha, \gamma)$ to $V_2$.

**Decision:** $V_2$ outputs $\rho(x, \alpha, \gamma, f_s(\alpha))$.

*Claim.* $(P_2, V_2)$ achieves AIZK. $(P_2, V_2)$ satisfies the efficient completeness, but doesn't satisfy the computational soundness.

*Proof.* The proof is essentially equivalent to the one of Claim 4.1.    □

The rest of the proof is equivalent to the corresponding one of Theorem 4 except that we use the argument in [GoKr96, Section 6.3] instead of the one in [GoKr96, Proof of Lemma 6.4]. The detail is omitted.    □

The theorem does not extend to the case of more than 3 rounds since we use the argument in [GoKr96, Section 6.3].

## 5  Concluding Remarks

In this paper, we have shown the gap between $Cl(AIZK)$ and $Cl(BSZK)$ is closely related to code obfuscation. We have focused on the following two statements: (1) There exists a constant round public-coin AIZK argument for a language outside of $\mathcal{BPP}$. (2) There exists a 3-round secret-coin AIZK argument for a language outside of $\mathcal{BPP}$. We have shown that if these statements are true, it implies negative results for code obfuscation. If the former is true, there exists no semantically secure code obfuscator for a PRFE. A similar negative result regarding the latter statement has also been shown. We don't know whether these non-existence results are reasonable or not. However, our result is a good reason to believe that any construction of constant round public-coin or 3-round secret-coin AIZK arguments for non-trivial languages essentially requires a computational assumption with a reverse-engineering property. Indeed, the 3-round AIZK argument constructed in [HT98] requires a computational assumption with a reverse-engineering property.

Dwork et al. showed that if there exits a secure bit commitment function resilient to selective decommitment then there exist 3-round public-coin ZK arguments for any $\mathcal{NP}$ language [DNRS99]. It is an open problem whether there exists a bit commitment function resilient to selective decommitment. They showed that in several special cases, such a bit commitment function exists in a weaker sense. However, the weaker resilience seems to be insufficient for the existence of 3-round public-coin ZK arguments for an $\mathcal{NP}$ language. Combining their result with Theorem 4, we can easily obtain the following relationship: Under the assumption that it does not hold that $\mathcal{NP} \subseteq \mathcal{BPP}$, if there exits a secure bit commitment function resilient to selective decommitment, then there is no semantically secure code obfuscator for PRFEs[6].

We considered only the setting of ZK arguments, but the problem studied in this paper really applies to any setting where we need a simulator [GM84][Ca00]. It is interesting to investigate what could be proven in other settings.

# References

[BrCr86]   G. Brassard and C. Crépeau, "Non-Transitive Transfer of Confidence : A Perfect Zero-Knowledge Interactive Protocol for SAT and Beyond, " Proceedings of 27th FOCS, 1986.

[Ca00]     R. Canetti, "Security and Composition of Multiparty Cryptographic Protocols," Journal of Cryptology, Vol.13, No. 1, pp.143-202, 2000.

[CTL97]    C. Collberg, C. Thomborson and D. Low, "A Taxonomy of Obfuscating Transformations," Technical Report 148, Department of Computer Science, University of Auckland, 1997.

[DNRS99]   C. Dwork, M. Naor, O. Reingold and L. Stockmeyer, "Magic Functions, " Proceedings of 40th FOCS, 1999.

[Go99]     O. Goldreich, "Modern Cryptography, Probabilistic Proofs and Pseudorandomness," Algorithms and Combinatorics Vol.17, Springer, 1999.

[GGM86]    O. Goldreich, S. Goldwasser, and S. Micali, "How to Construct Random Functions," Journal of the ACM, Vol.33, No.4, pp.792-807, 1986.

[GoKr96]   O. Goldreich and H. Krawczyk, "On the Composition of Zero-Knowledge Proof Systems," SIAM Journal on Computing, Vol.25, No.1, pp.169-192, 1996.

[GoOr94]   O. Goldreich and Y. Oren, "Definitions and Properties of Zero-Knowledge Proof Systems," Journal of Cryptology, Vol.7, No. 1, pp.1-32, 1994.

[GoOs96]   O. Goldreich and R. Ostrovsky, "Software Protection and Simulation on Oblivious RAMs," Journal of the ACM, Vol.43, No.3, pp.431-473, 1996.

[GM84]     S. Goldwasser and S. Micali, "Probabilistic Encryption," J. Comput. System Sci., 28, pp.270-299, 1984.

[GMR85]    S. Goldwasser, S. Micali, and C. Rackoff, "The Knowledge Complexity of Interactive Proofs," Proceedings of 17th STOC, pp.291-304, 1985.

[HT98]     S. Hada and T. Tanaka, "On the Existence of 3-Round Zero-Knowledge Protocols, " Proceedings of CRYPTO'98, pp. 408-423, 1998. A revised version is available as Theory of Cryptography Library: Record 99-9.

---

[6]   This requires the assumption regarding the $\mathcal{BPP}$ v.s. $\mathcal{NP}$ problem. We can remove it by directly observing the definitions of secure bit commitment functions resilient to selective decommitment and secure code obfuscator.

[HU79]     J. E. Hopcroft and J. D. Ullman, "Introduction to Automata Theory, Languages, and Computation," Addison-Wesley, 1979.

[Or87]     Y. Oren, "On the Cunning Power of Cheating Verifiers: Some Observations about Zero-Knowledge Proofs," Proceedings of 28th FOCS, pp. 462-471, 1987.

# A Note on Security Proofs in the Generic Model

Marc Fischlin

Fachbereich Mathematik (AG 7.2)
Johann Wolfgang Goethe-Universität Frankfurt am Main
Postfach 111932
60054 Frankfurt/Main, Germany
marc@mi.informatik.uni-frankfurt.de
http://www.mi.informatik.uni-frankfurt.de/

**Abstract.** A discrete-logarithm algorithm is called generic if it does not exploit the specific representation of the cyclic group for which it is supposed to compute discrete logarithms. Such algorithms include the well-known Baby-Step-Giant-Step procedure as well as the Pohlig-Hellman algorithm. In particular, these algorithms match a lower bound of Nachaev showing that generic discrete-log algorithms require exponentially many group operations. Building on this lower bound, Shoup and subsequently Schnorr and Jakobsson proved other discrete-log-based protocols to be intractable in the generic model. Here, we discuss pitfalls when applying the generic model to other schemes than the discrete-log problem and when interpreting such lower bounds as security proofs for these schemes.

## 1 Introduction

The Baby-Step-Giant-Step algorithm and the Pohlig-Hellman algorithm to compute discrete logarithms in cyclic groups operate representation-independent, i.e., they do not rely on the specific representation of the group, and thus work for any cyclic group. These examples match a lower bound of Nachaev [20] proving that such generic algorithms need $\Omega(\sqrt{q})$ group operations to compute discrete logarithms in a group of size $q$. The index calculus method, though, defeats this lower bound as it requires subexponential time for groups like $\mathbb{Z}_p^*$ with standard binary encoding. Yet, the index calculus is not known to work for arbitrary groups, e.g., it seems to be inapplicable to elliptic curves [26].

¿From a theoretical point of view, it is easy to see that security proofs in the generic model do not generally transfer to "the real world" when adding an encoding, because generic algorithms might cause an exponential blow-up in comparison to Turing machines: for the group $(\mathbb{Z}_q, +)$ the discrete logarithm for some element $x \in \mathbb{Z}_q$ with respect to the generator 1 is simply $x$. If we use the standard binary encoding of $\mathbb{Z}_q$ then it is easy to compute discrete logarithms for an algorithm that operates on bit strings. A generic algorithm, however, requires $\Omega(\sqrt{q})$ steps to find the discrete logarithm, because it cannot take advantage of the trivial encoding.

T. Okamoto (Ed.): ASIACRYPT 2000, LNCS 1976, pp. 458–469, 2000.

Shoup [25] and subsequently Schnorr and Jakobsson [24] extended the idea of Nachaev and proved schemes relying on the discrete-logarithm problem to be intractable for generic algorithms. Nonetheless, applying this model and its lower bound to other discrete-log-based protocols should not be viewed as providing the same security level as the fact that, currently, optimal discrete-log finders for appropriate groups like elliptic curves are generic. The aim of this work is to highlight some of these pitfalls when proceeding from the discrete-log problem to more sophisticated schemes in the generic model.

We present a simple explanatory example. Moving from a computational task like computing discrete logarithms to a decisional problem, e.g., distinguishing encryptions, without additional consideration is a dangerous step: the representation in decisional problems is related to the problem at a different level. For instance, for discrete-log-based algorithms allegedly producing pseudorandom strings an inappropriate encoding of group elements may cause the output to be easily distinguishable from random, even though the encoding does not help to compute discrete logarithms.

As another example consider the signed ElGamal encryption scheme in [24]. Informally, a signed ElGamal encryption consists of an ordinary ElGamal encryption [11] together with a Schnorr signature [23]. In [24] it is shown that the signed ElGamal scheme is secure against adaptive chosen-ciphertext attacks [22] in a combination of the generic model and the random oracle model. This proof relies on the fact that the adversary cannot generate a group element without knowing a representation of this value with respect to a set of given group elements[1] and that this representation is known to a simulator reducing an adaptive attack with decryption requests to one without such queries.

In Section 3 we present a three-round negligible-error zero-knowledge protocol in the generic model for all languages in NP. Our protocol, too, applies the property that a generic adversary cannot compute group elements without being aware of a representation, and that a simulator knows these representations; for a similar but more complicated protocol see [16]. In [13] it has been proved that three-round negligible-error *black-box* (i.e., observing only external behavior of parties) zero-knowledge proofs can only exist for languages in BPP. Since our protocol does not obey this black-box approach —we see internal data of the simulated adversary, specifically, the representations of the group elements chosen by the adversary— we simplify the problem to achieve something which we do not know how to do otherwise. The same trick enables [24] to prove the signed ElGamal scheme to be unbreakable in this model.

Another problem with viewing intractability results in the generic model as security proofs is the dependency of cryptographic primitives in this setting. Consider the well-known Schnorr signature scheme [23] in which a signature corresponds to a proof of knowledge for the secret key. The challenge for this proof of knowledge is generated by applying a suitable hash function to a group

---

[1] A representation of $X$ with respect to group elements $g_1, \ldots, g_n$ is a sequence $a_1, \ldots, a_n$ of integers such that $X = \prod g_i^{a_i}$. As for the special case $n = 1$ a representation corresponds to the discrete log of $X$ with respect to $g_1$.

element and the message. This suggests that in the generic model the hash function itself must be treated as a black box, because it does not solely operate on bit strings but partially on group elements. The hash function is therefore closely connected to the group. As for an actual implementation in practice, it is therefore necessary that one verifies that choosing a good hash function to a presumably strong group does not give origin to undesired problems. We elaborate on this in Section 4 by discussing the Schnorr signature scheme. Namely, we show that, although this signature scheme seems to be secure in the generic model for appropriate hash functions, depending on the choice of the group to the hash function, in the real world we either obtain a secure combination or we get an easy-to-forge scheme.

In contrast to the generic approach, a classical proof in cryptography is modular: once proved secure under certain properties of the primitives involved, one can take *any* instantiation of *any* of the primitives satisfying these properties, and it is guaranteed that the combined scheme is secure. Hence, in the generic case additional attention must be paid when implementing the schemes. Since the aforementioned problems could be subtle and hidden very well, this introduces a dangerous source of flaws.

In conclusion, *even if some cryptographic protocol is provably secure in the generic model, this does not necessarily give us the same confidence as the observation that nowadays optimal algorithms for the discrete-logarithm problem for groups like elliptic curves are generic.*

## 2    Preliminaries

Since the aim of this work is to highlight principle problems with security proofs in the generic model, the following discussion and all results are kept at a very informal level.

Generic algorithms and their application to cryptography have been introduced to the crypto community by Shoup [25] relying on a result by Nachaev [20]. Shoup models generic algorithms by oracle Turing machines. That is, choose a random encoding of group elements and give the oracle Turing machine access to a group operation oracle taking as input the random encodings of two group elements $X, Y$ and returning the random encoding of $XY$ or $XY^{-1}$. Schnorr and Jakobsson [24] use the a slightly different approach by dividing data into group data and non-group data. "Their" generic algorithms operate on non-group data as Turing machines[2] whereas for group data the algorithm is only allowed to compute the group element $\prod X_i^{a_i}$ for elements $X_1, \ldots, X_n$ and integers $a_1, \ldots, a_n$ in a single oracle step.

In [24] the signed ElGamal encryption scheme —introduced in [17,27]— has been analyzed in a combination of the random oracle and generic model. As mentioned before, basically, an encryption consists of an ElGamal encryption with a tag, a Schnorr signature. The system parameters are a group $G$ of prime

---

[2] Actually, Schnorr and Jakobsson [24] allow the generic algorithms to compute arbitrary functions on the non-group data and do not even restrict to recursive functions.

order $q$, a generator $g$ of $G$, and a random oracle $H : G^3 \rightarrow \mathbb{Z}_q$. The secret and the public key are given by $x \in \mathbb{Z}_q$ and $X = g^x$. In order to encrypt a message $m \in G$ select random $r, s \in \mathbb{Z}_q$, compute an ElGamal ciphertext $R = g^r, Y = mX^r$ and a Schnorr signature with $g^s$ and $c = H(g^s, R, Y), z = s + cr \bmod q$. Finally, output $(R, Y, c, z)$ as the ciphertext of $m$. To decrypt with the secret key $x$ first check the validity of the signature tag, i.e., that $c = H(g^z R^{-c}, R, Y)$, and if so return the message $Y/R^x$. The idea is that the Schnorr signature with secret key $r$ for message $(R, Y)$ guarantees that the adversary knows $r$ and thus the message $Y/X^r = m$.

A first formal proof that this combination of ElGamal encryption and Schnorr signature is indeed secure against adaptive chosen-ciphertext attacks has been given in [27], under the assumption that $H$ is a random oracle, that the decisional Diffie-Hellman problem [6] is intractable, and based on a somewhat strong assumption about the unforgeability of Schnorr signatures. In [24] the scheme has been proved secure in the generic model given that $H$ is a random oracle.

In order to show that certain approaches are possible when observing internal behavior, but are not known to be achievable otherwise, we present an example based on zero-knowledge proofs. Hence, we briefly discuss the definition of zero-knowledge proofs in the generic model. See [12] for a comprehensive treatment of zero-knowledge protocols. Informally, a zero-knowledge protocol [15] for a language $\mathcal{L}$ is an interactive proof system between an unbounded party, called the prover $P$, and a probabilistic polynomial-time machine, the verifier $V$, such that the following holds:

- completeness: if $P$ and $V$ both honestly follow the protocol then $V$ always accepts inputs $x \in \mathcal{L}$.
- soundness: for $x \notin \mathcal{L}$ the verifier $V$ only accepts with probability $\epsilon(|x|)$ for any malicious prover $P^*$ pretending to be $P$. The function $\epsilon$ is called the error of the protocol. Likewise, if $\epsilon$ is negligible then the protocol has negligible error.
- zero-knowledge: for any $x \in \mathcal{L}$, any possibly malicious verifier $V^*$ does not not learn anything useful beyond the fact that $x \in \mathcal{L}$ (in a computational sense) from the protocol execution with $P$. That is, for any verifier $V^*$ there exists a probabilistic (expected) polynomial-time simulator $S$ such that for $x \in \mathcal{L}$ the simulator's output $S(V^*, x)$ is computationally indistinguishable [14] from the random variable that describes the exchanged messages of a protocol execution between $P$ and $V^*$.

Basically, augmenting interactive proof systems by a group oracle means to provide all parties $P, P^*, V, V^*, S$ access to the same oracle. This, of course, implies that we have to transfer the instinguishability property to the generic model. Instead of demanding that any generic algorithm (with access to the same group oracle) cannot distinguish the prover's and the simulator's answers, we *afterwards* encode in both cases all group elements in a group with some encoding, like $\mathbb{Z}_p^*$ and the binary representation. Clearly, this leads to a conditional statement that the output of the zero-knowledge simulator is indistinguishable (in the standard sense) from the prover's answers *under the assumption that the group* *with the encoding is secure.* By this, we circumvent to introduce distinguishers in the generic model.

The zero-knowledge simulator which we present in Section 3 will not be a black-box simulator as it learns the queries of $V^*$ submitted to the group oracle, and hence observes some internal behavior of $V^*$. In fact, this is crucial for our three-round negligible-error zero-knowledge protocol in the next section. Furthermore, this is exactly what is done in [24] in order to prove the signed ElGamal encryption scheme to be secure against adaptive chosen-ciphertext attacks. There, it is demonstrated that the adversary essentially cannot create ciphertexts without knowing the message, and that the message can be extracted by looking at the adversary's oracle queries to the group oracle. Schnorr and Jakobsson [24] call this plaintext awareness (and thus implicitly suggest a definition in the generic model, although they do not present a formal definition). They refer to plaintext awareness as defined in [5] rather than to the refinement given in [2]. We are not aware if the signed ElGamal scheme is plaintext aware according to this refinement.

Finally, let us recall the three-round discrete-log-based oblivious transfer protocol of Bellare and Micali [3]. We apply this protocol as a tool to derive our zero-knowledge scheme. Informally, a chosen-one-out-of-two oblivious transfer scheme is a two-party protocol between a sender possessing messages $m_0, m_1$ and the receiver. The receiver would like to learn $m_b$ from the sender such that the sender does not learn the receiver's choice $b$. On the other hand, the sender is willing to reveal one of the messages to the receiver, but does not want to give away anything about the other message. Bellare and Micali introduce the following protocol in a group $G$ of prime order $q$ generated by $g$.[3]

- The sender generates a random pair $x \in \mathbb{Z}_q$, $X = g^x$ of private and public key and sends $X$ to the receiver.
- The receiver, trying to get $m_b$, randomly chooses $y \in \mathbb{Z}_q$, sets $Y_b = g^y$ and $Y_{b \oplus 1} = XY_b^{-1}$ and transmits $Y_0, Y_1$.
- The sender checks that $Y_0 Y_1 = X$. If so, it selects uniformly $a_0, a_1 \in \mathbb{Z}_q$ and computes the ElGamal encryptions $(A_i, B_i) = (g^{a_i}, Y_i^{a_i} m_i)$ for $i = 0, 1$ (where we presume for simplicity that the messages are in some way encoded as group elements). The sender transmits both pairs to the receiver.
- The receiver, knowing the discrete-log of $Y_b$, can decrypt $m_b$.

Intuitively, the receiver can only know one of the secret keys of $Y_0, Y_1$ and thus learns only a single message, i.e., the other message is computationally hidden under the decisional Diffie-Hellman assumption [3]. Conversely, the sender does not learn in an information-theoretical sense which of the messages the receiver

---

[3] In the generic model, the group is given, while in the real world it is generated by the sender in the first step, say, by selecting a subgroup $G$ of $\mathbb{Z}_p^*$. Since only the sender's privacy but not the receiver's depends on the intractability of the discrete-log in this group, and because the receiver can verify that a proper group of prime order has been generated, we can simply assume that even a malicious sender chooses the group honestly.

wants to retrieve, because the values $Y_0, Y_1$ are distributed independently of $b$. We remark that the same functionality can be accomplished with a less efficient scheme based solely on the computational Diffie-Hellman assumption and hardcore predicates [3].

# 3    Three-Round Negligible-Error Zero-Knowledge for NP

In this section we present our three-round negligible-erro zero-knowledge protocol for the NP-complete language $\mathcal{DHC}$, Directed Hamiltonian Cycle. The well-known atomic protocol with error $1/2$ works as follows (cf. [12]):

- The prover permutes the common input graph $H$ with permutation $\pi$ to obtain a graph $\pi(H)$. Then the prover sends a bit-wise commitment of the describing matrix of $\pi(H)$, and a commitment of $\pi$. We assume that the commitment scheme is non-interactive, computationally-hiding and unconditionally-binding; such commitment schemes exist for example under the discrete-log assumption (see [12]).
- The verifier chooses a random bit $b$ and asks the prover to reveal a Hamiltonian cycle in the permuted graph ($b = 0$), or to show that the commitment sequence really contains an isomorphic copy of $H$ ($b = 1$).
- The prover acts accordingly, i.e., for $b = 0$ opens $n$ committed 1-bits of the matrix of $\pi(H)$ describing a directed Hamiltonian cycle, and for $b = 1$ decommits to all of $\pi$ and $\pi(H)$.
- The verifier decides upon the opening.

Obviously, this protocol is complete. Soundness holds with error $1/2$ because for input $H \notin \mathcal{DHC}$ the prover can answer at most one of the two possible challenges correctly. The zero-knowledge simulator tries to guess the challenge prior to the commitment, i.e., commits to an arbitrary graph with a random Hamiltonian cycle if the guess is $b = 0$, and to a random permutation of $H$ for $b = 1$. Then it obtains the challenge $b^*$ of the verifier $V^*$ and if $b = b^*$ the simulator opens the commitments accordingly; else it restarts. The expected number of trials until the simulator successfully guesses $b^*$ is two.

Assume that instead of opening the parts of commitment according to the challenge in the atomic protocol, the prover executes both openings in parallel but encrypts each of both sequences of decommitments with an independent secret key. Additionally, the prover transfers one of these keys obliviously to the verifier (who decides at random which key) with the Bellare-Micali protocol. This technique has already been successfully applied in other works about zero-knowledge proofs (e.g. [18,9]). Completeness and soundness of the atomic protocol are preserved. But it is not clear that the zero-knowledge property still holds, because not knowing the right key the trial-and-error simulator above cannot check that its guess is correct. In the generic model, though, the simulator sees the verifier's group oracle queries and thus learns which key the verifier has chosen.

An important observation is that we can move the commitment step of the first round to the third round without affecting the soundness property. This leads to the protocol in Figure 3, where we use the notation $[\mathsf{OT}(K_0, K_1, b)]_i$ to denote the message in the $i$-th round ($i = 1, 2, 3$) of the OT scheme of Bellare-Micali with private input $K_0, K_1$ of the prover and random secret bit $b$ of the verifier. Additionally, we denote by $\mathsf{Enc}_K(\cdot)$ a semantically-secure [14] encryption scheme, e.g., the basic ElGamal encryption scheme under the decisional Diffie-Hellman assumption [27], or a hardcore-predicate-based bit encryption scheme based on the computational Diffie-Hellman assumption.

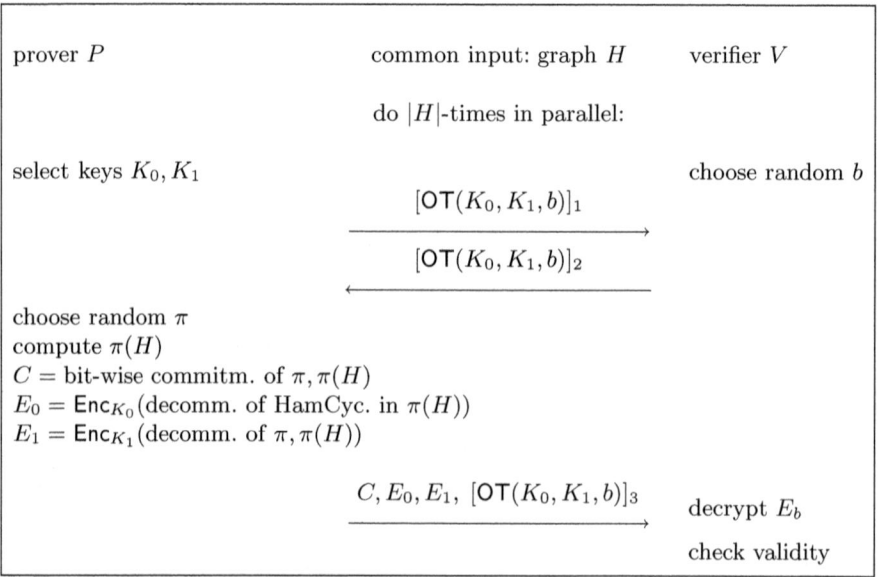

**Fig. 1.** Three-Round Negligible-Error Zero-Knowledge Proof of Hamiltonian Cycle

Apparently, our protocol is complete. A malicious prover $P^*$ can convince the verifier with probability at most $2^{-|H|}$ for inputs outside the language, because $P^*$ must then lie either about the permutation or about the Hamiltonian cycle, but does not know on which side he is checked (the verifier's choice $b$ is hidden information-theoretically in the Bellare-Micali protocol and the prover's commitments are unconditionally binding).

It remains to specify the zero-knowledge simulator $S$. In the generic model, $S$ knows which key the verifier in each parallel execution retrieves, because the simulator sees the internal group oracle queries of the verifier. That is, although the malicious verifier might generate $Y_0, Y_1$ in the oblivious transfer protocol different than the honest verifier, it always holds in the generic model that $Y_0 = g^a X^b$ and $Y_1 = X \cdot Y_0^{-1}$ for some $a, b$ which the simulator knows; thus, the verifier learns *at most* one of the keys $K_0, K_1$ and the simulator then knows which one (i.e., for $b \in \{0, 1\}$ the verifier knows the secret key to $Y_b$, and neither one for $b \notin \{0, 1\}$).

Our simulator imitates the simulator of the atomic protocol and chooses appropriate commitments and correct or dummy encryptions. That is, for each parallel execution our simulator selects keys $K_0, K_1$ and invokes in the oblivious transfer protocol with the verifier. The verifier answers with some group elements and we deduce which key (if any) the verifier wants to retrieve. If this is key $K_0$ we run the simulator of the atomic protocol to produce an appropriate commitment/decommitment for the guess $b = 0$; else, for $K_1$, we let this simulator generate a good instance for guess $b = 1$. We then correctly encrypt the opening with the chosen key and a produce dummy ciphertext of 0-bits with the other key. The fact that the views (afterwards encoded) are computationally indistinguishable under the decisional Diffie-Hellman assumption follows by standard techniques and is omitted. Also, if we use the less efficient variant of the Bellare-Micali protocol with hardcore predicates and the corresponding bit encryption scheme instead, then this scheme is computationally zero-knowledge under the computational Diffie-Hellman assumption.

Why does our result not contradict the lower bound in [13] for the round complexity of zero-knowledge proofs? The reason is that the BPP-algorithm in [13] relies on a black-box simulator observing merely the external behavior of the verifier. Here, the simulator sees some internal operations, namely, the queries to the group oracle. The same property has been used in [24]. Recall that a signed ElGamal encryption is a tuple $(R, Y, c, z) = (g^r, mX^r, c, z)$. In [24] it has been shown that submitting such a tuple to the decryption oracle can be simulated because the signature ensures that the adversary must have computed $g^r$ via the group oracle with very high probability; thus, $r$ is known to a simulator *that keeps track of the adversary's group oracle queries*. In particular, it follows that the answer $m = Y/X^r$ of the decryption oracle can be computed by the simulator without knowing the secret key to $X$. By this, any decryption requests can be simulated by a single group operation.

## 4    Instantiations of the Schnorr Signature Scheme

In this section we discuss problems when choosing bad combinations of instantiations of the primitives, although the constructed scheme is presumably secure in the generic model. Our demonstration example will be the Schnorr signature scheme [23]. First, let us briefly recall the system.

The scheme involves a group $G$ of prime order $q$ generated by some $g \in G$ and a hash function $H$; we will later discuss the properties of $H$. The secret key is a random element $x \in \mathbb{Z}_q$ and the public key is given by $X = g^x \in G$. To sign a message $m$ the signer chooses a random $r \in \mathbb{Z}_q$, computes $g^r$ and $c = H(g^r, m) \in \mathbb{Z}_q$ as well as $y = r + cx \bmod q$. The signature consists of the pair $(c, y)$. In order to verify a signature/message pair $(c, y), m$ the verifier calculates $Z = g^y X^{-c}$ and checks that $c = H(Z, m)$.

A potential attack on the Schnorr signature scheme is to reverse engineer the hash function, i.e., to choose a hash value $c$ beforehand and then to try to find $y \in \mathbb{Z}_q$ such that $H(g^y X^{-c}, m) = c$. Obviously, $(c, y)$ is then a valid signature. In practice, it is therefore assumed that $H$ is a collision-intractable hash function that cannot be reverse engineered.

We add the public parameters describing the group $G$ and $g, X$ to the hash evaluation process. That is, the hash value is computed as $H(\langle G \rangle, g, X, g^r, m)$ and as $H(\langle G \rangle, g, X, Z, m)$, respectively, where $\langle G \rangle$ denotes the group description. This is also suggested in [19] to prevent so-called adversarial hashing, and to best of our knowledge this does not weaken the Schnorr signature scheme. Nonetheless, it gives us the possibility to relate the hash function evaluation process to the underlying group.

We will consider two instantiations of the collision-intractable hash function and the group. Both instances use the same hash function, but each time a different cryptographically-strong group. One example will be completely insecure, whereas the other seems to be provide a secure signature scheme. By this, it follows that the choice of the group also affects the choice of the hash function and vice versa. As we will argue, both approaches conceivably provide a secure combination in the generic model. In contrast, a traditional security proof that a safe group and a collision-intractable hash function withstanding reverse engineering are sufficient would imply that any combination of, say, SHA-1 or MD5 with groups in $\mathbb{Z}_p^*$ or elliptic curves yields a secure scheme. Hence, security in the generic model does not support modular implementations in general.

For sake of clarity, we explain the example below for subgroups of $\mathbb{Z}_p^*$ of prime order $q$ with binary encoding. It also works for any other group, say, elliptic curves, if we hash down the binary representations of group elements to numbers between $0$ and $p - 1$. Let $h$ be a collision-intractable hash function that maps bit strings to the interval $[1, (q - 1)/2]$, viewed as a subset of $\mathbb{Z}_q$. Furthermore, let $h$ be secure against reverse engineering in the sense discussed above. Define the hash function $H$ for the signature scheme by dividing the input message $m$ into $m_1, m_2$ where $m_2 \in \{0, 1\}^{|p|}$ is interpreted as a group element in $\mathbb{Z}_p^*$. Set

$$H(p, q, g, X, R, m_1, m_2)$$
$$= \begin{cases} h(m_1) & \text{if } R \in [0, q) \text{ and } RX^{h(m_1)} = g \bmod p \\ & \text{and } g^R = m_2 \bmod p \\ h(R, m_1, m_2) + \frac{q-1}{2} \bmod q & \text{else} \end{cases}$$

It is easy to see that he derived hash function $H$ is collision-intractable for fixed $p, q, g, X$ and varying $(R, m_1, m_2)$.

The idea of the construction of $H$ is that its properties depend on the group. Specifically, assume that we choose $p = 2q + 1$. Then roughly half of the elements in $G \subseteq \mathbb{Z}_p^*$ fall into the interval $[0, q)$ (see [21]). If an adversary now picks $m_1$ at random and computes $R = gX^{-h(m_1)}$ then with probability approximately $1/2$ this value $R$ is less than $q$ as a natural number (assuming that the hash function $h$ distributes random values quite well). In this case, $RX^{h(m_1)} = g$ and for $m_2 = g^R \bmod p$ the hash funtion output equals $c = h(m_1)$. Thus, $(c, 1)$ is a valid signature for $(m_1, m_2)$ and the adversary easily succeeds in forging Schnorr signatures (without necessarily being able to compute discrete logarithms).

Now let $p = wq + 1$ for $w \gg q^2$. Assume for the moment that except for 1 none of the other $q - 1$ group elements lies in $[0, q]$. Unfortunately, we do not know whether this holds in general or not, and we are not aware of any results about the distribution of elements of this subgroup in $\mathbb{Z}_p^*$ (if the elements are almost uniformly in $\mathbb{Z}_p^*$ then this clearly follows from the choice of $w$). But again, we stress that this is an instructive example and we therefore admit this simplification. In this case, the hash function evaluation for $(R, m_1, m_2)$ can only result in $h(m_1)$ if $R = 1$. But then $RX^{h(m_1)} = X^{h(m_1)} = g$ is only possible for $\log_g X = 1/h(m_1) \bmod q$. This, in turn, is equivalent to computing the discrete-logarithm of $X$ to base $g$, and assuming the intractability of the discrete-log problem it is therefore very unlikely that this happens. Hence, given that the case above never occurs, the hash function evaluation merely results in values $h(R, m_1, m_2) + (q-1)/2 \bmod q$ and the scheme resembles to the original Schnorr system and is thus believed to be secure.

What happens in the generic model of Schnorr-Jakobsson? There, the adversary cannot interchange group data and non-group data. Hence, any hash function query cannot yield the answer $h(m_1)$ and the scheme is again conceivably secure. In other words, due to the generic model the hash function $H$ has the additional property of being immune against reverse engineering, although $H$ has not for the wrong choice of the group when implementing.

## 5    Conclusion

We have pointed out several pitfalls for security proofs in the generic model. Clearly, it is preferable to construct attractive protocols that are provably secure by classical methods. Yet, for some schemes used in practice like DSS such security proofs are still missing today (assuming that DSS can be proven secure at all). It is therefore a worthwhile effort to consider certain attacks on these schemes. But one should have in mind that it merely provides some evidence of hardness if these attacks fail. Also, the lack of proofs should incite researchers to find provably secure alternatives.

We remark that there are alternatives to the signed ElGamal scheme of [17,27,24] which are also discrete-log-based but require milder, yet still "non-standard" assumptions. One is the system based on the random oracle assumption and arbitrary trapdoor functions [5]. Another one is the DHAES scheme of Abdalla et al. [1] based on a potentially stronger assumption than the decisional Diffie-Hellmann assumption. The DHAES scheme seems to be at least as efficient as the signed ElGamal scheme: one exponentiation is traded for some private-key operations.

Finally, we remark that there is the ingenious encryption scheme of Cramer and Shoup [8] based only on the decisional Diffie-Hellman assumption; the

Cramer-Shoup scheme is only slightly less efficient than the signed ElGamal scheme.

## Acknowledgements

We thank Claus Schnorr for remarks on the generic model.

## References

1. M.ABDALLA, M.BELLARE, P.ROGAWAY: DHAES: An Encryption Scheme Based on the Diffie-Hellmann Problem, *available at* http://www-cse.ucsd.edu/users/mihir/, 1998.
2. M.BELLARE, A.DESAI, D.POINTCHEVAL, P.ROGAWAY: Relations Among Notions of Security for Public-Key Encryption Schemes, *Crypto '98, Lecture Notes in Computer Science, Vol. 1462, Springer-Verlag, pp. 26–45*, 1998.
3. M.BELLARE, S.MICALI: Non-Interactive Oblivious Transfer and Applications, *Crypto '89, Lecture Notes in Computer Science, Vol. 435, Springer-Verlag, pp. 547–559*, 1990.
4. M.BELLARE, P.ROGAWAY: Random Oracles are Practical: A Paradigm for Designing Efficient Protocols, *ACM Conference on Computer and Communication Security, pp. 62–73*, 1993.
5. M.BELLARE, P.ROGAWAY: Optimal Assymetric Encryption, *Eurocrypt '94, Lecture Notes in Computer Science, Vol. 950, Springer-Verlag, pp. 92–111*, 1994.
6. D.BONEH: The Decision Diffie-Hellman Problem, *Third Algorithmic Number Theory Symposium, Lecture Notes in Computer Science, Vol. 1423, Springer-Verlag, pp. 48–63*, 1998.
7. R.CANETTI, O.GOLDREICH, S.HALEVI: The Random Oracle Methodology, Revisited, *Proceedings of the 30th Annual ACM Symposium on the Theory of Computing, pp. 209–218*, 1998.
8. R.CRAMER, V.SHOUP: A Practical Public Key Cryptosystem Provably Secure Against Adaptive Chosen Ciphertext Attacks, *Crypto '98, Lecture Notes in Computer Science, Vol. 1462, Springer-Verlag, pp. 13–25*, 1998.
9. A.DE SANTIS, G.DI CRESCENZO, G.PERSIANO: Public-Key Cryptography and Zero-Knowledge Arguments, *Information and Computation, Vol. 121, No. 1, pp. 23–40*, 1995.
10. W.DIFFIE, M.HELLMAN: New Directions in Cryptography, *IEEE Transaction on Information Theory, Vol. 22, pp. 644–654*, 1976.
11. T.ELGAMAL: A Public Key Cryptosystem and a Signature Scheme Based on Discrete Logarithms, *IEEE Transaction on Information Theory, Vol. 31, pp. 469–472*, 1985.
12. O.GOLDREICH: Foundations of Cryptography (Fragments of a Book), *available at* http://www.wisdom.weizmann.ac.il/home/oded/public_html/index.html, 1998.
13. O.GOLDREICH, H.KRAWCZYK: On the Composition of Zero-Knowledge Proof Systems, *SIAM Journal on Computing, Vol. 25, pp. 169–192*, 1996.
14. S.GOLDWASSER, S.MICALI: Probabilistic Encryption, *Journal of Computer and System Sciences, Vol. 28(2), pp. 270–299*, 1984.

15. S.GOLDWASSSER, S.MICALI, C.RACKOFF: The Knowledge Complexity of Interactive Proof Systems, *SIAM Journal on Computing, Vol. 18, pp. 186–208*, 1989.
16. S.HADA, T.TANAKA: On the Existence of 3-Round Zero-Knowledge Protocols, *Crypto '98, Lecture Notes in Computer Science, Vol. 1462, Springer-Verlag, pp. 408–423*, 1998.
17. M.JAKOBSSON: A Practical Mix, *Eurocrypt '98, Lecture Notes in Computer Science, Vol. 1403, Springer-Verlag, pp. 448–461*, 1998.
18. J.KILIAN, S,MICALI, R.OSTROVSKY: Minimum Resource Zero-Knowledge Proofs, *Proceedings of the 30th IEEE Symposium on Foundations of Computer Science*, 1989.
19. S.MICALI, L.REYZIN: Signing with Partially Adversarial Hashing, *available at* `http://theory.lcs.mit.edu/~reyzin`, 2000.
20. V.NECHAEV: Complexity of a Determinate Algorithm for the Discrete Logarithm, *Mathematical Notes, Vol. 55, pp. 165–172*, 1994.
21. R.PERALTA: On the Distribution of Quadratic Residues and Non-Residues Modulo a Prime Number, *Mathematical Notes, Vol. 58*, 1995.
22. C.RACKOFF, D.SIMON: Non-Interactive Zero-Knowledge Proofs of Knowledge and Chosen Ciphertext Attacks, *Crypto '92, Lecture Notes in Computer Science, Vol. 576, Springer-Verlag, pp. 433–444*, 1998.
23. C.SCHNORR: Efficient Signature Generation for Smart Cards, *Journal of Cryptology, Vol. 4, pp. 161–174*, 1991.
24. C.SCHNORR, M.JAKOBSSON: Security of Signed ElGamal Encryption, *Asiacrypt 2000, Lecture Notes in Computer Science, Springer-Verlag*, 2000.
25. V.SHOUP: Lower Bounds for Discrete-Logarithm and Related Problems, *Eurocrypt '97, Lecture Notes in Computer Science, Vol. 1233, Springer-Verlag, pp. 256–266*, 1997.
26. J.SILVERMAN, J.SUZUKI: Elliptic Curve Discrete Logarithms and the Index Calculus, *Asiacrypt '98, Lecture Notes in Computer Science, Vol. 1514, Springer-Verlag, pp. 110–125*, 1998.
27. Y.TSIOUNIS, M.YUNG: On the Security of ElGamal Based Encryption, *PKC '98, Lecture Notes in Computer Science, Vol. 1431, Springer-Verlag, pp. 117–134*, 1998.

# On Relationships among Avalanche, Nonlinearity, and Correlation Immunity

Yuliang Zheng[1] and Xian-Mo Zhang[2]

[1] Monash University, Frankston, Melbourne, VIC 3199, Australia
yuliang.zheng@monash.edu.au,
http://www.netcomp.monash.edu.au/links/
[2] The University of Wollongong, Wollongong, NSW 2522, Australia
xianmo@cs.uow.edu.au

**Abstract.** We establish, for the first time, an explicit and simple lower bound on the nonlinearity $N_f$ of a Boolean function $f$ of $n$ variables satisfying the avalanche criterion of degree $p$, namely, $N_f \geq 2^{n-1} - 2^{n-1-\frac{1}{2}p}$. We also show that the lower bound is tight, and identify all the functions whose nonlinearity attains the lower bound. As a further contribution of this paper, we prove that except for very few cases, the sum of the degree of avalanche and the order of correlation immunity of a Boolean function of $n$ variables is at most $n-2$. These new results further highlight the significance of the fact that while avalanche property is in harmony with nonlinearity, it goes against correlation immunity.

**Key Words:**

Avalanche Criterion, Boolean Functions, Correlation Immunity, Nonlinearity, Propagation Criterion.

## 1 Introduction

Confusion and diffusion, introduced by Shannon [16], are two important principles used in the design of secret key cryptographic systems. These principles can be enforced by using some of the nonlinear properties of Boolean functions involved in a cryptographic transformation. More specifically, a high nonlinearity generally has a positive impact on confusion, whereas a high degree of avalanche enhances the effect of diffusion. Nevertheless, it is also important to note that some nonlinear properties contradict others. These motivate researchers to investigate into relationships among various nonlinear properties of Boolean functions.

One can consider three different relationships among nonlinearity, avalanche and correlation immunity, namely, nonlinearity and avalanche, nonlinearity and correlation immunity, and avalanche and correlation immunity. Zhang and Zheng [20] studied how avalanche property influences nonlinearity by establishing a number of upper and lower bounds on nonlinearity. Carlet [3] showed that one

T. Okamoto (Ed.): ASIACRYPT 2000, LNCS 1976, pp. 470–482, 2000.

may determine a number of different nonlinear properties of a Boolean function, if the function satisfies the avalanche criterion of a high degree. Zheng and Zhang [26] proved that Boolean functions satisfying the avalanche criterion in a hyper-space coincide with certain bent functions. They also established close relationships among plateaued functions with a maximum order, bent functions and the first order correlation immune functions [24]. Seberry, Zhang and Zheng were the first to research into relationships between nonlinearity and correlation immunity [14]. Very recently Zheng and Zhang have succeeded in deriving a new tight upper bound on the nonlinearity of high order correlation immune functions [25]. In the same paper they have also shown that correlation immune functions whose nonlinearity meets the tight upper bound coincide with plateaued functions introduced in [24,23]. All these results help further understand how nonlinearity and correlation immunity are at odds with each other.

The aim of this work is to widen our understanding of other connections among nonlinearity properties of Boolean functions, with a specific focus on relationships between nonlinearity and avalanche, and between avalanche and correlation immunity. We prove that if a function $f$ of $n$ variables satisfies the avalanche criterion of degree $p$, then its nonlinearity $N_f$ must satisfy the condition of $N_f \geq 2^{n-1} - 2^{n-1-\frac{1}{2}p}$. We also identify the cases when the equality holds, and characterize those functions that have the minimum nonlinearity. This result tells us that a high degree of avalanche guarantees a high nonlinearity.

In the second part of this paper, we look into the question of how avalanche and correlation immunity hold back each other. We prove that with very few exceptions, the sum of the degree of avalanche property and the order of correlation immunity of a Boolean function with $n$ variables is less than or equal to $n - 2$. This result clearly tells us that we cannot expect a function to achieve both a high degree of avalanche and a high order of correlation immunity.

## 2    Boolean Functions

We consider functions from $V_n$ to $GF(2)$ (or simply functions on $V_n$), where $V_n$ is the vector space of $n$ tuples of elements from $GF(2)$. The *truth table* of a function $f$ on $V_n$ is a $(0,1)$-sequence defined by $(f(\alpha_0), f(\alpha_1), \ldots, f(\alpha_{2^n-1}))$, and the *sequence* of $f$ is a $(1,-1)$-sequence defined by $((-1)^{f(\alpha_0)}, (-1)^{f(\alpha_1)}, \ldots, (-1)^{f(\alpha_{2^n-1})})$, where $\alpha_0 = (0, \ldots, 0, 0)$, $\alpha_1 = (0, \ldots, 0, 1)$, ..., $\alpha_{2^n-1} = (1, \ldots, 1, 1)$. A function is said to be *balanced* if its truth table contains $2^{n-1}$ zeros and an equal number of ones. Otherwise it called unbalanced.

The *matrix* of $f$ is a $(1,-1)$-matrix of order $2^n$ defined by $M = ((-1)^{f(\alpha_i \oplus \alpha_j)})$ where $\oplus$ denotes the addition in $V_n$.

Given two sequences $\tilde{a} = (a_1, \cdots, a_m)$ and $\tilde{b} = (b_1, \cdots, b_m)$, their *component-wise product* is defined by $\tilde{a} * \tilde{b} = (a_1 b_1, \cdots, a_m b_m)$. In particular, if $m = 2^n$ and $\tilde{a}, \tilde{b}$ are the sequences of functions $f$ and $g$ on $V_n$ respectively, then $\tilde{a} * \tilde{b}$ is the sequence of $f \oplus g$ where $\oplus$ denotes the addition in $GF(2)$.

Let $\tilde{a} = (a_1, \cdots, a_m)$ and $\tilde{b} = (b_1, \cdots, b_m)$ be two sequences or vectors, the *scalar product* of $\tilde{a}$ and $\tilde{b}$, denoted by $\langle \tilde{a}, \tilde{b} \rangle$, is defined as the sum of the component-wise multiplications. In particular, when $\tilde{a}$ and $\tilde{b}$ are from $V_m$, $\langle \tilde{a}, \tilde{b} \rangle = a_1 b_1 \oplus \cdots \oplus a_m b_m$, where the addition and multiplication are over $GF(2)$, and when $\tilde{a}$ and $\tilde{b}$ are $(1, -1)$-sequences, $\langle \tilde{a}, \tilde{b} \rangle = \sum_{i=1}^m a_i b_i$, where the addition and multiplication are over the reals.

An *affine* function $f$ on $V_n$ is a function that takes the form of $f(x_1, \ldots, x_n) = a_1 x_1 \oplus \cdots \oplus a_n x_n \oplus c$, where $a_j, c \in GF(2)$, $j = 1, 2, \ldots, n$. Furthermore $f$ is called a *linear* function if $c = 0$.

A $(1, -1)$-matrix $N$ of order $n$ is called a *Hadamard* matrix if $NN^T = nI_n$, where $N^T$ is the transpose of $N$ and $I_n$ is the identity matrix of order $n$. A Sylvester-Hadamard matrix of order $2^n$, denoted by $H_n$, is generated by the following recursive relation

$$H_0 = 1, \quad H_n = \begin{bmatrix} H_{n-1} & H_{n-1} \\ H_{n-1} & -H_{n-1} \end{bmatrix}, \quad n = 1, 2, \ldots.$$

Let $\ell_i$, $0 \le i \le 2^n - 1$, be the $i$ row of $H_n$. It is known that $\ell_i$ is the sequence of a linear function $\varphi_i(x)$ on $V_n$, defined by the scalar product $\varphi_i(x) = \langle \alpha_i, x \rangle$, where $\alpha_i$ is the binary representation of an integer $i$.

The *Hamming weight* of a $(0, 1)$-sequence $\xi$, denoted by $HW(\xi)$, is the number of ones in the sequence. Given two functions $f$ and $g$ on $V_n$, the *Hamming distance* $d(f, g)$ between them is defined as the Hamming weight of the truth table of $f(x) \oplus g(x)$, where $x = (x_1, \ldots, x_n)$.

## 3    Cryptographic Criteria of Boolean Functions

The following criteria for cryptographic Boolean functions are often considered: (1) **balance**, (2) **nonlinearity**, (3) **avalanche**, (4) **correlation immunity**, (5) **algebraic degree**, (6) absence of non-zero **linear structures**. In this paper we focus on avalanche, nonlinearity and correlation immunity.

Parseval's equation (Page 416 [8]) is a useful tool in this research: Let $f$ be a function on $V_n$ and $\xi$ denote the sequence of $f$. Then $\sum_{i=0}^{2^n-1} \langle \xi, \ell_i \rangle^2 = 2^{2n}$ where $\ell_i$ is the $i$th row of $H_n$, $i = 0, 1, \ldots, 2^n - 1$.

The *nonlinearity* of a function $f$ on $V_n$, denoted by $N_f$, is the minimal Hamming distance between $f$ and all affine functions on $V_n$, i.e.,

$$N_f = \min_{i=1,2,\ldots,2^{n+1}} d(f, \psi_i)$$

where $\psi_1, \psi_2, \ldots, \psi_{2^{n+1}}$ are all the affine functions on $V_n$. High nonlinearity can be used to resist a linear attack [9]. The following characterization of nonlinearity will be useful (for a proof see for instance [10]).

**Lemma 1.** *The nonlinearity of $f$ on $V_n$ can be expressed by*

$$N_f = 2^{n-1} - \frac{1}{2} \max\{|\langle \xi, \ell_i \rangle|, 0 \le i \le 2^n - 1\}$$

*where $\xi$ is the sequence of $f$ and $\ell_0, \ldots, \ell_{2^n-1}$ are the rows of $H_n$, namely, the sequences of linear functions on $V_n$.*

From Lemma 1 and Parseval's equation, it is easy to verify that $N_f \leq 2^{n-1} - 2^{\frac{1}{2}n-1}$ for any function $f$ on $V_n$. A function $f$ on $V_n$ is called a *bent function* if $\langle \xi, \ell_i \rangle^2 = 2^n$ for every $i$, $0 \leq i \leq 2^n - 1$ [13]. Hence $f$ is a bent function on $V_n$ if and only $N_f = 2^{n-1} - 2^{\frac{1}{2}n-1}$. It is known that a bent function on $V_n$ exists only when $n$ is even.

Let $f$ be a function on $V_n$. We say that $f$ satisfies the *avalanche criterion with respect to* $\alpha$ if $f(x) \oplus f(x \oplus \alpha)$ is a balanced function, where $x = (x_1, \ldots, x_n)$ and $\alpha$ is a vector in $V_n$. Furthermore $f$ is said to satisfy the *avalanche criterion of degree* $k$ if it satisfies the avalanche criterion with respect to every non-zero vector $\alpha$ whose Hamming weight is not larger than $k$. [1] From [13], a function $f$ on $V_n$ is bent if and only if $f$ satisfies the avalanche criterion of degree $n$. Note that the *strict avalanche criterion (SAC)* [18] is the same as the avalanche criterion of degree one.

Let $f$ be a function on $V_n$. For a vector $\alpha \in V_n$, denote by $\xi(\alpha)$ the sequence of $f(x \oplus \alpha)$. Thus $\xi(0)$ is the sequence of $f$ itself and $\xi(0) * \xi(\alpha)$ is the sequence of $f(x) \oplus f(x \oplus \alpha)$. Set $\Delta_f(\alpha) = \langle \xi(0), \xi(\alpha) \rangle$, the scalar product of $\xi(0)$ and $\xi(\alpha)$. $\Delta(\alpha)$ is called the auto-correlation of $f$ with a shift $\alpha$. We omit the subscript of $\Delta_f(\alpha)$ if no confusion occurs. Obviously, $\Delta(\alpha) = 0$ if and only if $f(x) \oplus f(x \oplus \alpha)$ is balanced, i.e., $f$ satisfies the avalanche criterion with respect to $\alpha$. In the case that $f$ does not satisfy the avalanche criterion with respect to a vector $\alpha$, it is desirable that $f(x) \oplus f(x \oplus \alpha)$ is almost balanced. Namely we require that $|\Delta_f(\alpha)|$ take a small value.

Let $f$ be a function on $V_n$. $\alpha \in V_n$ is called a *linear structure* of $f$ if $|\Delta(\alpha)| = 2^n$ (i.e., $f(x) \oplus f(x \oplus \alpha)$ is a constant). For any function $f$, we have $\Delta(\alpha_0) = 2^n$, where $\alpha_0$ is the zero vector on $V_n$. It is easy to verify that the set of all linear structures of a function $f$ form a linear subspace of $V_n$, whose dimension is called the *linearity of* $f$. A non-zero linear structure is cryptographically undesirable. It is also well-known that if $f$ has non-zero linear structures, then there exists a nonsingular $n \times n$ matrix $B$ over $GF(2)$ such that $f(xB) = g(y) \oplus \psi(z)$, where $x = (y, z)$, $y \in V_p$, $z \in V_q$, $g$ is a function on $V_p$ that has no non-zero linear structures, and $\psi$ is a linear function on $V_q$.

The following lemma is the re-statement of a relation proved in Section 2 of [4].

**Lemma 2.** *For every function $f$ on $V_n$, we have*

$$(\Delta(\alpha_0), \Delta(\alpha_1), \ldots, \Delta(\alpha_{2^n-1}))H_n = (\langle \xi, \ell_0 \rangle^2, \langle \xi, \ell_1 \rangle^2, \ldots, \langle \xi, \ell_{2^n-1} \rangle^2).$$

*where $\xi$ denotes the sequence of $f$, $\ell_i$ is the $i$th row of $H_n$, and $\alpha_i$ is the vector in $V_n$ that corresponds to the binary representation of $i$, $i = 0, 1, \ldots, 2^n - 1$.*

---

[1]  The avalanche criterion was called the propagation criterion in [12], as well as in all our earlier papers dealing with the subject. Historically, Feistel was apparently the first person who coined the term of "avalanche" and realized its importance in the design of a block cipher [6]. According to Coppersmith [5], a member of the team who designed DES, avalanche properties were employed in selecting the S-boxes used in the cipher, which contributed to the strength of the cipher against various attacks including differential [1] and linear [9] attacks.

The concept of correlation immune functions was introduced by Siegenthaler [17]. Xiao and Massey gave an equivalent definition [2,7]: A function $f$ on $V_n$ is called a *kth-order correlation immune function* if $\sum_{x \in V_n} f(x)(-1)^{\langle \beta, x \rangle} = 0$ for all $\beta \in V_n$ with $1 \le HW(\beta) \le k$, where in the the sum, $f(x)$ and $\langle \beta, x \rangle$ are regarded as real-valued functions. From Section 4.2 of [2], a correlation immune function can also be equivalently restated as follows: Let $f$ be a function on $V_n$ and let $\xi$ be its sequence. Then $f$ is called a *kth-order correlation immune function* if $\langle \xi, \ell \rangle = 0$ for every $\ell$, where $\ell$ is the sequence of a linear function $\varphi(x) = \langle \alpha, x \rangle$ on $V_n$ constrained by $1 \le HW(\alpha) \le k$. It should be noted that $\langle \xi, \ell \rangle = 0$, if and only if $f(x) \oplus \varphi(x)$ is balanced. Hence $f$ is a $k$th-order correlation immune function if and only if $f(x) \oplus \varphi(x)$ is balanced for each linear function $\varphi(x) = \langle \alpha, x \rangle$ on $V_n$ where $1 \le HW(\alpha) \le k$. Correlation immune functions are used in the design of running-key generators in stream ciphers to resist a correlation attack. Relevant discussions on correlation immune functions, and more generally on resilient functions, can be found in [22].

## 4    A Tight Lower Bound on Nonlinearity of Boolean Functions Satisfying Avalanche Criterion of Degree $p$

Let $(a_0, a_1, \ldots, a_{2^n-1})$ and $(b_0, b_1, \ldots, b_{2^n-1})$ be two real-valued sequences of length $2^n$, satisfying

$$(a_0, a_1, \ldots, a_{2^n-1})H_n = (b_0, b_1, \ldots, b_{2^n-1}) \tag{1}$$

Let $p$ be an integer with $1 \le p \le n-1$. Rewrite (1) as

$$(a_0, a_1, \ldots, a_{2^n-1})(H_{n-p} \times H_p) = (b_0, b_1, \ldots, b_{2^n-1}) \tag{2}$$

where $\times$ denotes the *Kronecker product* [19]. Let $e_j$ denote the $i$th row of $H_p$, $j = 0, 1, \ldots, 2^p - 1$. For any fixed $j$ with $0 \le j \le 2^p - 1$, comparing the $j$th, $(j + 2^p)$th, $\ldots$, $(j + (2^{n-p} - 1)2^p)$th terms in both sides of (2), we have

$$(a_0, a_1, \ldots, a_{2^n-1})(H_{n-p} \times e_j^T) = (b_j, b_{j+2^p}, b_{j+2 \cdot 2^p}, \ldots, b_{j+(2^{n-p}-1)2^p})$$

Write $(a_0, a_1, \ldots, a_{2^n-1}) = (\chi_0, \chi_1, \ldots, \chi_{2^{n-p}-1})$ where each $\chi_i$ is of length $2^p$. Then we have
$$(\langle \chi_0, e_j \rangle, \langle \chi_1, e_j \rangle, \ldots, \langle \chi_{2^{n-p}-1}, e_j \rangle)H_{n-p} = (b_j, b_{j+2^p}, b_{j+2 \cdot 2^p}, \ldots, b_{j+(2^{n-p}-1)2^p})$$
or equivalently,

$$2^{n-p}(\langle \chi_0, e_j \rangle, \langle \chi_1, e_j \rangle, \ldots, \langle \chi_{2^{n-p}-1}, e_j \rangle)$$
$$= (b_j, b_{j+2^p}, b_{j+2 \cdot 2^p}, \ldots, b_{j+(2^{n-p}-1)2^p})H_{n-p} \tag{3}$$

Let $\ell_i$ denote the $i$ row of $H_{n-p}$, where $j = 0, 1, \ldots, 2^{n-p} - 1$. In addition, write $(b_j, b_{j+2^p}, b_{j+2 \cdot 2^p}, \ldots, b_{j+(2^{n-p}-1)2^p}) = \lambda_j$, where $j = 0, 1, \ldots, 2^p - 1$. Comparing the $i$th terms in both sides of (3), we have $2^{n-p}\langle \chi_i, e_j \rangle = \langle \lambda_j, \ell_i \rangle$ where $\chi_i = (a_{i \cdot 2^p}, a_{1+i \cdot 2^p}, \ldots, a_{2^p-1+i \cdot 2^p})$. These discussions lead to the following lemma.

**Lemma 3.** *Let $(a_0, a_1, \ldots, a_{2^n-1})$ and $(b_0, b_1, \ldots, b_{2^n-1})$ be two real-valued sequences of length $2^n$, satisfying*

$$(a_0, a_1, \ldots, a_{2^n-1})H_n = (b_0, b_1, \ldots, b_{2^n-1})$$

*Let $p$ be an integer with $1 \leq p \leq n-1$. For any fixed $i$ with $0 \leq i \leq 2^{n-p}-1$ and any fixed $j$ with $0 \leq j \leq 2^p-1$, let $\chi_i = (a_{i \cdot 2^p}, a_{1+i \cdot 2^p}, \ldots, a_{2^p-1+i \cdot 2^p})$ and $\lambda_j = (b_j, b_{j+2^p}, b_{j+2 \cdot 2^p}, \ldots, b_{j+(2^{n-p}-1)2^p})$. Then we have*

$$2^{n-p}\langle \chi_i, e_j \rangle = \langle \lambda_j, \ell_i \rangle, \quad i = 0, 1, \ldots, 2^{n-p}-1, \quad j = 0, 1, \ldots, 2^p-1 \tag{4}$$

*where $\ell_i$ denotes the $i$th row of $H_{n-p}$ and $e_j$ denotes the $j$th row of $H_p$.*

Lemma 3 can be viewed as a refined version of the Hadamard transformation (1), and it will be a useful mathematical tool in proving the following two lemmas. These two lemmas will then play a significant role in proving the main results of this paper.

**Lemma 4.** *Let $f$ be a non-bent function on $V_n$, satisfying the avalanche criterion of degree $p$. Denote the sequence of $f$ by $\xi$. If there exists a row $L^*$ of $H_n$ such that $|\langle \xi, L^* \rangle| = 2^{n-\frac{1}{2}p}$, then $\alpha_{2^{t+p}+2^p-1}$ is a non-zero linear structure of $f$, where $\alpha_{2^{t+p}+2^p-1}$ is the vector in $V_n$ corresponding to the integer $2^{t+p}+2^p-1$, $t = 0, 1, \ldots, n-p-1$.*

*Proof.* First we note that $p > 0$. Since $f$ is not bent, $p \leq n-1$. Let us first rewrite the equality in Lemma 2 as follows

$$(\Delta(\alpha_0), \Delta(\alpha_1), \cdots, \Delta(\alpha_{2^n-1}))H_n = (\langle \xi, L_0 \rangle^2, \langle \xi, L_1 \rangle^2, \ldots, \langle \xi, L_{2^n-1} \rangle^2) \tag{5}$$

where $\alpha_i$ is the vector in $V_n$ corresponding to the integer $i$, and $L_i$ is the $i$th row of $H_n$, $i = 0, 1, \ldots, 2^n-1$. Set $i = 0$ in (4). Then we have $2^{n-p}\langle \chi_0, e_j \rangle = \langle \lambda_j, \ell_0 \rangle$. Since $f$ satisfies the avalanche criterion of degree $p$ and $HW(\alpha_j) \leq p$, $j = 1, \ldots 2^p-1$, we have

$$\Delta(\alpha_0) = 2^n, \quad \Delta(\alpha_1) = \cdots = \Delta(\alpha_{2^p-1}) = 0 \tag{6}$$

Applying $2^{n-p}\langle \chi_0, e_j \rangle = \langle \lambda_j, \ell_0 \rangle$ to (5), we obtain

$$2^{n-p}\Delta(\alpha_0) = \sum_{u=0}^{2^{n-p}-1} \langle \xi, L_{j+u \cdot 2^p} \rangle^2$$

or equivalently

$$\sum_{u=0}^{2^{n-p}-1} \langle \xi, L_{j+u \cdot 2^p} \rangle^2 = 2^{2n-p} \tag{7}$$

Since $L^*$ is a row of $H_n$, it can be expressed as $L^* = L_{j_0+u_0 \cdot 2^p}$, where $0 \leq j_0 \leq 2^p-1$ and $0 \leq u_0 \leq 2^{n-p}-1$. Set $j = j_0$ in (7), we have $\sum_{u=0}^{2^{n-p}-1} \langle \xi, L_{j_0+u \cdot 2^p} \rangle^2 = 2^{2n-p}$. From

$$\langle \xi, L_{j_0+u_0 \cdot 2^p} \rangle^2 = \langle \xi, L^* \rangle^2 = 2^{2n-p} \tag{8}$$

we have

$$\langle \xi, L_{j_0+u\cdot 2^p} \rangle = 0, \text{ for all } u, 0 \le u \le 2^{n-p} - 1, u \ne u_0 \tag{9}$$

Set $i = 2^t$ and $j = j_0$ in Lemma 3, where $0 \le t \le n - p - 1$, we have

$$2^{n-p}\langle \chi_{2^t}, e_{j_0} \rangle = \langle \lambda_{j_0}, \ell_{2^t} \rangle \tag{10}$$

where $\ell_{2^t}$ is the $2^t$th row of $H_{n-p}$ and $e_{j_0}$ is the $j_0$th row of $H_p$, $j = 0, 1, \ldots, 2^p - 1$. As $f$ satisfies the avalanche criterion of degree $p$ and $HW(\alpha_j) \le p$, $j = 2^{t+p}, 1 + 2^{t+p}, \ldots, 2^p - 2 + 2^{t+p}$, we have

$$\Delta(\alpha_{2^{t+p}}) = \Delta(\alpha_{1+2^{t+p}}) = \cdots = \Delta(\alpha_{2^p-2+2^{t+p}}) = 0 \tag{11}$$

Applying (10) to (5), and considering (8), (9) and (11), we have

$$2^{n-p}\Delta(\alpha_{2^p-1+2^{p+t}}) = \pm 2^{2n-p}$$

and thus

$$\Delta(\alpha_{2^p-1+2^{p+t}}) = \pm 2^n$$

This proves that $\alpha_{2^p-1+2^{p+t}}$ is indeed a non-zero linear structure of $f$, where $t = 0, 1, \ldots, n - p - 1$. □

**Lemma 5.** *Let $f$ be a non-bent function on $V_n$, satisfying the avalanche criterion of degree $p$. Denote the sequence of $f$ by $\xi$. If there exists a row $L^*$ of $H_n$, such that $|\langle \xi, L^* \rangle| = 2^{n-\frac{1}{2}p}$, then $p = n - 1$ and $n$ is odd.*

*Proof.* Since $|\langle \xi, L^* \rangle| = 2^{n-\frac{1}{2}p}$, $p$ must be even. Due to $p > 0$, we must have $p \ge 2$. We now prove the lemma by contradiction. Assume that $p \ne n - 1$. Since $p < n$, we have $p \le n - 2$. As $|\langle \xi, L^* \rangle| = 2^{n-\frac{1}{2}p}$, from Lemma 4, $\alpha_{2^{t+p}+2^p-1}$ is a non-zero linear structure of $f$, where $t = 0, 1, \ldots, n - p - 1$. Notice that $n - p - 1 \ge 1$. Set $t = 0, 1$. Thus both $\alpha_{2^p+2^p-1}$ and $\alpha_{2^{p+1}+2^p-1}$ are non-zero linear structures of $f$. Since all the linear structures of a function form a linear subspace, $\alpha_{2^p+2^p-1} \oplus \alpha_{2^{p+1}+2^p-1}$ is also a linear structure of $f$. Hence

$$\Delta(\alpha_{2^p+2^p-1} \oplus \alpha_{2^{p+1}+2^p-1}) = \pm 2^n \tag{12}$$

On the other hand, since $f$ satisfies the avalanche criterion of degree $p$ and $HW(\alpha_{2^p+2^p-1} \oplus \alpha_{2^{p+1}+2^p-1}) = 2 \le p$, we conclude that $\Delta(\alpha_{2^p+2^p-1} \oplus \alpha_{2^{p+1}+2^p-1}) = 0$. This contradicts (12). Thus we have $p > n - 2$. The only possible value for $p$ is $p = n - 1$. Since $p$ is even, $n$ must be odd. □

**Theorem 1.** *Let $f$ be a function on $V_n$, satisfying the avalanche criterion of degree $p$. Then*

*(i) the nonlinearity $N_f$ of $f$ satisfies $N_f \ge 2^{n-1} - 2^{n-1-\frac{1}{2}p}$,*
*(ii) the equality in (i) holds if and only if one of the following two conditions holds:*

(a) $p = n - 1$, $n$ is odd and $f(x) = g(x_1 \oplus x_n, \ldots, x_{n-1} \oplus x_n) \oplus h(x_1, \ldots, x_n)$, where $x = (x_1, \ldots, x_n)$, $g$ is a bent function on $V_{n-1}$, and $h$ is an affine function on $V_n$.

(b) $p = n$, $f$ is bent and $n$ is even.

*Proof.* Due to (7), i.e., $\sum_{u=0}^{2^{n-p}-1} \langle \xi, L_{j+u \cdot 2^p} \rangle^2 = 2^{2n-p}$, we have $\langle \xi, L_{j+u \cdot 2^p} \rangle^2 \leq 2^{2n-p}$. Since $u$ and $j$ are arbitrary, by using Lemma 1, we have $N_f \geq 2^{n-1} - 2^{n-1-\frac{1}{2}p}$. Now assume that

$$N_f = 2^{n-1} - 2^{n-1-\frac{1}{2}p} \tag{13}$$

From Lemma 1, there exists a row $L^*$ of $H_n$ such that $|\langle \xi, L^* \rangle| = 2^{n-\frac{1}{2}p}$. Two cases need to be considered: $f$ is non-bent and $f$ is bent. When $f$ is non-bent, thanks to Lemma 5, we have $p = n - 1$ and $n$ is odd. Considering Proposition 1 of [3], we conclude that $f$ must takes the form mentioned in (a). On the other hand, if $f$ is bent, then $p = n$ and $n$ is even. Hence (b) holds.

Conversely, assume that $f$ takes the form in (a). Applying a nonsingular linear transformation on the variables, and considering Proposition 3 of [11], we have $N_f = 2N_g$. Since $g$ is bent, we have $N_f = 2^{n-1} - 2^{\frac{1}{2}(n-1)}$. Hence (13) holds, where $p = n - 1$. On the other hand, it is obvious that (13) holds whenever (b) does. □

## 5  Relationships between Avalanche and Correlation Immunity

To prove the main theorems, we introduce two more results. The following lemma is part of Lemma 12 in [15].

**Lemma 6.** *Let $f_1$ be a function on $V_s$ and $f_2$ be a function on $V_t$. Then $f_1(x_1, \ldots, x_s) \oplus f_2(y_1, \ldots, y_t)$ is a balanced function on $V_{s+t}$ if $f_1$ or $f_2$ is balanced.*

Next we look at the structure of a function on $V_n$ that satisfies the avalanche criterion of degree $n - 1$.

**Lemma 7.** *Let $f$ be a function on $V_n$. Then*

(i) *$f$ is non-bent and satisfies the avalanche criterion of degree $n - 1$, if and only if $n$ is odd and $f(x) = g(x_1 \oplus x_n, \ldots, x_{n-1} \oplus x_n) \oplus c_1 x_1 \oplus \cdots \oplus c_n x_n \oplus c$, where $x = (x_1, \ldots, x_n)$, $g$ is a bent function on $V_{n-1}$, and $c_1, \ldots, c_n$ and $c$ are all constants in $GF(2)$,*

(ii) *$f$ is balanced and satisfies the avalanche criterion of degree $n - 1$, if and only if $n$ is odd and $f(x) = g(x_1 \oplus x_n, \ldots, x_{n-1} \oplus x_n) \oplus c_1 x_1 \oplus \cdots \oplus c_n x_n \oplus c$, where $g$ is a bent function on $V_{n-1}$, and $c_1, \ldots, c_n$ and $c$ are all constant in $GF(2)$, satisfying $\bigoplus_{j=1}^{n} c_j = 1$.*

*Proof.* (i) holds due to Proposition 1 of [3].

Assume that $f$ is balanced and satisfies the avalanche criterion of degree $n-1$. Since $f$ is balanced, it is non-bent. From (i) of the lemma, $f(x) = g(x_1 \oplus x_n, \ldots, x_{n-1} \oplus x_n) \oplus c_1 x_1 \oplus \cdots \oplus c_n x_n \oplus c$, where $x = (x_1, \ldots, x_n)$, $g$ is a bent function on $V_{n-1}$, and $c_1, \ldots, c_n$ and $c$ are all constant in $GF(2)$. Set $u_j = x_j \oplus x_n$, $j = 1, \ldots, n-1$. We have $f(u_1, \ldots, u_{n-1}, x_n) = g(u_1, \ldots, u_{n-1}) \oplus c_1 u_1 \oplus \cdots \oplus c_{n-1} u_{n-1} \oplus (c_1 \oplus \cdots \oplus c_n) x_n \oplus c$. Since $g(u_1, \ldots, u_{n-1}) \oplus c_1 x_1 \oplus \cdots \oplus c_{n-1} u_{n-1}$ is a bent function on $V_{n-1}$, it is unbalanced. On the other hand, since $f$ is balanced, we conclude that $\bigoplus_{j=1}^{n} c_j \neq 0$, namely, $\bigoplus_{j=1}^{n} c_j = 1$. This proves the necessity for (ii). Using the same reasoning as in the proof of (i), and taking into account Lemma 6, we can prove the sufficiency for (ii).     □

## 5.1   The Case of Balanced Functions

**Theorem 2.** *Let $f$ be a balanced $q$th-order correlation immune function on $V_n$, satisfying the avalanche criterion of degree $p$. Then we have $p + q \leq n - 2$.*

*Proof.* First we note that $q > 0$ and $p > 0$. Since $f$ is balanced, it cannot be bent. We prove the theorem in two steps. The first step deals with $p+q \leq n-2$, and the second step with $p + q \leq n - 1$.

We start with proving that $p + q \leq n - 1$ by contradiction. Assume that $p + q \geq n$. Set $i = 0$ and $j = 0$ in (4), we have $2^{n-p}\langle \chi_0, e_0 \rangle = \langle \lambda_0, \ell_0 \rangle$. Since $f$ satisfies the avalanche criterion of degree $p$ and $HW(\alpha_j) \leq p$, $j = 1, \ldots 2^p - 1$, we know that (6) holds. Note that $HW(\alpha_{u \cdot 2^p}) \leq n - p \leq q$ for all $u$, $0 \leq u \leq 2^{n-p} - 1$. Since $f$ is a balanced $q$th-order correlation immune function, we have

$$\langle \xi, L_0 \rangle = \langle \xi, L_{2^p} \rangle = \langle \xi, L_{2 \cdot 2^p} \rangle = \cdots = \langle \xi, L_{(2^{n-p}-1) \cdot 2^p} \rangle = 0 \tag{14}$$

Applying $2^{n-p}\langle \chi_0, e_0 \rangle = \langle \lambda_0, \ell_0 \rangle$ to (5), and noticing (6) and (14), we would have $2^{n-p}\Delta(\alpha_0) = 0$, i.e., $2^{2n-p} = 0$. This cannot be true. Hence we have proved that $p + q \leq n - 1$.

Next we complete the proof by showing that $p + q \leq n - 2$. Assume for contradiction that the theorem is not true, i.e., $p + q \geq n - 1$. Since we have already proved that $p + q \leq n - 1$, by assumption we should have $p + q = n - 1$. Note that $HW(\alpha_{u \cdot 2^p}) \leq n - p - 1 = q$ for all $u$ with $0 \leq u \leq 2^{n-p} - 2$, and $f$ is a balanced $q$th-order correlation immune function, where $q = n - p - 1$. Hence (14) still holds, with the exception that the actual value of $\langle \xi, L_{(2^{n-p}-1) \cdot 2^p} \rangle$ is not clear yet. Applying $2^{n-p}\langle \chi_0, e_0 \rangle = \langle \lambda_0, \ell_0 \rangle$ to (5), and noticing (6) and (14), we have $2^{n-p}\Delta(\alpha_0) = \langle \xi, L_{(2^{n-p}-1) \cdot 2^p} \rangle^2$. Thus we have $\langle \xi, L_{(2^{n-p}-1) \cdot 2^p} \rangle^2 = 2^{2n-p}$. Due to Lemma 5, we have $p = n - 1$. Since $q \geq 1$, we obtain $p + q \geq n$. This contradicts the inequality $p + q \leq n - 1$, that we have already proved. Hence $p + q \leq n - 2$ holds.     □

## 5.2   The Case of Unbalanced Functions

We turn our attention to unbalanced functions. A direct proof of the following Lemma can be found in [21].

**Lemma 8.** *Let $k \geq 2$ be a positive integer and $2^k = a^2 + b^2$, where both $a$ and $b$ are integers with $a \geq b \geq 0$. Then $a = 2^{\frac{1}{2}k}$ and $b = 0$ when $k$ is even, and $a = b = 2^{\frac{1}{2}(k-1)}$ otherwise.*

**Theorem 3.** *Let $f$ be an unbalanced $q$th-order correlation immune function on $V_n$, satisfying the avalanche criterion of degree $p$. Then*

*(i) $p + q \leq n$,*
*(ii) the equality in (i) holds if and only if $n$ is odd, $p = n - 1$, $q = 1$ and $f(x) = g(x_1 \oplus x_n, \ldots, x_{n-1} \oplus x_n) \oplus c_1 x_1 \oplus \cdots \oplus c_n x_n \oplus c$, where $x = (x_1, \ldots, x_n)$, $g$ is a bent function on $V_{n-1}$, $c_1, \ldots, c_n$ and $c$ are all constants in $GF(2)$, satisfying $\bigoplus_{j=1}^{n} c_j = 0$.*

*Proof.* Since $f$ is correlation immune, it cannot be bent. Once again we now prove (i) by contradiction. Assume that $p + q > n$. Hence $n - p < q$. We keep all the notations in Section 5.1. Note that $HW(\alpha_{u \cdot 2^p}) \leq n - p < q$ for all $u$ with $1 \leq u \leq 2^{n-p} - 1$. Since $f$ is an unbalanced $q$th-order correlation immune function, we have (14) again, with the understanding that $\langle \xi, L_0 \rangle \neq 0$. Applying $2^{n-p} \langle \chi_0, e_0 \rangle = \langle \lambda_0, \ell_0 \rangle$ to (5), and noticing (6) and (14) with $\langle \xi, L_0 \rangle \neq 0$, we have $2^{n-p} \Delta(\alpha_0) = \langle \xi, L_0 \rangle^2$. Hence $\langle \xi, L_0 \rangle^2 = 2^{2n-p}$ and $p$ must be even. Since $f$ is not bent, noticing Lemma 5, we can conclude that $p = n - 1$ and $n$ is odd. Using (ii) of Lemma 7, we have

$$f(x) = g(x_1 \oplus x_n, \ldots, x_{n-1} \oplus x_n) \oplus c_1 x_1 \oplus \cdots \oplus c_n x_n \oplus c$$

where $x = (x_1, \ldots, x_n)$, $g$ is a bent function on $V_{n-1}$, and $c_1, \ldots, c_n$ and $c$ are all constants in $GF(2)$, satisfying $\bigoplus_{j=1}^{n} c_j = 0$. One can verify that while $x_j \oplus f(x)$ is balanced, $j = 1, \ldots, n$, $x_j \oplus x_i \oplus f(x)$ is not if $j \neq i$. Hence $f$ is 1st-order, but not 2nd-order, correlation immune. Since $q > 0$, we have $q = 1$ and $p + q = n$. This contradicts the assumption that $p + q > n$. Hence we have proved that $p + q \leq n$.

We now prove (ii). Assume that $p + q = n$. Since $n - p = q$, we can apply $2^{n-p} \langle \chi_0, e_0 \rangle = \langle \lambda_0, \ell_0 \rangle$ to (5), and have (6) and (14) with $\langle \xi, L_0 \rangle \neq 0$. By using the same reasoning as in the proof of (i), we can arrive at the conclusion that (ii) holds. □

**Theorem 4.** *Let $f$ be an unbalanced $q$th-order correlation immune function on $V_n$, satisfying the avalanche criterion of degree $p$. If $p + q = n - 1$, then $f$ also satisfies the avalanche criterion of degree $p + 1$, $n$ is odd and $f$ must take the form mentioned in (ii) of Theorem 3.*

*Proof.* Let $p + q = n - 1$. Note that $HW(\alpha_{u \cdot 2^p}) \leq n - p - 1 = q$ for all $u$, $0 \leq u \leq 2^{n-p} - 2$. Since $f$ is unbalanced and $q$th-order correlation immune, we have (14), although once again $\langle \xi, L_0 \rangle \neq 0$ and the value of $\langle \xi, L_{(2^{n-p}-1) \cdot 2^p} \rangle$ is not clear yet. Applying $2^{n-p} \langle \chi_0, e_0 \rangle = \langle \lambda_0, \ell_0 \rangle$ to (5), noticing (6) and (14), with the understanding that $\langle \xi, L_0 \rangle \neq 0$ and $\langle \xi, L_{(2^{n-p}-1)\cdot 2^p} \rangle$ is not decided yet, we have $2^{n-p}\Delta(\alpha_0) = \langle \xi, L_0 \rangle^2 + \langle \xi, L_{(2^{n-p}-1)\cdot 2^p} \rangle^2$. That is

$$\langle \xi, L_0 \rangle^2 + \langle \xi, L_{(2^{n-p}-1)\cdot 2^p} \rangle^2 = 2^{2n-p} \tag{15}$$

There exist two cases to be considered: $p$ is even and $p$ is odd.

Case 1: $p$ is even and thus $p \geq 2$. Since $\langle \xi, L_0 \rangle \neq 0$, applying Lemma 8 to (15), we have $\langle \xi, L_0 \rangle^2 = 2^{2n-p}$ and $\langle \xi, L_{(2^{n-p}-1)\cdot 2^p} \rangle = 0$. Due to Lemma 5, $p = n - 1$. Since $q > 0$, we have $p + q \geq n$. This contradicts the assumption $p + q = n - 1$. Hence $p$ cannot be even.

Case 2: $p$ is odd. Applying Lemma 8 to (15), we obtain

$$\langle \xi, L_0 \rangle^2 = \langle \xi, L_{(2^{n-p}-1)\cdot 2^p} \rangle^2 = 2^{2n-p-1} \tag{16}$$

Set $i = 2^t$, $t = 0, 1, \ldots, n - p - 1$, where $n - p - 1 = q > 0$, and $j = 0$ in (4), we have

$$2^{n-p}\langle \chi_{2^t}, e_0 \rangle = \langle \lambda_0, \ell_{2^t} \rangle \tag{17}$$

where $\ell_{2^t}$ is the $2^t$th row of $H_{n-t}$ and $e_0$ is the all-one sequence of length $2^p$.

Since $f$ satisfies the avalanche criterion of degree $p$ and $HW(\alpha_j) \leq p$, $j = 2^{t+p}, 1 + 2^{t+p}, \ldots, 2^p - 2 + 2^{t+p}$, (11) holds.

Applying (17) to (5), noticing (11) and (14) with $\langle \xi, L_0 \rangle^2 = \langle \xi, L_{(2^{n-p}-1)\cdot 2^p} \rangle^2 = 2^{2n-p+1}$, we have $2^{n-p}\Delta(\alpha_{2^{t+p}+2^p-1}) = 2^{2n-p}$ or 0. In other words, $\Delta(\alpha_{2^{t+p}+2^p-1}) = 2^n$ or 0.

Note that $\ell_{2^t}$ is the sequence of a linear function $\psi$ on $V_{n-p}$ where $\psi(y) = \langle \beta_{2^t}, y \rangle$, $y \in V_{n-p}$, $\beta_{2^t} \in V_{n-p}$ corresponds to the binary representation of $2^t$. Due to (17), it is easy to verify that $\Delta(\alpha_{2^{t+p}+2^p-1}) = 2^n$ (or 0) if and only if $\langle \beta_{2^{n-p}-1}, \beta_{2^t} \rangle = 0$ (or 1) where $\beta_{2^{n-p}-1} \in V_{n-p}$ corresponds to the binary representation of $2^{n-p} - 1$. Note that $\beta_{2^{n-p}-1} = (0, \ldots, 0, 1, \ldots, 1)$ where the number of ones is equal to $n - p$. On the other hand $\beta_{2^t}$ can be written as $\beta_{2^t} = (0, \ldots, 0, 1, 0, \ldots, 0)$. Since $t \leq n - p - 1$, we conclude that $\langle \beta_{2^{n-p}-1}, \beta_{2^t} \rangle = 1$, for all $t$ with $0 \leq t \leq n - p - 1$. Hence $\Delta(\alpha_{2^{t+p}+2^p-1}) = 0$ for all such $t$.

Note that $HW(\alpha_{2^{t+p}+2^p-1}) = p + 1$. Permuting the variables, we can prove in a similar way that $\Delta(\alpha) = 0$ holds for each $\alpha$ with $HW(\alpha) = p + 1$. Hence $f$ satisfies the avalanche criterion of degree $p + 1$. Due to $p + q = n - 1$, we have $(p + 1) + q = n$. Using Theorem 3, we conclude that $n$ is odd and $f$ takes the form mentioned in (ii) of Theorem 3. $\square$

From Theorems 3 and 4, we conclude

**Corollary 1.** *Let $f$ be an unbalanced $q$th-order correlation immune function on $V_n$, satisfying the avalanche criterion of degree $p$. Then*

*(i) $p + q \leq n$, and the equality holds if and only if $n$ is odd, $p = n - 1$, $q = 1$ and $f(x) = g(x_1 \oplus x_n, \ldots, x_{n-1} \oplus x_n) \oplus c_1 x_1 \oplus \cdots \oplus c_n x_n \oplus c$, where $x = (x_1, \ldots, x_n)$, $g$ is a bent function on $V_{n-1}$, $c_1, \ldots, c_n$ and $c$ are all constants in $GF(2)$, satisfying $\bigoplus_{j=1}^n c_j = 0$,*

*(ii) $p + q \leq n - 2$ if $q \neq 1$.*

# 6   Conclusions

We have established a lower bound on nonlinearity over all Boolean functions satisfying the avalanche criterion of degree $p$. We have shown that the lower bound is tight. We have also characterized the functions that have the minimum nonlinearity. Furthermore, we have found a mutually exclusive relationship between the degree of avalanche and the order of correlation immunity.

There are still many interesting questions yet to be answered in this line of research. As an example, we believe that the upper bounds in Theorems 2 and 3 can be further improved, especially when $p$ and $q$ are neither too small, say close to 1, nor too large, say close to $n - 1$.

# Acknowledgment

The second author was supported by a Queen Elizabeth II Fellowship (227 23 1002).

# References

1. E. Biham and A. Shamir. Differential cryptanalysis of DES-like cryptosystems. *Journal of Cryptology*, Vol. 4, No. 1:3–72, 1991.
2. P. Camion, C. Carlet, P. Charpin, and N. Sendrier. On correlation-immune functions. In *Advances in Cryptology - CRYPTO'91*, volume 576 of *Lecture Notes in Computer Science*, pages 87–100. Springer-Verlag, Berlin, Heidelberg, New York, 1991.
3. C. Carlet and P. Codes. On the propagation criterion of degree $l$ and order $k$. In *Advances in Cryptology - EUROCRYPT'98*, volume 1403 of *Lecture Notes in Computer Science*, pages 462–474. Springer-Verlag, Berlin, Heidelberg, New York, 1998.
4. Claude Carlet. Partially-bent functions. *Designs, Codes and Cryptography*, 3:135–145, 1993.
5. D. Coppersmith. The development of DES, 2000. (Invited talk at CRYPTO2000).
6. H. Feistel. Cryptography and computer privacy. *Scientific American*, 228(5):15–23, 1973.
7. Xiao Guo-Zhen and J. L. Massey. A spectral characterization of correlation-immune combining functions. *IEEE Transactions on Information Theory*, 34(3):569–571, 1988.
8. F. J. MacWilliams and N. J. A. Sloane. *The Theory of Error-Correcting Codes*. North-Holland, Amsterdam, New York, Oxford, 1978.
9. M. Matsui. Linear cryptanalysis method for DES cipher. In *Advances in Cryptology - EUROCRYPT'93*, volume 765 of *Lecture Notes in Computer Science*, pages 386–397. Springer-Verlag, Berlin, Heidelberg, New York, 1994.
10. W. Meier and O. Staffelbach. Nonlinearity criteria for cryptographic functions. In *Advances in Cryptology - EUROCRYPT'89*, volume 434 of *Lecture Notes in Computer Science*, pages 549–562. Springer-Verlag, Berlin, Heidelberg, New York, 1990.

11. K. Nyberg. On the construction of highly nonlinear permutations. In *Advances in Cryptology - EUROCRYPT'92*, volume 658 of *Lecture Notes in Computer Science*, pages 92–98. Springer-Verlag, Berlin, Heidelberg, New York, 1993.

12. B. Preneel, W. V. Leekwijck, L. V. Linden, R. Govaerts, and J. Vandewalle. Propagation characteristics of boolean functions. In *Advances in Cryptology - EUROCRYPT'90*, volume 437 of *Lecture Notes in Computer Science*, pages 155–165. Springer-Verlag, Berlin, Heidelberg, New York, 1991.

13. O. S. Rothaus. On "bent" functions. *Journal of Combinatorial Theory*, Ser. A, 20:300–305, 1976.

14. J. Seberry, X. M. Zhang, and Y. Zheng. On constructions and nonlinearity of correlation immune functions. In *Advances in Cryptology - EUROCRYPT'93*, volume 765 of *Lecture Notes in Computer Science*, pages 181–199. Springer-Verlag, Berlin, Heidelberg, New York, 1994.

15. J. Seberry, X. M. Zhang, and Y. Zheng. Nonlinearity and propagation characteristics of balanced boolean functions. *Information and Computation*, 119(1):1–13, 1995.

16. C. E. Shannon. Communications theory of secrecy system. *Bell Sys. Tech. Journal*, Vol. 28:656–751, 1949.

17. T. Siegenthaler. Correlation-immunity of nonlinear combining functions for cryptographic applications. *IEEE Transactions on Information Theory*, IT-30 No. 5:776–779, 1984.

18. A. F. Webster and S. E. Tavares. On the design of S-boxes. In *Advances in Cryptology - CRYPTO'85*, volume 219 of *Lecture Notes in Computer Science*, pages 523–534. Springer-Verlag, Berlin, Heidelberg, New York, 1986.

19. R. Yarlagadda and J. E. Hershey. Analysis and synthesis of bent sequences. *IEE Proceedings (Part E)*, 136:112–123, 1989.

20. X. M. Zhang and Y. Zheng. Auto-correlations and new bounds on the nonlinearity of boolean functions. In *Advances in Cryptology - EUROCRYPT'96*, volume 1070 of *Lecture Notes in Computer Science*, pages 294–306. Springer-Verlag, Berlin, Heidelberg, New York, 1996.

21. X. M. Zhang and Y. Zheng. Characterizing the structures of cryptographic functions satisfying the propagation criterion for almost all vectors. *Design, Codes and Cryptography*, 7(1/2):111–134, 1996. special issue dedicated to Gus Simmons.

22. X. M. Zhang and Y. Zheng. Cryptographically resilient functions. *IEEE Transactions on Information Theory*, 43(5):1740–1747, 1997.

23. X. M. Zhang and Y. Zheng. On plateaued functions. *IEEE Transactions on Information Theory*, 2000. (accepted).

24. Y. Zheng and X. M. Zhang. Plateaued functions. In *Advances in Cryptology - ICICS'99*, volume 1726 of *Lecture Notes in Computer Science*, pages 284–300. Springer-Verlag, Berlin, Heidelberg, New York, 1999.

25. Y. Zheng and X. M. Zhang. Improved upper bound on the nonlinearity of high order correlation immune functions. In *Selected Areas in Cryptography, 7th Annual International Workshop, SAC2000*, volume xxxx of *Lecture Notes in Computer Science*, pages xxx–xxx. Springer-Verlag, Berlin, Heidelberg, New York, 2000. now in Preceedings pages 258-269.

26. Y. Zheng and X. M. Zhang. Strong linear dependence and unbiased distribution of non-propagative vectors. In *Selected Areas in Cryptography, 6th Annual International Workshop, SAC'99*, volume 1758 of *Lecture Notes in Computer Science*, pages 92–105. Springer-Verlag, Berlin, Heidelberg, New York, 2000.

# Cryptanalysis of the Yi-Lam Hash

David Wagner

University of California, Berkeley
daw@cs.berkeley.edu

**Abstract.** This paper analyzes the security of a hash mode recently proposed by Yi and Lam. Given a block cipher with $m$-bit block size and $2m$-bit key, they build a hash function with $2m$-bit outputs that can hash messages as fast as the underlying block cipher can encrypt. This construction was conjectured to have ideal security, i.e., to resist all collision attacks faster than brute force. We disprove this conjecture by presenting a collision attack that is substantially faster than brute force and which could even be considered practical for typical security parameters.

## 1 Introduction

The public cryptographic community has over 20 years of experience in building secure block ciphers. In contrast, the design of cryptographic hash functions has received only about half as many years of research. Yet hash functions are still a very important primitive to practitioners. Therefore, there is much interest in the problem of building a secure, fast hash function out of a secure block cipher.

This research program has been troubled by two major challenges. First, most existing block ciphers have a 64 bit block size, but a hash function with a 64 bit output cannot possibly resist collision search. Therefore, one must somehow securely double the width of the internal state, and this appears to be a non-trivial endeavor. Second, it is hard to maintain efficiency without sacrificing security. The critical figure of merit is the *rate* of the hash function, which is defined as the number of $m$-bit message blocks hashed per encryption, where $m$ is the block size of the underlying cipher. Many early proposals for building fast hash functions have been broken; in particular, Knudsen, Lai, and Preneel cryptanalyzed a large class of double-length hash functions of rate 1 [1,2,3,4,8].

In *ACISP '97*, Yi and Lam proposed a new construction for building a hash function from a block cipher (e.g., IDEA) with $m$-bit block width and $2m$-bit key size [10]. Typically, we will have $m = 64$. The Yi-Lam scheme has rate 1 and yields $2m$-bit outputs. With such high performance, it is an attractive candidate for building a fast hash function.

One crucial feature of the Yi-Lam design is the inclusion of incompatible group operations (XOR and addition modulo $2^m$) to combine internal state variables. Most previous work had used only XOR operations, so that the only non-linear component was the block cipher; however, Knudsen, Lai, and Preneel's

T. Okamoto (Ed.): ASIACRYPT 2000, LNCS 1976, pp. 483–488, 2000.

work broke all such hash functions of rate 1 using $m$-bit keys. The use of in-compatible group operations in the Yi-Lam hash is apparently intended to help frustrate those types of attacks.

Nonetheless, in this paper we still manage to find fast collision attacks on Yi and Lam's proposal. Our attacks work by controlling the effect of the "carry bits." This shows that the incompatibility of XOR and addition does not add much strength to the Yi-Lam hash mode.

Our primary results are as follows. We describe how to find full collisions for the Yi-Lam hash with $2^{.71m}$ work. Also, we give a free-start collision attack (also known as a pseudo-collision attack [6, Section 9.7.2]) that requires only $2^{m/2}$ steps of computation, when the adversary can choose the initial starting state of the hash function. Both attacks are very practical to implement for $m = 64$. Note that Satoh, Haga, and Kurosawa have previously shown that there are second-preimage attacks against this hash with complexity about $2^m$ [9], in contrast to its conjectured $2^{2m}$ security level. Thus, we may conclude that the Yi-Lam hash is neither strongly collision-free nor strongly one-way.

**Outline.** This paper is organized as follows. Section 2 describes Yi and Lam's new scheme briefly, for reference, and Section 3 shows how to find collisions in their proposal. Finally, we conclude the paper in Section 4. Appendix A includes a proof of our main lemma on the correlation between XOR and addition.

## 2    Description of the Yi-Lam Hash

A bit of notation is in order. We write $E(k, x)$ for the encryption of block $x$ under key $k$. The block cipher is assumed to be free of any weaknesses. Also, $x||y$ stands for the concatenation of two blocks $x$ and $y$. We let $x \oplus y$ stand for the XOR of $x$ and $y$, and write $x + y$ for the addition modulo $2^m$ of $x$ and $y$. Since Yi and Lam did not name their scheme, we simply call it the Yi-Lam hash in this paper.

The Yi-Lam hash operates as follows. We pad the message and divide the result into $k$ blocks of size $m$, denoted by $M_0, \ldots, M_{k-1}$. The $2m$-bit internal state is named $G||H$ and is initialized to a fixed public value $G_0||H_0$. We denote

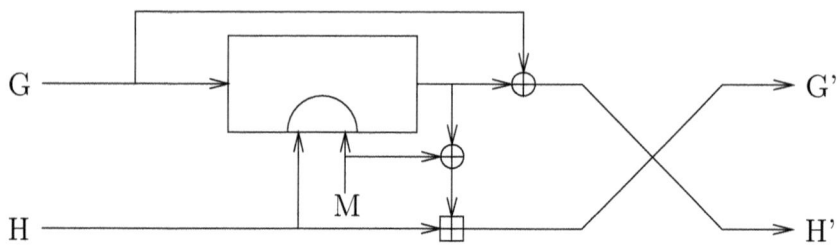

**Fig. 1.** The Yi-Lam hash.

the compression function by $h$, and define $h$ by $h(G||H, M) = G'||H'$, where $G', H'$ satisfy

$$t = E(H||M, G) \qquad G' = (t \oplus M) + H \bmod 2^m \qquad H' = t \oplus G.$$

The final hash digest is computed as $h(\ldots h(h(G_0||H_0, M_0), M_1)\ldots, M_{k-1})$. This construction is illustrated pictorially in Figure 1.

**Caution.** Beware: there are actually two hashing constructions which could be called "the Yi-Lam hash." One such scheme was proposed in *Electronics Letters* [11] (and subsequently cryptanalyzed [5]); another was published in *ACISP'97* [10]; and the two designs are quite different. In this paper, we focus on analysis of the *ACISP'97* proposal. To avoid confusion, phrases such as "the Yi-Lam hash" should be understood to refer to the *ACISP'97* proposal.

## 3   Collision Resistance

In this section, we explore the collision-resistance of the Yi-Lam hash. We exhibit a free-start collision attack with complexity $2^{m/2}$, and then we extend this to a full collision attack with complexity $2^{.71m}$. This is substantially lower than the conjectured security factor of $2^m$ for security against collision attacks. For the suggested parameters (i.e., $m = 64$), even the full collision attack is well within reach of most adversaries in practice, since it requires only $2^{.71\times64} < 2^{46}$ work.

**The free-start attack.** First, we describe a free-start collision attack on the Yi-Lam hash, as promised. Calculate

$$G_i'||H_i' = h(G_i||0, G_i) \qquad i = 1, \ldots, n,$$

for $n = 2^{m/2}$ values of $G_i$, observing that $G_i' = H_i'$ for all $i$. We search for a pair $i, j$ with $H_i' = H_j'$. Then $G_i' = H_i' = H_j' = G_j'$, so this produces a pseudo-collision for the compression function.

The total computational complexity of this attack is $2^{m/2}$ encryptions. A naive implementation of this attack might require $2^{m/2}$ units of storage, but we note that using Floyd's cycle-finding algorithm or any of its improvements [7] reduces the storage complexity of the attack to a very small constant.

**The full collision attack.** In the remainder of this section, we will analyze the security of the Yi-Lam hash against full collision attacks. Let $G_i||H_i$ (for $i = 1, \ldots, n$) be any $n$ values for the chaining variables. For instance, we could select $G_i||H_i = h(G_0||H_0, X_i)$ for $n$ different message blocks $X_i$.

To aid the intuition, let's first consider a variant of the Yi-Lam hash where all additions are replaced by XORs. Imagine calculating the values

$$G_i'||H_i' = h(G_i||H_i, G_i \oplus H_i) \qquad i = 1, \ldots, n,$$

for $n = 2^{m/2}$ arbitrary values of $G_i||H_i$ (which could be obtained by hashing $n$ different message prefixes). Then we would have the relation $G'_i = H'_i$ for all $i$, since

$$G'_i \oplus H'_i = t_i \oplus M_i \oplus H_i \oplus t_i \oplus G_i = 0. \tag{1}$$

After about $n = 2^{m/2}$ trial hashes, we would expect to find some $i \neq j$ such that $H'_i = H'_j$ by the birthday paradox. This would imply that $G'_i = H'_i = H'_j = G'_j$, so this algorithm would give a full collision in this Yi-Lam variant after $2^{m/2}$ trial hash computations.

Of course, the real Yi-Lam hash mixes addition with XOR to prevent these sorts of attacks. So we need to extend our analysis to handle the carry bits. We begin by establishing a simple lemma on the incompatibility of XOR and addition.

**Lemma 1.** *If $a, b, c$ are independently and uniformly distributed over all $m$-bit values, then*

$$\Pr[(a \oplus c) - (b \oplus c) \equiv a - b \bmod 2^m] = (3/4)^{m-1} > 2^{-.42m}.$$

*Proof.* See Appendix A.

For the collision attack on the full Yi-Lam hash, we suggest calculating the values

$$G'_i||H'_i = h(G_i||H_i, G_i \oplus (-H_i)) \qquad i = 1, \ldots, n,$$

for $n$ values of $G_i||H_i$. where $-H_i$ denotes the additive inverse of $H_i$ modulo $2^m$. The analysis goes as follows. Suppose that, for some pair $i \neq j$, we have $H'_i = H'_j$. Then $G'_i = (H'_i \oplus (-H_i)) + H_i$, and similarly $G'_j = (H'_j \oplus (-H_j)) + H_j = (H'_i \oplus (-H_j)) + H_j$. Now the lemma ensures that

$$(H'_i \oplus (-H_i)) - (H'_i \oplus (-H_j)) \equiv (-H_i) - (-H_j) \equiv H_j - H_i \bmod 2^m$$

holds with probability $> 2^{-.42m}$, so $G'_i = G'_j$ with the same probability. With $n = 2^{.71m}$ trials, we expect to see $2^{.42m-1}$ pairs $i, j$ such that $H'_i = H'_j$, which is enough that with non-negligible probability we expect to see one of them where $G'_i = G'_j$ also holds.

To summarize, this shows how to find a collision in the Yi-Lam hash with about $2^{.71m}$ offline hash computations. We expect that the storage complexity of the collision-finding attack will be negligible, if parallel collision search techniques are used to implement the attack [7].

## 4     Conclusions

We have shown that the Yi-Lam hash has serious flaws, and it is clear that the construction offers only minor benefits over traditional single-length hash functions.

The fundamental problem is that the Yi-Lam construction relied on the non-linearity of the carry bits found in modular addition. However, addition possesses only mild nonlinearity properties, and we have seen that in this case they were not enough to resist attack.

We considered a Yi-Lam variant where the addition operation is replaced by an XOR operation, and found a relation (see Equation 1) between the inputs and outputs of the compression function that is linear over XOR. This linear relation allows one to break the variant with a trivial attack. With the addition operation in place, the corresponding relation is no longer XOR-linear, but it is nearly so, and this permits a simple extension to the previous attack to handle the slight nonlinearity.

Therefore, our results provide some evidence to suggest that merely including an incompatible group operation in the hash design may not be sufficient to assure security. This does not rule out the possibility that mixing incompatible operations might improve the security of some cipher-based hash functions—at least in the case of the Yi-Lam hash, the inclusion of the addition operation did appear to frustrate certain trivial attacks—but we advise caution. We propose that in the future it would be wise for designers to avoid relying on addition modulo $2^m$ (instead of XOR) too much.

## 5    Acknowledgements

The author wishes to gratefully acknowledge a number of helpful comments from Lars Knudsen and from the anonymous reviewers of FSE'99 and ASIACRYPT 2000.

## References

1. L. Knudsen and X. Lai, "New attacks on all Double Block Length Hash Functions of Hash Rate 1, including the Parallel DM," *EUROCRYPT'94*, Springer Verlag, LNCS 950.
2. L.R. Knudsen, X. Lai, and B. Preneel, "Attacks on fast double block length hash functions," *Journal of Cryptology*, Winter 1998, vol.11, (no.1):59–72.
3. L. Knudsen and B. Preneel, "Hash Functions Based on Block Ciphers and Quaternary Codes," *ASIACRYPT'96*, Springer Verlag, LNCS 1163.
4. X. Lai and L. Knudsen, "Attacks on Double Block Length Hash Functions," *Fast Software Encryption '93*, Springer Verlag, LNCS 809.
5. K.M. Martin and C.J. Mitchell, "Analysis of hash function of Yi and Lam," *Electronics Letters*, vol.34, no. 24, 1998, pp.2327–2328.
6. A.J. Menezes, P.C. van Oorschot, and S.A. Vanstone, *Handbook of Applied Cryptography*, CRC Press, Boca Raton, 1997.
7. P.C. van Oorschot and M.J. Wiener, "Parallel Collision Search with Cryptanalytic Applications," *Journal of Cryptology*, vol.12, no. 1, 1999, pp.1–28.
8. B. Preneel, "Analysis and design of cryptographic hash functions," Doctoral dissertation, Katholieke Universiteit Leuven, 1993.
9. T. Satoh, M. Haga, and K. Kurosawa, "Towards Secure and Fast Hash Functions," *IEICE Trans. Fundamentals*, vol. E82-A, no. 1, Jan. 1999.
10. X. Yi and K.-Y. Lam, "A new hash function based on block cipher," *ACISP'97*, Second Australasian Conference on Information Security and Privacy, Springer, LNCS 1270.

11. X. Yi and K.-Y. Lam, "Hash function based on block cipher," *Electronics Letters*, 6 Nov 1997, vol.33, (no.23):1938–1940, IEEE.

# A  Proof of Lemma 1

**Lemma 1.** *If $a, b, c$ are independently and uniformly distributed over all $m$-bit values, then*

$$\Pr[(a \oplus c) - (b \oplus c) \equiv a - b \bmod 2^m] = (3/4)^{m-1} > 2^{-.42m}.$$

*Proof.* We rearrange the equation to become

$$(a \oplus c) + b \equiv a + (b \oplus c) \bmod 2^m.$$

Let $d$ stand for the left-hand carry bits, and $e$ the right-hand carry bits. Then $(a \oplus c) + b \equiv (a \oplus c) \oplus b \oplus d \bmod 2^m$, and $a + (b \oplus c) \equiv a \oplus (b \oplus c) \oplus e \bmod 2^m$, so our required equation becomes

$$(a \oplus c) \oplus b \oplus d \equiv a \oplus (b \oplus c) \oplus e \bmod 2^m.$$

In short, we need only calculate the probability that $d = e$. Let $d_j$ be the $j$-th bit position in $d$, for $j = 0, \ldots, m - 1$, where $d_0$ is the least significant bit of $d$. Define $e_j$ and $c_j$ similarly. We have $d_0 = e_0 = 0$ trivially. Now we proceed inductively. Suppose that $d_i = e_i$ for $i = 0, \ldots, j$ where $0 \leq j < 31$. We know that $d_{j+1}$ is computed as the majority function of $a_j \oplus c_j$, $b_j$, and $d_j$; also, $e_{j+1}$ is calculated as the majority of $a_j$, $b_j \oplus c_j$, and $e_j = d_j$. If $c_j = 0$, then clearly $d_{j+1} = e_{j+1}$. On the other hand, if $c_j = 1$, then $d_{j+1} = e_{j+1}$ holds with probability exactly $1/2$ (i.e., in the case that $a_j \neq b_j$). Moreover, the probabilities for each bit position are independent, so we can multiply them. Therefore, for any fixed value of $c$ the probability is $2^{-W(\bar{c})}$, where $W(\bar{c})$ denotes the Hamming weight of $\bar{c}$, and $\bar{c}$ denotes the $m - 1$ least significant bits of $c$.

Summing over $c$ and applying the binomial theorem, we get that the desired probability $p$ satisfies

$$p = 2^{-m} \sum_c 2^{-W(\bar{c})}$$
$$= 2^{-m+1} \sum_{c' < 2^{m-1}} 2^{-W(c')}$$
$$= 2^{-m+1} \sum_{w=0}^{m-1} \binom{m-1}{w} 2^{-w}$$
$$= 2^{-m+1} \cdot (1 + 1/2)^{m-1}$$
$$= 2^{-m+1} \cdot (3/2)^{m-1}$$
$$= (3/4)^{m-1}$$
$$\approx 2^{-.415(m-1)}$$
$$> 2^{-.42m}.$$

This completes the proof.                                                   □

# Power Analysis, What Is Now Possible...

Mehdi-Laurent Akkar[1][*], Régis Bevan[2], Paul Dischamp[2], and Didier Moyart[2]

[1] Bull CP8, 68 route de Versailles,
78431 Louveciennes, France.
ml.akkar@free.fr
[2] Oberthur Card systems,
25 rue Auguste Blanche, 92800 Puteaux, France.
{r.bevan, p.dischamp, d.moyart}@oberthurcs.com

**Abstract.** Since Power Analysis on smart-cards was introduced by Paul Kocher [KJJ98], the validity of the model used for smart-cards has not been given much attention. In this paper, we first describe and analyze some different possible models. Then we apply these models to real components and clearly define what can be detected by power analysis (simple, differential, code reverse engineering...). We also study, from a statistical point of view, some new ideas to exploit these models to attack the card by power analysis. Finally we apply these ideas to set up real attacks on cryptographic algorithms or enhance existing ones.

**Keywords:** Smart-cards, Power analysis, DPA, SPA.

## 1  Power Consuming Models for Smart-Cards

In cryptographic protocols, we normally assume that the attacker has at most the knowledge of the algorithm used and some input and/or output values. In the attack on smart-cards based on power-analysis, the situation is quite different: we assume the attacker has access to more than this, namely, see (within a certain limit) what is done during the computation. Therefore, we need to specify what kind of knowledge could be extracted from a card.

### 1.1  Sensitive Instructions

As opposed to modern computers, the smart-card processor is very limited in terms of capabilities, registers and memory. Usually, the arithmetic or the logical operations (xor, or, add...) are executed through a special register (e.g. the A-accumulator for Intel 8051 compatible family). Therefore, the programmer cannot load the variables to registers, execute the instructions on them and afterwards store them in their final positions in the memory. So, during the entire execution of a program, the micro-controller is always loading/saving the output of the calculation to/from memory. Moreover, the limited memory of the

---

[*] Research done while at Oberthur Card Systems.

T. Okamoto (Ed.): ASIACRYPT 2000, LNCS 1976, pp. 489–502, 2000.

devices often obliges the programmer to implement his code in a straightforward manner.

The limited set of instructions is another advantage for the attacker: usually, all the instructions involve at most two 8-bits variables for the input and one for the output. In most cases one or two of these variables is a special register. Taking advantage of this, the attacker can easily study nearly all the instruction set.

For these reasons, it seems reasonable to assume that a good point to attack a smart-card is when the processor is loading/saving a value.

## 1.2   Some Consumption Models

A smart-card, even if it is one of the most simple processors, consists of a lot of different blocks: the processor itself, the memory, the Bus ... This is why we need to consider a very general model to represent it correctly. All these blocks are perturbed by external parameters and so will react differently everytime (the power supply can fluctuate, so the clock ...).

Executing an instruction on a smart-card, like on most micro-controllers, takes several machine cycles. For example, the XOR of two values could be processed as follows by the CPU:

- analyze which instruction is to be executed (e.g. XOR);
- load the variables;
- execute the calculation;
- store the result.

Those operations can of course be pipelined and so are not serial in time.
This first (simple) analysis shows that, to model the power consumption of a smart-card, several things need to be taken into account:

- the instruction which is executed;
- the data involved in the calculation (input, output);
- the location in RAM/ROM of the instruction executed and the data;
- the instructions involved before and after the instruction considered;
- some random fluctuation.

Based on these considerations and some experimentations we consider here the following general model: the power consumption of an instruction $I$ (an array $P[I, 1..n]$ where n is the number of point during the acquisition of the instruction) could be represented as follows:

$$
\begin{aligned}
P[I] = & P_{gen}[I] \times N_{gen} \\
& + P_{in}[I, V_{in}] \times N_{in} \\
& + P_{out}[I, V_{in}, V_{out}] \times N_{out} \\
& + P_{last}[I', V'] \times N_{last}
\end{aligned}
$$

Where:

- $P_{gen}$ is the general consumption of the instruction and $N_{gen}$ the fluctuation of $P_{gen}$ (i.e. location of the code in ROM/EEPROM );
- $P_{in}[V_{in}]$ is the power consumption of the input operand (we shall see that in first approximation, the power consumption does not depend on the instruction executed) and $N_{in}$ the noise associated;
- $P_{out}[I, V_{in}, V_{out}]$ is the power consumption of the output operand (this time, of course, it depends on the nature of the instruction) and $N_{out}$ the noise associated;
- $P_{last}[I', V']$ represents the influence of the preceding instructions and their input/output and $N_{last/next}$ the associated noise.

## 1.3   Consumption Model for the Data

Intrinsic behaviour of the micro-controller hardware shows the following characteristics :

- there are some gates which commute when a value changes ($0 \rightarrow 1$ or $1 \rightarrow 0$);
- the bus drives the bit of information (so the value of a bit changes the consumption);
- some bits influence some other places in the micro-controller (e.g. the carry, the overflow);
- writing a 0 or a 1 does not consume the same power.

Taking these considerations into account, we can define the following different models:

- Global :   $P[x] = K_x$
- Linear :   $P[x] = \sum_{i=0}^{n_{bits}} x_i \times P_i$
- Flipping:   $P[x] = F(x, x_{last})$
- Quadratic: $P[x] = \sum_{i=0}^{n_{bits}} \sum_{j=0}^{n_{bits}} X_{00_{ij}} P_{00_{ij}} + X_{01_{ij}} P_{01_{ij}}$
  $$+ X_{10_{ij}} P_{10_{ij}} + X_{11_{ij}} P_{11_{ij}}$$

where:
- $X_{ab_{ij}} = 1$ iff $x_i = a$ and $x_j = b$
- $P_{ab_{ij}}$ is the associated consumption.
- $x_i$ corresponds to bit $i$ of $x$.

The global model is the most general one: it implies a specific consumption for every possible value.

The linear model assumes that every bit has a specific weight and is independent of the other bits. For example if you fix $P_i = 1$ for all $i \leq n$ you obtain a "hamming weight consumption" model.

The flipping model can represent the case where the last value influences the consumption. For example if you take $F = HW(data \oplus data_{last})$ - where $HW$ is the hamming weight - it represents the number of bit flips between the last data which has been manipulated and the actual one.

The quadratic is more powerful. It can represent component where the consumption depends on the value of the bits taken two at a time.

Now let us see how this model is correlated to reality.

## 2   Real Smart-Cards

All the results given are based on two different models of smart-card processors[1]: a low cost model and a more powerful one. But after further experimentation it seems that most smart-cards behave in a similar way to the defined consumption models.

We want to point out that some aspects have not been considered in the present paper:

- it was assumed that the instructions all had the same consumption model;
- the influence of past instructions were not taken in account.

### 2.1   Generalities

It appears that most of the processor consumption is due to the instruction being executed. The associated noise is relatively small and probably takes into account the previous executed instructions; the rest of the consumption comes from the values involved in the calculation (input and output).

| Instruction | Data | Last instructions/datas | Noise after filtering |
|---|---|---|---|
| 85% | 9% | 5% | 1% |

### 2.2   Validity of the Memory Models in Practice

Many people have concluded (cf. [KJJ98, CJRR99a, BS99]) that the consumption of the card directly depends on the Hamming weight, or the number of changes $0 \leftrightarrow 1$ in the binary value considered. It appears that the Hamming weight model is not adapted to the two smart-cards we studied.

The diagrams in figure 1 show the consumption associated with the storage of a value in RAM ordered by hamming weight, and this for the two different smart-cards. If this hypothesis was true we should see an increasing curve which is not the case. By ordered we mean that the values $x_1 \ldots x_{256}$ of the average consumption of the "store-$i$" instruction were taken and ordered by respecting the following: $x_i < x_j \iff HW(i) < HW(j)$ where $HW(i)$ represents the Hamming weight of $i$. If $HW(i) = HW(j)$ then we did order the value comparing their consumption.

One can see from figure 1 that for a given consumption, one would get different hamming weight values.

We need to find a more appropriate model. We have ordered the values of consumption to have a reference to the correct model. The goal was to find an order on $[0..255]$ which is close to the real order obtained by sorting the power consumptions. Due to the architecture of the micro-controller we have decomposed the value $n \in [0, 256]$ in its binary form $n_1, .., n_8$. The first idea was to find a "linear" order: this means assigning some weight to the different

---

[1] the exact models are obviously not given here

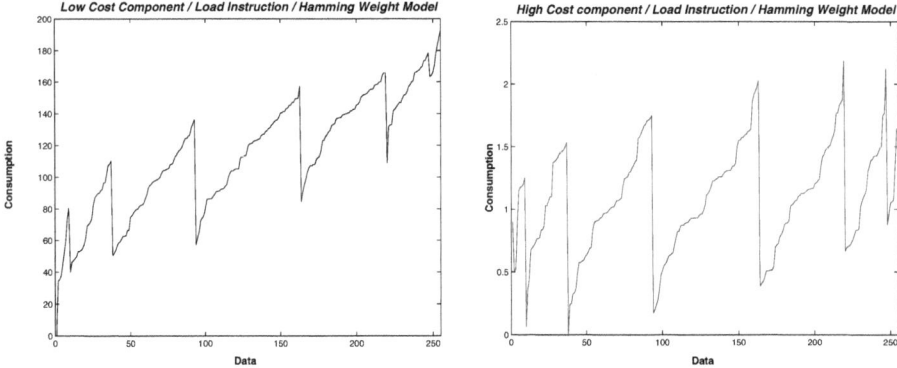

**Fig. 1.** Hamming Weight Model

bits of $n$. To compare $n$ to $m$, once the weights $p_1, ..., p_8$ have been chosen, you compare $n_1.p_1. + ... + n_8.p_8$ to $m_1.p_1 + ...m_8.p_8$. For example the hamming weight corresponds to $p_1 = p_2 = ... = p_8 = 1$ and the natural order to $p_i = 2^{8-i}$. Taking for $p_i$ the average difference of consumption of the bit $i$ (comparing the consumption between "$load(x)$" and "$load(y)$" where $x_j = y_j$ except for the bit $i$ (see below figure 3)) we obtain the following curves (figure 2)

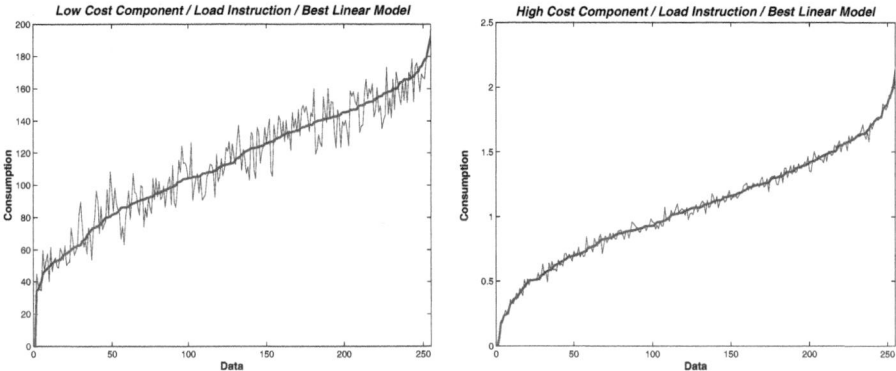

**Fig. 2.** Best Linear Model

We can clearly see that even if the results are satisfying, we are still far from the expected curve for the low end component (we add a "reference" curve : the curve obtained by sorting the consumption). To check if the "linear" model was correct, we computed for all values the difference caused by changing just one bit in the operand.

The table in fig. 3 presents all the results. The 'bit i' column is obtained by the following computation: the 128 values where the i'th bit was 0 were taken and for each individual value the i'th bit was turned on. The individual consumption changes were measured. Then, we extract the minimum and the maximum of the 128 values and compute the average and the standard deviation to sum up the results:

| Component | Consumption | bit 0 | bit 1 | bit 2 | bit 3 | bit 4 | bit 5 | bit 6 | bit 7 |
|---|---|---|---|---|---|---|---|---|---|
| low cost | average | 11.9 | 11.3 | 11.4 | 12.1 | 7.0 | 32.3 | 34.5 | 44.85 |
| | min | −12.1 | −25.6 | −15.1 | −15.4 | −30.0 | 1.2 | −8.0 | 7.2 |
| | max | 40.5 | 42.2 | 45.7 | 60.0 | 45.0 | 67.6 | 73.7 | 86.7 |
| | standard deviation | 10.3 | 11.7 | 10.8 | 13.6 | 14.1 | 11.9 | 16.9 | 16.3 |
| high cost | average | 12.1 | −21.9 | 16.1 | 10.7 | 14.1 | −20.1 | −5.0 | 10.3 |
| | min | 5.3 | −27.7 | 8.5 | 5.7 | 7.7 | −25.3 | −11.0 | 3.6 |
| | max | 17.0 | −15.5 | 24.0 | 18.1 | 19.7 | −14.0 | 2.4 | 16.9 |
| | standard deviation | 2.3 | 2.4 | 2.4 | 2.4 | 2.4 | 2.4 | 2.8 | 2.7 |

**Fig. 3.** Comparison in mV of the influence of one bit on low and high cost component

We can notice some interesting things that explain previous curves:

In both components, one can notice that flipping a bit from 0 to 1 does not affect the consumption by an equal difference. Moreover it can either increase or decrease the consumption. So it shows that in this case the hamming weight model is totally inadequate.

The influence of each bit is not increasing or decreasing with its position inside the byte.

For the low cost component, the variance is very important (i.e. for the bit 6, it can increase or decrease the consumption) explaining that the "best linear model" is not adequate. But, if we consider the following simplified quadratic model: we consider only the terms $b_i b_{i+1}$ (a bit only influences the bit just before and just after); then the curve is very close to the sorted one (not presented here).

For the high cost component the variance is significantly reduced. The silicon founder may have tried to separate the consumption of each bit, inducing this difference.

Now that we have a better model, let us see statistically how much information we can get from a card.

# 3   Attacks

In this section, we will consider that the following models are good enough to be as precise as the ideal one:

- simplified boolean polynomials of degree 2 for the low cost smart-card;
- best linear model for the high cost smartcard.

## 3.1   Existing Attacks

One of the most common attack by power-analysis is the DPA attack against DES. For more details, see [KJJ98]. The interesting aspect here is that this attack is based on the following assumption:

On average, the consumption of the card computing with values $x$ such as $x_i = 0$ [2] can be distinguished from the consumption where $x_i = 1$.

We need to prove this assumption. It would be obvious if the card was respecting the Hamming weight model, but, judging by our results, this is not clear. Table 4 presents some results about the efficiency of a DPA attack. We have computed the average consumption of the set of $x$ such as $x_i = 0$ (resp $x_i = 1$).

| Component | Consumption for | bit 0 | bit 1 | bit 2 | bit 3 | bit 4 | bit 5 | bit 6 | bit 7 |
|---|---|---|---|---|---|---|---|---|---|
| low | $x_i = 0$ | 0.474 | 0.475 | 0.475 | 0.473 | 0.485 | 0.429 | 0.424 | 0.402 |
| cost | $x_i = 1$ | 0.526 | 0.525 | 0.525 | 0.527 | 0.515 | 0.571 | 0.575 | 0.598 |
| high | $x_i = 0$ | 0.444 | 0.398 | 0.425 | 0.450 | 0.434 | 0.406 | 0.476 | 0.452 |
| cost | $x_i = 1$ | 0.556 | 0.602 | 0.575 | 0.550 | 0.566 | 0.593 | 0.524 | 0.548 |

**Fig. 4.** Efficiency of a DPA attack on low cost and high cost component

The table above confirms the experimental attack: the difference between the two distributions is quite important for every bit of data. But this attack being based on an imperfect model, we have tried to use a better model to enhance existing attacks and imagine new ones.

## 3.2   PODPA (Perhaps Optimal Differential Power Analysis )

**Theory:** Ordering the consumption, we compute the sets of value $A$ and $B$ such that:

- $A$ and $B$ are disjoint, $A \cup B = [0, ..255]$
- $|A|=|B|=128$
- $\forall(x, y) \in A \times B$, consumption$(x) <$ consumption$(y)$

---

[2] $x_i$ is the bit $i$ of x

Then we obtain a much higher difference between the consumption of subset $A$ and $B$ than in the usual DPA case:

- 36.7% / 63.3% for the low cost component;
- 34.3% / 65.7% for the high cost one.

In real attacks, this results in an improvement by a factor 5 (number of messages needed for a successful attack). Moreover, in critical situations (low quality acquisition, cards allowing few tries), it could result in a sufficient improvement to make an otherwise unsuccessful attack work.

**Practical Scenario:** we assume for this example that the scheme of DES is known. As in the classical DPA attack, we repeat the following operations on several DES acquisitions with the message $M$:

- apply IP (DES Initial Permutation) to $M$;
- apply the expansion permutation to the 32 right bits of $IP(M)$;
- guess 6 bits of the key (output of the first round key scheduling) and XOR it to the corresponding part of the message;
- those 6 bits will be the input of the S-Box table look-up.

In the classical attack we consider one bit of the output of the S-Boxes. But here, we will just separate the set $[0, 63]$ (6 input bits of the S-Boxes) in two optimal subsets by the method explained in the last section. Then, by analyzing all the 6 bits guesses, we can determine the single correct guess. Indeed, by definition of the subsets, it will be the difference curve with the biggest peak.

**Remarks:**

To realize this attack, we need to know the optimal linear model for the consumption of the smart-card: in reality, this information is not obtained easily[3]. So we have to find an experimental method to determine the consumption model. The easiest method is to implement the "usual" DPA attack to determine a key on one component. Next, to analyze the leaking instruction (in the DPA attack) for every possible input; one can do this knowing the message and the key. And then the results can be applied to every implementation of crypto-algorithms on the same component.

Even if this example is specific, it highlights under which conditions this attack could be applied: one has to go through the algorithm with a known message until a part of the key is involved and mixed with the message. This situation is quite general for secret and public key algorithms. (i.e. AES candidates, RSA, ECC ...)

---

[3] Such information is considered proprietary by the silicon manufacturers

### 3.3   BPA: Binary Power Analysis

We will now study another type of attack, the Binary Power Analysis.

**Theory:** In many secret key algorithms (like DESX, cellular phone algorithms...), the "whitening" technique is used to increase the size of the key without modifying the algorithm used or using multiple encryption. The method used is the following: before (respectively after) the algorithms core, the message (the result) is xored (whitening operation) with a part of the key.

**e.g.:** whitening a 8-bit message $M_0 \ldots M_7$ with a 8-bit key $K = K_0 \ldots K_7$

- for(i=0..8) do $M_i = M_i \oplus K_i$
- C = Encrypt(M)
- for(i=0..8) do $C_i = C_i \oplus K_i$

This method, from a power analysis point of view, prevents a DPA-like simulation on the message. Unfortunately, it is quite easy to attack this operation by power analysis and then use the usual attack on the algorithm core. This is based on the fact that if a bit of the key is 0 the associated bit in the message will remain unaltered.

Here are the BPA attack operations :

- Obtain $n$ curves $\{T_k\}_{k=1..n}$ associated with the messages $M_k$
- for(i=0..8) do (if there are 8 bytes)
  - for(j=0..8) (8 bits / byte)
    - Separate the curves in 2 subsets according to the bit $j$ of byte $i$ of all messages $M_k$
    - Subtract the average of the 2 subsets
- Process the results to determine the whitening key possibilities

The last step needs more explanations to understand how the separation is done in practice.

**Practical Scenario:** The graphic 5 presents the results of the analysis of a whitening operation (just one byte).

In this case, the analysis is quite easy. The XOR operation is executed in 3 steps: first the input values are loaded, then the xor is executed, lastly the result is stored. So, when attacking a special bit, on average the consumption will not change when the key bit is 0 and will when the key bit is 1. In this case, we can quite easily recover the key byte by comparing the beginning and the end of the instruction: $K = 00001001$ [4]

One can find more details about the attack in [CJRR99a].

---

[4]   In this case the second peak of each part is representative of the key bit.

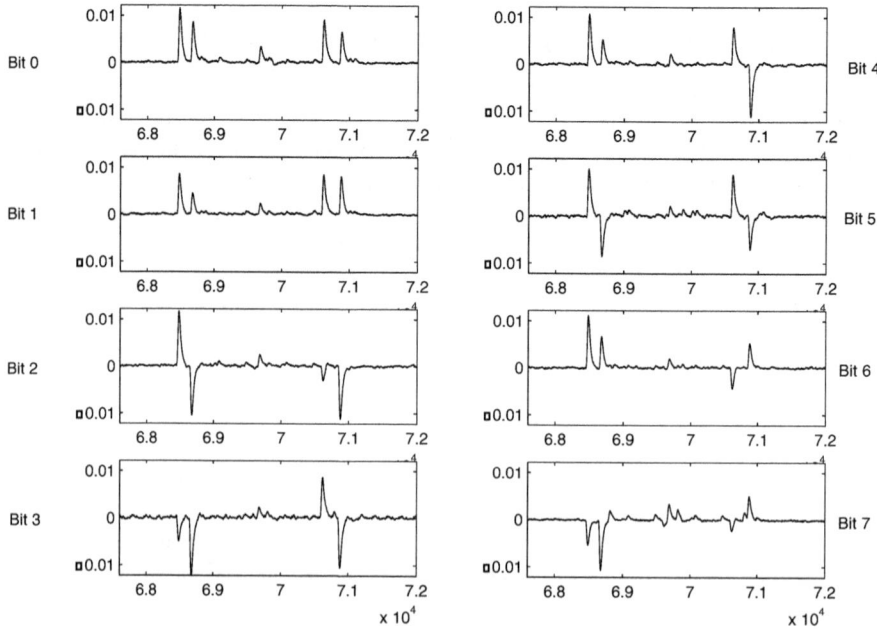

**Fig. 5.** Whitening operation on a low cost component

### 3.4   DiPA: Direct Power Analysis

If the consumption model of a card is well known, some very powerful attacks can be set up, avoiding two restriction of DPA like attacks:

– Why just separate the consumption in only two sets?
– Why only do an average differential analysis?

Considering the consumption distribution, it is surely possible to distinguish more than two parts in the consumption curves and build more sophisticated attacks. For examples, one can construct boolean or linear formulas on key-bits or key-bytes like:

– The $i$-byte of the key is one of these values...
– The bit $j$ and $k$ are equal...
– If bit $j$ and $i$ are different then...
– $b_i+b_j+b_k=2$...

One can then extract some bits of the key by appropriate computation (involving boolean solving methods, Gröbner basis...).

Moreover, if you have a reasonable model for the card, the important point is that all these equations can be obtained with separate acquisitions and does not involve any averaging or differential. You can extract information on each individual computation done by the card.

For example, with the DES, you can extract a lot of information: you can have a very precise idea of the key if you can (and this is possible in reality) localize operations during the end phase of the key-scheduling (just before the xor preceding the S-box operation). You obtain 16 sets of boolean equations on the 6 bits values of the keys. If the accuracy of the model is good (number of subsets in $[0, 255] > 8$) you can recover the entire key with just one acquisition and even without knowing which message is being encrypted.

## 3.5   Countermeasures

In reality, many countermeasures exist to prevent power analysis attacks on the smart-card:

- desynchronisation of the measures: dummy operations, random frequency clock...;
- randomization of the operations: i.e. random permutations;
- transformations of the data: i.e. public key blinding, Duplication method [GP99], masking methods[Mes00, CJRR99b].

## 3.6   Using Previous Results

We summarize some generals "counter-countermeasures" to these countermeasures:

  **- Pattern Recognition** (to localize interesting instructions from dummy operations): it appears that the general power comsumption of an algorithm does not change from one acquisition to the other. Moreover it could be that every type of instruction (arithmetic, boolean, load and store ...) and every type of adressing mode has a particular consumption profile. Hence, it might be possible to classify every assembly instruction by its power consumption. Such a study has been undertaken with some success. Knowing the instruction profile, one could disassemble part of the code or at least retrieve some instruction class. Consumption profile helps in synchronising the curves when an external glitch is not provided. Last but not least, an attacker could see what type of countermeasures have been implemented in the code (use of random parameters). The signals are first filtered in order to remove any local noise. Then, the adequate algorithms (classical matching algorithms) are used to retrieve a given instruction in a set of power consumption curves.

  **- Synchronizing a randomized clock:**  a hardware counter-measure against waveform attacks consists in using a random clock. Experimentally, it is possible to quite quickly rebuild the synchronous signals from the randomized signals. Once again, the signals are processed with an adequate filter in order to remove any local variations. As a result, it is easy to recognize (studying the first

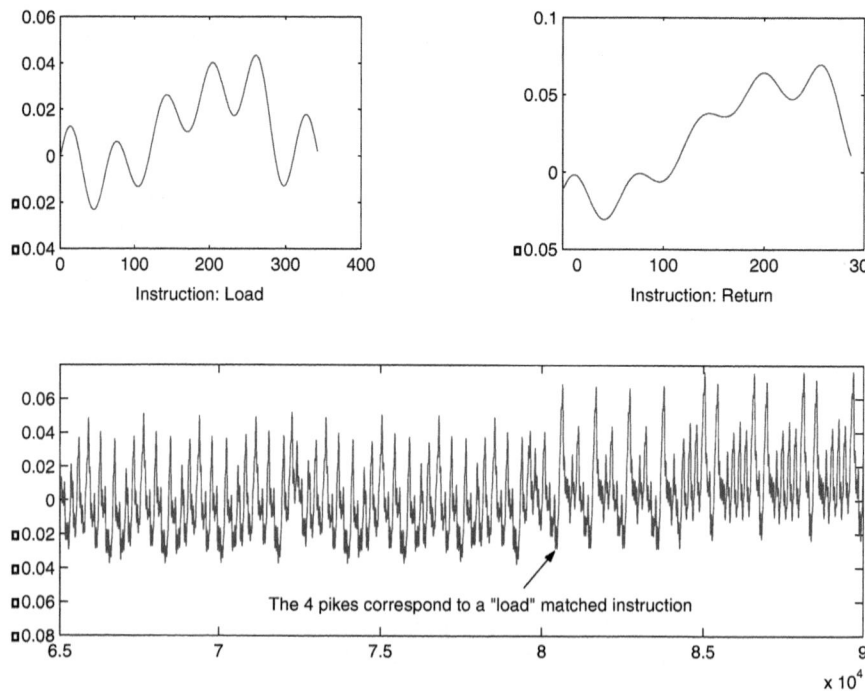

**Fig. 6.** Pattern Recognition example

and second derivate) the clock cycles. Then, according (or not) to a reference curve, one can expand or shrink every single clock cycle of a given curve. The procedure is then repeated for every clock cycle of the signal. One can then re-use normal DPA attacks.

- **Transformations/Masking Methods:** we will just present here an example. Often in the DES (due to its structure), the card computes randomly with the data or with the flipped data ($M = M \oplus 0xFF..F$) to protect the card against a bit prediction of the data. But using the PODPA method, we can neutralize this countermeasure. Knowing the consumption distribution of the component, we proceed as follows:

- construct the 128 pairs $(x, x \oplus 255)$ for $x < 128$ with its average consumption $(c(x) + c(x \oplus 255))/2$;
- order the pairs by consumption;
- construct two subsets with the 64 lower/higher consumption pairs;
- proceed a PODPA attack with this distribution.

This attack will work because it does not distinguish $x$ and $x \oplus 255$ for all $x \in [0, 255]$ (there are in the same subset). Moreover, in most components, the consumption of these two subsets is quite different (similar to the usual DPA bit selection subsets). For example, this attack is very efficient against the counter-measure explained in [Mes00]. It exploits a small leakage when transforming an "arithmetic" mask into a "logical" mask. However we want to point out the fact that the PODPA attack does not defeat "arithmetic" or "logical" masks, just the transformation from one to the other.

Unfortunately, DiPA attacks can counteract masking countermeasures men-tioned in ([Mes00, CJRR99b]). The reason is the following: in their security proof, they are assuming that the information is extracted from several mes-sages. But, with our attack some information about the key is extracted from each computation and not from the comparison of different acquisitions.

## 4   Summary of an Effective Attack

- Acquire sufficient signals to obtain general information.
- Apply adequate DSP[5] to determine the type of clock and retrieve the struc-ture of the program (rounds, countermeasures ...).
- By statistical test (variance) check if the execution is deterministic.
- Obtain sufficient curves of a specific part of the algorithm.
- Rescale if needed (noise, random clock...).
- By usual methods extract one key.
- Use SPA analysis to obtain more information about the card consumption.
- Now one is able to use PODPA, POSPA or DiPA attacks to "easily" break the card.
- One can directly attack other algorithms using the same card model!

## 5   Conclusions

We have shown that there are several potential attack scenarios which need to be further explored. These attacks require a more detailed study of the com-ponent than a classical DPA. A better knowledge of the behaviour of the chip enables to conduct powerful attacks even with little knowledge of the algorithm implementation.

## References

[BS99]      E. Biham and A. Shamir. Power analysis of the key scheduling of the AES candidates. *Second AES Candidate Conference*, 1999.

[CJRR99a]   S. Chari, C. Jutla, J.R. Rao, and P. Rohatgi. A cautionary note regarding evaluation of AES candidates on smart-cards. *CHES*, 1999.

[CJRR99b]   S. Chari, C. Jutla, J.R. Rao, and P. Rohatgi. Towards sound approaches to counteract power-analysis attacks. *Crypto*, 1999.

[GP99]      L. Goubin and J. Patarin. DES and differential power analysis, the du-plication method. *CHES*, 1999.

---

[5] Digital Signal Processing

[KJJ98]    Paul Kocher, Joshua Jaffe, and Benjamin Jun. Differential power analysis. *Web Site: www.cryptography.com/dpa*, 1998.
[Mes00]    T.S. Messerges. Securing the AES finalists against power analysis attacks. *FSE*, 2000.

# Concrete Security Characterizations of PRFs and PRPs: Reductions and Applications

Anand Desai[1] and Sara Miner[2]

[1] Bell Labs Research Silicon Valley, 3180 Porter Drive, Palo Alto, CA 94304, USA.
emailadesai@research.bell-labs.com

[2] Dept. of Computer Science & Engineering, University of California at San Diego,
9500 Gilman Drive, La Jolla, CA 92093, USA.
sminer@cs.ucsd.edu

**Abstract.** We investigate several alternate characterizations of pseudorandom functions (PRFs) and pseudorandom permutations (PRPs) in a concrete security setting. By analyzing the concrete complexity of the reductions between the standard notions and the alternate ones, we show that the latter, while equivalent under polynomial-time reductions, are weaker in the concrete security sense. With these alternate notions, we argue that it is possible to get better concrete security bounds for certain PRF/PRP-based schemes. As an example, we show how using an alternate characterization of a PRF could result in tighter security bounds for some types of message authentication codes. We also use this method to give a simple concrete security analysis of the counter mode of encryption. In addition, our results provide some insight into how injectivity impacts pseudorandomness.

## 1 Introduction

Pseudorandom functions (PRFs) and pseudorandom permutations (PRPs) are extremely useful and widely used tools in cryptographic protocol design, particularly in the setting of private-key cryptography. In this paper, we study several different notions of security for these objects. Specifically, we study these notions in a concrete security framework, and we show how different characterizations may be used to derive better security bounds for some commonly used private-key cryptographic protocols.

### 1.1 Descriptions of Notions

The notion of a PRF family was proposed by Goldreich, Goldwasser and Micali [8]. In such a family, each function is specified by a short key, and can be easily computed given the key. Yet it has the property that telling apart a function sampled from the PRF family and one from a random function family, given adaptive access to the function as a black-box, is computationally infeasible (for someone who does not know the key). This is the standard notion of a PRF, and (to distinguish it from alternate notions) we refer to it in this paper as the

T. Okamoto (Ed.): ASIACRYPT 2000, LNCS 1976, pp. 503–516, 2000.
© Springer-Verlag Berlin Heidelberg 2000

PRF notion. Luby and Rackoff extended the above to permutation families by introducing the notion of a PRP family [11]. The reference family for defining the security of a PRP family can be that of random functions, as in [11], or that of random *permutations*, a practice started by Bellare, Kilian and Rogaway [5]. We adopt the definition of [5] and refer to it here as the PRP notion.

ALTERNATE CHARACTERIZATIONS. PRFs may be characterized in several ways other than the standard notion. We are particularly interested in one way suggested in the very paper that introduced the standard notion [8]. This alternate notion can be described informally through the following interactive protocol: a distinguisher who is given adaptive oracle access to the function obtains the output of the function on some points of its choice through oracle queries. It then outputs a point that has not yet been queried and gets back, based on a hidden coin flip, either the output of the function on that point or a uniformly distributed point in the range of the function. It should be computationally infeasible for the distinguisher to guess which of the two possibilities it was presented. We call this notion indistinguishable-uniform functions or IUF, to distinguish it from the standard notion PRF. A similar notion may be defined for permutation families, and we call this IUP for indistinguishable-uniform permutations.

We also consider another notion that is normally associated with the security of encryption schemes. In this notion too, the distinguisher is given adaptive oracle access to the function. It then outputs *two* new points and, based on a hidden coin flip, is presented with the output of the function on one of them. We require that a computationally-restricted distinguisher have negligible success in telling apart the two cases. In this paper, we refer to this notion as IPF, for indistinguishable-point functions. We show that this notion does not imply pseudorandomness for functions. However, when we consider the analogous notion for permutations, which we call IPP (indistinguishable-point permutations), we find that pseudorandomness is captured.

## 1.2    Concrete Security and Reductions Among the Notions

Making a break from the traditional approach of presenting PRF families in an asymptotic way, Bellare, Kilian and Rogaway began the practice of explicitly specifying the resources determining security and paying particular attention to the quality of security reductions [5]. This approach forms the basis of concrete security analysis and has been used in many subsequent works [4,2,3]. One benefit of this approach is that it enables the comparison (and classification as weaker or stronger) of polynomially-equivalent notions in cryptography. Paying attention to the concrete complexity of reductions between notions is important in practice, as inefficient reductions translate to a penalty either in security assurance or in running time.

REDUCTIONS AMONG THE NOTIONS. Under polynomial-time reductions, the equivalence between the notions of PRF and IUF has been established by Goldreich et al [8]. (In fact, the concrete security bounds we derive in our reductions between these notions are implicit in theirs.) We establish that our reductions

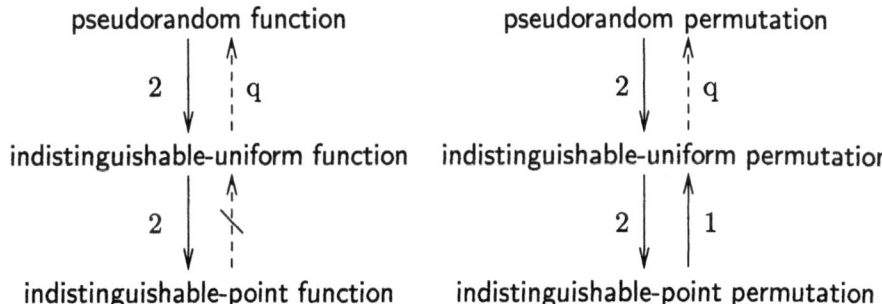

**Fig. 1.** *Relating the notions. A solid arrow from notion A to notion B means that there is a security-preserving reduction from A to B. A broken arrow indicates a reduction that is not security-preserving. The arrows are labeled by the loss-factor of the reduction. A hatched arrow means that there is no polynomial-time reduction.*

are tight. Additionally, we relate the notions of PRP and IUP. The reductions between these two permutation notions are the same as those between the corresponding notions for functions.

Furthermore, we show that IUP and IPP are equivalent, up to a small constant factor in the reduction. However, as mentioned above, a different picture emerges when we look at the corresponding notions for functions. It turns out that IPF and IUF (or PRF) are not equivalent, even in just an asymptotic sense. We show that IPF is a strictly weaker notion, in that there are function families which are secure in the IPF sense, but completely insecure in the IUF sense. A summary of the reductions is given in Figure 1.

### 1.3    Motivation: Tighter Security Analyses

Our demonstration that the alternate notions we consider here are weaker in the concrete security sense than the standard notions might be seen as an argument *against* using any of them. Yet we will recommend their use in certain circumstances (to complement, rather than replace the standard notions).

In a concrete security analysis of a protocol which is based on a particular primitive, the security of the protocol is related to that of the underlying primitive in a precise way. If we know the concrete security of a protocol in terms of the security of the underlying primitive under one notion, it is easy to translate this to the security of the protocol in terms of the security of the primitive under a weaker notion. We simply use the appropriate security reduction between the notions. We then see a drop in the translated security, reflecting the gap in the reduction between the notions. However, we show that it is sometimes possible to directly reduce the security of the protocol to that of the underlying primitive under a weaker notion *without* the expected drop in security. Such a situation exists when the weaker notion somehow "meshes" better with the notion of security for the protocol.

We make the above discussion more concrete with two examples: message authentication codes and symmetric encryption schemes. The security of (deterministic) message authentication codes (MACs) is captured by the notion of unpredictable functions [10,1,12]. In the context of MACs, this means that an adversary who is given valid MACs on some messages of its choice will be unlikely to succeed in outputting a "new" message (that is, one different from those whose MACs it was been given) along with a valid MAC on that message. It is well-known that any PRF is unpredictable (i.e. a secure MAC) [8]. Moreover, the reduction from unpredictable functions to PRFs is almost tight [5]. We show that using a direct reduction from unpredictable functions to IUF, one can obtain *exactly* the same bounds. This represents a tightening of the analysis, as we expect security of a PRF in the IUF sense to be smaller than the security in the standard PRF sense. (Our reductions show that the security in the IUF sense will never be more than a constant factor of 2 greater than the security in the PRF sense and will typically be a quantitative factor less.)

Now let us examine in what sense IUF "meshes" better with the notion of unpredictable functions. The quantitative drop in security in the reduction from PRF to IUF can be traced to the fact that under IUF the distinguisher must decide given *one* challenge whereas, under PRF, every response to a query potentially constitutes a "challenge". Like IUF, the notion of unpredictable functions also has a single distinguished challenge. In the reduction to PRF, however, we cannot really take any advantage of the source of the strength of this notion, and hence the bounds derived are not as tight as what could be achieved otherwise.

Another example of a notion with a distinguished challenge phase is the standard indistinguishability of encryptions notion of security for encryption schemes [9,3]. Here again, using the notion of IUF instead of the standard PRF, we can hope to tighten analysis of PRF-based encryption schemes. We do this for the *counter mode* of encryption.

## 1.4   Related Work

We have already mentioned the foundational work on PRFs and PRPs [8,11] and the concrete security analysis of these objects [5,4]. Our approach in this work follows that of Bellare et al [3], who compared and classified notions of security for symmetric encryption schemes according to the concrete complexity of reductions. A concrete security analysis of various symmetric encryption schemes, including the counter mode, is given in that paper. Naor and Reingold have explored the relationship between unpredictable functions and PRFs under different attack models [12].

## 2   Definitions and Notation

We describe different notions of security for (finite) function families in this section. A function family is a keyed multi-set $F$ of functions where all functions have the same domain and range. To pick a function $f$ from family $F$ means to pick a key $a$, uniformly from key space $\mathsf{Keys}(F)$ of $F$, and let $f = F_a$. A family $F$ has input length $l$ and output length $L$ if each $f \in F$ maps $\{0,1\}^l$ to $\{0,1\}^L$.

We let $R_{l,L}$ denote the function family consisting of all functions with input length $l$ and output length $L$. Similarly, we let $P_l$ denote the set of all permutations on $l$-bit strings.

A function family $F$ is pseudorandom if the input-output behavior of $F_a$ is indistinguishable from the behavior of a random function of the same domain and range. This is formalized via the notion of statistical tests of Goldreich et al [8]. Our concrete security formalizations follow those of Bellare et al [5].

We first informally describe the two additional notions (IUF and IPF) for function families considered in this paper. The corresponding notions for permutation families (IUP and IPP) are analogous to these, and so we skip their description. At the end of this section, we formally define all these notions (for both function and permutation families).

INDISTINGUISHABLE-UNIFORM FUNCTIONS. This is an adaptation of a notion given by Goldreich et al [8]. The idea is that a distinguisher should not be able to distinguish the output of the PRF from a uniformly distributed value in the range of the function. The formalization considers two different experiments. In both experiments we start by choosing a random key $a \leftarrow \mathsf{Keys}(F)$, specifying a function $F_a$. In the first phase, the distinguisher is given an oracle for $F_a$ and allowed to query this oracle on points of its choice. It then outputs a point $x$ that has not been queried yet and some state information $s$ that it may want to preserve for use during the second phase. In one experiment, it receives in response the value $F_a(x)$. In the other experiment, it receives a uniformly distributed value in the range of $F$. The PRF family is "good" if no "reasonable" distinguisher can obtain significant advantage in distinguishing the two experiments.

INDISTINGUISHABLE-POINT FUNCTIONS. This is an adaptation of the indistinguishability of encryptions notion of security for encryption schemes. Here again we imagine a distinguisher $A$ that runs in two phases. In the find phase, given adaptive access to an oracle for the function, it comes up with a pair of points $x_0, x_1$ that it has not queried yet and some state information $s$. In the guess phase, given the output of the function $y$ on one of these points and $s$, it must identify which of the two points goes with $y$.

It is interesting that the notion IPP does capture pseudorandomness for permutation families. For most other primitives, we find that an indistinguishable-point-based characterization is weaker than an indistinguishable-uniform-based characterization. This is true for encryption schemes and turns out to be true for function families, as well. Observe that, for encryption schemes, we are usually concerned with this weaker characterization, because it captures the desired security requirements.

FORMAL DEFINITIONS. For each of the six notions we consider in this paper, we give definitions using the experiments defined in Figure 2. First, we consider the function family notions: PRF, IUF, and IPF.

| PRF: $\mathrm{Exp}_F^{\mathrm{PRF}}(A, b)$ | PRP: $\mathrm{Exp}_F^{\mathrm{PRP}}(A, b)$ |
|---|---|
| $\quad a \leftarrow \mathrm{Keys}(F)$ <br> $\quad \mathcal{O}_0 \leftarrow F_a; \mathcal{O}_1 \leftarrow R_{l,L}$ <br> $\quad d \leftarrow A^{\mathcal{O}_b}$ <br> $\quad$ return $d$ | $\quad a \leftarrow \mathrm{Keys}(F)$ <br> $\quad \mathcal{O}_0 \leftarrow F_a; \mathcal{O}_1 \leftarrow P_l$ <br> $\quad d \leftarrow A^{\mathcal{O}_b}$ <br> $\quad$ return $d$ |
| IUF: $\mathrm{Exp}_F^{\mathrm{IUF}}(A, b)$ | IUP: $\mathrm{Exp}_F^{\mathrm{IUP}}(A, b)$ |
| $\quad a \leftarrow \mathrm{Keys}(F)$ <br> $\quad (x, s) \leftarrow A^{F_a}(\mathsf{find})^{\dagger}$ <br> $\quad y_0 \leftarrow F_a(x); y_1 \xleftarrow{R} \{0,1\}^L$ <br> $\quad d \leftarrow A(\mathsf{guess}, y_b, s)$ <br> $\quad$ return $d$ | $\quad a \leftarrow \mathrm{Keys}(F)$ <br> $\quad (x, s) \leftarrow A^{F_a}(\mathsf{find})^{\dagger}$ <br> $\quad y_0 \leftarrow F_a(x); y_1 \xleftarrow{R} \{0,1\}^l$ <br> $\quad d \leftarrow A(\mathsf{guess}, y_b, s)$ <br> $\quad$ return $d$ |
| $^{\dagger}$ $x$ not queried to $F_a$ | $^{\dagger}$ $x$ not queried to $F_a$ |
| IPF: $\mathrm{Exp}_F^{\mathrm{IPF}}(A, b)$ | IPP: $\mathrm{Exp}_F^{\mathrm{IPP}}(A, b)$ |
| $\quad a \leftarrow \mathrm{Keys}(F)$ <br> $\quad (x_0, x_1, s) \leftarrow A^{F_a}(\mathsf{find})^{\dagger}$ <br> $\quad y \leftarrow F_a(x_b)$ <br> $\quad d \leftarrow A(\mathsf{guess}, y, s)$ <br> $\quad$ return $d$ | $\quad a \leftarrow \mathrm{Keys}(F)$ <br> $\quad (x_0, x_1, s) \leftarrow A^{F_a}(\mathsf{find})^{\dagger}$ <br> $\quad y \leftarrow F_a(x_b)$ <br> $\quad d \leftarrow A(\mathsf{guess}, y, s)$ <br> $\quad$ return $d$ |
| $^{\dagger}$ $x_0, x_1$ not queried to $F_a$ | $^{\dagger}$ $x_0, x_1$ not queried to $F_a$ |

**Fig. 2.** Experiments defining each of the notions considered in this paper.

**Definition 1.** *For each notion* $\mathsf{N} \in \{\mathrm{PRF}, \mathrm{IUF}, \mathrm{IPF}\}$, *let* $F: \mathrm{Keys}(F) \times \{0,1\}^l \to \{0,1\}^L$ *be a finite function family. For an adversary* $A$ *and* $b = 0, 1$ *define the experiment* $\mathrm{Exp}_F^{\mathsf{N}}(A, b)$, *as given in Figure 2. Define the advantage of* $A$ *and the advantage function of* $F$, *respectfully, as follows. For any integers* $t, q \geq 0$,

$$\mathrm{Adv}_F^{\mathsf{N}}(A) = \Pr[\, \mathrm{Exp}_F^{\mathsf{N}}(A, 0) = 0 \,] - \Pr[\, \mathrm{Exp}_F^{\mathsf{N}}(A, 1) = 0 \,]$$
$$\mathrm{Adv}_F^{\mathsf{N}}(t, q) = \max_A \{\mathrm{Adv}_F^{\mathsf{N}}(A)\}$$

*where the maximum is over all* $A$ *with time complexity* $t$, *making* $\leq q$ *queries.* ∎

Here the "time-complexity" is the worst-case total execution time of the experiment, plus the size of the code of the adversary, in some fixed RAM model of computation. This convention is used for all definitions in this paper.

Next, we turn our attention to the definitions for the corresponding permutation family notions: PRP, IUP, and IPP.

**Definition 2.** *For each notion* $\mathsf{N} \in \{\mathrm{PRP}, \mathrm{IUP}, \mathrm{IPP}\}$, *let* $F: \mathrm{Keys}(F) \times \{0,1\}^l \to \{0,1\}^l$ *be a finite permutation family. For an adversary* $A$ *and* $b = 0, 1$ *define* *the experiment* $\mathsf{Exp}_F^N(A, b)$, *as given in Figure 2. Define the advantage of A and the advantage function of F, respectfully, as follows. For any integers* $t, q \geq 0$,

$$\mathsf{Adv}_F^N(A) = \Pr[\,\mathsf{Exp}_F^N(A, 0) = 0\,] - \Pr[\,\mathsf{Exp}_F^N(A, 1) = 0\,]$$
$$\mathsf{Adv}_F^N(t, q) = \max_A \{\mathsf{Adv}_F^N(A)\}$$

*where the maximum is over all A with time complexity t, making* $\leq q$ *queries.* ∎

## 3    Reductions Among the Notions

In this section, we formally state the relations shown in Figure 1. The proofs for these results are given in the full version of this paper [7].

We use the notation $A \Rightarrow B$ to indicate a security-preserving reduction from notion A to notion B. $A \rightarrow B$ indicates a reduction (not necessarily security-preserving) from A to B. $A \not\Rightarrow B$ and $A \not\rightarrow B$ are the natural interpretations given the above. This convention is followed for all reductions given in this paper.

### 3.1    Function Family Notions

The first theorem says that if a function family has certain security in the standard PRF sense, then it has essentially the same security in the IUF sense.

**Theorem 1.** [PRF $\Rightarrow$ IUF] *For any function family F and integers* $t, q \geq 1$,

$$\mathsf{Adv}_F^{\mathrm{IUF}}(t, q) \leq 2 \cdot \mathsf{Adv}_F^{\mathrm{PRF}}(t', q)$$

*where* $t' = t + O(l + L)$. ∎

Our next theorem says that if a function family is secure in the IUF sense, then it is also secure in the PRF sense, but the security is quantitatively lower.

**Theorem 2.** [IUF $\rightarrow$ PRF] *For any function family F and integers* $t, q \geq 1$,

$$\mathsf{Adv}_F^{\mathrm{PRF}}(t, q) \leq q \cdot \mathsf{Adv}_F^{\mathrm{IUF}}(t', q)$$

*where* $t' = t + O(l + L)$. ∎

The following proposition establishes that the drop in security in the previous theorem was not due to any weakness of our reduction but is, in fact, intrinsic to the notions. We give a concrete example of a function family that has higher security in the PRF sense, with a gap of the same order as in Theorem 2.

**Proposition 1.** [IUF $\not\Rightarrow$ PRF] *There exists a function family F such that*

$$\mathsf{Adv}_F^{\mathrm{PRF}}(t, q) \geq \frac{1}{2} \text{ and } \mathsf{Adv}_F^{\mathrm{IUF}}(t, q) \leq \frac{1}{q}$$

*for any integers* $t \geq 1$ *and* $1 \leq q \leq 2^{L-1}$. ∎

Our next two results demonstrate that the IPF notion is weaker than the other two notions we have considered, and hence does not capture pseudorandomness.

**Theorem 3.** [IUF $\Rightarrow$ IPF] *For any function family $F$ and integers $t, q \geq 1$,*

$$\mathsf{Adv}_F^{\mathrm{IPF}}(t, q) \leq 2 \cdot \mathsf{Adv}_F^{\mathrm{IUF}}(t', q)$$

*where $t' = t + O(l + L)$.* ∎

**Proposition 2.** [IPF $\not\Rightarrow$ IUF] *There exists a function family $F$ such that,*

$$\mathsf{Adv}_F^{\mathrm{IUF}}(t, q) \geq 1 - 2^{-qL} \text{ and } \mathsf{Adv}_F^{\mathrm{IPF}}(t, q) = 0$$

*for any integers $t, q \geq 1$.* ∎

### 3.2   Permutation Family Notions

We first give the reductions between PRP and IUP. Our next three claims show that the security bounds we had derived between the notions for function families also hold between the corresponding notions for permutation families.

**Theorem 4.** [PRP $\Rightarrow$ IUP] *For any permutation family $F$ and integers $t, q \geq 1$,*

$$\mathsf{Adv}_F^{\mathrm{IUP}}(t, q) \leq 2 \cdot \mathsf{Adv}_F^{\mathrm{PRP}}(t', q)$$

*where $t' = t + O(l)$.* ∎

**Theorem 5.** [IUP $\rightarrow$ PRP] *For any permutation family $F$ and integers $t, q \geq 1$,*

$$\mathsf{Adv}_F^{\mathrm{PRP}}(t, q) \leq q \cdot \mathsf{Adv}_F^{\mathrm{IUP}}(t', q)$$

*where $t' = t + O(l)$.* ∎

**Proposition 3.** [IUP $\not\Rightarrow$ PRP] *There exists a permutation family $F$ such that*

$$\mathsf{Adv}_F^{\mathrm{PRP}}(t, q) \geq \frac{1}{2} \text{ and } \mathsf{Adv}_F^{\mathrm{IUP}}(t, q) \leq \frac{1}{q}$$

*for any integers $t \geq 1$ and $1 \leq q \leq 2^{L-1}$.* ∎

Next, we establish that IUP and IPP are of essentially equivalent strength. Note that this is a departure from the relationship that exists between the corresponding function family notions.

**Theorem 6.** [IUP $\Rightarrow$ IPP] *For any permutation family $F$ and integers $t, q \geq 1$,*

$$\mathsf{Adv}_F^{\mathrm{IPP}}(t, q) \leq 2 \cdot \mathsf{Adv}_F^{\mathrm{IUP}}(t', q)$$

*where $t' = t + O(l)$.* ∎

**Theorem 7.** [IPP $\Rightarrow$ IUP] *For any permutation family $F$ and integers $t, q \geq 1$,*

$$\mathsf{Adv}_F^{\mathrm{IUP}}(t, q) \leq \mathsf{Adv}_F^{\mathrm{IPP}}(t', q)$$

*where $t' = t + O(l)$.* ∎

# 4  Applications

Here, we give some motivation for the use of the IUF characterization of PRF families. As discussed in Section 1, use of this notion gives tighter security bounds for certain cryptographic protocols. We give two such examples in this section.

## 4.1  The Case of Message Authentication Codes

A message authentication code (MAC) enables two parties who share a secret key to authenticate their transmissions. To be secure, MACs must resist existential forgery under chosen-message attacks [10,5]. For deterministic MACs, this notion matches that of unpredictable functions (UPF) [1,12].

Formally, the notion is captured by allowing a distinguisher $A$ to query a MAC oracle, $F_a$, where $F$ is a function family and $a$ is a random MAC key. $A$ must then output a point $x$ that has not been queried yet, along with its prediction $y$ for the value of $F_a(x)$.

**Definition 3.** [Message authentication security: UPF] *Let $F$:* $\mathsf{Keys}(F) \times \{0,1\}^l \to \{0,1\}^L$ *be a MAC. For an adversary $A$ define the following experiment:*

Experiment $\mathsf{Exp}_F^{\mathrm{UPF}}(A)$
   $a \leftarrow \mathsf{Keys}(F);\ (x,y) \leftarrow A^{F_a}$ //where $x$ is a point that $A$ has not queried
   If $y = F_a(x)$ then $d \leftarrow 0$ else $d \leftarrow 1$;   Return $d$.

*Define the advantage of $A$ and the advantage function of $F$, respectfully, as follows. For any integers $t, q \geq 0$,*

$$\mathsf{Adv}_F^{\mathrm{UPF}}(A) = \Pr[\,\mathsf{Exp}_F^{\mathrm{UPF}}(A) = 0\,]$$

$$\mathsf{Adv}_F^{\mathrm{UPF}}(t, q) = \max_A \{\mathsf{Adv}_F^{\mathrm{UPF}}(A)\}$$

*where the maximum is over all $A$ with time complexity $t$, making $\leq q$ queries.* ∎

PRF families are more well-studied than unpredictable function families and, moreover, are widely available. Hence, the observation that a PRF family constitutes a secure MAC [8] has proven very useful in practice. The following exact security reduction is already known [5].

**Proposition 4.** [PRF $\Rightarrow$ UPF] *For any function family $F$ and integers $t, q \geq 1$,*

$$\mathsf{Adv}_F^{\mathrm{UPF}}(t, q) \leq \mathsf{Adv}_F^{\mathrm{PRF}}(t', q) + 2^{-L}$$

*where $t' = t + O(l + L)$.* ∎

The reduction is almost tight. Consider now translating the above, to get security as a MAC in terms of the security as a PRF family in the IUF sense. Using Theorem 2 will lead to a drop in security by a factor $q$. However, by applying a direct reduction, we avoid this expected loss.

**Proposition 5.** [IUF $\Rightarrow$ UPF] *For any function family $F$ and integers $t, q \geq 1$,*

$$\mathsf{Adv}_F^{\mathrm{UPF}}(t, q) \leq \mathsf{Adv}_F^{\mathrm{IUF}}(t', q) + 2^{-L}$$

*where $t' = t + O(l + L)$.*

*Proof.* The reduction is standard. Let $A$ be a forger attacking the MAC $F$, making at most $q$ oracle queries and running in time at most $t$, in the experiment $\mathsf{Exp}_F^{\mathrm{UPF}}(A)$. We construct a distinguisher $A'$, making at most $q$ queries and running in time at most $t'$, using the forger $A$ as a subroutine.

Let $\mathcal{O}_f$ be $A$'s oracle. $A'^{\mathcal{O}_f}$ will run $A$ using $\mathcal{O}_f$ to provide an appropriate simulation of $A$'s oracle, as indicated below.

Algorithm $A'^{\mathcal{O}_f}$

(1)    Run $A$, answering any query $u$ with $\mathcal{O}_f(u)$.
(2)    Let $(x, y) \leftarrow A$.
(3)    Output $(x, y)$ and receive $y'$ as the challenge.
(4)    If $y' = y$ then output 0, else output 1.

For simplicity, we assume that $A$ makes exactly $q$ queries in $\mathsf{Exp}_F^{\mathrm{UPF}}(A)$. It is easy to check that the time and query complexity are as claimed. Next, we compute the advantage of $A'$.

$$\mathsf{Adv}_F^{\mathrm{IUF}}(A') = \Pr[\mathsf{Exp}_F^{\mathrm{IUF}}(A', 0) = 0] - \Pr[\mathsf{Exp}_F^{\mathrm{IUF}}(A', 1) = 0]$$
$$= \Pr[\mathsf{Exp}_F^{\mathrm{UPF}}(A) = 0] - 2^{-L} \; = \; \mathsf{Adv}_F^{\mathrm{UPF}}(A) - 2^{-L}$$

Given that $A$ was any arbitrary forger, the claimed relation follows. ∎

We say that Proposition 5 represents a tightening of the security bounds given in Proposition 4 since, from Theorems 1 and 2, we know that $\mathsf{Adv}_F^{\mathrm{IUF}}(t', q)$ is at most $2 \cdot \mathsf{Adv}_F^{\mathrm{PRF}}(t', q)$ and can be as small as $\frac{1}{q} \cdot \mathsf{Adv}_F^{\mathrm{PRF}}(t', q)$.

### 4.2    The Case of Symmetric Encryption Schemes

In the following discussion, we use the standard syntax and notion of security for encryption schemes given by Bellare et al [3], which is an adaptation of one given by Goldwasser and Micali [9]. In the indistinguishability of encryptions under chosen-plaintext attack (IND) notion, the adversary $A$ is imagined to run in two phases. In the find phase, given adaptive access to an encryption oracle, $A$ produces a pair of equal-length messages $x_0, x_1$, along with some state information $s$. In the guess phase, given the encryption $y$ of one of the messages and $s$, it must identify which of the two messages goes with $y$.

**Definition 4.** [Symmetric encryption security: IND] *Let $\Pi = (\mathcal{K}, \mathcal{E}, \mathcal{D})$ be an encryption scheme. For an adversary $A$ and $b = 0, 1$ define the experiment:*

Experiment $\mathsf{Exp}_\Pi^{\mathrm{IND}}(A, b)$
   $a \leftarrow \mathcal{K}; \; (x_0, x_1, s) \leftarrow A^{\mathcal{E}_a}(\mathsf{find}); \; y \leftarrow \mathcal{E}_a(x_b); \; d \leftarrow A^{\mathcal{E}_a}(\mathsf{guess}, y, s); \; \text{Return } d.$

*It is mandated that* $|x_0| = |x_1|$ *above. Define the advantage of A and the advantage function of* $\Pi$, *respectfully, as follows. For any integers* $t, q, \mu \geq 0$,

$$\mathsf{Adv}_\Pi^{\mathrm{IND}}(A) = \Pr[\,\mathsf{Exp}_\Pi^{\mathrm{IND}}(A, 0) = 0\,] - \Pr[\,\mathsf{Exp}_\Pi^{\mathrm{IND}}(A, 1) = 0\,]$$
$$\mathsf{Adv}_\Pi^{\mathrm{IND}}(t, q, \mu) = \max_A \{\mathsf{Adv}_\Pi^{\mathrm{IND}}(A)\}$$

*where the maximum is over all A with time complexity t, making* $\leq q$ *oracle queries which total* $\leq \mu$ *bits.* ∎

We analyze the counter mode of encryption based on a finite PRF. In practice, the finite PRF may be instantiated by a block cipher. For a finite PRF $F$, the counter mode $\mathrm{CTR}(F) = (\mathcal{E}\text{-}\mathrm{CTR}, \mathcal{D}\text{-}\mathrm{CTR}, \mathcal{K}\text{-}\mathrm{CTR})$ can be described as follows. The key generation algorithm $\mathcal{K}\text{-}\mathrm{CTR}$ outputs a random key $a$ for the underlying PRF family $F$, thereby specifying a function $f = F_a$ of $l$-bits to $L$-bits. The sender maintains a $l$ bit counter $ctr$ that is initially $-1$ and is incremented after each encryption by the number of blocks encrypted. The message $x$ to be encrypted is regarded as a sequence of $L$-bit blocks (padding is done first, if necessary), $x = x_1 \cdots x_n$. We define $\mathcal{E}\text{-}\mathrm{CTR}_a(x, ctr) = \mathcal{E}\text{-}\mathrm{CTR}^{F_a}(x, ctr)$ and $\mathcal{D}\text{-}\mathrm{CTR}_a(z) = \mathcal{D}\text{-}\mathrm{CTR}^{F_a}(z)$, where:

| Algorithm $\mathcal{E}\text{-}\mathrm{CTR}^f(x, ctr)$ | Algorithm $\mathcal{D}\text{-}\mathrm{CTR}^f(z)$ |
|---|---|
| for $i = 1, \ldots, n$ do | Parse $z$ as $ctr, y_1 \cdots y_n$ |
| $\quad y_i = f(ctr + i) \oplus x_i$ | for $i = 1, \ldots, n$ do |
| $ctr \leftarrow ctr + n$ | $\quad x_i = f(ctr + i) \oplus y_i$ |
| return $(ctr, y_1 y_2 \cdots y_n)$ | return $x = x_1 \cdots x_n$ |

We show that $\mathrm{CTR}(F)$ is secure in the IND sense if $F$ is secure in the IUF sense. As with our previous example, the reduction achieves the same concrete security bounds as those possible using the standard notion of PRF families.

**Theorem 8. [Security of CTR using an IUF function family]** *For any function family $F$ and integers $t, q \geq 1$ and $L \leq \mu \leq L2^l$,*

$$\mathsf{Adv}_{\mathrm{CTR}(F)}^{\mathrm{IND}}(t, q, \mu) \leq 2 \cdot \mathsf{Adv}_F^{\mathrm{IUF}}(t', q')$$

*where* $t' = t + O(\frac{\mu}{L}(l + L))$ *and* $q' = \frac{\mu}{L}$.

*Proof.* We want to show that if $\mathrm{CTR}(F)$ is not secure in the IND sense, then it must be the case that $F$ is not secure in the IUF sense. Let $A$ be an adversary attacking the $\mathrm{CTR}(F)$, running in time at most $t$ and making at most $q$ oracle queries, these totalling at most $\mu$ bits, in the experiment $\mathsf{Exp}_{\mathrm{CTR}(F)}^{\mathrm{IND}}(A)$. We construct a distinguisher $A'$, making at most $q'$ queries and running in time at most $t'$, using the adversary $A$ as a subroutine.

Let $\mathcal{O}_f$ be $A'$'s oracle. $A'^{\mathcal{O}_f}$ will run $A$ using $\mathcal{O}_f$ to provide an appropriate simulation of $A$'s encryption oracle. We assume, for the sake of simplicity of the exposition, that the two messages $A$ outputs at the end of its first phase are exactly $L$ bits in length (i.e. of the size of one block). In the following, $\mu_G < \mu$, is the amount of ciphertext $A$ needs to see in its **guess** phase.

Algorithm $A'^{\mathcal{O}_f}$

(1)   Initialize counter: $ctr \leftarrow -1$.

(2)   Run $A(\text{find})$, answering any query $u$ with $\mathcal{E}\text{-CTR}^{\mathcal{O}_f}(u)$.

(3)   Let $(x_0, x_1, s) \leftarrow A(\text{find})$.

(4)   Let the current value of the counter be $ctr_0$.

(5)   Compute $\mathcal{F} = \{\mathcal{O}_f(ctr_0 + i) : 1 \leq i \leq \frac{\mu_G}{L}\}$.

(6)   Let $s' = (s, x_0, x_1, ctr_0, \mathcal{F})$.

(7)   Output $(ctr_0, s')$ and receive $y$ as the challenge.

(8)   Let $d \leftarrow \{0, 1\}$.

(9)   Run $A(\text{guess}, y \oplus x_d, s)$, answering any query $u$, using $\mathcal{F}$, with $\mathcal{E}\text{-CTR}(u)$.

(10)  Let $d' \leftarrow A(\text{guess}, y \oplus x_d, s)$.

(11)  If $d = d'$ then output 0, else output 1.

In the reduction above, $A'$ maintains the counter $ctr$, incrementing it appropriately. It is important here that $A'$ can implement $\mathcal{E}\text{-CTR}^f(\cdot, ctr)$ given an oracle for $f$. At the end of the find phase queries of $A$, it picks the current value of counter $ctr_0$ to be the output of its own find phase, along with the state information. A slight problem that comes up here is that $A'$ does not have access to $\mathcal{O}_f$ in its **guess** phase but it will still need to provide a simulation of the encryption oracle during $A$'s guess phase queries. We get around this by having $A'$ pre-compute the value of $\mathcal{O}_f$ on as many points as necessary, starting from $ctr_0 + 1$, to answer all of $A$'s **guess** phase encryption oracle queries. These pre-computed values are in the set $\mathcal{F}$ which is passed to $A''$s **guess** phase via state information $s$. Note that it is important that $A'$ did not query $\mathcal{O}_f$ with $ctr_0$, since otherwise it could not output $ctr_0$ as the point on which it gets its challenge. The counter mode guarantees that, as long as fewer than $\frac{\mu}{L}$ queries are made (i.e the counter does not loop around), the function will always be invoked on a new point.

The total number of oracle queries made by $A'$ is at most $\frac{\mu}{L}$, which by assumption is $q'$. Given this, one can check that the running time of $A'$ is as claimed. The advantage of $A'$ is given by,

$$\mathsf{Adv}_F^{\mathrm{IUF}}(A') = \Pr[\mathsf{Exp}_F^{\mathrm{IUF}}(A', 0) = 0] - \Pr[\mathsf{Exp}_F^{\mathrm{IUF}}(A', 1) = 0]$$

$$= \Pr[\mathsf{Exp}_{\mathrm{CTR}(F)}^{\mathrm{IND}}(A, 0) = 0] + \Pr[\mathsf{Exp}_{\mathrm{CTR}(F)}^{\mathrm{IND}}(A, 1) = 1] - \frac{1}{2}$$

$$= \frac{1}{2}(1 + \mathsf{Adv}_{\mathrm{CTR}(F)}^{\mathrm{IND}}(A)) - \frac{1}{2} \;\; = \;\; \frac{1}{2} \cdot \mathsf{Adv}_{\mathrm{CTR}(F)}^{\mathrm{IND}}(A)$$

Given that $A$ is an arbitrary adversary, the claimed relation follows.  ∎

# 5   Discussion

We stress that the benefits of tighter security analyses, such as those we have presented here, are real. For example, using the standard notion of a PRF, the security of a protocol may appear to be marginal, prompting the use of a larger security parameter. However, using a tighter characterization, such as IUF, the security might have been determined to be adequate.

In criticism to our approach to getting tighter bounds for MACs and symmetric encryption schemes, one may suggest that we are looking at the wrong notions of security for these protocols. Indeed, there are alternate notions for which our gains would disappear. However, the notions of security we consider for both MACs and symmetric encryption are, in practice, the notions which are most widely used.

FUTURE DIRECTIONS. Unlike the case with the counter mode of encryption, in our first example we view the entire MAC as being the primitive, when in fact it too may be built on a PRF (for example, the CBC-MAC based on a block cipher). While it seems unlikely that we can achieve a tighter security analysis for the CBC-MAC scheme using the same approach, it may be possible for other message authentication schemes. Then there are other schemes, besides those for message authentication and symmetric encryption, to which our techniques could be applied. For example, it may be possible to improve the security bounds of variable-length input pseudorandom functions (VI-PRFs) [2] and variable-input-length ciphers [6].

Using similar techniques as above, we can also get tighter bounds for PRP-based protocols. In a sense, this is more interesting, given that PRP families provide a more natural model for block ciphers [5]. Viewing a block cipher as a PRP family rather than a PRF family itself can lead to tighter security bounds. However, our examples were motivated by the fact that analysis of a block-cipher-based scheme is, as far as possible, done modeling the block cipher as a PRF. This is because the analysis using PRFs is usually significantly simpler.

We remark that it seems somewhat significant that, in the indistinguishability of points characterization, there is a difference between function and permutation families. This seems to be the first such distinction, as far as we know, when asymptotic measures are used. It may be interesting to investigate further the impact of injectivity upon pseudorandomness.

# Acknowledgements

We are grateful to Mihir Bellare for his advice and assistance with this work. We also thank the Asiacrypt 2000 program committee for their helpful comments.

This work was completed while the first author was a student at the University of California at San Diego, USA. Both authors were supported in part by Mihir Bellare's 1996 Packard Foundation Fellowship in Science and Engineering and NSF CAREER Award CCR-9624439.

# References

1. M. BELLARE, R. CANETTI AND H. KRAWCZYK, "Keying hash functions for message authentication," *Advances in Cryptology - Crypto '96*, LNCS Vol. 1109, N. Koblitz ed., Springer-Verlag, 1996.

2. M. BELLARE, R. CANETTI AND H. KRAWCZYK, "Pseudorandom functions revisited: The cascade construction and its concrete security," *Proceedings of the 37th Symposium on Foundations of Computer Science*, IEEE, 1996.

3. M. BELLARE, A. DESAI, E. JOKIPII AND P. ROGAWAY, "A concrete security treatment of symmetric encryption: Analysis of the DES modes of operation," *Proceedings of the 38th Symposium on Foundations of Computer Science*, IEEE, 1997.

4. M. BELLARE, R. GUÉRIN AND P. ROGAWAY, "XOR MACs: New methods for message authentication using finite pseudorandom functions," *Advances in Cryptology - Crypto '95*, LNCS Vol. 963, D. Coppersmith ed., Springer-Verlag, 1995.

5. M. BELLARE, J. KILIAN AND P. ROGAWAY, "The security of the cipher block chaining message authentication code," *Advances in Cryptology - Crypto '94*, LNCS Vol. 839, Y. Desmedt ed., Springer-Verlag, 1994.

6. M. BELLARE AND P. ROGAWAY, "On the construction of variable-input-length ciphers," *Proceedings of the Sixth Workshop on Fast Software Encryption*, L. Knudsen ed., 1999.

7. A. DESAI AND S. MINER, "Concrete security characterizations of PRFs and PRPs: Reductions and applications," Full version of this paper, available via: http://www-cse.ucsd.edu/users/sminer/.

8. O. GOLDREICH, S. GOLDWASSER AND S. MICALI, How to construct random functions. *Journal of the ACM*, 33(4): 792-807, 1986.

9. S. GOLDWASSER AND S. MICALI, "Probabilistic encryption," *Journal of Computer and System Science*, Vol. 28, pp. 270-299, 1984.

10. S. GOLDWASSER, S. MICALI AND R. RIVEST, "A digital signature signature scheme secure against adaptive chosen-message attacks," *SIAM J. of Computing*, 17(2): 281-308, April 1988.

11. M. LUBY AND C. RACKOFF, "How to construct pseudorandom permutations from pseudorandom functions," *SIAM J. Computing*, 17(2), April 1988.

12. M. NAOR AND O. REINGOLD, "From unpredictability to indistinguishability: A simple construction of PRFs from MACs," *Advances in Cryptology - Crypto '98*, LNCS Vol. 1462, H. Krawczyk ed., Springer-Verlag, 1998.

# The Security of Chaffing and Winnowing

Mihir Bellare and Alexandra Boldyreva

Dept. of Computer Science & Engineering, University of California at San Diego,
9500 Gilman Drive, La Jolla, California 92093, USA.
{mihir, aboldyre}@cs.ucsd.edu
www-cse.ucsd.edu/users/{mihir, aboldyre}

**Abstract.** This paper takes a closer look at Rivest's chaffing-and-winnowing paradigm for data privacy. We begin with a definition which enables one to clearly determine whether a given scheme qualifies as "chaffing-and-winnowing." We then analyze Rivest's schemes to see what quality of data privacy they provide. His bit-by-bit scheme is easily proven secure but is inefficient. His more efficient scheme —based on all-or-nothing transforms (AONTs)— can be attacked under Rivest's definition of security of an AONT, and even under stronger notions does not appear provable. However we show that by using OAEP as the AONT one can prove security, and also present a different scheme, still using AONTs, that is equally efficient and easily proven secure even under a relatively weak notion of security of AONTs.

## 1 Introduction

Rivest presents a number of methods to achieve data privacy based on a paradigm he calls "chaffing and winnowing" [11]. In this paper we provide a *definition* of chaffing and winnowing; assess whether the schemes of [11] can be *proven* to meet standard data-privacy goals, and, if so, under what kinds of *assumptions* on the underlying primitives; and suggest more efficient schemes and analyze their security. Let us first provide some background and motivation, and see what are the basic questions. Then we discuss our contributions in more detail.

### 1.1 Background, Motivation, and Questions

Chaffing and winnowing uses a message authentication code (MAC) to provide privacy. However Rivest notes that in order to have privacy the MAC must be a pseudorandom function. (Any PRF is a good MAC [9,3] but not vice-versa.) Of course, there are many well-known ways to use a PRF to provide privacy; the interest of chaffing and winnowing arises from the particular manner in which the MAC is used, which is roughly the following. Each data block is authenticated so that one has a sequence of data-MAC pairs. Then "chaff" is interspaced, this consisting of pairs, each being a block with a random tag. The receiver can discard blocks with invalid tags —this is called "winnowing"— thereby recovering the data. (Within this framework, many specific methods are possible.) Privacy requires that it be computationally infeasible for an adversary to tell valid MACs from random tags. (But is also very sensitive to the manner in which chaff is interspaced.) Rivest argues that the use of the MAC here stays within

T. Okamoto (Ed.): ASIACRYPT 2000, LNCS 1976, pp. 517–530, 2000.

its functionality as an authentication mechanism, and thereby makes moot a policy that restricts "encryption" while allowing authentication.

Chaffing-and-winnowing has received a lot of attention on the political front, but little on the technical front barring that in the initial paper. It merits more serious attention from cryptographers. One is that a better technical understanding leads to a better understanding of the implications for the debate on cryptographic policy. Another reason is foundational. As Rivest notes in introducing his idea, there are very few paradigms for cryptography's main goal, namely data privacy. A new paradigm such as the one he presents should be explored in order to assess its potential.

The description and examples in [11] suffice to get the gist of the idea, but as we consider more complex mechanisms it is sometimes difficult to decide whether they obey the "rules of the game." A definition is needed to settle such questions, and also for rigorous security analysis. Accordingly, we begin there.

Chaffing and winnowing purports to provide data-privacy. A basic question is about the quality of privacy that it can provide. Specifically we want to know how it compares to standard mechanisms such as modes of operation of a block cipher, which have been proven to meet strong, well-defined notions of privacy under appropriate assumptions [2]. Can chaffing-and-winnowing based schemes provide the same level of privacy, and, if so, can this be proven, and under what assumptions?

## 1.2   Defining Chaffing and Winnowing

The security goal of a chaffing-and-winnowing scheme is to provide privacy in a symmetric setting. Accordingly, from the security point of view, it is —in fact, must be— treated simply as a symmetric encryption scheme. There is some "encryption" process that takes a message and creates a "ciphertext", and some "decryption" process that takes the ciphertext and recovers the message, both operating under a common secret key. (This is the key for the MAC.) These processes are not implemented in "usual" ways, but, abstractly, they must exist, else it is moot to talk of achieving privacy. Once this is understood, security can be measured using any of several well-known notions in the literature. (We adopt the simplest, namely the "find-then-guess" notion of [2], which is the most direct extension to the symmetric case of the notion of indistinguishability of [10].)

Thus what defines chaffing and winnowing as a "notion" is not some novel security property but rather a novel set of restrictions on the processes (namely encryption and decryption) directed at achieving a standard security property (namely data privacy). The crux of the definition is to pin down these restrictions. We view a chaffing-and-winnowing based encryption scheme as arising by the use of an *authentication to privacy transform* (ATPT) over a *MAC-based authentication channel*.

The channel captures the manner in which the parties have access to the MAC function $\text{MAC}(K, \cdot)$. An application on the sender side can pass data down to be MACed, thereby creating a packet (data-MAC pair) which is transmitted over the channel. (The application has no direct access to the $\text{MAC}(K, \cdot)$ function let alone to the underlying key $K$.) At the receiving end, packets with invalid MACs are dropped and the data from valid packets is passed up to the receiving application. (The latter sees no MACs and does not even know of the existence of the dropped packets.) See Figure 1 and Definition 4.

The ATPT consists of three algorithms whose important feature is that they are all entirely *keyless*. The sending application applies a MakeWheat algorithm to the plaintext to turn it into a sequence of data blocks to be passed to the authentication channel. An AddChaff procedure is responsible for interspacing chaff packets into the stream of valid packets output by the authentication channel. Finally the receiving application applies a Recover algorithm to the received data blocks to get back the plaintext. See Definition 5.

We stress again that the algorithms in the ATPT are keyless, so that on their own they cannot be used to provide privacy. The chaff and winnow based encryption scheme is realized by coupling these algorithms with the authentication channel. This is illustrated in Figure 2.

The definition (provided in Section 3) can help clarify the contribution of chaffing and winnowing to the debate on cryptographic policy by providing a means to evaluate whether a particular method qualifies as "legal encryption based on authentication." If one scheme meeting the definition qualifies, so do the rest, even if their implementation is more complex.

### 1.3   Security of Rivest's Schemes

Rivest notes that his first few examples will not provide a high level of privacy. (In particular they will not meet a notion of privacy such as find-then-guess.) The first serious candidate is the bit-by-bit scheme.

BIT-BY-BIT SCHEME. Here the MakeWheat procedure splits the plaintext into bits and appends a counter or nonce to each bit. These data blocks are MACed. The AddChaff procedure inserts, for every valid packet, an invalid packet with the opposite bit value and an appended nonce, together with a random value for the tag. We prove that this scheme provides privacy in the find-then-guess sense assuming the MAC is a pseudorandom function. The concrete security analysis is provided in Theorem 1.

This indicates that chaffing and winnowing can provide privacy of as high a quality as standard encryption schemes, and furthermore with provable guarantees based on the same assumption —namely a pseudorandom function— used to prove the security of popular block cipher modes of operation [2]. There is however a high cost in bandwidth: two nonces and two tags are needed per bit of plaintext.

SCATTERING SCHEMES. In order to reduce the bandwidth, Rivest suggests an alternative paradigm. First apply an all-or-nothing transform (AONT) [12] to the plaintext. (This is a keyless, invertible transform with the property that inversion is hard if any block of the output is missing.) Each block of the output of the AONT is MACed, resulting in a stream of valid packets. Then $s'$ chaff packets are inserted into random positions in this stream. Intuitively, an adversary must guess the positions of all $s'$ chaff packets in order to decipher. (Accordingly it is suggested that security will be provided for a value of $s'$ that does not depend on the length of the plaintext, eg. $s' = 128$, so this method is cost-effective for long plaintexts.) Upon closer examination, however, the security provided by this paradigm is unclear. We note first that under the original definition of an AONT provided in [12] the scheme is insecure. (We show that there are example AONTs that meet the definition of [12] but for which there are attacks compromising the privacy of the chaffing-and-winnowing scheme.) It is natural to then try to use Boyko's stronger definition of security for an AONT [6]. In that case the analysis is inconclusive: the stronger property of an AONT still does not appear to suffice to prove security of the chaffing-and -winnowing scheme, but neither do we exhibit a counter-example that confirms this. We would prefer a provably-secure scheme.

SCATTERING WITH OAEP. The above-mentioned analyses indicate that in general an AONT seems neither necessary nor sufficient as the initial transform to provide privacy of the scattering based chaffing-and-winnowing scheme. We show however that if the OAEP transform of [5] is used as the AONT, then privacy can be proved. (This, like all security proofs involving OAEP, is in a random oracle model [4]). The concrete security analysis provided in Theorem 2 supports the intuition regarding the scattering scheme provided in [11]— the probability of breaking the scheme is inversely proportional to $\binom{s+s'}{s}$ where $s$ is the number of blocks in the output of OAEP and $s'$ is the number of chaff blocks.

Note that OAEP has been shown to be a secure AONT [6], but given the above we cannot exploit this here. Instead, our proof is direct, based on techniques from [5,6]. It is an open question whether other specific constructions of AONTs such as that of [7] suffice to prove security of the scheme.

## 1.4   New Schemes

We point out that there is an alternative to the scattering scheme that is simpler, can be proven secure, and is equally cost-effective. It too makes use of AONTs and can be proven secure for any AONT meeting a notion of security that is actually weaker than that of [6]. (In particular one can use OAEP or use the construct of [7] and avoid random oracles but there may be other more efficient instantiations.) The construction applies the AONT to the plaintext as before. Rather than scattering chaff into the output blocks, however, it simply treats a prefix of this output as the plaintext for the bit-by-bit chaffing-and-winnowing scheme and applies the latter. Theorem 3 provides a concrete security analysis of the final chaffing-and-winnowing scheme.

## 1.5   Is Chaffing and Winnowing "Encryption"?

We view chaffing-and-winnowing schemes as (special kinds of) symmetric encryption schemes, the key for encryption and decryption being that of the MAC function. This might at first seem to contradict Rivest's view [11]. He says that the process of chaffing and winnowing is "not encryption" and that there is no "decryption key." These views are not at odds with each other; the difference is purely in terminology. We are using the technical terminology of cryptographers which is more suited to security analysis, while Rivest uses the terminology of cryptographic policy discussion. (The convention in modern cryptography, which we are following here, is to use the term "encryption scheme" for any mechanism whose goal is to provide privacy. Under this convention, the key for the MAC is, by definition, a decryption key, since it enables recovery of the plaintext from the ciphertext. In cryptographic policy, "encryption" seems to refer to certain mechanisms rather than a goal. Actually, exactly what it refers to is unclear, which is part of the point made in [11].)

## 2   Symmetric Encryption, PRFs, and AONTs

SYMMETRIC ENCRYPTION. A symmetric encryption scheme $\mathsf{SE} = (\mathcal{K}, \mathcal{E}, \mathcal{D})$ consists of a (randomized) key generation algorithm $\mathcal{K}$ (returning a key $K$), a (randomized or stateful) encryption algorithm $\mathcal{E}$ (taking $K$ and a message $M \in \{0,1\}^*$ to return a ciphertext $C$) and a decryption algorithm $\mathcal{D}$ (taking $K$ and a ciphertext and returning a message). We require that $\mathcal{D}_K(\mathcal{E}_K(M)) = M$ for all $M \in \{0,1\}^*$. In the "find-then-guess" model [2] an adversary is given an oracle for encryption under key $K$ and wins if it can find two equal-length messages whose ciphertexts it can later distinguish. Below we associate to any adversary an "advantage" which measures its winning probability, and then use as security measure of the scheme the maximum possible advantage subject to stated resource restrictions on the adversary.

**Definition 1. (Find-then-guess security of encryption, [2])** Let $\mathsf{SE} = (\mathcal{K}, \mathcal{E}, \mathcal{D})$ be a symmetric encryption scheme. For an adversary $A$ and $b = 0, 1$ define the experiment

Experiment $\mathbf{Exp}_{\mathsf{SE}}^{\mathrm{priv}}(A, b)$
$\quad K \xleftarrow{R} \mathcal{K} \; ; \; (M_0, M_1, St) \leftarrow A^{\mathcal{E}_K(\cdot)}(\mathsf{find}) \; ; \; C \leftarrow \mathcal{E}_K(M_b) \; ;$
$\quad d \leftarrow A^{\mathcal{E}_K(\cdot)}(\mathsf{guess}, C, St) \; ; \; \mathsf{Return} \; d$

Here $St$ is some state information that the adversary may want to preserve to help it later. It is mandated that $|M_0| = |M_1|$ above. Now define the *advantage* of $A$ and the *advantage function* of the scheme respectively, as follows:

$$\mathbf{Adv}_{\mathsf{SE}}^{\mathrm{priv}}(A) = \Pr\left[\, \mathbf{Exp}_{\mathsf{SE}}^{\mathrm{priv}}(A, 0) = 0 \,\right] - \Pr\left[\, \mathbf{Exp}_{\mathsf{SE}}^{\mathrm{priv}}(A, 1) = 0 \,\right]$$

$$\mathbf{Adv}_{\mathsf{SE}}^{\mathrm{priv}}(t, q, \mu) = \max \{ \, \mathbf{Adv}_{\mathsf{SE}}^{\mathrm{priv}}(A) \, \}$$

where the maximum is over all $A$ with "time-complexity" $t$, making at most $q$ encryption oracle queries, these totalling at most $\mu$ bits. ∎

In this paper for simplicity we assume that all messages encrypted have the same length, usually denoted $m$. This means that $\mu = mq$. We also assume that the length of each of the challenge messages is $m$. The "time-complexity" refers to the worst case execution time of experiment $\mathbf{Exp}_{\mathsf{SE}}^{\mathrm{priv}}(A)$ plus the size of the code of $A$, in some fixed RAM model of computation. We are considering only chosen-plaintext attacks, not chosen-ciphertext attacks.

PSEUDORANDOM FUNCTIONS. Consider a map $F \colon \{0,1\}^k \times S \to \{0,1\}^l$ which takes a key $K \in \{0,1\}^k$ and an input $x$ from the domain $S$ to return an output $y = F(K, x)$. The domain $S$ is for convenience $\{0,1\}^*$, or at least the set of all strings of length up to some very large maximum length. The notation $g \xleftarrow{R} F$ is shorthand for $K \xleftarrow{R} \{0,1\}^k \; ; \; g \leftarrow F(K, \cdot)$. We let $R$ denote the family of all functions of $S$ to $\{0,1\}^l$ so that $g \xleftarrow{R} R$ denotes the operation of selecting at random a function of $S$ to $\{0,1\}^l$. A distinguisher $D$ is an algorithm that takes an oracle for a function $g \colon S \to \{0,1\}^l$, and after computing with this oracle returns a bit. The following is the notion of [9] concretized as per [3].

**Definition 2.** Let $F, R$ be as above, let $D$ be a distinguisher, and suppose $t, q, \mu \geq 0$. Define the *advantage of D*, and the *advantage function of F*, respectively, as

$$\mathbf{Adv}_F^{\mathrm{prf}}(D) = \Pr\left[ D^g = 1 \; : \; g \overset{R}{\leftarrow} F \right] - \Pr\left[ D^g = 1 \; : \; g \overset{R}{\leftarrow} R \right]$$

$$\mathbf{Adv}_F^{\mathrm{prf}}(t, q, \mu) = \max\left\{ \mathbf{Adv}_F^{\mathrm{prf}}(D) \right\}.$$

where the maximum is over all $D$ with time-complexity at most $t$, making at most $q$ oracle queries, these totalling at most $\mu$ bits. ∎

ALL-OR-NOTHING TRANSFORMS. An all-or-nothing transform is an efficiently computable, keyless, randomized transformation AONT which maps a message to a sequence of blocks such that given the AONT of some message, one can easily compute the original message [12]. The (deterministic) inverse transformation permitting recovery of the message from the output is denoted AONT$^{-1}$. Security pertains to the question of what information you can compute about the message if you are given all but one of the output blocks, and several notions have been suggested [12,6,8]. We provide our formalization and compare it to the others later.

We assume for simplicity that the AONT takes input messages of length $m$ and has outputs of length $sn$. The attack allowed is non-adaptive, meaning the adversary fixes beforehand the position of the output block that will be omitted. Denote this by $L \in \{1, \ldots, s\}$. During the ❚nd stage the adversary comes up with a pair of messages $M_0$ and $M_1$, both of length $m$. In its guess stage it is given a AONT for one of the plaintexts $M_0, M_1$, with block $L$ missing. The adversary wins if it correctly guesses which message goes with the challenge AONT. If $X \in \{0,1\}^{sn}$ is a string of $s$ blocks, each $n$-bits long, then we let $X[1, \ldots, L-1, L+1, \ldots, s]$ denote the string consisting of blocks $1, \ldots, L-1, L+1, \ldots, s$ of $X$, meaning all but block $L$.

**Definition 3.** Let AONT be a (randomized) algorithm taking an input of length $m$ and returning an output of length $sn$. Let $L \in \{1, \ldots, s\}$ be a block number. $St$ denotes some state information. For $b = 0, 1$ define the experiment

Experiment $\mathbf{Exp}_{\mathrm{AONT}, L}^{\mathrm{aont}}(A, b)$
    $(M_0, M_1, St) \leftarrow A(\mathsf{find})$ ; $C \leftarrow \mathrm{AONT}(M_b)[1, \ldots, L-1, L+1, \ldots, s]$ ;
    $d \leftarrow A(\mathsf{guess}, C, St)$ ; Return $d$

Now define the *advantage* of $A$ and the *advantage function* of AONT, respectively, as follows:

$$\mathbf{Adv}_{\mathrm{AONT}, L}^{\mathrm{aont}}(A) = \Pr\left[ \mathbf{Exp}_{\mathrm{AONT}, L}^{\mathrm{aont}}(A, 0) = 0 \right] - \Pr\left[ \mathbf{Exp}_{\mathrm{AONT}, L}^{\mathrm{aont}}(A, 1) = 0 \right]$$

$$\mathbf{Adv}_{\mathrm{AONT}, L}^{\mathrm{aont}}(t) = \max\left\{ \mathbf{Adv}_{\mathrm{AONT}, L}^{\mathrm{aont}}(A) \right\}$$

where the maximum is over all $A$ with "time-complexity" $t$. ∎

We now compare this to other notions, in all cases considering an adversary having a string $C$ consisting of all but one block of the output. Rivest [12] asks that given $C$ it be computationally infeasible to get any non-trivial information about any block of the message. Our definition is stronger than his, meaning any AONT secure in our

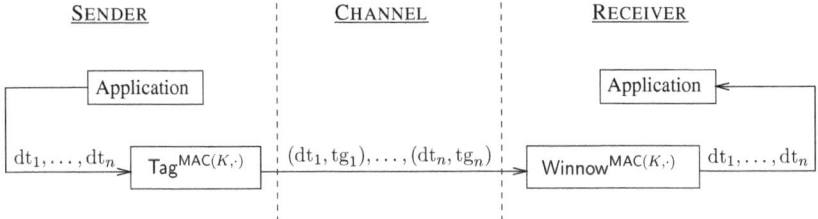

**Fig. 1. Authentication channel**

sense is secure in his sense. Boyko's definition of security [6] asks that given $C$ it be computationally infeasible to get any non-trivial information about the message as a whole, not just individual blocks. Desai [8] asks that $C$ be indistinguishable from a random string of the same length. Our definition is weaker than either of these, in the sense that any AONT secure in their sense is secure in our sense.

## 3   Defining Chaffing and Winnowing

Fix a map MAC: $\{0,1\}^k \times \{0,1\}^* \to \{0,1\}^l$ to be used as a message authentication code. (Security assessments will assume that this map is a pseudorandom function, but discussions and constructions will refer to it as a MAC.) A *packet* is a pair $(\mathrm{dt},\mathrm{tg})$ consisting of *data* dt and a *tag* tg where the length of tg is $l$-bits, the length of the output of MAC. A packet $(\mathrm{dt},\mathrm{tg})$ is *valid* with respect to $\mathrm{MAC}_K$ —where $K \in \{0,1\}^k$ is some key for the MAC— if $\mathrm{MAC}_K(\mathrm{dt}) = \mathrm{tg}$, and *invalid* with respect to $\mathrm{MAC}_K$ otherwise. When $\mathrm{MAC}_K$ is understood, we simply talk of valid and invalid packets.

The sender and receiver have an authenticated channel of communication based on the MAC. Each party has a module responsible for authentication. These modules hold in common a key $K \in \{0,1\}^k$ for the MAC. When the sender wants to send data dt to the receiver in an authenticated way, the sender passes dt to its authentication module, which creates the (valid) packet $\mathrm{Pkt} = (\mathrm{dt}, \mathrm{MAC}(K,\mathrm{dt}))$. This packet is sent to the receiver. We call this the "tag" procedure. The packet is received by the receiver's authentication module, which verifies the tag. If the tag is valid, it passes the data "up" to the receiver. If the tag is not valid, the packet is simply discarded; nothing is passed up to the receiver. The receiving module thus acts as a "filter", separating "wheat" (valid) packets from "chaff" (invalid) packets, and passing to the receiver only the data from the valid packets. This is what [11] calls the "winnow" procedure. The two procedures are specified in detail below, and the channel is depicted in Figure 1.

**Definition 4. [MAC-based tag and winnow procedures]** We associate to a MAC function MAC: $\{0,1\}^k \times \{0,1\}^* \to \{0,1\}^l$ the following *tag and winnow procedures*. The tag procedure produces a valid packet from the input data. The winnow procedure takes as input a stream of packets and returns the data of the valid packets:

| Algorithm $\mathsf{Tag}^{\mathsf{MAC}(K,\cdot)}$ | Algorithm $\mathsf{Winnow}^{\mathsf{MAC}(K,\cdot)}(\mathrm{Pkt}'_1,\ldots,\mathrm{Pkt}'_{n'})$ |
|---|---|
| For $i = 1,\ldots,n$ do | For $i = 1,\ldots,n'$ do |
| $\quad \mathrm{tg}_i \leftarrow \mathsf{MAC}(K,\mathrm{dt}_i)$ | $\quad$ Parse $\mathrm{Pkt}'_i$ as $(\mathrm{dt},\mathrm{tg})$ |
| $\quad$ Return $(\mathrm{dt}_i,\mathrm{tg}_i)$ | $\quad$ If $\mathsf{MAC}(K,\mathrm{dt}) = \mathrm{tg}$ then return $\mathrm{dt}$ |
| EndFor | EndFor |

Here $K \in \{0,1\}^k$ is a key for the MAC and $n$ is the number of packets in the input stream. ∎

The receiver has no direct access to the packets or their MACs, no access to (or even knowledge of) the invalid packets, which are simply discarded by the winnow procedure. The receiver only gets, in order, the data part of the valid packets.

**Definition 5. [ATPT]** An *authentication to privacy transform* (ATPT) with tag length $l$ is a triple ATPT = (MakeWheat, AddChaff, Recover) of algorithms, where

- MakeWheat takes as input a message $M$ and returns a sequence $w_1,\ldots,w_n$ of strings called the *wheat strings*
- AddChaff takes as input a sequence $\mathrm{Pkt}_1,\ldots,\mathrm{Pkt}_n$ of packets called the *wheat packets* and returns another sequence $\mathrm{Pkt}'_1,\ldots,\mathrm{Pkt}'_{n'}$ of packets
- Recover takes as input strings $w_1,\ldots,w_n$ and returns a message $M$.

The first two algorithms can be probabilistic or stateful (accessing a global state variable such as a counter). The last algorithm is usually deterministic and stateless. ∎

An ATPT above is used in combination with an authentication channel to provide confidentiality. The way the process works is depicted and explained in Figure 2. Our interest is in the security of this entire procedure viewed as a symmetric encryption scheme. For this purpose it is convenient to think of it more as a standard symmetric encryption scheme, consisting of a key generation, encryption and decryption procedure. (The fact that it works by chaff and winnow is irrelevant to the security, although of course crucial to policy debate.) Below, we specify the symmetric encryption scheme that results from running a given ATPT over a given authentication channel, by specifying the three constituent algorithms.

**Definition 6.** Let ATPT = (MakeWheat, AddChaff, Recover) be an ATPT with tag length $l$, and let MAC: $\{0,1\}^k \times \{0,1\}^* \to \{0,1\}^l$ be a MAC. Associated to them is a *canonical encryption scheme* $(\mathcal{K},\mathcal{E},\mathcal{D})$. The key generation algorithm $\mathcal{K}$ is the same as that of the MAC, namely it outputs a random $k$-bit key $K$, and the encryption and decryption algorithms are as follows:

| Algorithm $\mathcal{E}_K(M)$ | Algorithm $\mathcal{D}_K(\mathrm{Pkt}'_1,\ldots,\mathrm{Pkt}'_{n'})$ |
|---|---|
| $(w_1,\ldots,w_n) \leftarrow \mathsf{MakeWheat}(M)$ | $(\mathrm{dt}_1,\ldots,\mathrm{dt}_n) \leftarrow$ |
| For $i = 1,\ldots,n$ do | $\quad\quad \mathsf{Winnow}^{\mathsf{MAC}(K,\cdot)}(\mathrm{Pkt}'_1,\ldots,\mathrm{Pkt}'_{n'})$ |
| $\quad \mathrm{Pkt}_i \leftarrow (w_i, \mathsf{MAC}_K(w_i))$ | $M \leftarrow \mathsf{Recover}(\mathrm{dt}_1,\ldots,\mathrm{dt}_n)$ |
| EndFor | Return $M$ |
| $(\mathrm{Pkt}'_1,\ldots,\mathrm{Pkt}'_{n'}) \leftarrow$ | |
| $\quad\quad \mathsf{AddChaff}(\mathrm{Pkt}_1,\ldots,\mathrm{Pkt}_n)$ | |
| Return $\mathrm{Pkt}'_1,\ldots,\mathrm{Pkt}'_{n'}$ | |

We require that $\mathcal{D}_K(\mathcal{E}_K(M)) = M$ for all $M \in \{0,1\}^*$. ∎

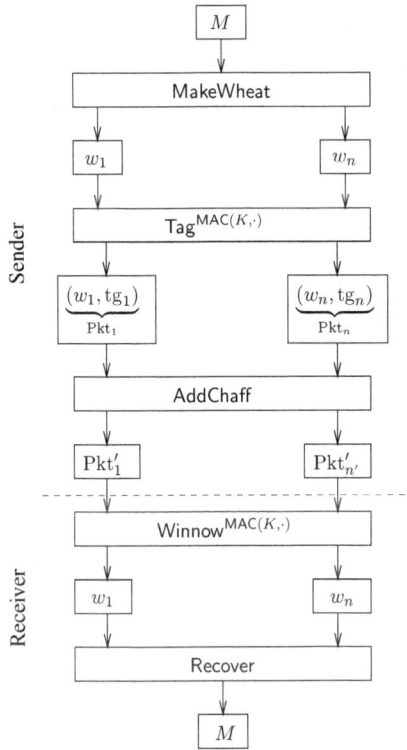

The (plaintext) message $M$ is first processed by a (keyless) transform MakeWheat to yield a sequence $w_1, \ldots, w_n$ of strings, each of which is MACed to yield a stream $\text{Pkt}_1 = (w_1, \text{tg}_1), \ldots, \text{Pkt}_n = (w_n, \text{tg}_n)$ of valid packets, where $\text{tg}_i = \text{MAC}(K, w_i)$ for $i = 1, \ldots, n$. The (keyless) AddChaff procedure adds chaff packets to produce a new stream $\text{Pkt}'_1, \ldots, \text{Pkt}'_{n'}$ of packets (the ciphertext) which is sent to the receiver. They hit the receiver's winnow (cf. Definition 4) which discards packets with invalid MACs, and passes up to the receiver the data from the valid packets. A (keyless) Recover procedure now puts this data together to get back the original message $M$. The three keyless algorithms MakeWheat, AddChaff, and Recover comprise what we call the ATPT (authentication to privacy transform)— they enable the possibility of obtaining confidentiality via an existing authentication channel without the addition of any extra cryptographic elements.

**Fig. 2. Chaff-and-winnow based "encryption".**

The last requirement is made so that this is a valid symmetric encryption scheme, meaning correctly encrypted data can be decrypted by a receiver that knows the secret key.

In the sequel, we will specify chaff-and-winnow based encryption schemes directly as standard symmetric encryption schemes, because this is more conducive to security assessments. Accordingly it is useful to have the following terminology.

**Definition 7.** Let $\text{SE} = (\mathcal{K}, \mathcal{E}, \mathcal{D})$ be a symmetric encryption scheme. We say that SE is a *chaff-and-winnow based encryption scheme* if there exists an ATPT transform ATPT and a MAC MAC: $\{0,1\}^k \times \{0,1\}^* \rightarrow \{0,1\}^l$ such that SE is exactly the canonical confidentiality procedure associated to ATPT and MAC as per Definition 6. ∎

## 4    Analysis of Rivest's Schemes

As above, MAC: $\{0,1\}^k \times \{0,1\}^* \rightarrow \{0,1\}^l$ is a message authentication code. In the bit-by-bit scheme, the sender maintains a counter $ctr$ that is initially zero. The encryption procedure (more precisely, the MakeWheat algorithm) increments this counter upon each invocation. Assume all messages to be encrypted have length $m$.

**Scheme 1. [Bit-by-bit CW]** The key generation algorithm $\mathcal{K}$ of this symmetric encryption scheme returns a random $k$-bit key $K$ for the MAC, and the encryption and decryption algorithms are as follows:

Algorithm $\mathcal{E}_K(M)$
　Break $M$ into bits, $M = b_1 \dots b_m$
　For $i = 1, \dots, m$ do
　　$tg[i, b_i] \leftarrow \mathrm{MAC}(K, b_i \| \langle ctr + i \rangle)$
　　$tg[i, \bar{b}_i] \stackrel{R}{\leftarrow} \{0, 1\}^l$
　　$\mathrm{Pkt}[i, 0] \leftarrow (0 \| \langle ctr + i \rangle, tg[i, 0])$
　　$\mathrm{Pkt}[i, 1] \leftarrow (1 \| \langle ctr + i \rangle, tg[i, 1])$
　EndFor
　$ctr \leftarrow ctr + m$
　Return　$\mathrm{Pkt}[1, 0], \mathrm{Pkt}[1, 1], \dots,$
　　　　　$\mathrm{Pkt}[m, 0], \mathrm{Pkt}[m, 1]$

Algorithm $\mathcal{D}_K(\mathrm{Pkt}_1, \dots, \mathrm{Pkt}_{2m})$
　For $i = 1, \dots, 2m$ do
　　Parse $\mathrm{Pkt}_i$ as $(dt, tg)$
　　If $\mathrm{MAC}(K, dt) = tg$
　　then return ﬁrst bit of $dt$
　EndFor

Here $\bar{b}$ denotes the complement bit of $b$ and $\langle i \rangle$ denotes the binary representation of integer $i$ as a binary string of some fixed, predefined length $p$. The "wheat" packets are $\mathrm{Pkt}[i, b_i]$ for $i = 1, \dots, m$ and the "chaff" packets are $\mathrm{Pkt}[i, \bar{b}_i]$ for $i = 1, \dots, m$. ∎

In the full version of this paper [1] we show formally that the above (and other schemes of this paper) are chaff-and-winnow based encryption schemes by saying what are the algorithms MakeWheat, AddChaff, and Recover.

　The following theorem shows that this scheme meets the "find-then-guess" notion of privacy under the assumption that MAC is a PRF. The reduction is almost tight. The proof is in [1].

**Theorem 1.** *Let* MAC: $\{0, 1\}^k \times \{0, 1\}^* \to \{0, 1\}^l$ *be a pseudorandom function and let* SE $= (\mathcal{K}, \mathcal{E}, \mathcal{D})$ *be the bit-by-bit chaff-and-winnow based encryption scheme of Scheme 1. Assume the counter is p-bits long. Then for any $t, q, \mu$ with $\mu < 2^p-$*

$$\mathbf{Adv}_{\mathrm{SE}}^{\mathrm{priv}}(t, q, \mu) \leq 2 \cdot \mathbf{Adv}_{\mathrm{MAC}}^{\mathrm{prf}}(t, q', \mu') \, ,$$

*where $q' = \mu$ and $\mu' = (1 + p)\mu$.* ∎

If the counter is allowed to wrap around the scheme is obviously insecure. It is possible to use randomness instead of a counter. In this case each bit of the message is concatenated with random value represented as a string of some fixed predefined length. This value is drawn at random for each bit of the message. The analysis is analogous but the concrete security is worse due to birthday attacks.

　The security of these schemes comes with a price. They are very inefficient since they have large data expansion: if the message is $m$ bits long then $2m(1 + p + l)$ bits are transmitted, where $p$ is the length of a counter and $l$ is the length of the output of the MAC. Bleichenbacher suggested that it is possible to reduce the communication cost by a factor of two by selecting at random and sending just one packet for each bit, either a chaff or a wheat packet. The receiver checks the validity of the data packet and complements the bit if the packet is invalid. Bleichenbacher further suggested that it is possible to reduce the communication even more if the sender authenticates each byte of the message and transmits only the computed MAC, but not the byte itself. The receiver then has to compute MACs for all possible bytes and take that for which the MACs match.

Another scheme mentioned in [11] authenticated message blocks of some length rather than single bits, but the author already indicated that it was insecure under stringent notions of privacy such as the one we use here. We go on to the scattering scheme.

As before let MAC: $\{0,1\}^k \times \{0,1\}^* \to \{0,1\}^l$ be our message authentication code. In the scattering scheme, the output of an AONT is viewed as a sequence of $s$ blocks, each $n$ bits long. The ciphertext will contain $s$ wheat packets interspaced with $s' > 0$ chaff packets, where $s'$ is a parameter of the scheme. The wheat packet positions are a random subset of $\{1, \ldots, s + s'\}$. The full description follows.

**Scheme 2. [Scattering scheme]** We fix an all-or-nothing transform AONT: $\{0,1\}^m \to \{0,1\}^{sn}$. We assume that all messages to be encrypted have length $m$. The key generation algorithm $\mathcal{K}$ of this symmetric encryption scheme returns a random $k$-bit key $K$ for the MAC, and the encryption and decryption algorithms are as follows:

Algorithm $\mathcal{E}_K(M)$
   $M' \leftarrow \mathsf{AONT}(M)$
   Parse $M'$ as $m_1\|m_2\|\cdots\|m_s$ where $|m_i| = n$
   Pick $S \subseteq \{1, \ldots, s + s'\}$ at random
   subject to $|S| = s$
   $j \leftarrow 0$
   For $i = 1, \ldots, s + s'$ do
      If $i \in S$ then
         $j \leftarrow j + 1$
         $\mathrm{tg}[i] \leftarrow \mathsf{MAC}(K, m_j)$
         $\mathrm{Pkt}[i] \leftarrow (m_j, \mathrm{tg}[i])$
      else
         $\mathrm{dt}[i] \stackrel{R}{\leftarrow} \{0,1\}^n$
         $\mathrm{tg}[i] \stackrel{R}{\leftarrow} \{0,1\}^l$
         $\mathrm{Pkt}[i] \leftarrow (\mathrm{dt}[i], \mathrm{tg}[i])$
      EndIf
   EndFor
   Return $\mathrm{Pkt}[1], \mathrm{Pkt}[2], \ldots \mathrm{Pkt}[s + s']$

Algorithm $\mathcal{D}_K(\mathrm{Pkt}_1, \ldots, \mathrm{Pkt}_{s+s'})$
   For $i = 1, \ldots, s + s'$ do
      Parse $\mathrm{Pkt}_i$ as $(\mathrm{dt}, \mathrm{tg})$
      If $\mathsf{MAC}(K, \mathrm{dt}) = \mathrm{tg}$
      then $m_i \leftarrow \mathrm{dt}$
   EndFor
   $M \leftarrow \mathsf{AONT}^{-1}(m_1\|m_2\|\cdots\|m_s)$
   Return $M$

The most obvious attack is to test each group of $s$ packets to see whether they are the wheat packets. The adversary goes through all size $s$ subsets of the packets. In each case it forms a candidate output of AONT and applies $\mathsf{AONT}^{-1}$. Assuming it knows some partial information about the message, it can tell when it got the choice of the subset right. The time taken by this attack is proportional to $\binom{s+s'}{s}$.

The intuition for security given in [11] is that this is the best possible attack. The complexity is large as long as both $s$ and $s'$ are above some minimal threshold; for example, both more than 128. Accordingly we could set $s' = 128$ and choose the AONT so that its output always had at least 128 blocks.

A closer look reveals however that security is not so straightforward. For example, another thing to consider is the effect of equal data blocks. If the data blocks in two packets are equal, an adversary can get some information by looking at their tags: if the tags are unequal, they cannot both be wheat packets, because the MAC is deterministic. This can reduce the complexity of an attack, indicating that the time-complexity of an attack must also be a function of the block size $n$ of the output.

There are other such considerations, but more importantly, we claim that Rivest's notion of security for the AONT can be shown to be insufficient to make this scheme secure. An example illustrating this is to consider an AONT each of whose output blocks has the property that the first few bits are 0. (One can show that if any AONT meeting Rivest's definition exists , then so does one with this property.) But with this AONT, Scheme 2 can be broken because wheat packets can be distinguished from chaff packets: the wheat packets are the ones whose data has first few bits zero. The same counterexample shows that Definition 3 is also not enough.

The example AONT we constructed above does not however meet Boyko's stronger notion of security for AONTs [6], so the next question is whether Scheme 2 could be proven secure under this stronger notion. However even with this stronger notion it is unclear one can prove security. The reason is that the ciphertext contains the complete output of the AONT, while the security property of the AONT pertains to a setting where the adversary has no information about at least one block of the output of the AONT. This makes it unclear how to do a reduction. Indeed, the security property of an AONT does not seem to mesh well with what is required to prove security of Scheme 2. We will see next that a positive statement can be made by considering a particular AONT, namely the OAEP transform of Bellare and Rogaway [5]. But in general, as the transform used in the initial step, an AONT seems to be neither sufficient nor necessary for the security of Scheme 2.

The OAEP transform appeals to random oracles $G: \{0,1\}^n \to \{0,1\}^m$ and $H: \{0,1\}^m \to \{0,1\}^n$ where $n$ is the length of the OAEP seed and $m$ as usual is the message length. It takes as input an $m$-bit string $M$ and proceeds as follows–

**Algorithm OAEP$^{G,H}(M)$**

    $r \xleftarrow{R} \{0,1\}^n$ ; $y \leftarrow G(r) \oplus M$ ; $w \leftarrow H(y) \oplus r$ ; **Return** $w\|y$

Boyko showed that OAEP is an AONT, but this will not help us here given the above discussion. Instead, we go back to the transform itself and prove the security of Scheme 2 when AONT is set to OAEP.

As with any proof concerning OAEP, we work in the random oracle model of [4]. We must "lift" our definitions to allow all algorithms and parties, including the adversary, access to the random oracles $G, H$. Briefly, modify $\mathbf{Exp}^{\mathrm{priv}}_{\mathsf{SE}}(A, b)$ in Definition 1 to begin by picking $G, H$ randomly. Allow $\mathcal{E}_K$ and $A$ oracle access to $G, H$. Allow the scheme advantage to take extra parameters, $\mathbf{Adv}^{\mathrm{priv}}_{\mathsf{SE}}(t, q, \mu; q_G, q_H)$, these being bounds on the number of queries made by the adversary to the oracles in question.

The bound below reflects the above intuition: it is inversely proportional to $\binom{s+s'}{s}$ and also to $2^n$. This shows that for OAEP the security is what one would have liked it to be for a "good" AONT. The proof of Theorem 2 can be found in [1].

**Theorem 2.** *Let $n, m$ be integers with $m$ a multiple of $n$. Let* MAC: $\{0,1\}^k \times \{0,1\}^* \to \{0,1\}^l$ *be a pseudorandom function. Let* SE $= (\mathcal{K}, \mathcal{E}, \mathcal{D})$ *be Scheme 2 using* OAEP *as the* AONT, *with parameters $n, s, s'$ where $s = m/n + 1$. For any $t, q$ we let $\mu = nq(s - 1)$. Assume $q_H \leq 2^n/2$ and $q_G \leq (1/2) \cdot \binom{s+s'}{s}$. Then*

$$\mathbf{Adv}_{\mathsf{SE}}^{\mathrm{priv}}(t, q, \mu; q_G, q_H) \leq$$

$$\frac{2q_H}{\binom{s+s'}{s-1}} + \frac{2q_G + (q+1)^2 \cdot [(s+s')^2 + 1]}{2^n} + \mathbf{Adv}_{\mathsf{MAC}}^{\mathrm{prf}}(t', q', \mu')$$

*where $t' = t$, $q' = q(s + s')$ and $\mu' = nq(s + s')$.*

## 5   A New Chaffing-and-Winnowing Scheme

Here we suggest an alternative scheme that has low data expansion and analyze its advantage function. It returns to much more "standard" paradigms of encryption than the scattering scheme. Simply apply an AONT to the message and then encrypt the first block of the message. If the last encryption is done by chaffing-and-winnowing, say using the bit-by-bit scheme, the whole scheme is also a chaffing-and-winnowing scheme, since the AONT is keyless. The savings in bandwidth comes from the fact that the number of bits encrypted using the bit-by-bit scheme is independent of the length of the message.

**Scheme 3.** Let AONT be an all-or-nothing transform taking input messages of length $m$ and returning outputs of length $sn$. The output is viewed as a sequence of $n$-bit blocks. Let se $= (\mathcal{K}, \mathsf{e}, \mathsf{d})$ be the bit-by-bit scheme of Scheme 1 with message space $\{0,1\}^n$. The new scheme is SE $= (\mathcal{K}, \mathcal{E}, \mathcal{D})$ where

| Algorithm $\mathcal{E}_K(M)$ | Algorithm $\mathcal{D}_K(C_1 \| (m'', \tau))$ |
|---|---|
| $M' \leftarrow$ AONT$(M)$ | $m' \leftarrow \mathsf{d}_K(C_1)$ |
| Let $m'$ be the first block of $M'$ and $m''$ the rest | $M' \leftarrow m' \| m''$ |
| $C_1 \leftarrow \mathsf{e}_K(m')$ | $M \leftarrow$ AONT$^{-1}(M')$ |
| Return $C_1 \| (m'', \mathrm{MAC}(K, m''))$ | Return $M$ |

Note that the MAC attached to $m''$ is irrelevant to security; it is only there in order to make the final scheme a chaffing-and-winnowing scheme. ∎

We now analyze the security of Scheme 3. Refer to Definition 3 for the definition of the advantage function of AONT and note that $L = 1$ in this case, meaning we are requiring security only in the case where the first block is the one not provided to the adversary. A proof of the following theorem is in [1].

**Theorem 3.** *Let* MAC: $\{0,1\}^k \times \{0,1\}^* \to \{0,1\}^l$ *be a pseudorandom function and let* AONT *be an all-or-nothing transform with input length $m$, output length $sn$. Let* SE $= (\mathcal{K}, \mathcal{E}, \mathcal{D})$ *be Scheme 3 using* AONT *as the all-or-nothing transform and Scheme 1 as* se. *Assume the counter in the latter is $p$-bits long. Then for any $t, q, \mu$ with $\mu = qm$ and $qn < 2^p-$*

$$\mathbf{Adv}_{\mathsf{SE}}^{\mathrm{priv}}(t, q, \mu) \leq 2 \cdot \mathbf{Adv}_{\mathsf{MAC}}^{\mathrm{prf}}(t, q_1, \mu_1) + \mathbf{Adv}_{\mathsf{AONT},1}^{\mathrm{aont}}(t)$$

*where $q_1 = q(n + 1)$ and $\mu_1 = qn(s + p)$.* ∎

A concrete instantiation can be obtained by using OAEP in the role of the AONT. The security of this instantiation relies on the fact that OAEP is a secure AONT [6], and the concrete security can be obtained by combining the above with the results in [6]. (In that case we would have to lift all of the above to the random oracle model, but this is easily done.)

## Acknowledgments

We thank Ron Rivest and Daniel Bleichenbacher for helpful discussions, and the anonymous referees for their constructive comments.

The authors are supported in part by a 1996 Packard Foundation Fellowship in Science and Engineering and NSF CAREER Award CCR-9624439.

## References

1. M. BELLARE AND A. BOLDYREVA, "The security of chaffing and winnowing," Full version of this paper, available via http://www-cse.ucsd.edu/users/mihir.
2. M. BELLARE, A. DESAI, E. JOKIPII AND P. ROGAWAY, "A concrete security treatment of symmetric encryption: Analysis of the DES modes of operation," *Proceedings of the 38th Symposium on Foundations of Computer Science*, IEEE, 1997.
3. M. BELLARE, J. KILIAN AND P. ROGAWAY, "The security of cipher block chaining," *Advances in Cryptology – Crypto '94*, Lecture Notes in Computer Science Vol. 839, Y. Desmedt ed., Springer-Verlag, 1994.
4. M. BELLARE AND P. ROGAWAY, "Random oracles are practical: a paradigm for designing efficient protocols," *Proceedings of the 1st Annual Conference on Computer and Communications Security*, ACM, 1993.
5. M. BELLARE, P. ROGAWAY, "Optimal asymmetric encryption - How to encrypt with RSA," *Advances in Cryptology – Eurocrypt '94*, Lecture Notes in Computer Science Vol. 950, A. De Santis ed., Springer-Verlag, 1994.
6. V. BOYKO, "On the security properties of OAEP as an all-or-nothing transform,"*Advances in Cryptology – Crypto '99*, Lecture Notes in Computer Science Vol. 1666, M. Wiener ed., Springer-Verlag, 1999.
7. R. CANETTI, Y. DODIS, S. HALEVI, E. KUSHILEVITZ, A. SAHAI, "Exposure-resilient functions and all-or-nothing transforms," *Advances in Cryptology – Eurocrypt '00*, Lecture Notes in Computer Science Vol. 1807, B. Preneel ed., Springer-Verlag, 2000.
8. A. DESAI, "The security of all-or-nothing encryption: protecting against exhaustive key search,"*Advances in Cryptology – Crypto '00*, Lecture Notes in Computer Science Vol. 1880, M. Bellare ed., Springer-Verlag, 2000.
9. O. GOLDREICH, S. GOLDWASSER AND S. MICALI, "How to construct random functions,"*Journal of the ACM*, Vol. 33, No. 4, 210–217, (1986).
10. S. GOLDWASSER AND S. MICALI, "Probabilistic encryption," *Journal of Computer and System Science*, Vol. 28, 1984, pp. 270–299.
11. R. RIVEST, "Chaffing and winnowing: Confidentiality without encryption," http://theory.lcs.mit.edu/~rivest/publications.html.
12. R. RIVEST, "all-or-nothing encryption and the package transform," *Proceedings of the 4th Workshop on Fast Software Encryption*, Lecture Notes in Computer Science Vol. 1267, Springer-Verlag, 1997.

# Authenticated Encryption: Relations among Notions and Analysis of the Generic Composition Paradigm

Mihir Bellare and Chanathip Namprempre

Dept. of Computer Science & Engineering, University of California at San Diego,
9500 Gilman Drive, La Jolla, California 92093, USA.
{mihir, cnamprem}@cs.ucsd.edu
www-cse.ucsd.edu/users/{mihir, cnamprem}

**Abstract.** We consider two possible notions of authenticity for symmetric encryption schemes, namely integrity of plaintexts and integrity of ciphertexts, and relate them to the standard notions of privacy for symmetric encryption schemes by presenting implications and separations between all notions considered. We then analyze the security of authenticated encryption schemes designed by "generic composition," meaning making black-box use of a given symmetric encryption scheme and a given MAC. Three composition methods are considered, namely *Encrypt-and-MAC plaintext*, *MAC-then-encrypt*, and *Encrypt-then-MAC*. For each of these, and for each notion of security, we indicate whether or not the resulting scheme meets the notion in question assuming the given symmetric encryption scheme is secure against chosen-plaintext attack and the given MAC is unforgeable under chosen-message attack. We provide proofs for the cases where the answer is "yes" and counter-examples for the cases where the answer is "no."

## 1 Introduction

We use the term *authenticated encryption scheme* to refer to a shared-key based transform whose goal is to provide *both* privacy *and* authenticity of the encapsulated data. In such a scheme the *encryption* process applied by the sender takes the key and a plaintext to return a ciphertext, while the *decryption* process applied by the receiver takes the same key and a ciphertext to return either a plaintext or a special symbol indicating that it considers the ciphertext invalid or unauthentic.

The design of such schemes has attracted a lot of attention historically. The early schemes were typically based on adding "redundancy" to the message before CBC encrypting, and many of these schemes were broken. Today authenticated encryption schemes continue to be the target of design and standardization efforts. A popular modern design paradigm is to combine MACs with standard block cipher modes of operation.

The goal of symmetric encryption is usually viewed as privacy, but an authenticated encryption scheme is simply a symmetric encryption scheme meeting additional authenticity goals. The first part of this paper formalizes several different possible notions of authenticity for symmetric encryption schemes, and integrates them into the existing mosaic of notions by relating them to the main known notions of privacy for

T. Okamoto (Ed.): ASIACRYPT 2000, LNCS 1976, pp. 531–545, 2000.
© Springer-Verlag Berlin Heidelberg 2000

symmetric encryption, via implications and separations in the style of [3]. The second part of this paper is motivated by emerging standards such as [16] which design authenticated encryption schemes by what we call "generic composition" of encryption and MAC schemes. We analyze, with regard to meeting the previous notions, several generic composition methods. Let us now look at these items in more detail.

## 1.1    Relations among Notions

Privacy goals for symmetric encryption schemes include indistinguishability and non-malleability, each of which can be considered under either chosen-plaintext or (adaptive) chosen-ciphertext attack, leading to four notions of security we abbreviate IND-CPA, IND-CCA, NM-CPA, NM-CCA. (The original definitions were in the asymmetric setting [12,10,18] but can be "lifted" to the symmetric setting using the encryption oracle based template of [2]). The relations among these notions are well-understood [3,11]. (These papers state results for the asymmetric setting, but as noted in [3] it is an easy exercise to transfer them to the symmetric setting.)

We consider two notions of integrity (we use the terms authenticity and integrity interchangeably) for symmetric encryption schemes. INT-PTXT (integrity of plaintexts) requires that it be computationally infeasible to produce a ciphertext decrypting to a message which the sender had never encrypted, while INT-CTXT (integrity of ciphertexts) requires that it be computationally infeasible to produce a ciphertext not previously produced by the sender, regardless of whether or not the underlying plaintext is "new." (In both cases, the adversary is allowed a chosen-message attack.) The first of these notions is the more natural security requirement while the interest of the second, stronger notion is perhaps more in the implications we discuss below.

These notions of authenticity are by themselves quite disjoint from the notions of privacy; for example, sending the message in the clear with an accompanying (strong) MAC achieves INT-CTXT but no kind of privacy. To make for useful comparisons, we consider each notion of authenticity coupled with IND-CPA, the weakest notion of privacy; namely the notions on which we focus for comparison purposes are INT-PTXT ∧ IND-CPA and INT-CTXT ∧ IND-CPA. (Read "∧" as "and".)

Figure 1 shows the graph of relations between these notions and the above-mentioned older ones in the style of [3]. An "implication" $\mathbf{A} \to \mathbf{B}$ means that every symmetric encryption scheme meeting notion $\mathbf{A}$ also meets notion $\mathbf{B}$. A "separation" $\mathbf{A} \not\to \mathbf{B}$ means that there exists a symmetric encryption scheme meeting notion $\mathbf{A}$ but not notion $\mathbf{B}$. (This under the minimal assumption that some scheme meeting notion $\mathbf{A}$ exists since otherwise the question is moot.) Only a minimal set of relations is explicitly indicated; the relation between any two notions can be derived from the shown ones. (For example, IND-CCA does not imply INT-CTXT ∧ IND-CPA because otherwise, by following arrows, we would get IND-CCA → INT-PTXT ∧ IND-CPA contradicting a stated separation.) The dotted lines are reminders of existing relations while the numbers annotating the dark lines are pointers to Propositions or Theorems in this paper.

A few points may be worth highlighting. Integrity of ciphertexts —even when coupled only with the weak privacy requirement IND-CPA— emerges as the most powerful notion. Not only does it imply security against chosen-ciphertext attack, but it is strictly stronger than this notion. Non-malleability —whether under chosen-plaintext or

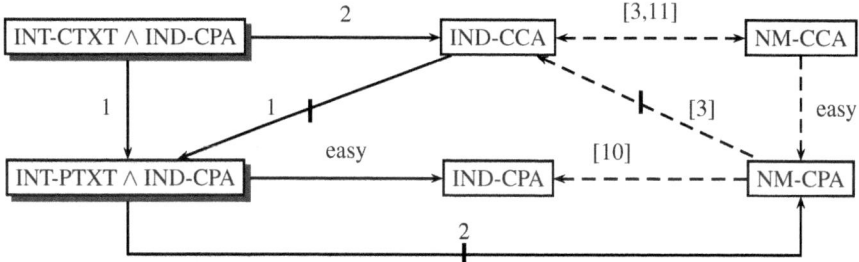

**Fig. 1. Relations among notions of symmetric encryption:** An arrow denotes an implication while a barred arrow denotes a separation. The full arrows are relations proved in this paper, annotated with the number of the corresponding Proposition or Theorem, while dotted arrows are reminders of existing relations, annotated with citations to the papers establishing them.

chosen-ciphertext attack— does not imply any type of integrity. The intuitive reason is that non-malleability only prevents the generation of ciphertexts whose plaintexts are meaningfully related to those of some challenge ciphertexts, while integrity requires it to be hard to generate ciphertexts of new plaintexts even if these are unrelated to plaintexts underlying any existing ciphertexts. Finally, INT-PTXT ∧ IND-CPA does not imply INT-CTXT ∧ IND-CPA.

## 1.2 Analysis of Generic Composition

There are many possible ways to design authenticated encryption schemes. We focus in this paper on "generic composition:" simply combine a standard symmetric encryption scheme with a MAC in some way. There are a few possible ways to do it, and our goal is to analyze and compare their security. (The motivation, as we will argue, is that these "obvious" methods, as often the case in practice, remain the most pragmatic from the point of view of performance and security architecture design.)

GENERIC COMPOSITION. Assume we are given a symmetric encryption scheme $\mathcal{SE}$ specified by an encryption algorithm $\mathcal{E}$ and a decryption algorithm $\mathcal{D}$. (Typically this will be a block cipher mode of operation.) Also assume we are given a message authentication scheme $\mathcal{MA}$ specified by a tagging algorithm $\mathcal{T}$ and a tag verifying algorithm $\mathcal{V}$ and meeting some appropriate notion of unforgeability under chosen-message attack. (Possibilities include the CBC-MAC, HMAC [1], or UMAC [8]). We consider the following methods of "composing" these schemes in order to create an authenticated encryption scheme meeting either INT-CTXT ∧ IND-CPA or INT-PTXT ∧ IND-CPA. We call them "generic" because the algorithms of the authenticated encryption scheme appeal to the given ones as black-boxes only:

| Composition Method | Privacy | | | Integrity | |
|---|---|---|---|---|---|
| | IND-CPA | IND-CCA | NM-CPA | INT-PTXT | INT-CTXT |
| Encrypt-and-MAC plaintext | insecure | insecure | insecure | secure | insecure |
| MAC-then-encrypt | secure | insecure | insecure | secure | insecure |
| Encrypt-then-MAC | secure | insecure | insecure | secure | insecure |

**Fig. 2.** Summary of security results for the composed authenticated encryption schemes under the assumption that the given encryption scheme is IND-CPA and the given MAC is weakly unforgeable.

| Composition Method | Privacy | | | Integrity | |
|---|---|---|---|---|---|
| | IND-CPA | IND-CCA | NM-CPA | INT-PTXT | INT-CTXT |
| Encrypt-and-MAC plaintext | insecure | insecure | insecure | secure | insecure |
| MAC-then-encrypt | secure | insecure | insecure | secure | insecure |
| Encrypt-then-MAC | secure | secure | secure | secure | secure |

**Fig. 3.** Summary of security results for the composed authenticated encryption schemes under the assumption that the given encryption scheme is IND-CPA and the given MAC is strongly unforgeable.

---

— Encrypt-and-MAC plaintext: $\overline{\mathcal{E}}_{K_e,K_m}(M) = \mathcal{E}_{K_e}(M)\|\mathcal{T}_{K_m}(M)$.[1] Namely, encrypt the plaintext and append a MAC of the plaintext. "Decrypt+verify" is performed by first decrypting to get the plaintext and then verifying the tag.

— MAC-then-encrypt: $\overline{\mathcal{E}}_{K_e,K_m}(M) = \mathcal{E}_{K_e}(M\|\mathcal{T}_{K_m}(M))$. Namely, append a MAC to the plaintext and then encrypt them together. "Decrypt+verify" is performed by first decrypting to get the plaintext and candidate tag, and then verifying the tag.

— Encrypt-then-MAC: $\overline{\mathcal{E}}_{K_e,K_m}(M) = C\|\mathcal{T}_{K_m}(C)$ where $C = \mathcal{E}_{K_e}(M)$. Namely, encrypt the plaintext to get a ciphertext $C$ and append a MAC of $C$. "Decrypt+verify" is performed by first verifying the tag and then decrypting $C$. This is the method of Internet RFC [16].

Here $\overline{\mathcal{E}}$ is the encryption algorithm of the authenticated encryption scheme while the "decrypt+verify" process specifies a decryption algorithm $\overline{\mathcal{D}}$. The latter will either return a plaintext or a special symbol indicating that it considers the ciphertext unauthentic.

SECURITY RESULTS. Figure 2 and Figure 3 summarize the security results for the three composite authenticated encryption schemes. (We omit NM-CCA since it is equivalent to IND-CCA). Figure 2 shows the results assuming that the base MAC is weakly unforgeable while Figure 3 shows the results assuming that the MAC is strongly unforge-

---

[1]  Here (and everywhere in this paper) "$\|$" denotes an operation that combines several strings into one in such a way that the constituent strings are uniquely recoverable from the final one. (If lengths of all strings are fixed and known, concatenation will serve the purpose.)

able. Weak unforgeability is the standard notion [4]— it should be computationally infeasible for the adversary to find a message-tag pair in which the message is "new," even after a chosen-message attack. Strong unforgeability requires that it be computationally infeasible for the adversary to find a new message-tag pair even after a chosen-message attack. (The message does not have to be new as long as the output tag was not previously attached to this message by the legitimate parties.) We note that any pseudorandom function is a strongly unforgeable MAC, and most practical MACs seem to be strongly unforgeable. Therefore, analyzing the composition methods under this notion is a realistic and useful approach. Entries in the above tables have the following meaning:

— *Secure:* The composite encryption scheme in question is proven to meet the security requirement in question, assuming only that the component encryption scheme meets IND-CPA and the message authentication scheme is unforgeable under chosen-message attack.

— *Insecure:* There exists *some* IND-CPA secure symmetric encryption and some message authentication scheme unforgeable under chosen-message attack such that the composite scheme based on them does not meet the security requirement in question.

As we can see from Figure 3, the *encrypt-then-MAC* method of [16] is secure from all points of view, making it a good choice for a standard.

The use of a generic composition method secure in the sense above is advantageous from the point of view both of performance and of security architecture. The performance benefit arises from the presence of fast MACs such as HMAC [1] and UMAC [8]. The architectural benefits arise from the stringent notion of security being used. To be secure, the composition must be secure for *all* possible secure instantiations of its constituent primitives. (If it is secure for some instantiations but not others, we declare it insecure.) An application can thus choose a symmetric encryption scheme and a message authentication scheme independently (these are usually already supported by existing security analyses) and then appeal to some fixed and standard composition technique to combine them. No tailored security analysis of the composed scheme is required.

In Section 4 we state formal theorems to support the above claims, providing quantitative bounds for the positive results, and counter-examples with attacks for the negative result. For brevity, we provide theorems and proofs for only the results in Figure 3 (i.e. the strong MAC case).

QUANTITATIVE RESULTS AND COMPARISONS. Above we have discussed our results at a qualitative level. Each result also has a quantitative counterpart; these are what our theorems actually state and prove. These "concrete security" analyses enable a designer to estimate the security of the authenticated encryption scheme in terms of that of its components. All the reductions in this paper are tight, meaning there is little to no loss of security.

### 1.3   Related Work

The notions IND-CCA, NM-CCA were denoted IND-CCA2 and NM-CCA2, respectively, in [3]. The chosen-ciphertext attacks here are the adaptive kind [18]. Consideration of non-adaptive chosen-ciphertext attacks [17] leads to two more notions, denoted IND-CCA1 and NM-CCA1 by [3], who worked out the relations between six notions of privacy, these two and the four we consider here. (Their results hold for both the asymmetric and the symmetric settings, as mentioned before.) Three additional notions of privacy are considered and related to these six by [14]. In this paper, we have for simplicity avoided consideration of all the possible notions of privacy, focusing instead on what we consider the (four) main ones and their relations to the notions of authenticity. Relations of the remaining notions of privacy to the notions of authenticity considered here can be easily worked out.

Authenticity of an encryption scheme has been understood as a goal by designers for many years. The first formalization of which we are aware is that of [6]. (Early versions of their work date to 1998.) The notion they formalized was INT-CTXT. The formalization of INT-PTXT we use here seems to be new. In independent and concurrent work (both papers were submitted to FSE00) Katz and Yung [15] formalize INT-CTXT plus two other notions of authenticity not considered here. They also observe the implication INT-CTXT $\wedge$ IND-CPA $\to$ IND-CCA.

Generic composition is one of many approaches to the design of authenticated encryption schemes. Two more general approaches are "encryption with redundancy" — append redundancy to the message before encrypting, the latter typically with some block cipher mode of operation— and "encode then encipher" [6] —add randomness and redundancy and then encipher rather than encrypt. As indicated above, attacks have been found on many encrypt with redundancy schemes. Encode then encipher, however, can be proven to work [6] —meaning yields schemes achieving INT-CTXT $\wedge$ IND-CPA— but requires a variable-input length pseudorandom permutation, which can be relatively expensive to construct. In addition, there are many specific schemes. One such scheme is the RPC mode of [15] but it is computation and space inefficient compared to the generic composition methods. (Processing an $n$-block plaintext requires $(1 + c)n$ block cipher computations and results in a ciphertext of this many blocks, where $c \geq 0.3$.) Another scheme is the elegant IACBC mode of Jutla [13] which uses $n + O(\log n)$ block cipher operations to process an $n$-block plaintext. Implementation and testing would be required to compare its speed with that of generic composition methods that use fast MACs (cf. [1,8]).

Authenticated encryption is not the only approach to achieving security against chosen-ciphertext attacks. Direct approaches yielding more compact schemes have been provided by Desai [9].

## 2   Definitions

We present definitions for symmetric encryption following [2], first specifying the *syntax* —meaning what kinds of algorithms make up the scheme— and then specifying formal security measures. Associated with each scheme, each notion of security and each adversary is an advantage function that measures the success probability of this adversary as a function of the security parameter. We define asymptotic notions of security result by asking this function to be negligible for adversaries of time complexity polynomial in the security parameter. Concrete security assessments are made by associating to the scheme another advantage function that for each value of the security parameter and given resources for an adversary returns the maximum, over all adversaries limited to the given resources, of the success probability.

The concrete security assessments are important in practical applications— block cipher based schemes have no associated asymptotics. Hence, we provide concrete security assessments for all positive results (implications or proofs that composition methods meet some notion of security). For simplicity, however, negative results (separations or counter-examples) are phrased in the asymptotic style. (Concrete security statements are, however, easily derived from the proofs.)

SYNTAX OF (SYMMETRIC) ENCRYPTION SCHEMES. A *(symmetric) encryption scheme* $\mathcal{SE} = (\mathcal{K}, \mathcal{E}, \mathcal{D})$ consists of three algorithms. The randomized *key generation* algorithm $\mathcal{K}$ takes input a security parameter $k \in \mathbb{N}$ and returns a key $K$; we write $K \stackrel{R}{\leftarrow} \mathcal{K}(k)$. The *encryption* algorithm $\mathcal{E}$ could be randomized or stateful. It takes the key $K$ and a *plaintext* $M$ to return a *ciphertext* $C$; we write $C \stackrel{R}{\leftarrow} \mathcal{E}_K(M)$. (If randomized, it flips coins anew on each invocation. If stateful, it uses and then updates a state that is maintained across invocations.) The *decryption* algorithm $\mathcal{D}$ is deterministic and stateless. It takes the key $K$ and a string $C$ to return either the corresponding plaintext $M$ or the symbol $\perp$; we write $x \leftarrow \mathcal{D}_K(C)$ where $x \in \{0,1\}^* \cup \{\perp\}$. We require that $\mathcal{D}_K(\mathcal{E}_K(M)) = M$ for all $M \in \{0,1\}^*$. An authenticated encryption scheme is syntactically identical to an encryption scheme as defined above; we will use the term only to emphasize cases where we are targeting authenticity goals.

PRIVACY. We measure indistinguishability via the "left-or-right" model of [2]. Define the *left-or-right* oracle $\mathcal{E}_K(\mathcal{LR}(\cdot, \cdot, b))$, where $b \in \{0,1\}$, to take input $(x_0, x_1)$ and do the following: if $b = 0$ it computes $C \leftarrow \mathcal{E}_K(x_0)$ and returns $C$; else it computes $C \leftarrow \mathcal{E}_K(x_1)$ and returns $C$. The adversary makes oracle queries of the form $(x_0, x_1)$ consisting of two equal length messages and must guess the bit $b$. To model chosen-ciphertext attacks we allow the adversary to also have access to a decryption oracle.

**Definition 1. (Indistinguishability of a Symmetric Encryption Scheme [2])** Let $\mathcal{SE} = (\mathcal{K}, \mathcal{E}, \mathcal{D})$ be a symmetric encryption scheme. Let $b \in \{0,1\}$ and $k \in \mathbb{N}$. Let $A_{\mathrm{cpa}}$ be an adversary that has access to the oracle $\mathcal{E}_K(\mathcal{LR}(\cdot, \cdot, b))$ and let $A_{\mathrm{cca}}$ be an adversary that has access to the oracles $\mathcal{E}_K(\mathcal{LR}(\cdot, \cdot, b))$ and $\mathcal{D}_K(\cdot)$. Now, we consider the following experiments:

$$
\begin{array}{l|l}
\text{Experiment } \mathbf{Exp}^{\text{ind-cpa-}b}_{\mathcal{SE}, A_{\text{cpa}}}(k) & \text{Experiment } \mathbf{Exp}^{\text{ind-cca-}b}_{\mathcal{SE}, A_{\text{cca}}}(k) \\
\quad K \stackrel{R}{\leftarrow} \mathcal{K}(k) & \quad K \stackrel{R}{\leftarrow} \mathcal{K}(k) \\
\quad x \leftarrow A_{\text{cpa}}^{\mathcal{E}_K(\mathcal{LR}(\cdot,\cdot,b))}(k) & \quad x \leftarrow A_{\text{cca}}^{\mathcal{E}_K(\mathcal{LR}(\cdot,\cdot,b)),\mathcal{D}_K(\cdot)}(k) \\
\quad \text{Return } x & \quad \text{Return } x
\end{array}
$$

Above it is mandated that $A_{\mathrm{cca}}$ never queries $\mathcal{D}_K(\cdot)$ on a ciphertext $C$ output by the $\mathcal{E}_K(\mathcal{LR}(\cdot, \cdot, b))$ oracle, and that the two messages queried of $\mathcal{E}_K(\mathcal{LR}(\cdot, \cdot, b))$ always have equal length. We define the *advantages* of the adversaries via

$$\mathbf{Adv}^{\text{ind-cpa}}_{\mathcal{SE},A_{\text{cpa}}}(k) = \Pr\left[\,\mathbf{Exp}^{\text{ind-cpa-1}}_{\mathcal{SE},A_{\text{cpa}}}(k) = 1\,\right] - \Pr\left[\,\mathbf{Exp}^{\text{ind-cpa-0}}_{\mathcal{SE},A_{\text{cpa}}}(k) = 1\,\right]$$

$$\mathbf{Adv}^{\text{ind-cca}}_{\mathcal{SE},A_{\text{cca}}}(k) = \Pr\left[\,\mathbf{Exp}^{\text{ind-cca-1}}_{\mathcal{SE},A_{\text{cca}}}(k) = 1\,\right] - \Pr\left[\,\mathbf{Exp}^{\text{ind-cca-0}}_{\mathcal{SE},A_{\text{cca}}}(k) = 1\,\right].$$

We define the *advantage functions of the scheme* as follows. For any integers $t, q_e, q_d, \mu$,

$$\mathbf{Adv}^{\text{ind-cpa}}_{\mathcal{SE}}(k, t, q_e, \mu) = \max_{A_{\text{cpa}}}\{\mathbf{Adv}^{\text{ind-cpa}}_{\mathcal{SE},A_{\text{cpa}}}(k)\}$$

$$\mathbf{Adv}^{\text{ind-cca}}_{\mathcal{SE}}(k, t, q_e, q_d, \mu) = \max_{A_{\text{cca}}}\{\mathbf{Adv}^{\text{ind-cca}}_{\mathcal{SE},A_{\text{cca}}}(k)\}$$

where the maximum is over all $A_{\text{cpa}}, A_{\text{cca}}$ with "time complexity" $t$, each making at most $q_e$ queries to the $\mathcal{E}_K(\mathcal{LR}(\cdot, \cdot, b))$ oracle, totaling at most $\mu$ bits, and, in the case of $A_{\text{cca}}$, also making at most $q_d$ queries to the $\mathcal{D}_K(\cdot)$ oracle. The scheme $\mathcal{SE}$ is said to be *IND-CPA secure* (resp. *IND-CCA secure*) if the function $\mathbf{Adv}^{\text{ind-cpa}}_{\mathcal{SE},A}(\cdot)$ (resp. $\mathbf{Adv}^{\text{ind-cca}}_{\mathcal{SE},A}(\cdot)$) is negligible for any adversary $A$ whose time complexity is polynomial in $k$. ∎

The "time complexity" is the worst case total execution time of the experiment, plus the size of the code of the adversary, in some fixed RAM model of computation. We stress that the the total execution time of the experiment includes the time of *all* operations in the experiment, including the time for key generation and the computation of answers to oracle queries. Thus, when the time complexity is polynomially bounded, so are all the other parameters. This convention for measuring time complexity and other resources of an adversary is used for all definitions in this paper. The advantage function is the maximum probability that the security of the scheme $\mathcal{SE}$ can be compromised by an adversary using the indicated resources, and is used for concrete security analyses.

We will not use definitions of non-malleability as per [10,3] but instead use the equivalent indistinguishability under parallel chosen-ciphertext attack characterization of [7]. This facilitates our proofs and analyses and also facilitates concrete security measurements. The notation $\mathcal{D}_K(\cdot)$ denotes the algorithm which takes input a vector $\mathbf{c} = (c_1, \dots, c_n)$ of ciphertexts and returns the corresponding vector $\mathbf{p} = (\mathcal{D}_K(c_1), \dots, \mathcal{D}_K(c_n))$ of plaintexts.

**Definition 2. (Non-Malleability of a Symmetric Encryption Scheme [7])** Let $\mathcal{SE} = (\mathcal{K}, \mathcal{E}, \mathcal{D})$ be a symmetric encryption scheme. Let $b \in \{0, 1\}$ and $k \in \mathbb{N}$. Let $A_{\text{cpa}} = (A_{\text{cpa}_1}, A_{\text{cpa}_2})$ be an adversary that has access to the oracle $\mathcal{E}_K(\mathcal{LR}(\cdot, \cdot, b))$ and let $A_{\text{cca}} = (A_{\text{cca}_1}, A_{\text{cca}_2})$ be an adversary that has access to the oracles $\mathcal{E}_K(\mathcal{LR}(\cdot, \cdot, b))$ and $\mathcal{D}_K(\cdot)$. Now, we consider the following experiments:

| Experiment $\mathbf{Exp}^{\text{nm-cpa-}b}_{\mathcal{SE},A_{\text{cpa}}}(k)$ | Experiment $\mathbf{Exp}^{\text{nm-cca-}b}_{\mathcal{SE},A_{\text{cca}}}(k)$ |
|---|---|
| $K \stackrel{R}{\leftarrow} \mathcal{K}(k)$ | $K \stackrel{R}{\leftarrow} \mathcal{K}(k)$ |
| $(\mathbf{c}, s) \leftarrow A^{\mathcal{E}_K(\mathcal{LR}(\cdot,\cdot,b))}_{\text{cpa}_1}(k)$ | $(\mathbf{c}, s) \leftarrow A^{\mathcal{E}_K(\mathcal{LR}(\cdot,\cdot,b)),\mathcal{D}_K(\cdot)}_{\text{cca}_1}(k)$ |
| $\mathbf{p} \leftarrow \mathcal{D}_K(\mathbf{c})$ | $\mathbf{p} \leftarrow \mathcal{D}_K(\mathbf{c})$ |
| $x \leftarrow A_{\text{cpa}_2}(\mathbf{p}, \mathbf{c}, s)$ | $x \leftarrow A_{\text{cca}_2}(\mathbf{p}, \mathbf{c}, s)$ |
| Return $x$ | Return $x$ |

Above it is mandated that the vector $c$ output by $A_{\mathrm{cpa}_1}$ does not contain any of the ciphertexts output by the $\mathcal{E}_K(\mathcal{LR}(\cdot,\cdot,b))$ oracle, and that the pairs of messages queried of $\mathcal{E}_K(\mathcal{LR}(\cdot,\cdot,b))$ are always of equal length. We define the *advantages* of the adversaries via

$$\mathbf{Adv}^{\mathrm{nm\text{-}cpa}}_{\mathcal{SE},A_{\mathrm{cpa}}}(k) = \Pr\left[\mathbf{Exp}^{\mathrm{nm\text{-}cpa\text{-}1}}_{\mathcal{SE},A_{\mathrm{cpa}}}(k) = 1\right] - \Pr\left[\mathbf{Exp}^{\mathrm{nm\text{-}cpa\text{-}0}}_{\mathcal{SE},A_{\mathrm{cpa}}}(k) = 1\right]$$

$$\mathbf{Adv}^{\mathrm{nm\text{-}cca}}_{\mathcal{SE},A_{\mathrm{cca}}}(k) = \Pr\left[\mathbf{Exp}^{\mathrm{nm\text{-}cca\text{-}1}}_{\mathcal{SE},A_{\mathrm{cca}}}(k) = 1\right] - \Pr\left[\mathbf{Exp}^{\mathrm{nm\text{-}cca\text{-}0}}_{\mathcal{SE},A_{\mathrm{cca}}}(k) = 1\right].$$

We define the *advantage functions of the scheme* as follows. For any integers $t, q_e, q_d, \mu$,

$$\mathbf{Adv}^{\mathrm{nm\text{-}cpa}}_{\mathcal{SE}}(k, t, q_e, \mu) = \max_{A_{\mathrm{cpa}}}\{\mathbf{Adv}^{\mathrm{nm\text{-}cpa}}_{\mathcal{SE},A_{\mathrm{cpa}}}(k)\}$$

$$\mathbf{Adv}^{\mathrm{nm\text{-}cca}}_{\mathcal{SE}}(k, t, q_e, q_d, \mu) = \max_{A_{\mathrm{cca}}}\{\mathbf{Adv}^{\mathrm{nm\text{-}cca}}_{\mathcal{SE},A_{\mathrm{cca}}}(k)\}$$

where the maximum is over all $A_{\mathrm{cpa}}, A_{\mathrm{cca}}$ with time complexity $t$, each making at most $q_e$ queries to the $\mathcal{E}_K(\mathcal{LR}(\cdot,\cdot,b))$ oracle, totaling at most $\mu$ bits, and, in the case of $A_{\mathrm{cca}}$, also making at most $q_d$ queries to the $\mathcal{D}_K(\cdot)$ oracle. The scheme $\mathcal{SE}$ is said to be *NM-CPA secure* (resp. *NM-CCA secure*) if the function $\mathbf{Adv}^{\mathrm{nm\text{-}cpa}}_{\mathcal{SE},A}(\cdot)$ (resp. $\mathbf{Adv}^{\mathrm{nm\text{-}cca}}_{\mathcal{SE},A}(\cdot)$) is negligible for any adversary $A$ whose time complexity is polynomial in $k$. ∎

INTEGRITY. Now we specify security definitions for integrity (authenticity) of a symmetric encryption scheme $\mathcal{SE} = (\mathcal{K}, \mathcal{E}, \mathcal{D})$. It is convenient to define an algorithm $\mathcal{D}^*_K(\cdot)$ as follows: If $\mathcal{D}_K(C) \neq \perp$, then return 1 Else return 0. We call this the *verification algorithm* or *verification oracle*. The adversary is allowed a chosen-message attack on the scheme, modeled by giving it access to an encryption oracle $\mathcal{E}_K(\cdot)$. It is successful if it makes the verification oracle accept a ciphertext that was not "legitimately produced." Different interpretations of the latter give rise to different notions.

**Definition 3. (Integrity of an Authenticated Encryption Scheme)** Let $\mathcal{SE} = (\mathcal{K}, \mathcal{E}, \mathcal{D})$ be a symmetric encryption scheme. Let $k \in \mathsf{N}$, and let $A_{\mathrm{ptxt}}$ and $A_{\mathrm{ctxt}}$ be adversaries each of which has access to two oracles: $\mathcal{E}_K(\cdot)$ and $\mathcal{D}^*_K(\cdot)$. Consider these experiments.

Experiment $\mathbf{Exp}^{\mathrm{int\text{-}ptxt}}_{\mathcal{SE},A_{\mathrm{ptxt}}}(k)$

   $K \xleftarrow{R} \mathcal{K}(k)$
   If $A^{\mathcal{E}_K(\cdot),\mathcal{D}^*_K(\cdot)}_{\mathrm{ptxt}}(k)$ makes a query $C$ to
   the oracle $\mathcal{D}^*_K(\cdot)$ such that
     $- \mathcal{D}^*_K(C)$ returns 1, and
     $- M \stackrel{\mathrm{def}}{=} \mathcal{D}_K(C)$ was never a query to $\mathcal{E}_K(\cdot)$
   then return 1 else return 0.

Experiment $\mathbf{Exp}^{\mathrm{int\text{-}ctxt}}_{\mathcal{SE},A_{\mathrm{ctxt}}}(k)$

   $K \xleftarrow{R} \mathcal{K}(k)$
   If $A^{\mathcal{E}_K(\cdot),\mathcal{D}^*_K(\cdot)}_{\mathrm{ctxt}}(k)$ makes a query $C$ to
   the oracle $\mathcal{D}^*_K(\cdot)$ such that
     $- \mathcal{D}^*_K(C)$ returns 1, and
     $- C$ was never a response of $\mathcal{E}_K(\cdot)$
   then return 1 else return 0.

We define the *advantages* of the adversaries via

$$\mathbf{Adv}^{\mathrm{int\text{-}ptxt}}_{\mathcal{SE},A_{\mathrm{ptxt}}}(k) = \Pr\left[\mathbf{Exp}^{\mathrm{int\text{-}ptxt}}_{\mathcal{SE},A_{\mathrm{ptxt}}}(k) = 1\right]$$

$$\mathbf{Adv}^{\mathrm{int\text{-}ctxt}}_{\mathcal{SE},A_{\mathrm{ctxt}}}(k) = \Pr\left[\mathbf{Exp}^{\mathrm{int\text{-}ctxt}}_{\mathcal{SE},A_{\mathrm{ctxt}}}(k) = 1\right]$$

We define the *advantage functions of the scheme* as follows. For any integers $t, q_e, q_d, \mu$,

$$\mathbf{Adv}^{\text{int-ptxt}}_{\mathcal{SE}}(k, t, q_e, q_d, \mu) = \max_{A_{\text{ptxt}}}\{\mathbf{Adv}^{\text{int-ptxt}}_{\mathcal{SE}, A_{\text{ptxt}}}(k)\}$$

$$\mathbf{Adv}^{\text{int-ctxt}}_{\mathcal{SE}}(k, t, q_e, q_d, \mu) = \max_{A_{\text{ctxt}}}\{\mathbf{Adv}^{\text{int-ctxt}}_{\mathcal{SE}, A_{\text{ctxt}}}(k)\}$$

where the maximum is over all $A_{\text{ptxt}}, A_{\text{ctxt}}$ with time complexity $t$, each making at most $q_e$ queries to the oracle $\mathcal{E}_K(\cdot)$ and at most $q_d$ queries to $\mathcal{D}^*_K(\cdot)$ such that the sum of the lengths of all oracle queries is at most $\mu$ bits. The scheme $\mathcal{SE}$ is said to be *INT-PTXT secure* (resp. *INT-CTXT secure*) if the function $\mathbf{Adv}^{\text{int-ptxt}}_{\mathcal{SE}, A}(\cdot)$ (resp. $\mathbf{Adv}^{\text{int-ctxt}}_{\mathcal{SE}, A}(\cdot)$) is negligible for any adversary $A$ whose time complexity is polynomial in $k$. ∎

MESSAGE AUTHENTICATION SCHEMES. A *message authentication scheme* $\mathcal{MA} = (\mathcal{K}, \mathcal{T}, \mathcal{V})$ consists of three algorithms. The randomized *key generation* algorithm $\mathcal{K}$ takes input a security parameter $k \in \mathsf{N}$ and returns a key $K$; we write $K \xleftarrow{R} \mathcal{K}(k)$. The *tagging* algorithm $\mathcal{T}$ could be either randomized or stateful. It takes the key $K$ and a message $M$ to return a *tag* $\sigma$; we write $\sigma \xleftarrow{R} \mathcal{T}_K(M)$. The *verification* algorithm $\mathcal{V}$ is deterministic. It takes the key $K$, a message $M$, and a candidate tag $\sigma$ for $M$ to return a bit $v$; we write $v \leftarrow \mathcal{V}_K(M, \sigma)$. We require that $\mathcal{V}_K(M, \mathcal{T}_K(M)) = 1$ for all $M \in \{0,1\}^*$. The scheme is said to be deterministic if the tagging algorithm is deterministic and verification is done via tag re-computation. We sometimes call a message authentication scheme a MAC, and also sometimes call the tag $\sigma$ a MAC.

Security for message authentication considers an adversary $F$ who is allowed a chosen-message attack, modeled by allowing it access to an oracle for $\mathcal{T}_K(\cdot)$. $F$ is "successful" if it can make the verifying oracle $\mathcal{V}_K(\cdot, \cdot)$ accept a pair $(M, \sigma)$ that was not "legitimately produced." There are two possible conventions with regard to what "legitimately produced" can mean, leading to two measures of advantage. In the following definition, we use the acronyms WUF-CMA and SUF-CMA respectively for weak and strong unforgeability against chosen-message attacks.

**Definition 4. (Message Authentication Scheme Security)** Let $\mathcal{MA} = (\mathcal{K}, \mathcal{T}, \mathcal{V})$ be a message authentication scheme. Let $k \in \mathsf{N}$, and let $F_{\text{w}}$ and $F_{\text{s}}$ be adversaries that have access to two oracles: $\mathcal{T}_K(\cdot)$ and $\mathcal{V}_K(\cdot, \cdot)$. Consider the following experiment:

| Experiment $\mathbf{Exp}^{\text{wuf-cma}}_{\mathcal{MA}, \mathcal{F}_{\text{w}}}(k)$ | Experiment $\mathbf{Exp}^{\text{suf-cma}}_{\mathcal{MA}, \mathcal{F}_{\text{s}}}(k)$ |
|---|---|
| $K \xleftarrow{R} \mathcal{K}(k)$ | $K \xleftarrow{R} \mathcal{K}(k)$ |
| If $F_{\text{w}}^{\mathcal{T}_K(\cdot), \mathcal{V}_K(\cdot, \cdot)}(k)$ makes a query $(M, \sigma)$ | If $F_{\text{s}}^{\mathcal{T}_K(\cdot), \mathcal{V}_K(\cdot, \cdot)}(k)$ makes a query $(M, \sigma)$ |
| to the oracle $\mathcal{V}_K(\cdot, \cdot)$ such that | to the oracle $\mathcal{V}_K(\cdot, \cdot)$ such that |
| – $\mathcal{V}_K(M, \sigma)$ returns 1, and | – $\mathcal{V}_K(M, \sigma)$ returns 1, and |
| – $M$ was never queried to | – $\sigma$ was never returned by the |
| the oracle $\mathcal{T}_K(\cdot)$, | oracle $\mathcal{T}_K(\cdot)$ in response to query $M$, |
| then return 1 else return 0. | then return 1 else return 0. |

We define the *advantages* of the forgers via

$$\mathbf{Adv}^{\text{wuf-cma}}_{\mathcal{MA}, F_{\text{w}}}(k) = \Pr\left[\mathbf{Exp}^{\text{wuf-cma}}_{\mathcal{MA}, \mathcal{F}_{\text{w}}}(k) = 1\right]$$

$$\mathbf{Adv}^{\text{suf-cma}}_{\mathcal{MA}, F_{\text{s}}}(k) = \Pr\left[\mathbf{Exp}^{\text{suf-cma}}_{\mathcal{MA}, \mathcal{F}_{\text{s}}}(k) = 1\right]$$

We define the *advantage functions of the scheme* as follows. For any integers $t, q_t, q_v, \mu$,

$$\mathbf{Adv}_{\mathcal{MA}}^{\text{wuf-cma}}(k, t, q_t, q_v, \mu) = \max_{F_{\text{w}}}\{\mathbf{Adv}_{\mathcal{MA}, F_{\text{w}}}^{\text{wuf-cma}}(k)\}$$

$$\mathbf{Adv}_{\mathcal{MA}}^{\text{suf-cma}}(k, t, q_t, q_v, \mu) = \max_{F_{\text{s}}}\{\mathbf{Adv}_{\mathcal{MA}, F_{\text{s}}}^{\text{suf-cma}}(k)\}$$

where the maximum is over all $F_{\text{w}}, F_{\text{s}}$ with time complexity $t$, making at most $q_t$ oracle queries to $\mathcal{T}_K(\cdot)$ and at most $q_v$ oracle queries to $\mathcal{V}_K(\cdot, \cdot)$ such that the sum of the lengths of all oracle queries is at most $\mu$ bits. The scheme $\mathcal{MA}$ is said to be *WUF-CMA secure* (resp. *SUF-CMA secure*) if the function $\mathbf{Adv}_{\mathcal{MA}, F}^{\text{wuf-cma}}(\cdot)$ (resp. $\mathbf{Adv}_{\mathcal{MA}, F}^{\text{suf-cma}}(\cdot)$) is negligible for any forger $F$ whose time complexity is polynomial in $k$. ∎

## 3   Relations among Notions

In this section, we state the formal versions of the results summarized in Figure 1. We begin with the implications and then move to the separations. All proofs are in the full version of this paper [5]. The first implication, below, is a triviality:

**Theorem 1.** *(INT-CTXT → INT-PTXT) Let $\mathcal{SE}$ be an encryption scheme. If $\mathcal{SE}$ is INT-CTXT secure, then it is INT-PTXT secure as well. Concretely:*

$$\mathbf{Adv}_{\mathcal{SE}}^{\text{int-ptxt}}(k, t, q_e, q_d, \mu) \leq \mathbf{Adv}_{\mathcal{SE}}^{\text{int-ctxt}}(k, t, q_e, q_d, \mu) . \quad ∎$$

The next implication is more interesting:

**Theorem 2.** *(INT-CTXT ∧ IND-CPA → IND-CCA) Let $\mathcal{SE}$ be an encryption scheme. If $\mathcal{SE}$ is INT-CTXT secure and IND-CPA secure, then it is IND-CCA secure. Concretely:*

$$\mathbf{Adv}_{\mathcal{SE}}^{\text{ind-cca}}(k, t, q_e, q_d, \mu) \leq 2 \cdot \mathbf{Adv}_{\mathcal{SE}}^{\text{int-ctxt}}(k, t, q_e, q_d, \mu) + \mathbf{Adv}_{\mathcal{SE}}^{\text{ind-cpa}}(k, t, q_e, \mu) . \quad ∎$$

Next we have the formal statements of the separation results.

**Proposition 1.** *(IND-CCA ↛ INT-PTXT) Given a symmetric encryption scheme $\mathcal{SE}$ which is IND-CCA secure, we can construct a symmetric encryption scheme $\mathcal{SE}'$ which is also IND-CCA secure but is not INT-PTXT secure.* ∎

**Proposition 2.** *(INT-PTXT ∧ IND-CPA ↛ NM-CPA) Given a symmetric encryption scheme $\mathcal{SE}$ which is both INT-PTXT secure and IND-CPA secure, we can construct a symmetric encryption scheme $\mathcal{SE}'$ which is also both INT-PTXT secure and IND-CPA secure but is not NM-CPA secure.* ∎

## 4   Security of the Composite Schemes

We now present the formal security results for the composite schemes as summarized in Figure 3. The proofs can be found in the full version of this paper [5]. Proofs for the results of Figure 2 are omitted.

Throughout this section, $\mathcal{SE} = (\mathcal{K}_e, \mathcal{E}, \mathcal{D})$ is a given symmetric encryption scheme which is IND-CPA secure, $\mathcal{MA} = (\mathcal{K}_m, \mathcal{T}, \mathcal{V})$ is a given message authentication scheme which is SUF-CMA secure, and $\overline{S\mathcal{E}} = (\overline{\mathcal{K}}, \overline{\mathcal{E}}, \overline{\mathcal{D}})$ is a composite scheme according to one of the three methods we are considering. The presentation below is method by method, and in each case we begin by specifying the method in more detail.

We make the simplifying assumption that $\mathcal{D}$ never returns $\perp$. It can take any string as input, and the output is always some string. (This is without loss of generality because we can modify $\mathcal{D}$ so that instead of returning $\perp$ it just returns some default message. Security under chosen-plaintext attack is unaffected.) However, $\overline{\mathcal{D}}$ can and will return $\perp$ at times, and this is crucial for integrity.

ENCRYPT-AND-MAC PLAINTEXT. The composite scheme is defined as follows:

| Algorithm $\overline{\mathcal{K}}(k)$ | Algorithm $\overline{\mathcal{E}}_{\langle K_e, K_m \rangle}(M)$ | Algorithm $\overline{\mathcal{D}}_{\langle K_e, K_m \rangle}(C)$ |
|---|---|---|
| $K_e \stackrel{R}{\leftarrow} \mathcal{K}_e(k)$ | $C' \leftarrow \mathcal{E}_{K_e}(M)$ | Parse $C$ as $C' \| \tau$ |
| $K_m \stackrel{R}{\leftarrow} \mathcal{K}_m(k)$ | $\tau \leftarrow \mathcal{T}_{K_m}(M)$ | $M \leftarrow \mathcal{D}_{K_e}(C')$ |
| Return $\langle K_e, K_m \rangle$ | $C \leftarrow C' \| \tau$ | $v \leftarrow \mathcal{V}_{K_m}(M, \tau)$ |
| | Return $C$ | If $v = 1$, return $M$ |
| | | else return $\perp$. |

This composition method does not preserve privacy because the MAC could reveal information about the plaintext.

**Proposition 3. (Encrypt-and-MAC plaintext method is not IND-CPA secure)** *Given a IND-CPA secure symmetric encryption scheme $S\mathcal{E}$ and a SUF-CMA secure message authentication scheme $\mathcal{MA}$, we can construct a message authentication scheme $\mathcal{MA}'$ such that $\mathcal{MA}'$ is SUF-CMA secure, but the composite scheme $\overline{S\mathcal{E}}$ formed by the encrypt-and-MAC plaintext composition method based on $S\mathcal{E}$ and $\mathcal{MA}'$ is not IND-CPA secure.* ∎

Since both IND-CCA and NM-CPA imply IND-CPA, this means that this composition method is also *neither* IND-CCA *nor* NM-CPA secure.

The *encrypt-and-MAC plaintext* composition method, however, inherits the integrity of the MAC in a direct way:

**Theorem 3. (Encrypt-and-MAC plaintext method is INT-PTXT secure)** *Let $S\mathcal{E}$ be a symmetric encryption scheme, let $\mathcal{MA}$ be a message authentication scheme, and let $\overline{S\mathcal{E}}$ be the encryption scheme obtained from $S\mathcal{E}$ and $\mathcal{MA}$ via the encrypt-and-MAC plaintext composition method. Then, if $\mathcal{MA}$ is SUF-CMA secure, then $\overline{S\mathcal{E}}$ is INT-PTXT secure. Concretely:*

$$\mathbf{Adv}_{\overline{S\mathcal{E}}}^{\text{int-ptxt}}(k, t, q_e, q_d, \mu) \leq \mathbf{Adv}_{\mathcal{MA}}^{\text{suf-cma}}(k, t, q_e, q_d, \mu) .$$ ∎

However, this composition method fails in general to provide integrity of ciphertexts. This is because there are secure encryption schemes with the property that a ciphertext can be modified without changing its decryption. When such an encryption scheme is used as the base symmetric encryption scheme, an adversary can query the encryption oracle, modify part of the response, and still submit the result to the verification oracle as a valid ciphertext. The following proposition states this result.

**Proposition 4. (Encrypt-and-MAC plaintext method is not INT-CTXT secure)** *Given a IND-CPA secure symmetric encryption scheme $S\mathcal{E}$ and a SUF-CMA secure message* *authentication scheme* $\mathcal{MA}$, *we can construct a symmetric encryption scheme* $\mathcal{SE}'$ *such that* $\mathcal{SE}'$ *is IND-CPA secure, but the composite scheme* $\overline{\mathcal{SE}}$ *formed by the* encrypt-and-MAC *plaintext composition method based on* $\mathcal{SE}'$ *and* $\mathcal{MA}$ *is not INT-CTXT secure.* ∎

MAC-THEN-ENCRYPT. The composite scheme is defined as follows:

| Algorithm $\overline{\mathcal{K}}(k)$ | Algorithm $\overline{\mathcal{E}}_{\langle K_e, K_m \rangle}(M)$ | Algorithm $\overline{\mathcal{D}}_{\langle K_e, K_m \rangle}(C)$ |
|---|---|---|
| $K_e \xleftarrow{R} \mathcal{K}_e(k)$ | $\tau \leftarrow \mathcal{T}_{K_m}(M)$ | $M' \leftarrow \mathcal{D}_{K_e}(C)$ |
| $K_m \xleftarrow{R} \mathcal{K}_m(k)$ | $C \leftarrow \mathcal{E}_{K_e}(M\|\tau)$ | Parse $M'$ as $M\|\tau$ |
| Return $\langle K_e, K_m \rangle$ | Return $C$ | $v \leftarrow \mathcal{V}_{K_m}(M, \tau)$ |
| | | If $v = 1$, return $M$ |
| | | else return $\perp$. |

The *MAC-then-encrypt* composition method preserves both privacy against chosen-plaintext attack and integrity of plaintexts, as stated in the following theorem.

**Theorem 4. (MAC-then-encrypt method is both INT-PTXT and IND-CPA secure)** *Let* $\mathcal{MA}$ *be a message authentication scheme, and let* $\mathcal{SE}$ *be a symmetric encryption scheme secure against chosen-plaintext attacks. Let* $\overline{\mathcal{SE}}$ *be the encryption scheme obtained from* $\mathcal{SE}$ *and* $\mathcal{MA}$ *via the MAC-then-encrypt composition method. Then, if* $\mathcal{MA}$ *is SUF-CMA secure, then* $\overline{\mathcal{SE}}$ *is INT-PTXT secure. Furthermore, if* $\mathcal{SE}$ *is IND-CPA secure, then so is* $\overline{\mathcal{SE}}$. *Concretely:*

$$\mathbf{Adv}_{\overline{\mathcal{SE}}}^{\text{int-ptxt}}(k, t_i, q_e, q_d, \mu_i) \leq \mathbf{Adv}_{\mathcal{MA}}^{\text{suf-cma}}(k, t_i, q_e, q_d, \mu_i)$$
$$\mathbf{Adv}_{\overline{\mathcal{SE}}}^{\text{ind-cpa}}(k, t_p, q, \mu_p) \leq \mathbf{Adv}_{\mathcal{SE}}^{\text{ind-cpa}}(k, t_p, q, \mu_p) . \quad ∎$$

However, the base encryption scheme might be malleable, and this will be inherited by the composite scheme.

**Proposition 5. (MAC-then-encrypt method is not NM-CPA secure)** *Given a IND-CPA secure symmetric encryption scheme* $\mathcal{SE}$ *and a SUF-CMA secure message authentication scheme* $\mathcal{MA}$, *we can construct a symmetric encryption scheme* $\mathcal{SE}'$ *such that* $\mathcal{SE}'$ *is IND-CPA secure, but the composite scheme* $\overline{\mathcal{SE}}$ *formed by the MAC-then-encrypt composition method based on* $\mathcal{SE}'$ *and* $\mathcal{MA}$ *is not NM-CPA secure.* ∎

Since IND-CCA implies NM-CPA, this composition method is also *not* IND-CCA secure. Furthermore, the fact that it is IND-CPA secure but not NM-CPA secure implies that it is not INT-CTXT secure.

ENCRYPT-THEN-MAC. The composite scheme is defined as follows:

| Algorithm $\overline{\mathcal{K}}(k)$ | Algorithm $\overline{\mathcal{E}}_{\langle K_e, K_m \rangle}(M)$ | Algorithm $\overline{\mathcal{D}}_{\langle K_e, K_m \rangle}(C)$ |
|---|---|---|
| $K_e \xleftarrow{R} \mathcal{K}_e(k)$ | $C' \leftarrow \mathcal{E}_{K_e}(M)$ | Parse $C$ as $C'\|\tau'$ |
| $K_m \xleftarrow{R} \mathcal{K}_m(k)$ | $\tau' \leftarrow \mathcal{T}_{K_m}(C')$ | $M \leftarrow \mathcal{D}_{K_e}(C')$ |
| Return $\langle K_e, K_m \rangle$ | $C \leftarrow C'\|\tau'$ | $v \leftarrow \mathcal{V}_{K_m}(C', \tau')$ |
| | Return $C$ | If $v = 1$, return $M$ |
| | | else return $\perp$. |

The following theorem implies that the *encrypt-then-MAC* composition method is IND-CPA, IND-CCA, NM-CPA, INT-PTXT and INT-CTXT secure.

**Theorem 5.** *(Encrypt-then-MAC method is INT-CTXT, IND-CPA, and IND-CCA se-*
*cure) Let $\mathcal{SE}$ be a symmetric encryption scheme, and let $\mathcal{MA}$ be a message authenti-*
*cation scheme. Let $\overline{\mathcal{SE}}$ be the authenticated encryption scheme obtained from $\mathcal{SE}$ and*
*$\mathcal{MA}$ via the encrypt-then-MAC composition method. Then, if $\mathcal{MA}$ is SUF-CMA se-*
*cure, then $\overline{\mathcal{SE}}$ is INT-CTXT secure. If $\mathcal{SE}$ is IND-CPA secure, then so is $\overline{\mathcal{SE}}$. And if we*
*have both of the previous conditions, then $\overline{\mathcal{SE}}$ is IND-CCA secure. Concretely:*

$$\mathbf{Adv}_{\overline{\mathcal{SE}}}^{\text{int-ctxt}}(k, t_2, q_2, q_2', \mu_2) \leq \mathbf{Adv}_{\mathcal{MA}}^{\text{suf-cma}}(k, t_2, q_2, q_2', \mu_2)$$

$$\mathbf{Adv}_{\overline{\mathcal{SE}}}^{\text{ind-cpa}}(k, t_3, q_3, \mu_3) \leq \mathbf{Adv}_{\mathcal{SE}}^{\text{ind-cpa}}(k, t_3, q_3, \mu_3)$$

*and*

$$\mathbf{Adv}_{\overline{\mathcal{SE}}}^{\text{ind-cca}}(k, t_4, q_4, q_4', \mu_4) \leq$$
$$2 \cdot \mathbf{Adv}_{\mathcal{MA}}^{\text{suf-cma}}(k, t_4, q_4, q_4', \mu_4) + \mathbf{Adv}_{\mathcal{SE}}^{\text{ind-cpa}}(k, t_4, q_4, \mu_4) . \ \blacksquare$$

## Acknowledgments

The authors are supported in part by a 1996 Packard Foundation Fellowship in Science and Engineering and NSF CAREER Award CCR-9624439.

## References

1. M. BELLARE, R. CANETTI AND H. KRAWCZYK, "Keying hash functions for message authentication," *Advances in Cryptology – Crypto '96*, LNCS Vol. 1109, N. Koblitz ed., Springer-Verlag, 1996.

2. M. BELLARE, A. DESAI, E. JOKIPII AND P. ROGAWAY, "A concrete security treatment of symmetric encryption: Analysis of the DES modes of operation," *Proc. of the 38th* IEEE FOCS, IEEE, 1997.

3. M. BELLARE, A. DESAI, D. POINTCHEVAL AND P. ROGAWAY, "Relations among notions of security for public-key encryption schemes," *Advances in Cryptology – Crypto '98*, LNCS Vol. 1462, H. Krawczyk ed., Springer-Verlag, 1998.

4. M. BELLARE, J. KILIAN, P. ROGAWAY, "The security of the cipher block chaining message authentication code," *Advances in Cryptology – Crypto '94*, LNCS Vol. 839, Y. Desmedt ed., Springer-Verlag, 1994.

5. M. BELLARE, C. NAMPREMPRE, "Authenticated Encryption: Relations among notions and analysis of the generic composition paradigm," Full version of this paper, available via http://www-cse.ucsd.edu/users/mihir.

6. M. BELLARE AND P. ROGAWAY, "Encode-then-encipher encryption: How to exploit nonces or redundancy in plaintexts for efficient cryptography," *Advances in Cryptology – ASIACRYPT '00*, LNCS Vol. ??, T. Okamoto ed., Springer-Verlag, 2000.

7. M. BELLARE AND A. SAHAI, "Non-Malleable Encryption: Equivalence between Two Notions, and an Indistinguishability-Based Characterization," *Advances in Cryptology – Crypto '99*, LNCS Vol. 1666, M. Wiener ed., Springer-Verlag, 1999.

8. J. BLACK, S. HALEVI, H. KRAWCZYK, T. KROVETZ AND P. ROGAWAY, "UMAC: Fast and secure message authentication," *Advances in Cryptology – Crypto '99*, LNCS Vol. 1666, M. Wiener ed., Springer-Verlag, 1999.

9.  A. DESAI, "New paradigms for constructing symmetric encryption schemes secure against chosen ciphertext attack," *Advances in Cryptology – Crypto '00*, LNCS Vol. 1880, M. Bellare ed., Springer-Verlag, 2000.

10. D. DOLEV, C. DWORK, AND M. NAOR, "Non-malleable cryptography," *Proc. of the 23rd* ACM STOC, ACM, 1991.

11. D. DOLEV, C. DWORK, AND M. NAOR, "Non-malleable cryptography," to appear in *SIAM J. Comput.*

12. S. GOLDWASSER AND S. MICALI, "Probabilistic encryption," *Journal of Computer and System Science*, Vol. 28, 1984, pp. 270-299.

13. C. JUTLA, "Encryption modes with almost free message integrity," Report 2000/039, *Cryptology ePrint Archive,* http://eprint.iacr.org/, August 2000.

14. J. KATZ AND M. YUNG, "Complete characterization of security notions for probabilistic private-key encryption," *Proc. of the 32nd* ACM STOC, ACM, 2000.

15. J. KATZ AND M. YUNG, "Unforgeable Encryption and Adaptively Secure Modes of Operation," *Fast Software Encryption '00*, LNCS Vol. ??, B. Schneier ed., Springer-Verlag, 2000.

16. S. KENT AND R. ATKINSON, "IP Encapsulating Security Payload (ESP)," Request for Comments 2406, November 1998.

17. M. NAOR AND M. YUNG, "Public-key cryptosystems provably secure against chosen ciphertext attacks," *Proc. of the 22nd* ACM STOC, ACM, 1990.

18. C. RACKOFF AND D. SIMON, "Non-Interactive zero-knowledge proof of knowledge and chosen ciphertext attack," *Advances in Cryptology – Crypto '91*, LNCS Vol. 576, J. Feigenbaum ed., Springer-Verlag, 1991.

# Increasing the Lifetime of a Key: A Comparative Analysis of the Security of Re-keying Techniques

Michel Abdalla and Mihir Bellare

Dept. of Computer Science & Engineering, University of California at San Diego,
9500 Gilman Drive, La Jolla, California 92093, USA.
{mabdalla, mihir}@cs.ucsd.edu
www-cse.ucsd.edu/users/{mabdalla, mihir}

**Abstract.** Rather than use a shared key directly to cryptographically process (e.g. encrypt or authenticate) data one can use it as a master key to derive subkeys, and use the subkeys for the actual cryptographic processing. This popular paradigm is called re-keying, and the expectation is that it is good for security. In this paper we provide concrete security analyses of various re-keying mechanisms and their usage. We show that re-keying does indeed "increase" security, effectively extending the lifetime of the master key and bringing significant, provable security gains in practical situations. We quantify the security provided by different re-keying processes as a function of the security of the primitives they use, thereby enabling a user to choose between different re-keying processes given the constraints of some application.

## 1 Introduction

Re-keying (also called key-derivation) is a commonly employed paradigm in computer security systems, about whose security benefits users appear to have various expectations. Yet the security of these methods has not been systematically investigated. Let us begin with some examples that illustrate usage, commonly employed implementations, and motivation for re-keying, and see what security issues are raised. We then go on to our results.

RE-KEYED ENCRYPTION. Say two parties share a key $K$, and want to encrypt data they send to each other. They will use some block cipher based mode of operation, say CBC. The straightforward approach is to use $K$ directly to encrypt the data. An often employed alternative is re-keyed encryption. The key $K$ is not used to encrypt data but rather viewed as a master key. Subkeys $K_1, K_2, K_3, \ldots$ are derived from $K$, by some process called the re-keying process. A certain number $l$ of messages are encrypted using $K_1$ and then the parties switch to $K_2$. Once $l$ messages have been encrypted under $K_2$ they switch to $K_3$ and so on.

EXAMPLES OF RE-KEYING METHODS. Many different re-keying methods are possible. Let us outline two of them. In each case $F(\cdot, \cdot)$ is a map that takes a $k$-bit key $\kappa$ and $k$-bit input $x$ to a $k$-bit output $F(\kappa, x)$. (This might be implemented via a block cipher or a keyed hash function.) The *parallel* method consists of

T. Okamoto (Ed.): ASIACRYPT 2000, LNCS 1976, pp. 546–559, 2000.

setting $K_i = F(K, i)$ for $i = 1, 2, \ldots$ The *serial* method sets $k_0 = K$ and then sets $K_i = F(k_{i-1}, 0)$ and $k_i = F(k_{i-1}, 1)$ for $i = 1, 2, \ldots$ Many other methods are possible, including hybrids of these two such as tree-based re-keying [1].

WHY RE-KEY? Common attacks base their success on the ability to get lots of encryptions under a single key. For example differential or linear cryptanalysis [10,17] will recover a DES key once a certain threshold number of encryptions have been performed using it. Furthermore, most modes of operation are subject to birthday attacks [3], leading to compromise of the privacy of a scheme based on a block cipher with block size $k$ once $2^{k/2}$ encryptions are performed under the same key. Typically, the birthday threshold is lower than that of the cryptanalytic attacks.

Thus, if encryption is performed under a single key, there is a certain maximum threshold number of messages that can be safely encrypted. Re-keying protects against attacks such as the above by changing the key before the threshold number of encryptions permitting the attack is reached. It thus effectively extends the lifetime of the (master) key, increasing the threshold number of encryptions that can be performed without requiring a new exchange of keys.

QUESTIONS. Although re-keying is common practice, its security has not been systematically investigated. We are interested in the following kinds of questions. Does re-keying really work, in the sense that there is some *provable* increase in security of an application like re-keyed encryption described above? That is, can one prove that the encryption threshold —number of messages of some fixed length that can be safely encrypted— increases with re-keying? How do different re-keying processes compare in terms of security benefits? Do some offer more security than others? How frequently should the key be changed, meaning how should one choose the parameter $l$ given the parameters of a cryptographic system?

HIGH LEVEL ANSWERS. At the highest level, our answer to the most basic question (does re-keying increase security?) is "YES." We are able to justify the prevailing intuition with concrete security analyses in the provable security framework and show that re-keying, properly done, brings significant security gains in practical situations, including an increase in the encryption threshold. Seen from closer up, our results give more precise and usable information. We quantify the security provided by different re-keying processes as a function of the security of the primitives they use. This enables comparison between these processes. Thus, say a user wants to encrypt a certain amount of data with a block cipher of a certain strength: our results can enable this user to figure out which re-keying scheme to use, with what parameters, and what security expectations.

RE-KEYED CBC ENCRYPTION. As a sample of our results we discuss CBC encryption. Suppose we CBC encrypt with a block cipher $F$ having key-length and block-length $k$. Let's define the encryption threshold as the number $Q$ of $k$-bit messages that can be safely encrypted. We know from [3] that this value is $Q \approx 2^{k/2}$ for the single-key scheme. We now consider re-keyed CBC encryption under the parallel or serial re-keying methods discussed above where we use the same block cipher $F$ as the re-keying function. We show that by re-keying every $2^{k/3}$ encryptions —i.e. set the subkey lifetime $l = 2^{k/3}$— the encryption threshold increases to $Q \approx 2^{2k/3}$. That is, one can safely encrypt significantly more data by using re-keying. The analysis can be found in Section 3.

OVERVIEW OF APPROACH AND RESULTS. Re-keying can be used in conjunction with any shared-key based cryptographic data processing. This might be data encryption, under any of the common modes of operation; it might be data authentication using some MAC; it might be something else. We wish to provide tools that enable the analysis of any of these situations. So rather than analyze each re-keyed application independently, we take a modular approach. We isolate the re-keying process, which is responsible for producing subkeys based on a master key, from the application which uses the subkeys. We then seek a general security attribute of the re-keying process which, if present, would enable one to analyze the security of any re-keying based application. We suggest that this attribute is pseudorandomness. We view the re-keying process as a stateful pseudorandom bit generator and adopt a standard notion of security for pseudorandom bit generators [11,18]. We measure pseudorandomness quantitatively, associating to any re-keying process (stateful generator) $\mathcal{G}$ an advantage function $\mathsf{Adv}^{\mathrm{prg}}_{\mathcal{G},n}(t)$, which is the maximum probability of being able to distinguish $n$ output blocks of the generator from a random string of the same length when the distinguishing adversary has running time at most $t$. We then analyze the parallel and serial generators, upper bounding their advantage functions in terms of an advantage function associated to the underlying primitive $F$. See Section 2.

To illustrate an application, we then consider re-keyed symmetric encryption. We associate a re-keyed encryption scheme to any base symmetric encryption scheme (e.g. CBC) and any generator. We show how the advantage function of the re-keyed encryption scheme can be bounded in terms of the advantage function of the base scheme and the advantage function of the generator. (The advantage function of an encryption scheme, whether the base or re-keyed one, measures the breaking probability as a function of adversary resources under the notion of left-or-right security of [3].) Coupling our results about the parallel and serial generators with known analyses of CBC encryption [3] enables us to derive conclusions about the encryption threshold for CBC as discussed above. See Section 3.

SECURITY OF THE PARALLEL AND SERIAL GENERATORS. Our analysis of the parallel and serial generators as given by Theorems 1 and 2 indicates that their advantage functions depend differently on the advantage function of the underlying primitive $F$. (We model the latter as a pseudorandom function [13] and associate an advantage function as per [5].) In general, the parallel generator provides better security. This is true already when $F$ is a block cipher but even more strikingly the case when $F$ is a non-invertible PRF. This should be kept in mind when choosing between the generators for re-keying. However, whether or not it eventually helps depends also on the application. For example, with CBC encryption, there is no particular difference in the quantitative security providing by parallel and serial re-keying (even though both provide gains over the single-key scheme). This is due to the shape of the curve of the advantage function of the base CBC encryption function as explained in Section 3.

FORWARD SECURITY. Another possible motivation for re-keying is to provide forward security. The goal here is to minimize the amount of damage that might be caused by key exposure due, for instance, to compromise of the security of the underlying system storing the secret key. (Forward security was first considered for session keys [15,12] and then for digital signatures [7].) Under re-keying, the adversary would only get the current subkey and state of the system. It could certainly figure out all future subkeys, but what about past ones? If the re-keying process is appropriately designed, it can have forward security: the past subkeys will remain computationally infeasible for the adversary to derive even given the current subkey and state, and thus ciphertexts that were formed under them will not be compromised. It is easy to see that the parallel generator does not provide forward security. It can be shown however that the serial one does. A treatment of forward security in the symmetric setting, including a proof of the forward security of the serial generator and the corresponding re-keyed encryption scheme, can be found in [9].

RELATED WORK. Another approach to increasing the encryption threshold, discussed in [6], is to use a mode of encryption not subject to birthday attack (e.g. CTR rather than CBC) and implement this using a non-invertible, high security PRF rather than a block cipher. Constructions of appropriate PRFs have been provided in [6,16]. Re-keying is cheaper in that one can use the given block cipher and a standard mode like CBC and still push the encryption threshold well beyond the birthday threshold.

Re-keying requires that parties maintain state. Stateless methods of increasing security beyond the birthday bound are discussed in [4].

## 2    Re-keying Processes as Pseudorandom Generators

The subkeys derived by a re-keying process may be used in many different ways: data encryption or authentication are some but not all of these. To enable modular analysis, we separate the subkey generation from the application that uses the subkeys. We view the re-keying process —which generates the subkeys— as a stateful pseudorandom bit generator. In this section we provide quantitative assessments of the security of various re-keying schemes with regard to notions of security for pseudorandom generators. These application independent results are used in later sections to assess the security of a variety of different applications under re-keying.

STATEFUL GENERATORS. A stateful generator $\mathcal{G} = (\mathcal{K}, \mathcal{N})$ is a pair of algorithms. The probabilistic *key generation* algorithm $\mathcal{K}$ produces the initial state, or seed, of the generator. The deterministic *next step* algorithm $\mathcal{N}$ takes the current state as input and returns a block, viewed as the output of this stage, and an updated state, to be stored and used in the next invocation. A sequence $Out_1, Out_2, \ldots$ of pseudorandom blocks is defined by first picking an initial seed $St_0 \leftarrow \mathcal{K}$ and then iterating: $(Out_i, St_i) \leftarrow \mathcal{N}(St_{i-1})$ for $i \geq 1$. (When the generator is used for re-keying, these are the subkeys. Thus $Out_i$ was denoted $K_i$ in Section 1). We assume all output blocks are of the same length and call this the block length.

We now specify two particular generators, the parallel and serial ones. We fix a PRF $F$: $\{0,1\}^k \times \{0,1\}^k \rightarrow \{0,1\}^k$. (As the notation indicates, we are making the simplifying assumption that the key length, as well as the input and output lengths of each individual function $F(K, \cdot)$ are all equal to $k$.) In practice, this might be instantiated via a block cipher or via a keyed hash function such as HMAC [2]. (For example, if DES is used, then we set $k = 64$ and define $F(K, \cdot)$ to be DES$(K[1..56], \cdot)$.)

**Construction 1. (Parallel generator)** The $F$-based parallel generator $\mathcal{PG}[F] = (\mathcal{K}, \mathcal{N})$ is defined by

$$
\begin{array}{c|c}
\text{Algorithm } \mathcal{K} & \text{Algorithm } \mathcal{N}(\langle i, K \rangle) \\
K \stackrel{R}{\leftarrow} \{0,1\}^k & Out \leftarrow F(K, i) \\
\text{Return } \langle 0, K \rangle & \text{Return } (Out, \langle i+1, K \rangle)
\end{array}
$$

The state has the form $\langle i, K \rangle$ where $K$ is the initial seed and $i$ is a counter, initially zero. In the $i$-th stage, the output block is obtained by applying the $K$-keyed PRF to the ($k$-bit binary representation of the integer) $i$, and the counter is updated. This generator has block length $k$. ∎

**Construction 2. (Serial generator)** The $F$-based serial generator $\mathcal{SG}[F] = (\mathcal{K}, \mathcal{N})$ is defined by

$$
\begin{array}{c|c}
\text{Algorithm } \mathcal{K} & \text{Algorithm } \mathcal{N}(K) \\
K \stackrel{R}{\leftarrow} \{0,1\}^k & Out \leftarrow F(K, 0) \\
\text{Return } K & K \leftarrow F(K, 1) \\
& \text{Return } (Out, K)
\end{array}
$$

The state is a key $K$. In the $i$-th stage, the output block is obtained by applying the $K$-keyed PRF to the ($k$-bit binary representation of the integer) 0, and the new state is a key generated by applying the $K$-keyed PRF to the ($k$-bit binary representation of the integer) 1. This generator has block length $k$. ∎

PSEUDORANDOMNESS. The standard desired attribute of a (stateful) generator is pseudorandomness of the output sequence. We adopt the notion of [11,18] which formalizes this by asking that the output of the generator on a random seed be computationally indistinguishable from a random string of the same length. Below, we concretize this notion by associating to any generator an advantage function which measures the probability that an adversary can detect a deviation in pseudorandomness as a function of the amount of time invested by the adversary.

**Definition 1. (Pseudorandomness of a stateful generator)** Let $\mathcal{G} = (\mathcal{K}, \mathcal{N})$ be a stateful generator with block length $k$, let $n$ be an integer, and let $A$ be an adversary. Consider the experiments

Experiment $\mathbf{Exp}_{\mathcal{G},n}^{\text{prg-real}}(A)$

   $St_0 \leftarrow \mathcal{K}$ ; $s \leftarrow \varepsilon$
   for $i = 1, \ldots, n$ do
      $(Out_i, St_i) \leftarrow \mathcal{N}(St_{i-1})$ ; $s \leftarrow s \parallel Out_i$
   $g \leftarrow A(s)$
   return $g$

Experiment $\mathbf{Exp}_{\mathcal{G},n}^{\text{prg-rand}}(A)$

   $s \leftarrow \{0,1\}^{n \cdot k}$
   $g \leftarrow A(s)$
   return $g$

Now define the *advantage* of $A$ and the *advantage function of the generator*, respectively, as follows:

$$\mathsf{Adv}_{\mathcal{G},n}^{\text{prg}}(A) = \Pr[\,\mathbf{Exp}_{\mathcal{G},n}^{\text{prg-real}}(A) = 1\,] - \Pr[\,\mathbf{Exp}_{\mathcal{G},n}^{\text{prg-rand}}(A) = 1\,]$$
$$\mathsf{Adv}_{\mathcal{G},n}^{\text{prg}}(t) = \max_A \{\, \mathsf{Adv}_{\mathcal{G},n}^{\text{prg}}(A) \,\} ,$$

where the maximum is over all $A$ with "time-complexity" $t$. ∎

Here "time-complexity" is the maximum of the execution times of the two experiments plus the size of the code for $A$, all in some fixed RAM model of computation. (Note that the execution time refers to that of the entire experiment, not just the execution time of the adversary.) The advantage function is the maximum likelihood of the security of the pseudorandom generator $\mathcal{G}$ being compromised by an adversary using the indicated resources.

SECURITY MEASURE FOR PRFS. Since the security of the above constructions depends on that of the underlying PRF $F$: $\{0,1\}^k \times \{0,1\}^k \to \{0,1\}^k$, we recall the measure of [5], based on the notion of [13]. Let $R^k$ denote the family of all functions mapping $\{0,1\}^k$ to $\{0,1\}^k$, under the uniform distribution. If $D$ is a distinguisher having an oracle, then

$$\mathsf{Adv}_F^{\text{prf}}(D) \;=\; \Pr[\, D^{F(K,\cdot)} = 1 \;:\; K \xleftarrow{R} \{0,1\}^k \,] - \Pr[\, D^{f(\cdot)} = 1 \;:\; f \xleftarrow{R} R^k \,]$$

is the advantage of $D$. The advantage function of $F$ is

$$\mathsf{Adv}_F^{\text{prf}}(t, q) \;=\; \max_D \{\, \mathsf{Adv}_F^{\text{prf}}(D) \,\} ,$$

where the maximum is over all $A$ with "time-complexity" $t$ and making at most $q$ oracle queries. The time-complexity is the execution time of the experiment $K \xleftarrow{R} \{0,1\}^k$ ; $v \leftarrow D^{F(K,\cdot)}$ plus the size of the code of $D$, and, in particular, includes the time to compute $F_K(\cdot)$ and reply to oracle queries of $D$.

PSEUDORANDOMNESS OF THE PARALLEL AND SERIAL GENERATORS. The following theorems, whose proofs can be found in Appendices A and B, show how the pseudorandomness of the two generators is related to the security of the underlying PRF.

**Theorem 1.** *Let $F$: $\{0,1\}^k \times \{0,1\}^k \to \{0,1\}^k$ be a PRF and let $\mathcal{PG}[F]$ be the $F$-based parallel generator defined in Construction 1. Then*

$$\mathsf{Adv}_{\mathcal{PG}[F],n}^{\text{prg}}(t) \leq \mathsf{Adv}_F^{\text{prf}}(t, n) .\;∎$$

**Theorem 2.** *Let $F \colon \{0,1\}^k \times \{0,1\}^k \to \{0,1\}^k$ be a PRF and let $\mathcal{SG}[F]$ be the F-based parallel generator defined in Construction 2. Then*

$$\mathsf{Adv}^{\mathrm{prg}}_{\mathcal{SG}[F],n}(t) \le n \cdot \mathsf{Adv}^{\mathrm{prf}}_{F}(t + \log n, 2) . \blacksquare$$

The qualitative interpretation of the two theorems is the same: both the parallel and the serial generator are secure pseudorandom bit generators if the PRF is secure. The quantitative statements show however that the pseudorandomness of $n$ output blocks depends differently on the security of the PRF in the two cases. For the parallel generator, it depends on the security of the PRF under $n$ queries. For the serial generator, it depends on the security of the PRF against only a constant number of queries, but this term is multiplied by the number of output blocks. Comparing the functions on the right hand side in the two theorems will tell us which generator is more secure.

EXAMPLES. As an example, assume $F$ is a block cipher. Since $F$ is a cipher, each map $F(K, \cdot)$ is a permutation, and birthday attacks can be used to distinguish $F$ from the family of random functions with a success rate growing as $q^2/2^k$ for $q$ queries (c.f.. [5, Proposition 2.4]). Let us make the (heuristic) assumption that this is roughly the best possible, meaning

$$\mathsf{Adv}^{\mathrm{prf}}_{F}(t, q) \approx \frac{q^2 + t}{2^k} \tag{1}$$

for $t$ small enough to prevent cryptanalytic attacks. Now the above tells us that the advantage functions of the two generators grow as follows:

$$\mathsf{Adv}^{\mathrm{prg}}_{\mathcal{PG}[F],n}(t) \approx \frac{n^2 + t}{2^k} \quad \text{and} \quad \mathsf{Adv}^{\mathrm{prg}}_{\mathcal{SG}[F],n}(t) \approx \frac{nt}{2^k} .$$

Since $t \ge n$, the two functions are roughly comparable, but in fact the first one has a somewhat slower growth because we would expect that $t \gg n$. So, in this case, the parallel generator is somewhat better.

Now assume $F$ is not a block cipher but something that better approximates a random function, having security beyond the birthday bound. Ideally, we would like something like

$$\mathsf{Adv}^{\mathrm{prf}}_{F}(t, q) \approx \frac{q + t}{2^k} \tag{2}$$

for $t$ small enough to prevent cryptanalytic attacks. This might be achieved by using a keyed hash function based construction, or by using PRFs constructed from block ciphers as per [6,16]. In this case we would get

$$\mathsf{Adv}^{\mathrm{prg}}_{\mathcal{PG}[F],n}(t) \approx \frac{n + t}{2^k} \quad \text{and} \quad \mathsf{Adv}^{\mathrm{prg}}_{\mathcal{SG}[F],n}(t) \approx \frac{nt}{2^k} .$$

Thinking of $t \approx n$ (it cannot be less but could be more, so this is an optimistic choice), we see that the first function has linear growth and the second has quadratic growth, meaning the parallel generator again offers better security, but this time in a more decisive way.

These examples illustrate how the quantitative results of the theorems can be coupled with cryptanalytic knowledge or assumptions about the starting primitive $F$ to yield information enabling a user to choose between the generators.

# 3   Re-keyed Symmetric Encryption

We fix a *base encryption scheme*. (For example, CBC mode encryption based on some block cipher.) We wish to encrypt data using this scheme, but with re-keying. Two things need to be decided. The first is how the re-keying is to be done, meaning how the subkeys will be computed. This corresponds to making a choice of stateful generator to generate the subkey sequence. The second is the lifetime of each subkey, meaning how many encryptions will be done with it. This corresponds to choosing an integer parameter $l > 0$ which we call the *subkey lifetime*. Associated to a base scheme, generator and subkey lifetime, is a particular *re-keyed encryption scheme*. We are interested in comparing the security of the re-keyed encryption scheme across different choices of re-keying processes (i.e. generators), keeping the base scheme and subkey lifetime fixed. In particular, we want to compare the use of the parallel and serial generators.

Our analysis takes a modular approach. Rather than analyzing separately the re-keyed encryption schemes corresponding to different choices of generators, we first analyze the security of a re-keyed encryption scheme with an arbitrary generator, showing how the advantage of the encryption scheme can be bounded in terms of that of the generator and the base scheme. We then build on results of Section 2 to get results for re-keyed encryption with specific generators. We begin by specifying in more detail the re-keyed encryption scheme and saying how we measure security of symmetric encryption schemes.

RE-KEYED ENCRYPTION SCHEMES. Let $\mathcal{SE} = (\mathcal{K}_e, \mathcal{E}, \mathcal{D})$ be the base (symmetric) encryption scheme, specified by its key generation, encryption and decryption algorithms [3]. Let $\mathcal{G} = (\mathcal{K}_g, \mathcal{N})$ be a stateful generator with block size $k$, where $k$ is the length of the key of the base scheme. Let $l > 0$ be a subkey lifetime parameter. We associate to them a *re-keyed encryption scheme* $\overline{\mathcal{SE}}[\mathcal{SE}, \mathcal{G}, l] = (\overline{\mathcal{K}}, \overline{\mathcal{E}}, \overline{\mathcal{D}})$. This is a stateful encryption scheme which works as follows. The initial state of the encryption scheme includes the initial state of the generator, given by $St_0 \overset{R}{\leftarrow} \mathcal{K}_g$. Encryption is divided into stages $i = 1, 2, \ldots$. Stage $i$ begins with the generation of a new key $K_i$ using the generator: $(K_i, St_i) \leftarrow \mathcal{N}(St_{i-1})$. In stage $i$ encryption is done using the encryption algorithm of the base scheme with key $K_i$. An *encryption counter* is maintained, and when $l$ encryptions have been performed, this stage ends. The encryption counter is then reset, the stage counter is incremented, and the key for the next stage is generated. If the base scheme is stateful, its state is reset whenever the key changes.

Formally, the key generation algorithm $\overline{\mathcal{K}}$ of the re-keyed scheme is run once, at the beginning, to produce an initial state which is shared between sender and receiver and includes $St_0$. The encryption algorithm $\overline{\mathcal{E}}$ takes the current state (which includes $K_i, St_i$, a stage counter, the encryption counter, and a state for the base scheme if the latter happens to be stateful) and the message $M$ to be encrypted, and returns ciphertext $C \leftarrow \mathcal{E}_{K_i}(M)$. It also returns an updated state which is stored locally. It is advisable to include with the ciphertext the number $i$ of the current stage, so that the receiver can maintain decryption capability even if messages are lost in transit. The $\overline{\mathcal{D}}$ algorithm run by the receiver can be stateless in this case. (This is true as long as the goal is privacy against chosen-plaintext attacks as we consider here, but if active attacks are considered, meaning we want privacy against chosen-ciphertext attacks or authenticity, the receiver will have to maintain state as well.)

SECURITY MEASURES FOR ENCRYPTION SCHEMES. Several (polynomial-time equivalent) definitions for security of a symmetric encryption scheme under chosen-plaintext attack were given in [3]. We use one of them, called left-or-right security. The game begins with a random bit $b$ being chosen. The adversary then gets access to an oracle which can take as input any two equal-length messages $(x_0, x_1)$ and responds with a ciphertext formed by encrypting $x_b$. The adversary wins if it can eventually guess $b$ correctly. We can associate to any adversary an advantage measuring the probability it wins. We then associate to the base encryption scheme —respectively, the re-keyed encryption scheme— an advantage function $\mathsf{Adv}_{\mathcal{SE}}^{\mathrm{ind\text{-}cpa}}(t, q, m)$ —respectively $\mathsf{Adv}_{\overline{\mathcal{SE}}}^{\mathrm{ind\text{-}cpa}}(t, q, m)$— which measures the maximum probability of the scheme being compromised by an adversary running in time $t$ and allowed $q$ oracle queries each consisting of a pair of $m$-bit messages. Intuitively, this captures security against a chosen-plaintext attack of $q$ messages. (The usual convention [3] is to allow messages of different lengths and count the sum of the lengths of all messages but for simplicity we ask here that all messages have the same length. Note that for the base encryption scheme, all encryption is done using a single, random key. For the re-keyed scheme, it is done as the scheme specifies, meaning with the key changing every $l$ encryptions. We omit details here, but precise definitions with this type of notation can be found for example in [8].)

SECURITY OF RE-KEYED ENCRYPTION. The qualitative interpretation of the following theorem is that if the generator and base encryption scheme are secure then so is the re-keyed encryption scheme. It is the quantitative implications however on which we focus. The theorem says that the security of encrypting $ln$ messages with the re-keyed scheme relates to the pseudorandomness of $n$ blocks of the generator output and the security of encrypting $l$ messages under the base scheme with a single random key. The $\mathsf{Adv}_{\mathcal{SE}}^{\mathrm{ind\text{-}cpa}}(t, l, m)$ term is multiplied by $n$, yet there is a clear gain, in that the security of the base encryption scheme relates to encrypting only $l$ messages. The proof of Theorem 3 can be found in the full version of this paper [1].

**Theorem 3. (Security of re-keyed encryption)** *Let $\mathcal{SE}$ be a base encryption scheme with key size $k$, let $\mathcal{G}$ be a stateful generator with blocksize $k$, and let $l > 0$ be a subkey lifetime. Let $\overline{\mathcal{SE}} = \overline{\mathcal{SE}}[\mathcal{SE}, \mathcal{G}, l]$ be the associated re-keyed encryption scheme. Then*

$$\mathsf{Adv}_{\overline{\mathcal{SE}}}^{\mathrm{ind\text{-}cpa}}(t, ln, m) \ \leq \ \mathsf{Adv}_{\mathcal{G}, n}^{\mathrm{prg}}(t) + n \cdot \mathsf{Adv}_{\mathcal{SE}}^{\mathrm{ind\text{-}cpa}}(t, l, m) . \ \blacksquare$$

RE-KEYED ENCRYPTION WITH THE PARALLEL AND SERIAL GENERATORS. Combining Theorem 3 with Theorems 1 and 2 gives us information about the security of re-keyed encryption under the parallel and serial generators.

**Corollary 1. (Security of re-keyed encryption with the parallel generator)** *Let $\mathcal{SE}$ be a base encryption scheme, let $F\colon \{0,1\}^k \times \{0,1\}^k \to \{0,1\}^k$ be a PRF, let $\mathcal{PG}[F]$ be the $F$-based parallel generator defined in Construction 1, and let $l > 0$ be a subkey lifetime. Let $\overline{\mathcal{SE}} = \overline{\mathcal{SE}}[\mathcal{SE}, \mathcal{PG}[F], l]$ be the associated re-keyed encryption scheme. Then*

$$\mathsf{Adv}_{\overline{\mathcal{SE}}}^{\text{ind-cpa}}(t, ln, m) \leq \mathsf{Adv}_F^{\text{prf}}(t, n) + n \cdot \mathsf{Adv}_{\mathcal{SE}}^{\text{ind-cpa}}(t, l, m) \ . \blacksquare$$

**Corollary 2. (Security of re-keyed encryption with the serial generator)** *Let $\mathcal{SE}$ be a base encryption scheme, let $F\colon \{0,1\}^k \times \{0,1\}^k \to \{0,1\}^k$ be a PRF, let $\mathcal{SG}[F]$ be the $F$-based serial generator defined in Construction 2, and let $l > 0$ be a subkey lifetime. Let $\overline{\mathcal{SE}} = \overline{\mathcal{SE}}[\mathcal{SE}, \mathcal{SG}[F], l]$ be the associated re-keyed encryption scheme. Then*

$$\mathsf{Adv}_{\overline{\mathcal{SE}}}^{\text{ind-cpa}}(t, ln, m) \leq n \cdot \mathsf{Adv}_F^{\text{prf}}(t + \log n, 2) + n \cdot \mathsf{Adv}_{\mathcal{SE}}^{\text{ind-cpa}}(t, l, m) \ . \blacksquare$$

EXAMPLE. For the base encryption scheme, let us use CBC with some block cipher $B\colon \{0,1\}^k \times \{0,1\}^b \to \{0,1\}^b$ having block length $b$. We wish to compare the security of encrypting $q$ messages directly with one key; doing this with re-keying using the parallel generator; and doing this with re-keying using the serial generator. The re-keying is based on a PRF $F\colon \{0,1\}^k \times \{0,1\}^k \to \{0,1\}^k$ having block length $k$. Note that $B$ and $F$ can but need not be the same. In particular $B$ must be a cipher (i.e. invertible) in order to enable CBC decryption, but we have seen that better security results for the re-keying schemes by choosing $F$ to be non-invertible and might want to choose $F$ accordingly.

Let $\mathcal{CBC}$ denote the base encryption scheme. Let $\mathcal{PCBC}$ denote the re-keyed encryption scheme using $\mathcal{CBC}$ as the base scheme, the $F$-based parallel generator, and subkey lifetime parameter $l$. Let $\mathcal{SCBC}$ denote the re-keyed encryption scheme using $\mathcal{CBC}$ as the base scheme, the $F$-based serial generator, and subkey lifetime parameter $l$. Since $B$ is a cipher we take its advantage to be

$$\mathsf{Adv}_B^{\text{prf}}(t, q) \approx \frac{q^2}{2^b} + \frac{t}{2^k} \ . \tag{3}$$

We know from [3] that

$$\mathsf{Adv}_{\mathcal{CBC}}^{\text{ind-cpa}}(t, q, m) \approx \frac{q^2 m^2}{b^2 2^b} + 2 \cdot \mathsf{Adv}_B^{\text{prf}}(t, qm/b) \approx \frac{3q^2 m^2}{b^2 2^b} + \frac{2t}{2^k} \ .$$

For simplicity we let the message length be $m = b$. Thus if $q = ln$ messages of length $m$ are CBC encrypted we have

$$\mathsf{Adv}_{\mathcal{CBC}}^{\text{ind-cpa}}(t, ln, b) \approx \frac{3l^2 n^2}{2^b} + \frac{2t}{2^k}$$

$$\mathsf{Adv}_{\mathcal{PCBC}}^{\text{ind-cpa}}(t, ln, b) \approx \mathsf{Adv}_F^{\text{prf}}(t, n) + \frac{3l^2 n}{2^b} + \frac{2nt}{2^k}$$

$$\mathsf{Adv}_{\mathcal{SCBC}}^{\text{ind-cpa}}(t, ln, b) \approx n \cdot \mathsf{Adv}_F^{\text{prf}}(t + \log n, 2) + \frac{3l^2 n}{2^b} + \frac{2nt}{2^k} \ .$$

The first corresponds to encryption with a single key, the second to re-keying with the parallel generator, and the third to re-keying with the serial generator.

Suppose we let $F$ be a block cipher. (This is the easiest choice in practice.) We can simply let $F = B$. In that case $F$ obeys Equation (1) and we get

$$\mathsf{Adv}_{\mathcal{CBC}}^{\mathrm{ind\text{-}cpa}}(t, ln, m) \approx \frac{3l^2n^2 + 2t}{2^k}$$

$$\mathsf{Adv}_{\mathcal{PCBC}}^{\mathrm{ind\text{-}cpa}}(t, ln, m) \approx \frac{3l^2n + n^2 + 2nt}{2^k}$$

$$\mathsf{Adv}_{\mathcal{SCBC}}^{\mathrm{ind\text{-}cpa}}(t, ln, m) \approx \frac{3l^2n + 2nt + t}{2^k}.$$

The two generators deliver about the same advantage. To gauge the gains provided by the re-keying schemes over the single-key scheme, let us define the *encryption threshold* of a scheme to be the smallest number of messages $Q = ln$ that can be encrypted before the advantage hits one. (Roughly speaking, this is the number of messages we can safely encrypt.) We want it to be as high as possible. Let's take $t \approx nl$. (It cannot be less but could be more so this is an optimistic choice). In the single-key scheme $Q \approx 2^{k/2}$. In the re-keyed schemes let us set $l = 2^{k/3}$. (This is the optimal choice.) In that case $Q \approx 2^{2k/3}$. This is a significant increase in the encryption threshold, showing that re-keying brings important security benefits.

We could try to set $F$ to be a non-invertible PRF for which Equation (2) is true. (In particular $F$ would not be $B$.) Going through the calculations shows that again the two generators will offer the same advantage, but this would be an improvement over the single-key scheme only if $k > b$. (Setting $k = 2b$ yields an encryption threshold of $2^b$ for the re-keyed schemes as compared to $2^{b/2}$ for the single-key scheme.)

We saw in Section 2 that the parallel generator offered greater security than the serial one. We note that this did not materialize in the application to re-keyed CBC encryption: here, the advantage functions arising from re-keying under the two generators are the same. This is because the term corresponding to the security of the base scheme in Corollaries 1 and 2 dominates when the base scheme is CBC.

In summary we wish to stress two things: that security increases are possible, and that our results provide general tools to estimate security in a variety of re-keyed schemes and to choose parameters to minimize the advantage functions of the re-keyed schemes.

## Acknowledgments

Michel Abdalla was supported in part by CAPES under Grant BEX3019/95-2. Mihir Bellare was supported in part by a 1996 Packard Foundation Fellowship in Science and Engineering and NSF CAREER Award CCR-9624439.

## References

1. M. ABDALLA AND M. BELLARE, "A comparative analysis of the security of re-keying techniques," Full version of this paper, available via http://www-cse.ucsd.edu/users/mihir.

2. M. BELLARE, R. CANETTI AND H. KRAWCZYK, "Keying hash functions for message authentication," *Advances in Cryptology – Crypto '96*, LNCS Vol. 1109, N. Koblitz ed., Springer-Verlag, 1996.

3. M. BELLARE, A. DESAI, E. JOKIPII AND P. ROGAWAY, "A concrete security treatment of symmetric encryption: Analysis of the DES modes of operation," *Proc. of the 38th* IEEE FOCS, IEEE, 1997.

4. M. BELLARE, O. GOLDREICH AND H. KRAWCZYK, "Stateless evaluation of pseudorandom functions: Security beyond the birthday barrier," *Advances in Cryptology – Crypto '99*, LNCS Vol. 1666, M. Wiener ed., Springer-Verlag, 1999.

5. M. BELLARE, J. KILIAN AND P. ROGAWAY, "The security of cipher block chaining," available via `http://www-cse.ucsd.edu/users/mihir`. Preliminary version in *Advances in Cryptology – Crypto '94*, LNCS Vol. 839, Y. Desmedt ed., Springer-Verlag, 1994.

6. M. BELLARE, T. KROVETZ AND P. ROGAWAY, "Luby-Rackoff backwards: Increasing security by making block ciphers non-invertible," *Advances in Cryptology – Eurocrypt '98*, LNCS Vol. 1403, K. Nyberg ed., Springer-Verlag, 1998.

7. M. BELLARE AND S. MINER, "A forward-secure digital signature scheme," *Advances in Cryptology – Crypto '99*, LNCS Vol. 1666, M. Wiener ed., Springer-Verlag, 1999.

8. M. BELLARE AND C. NAMPREMPRE, "Authenticated Encryption: Relations among notions and analysis of the generic composition paradigm," *Advances in Cryptology – ASIACRYPT '00*, LNCS Vol. ??, T. Okamoto ed., Springer-Verlag, 2000. Available via `http://www-cse.ucsd.edu/users/mihir`.

9. M. BELLARE AND B. YEE, "Forward security in private-key cryptography," Manuscript, 1998.

10. E. BIHAM AND A. SHAMIR, "Differential cryptanalysis of the Full 16-round DES," *Advances in Cryptology – Crypto '92*, LNCS Vol. 740, E. Brickell ed., Springer-Verlag, 1992.

11. M. BLUM AND S. MICALI, "How to generate cryptographically strong sequences of pseudo-random bits," *SIAM Journal on Computing*, Vol. 13, No. 4, 850-864, November 1984.

12. W. DIFFIE, P. VAN OORSCHOT AND M. WIENER, "Authentication and authenticated key exchanges," *Designs, Codes and Cryptography*, 2, 107–125, 1992.

13. O. GOLDREICH, S. GOLDWASSER AND S. MICALI, "How to construct random functions," *Journal of the ACM*, Vol. 33, No. 4, 1986, pp. 210–217.

14. S. GOLDWASSER AND S. MICALI, "Probabilistic encryption," *Journal of Computer and System Sciences*, Vol. 28, 1984, pp. 270–299.

15. C. GÜNTHER, "An identity-based key-exchange protocol," *Advances in Cryptology – Eurocrypt '89*, LNCS Vol. 434, J-J. Quisquater, J. Vandewille ed., Springer-Verlag, 1989.

16. C. HALL, D. WAGNER, J. KELSEY AND B. SCHNEIER, "Building PRFs from PRPs," *Advances in Cryptology – Crypto '98*, LNCS Vol. 1462, H. Krawczyk ed., Springer-Verlag, 1998.

17. M. MATSUI, "The first experimental cryptanalysis of the Data Encryption Standard," *Advances in Cryptology – Crypto '94*, LNCS Vol. 839, Y. Desmedt ed., Springer-Verlag, 1994.

18. A. YAO, "Theory and applications of trapdoor functions," *Proc. of the 23rd* IEEE FOCS, IEEE, 1982.

# A   Proof of Theorem 1

Let $A$ be an adversary attacking the pseudorandomness of $\mathcal{PG}[F]$ and $t$ be the maximum of the running times of $\mathbf{Exp}_{\mathcal{PG}[F],n}^{\text{prg-real}}(A)$ and $\mathbf{Exp}_{\mathcal{PG}[F],n}^{\text{prg-rand}}(A)$. We want to upper bound $\mathsf{Adv}_{\mathcal{PG}[F],n}^{\text{prg}}(A)$. We do so by constructing a distinguisher $D$ for $F$ and relating its advantage to that of $A$. $D$ has access to an oracle $\mathcal{O}$. It simply computes $s = \mathcal{O}(1) \| \ldots \| \mathcal{O}(n)$ and outputs the same guess as $A$ on input $s$. We can see that when the oracle $\mathcal{O}$ is drawn at random from the family $F$, the probability that $D$ returns 1 equals the probability that the experiment $\mathbf{Exp}_{\mathcal{PG}[F],n}^{\text{prg-real}}(A)$ returns 1. Likewise, the probability that the experiment $\mathbf{Exp}_{\mathcal{PG}[F],n}^{\text{prg-rand}}(A)$ returns 1 equals that of $D$ returning 1 when $\mathcal{O}$ is drawn at random from the family of random functions $R^k$. As $D$ runs in time at most $t$ and makes exactly $n$ queries to its oracle, we get that

$$\mathsf{Adv}_{\mathcal{PG}[F],n}^{\text{prg}}(A) \leq \mathsf{Adv}_{F}^{\text{prf}}(t, n) .$$

Since $A$ was an arbitrary adversary and the maximum of the running times of experiments $\mathbf{Exp}_{\mathcal{PG}[F],n}^{\text{prg-real}}(A)$ and $\mathbf{Exp}_{\mathcal{PG}[F],n}^{\text{prg-rand}}(A)$ is $t$, we obtain the conclusion of the theorem.

# B   Proof of Theorem 2

Let $A$ be an adversary attacking the pseudorandomness of $\mathcal{SG}[F]$ and $t$ be the maximum of the running times of $\mathbf{Exp}_{\mathcal{SG}[F],n}^{\text{prg-real}}(A)$ and $\mathbf{Exp}_{\mathcal{SG}[F],n}^{\text{prg-rand}}(A)$. We want to upper bound $\mathsf{Adv}_{\mathcal{SG}[F],n}^{\text{prg}}(A)$. We begin by defining the following sequence of hybrid experiments, where $j$ varies between 0 and $n$.

Experiment $\mathbf{Hybrid}(A, j)$

> $St \xleftarrow{R} \{0,1\}^k$ ; $s \leftarrow \varepsilon$
> for $i = 1, \ldots, n$ do
> > if $i \leq j$ then $Out_i \xleftarrow{R} \{0,1\}^k$
> > else $(Out_i, St) \leftarrow N(St)$
> > $s \leftarrow s \| Out_i$
> $g \leftarrow A(s)$
> return $g$

Let $P_j$ be the probability that experiment $\mathbf{Hybrid}(A, j)$ returns 1, for $j = 0, \ldots, n$. Note that the experiments $\mathbf{Exp}_{\mathcal{SG}[F],n}^{\text{prg-real}}(A)$ and $\mathbf{Exp}_{\mathcal{SG}[F],n}^{\text{prg-rand}}(A)$ are identical to $\mathbf{Hybrid}(A, 0)$ and $\mathbf{Hybrid}(A, n)$, respectively. (Not syntactically, but semantically.) This means that $P_0 = \Pr[\mathbf{Exp}_{\mathcal{SG}[F],n}^{\text{prg-real}}(A) = 1]$ and $P_n = \Pr[\mathbf{Exp}_{\mathcal{SG}[F],n}^{\text{prg-rand}}(A) = 1]$. Putting it all together, we have

$$\mathsf{Adv}_{\mathcal{SG}[F],n}^{\text{prg}}(A) = \Pr[\mathbf{Exp}_{\mathcal{SG}[F],n}^{\text{prg-real}}(A) = 1] - \Pr[\mathbf{Exp}_{\mathcal{SG}[F],n}^{\text{prg-rand}}(A) = 1]$$
$$= P_0 - P_n . \tag{4}$$

We now claim that

$$\mathsf{Adv}^{\mathrm{prg}}_{\mathcal{SG}[F],n}(A) = P_0 - P_n \leq n \cdot \mathsf{Adv}^{\mathrm{prf}}_F(t + \log n, 2) . \tag{5}$$

Since $A$ was an arbitrary adversary, we obtain the conclusion of the theorem. It remains to justify Equation (5). We will do this using the advantage function of $F$. Consider the following distinguisher for $F$.

```
Algorithm D^O
    j ←R {1,...,n} ; s ← ε
    for i = 1,...,n do
        if i < j then Out_i ←R {0,1}^k
        if i = j then Out_i ← O(0) ; St ← O(1)
        if i > j then (Out_i, St) ← N(St)
        s ← s || Out_i
    g ← A(s)
    return g
```

Suppose the oracle given to $D$ was drawn at random from the family $F$. Then, the probability that it returns 1 equals the probability that the expirement **Hybrid**$(A, j-1)$ returns 1, where $j$ is the value chosen at random by $D$ in its first step. Similarly, if the given oracle is drawn at random from the family of random functions $R^k$, then the probability that $D$ returns 1 equals the probability that the experiment **Hybrid**$(A, j)$ returns 1, where $j$ is the value chosen at random by $D$ in its first step. Hence,

$$\Pr\left[ D^O \mid O \xleftarrow{R} F \right] = \tfrac{1}{n}\textstyle\sum_{j=1}^{n} P_{j-1}$$

$$\Pr\left[ D^O \mid O \xleftarrow{R} R^k \right] = \tfrac{1}{n}\textstyle\sum_{j=1}^{n} P_{j} .$$

Subtract the second sum from the first and exploit the collapse to get

$$\frac{P_0 - P_n}{n} = \tfrac{1}{n}\textstyle\sum_{j=1}^{n} P_{j-1} - \tfrac{1}{n}\textstyle\sum_{j=1}^{n} P_j = \mathsf{Adv}^{\mathrm{prf}}_F(D) .$$

Note that $D$ runs in time at most $t + O(\log n)$ and makes exactly 2 queries to its oracle, whence we get Equation (5). This concludes the proof of the theorem.

# Proofs of Security for the Unix Password Hashing Algorithm

David Wagner and Ian Goldberg

University of California, Berkeley
{daw,iang}@cs.berkeley.edu

**Abstract.** We give the first proof of security for the full Unix password hashing algorithm (rather than of a simplified variant). Our results show that it is very good at extracting almost all of the available strength from the underlying cryptographic primitive and provide good reason for confidence in the Unix construction.

## 1   Introduction

This paper examines the security of the Unix password hashing algorithm, the core of the Unix password authentication protocol [14]. Although the algorithm has been conjectured cryptographically secure, after two decades and deployment in millions of systems worldwide it still has not been proven to resist attack. In this paper, we provide the first practical proof of security (under some reasonable cryptographic assumptions) for the Unix algorithm.

The hashing algorithm is a fairly simple application of DES, perhaps the best-known block cipher available to the public. Since DES has seen many man-years of analysis, in an ideal world we might hope for a proof (via some reduction) that the Unix password hash is secure if DES is. However, so far no such proof has appeared in the literature.

In earlier work, Luby and Rackoff presented strong theoretical evidence that the basic approach found in the Unix algorithm is likely to be sound, by presenting proofs of security for a simplified variant of the Unix hash [12,13]. However, their proofs have three serious limitations: the abstract model they analyze omits some important features of the real algorithm (they analyze the variant $k \mapsto E_k(0)$ rather than the full iterated construction $k \mapsto E_k^{25}(0)$); their proofs of security are asymptotic, and so do not directly apply to real (necessarily finite) instantiations of the construction; and they assume a uniform distribution on passwords. Therefore, we feel that, from a practical point of view, the security of the real Unix password hash remains an open question.

In the first half of this paper, we take a further step towards justifying the design of the Unix password hash by removing the first two limitations mentioned above (we also make some progress towards removing the final limitation in the second half of this paper, as will be discussed below). Our primary contribution is that we show how to analyze the *full* Unix construction, removing the need to abstract away features of the algorithm. This removes the gap between what

T. Okamoto (Ed.): ASIACRYPT 2000, LNCS 1976, pp. 560–572, 2000.
© Springer-Verlag Berlin Heidelberg 2000

has been analyzed and what is currently in use. In particular, we extend Luby and Rackoff's proof techniques to handle the iterated construction $k \mapsto (E_k \circ E_k \circ \cdots \circ E_k)(0)$ found in the real Unix hash.

We also provide explicit, quantitative security measures for the Unix hash (instead of asymptotic estimates), and as a result, our proofs can be directly applied to the real (finite) Unix algorithm. We make no claims about the novelty of this calculation—it is a straightforward technical exercise—but the concrete bounds are important if we are to assess the practical security of the Unix hash construction in real life.

In practice the Unix password hash suffers from the same limitation as DES: both algorithms appear to be very well designed, but their short key size limits the attainable security level. Nonetheless, we show here that the construction used in the Unix password hash is cryptographically sound and does a very good job of extracting almost all of the available strength from the underlying cryptographic primitive.

We also show a result that may be of independent interest: every pseudo-random generator forms a one-way function, and this construction is simultaneously very efficient and strongly security-preserving. See Theorem 1 for a surprisingly tight reduction (our concrete security parameters are better than those obtained from the apparently-standard analysis of this construction [6, Proposition 3.3.6]). This theorem was effectively the main tool in the Luby-Rackoff proof, but it was never separated out explicitly.

The main practical shortcoming of the proof techniques discussed above is that, for best results, we must assume that the passwords are uniformly distributed. To remedy this shortcoming, we also present some initial progress towards handling the non-uniform case as well.

In general, the security issues associated with non-uniformly distributed keying material appear to be under-represented in the literature. A second contribution of this paper is that we make some initial progress on this problem, presenting a formal model that we hope may serve as a foundation for future exploration in this area. Using this model, we are able to show relatively good lower bounds on the security of the Unix algorithm when used with non-uniformly distributed passwords. These techniques provide practically useful results for the special case of the Unix hash function, but in general the results that can be obtained via these methods are not as strong as we would like, and so we leave this as an open question for further research. See Section 7.

This paper is structured as follows. We recall the definition of the Unix password hash in Section 2 and then summarize the results of our analysis in Section 3. The remainder of the paper is dedicated to the theoretical analysis: Section 4 outlines the main ideas from a high level, Section 5 gives some important definitions, and Section 6 dives into the details of the proofs. Finally, Section 7 gives a formal model for security with non-uniformly distributed passwords and presents some initial results in this area.

## 2     The Unix Algorithm

We briefly recount the definition of the Unix password hashing function. The function—let us call it $H$—is built out of a 25-fold iterated version of DES, in the following way. Let $\mathrm{DES}_k(x)$ denote the DES encryption of plaintext $x$ under key $k$ and $\mathrm{DES}_k^n(x) = \mathrm{DES}_k(\mathrm{DES}_k^{n-1}(x))$ denote the $n$-fold iteration of $\mathrm{DES}_k$. Then the hash $H(k)$ of the 8-character password $k$ may be defined as

$$H(k) = \mathrm{DES}_k^{25}(0).$$

When a new user account is created, the hash $H(k)$ of the user's initial password is stored with the user's id in the (world-readable) system password file /etc/passwd. When the user attempts to log on with password $k'$, the system computes $H(k')$ and compares the result to the value stored in the password file, allowing the user to log on only if $H(k') = H(k)$.

Our description of the Unix password scheme omits one important feature of the construction: the salt. In fact, when the user register his password $k$ for the first time, a random 12-bit salt $s$ is generated, and the system computes a salted hash $H_s(k)$ from $k$ and $s$. We do not analyze the effect of the salt in this paper.

## 3     Results

The main consequence of our analysis is the following informal result:

> If DES is a $(t, 25, e)$-secure block cipher, then the Unix password hashing function is a $(t', p)$-secure password hashing function, where $t' \approx t$ and $p \approx (1 + 1/255)e$.

Some interpretation of this analytical result is clearly in order. Formal definitions of $(t, q, e)$-security for block ciphers and $(t', p)$-security for hash functions will be provided later in Section 5, but for now we just sketch the intuition. Roughly speaking, the theorem says that if DES is secure against all attacks using at most 25 chosen plaintexts, and if the password is chosen uniformly at random, then the Unix construction is secure against password guessing attacks.

Note that our security proofs require only very mild assumptions on the properties of DES. To break the Unix algorithm, the adversary must have some way to break DES with only 25 chosen plaintexts, which is likely to be a very difficult task. Furthermore, even the existence of such an attack on DES is no guarantee of success at breaking the Unix hash function, since it seems to be very difficult to control the internal values of the hash computation. Therefore, we expect that the Unix hash function is likely to be even stronger than our lower bounds would suggest.

AN EXAMPLE. Let us try to estimate the resources needed to reverse the Unix hash function. We start by estimating the concrete security level afforded by DES. The best attack reported in the literature for breaking DES with 25 chosen plaintexts is exhaustive keysearch; differential and linear cryptanalysis do not help with such a small number of chosen texts. If we mount a partial exhaustive keysearch, searching over $t$ keys, we obtain an attack with time complexity $t$ and success probability[1] $e \leq t/2^{55}$. Therefore, if the cryptanalytic results reported in the literature are representative and this is indeed the best available attack, we may conclude that DES forms a $(t, 25, t/2^{55})$-secure block cipher. Theorem 1 then says that the Unix scheme is $(t, p)$-secure for $p \approx (1 + 1/255)t/2^{55}$, which is only larger than the corresponding success probability for attacking DES by the small multiplicative factor of $1 + 1/255$. To summarize:

> For an adversary with a given set of resources, the chances of breaking the Unix password hash are at most only slightly higher—less than 1% higher—than the chances of breaking DES with the same resources.

This illustrates that the reduction is nearly tight: our analysis requires only very weak assumptions of security for DES, and as a result, our results will still be relevant even if DES is found to have some small weakness. In other words, the Unix construction is robust: any small imperfections that might exist in DES are guaranteed not to be magnified by the Unix construction into a fatal flaw for a hashing function.

LIMITATIONS. There are several important technical limitations to our work. First, we do not analyze the salt, so we do not consider attacks on many passwords in parallel. Second, although we are for the first time able to show that the iteration in the Unix hash does not harm security, we were not able to prove that iteration actually improves security, as one would intuitively expect. Third, our results for the non-uniform distribution are not as strong as we would like, as is discussed in more detail elsewhere in this paper.

In practice probably the most significant vulnerabilities in the Unix password hash function are that real passwords often do not contain enough entropy to resist dictionary attacks [3,5,9,11,15], that the 56-bit keysize of DES is too short to resist exhaustive keysearch attacks [4], and that cleartext passwords are inappropriate for use in a networked environment. However, our results show that the Unix password hashing construction attains about as much cryptographic strength as possible, given these unavoidable limits on its security.

## 4   An Outline of the Analysis

Our analysis of the Unix hash uses essentially only one new idea[2]: an observation about close ties between the Unix hash and the CBC-MAC construction. In the remainder of this section, we give a high-level sketch of these two fundamental observations.

---

[1] We assume the adversary exploits the DES complementation property, and thus $e = t/2^{55}$, not $t/2^{56}$ as one might naively expect.

[2] We also give what we believe to be a simpler presentation of Luby and Rackoff's proof that $k \mapsto E_k(0)$ is a good one-way function if $E$ is a good block cipher, but the result itself is not new.

RELEVANCE OF THE CBC-MAC. First, we show that the Unix password hashing algorithm is just a special case of the more general and better-studied DES-CBC-MAC construction [1]. Consequently, we can take advantage of well-known results on the security of DES-CBC-MAC.

Let $f$-CBC-MAC$(x)$ denote the CBC-MAC of the message $x$ under the function $f$. Recall that the $f$-CBC-MAC of a $n$-block message $x$ under function $f$ is defined as

$$f\text{-CBC-MAC}(x) = f(x_n \oplus \cdots f(x_2 \oplus f(x_1)) \cdots).$$

Then it is not hard to see that we get a close relation between $n$-fold iterated encryption and the CBC-MAC on a $n$-block message:

$$f^n(x) = f\text{-CBC-MAC}(\langle x, 0, 0, \dots, 0 \rangle).$$

This observation may be of independent interest, because it gives a simple and powerful way to analyze iterated encryption.

Using this trick, we observe that the Unix password hashing algorithm can be related to the DES-CBC-MAC by

$$\text{Unix-hash}(k) = \text{DES-CBC-MAC}_k(\langle 0, \dots, 0 \rangle).$$

This is the basis for our treatment of iteration in the Unix password hash.

## 5 Definitions

Definitions of concrete security are parametrized by a measure of the resources needed to break the cryptographic primitive. In general, we say that an attack $R$-breaks a crypto primitive if the algorithm succeeds in breaking the primitive with resources specified by $R$, and we say that a crypto primitive is $R$-secure if there is no algorithm[3] to $R$-break it. In the definitions to follow, we elaborate on the measure of an adversary's resources.

First, we formally define the concept of a pseudorandom function (PRF). Let $F : \mathcal{K} \times \mathcal{X} \to \mathcal{Y}$ be a keyed function. We say that the oracle algorithm $A$ is an adversary which $(t, q, e)$-breaks the alleged-PRF $F$ if $A$ runs in time $t$, makes at most $q$ queries to its oracle, and has advantage $\text{Adv}\, A = e$. The adversary's advantage $\text{Adv}\, A$ is defined to be

$$\text{Adv}\, A = |\Pr[A^{F_k} = 1] - \Pr[A^R = 1]|,$$

where the probability is taken over the choice of $k$ and $R$, and where the random variable $k$ is drawn from the uniform distribution over $\mathcal{K}$ and $R : \mathcal{X} \to \mathcal{Y}$ is a random function. We say that $F$ is a $(t, q, e)$-secure PRF if there is no adversary which $(t, q, e)$-breaks $F$.

---

[3] We may assume without loss of generality that all adversarial algorithms behave deterministically, since any probabilistic adversary can be de-randomized using standard techniques.

A $(t, q, e)$-secure "super" pseudorandom permutation (PRP) $E : \mathcal{K} \times \mathcal{Y} \to \mathcal{Y}$ is a family of permutations with the property that $E_k$ is indistinguishable from a random permutation $\pi : \mathcal{Y} \to \mathcal{Y}$ chosen uniformly at random from the set of all permutations on $\mathcal{Y}$, when $k$ is drawn uniformly at random from $\mathcal{K}$. The advantage of an adversary $A$ is defined as $\text{Adv } A = \left| \Pr[A^{E_k, E_k^{-1}} = 1] - \Pr[A^{\pi, \pi^{-1}} = 1] \right|$. Note that we typically omit the "super" prefix for brevity.

A pseudorandom generator (PRG) is a function $g : \mathcal{K} \to \mathcal{Y}$ which stretches a short seed (from $\mathcal{K}$) into a long, random-looking output. The advantage of an adversary for $g$ is defined to be $\text{Adv } A = |\Pr[A(g(k)) = 1] - \Pr[A(u_Y) = 1]|$, where the random variables $k$ and $u_Y$ are chosenly randomly according to the uniform distributions on $\mathcal{K}$ and $\mathcal{Y}$ (resp.). In the case of pseudorandom generators, normally one insists that the output be longer than the seed, i.e., that $|\mathcal{Y}| > |\mathcal{K}|$.

Also, it is useful to have the concept of a one-way function (OWF). Let $h : \mathcal{K} \to \mathcal{Y}$ be an unkeyed function. An adversary $B$ attacking $h$ is an algorithm with input $y \in \mathcal{Y}$ which outputs a symbol in $\mathcal{K} \cup \{\bot\}$, and which is *correct*; $B$ is correct when $y \in \mathcal{Y}, B(y) \neq \bot$ implies $h(B(y)) = y$. We say that an adversary $B$ $(t, p)$-breaks the alleged-OWF $h$ if $B$ runs in time $t$ and succeeds with probability $p = \Pr[B(h(k)) \neq \bot]$, where the probability is taken over the choice of $k \in \mathcal{K}$, and the random variable $k$ is drawn from the uniform distribution. Finally, we say that $g$ is a $(t, p)$-secure OWF if there is no adversary which $(t, p)$-breaks it.

Note that the notion of a one-way function exactly captures the security properties we need from a password hashing function. In particular, if $g$ is a $(t, p)$-secure OWF, then the success probability of any adversary running in time $t$ is at most $p$.

## 6   Analysis

The main result is a proof that any pseudorandom generator is a good one-way function. This is an version of Luby and Rackoff's result [12,13], adapted to the concrete security model.

**Theorem 1.** *Let* $g : \mathcal{K} \to \mathcal{Y}$ *be a* $(t, e)$-*secure pseudorandom generator, with* $|\mathcal{Y}| > |\mathcal{K}|$. *Then* $g$ *is a* $(t', p)$-*secure one-way function, where* $p = e/(1 - |\mathcal{K}|/|\mathcal{Y}|)$ *and* $t' \approx t$.

*Remark 1.* To be more precise, we show that $g$ is $(t', p)$-secure, where $t' = t - c$ and $c$ is a universal constant which depends only on the machine model. However, in practice $c$ is extremely small compared to $t$, so for simplicity of exposition in this paper we omit these tiny constants and summarize the situation by writing $t' \approx t$.

*Proof.* We prove the contrapositive. Let $h = g$ be our alleged one-way function. Suppose that there is an adversary $B$ which $(t, p)$-breaks $h$ (viewed as a one-way function). We construct an adversary $A$ against $g$ (viewed as a PRG), defined by

$$A(y) = \begin{cases} 1 & \text{if } B(y) \neq \perp \\ 0 & \text{otherwise.} \end{cases}$$

Our claim is that $A$ $(t,e)$-breaks $g$ (the pseudorandom generator), i.e., that Adv $A \geq (1 - |\mathcal{K}|/|\mathcal{Y}|) \cdot p$. A bit of notation: we let $k$ stand for a random variable uniformly distributed over $\mathcal{K}$, and $u_Y$ for a r.v. that is uniform over $\mathcal{Y}$. All probabilities are calculated with respect to $k$.

Let $V = \{y \in \mathcal{Y} : B(y) \neq \perp\}$ be the set of outputs of $h$ where $B$ succeeds. Also, let $W = \{k \in \mathcal{K} : h(k) \in V\}$ be the set of inputs to $h$ which are not secure against $B$. We see that $p = \Pr[h(k) \in V] = \Pr[k \in W]$.

Next, we observe that $|V| \leq |W|$. The argument goes like this. We may view $B$ as a deterministic function (by standard de-randomization results). We examine $B'$, the restriction of $B$ to the domain $V$. This restriction is well-defined, since when $v \in V$, $B(v)$ is a well-defined element of $K$. Moreover, using the correctness of $B$, we have $g(B(v)) = v \in V$ for all $v \in V$, so that $B(v) \in W$ for all $v \in V$. Thus we may consider $B'$ as a function with signature $V \to W$. Also, if $v, v' \in V$ and $B(v) = B(v')$, we find that $v = g(B(v)) = g(B(v')) = v'$; therefore, $B'$ is one-to-one. In summary, we have exhibited a one-to-one function from $V$ to $W$, which demonstrates that $|V| \leq |W|$.

Finally, we are ready to calculate the advantage of the adversary $A$. First,

$$\Pr[A(g(k)) = 1] = \Pr[B(h(k)) \neq \perp] = \Pr[h(k) \in V] = p.$$

Also $|W| = |\mathcal{K}| \cdot \Pr[k \in W] = |\mathcal{K}| \cdot p$ and $|V| \leq |W|$, so

$$\Pr[A(u_Y) = 1] = \Pr[u_Y \in V] = |V|/|\mathcal{Y}| \leq |W|/|\mathcal{Y}| = |\mathcal{K}|/|\mathcal{Y}| \cdot p.$$

Plugging into the definition of Adv $A$ gives

$$\text{Adv } A \geq |p - |\mathcal{K}|/|\mathcal{Y}| \cdot p| = (1 - |\mathcal{K}|/|\mathcal{Y}|) \cdot p = e.$$

To recap, under the assumption that there is an adversary $B$ which $(t,p)$-breaks $h$, we obtain an adversary $A$ which shows that $g$ is not $(t,e)$-secure, and this is the desired result. □

**Lemma 1.** A $(t,q,e)$-secure PRP $E : \mathcal{K} \times \mathcal{Y} \to \mathcal{Y}$ is a $(t,q,e+q^2/2|\mathcal{Y}|)$-secure PRF.

*Proof.* This lemma is a well-known consequence of the birthday paradox. For a full proof, see, e.g., [1]. □

**Lemma 2.** If $F : \mathcal{K} \times \mathcal{Y} \to \mathcal{Y}$ is a $(t,q,e)$-secure PRF, then $F^n$ is a $(t',q/n,e')$-secure PRF, where $t' = t - q \log_2 |\mathcal{Y}|$ and $e' = e + 1.5q^2/|\mathcal{Y}|$.

*Proof.* Recall that $F_k^n(x) = F_k(\cdots(F_k(x))\cdots)$ is a $F_k$-CBC-MAC on the message $\langle x, 0, 0, \ldots, 0\rangle$, as noted in Section 4. Now invoke [1, Lemma 4.1] to show that the $F_k$-CBC-MAC is a secure PRF. □

**Lemma 3.** *If $F$ is a $(t, 1, e)$-secure PRF, then $g(k) = F_k(0)$ is a $(t, e)$-secure PRG.*

*Proof.* Immediate from the definitions.    □

**Theorem 2.** *If DES is a $(t, 25, e)$-secure pseudorandom permutation, then the Unix construction is a $(t', p)$-secure password hashing function, where $p = (1 + 1/255)e + (1 + 1/255) \cdot 25^2/2^{63} \approx (1 + 1/255)e$ and $t' \approx t$.*

*Proof.* Applying Lemmas 1 and 2, we see that $x \mapsto \mathrm{DES}_k^{25}(x)$ is a $(\tau, 1, \epsilon)$-secure PRF, where $\tau = t - 25 \times 64$ and $\epsilon = e + 2q^2/|\mathcal{Y}|$. Lemma 3 then shows that the Unix algorithm is a $(\tau, \epsilon)$-secure PRG. Finally, Theorem 1 assures us that the Unix construction is a $(t', p)$-secure one-way function, where $t' = \tau - c = t - 25 \times 64 - c \approx t$ and $p = \epsilon/(1 - 2^{-8}) = (1 + 1/255)\epsilon$. As discussed above, this is exactly the notion needed to show that the Unix password hashing algorithm is secure when used with uniformly-distributed passwords.    □

## 7    Non-uniformly Distributed Secrets

So far our proofs of security have assumed that all passwords are uniformly distributed. In practice, though, such an assumption is often far from the mark [3,5,9,11,15]. This section tackles the issue of security for non-uniform distributions.

In this section, we introduce a new security model, *passphrase-based cryptography*, where keying material and other cryptographic secrets are derived from human-entered passphrases and thus are likely to have a highly non-uniform distribution. This is a significant departure from the standard model, where the very definitions of security assume a uniform distribution on the keys. A second important difference is that passphrases are typically relatively short, so the secret entropy in them is a scarce resource which we must not waste. We show that the standard approaches to smoothing non-uniform distributions are unsuitable for practical use because they waste too much entropy. Therefore, new techniques are needed.

Let us start by developing formal definitions of security for passphrase-based cryptography. We need a small amount of background. Let $D$ be a distribution on $\mathcal{K}$ which assigns the probability $D(k)$ to each $k \in \mathcal{K}$, and let $D(S) = \sum_{k \in S} D(k)$ for all $S \subseteq \mathcal{K}$.

We define the notion of a *one-way function secure for $D$* as in Section 5, except that the success probability is now calculated when inputs are distributed according to $D$ rather than the uniform distribution. (We assume that the distribution is fixed in advance, so that the attack algorithm may depend on $D$.) Let $f : \mathcal{K} \to \mathcal{Y}$ be an unkeyed function and let $B : \mathcal{Y} \to \mathcal{K} \cup \{\bot\}$ be an adversary against $f$ that is *correct* (i.e., $B(y) \neq \bot$ implies $f(B(y)) = y$). We say that the algorithm $B$ $(t, p)$-*breaks* $f$ (for $D$) if $B$ runs in time $t$ and succeeds with probability $p = \Pr[B(f(k)) \neq \bot]$, where $k$ is chosen from $\mathcal{K}$ according to the distribution $D$ and the probability is taken with respect to the choice of $k$. Finally, the one-way function $f$ is $(t, p)$-secure for $D$ if there is no adversary which $(t, p)$-breaks $f$ for $D$.

In this paper, we define

$$\chi_D(t) = \max\{D(S) : S \subseteq \mathcal{K}, |S| \leq t\}.$$

This definition is motivated by the following upper bound on the security of hashing inputs with distribution $D$:

**Theorem 3.** *For all one-way hash functions $f$, all distributions $D$, and all time bounds $t$, there is a generic attack, called the* dictionary attack, *which $(t, \chi_D(t))$-breaks $f$.*

*Proof.* The dictionary attack proceeds by trying the $t$ elements of $D$ with the $t$ largest probabilities. (Each guess can be easily checked with a single computation of $f$.) If we write the $D$-probabilities in decreasing order, $d_1 \geq d_2 \geq d_3 \geq \cdots$, we can see that the success probability of the dictionary attack is $d_1 + d_2 + \cdots + d_t$; furthermore, this quantity is precisely $\chi_D(t)$.  □

Therefore, $\chi_D$ describes the effectiveness of the optimal dictionary-search attack against $D$: no matter what we do, *every* one-way hash function with inputs chosen according to $D$ can be broken with probability $\chi_D(t)$ and time $t$.

There is no way to avoid the dictionary attack. This motivates our definition of security for a one-way hash function that operates on inputs with a non-uniform distribution:

**Definition 1.** *We say that the one-way function $f$ is* ideally-secure *for distribution $D$ if $f$ is $(t, \chi_D(t))$-secure (for all $t$) when its inputs are distributed according to $D$.*

Intuitively speaking, a one-way function is ideally-secure if the dictionary attack is the best attack.

We are able to show that any one-way function that is sufficiently strong for uniformly distributed inputs will also be relatively good for other distributions.

**Theorem 4.** *Let $f$ be a one-way function that is $(t, p)$-secure for uniformly-distributed inputs. Then, for every distribution $D$ on $\mathcal{K}$, $f$ is a $(t, \chi_D(p|\mathcal{K}|))$-secure one-way hash function for $D$.*

*Proof.* Let $A$ be an adversary which $(t, p')$-breaks $f$ for $D$, where $p' \geq \chi_D(p|\mathcal{K}|)$. We will show that $A$ also $(t, p)$-breaks $f$ (for uniformly distributed inputs), and then taking the contrapositive will yield the desired result.

Let $S = \{k \in \mathcal{K} : A(f(k)) \neq \perp\}$ be the set of $f$-inputs which are not safe against $A$. Note that $p' = D(S)$, and moreover that $\chi_D(|S|) \geq D(S)$ (by the definition of $\chi_D$), so we have

$$\chi_D(|S|) \geq D(S) = p' \geq \chi_D(p|\mathcal{K}|).$$

Since $\chi_D(t)$ is a monotonically increasing function of $t$, we may conclude that $|S| \geq p|\mathcal{K}|$.

Now we may prove that $A$ indeed works well, not just for the distribution $D$, but also for the uniform distribution. Note that

$$\Pr[A(f(k)) \neq \perp] = \Pr[k \in S] = |S|/|\mathcal{K}| \geq p$$

when $k$ is drawn from the uniform distribution on $\mathcal{K}$. Therefore, $A$ is an adversary that $(t, p)$-breaks $f$ (for the uniform distribution), as claimed, and the theorem follows.                                                                □

**Corollary 1.** *If the one-way function $f$ is ideally-secure for the uniform distribution, then it is also ideally-secure for all other distributions as well.*

*Proof.* For the uniform distribution $U$ on $\mathcal{K}$, we have $\chi_U(t) = t/|\mathcal{K}|$, so by assumption $f$ is $(t, t/|\mathcal{K}|)$-secure for all $t$. Now Theorem 4 assures us that $f$ is $(t, \chi_D(t))$-secure for all distributions $D$, since $\chi_D((t/|\mathcal{K}|) \cdot |\mathcal{K}|) = \chi_D(t)$.                □

APPLICATIONS TO UNIX PASSWORD HASHING. We can show that the Unix hash is good at hashing even non-uniformly distributed passwords, under assumptions on DES that appear to be reasonable (albeit stronger than one might like).

In Section 3, we argued that DES appears to be $(t, 25, t/2^{55})$-secure, if the cryptanalysis results reported in the literature do indeed represent the best attacks on DES (as many researchers believe). This assumption implies that the Unix hash $H$ is a $(t', t/2^{56})$-secure one-way function when its inputs are uniformly distributed, where $t' = (1 - 2^{-8})(t/2 - 25^2/2^8) \approx (1 + 2^{-8})t/2$. Thus, Theorem 4 allows us to conclude that the Unix hash is $(t', \chi_D(t))$-secure—i.e., nearly $(t/2, \chi_D(t))$-secure—for every distribution $D$.

This lower bound only differs from Theorem 3's upper bound by a factor of about two[4]. Roughly speaking, this means that the Unix hash appears to be nearly ideally-secure for all distributions $D$: no shortcut attack can do much better than the dictionary attack.

Whether this result is useful in practice will depend on several factors. One disadvantage is that the approach requires relatively strong assumptions about DES—that there are no shortcut attacks on DES that reduce the workfactor of exhaustive keysearch by more than a small factor when the key is uniformly distributed—and as a result, the result is not as robust as we would like. For example, if small weaknesses are present in DES, our proof techniques cannot rule out the possibility that these weaknesses might be greatly magnified when one uses DES with patterned passwords, even though such a worst-case scenario is considered unlikely by practitioners.

However, it is interesting to point out that we obtain a proof of security for the Unix hash of patterned passwords starting only with the assumption

---

[4] If we consider that the Unix hash internally iterates DES 25 times and thus costs 25 times as much to compute as does a single DES trial encryption, the gap between the upper and lower bounds becomes a factor of about 50, which is still quite small.

that DES is secure for uniformly-distributed keys. In particular, we make no assumptions whatsoever about the behavior of DES when keyed from a non-uniform distribution. Consequently, we can take advantage of the decades of analysis on DES (which has all been premised on the assumption of uniformly-distributed keys) to gain confidence in the security of the Unix algorithm.

APPLICATIONS TO OTHER CRYPTO PRIMITIVES. It is also worth noting that Theorem 4 can also be generalized to many other keyed cryptographic primitives, such as block ciphers, stream ciphers, and PRF's, using the same style of proof.

TIGHTNESS. One can show that our lower bound (given in Theorem 4) on the security of $f$ for non-uniform distributions is essentially tight. In other words, it is unlikely that one can do much better without either making additional assumptions on $f$ or finding a better construction.

The following simple example is due to David Zuckerman [16]. Let $g : \mathcal{K}_1 \to \mathcal{Y}_1$ be an ideally-secure one-way function with keyspace $\mathcal{K}_1$ and output space $\mathcal{Y}_1$. We construct $f : \mathcal{K} \to \mathcal{Y}$ as $f(\langle x,y \rangle) = \langle g(x),y \rangle$, where $\mathcal{K} = \mathcal{K}_1 \times \mathcal{K}_2$ and $\mathcal{Y} = \mathcal{Y}_1 \times \mathcal{K}_2$. Note that $f$ is $(t, t/|\mathcal{K}_1|)$-secure (for all $t$) for the uniform distribution on $\mathcal{K}$.

Next consider the uniform distribution $D$ on $S \times \mathcal{K}_2$ for some $S \subseteq \mathcal{K}_1$, i.e., $D(\langle x,y \rangle) = 1/(|S| \cdot |\mathcal{K}_2|)$ for all $\langle x,y \rangle \in S \times \mathcal{K}_2$ and $D(\langle x,y \rangle) = 0$ for $x \notin S$. Theorem 4 implies $f$ is $(t, p)$-secure for $D$, where

$$p = \chi_D \left( \frac{t}{|\mathcal{K}_1|} |\mathcal{K}| \right) = \chi_D(t|\mathcal{K}_2|) = \frac{t \cdot |\mathcal{K}_2|}{|S| \cdot |\mathcal{K}_2|} = t/|S|.$$

At the same time, one may clearly $(t, t/|S|)$-break $f$ using a dictionary attack when its inputs are distributed according to $D$ (see Theorem 3). This shows that Theorem 4 is tight.

THE POWER OF STRONGER ASSUMPTIONS. One alternative approach is to start from the assumption that DES is secure (up to the possibility of dictionary attacks) no matter what distribution the key is drawn from. Then we may attempt to prove that the Unix hash is secure for passwords with distribution $D$ if DES is secure for keys with distribution $D$.

The following theorem, which forms a nice example of this approach, is due to Bellare (and was stated as a homework problem in [7]):

**Theorem 5.** *If $g : \mathcal{K} \to \mathcal{Y}$ is a $(t, e)$-secure pseudorandom generator for seeds distributed according to $D$, then $g$ is a $(t, p)$-secure one-way function for $D$, where $p = e + |\mathcal{K}|/|\mathcal{Y}|$.*

*Remark 2.* Of course, we may take $D$ to be the uniform distribution in the above; however, this gives strictly weaker bounds than Theorem 1's dedicated analysis.

*Proof.* Use the same notation as in the proof of Theorem 1, and define the adversary $A$ in the same way. Note that $\Pr[A(u_Y) = 1] \leq |\{g(k) : k \in \mathcal{K}\}|/|\mathcal{Y}| \leq |\mathcal{K}|/|\mathcal{Y}|$, and $\Pr[A(g(k)) = 1] = p$ as before, so we get

$$\text{Adv } A = |\Pr[A(g(k)) = 1] - \Pr[A(u_Y) = 1]| \geq |p - |\mathcal{K}|/|\mathcal{Y}|| = e.$$

In other words, if there is an adversary $B$ to $(t, p)$-break $g$ as a one-way function, then there is another adversary $A$ to $(t, e)$-break $g$ as a pseudorandom generator, and the theorem follows. □

While this result may be useful in some contexts, it doesn't give terribly useful lower bounds for the security of the Unix hash. For the Unix algorithm, we have $|\mathcal{K}|/|\mathcal{Y}| = 2^{-8}$, so we won't be able to rule out the possibility that there exists an algorithm that succeeds in breaking $1/256$ of all passwords in constant time. Such a result is not very reassuring.

One could attempt to repair the flaw by defining a new hash construction, e.g., New-hash$(k) = \langle \text{DES}_k(0 \ldots 00), \text{DES}_k(0 \ldots 01) \rangle$. Such an approach would work—if one is willing to deploy an updated implementation of the password hashing algorithm on millions of machines around the world!—but it would still require strong assumptions about the security of DES when used with non-uniformly distributed keys. Since DES has not received as much scrutiny in this setting (where the key is non-uniformly distributed), it becomes harder to gain much confidence that the necessary assumptions are indeed satisfied.

Therefore, we conclude that this approach does not seem to yield security bounds that are as meaningful as those that can be achieved with Theorem 4.

COMPARISON TO ENTROPY SMOOTHING. Another alternative approach to dealing with patterned passwords is to smooth out the non-uniformity in the distribution. A well-known result called the *leftover hash lemma* [8,10] shows that universal hash functions are good at entropy smoothing: if $h$ is selected uniformly at random from a family of universal hash functions with $m$-bit outputs, and if $k$ is drawn from a distribution with at least $3m$ bits of Renyi entropy, the random variable $\langle h, h(k) \rangle$ will be approximately uniformly distributed.

The disadvantage with the leftover hash lemma is that it wastes at least two-thirds of the entropy of the password $k$: if we want to feed the smoothed bits into the Unix hash function (e.g., New-hash$(k) = \langle h, \text{Unix-hash}(h(k)) \rangle$), we need a passphrase with at least $3 \times 56 = 168$ bits of entropy. This would require that passphrases consist of hundreds of characters, which is too difficult for most mere mortals to memorize. When we consider that, in the real world, one is lucky to find a password with more than 25–35 bits of entropy [3,5,9,11,15], it becomes clear that the leftover hash lemma is thoroughly unsuitable for practical use.

The problem is that universal hash functions (and their generalizations, e.g., extractors) are designed for use in de-randomization, where the scarce resource is uniformly-distributed randomness, and where non-uniformly distributed bits are very cheap. In contrast, for passphrase-based cryptography, *secret randomness* (e.g., passwords, passphrases, etc.) should be considered a very precious resource that must be conserved at all costs, whereas *public randomness* (even uniformly-distributed public randomness) is nearly free. This suggests that new approaches may be required, and we leave this as an interesting challenge for further study.

## 8   Acknowledgements

We would like to gratefully acknowledge the contributions of Mihir Bellare and Phil Rogaway (for their helpful observations and careful commentary), Dan Boneh (who pointed out the relevance of the leftover hash lemma), David Zuckerman (who contributed the simplified example used in showing the tightness of Theorem 4), Umesh Vazirani (for helpful discussions on extractors and entropy smoothing), and the anonymous referees for CRYPTO 2000 and ASIACRYPT 2000.

## References

1. M. Bellare, J. Kilian, P. Rogaway, "The Security of the Cipher Block Chaining Message Authentication Code," *CRYPTO '94*, Springer-Verlag, 1994.
2. M. Bellare, R. Canetti, H. Krawczyk, "Keying hash functions for message authentication,", *CRYPTO '96*, Springer-Verlag, 1996.
3. M. Bishop, D.V. Klein, "Improving system security via proactive password checking," *Computers & Security*, vol.14, (no.3), 1995, pp.233–249.
4. Electronic Frontier Foundation, *Cracking DES: Secrets of Encryption Research, Wiretap Politics, and Chip Design*, O'Reilly, 1998.
5. D.C. Feldmeier and P.R. Karn, "UNIX password security—ten years later." *CRYPTO '89*, Springer-Verlag, 1990, pp.44–63.
6. O. Goldreich, "Foundations of Cryptography (Fragments of a Book)," Chapter 3.
7. S. Goldwasser, M. Bellare, "Lecture Notes on Cryptography," available online from `http://www-cse.ucsd.edu/users/mihir/papers/gb.html`.
8. J. Hastad, R. Impagliazzo, L.A. Levin, M. Luby, "A pseudorandom generator from any one-way function," *SIAM Journal on Computing*, vol.28 no.4, 1999.
9. J. Hietaniemi, "ipasswd—proactive password security," *Proc. 6th Systems Administration Conf. (LISA VI)*, USENIX Association, 1992, pp.105–114.
10. R. Impagliazzo and D. Zuckerman, "How to Recycle Random Bits," *FOCS '89*, IEEE Press, 1989.
11. D.V. Klein, "Foiling the cracker: a survey of, and improvements to, password security," *USENIX Workshop Proceedings: UNIX Security II*, USENIX Assoc., 1990.
12. M. Luby, C. Rackoff, "A study of password security," *CRYPTO '87*, Springer-Verlag, 1988.
13. M. Luby, C. Rackoff, "A study of password security," *J. Cryptology*, vol. 1 no. 3, 1989.
14. R. Morris, K. Thompson, "Password security: a case history," *Communications of the ACM*, vol. 22, no. 11, Nov. 1979.
15. T. Wu, "A real-world analysis of Kerberos password security," *Proc. 1999 Network and Distributed System Security Symp.*, Internet Soc., 1999, pp.13–22.
16. D. Zuckerman, personal communication, July 1999.

# Trapdooring Discrete Logarithms on Elliptic Curves over Rings

Pascal Paillier

Cryptography and Security Group, Gemplus Card International
34 rue Guynemer, F-92447 Issy-Les-Moulineaux
pascal.paillier@gemplus.com

**Abstract.** This paper introduces three new probabilistic encryption schemes using elliptic curves over rings. The cryptosystems are based on three specific trapdoor mechanisms allowing the recipient to recover discrete logarithms on different types of curves. The first scheme is an embodiment of Naccache and Stern's cryptosystem and realizes a discrete log encryption as originally wanted in [23] by Vanstone and Zuccherato. Our second scheme provides an elliptic curve version of Okamoto and Uchiyama's probabilistic encryption, thus answering a question left open in [10] by the same authors. Finally, we introduce a Paillier-like encryption scheme based on the use of twists of anomalous curves. Our contributions provide probabilistic, homomorphic and semantically secure cryptosystems that concretize all previous research works on discrete log encryption in the elliptic curve setting.

**Keywords.** Elliptic Curve Cryptosystems, Discrete Logarithm Encryption, Homomorphic Encryption, Naccache-Stern, Okamoto-Uchiyama, Paillier.

## 1 Introduction

At the present time, one of the most challenging open problems in cryptography is certainly the realization of a trapdoor in the discrete logarithm problem. A discrete-log (DL) encryption scheme over a group G intends to encrypt a plaintext $m$ by simply raising some base element $g \in$ G to the power $m$, while decryption recovers $m$ up to a public bound[1]. Motivations for this may be diverse. The main advantage in comparison to other public-key techniques such as RSA or El-Gamal comes from the *additive* homomorphic property of ciphertexts (the group product of encryptions of $m_1$ and $m_2$ yields an encryption of $m_1 + m_2$). This property constitutes the necessary condition for many cryptographic protocols to exist in fields like electronic voting [4], key escrow [13] or group signatures, to quote a few. Clearly, discovering novel discrete-log encryption techniques has a crucial positive impact on these research domains. In contrast, direct applications of these for simple encryption purposes may be of more moderate interest

---

[1] the decryption is only expected to retrieve $m$ modulo the given bound, *i.e.* the trapdoor is partial.

T. Okamoto (Ed.): ASIACRYPT 2000, LNCS 1976, pp. 573–584, 2000.

as malleability destroys chosen-ciphertext security anyway[2]. Without considering all potential applications, this paper focuses on providing and analyzing new discrete log trapdoors and comparing their properties with the ones recently discovered in [8,9,11].

High degree residuosity was introduced by Benaloh [1] as an algebraic framework extending the properties of quadratic residuosity to prime degrees greater than two. Since then, successive works have considerably improved the efficiency of residuosity-based encryption. Naccache and Stern [8], utilizing a smooth degree modulo $n = pq$, increased Benaloh's encryption rate up to $\approx 1/5$. More recently, Okamoto and Uchiyama [9] and Paillier [11] came up with modulus-independent encryption rates of $1/3$ and $1/2$ respectively, basing trapdoorness on a joint use of Fermat quotients and clever parameter choices. Interestingly, these three cryptosystems only stand in the multiplicative groups $\mathbb{Z}_n^*$ where $n = pq$, $p^2 q$ or $p^2 q^2$ and $p$, $q$ are large prime numbers.

There have been several attempts, in the meantime, to realize discrete-log encryption over elliptic curves instead of standard groups. This was motivated by the fact that no subexponential time algorithm for extracting discrete logarithms is known so far, at least for most elliptic curves[3]. As a matter of fact, all such design proposals have revealed themselves unsuccessful. Vanstone and Zuccherato [23] proposed a deterministic DL encryption scheme that was shown to be insecure a few months later by McKee and Pinch [6] and Coppersmith [2]. Independently, Okamoto and Uchiyama failed in attempting to design DL encryption over composite anomalous curves [10].

This paper introduces cryptosystems successfully answering the quests of [23] and [10] respectively, with guaranteed semantic security relatively to well identified computational problems. The first scheme is an embodiment of Naccache and Stern's cryptosystem on curves defined over $\mathbb{Z}_n$ ($n = pq$) which realizes a discrete-log encryption as originally imagined by Vanstone and Zuccherato. Probabilistic, the scheme is also provably semantically secure relatively to the so-called high-degree residuosity problem. Our second cryptosystem relates to the $p$-residuosity of a well-chosen curve over the ring $\mathbb{Z}_{p^2 q}$, that is, provides an elliptic curve instance of Okamoto and Uchiyama's encryption scheme. Finally, we show how to extend the same design framework to Paillier encryption [11], while preserving all security and efficiency properties inherent to the original cryptosystem. All three schemes are reasonably efficient, simple to understand, additively homomorphic, probabilistic and provably secure against chosen plaintext attacks (IND-CPA) in the standard model. We believe our cryptosystems to be the only ones that verify these properties.

Due to space limitations, we do not recall here the basics of high-degree residuosity (neither do we give the description of the encryption schemes we

---

[2] like for other cryptosystems however, security improvements are possible to reach resistance against active adversaries, see [12].

[3] it is known that there exist subexponential algorithms for curves of trace zero over $\mathbb{F}_p$ for $p$ prime. The discrete-log problem happens to be trivially polynomial in the case of trace one, see [20].

work with), referring the reader to the bibliography for further information when needed.

## 2    Elliptic Curve Naccache-Stern Encryption

The first encryption scheme that we describe here is a variant of Naccache and Stern's encryption scheme [8] where the working group is an elliptic curve over the ring $\mathbb{Z}_n$. The construction of such a curve is similar in spirit to the work of Koyama, Maurer, Okamoto and Vanstone [5] that allowed to export factoring-based cryptosystems like RSA [15] and Rabin [14] on a particular family of curves over the ring $\mathbb{Z}_n$ (KMOV). We now describe briefly their construction.

In the sequel, $p$ and $q$ denote distinct large primes of product $n$. Recall that for any integer $k$, $E_k(a, b)$ is defined as the set of points $(x, y) \in \mathbb{Z}_k \times \mathbb{Z}_k$ such that

$$y^2 = x^3 + ax + b \mod k ,$$

together with a special element $\mathcal{O}_k$ called the point at infinity. It is known that given a composite integer $k$, a curve $E_k(a, b)$ defined over the ring $\mathbb{Z}_k$ has no reason to be a group. This problem, however, does not have real consequences in practice when $k = n$ because exhibiting a litigious addition leads to factor $n$ and this event remains of negligible probability. Furthermore, projections of $E_n(a, b)$ over $\mathbb{F}_p$ and $\mathbb{F}_q$ (namely, $E_p(a, b)$ and $E_q(a, b)$) being finite abelian groups, the Chinese remainder theorem easily conducts to the following statement:

**Lemma 1 (Koyama** *et al.* **[5]).** *Let* $E_n(a, b)$ *be an elliptic curve where* $n = pq$ *is the product of two primes such that* $\gcd(4a^3 + 27b^2, n) = 1$. *Let us define the order of* $E_n(a, b)$ *as*

$$|E_n(a, b)| = \mathrm{lcm}(|E_p(a, b)|, |E_q(a, b)|) .$$

*Then, for any point* $P \in E_n(a, b)$,

$$|E_n(a, b)| \cdot P = \mathcal{O}_n$$

*where* $\mathcal{O}_n$ *denotes the point at infinity of* $E_n(a, b)$.

Although not being a group in a strict sense, the structure of the curve $E_n(a, b)$ complies to Lagrange's theorem and, from this standpoint, can be used as a group. Koyama *et al.* take advantage of this feature by focusing on curves of the following specific form:

$$E_n(0, b) \ : \ y^2 = x^3 + b \mod n \quad \text{for } b \in \mathbb{Z}_n^* ,$$

with $p \equiv q \equiv 2 \pmod 3$. This is motivated by the fact that the projected curves $E_p(0, b)$ and $E_q(0, b)$ happen to be of trace of Frobenius equal to zero. More specifically,

**Lemma 2.** *Let $p$ be an odd prime satisfying $p \equiv 2$ (mod 3). Then, for all $b \in [1, p-1]$, $E_p(0, b)$ is a cyclic group of order*

$$|E_p(0, b)| = p + 1 .$$

Subsequently, the problem of recovering $|E_n(0, b)| = \mathrm{lcm}(p+1, q+1)$ from $n$ is equivalent to factoring $n$ when $p \equiv q \equiv 2$ (mod 3). Note that another possible choice of parameters are curves $E_n(a, 0)$ for $a \in \mathbb{Z}_n^*$ and $p \equiv q \equiv 3$ (mod 4). We refer the reader to [5] for further details.

## 2.1   Our Setting

Just as above, for some $b \in \mathbb{Z}_n^*$, we will be considering the curve $E_n(0, b)$ as a finite abelian group of order

$$\mu = |E_n(0, b)| = \mathrm{lcm}(p+1, q+1) .$$

In our setting, the prime factors $p$ and $q$ are both chosen congruent to 2 modulo 3 so that, by virtue of lemma 2, the two curves $E_p(0, b)$ and $E_q(0, b)$ are cyclic groups of respective orders $p+1$ and $q+1$. We also impose

$$p + 1 = 6 \cdot u \cdot p' \quad \text{where} \quad u = \prod p_i^{\delta_i} \quad \text{and} \tag{1}$$

$$q + 1 = 6 \cdot v \cdot q' \quad \text{where} \quad v = \prod p_j^{\delta_j} , \tag{2}$$

for some $B$-smooth integers $u$ and $v$ of (roughly) equal bitsize such that

$$\gcd(6, u, v, p', q') = 1$$

and $B = O(\log n)$. Integers $p'$ and $q'$ are taken prime. The whole construction is closely related to Naccache and Stern's encryption scheme [8]. In our case, we focus on base points of $E_n(0, b)$ of order a multiple of $\sigma = uv$. If $G$ is such a point, then one could envision to encrypt some plaintext $m \in \mathbb{Z}_\sigma$ by

$$m \longmapsto m \cdot G + \sigma \cdot R \quad \text{where} \quad R \in_R E_n(0, b) , \tag{3}$$

and decrypt by computing the residuosity class with respect to $G$. Because $\sigma$ was chosen to be smooth, computing discrete logarithms for a base of degree $\sigma$ can be efficiently done using the baby-step giant-step algorithm combined with Pohlig and Hellman's method. Thus, one can compute residuosity classes on $E_n(0, b)$ in polynomial time provided that $\mu$ is known, *i.e.* knowing the factors of $n$. There still remains the problem of randomly choosing an element $R \in_R E_n(0, b)$ during encryption: the spontaneous creation of an arbitrary point seems to require either the computation of a quadratic root of $R_x^3 + b$ with $R_x \in_R \mathbb{Z}_n$ (equivalent to the knowledge of the factors), or the computation of $\sqrt[3]{R_y^2 - b}$ with $R_y \in_R \mathbb{Z}_n$

(equivalent to RSA on $\mathbb{Z}_n^*$ with $e = 3$). An elegant solution[4] consists in modifying the encryption function so that $m \in \mathbb{Z}_\sigma$ is now encrypted as

$$m \longmapsto C = (m + \sigma r) \cdot G \quad \text{with} \quad r \in_R \mathbb{Z}_n ,$$

and decryption necessitates to compute the discrete logarithm of $(\mu/\sigma) \cdot C$ with respect to the base $G' = (\mu/\sigma) \cdot G$, which is done as previously discussed since $G'$ is of smooth order $\sigma$. The so-obtained probabilistic encryption scheme is described more precisely hereafter.

Our parameter generation process is very similar to Naccache and Stern's. One chooses two $B$-smooth integers $u$ and $v$ of product $\sigma$ such that $\log \sigma = O(\log^\varepsilon n)$ with $\varepsilon > 0$. For practical use, one sets as in [8] $\lceil \log_2 \sigma \rceil = 160$ and $B \approx 2^{10}$. Prime numbers $p$ and $q$ are then generated according to equations 1 and 2. The choice of $b$ is arbitrary in $\mathbb{Z}_n^*$: we recommend a small constant value such as $b = 1$ which renders point additions easier. The base point $G$ can be chosen of maximal order $\mu = \text{lcm}(p + 1, q + 1)$, computed separately mod $p$ and mod $q$, and recombined at the very end by Chinese remaindering.

**Public key**     $n, b, \sigma, G$.

**Private key**    $(p, q)$ or $\mu = \text{lcm}(p + 1, q + 1)$.

**Encryption**     plaintext $m \in \mathbb{Z}_\sigma$,
                   pick a random $r < n$,
                   ciphertext $C = (m + \sigma r) \cdot G$.

**Decryption**     compute $u = (\mu/\sigma) \cdot C = m \cdot G'$.
                   Use Pohlig-Hellman and baby-step giant-step
                   to compute the discrete log of $u$ in base $G'$.

Decryption can also be performed over $E_p(0, b)$ and $E_q(0, b)$: in this case, one separately computes $m \mod u$ and $m \mod v$. The plaintext $m$ is then recovered modulo $\sigma$ by Chinese remaindering.

## 2.2  Security Analysis

Clearly, inverting the encryption function of our scheme is equivalent to computing residuosity classes on $E_n(0, b)$, and the semantic security is equivalent to the decisional version of the same problem. By analogy with [8], we conjecture that these two problems are actually intractable.

Note also that the scheme can be made deterministic by setting $r = 0$ in the encryption function. We therefore have $C = m \cdot G$ like in Vanstone and Zuccherato's cryptosystem [23]. This variant is of moderate interest as it looses semantic security.

---

[4] alternatively, one can pick random coordinates for $R$ and then select the coefficient $b$ as $b = R_y^2 - R_x^3 \mod n$. During decryption, $b$ is recovered by $b = C_y^2 - C_x^3$. In this event, the scheme relies on a *family* of curves, see [5].

## 2.3   Implementation Aspects

We analyze briefly the performances of our encryption scheme. Note first that since $E_p(0, b)$ and $E_q(0, b)$ are cyclic and $G$ chosen of maximal order, the ciphertext space is $E_n(0, b)$ itself. The expansion rate is therefore $\rho = 2\lceil \log_2 n \rceil / \lceil \log_2 \sigma \rceil$ i.e. twice the one of Naccache and Stern's cryptosystem. This is due to the fact that the ciphertext has two coordinates modulo $n$. For instance, we have $\rho \approx 10$ when $\lceil \log_2 n \rceil = 768$ and $\lceil \log_2 \sigma \rceil = 160$. One way to increase the encryption bandwidth is to transmit only one ciphertext coordinate. Transmitting $C_y$, $C_x$ is recovered before decryption by extracting the cubic root of $C_y^2 - b \mod n$. Transmitting $C_x$, decryption leads to exactly four message solutions: necessarily, 2 redundant bits have then to be included in the plaintext to eradicate any decryption ambiguity. This is similar to Rabin encryption [14].

## 3   Elliptic Curve Okamoto-Uchiyama Encryption

In this section, we show how to extend the setting defined in [9] to the one of elliptic curves. In particular, the technique we suggest addresses an open question described in [10].

It is known that curves $E_p(\bar{a}, \bar{b})$ over $\mathbb{F}_p$ which have trace of Frobenius one (they are said to be anomalous) present the property that computing discrete logarithms on them is *very easy*. To be more precise, such an extraction requires a linear number of field operations over $\mathbb{F}_p$, i.e. $O(\log^3 p)$ bit operations. This was studied by several authors [20,19,22]. Okamoto *et al.* [10] attempted to take advantage of this feature to design an identity-based cryptosystem, but due to $|E_p(\bar{a}, \bar{b})| = p$, we believe that this property can hardly be captured so directly into a properly secure encryption scheme. Instead, we extend the discrete logarithm recoverability property to a $p$-subgroup of $E_{p^2}(a, b)$ so that the projection onto $\mathbb{F}_p$ gives the *twist* of an anomalous curve. This is done as follows. We begin by stating a few useful facts that derive from Hasse's theorem.

**Lemma 3.** *Let $E_p(\bar{a}, \bar{b}) : y^2 = x^3 + \bar{a}x + \bar{b} \mod p$ be an elliptic curve of order $|E_p(\bar{a}, \bar{b})| = p + 1 - t$ where $|t| \leq 2\sqrt{p}$. Then for any integers $a, b$ such that $a = \bar{a} \mod p$ and $b = \bar{b} \mod p$, we have*

$$|E_{p^2}(a, b)| = (p + 1 - t)(p + 1 + t) .$$

The curve $E_{p^2}(a, b)$ is usually said to be a lift of $E_p(\bar{a}, \bar{b})$ to $\mathbb{F}_{p^2}$. One consequence of lemma 3 is that if $E_p(\bar{a}, \bar{b})$ has $p + 2$ points, then any lift $E_{p^2}(a, b)$ must be of order $p(p + 2)$.

**Lemma 4.** *Let $E_p(\bar{a}, \bar{b})$ be an elliptic curve over $\mathbb{F}_p$ of order $p + 2$. Provided that $p \equiv 2 \pmod 3$, any lift $E_{p^2}(a, b)$ of $E_p(\bar{a}, \bar{b})$ to $\mathbb{F}_{p^2}$ is cyclic.*

*Proof.* Let $E_{p^2}(a, b)$ be a non-cyclic lift of $E_p(\bar{a}, \bar{b})$. From Rück's theorem [17], we know that $E_{p^2}(a, b) = \mathbb{Z}_{d_1} \times \mathbb{Z}_{d_2}$ with $d_1 | d_2$, $d_1 > 1$ and $d_1 | p^2 - 1$. By virtue of lemma 3, we must have $d_1 d_2 = p(p + 2)$. Therefore, $d_1$ divides

$$\gcd(p + 2, p^2 - 1) = \gcd(p + 2, p - 1) = \gcd(3, p - 1),$$

which implies $d_1 = 3$ or $1$. Since $d_1 \neq 1$ and $p = 2 + 3\eta$ for some integer $\eta$, we get the contradiction $3|(1 + 3\eta)$. Hence $E_{p^2}(a, b)$ must be cyclic.    □

In what follows, $p$ denotes a large prime verifying $p \equiv 2 \pmod{3}$, $E_p(\bar{a}, \bar{b})$ stands for a curve of order $p + 2$ and $E_{p^2}(a, b)$ is some lift of $E_p(\bar{a}, \bar{b})$ to $\mathbb{F}_{p^2}$. We note

$$E[p] = (p + 2) \cdot E_{p^2}(a, b)$$

the (cyclic) $p$-torsion subgroup formed by the points of order dividing $p$, i.e. points of order $p$ together with the point at infinity $\mathcal{O}_{p^2}$ of $E_{p^2}(a, b)$. We state:

**Theorem 1.** *There exist a polynomial time algorithm that computes discrete logarithms on $E[p]$ with complexity at most $O(\log^3 p)$.*

*Proof.* Since $E[p]$ is the group of $p$-torsion points of $E_{p^2}(a, b)$, we could apply Semaev's algorithm [20] *stricto sensu*. We rather rely on a (simpler) elliptic-log-based approach similar to Smart's [22] as follows. Observe that any point $P$ belongs to $E[p]$ if (and only if) it is a lift of $\mathcal{O}_p \in E_p(\bar{a}, \bar{b})$, wherefrom $E[p]$ is the kernel of the reduction map $P \mapsto P \bmod p$. Hence the $p$-adic elliptic logarithm (see [21, p. 175])

$$\psi_p(x, y) = -\frac{x}{y} \bmod p^2$$

is well-defined and can be applied on any point of $E[p]$. $\psi_p$ being actually a morphism, if $P = m \cdot G$ stands for any arbitrary points $P, G \in E[p]$, we have

$$m = \frac{\psi_p(P)}{\psi_p(G)} \bmod p,$$

provided that $G \neq \mathcal{O}_{p^2}$. The main computational workload stands in the modular divisions which require at most $O(\log^3 p)$ bit operations.    □

Note that other approaches such as Satoh and Araki's [19] or Rück [16], in application to our case, would have led to somehow equivalent computation methods.

## 3.1   Description

This section shows how to realize an analogue of Okamoto and Uchiyama's encryption scheme [9] on elliptic curves, in the sense wanted by the same authors in [10]. We make use of our previous results as follows.

One first chooses two large primes $p$ (with $p \equiv 2 \pmod{3}$) and $q$ of bitsize $k$, and sets $n = p^2 q$. The user then picks integers $\bar{a}_p, \bar{b}_p \in \mathbb{F}_p$ such that $E_p(\bar{a}_p, \bar{b}_p)$ is of order $p+2$, using techniques such as [7]. He then chooses some lift $E_{p^2}(a_p, b_p)$ of $E_p(\overline{a_p}, \overline{b_p})$ to $\mathbb{F}_{p^2}$, as well as a random curve $E_q(\overline{a_q}, \overline{b_q})$ defined over $\mathbb{F}_q$. Using Chinese remaindering, the user combines $E_{p^2}(a_p, b_p)$ and $E_q(\overline{a_q}, \overline{b_q})$ to get the curve $E_n = E_n(a, b)$ where $a, b \in \mathbb{Z}_n$. Finally, the user picks a point $G \in E_n$ of maximal order $\mathrm{lcm}(|E_{p^2}|, |E_q|)$ and sets $H = n \cdot G$. Our cryptosystem is as depicted below.

**Public key**   $n = p^2 q$, $E_n$, $G$ of maximal order, $H = n \cdot G$.

**Private key**   $p$.

**Encryption**   plaintext $m < 2^{k-1}$,
pick a random $r < 2^{2k}$,
ciphertext $C = m \cdot G + r \cdot H$

**Decryption**   compute $m = \dfrac{\psi_p((p+2) \cdot C)}{\psi_p((p+2) \cdot G)} \mod p$ .

Our system is very similar in spirit to Okamoto and Uchiyama's encryption as originally discovered. For this reason, most properties of their scheme still apply to ours: in particular, chosen ciphertext security can be easily shown equivalent to factoring $n = p^2 q$. The proof of this fact is a straightforward adaptation of Okamoto and Uchiyama's, see [9]. Besides, one-wayness and semantic security remain effective, except that they rely on problems related to high ($p$-degreed) residuosity on $E_n$ instead of $\mathbb{Z}_n^*$. The scheme also features additive homomorphic properties for short messages.

## 4   Elliptic Curve Paillier Encryption

In this section, we refine the previous encryption technique to meet more advanced security requirements: we show how to construct an efficient yet natural embodiment of Paillier's cryptosystem [11] on elliptic curves. We first extend the setting of section 3 to curves defined over $\mathbb{Z}_{n^2}$ where $n = pq$. Suppose $E_{p^2}(a_p, b_p)$ (resp. $E_{q^2}(a_q, b_q)$) is some lift of a curve of trace $p+2$ (resp. $q+2$) defined over $\mathbb{F}_p$ (resp. $\mathbb{F}_q$). Considering $E_{n^2}(a, b)$ as the Chinese remaindering of $E_{p^2}(a_p, b_p)$ and $E_{q^2}(a_q, b_q)$ (hence it is defined over the ring $\mathbb{Z}_{n^2}$), it is easily seen that $E_{n^2}(a, b)$ is of order $n\mu$ where

$$\mu = \mu(n) = \mathrm{lcm}(p + 2, q + 2) .$$

We extend theorem 1 up to the present setting as follows. Noting

$$E[n] = \mu \cdot E_{n^2}(a, b) ,$$

we state:

**Corollary 1 (of theorem 1).** *There exist a polynomial time algorithm that computes discrete logarithms on $E[n]$ with complexity $O(\log^3 n)$.*

*Proof.* This is easily proven, either by applying theorem 1 twice on curves $E[p] \simeq E[n] \mod p^2$ and $E[q] \simeq E[n] \mod q^2$ and then by Chinese remaindering local logarithms, or more compactly by defining over $E[n]$ an $n$-adic elliptic logarithm

$$\psi_n(x, y) = -\frac{x}{y} \mod n^2 .$$

Provided that $P = m \cdot G$ for $P, G \in E[n]$ and $G \neq \mathcal{O}_{n^2}$, we retrieve $m$ by computing

$$m = \frac{\psi_n(P)}{\psi_n(G)} \mod n .$$

$\square$

Here is how the cryptosystem is initialized: the user chooses two large primes $p$ and $q$ (with $p \equiv q \equiv 2 \pmod{3}$) and sets $n = pq$. He then picks up integers $\overline{a_p}$, $\overline{b_p} \in \mathbb{F}_p$ and $\overline{a_q}, \overline{b_q} \in \mathbb{F}_q$ such that $E_p(\overline{a_p}, \overline{b_p})$ is of order $p+2$ and $E_q(\overline{a_q}, \overline{b_q})$ is of order $q + 2$. Lifted curves $E_{p^2}(a_p, b_p)$ and $E_{q^2}(a_q, b_q)$ are chosen and combined to get $E_{n^2} = E_{n^2}(a, b)$. Finally, a base point $G \in E_{n^2}$ is chosen of order divisible by $n$, possibly of maximal order $n\mu$.

**Public key**   $n = pq,\ E_{n^2},\ G$.

**Private key**   $\mu = \operatorname{lcm}(p + 2, q + 2)$ or equivalently $(p, q)$.

**Encryption**   plaintext $m \in \mathbb{Z}_n$,
                 pick a random $r < n$,
                 ciphertext $C = (m + nr) \cdot G$

**Decryption**   compute $m = \dfrac{\psi_n(\mu \cdot C)}{\psi_n(\mu \cdot G)} \mod n$ .

Note that, due to lemma 4, the ciphertext space covers the entire curve $E_{n^2}$ i.e. , any point of $E_{n^2}$ is the image of some plaintext. We therefore have a maximal encryption bandwidth. This is obtained thanks to the fact that all curves we work with are cyclic.

## 4.1   Security Analysis

Here again, the very high resemblance of our encryption scheme with [11] implies that most cryptographic features happen to be identical in the two cases. The one-wayness of our scheme is equivalent to the problem of computing residuosity classes over $E_{n^2}$ which, provided that $n$ is hard to factor, we conjecture to be intractable[5]. Similarly, semantic security relates to the indistinguishability of $n$-residues of $E_{n^2}$, i.e. points belonging to $E[\mu] = n \cdot E_{n^2}$, from other points of the curve. We conjecture this problem to be intractable as well.

Our scheme is clearly malleable, and as such, does not resist adaptive chosen-ciphertext attacks. We believe, however, that security enhancement techniques such as [12] could be applied *mutatis mutandis* to meet provable security at the strongest level NM-CCA2.

---

[5] this is similar to the Composite Residuosity Assumption over $\mathbb{Z}_{n^2}^*$, see [11,12].

## 4.2   Implementation Aspects

Slight modifications of our encryption scheme may allow significant cost savings: a typical implementation speed-up is obtained by choosing a base point $G$ of order $n\alpha$ with $\alpha = \alpha_p \alpha_q$, where

$$\alpha_p \,|\, p+2, \quad \alpha_p \nmid q+2, \quad \alpha_q \,|\, q+2, \quad \alpha_q \nmid p+2 \,,$$

and $\lceil \log_2 \alpha \rceil$ is fixed to 160 for practical use. The decryption process is then advantageously replaced by

$$m = \frac{\psi_n(\alpha \cdot C)}{\psi_n(\alpha \cdot G)} \mod n$$

where the main computational workload is now a single scalar multiplication[6] by a short 160-bit constant. Chinese remaindering can also be used during decryption.

## 4.3   Homomorphic Properties

Our encryption scheme is $(+, +)$-homomorphic, *i.e.* an elliptic curve addition of two or several ciphertexts induces the implicit modular addition of the corresponding plaintexts. It also allows self-blinding, that is, provides the ability to publicly randomize a given ciphertext while conserving the correspondence with the initial plaintext. Finally, just like other known one-way trapdoor morphisms, the scheme provides random self-reducible encryption [3,18].

## 5   Conclusions

This paper introduced three new probabilistic encryption schemes on elliptic curves over rings. The cryptosystems are based on three specific trapdoor mechanisms allowing the recipient to recover discrete logarithms on different types of curves. More specifically, we showed how to design embodiments of Naccache-Stern, Okamoto-Uchiyama and Paillier discrete-log encryption schemes. Each provided cryptosystem is probabilistic and semantically secure relatively to the high residuosity problem associated with its curve type. We believe our work positively concretizes all previous research works on discrete log encryption in the elliptic curve setting.

## 6   Acknowledgements

The author is very grateful to Marc Joye who guided much of this research and to Moti Yung for various thoughts and discussions. I also thank the anonymous referees for their helpful and detailed comments.

---

[6] the value of $\psi_p(\alpha \cdot G)^{-1} \mod n$ can be pre-computed and stored before decryption takes place.

# References

1. J. C. Benaloh. *Verifiable Secret-Ballot Elections*. PhD Thesis, Yale University, 1988.

2. D. Coppersmith. *Specialized Integer Factorization*. In Advances in Cryptology, Proceedings of Eurocrypt'98, LNCS 1403, Springer-Verlag, pp. 542–545, 1992.

3. J. Feigenbaum, S. Kannan and N. Nisan. *Lower Bounds on Random-Self-Reducibility*. In Proceedings of Structures 1990, 1990.

4. P-A. Fouque, G. Poupard, and J. Stern. *Sharing Decryption in the Context of Voting or Lotteries*. In Proceedings of Financial Cryptography '00, LNCS, Springer-Verlag, 2000.

5. K. Koyama, U. Maurer, T. Okamoto and S. Vanstone. *New Public-Key Schemes based on Elliptic Curves over the ring $\mathbb{Z}_n$*. In Advances in Cryptology, Proceedings of Crypto'91, LNCS 576, Springer-Verlag, pp. 252–266, 1992.

6. J. McKee and R. Pinch. *On a Cryptosystem of Vanstone and Zuccherato*. Preprint, 1998.

7. A. Miyaji. *Elliptic Curves over $\mathbb{F}_p$ Suitable for Cryptosystems*. In Advances in Cryptology, Proceedings of Auscrypt'92, LNCS 718, Springer-Verlag, pp. 479–491, 1993.

8. D. Naccache and J. Stern. *A New Cryptosystem based on Higher Residues*. In Proceedings of the 5th CCCS, ACM Press, pp. 59–66, 1998.

9. T. Okamoto and S. Uchiyama. *A New Public Key Cryptosystem as Secure as Factoring*. In Advances in Cryptology, Proceedings of Eurocrypt '98, LNCS 1403, Springer Verlag, pp. 308–318, 1998.

10. T. Okamoto and S. Uchiyama. *Security of an Identity-Based Cryptosystem and the Related Reductions*. In Advances in Cryptology, Eurocrypt'98, LNCS 1403, pp. 546–560, Springer Verlag, 1998.

11. P. Paillier. *Public-Key Cryptosystems Based on Composite-Degree Residuosity Classes*. In Advances in Cryptology, Eurocrypt'99, LNCS 1592, pp. 223–238, Springer Verlag, 1999.

12. P. Paillier and D. Pointcheval. *Efficient Public-Key Cryptosystems Provably Secure Against Active Adversaries*. In Advances in Cryptology, Asiacrypt'99, LNCS 1716, pp. 165–179, Springer Verlag, 1999.

13. G. Poupard and J. Stern. *Fair Encryption of RSA Keys*. In Advances in Cryptology, Eurocrypt'00, LNCS 1807, Springer Verlag, 2000.

14. M. O. Rabin. *Digital Signatures and Public-Key Encryptions as Intractable as Factorization*. MIT Technical Report No 212, 1979.

15. R. Rivest, A. Shamir, and L. Adleman. *A Method for Obtaining Digital Signatures and Public Key Cryptosystems*. Communications of the ACM, vol. 21, no. 2, pp. 120–126, 1978.

16. H.-G. Rück. *On the Discrete Logarithm in the Divisor Class Group of Curves*. Math. Comp, vol. 68, no. 226, pp. 805–806, 1999.

17. H.-G. Rück. *A Note on Eliiptic Curves over Finite Fields.* Math. Comp, vol. 49, no. 179, pp. 301–304, 1987.

18. T. Sander, A. Young and M. Yung. *Non-Interactive CryptoComputing for $NC^1$.* IEEE FOCS'99, 1999.

19. T. Satoh and K. Araki. *Fermat Quotient and the Polynomial Time Discrete Log Algorithm for Anomalous Elliptic Curves.* Preprint, 1997.

20. I. A. Semaev. *Evaluation of Discrete Logarithms in a Group of p-Torsion Points of an Elliptic Curve in Characteristic p.* Math. Comp., vol. 67, pp. 353-356, 1998.

21. J. H. Silverman. *The Arithmetic of Elliptic Curves.* Springer-Verlag, GTM 106, 1986.

22. N. Smart. *The Discrete Logarithm Problem on Elliptic Curves of Trace One.* Journal of Cryptology, vol. 12, no. 3, pp. 193–196, 1999.

23. S. Vanstone and R. Zuccherato. *Elliptic Curve Cryptosystem Using Curves of Smooth Order Over the Ring $Z_n$.* In IEEE Trans. Inf. Theory, vol. 43, no. 4, 1997.

# Strengthening McEliece Cryptosystem

Pierre Loidreau

Project CODES, INRIA Rocquencourt Research Unit
B.P. 105 - 78153 Le Chesnay - Cedex France
Pierre.Loidreau@inria.fr

**Abstract.** McEliece cryptosystem is a public-key cryptosystem based on error-correcting codes. It constitutes one of the few alternatives to cryptosystems relying on number theory. We present a modification of the McEliece cryptosystem which strengthens its security without increasing the size of the public key. We show that it is possible to use some properties of the automorphism groups of the codes to build decodable patterns of large weight errors. This greatly strengthens the system against the decoding attacks.

## 1 Introduction

Since public-key cryptography was introduced in 1977 in the fundamental paper of Diffie and Hellman, it has taken an increasing importance in research as well as application fields. Many public-key ciphers have been proposed during the last twenty years; they rely on various difficult problems such as factoring numbers, computing discrete logarithms, solving knapsack problems... However, the conjugate development of computing power and efficient algorithms have made many of them insecure. A common point between the non-yet broken systems is that they remain dangerously linked with only two problems of number theory – the difficulty of factoring an integer and the difficulty of computing a discrete logarithm – and we are not protected from a theoretical breakthrough.

McEliece proposed an alternative to such systems in 1978 [McE78]. It consists in a public-key cryptosystem based on error-correcting codes. Together with its Niederreiter [Nie86] version - of equivalent security [LDW94] - the original system based on the family of Goppa codes still resists cryptanalysis. The general security of the scheme relies on the inherent intractability of decoding a random code up to its error-correcting capability. The great advantage of systems based on error-correcting codes is the extremely low cost of their encryption and decryption procedures. It approaches the complexity of secret key encryption schemes. Furthermore, if by chance major breakthroughs were made in number theory problems, such systems would constitute one of the few possible alternatives; therefore the study of their security is essential. The cost of a general decoding attack on these systems depends on the size of the chosen code and its error-correcting capability. The best known algorithm based on this method points out [CS98] that the size of the public key of the original system is becoming short regarding the increasing power of the computers. A safer step should be

T. Okamoto (Ed.): ASIACRYPT 2000, LNCS 1976, pp. 585–598, 2000.

to take a larger key. Yet, in that case the huge size of key (more than 880kbytes) would be a major disadvantage for implementation on limited resource systems. In this paper we present a modification of McEliece cryptosystem which makes all decoding attacks infeasible without increasing the size of the public-key.

The underlying idea results from a trade-off between the strong security of the system against structural attack and its much weaker security regarding decoding attacks. We allow ourselves to reduce the size of the space of public-keys weakening the system against structural attacks to increase the security of the system regarding the decoding attacks. This can be done by using to some property of the automorphism group of Goppa codes. Namely, whenever the Frobenius automorphism lies in the automorphism group of the code we can generate large sets of decodable error-words of a larger weight than the constructed error-correcting capability of the code. We show that whenever such sets are used in the system, the cost decoding attacks is significantly increased.

## 2    McEliece Public-Key Cryptosystem

In the family of public-key cryptosystems based on coding theory the one proposed by McEliece is the most widely considered. Namely, it is not only the first encryption scheme using coding theory ever proposed but it has also ever resisted to the attacks attempting to recover the secret key.

Other McEliece-like systems using different families of codes have been structurally cryptanalysed [SS92]. The credit for its resistance can thus be given to the family of Goppa codes taken as the secret key space. Their poor structure prevents an attacker to find a way to reduce significantly the size of the key-space. However, the size of the public key has to be significantly large to avoid general decoding attacks. Even with such a constraint both encryption and decryption procedures for the system remain much faster than for RSA.

### 2.1    Description of the Cryptosystem

A linear binary code of length $n$ and dimension $k$ is a linear subspace of $\mathbf{F}_2^n$. It can be represented by a $k \times n$ binary matrix called generating matrix. Two codes $C_1$ and $C_2$ of length $n$ are said to be equivalent if there exists a permutation of the $n$ coordinate places changing $C_1$ into $C_2$.

The permutations of coordinate places sending a code $C$ into itself form the automorphism group of the code $C$.

*Irreducible Goppa Codes* The secret key space is a family of irreducible Goppa codes [MS77] pp. 338. The receiver must thus consider some notions of finite fields algebra. Namely, in the construction of a Goppa code $\Gamma(L, g)$ [Gop70], we use

1. a finite field $\mathbf{F}_{2^m}$ with $2^m$ elements. $\mathbf{F}_{2^m}$ is the support field of the code,
2. a labeling $L$ of $\mathbf{F}_{2^m}$. $L$ is called generating vector of the code,
3. an irreducible polynomial $g$ over $\mathbf{F}_{2^m}$ of degree $t$. $g$ is called generating polynomial of the code.

*Properties* Every irreducible Goppa codes $\Gamma(L, g)$ has a fast polynomial time decoding algorithm [Pat75] up to its constructed error-correcting capability. The error-correcting capability of the codes is lower bounded by $t$, the degree of the generating polynomial, that is any error of weight less than $t$ occurring on a codeword can be corrected.

*Key Space* To construct the scheme one takes the family $\mathcal{G}$ of irreducible Goppa codes of length $n = 2^m$, dimension $k$ and error-correcting capability $t$. The cardinality of $\mathcal{G}$ is almost always equal to the number of irreducible polynomials of degree $t$ over the finite field with $2^m$ elements that is approximately $2^m/t$. With the original parameters, $n = 1024$, $k = 524$, $t = 50$, the size of the space is around $2^{496}$.

*Cryptosystem* It has the following form:

1. Private key: a Goppa code $\Gamma(L, g)$ randomly picked up in the family $\mathcal{G}$, a random $k \times k$ non-singular binary matrix $S$, and a random $n \times n$ permutation matrix $P$.
2. Public key: the product $G' = SGP$, where $G$ is a generating matrix for $\Gamma(L, g)$.
3. Encryption: let $x$ be the $k$-bit message to be encrypted, the sender computes $x' = xG' + e$ where $e$ is a $n$-bit error-vector of weight $t$.
4. Decryption: the receiver computes $x'P^{-1} = xSG + eP^{-1}$, and then recovers $xS$ by using the fast decoding algorithm of $\Gamma(L, g)$. Since $S$ is non-singular the receiver recovers $x$.

The security of the system depends on the difficult problem of decoding a code up to its error-correcting capability.

*Complexity of Encryption-Decryption* This encryption scheme has an extremely low complexity compared to the RSA. Namely, [Can96]

- in the encryption procedure we can take for granted that the cost of generating a random word of length $n$ and weight $t$ is negligible compared to the cost of a matricial product. Hence the work factor for encryption is

$$W^C = nk/2$$

- by using the Euclidian algorithm –which is not the most efficient but whose complexity is the easiest to evaluate – to make the decoding the work factor for decryption is:

$$W^D \approx \underbrace{3mnt + 4m^2t^2}_{\text{decoding algorithm}} + k^2/2$$

Originally Goppa codes of length $2^{10}$ dimension 524, and degree of the generator 50 are taken. This gives:

- Number of binary operations per information bit for encryption:
  $W^C/k = 512$, which is smaller than the 2402.7 binary operations per information bit required in the RSA-1024 encryption procedure.
- Number of binary operations per information bit for the decryption:
  $W^D/k = 5101.7$, which is much smaller than the 738 112.5 binary operations per information bit required in the RSA-1024 decryption procedure.

With such parameters the system runs more than 100 times faster for decryption than the RSA-1024 [CS98].

However the system has three main drawbacks:

1. the transmission rate is low: $k/n$ that is 51 percent in this case. Some attempts have been made to increase the transmission rate.
2. the size of the public key has to be huge: $kn$ bits, approximately $500Kbits$. If keys are smaller the scheme does not resist to decoding attacks.
3. encrypting the same message twice is recognizable and the plaintext can be recovered straightforward.

Note that by using the Niederreiter variant [Nie86] of the system, we can completely eradicate the problem of encrypting the same message twice. Moreover it allows to increase the transmission rate and to halve the size of the public-key without reducing the security of the system [LDW94]

## 2.2   Attacks on the System

There are two main approaches to cryptanalyse the system. They rely on two separate difficult problems.

1. The first one consists reconstructing a decoder for the code generated by the public-key $G'$ by studying its structure. A such approach is denoted structural attack.
   From the very construction of the system, the code $C'$ generated by the public key $G'$ is equivalent to $\Gamma(L, g)$. The attack consists in enumerating the codes in the family $\mathcal{G}$ to find a code $\Gamma \in \mathcal{G}$ which is equivalent to $C'$. Since equivalence classes of Goppa codes are constructible [Gib91] one can reduce the cost of the attack by examining a single element in each equivalence class. Yet the equivalence classes have a too small cardinality to decrease significantly the cost of the attack. For instance if we take the original parameters, $- t = 50$, and $n = 1024$ – there are $\approx 2^{496}$ irreducible polynomials of degree 50 over $\mathbf{F}_{2^{10}}$, and the equivalence classes have at most $2^{30}$ elements. Finding a code equivalent to $C'$ implies thus to explore on average more than $2^{466}$ codes. This remains largely beyond the capabilities of the most powerful computers.
   Once $\Gamma$ equivalent to $C'$ has been found, one recovers the permutation between $\Gamma$ and $C'$ by applying for instance the Support Splitting Algorithm [Sen99].

2. The second approach consists in decoding the intercepted ciphertexts $m'$ relatively to the public code $C'$ generated by the public-key. It is called decoding attack.

Since $\Gamma(L, g)$ is equivalent to $C'$ both codes have the same error-correcting capability $t$ and the equation $x' = x + e$, $x \in C'$ has a unique solution $(x, e)$ with $e$ of weight less than $t$. The cost of the attack depends only on the parameters of $C'$, its length, its dimension and its error-correcting capability. It implies that the parameters of the system have to be chosen very carefully and large enough. For this reason, the original parameters given by McEliece (length 1024, dimension 524, error-correcting capability $t = 50$) are becoming rather small for the state of art [CS98]: decoding one word takes on average $2^{64}$ binary operations. The next "safer" step would be to take $n = 2048$ for the code length. However the size of the key would become really prohibitive, for implementation on limited resource systems.

Whereas efficient decoding attacks were developed, the investigations concerning the reconstruction of a decoder remain rather scarce. In the general instance of the system there is no better way than exhaustive search on the key space - reduced modulo the equivalence relation -, testing the equivalence of each code with the code generated by the public-key.

One could replace the Goppa codes by any other family of codes with a fast polynomial-time decoding algorithm. Many codes are better than Goppa code regarding the decoding attacks. However the structure of theses families make the system insecure against structural attacks. For instance if one replaces the family of Goppa codes with the family of generalized Reed-Solomon codes or the family of concatenated codes, the recovering of a decoder can be done straightforward. [SS92, Sen98].

## 3    Tower Decodable Patterns

Taken randomly Goppa codes have a similar structure to random codes. In particular their automorphism group is usually trivial. Yet, Goppa codes with non trivial automorphism group are constructible: if the generating polynomial has coefficients in a subfield $\mathbf{F}_{2^s}$ of support field $\mathbf{F}_{2^m}$, then the automorphism group of the code is generated by the Frobenius automorphism. The attacker can detect this property by applying the Support Splitting Algorithm to the public key. This property was used to derive an almost realistic structural attack on the McEliece parameters, whenever the generating polynomial has binary coefficients [LS98].

Although such a property weakens the system against structural attack by reducing the size of the secret key space, we show that it can equally be used to strengthen the system against decoding attacks. By using properties of the automorphism group the conceiver can build sets of decodable patterns of large weight.

Moreover, from a cryptographic standpoint this set should satisfies some preliminary conditions: it must be large enough to avoid exhaustive search, the error words must have a weight larger than the error-correcting capability of the code. If such sets are used in place of the error vectors added in the original system, the cost of decoding attacks is greatly increased without changing the size of the public key.

## 3.1    Automorphism Group of Goppa Codes

Suppose the support field is $\mathbf{F}_{2^m}$, and let $L = (\alpha_1, \dots, \alpha_n)$ be a labeling of the support field. Let us consider the Goppa code $\Gamma(L, g)$ where the generating polynomial $g$ has coefficients in a subfield $\mathbf{F}_{2^s}$ of $\mathbf{F}_{2^m}$. Then we have

**Proposition 1.** *The automorphism group of $\Gamma(L, g)$ contains the group generated by the Frobenius automorphism $\sigma : z \mapsto z^{2^s}$ of $\mathbf{F}_{2^m}/\mathbf{F}_{2^s}$.*

The proof can be derived from Moreno's theorem [MS77] pp 347.

This means that the code $\Gamma(L, g)$ is invariant under the action of the Frobenius automorphism. If any word $c$ of length $n$ is labeled by $L$, we have

$$\forall\, c = (c_{\alpha_1}, \dots, c_{\alpha_n}) \in \Gamma(L, g), \quad \sigma(c) = (c_{\sigma(\alpha_1)}, \dots, c_{\sigma(\alpha_n)}) \in \Gamma(L, g)$$

## 3.2    t-Tower Decodability

**Definition 1.** *Let $\mathcal{E}$ be a set of words of length $n = 2^m$, let $\mathbf{F}_{2^s}$ be a subfield of $\mathbf{F}_{2^m}$ and $\sigma : z \mapsto z^{2^s}$ the Frobenius automorphism of the extension field. We say that $\mathcal{E}$ is t-tower decodable if*

1. *for all $e \in \mathcal{E}$, there exists a linear combination*

$$E = \sum_{i=0}^{m/s-1} \epsilon_i \sigma^i(e), \quad \epsilon_i \in \mathbf{F}_2$$

   *having a Hamming weight less than $t$, where $\sigma(e)$ denotes the action of the Frobenius on the word $e$,*
2. *the knowledge of $E$ enables the receiver to recover $e$ in $\mathcal{E}$ in a unique way.*

In other words $\mathcal{E}$ is a t-tower decodable set if there exists a linear combination of the powers of the Frobenius automorphism $\sigma$ that is a one-to-one mapping from $\mathcal{E}$ into the vectors of length $n$ and weight less than the correcting capability of the Goppa code.

The second condition in the definition is fundamental. It ensures that given a pattern we can invert all the operations to recover the original vector $e$.

The first condition is simple to achieve: Let us take $\mathcal{E}$, the set of all the binary words $e$ of length $n$ satisfying

$$\sum_{i=0}^{m/s-1} \sigma^i(e) = 0$$

However it does not satisfy the second condition. Namely every word in $\mathcal{E}$ is mapped onto the null word.

$t$-tower decodability is intimately linked with classical decodability up to $t$ in the family of Goppa codes with a non-trivial automorphism group :

**Proposition 2.** *Let $\Gamma(L, g)$ be a Goppa code with generating vector of degree $t$ over a subfield $\mathbf{F}_{2^s}$ of the support field $\mathbf{F}_{2^m}$, then any error vector of a $t$-tower decodable set $\mathcal{E}$ is correctable in $\Gamma(L, g)$.*

*Proof.* Let $x' = x + e$ where $x$ is a codeword in $\Gamma(L, g)$ and $e \in \mathcal{E}$. By definition of $\mathcal{E}$ there exist a linear combination of the power of the Frobenius $E = \sum_{i=0}^{m/s-1} \epsilon_i \sigma^i(e)$ having weight less than $t$.

From Sect. 3.1 the automorphism group of $\Gamma(L, g)$ contains $\sigma$. Thus, the linear combination $x' = \sum_{i=0}^{m/s-1} \epsilon_i \sigma^i(x)$ is also in the code $\Gamma(L, g)$.

Since $\sum_{i=0}^{m/s-1} \epsilon_i \sigma^i(x) = x' + E$, by applying the decoding algorithm of $\Gamma(L, g)$ one recovers $E$. From Definition 1, the error-vector $e$ can be recovered in a unique manner.    □

## 3.3   Modified Cryptosystem

*Space of Secret Keys* Let $g_1$ be an irreducible polynomial of degree $t_1$ over $\mathbf{F}_{2^m}$. $g_1$ is called hiding polynomial. Let $\mathcal{G}$ be the family of the Goppa codes $\Gamma(L, g_1 g)$ where $g$ describes the family of irreducible polynomials of degree $t$ over a subfield $\mathbf{F}_{2^s}$ of $\mathbf{F}_{2^m}$.

*Private Key* Not changing from the original scheme, it is made of 3 parts:

  - a $k \times n$-generating matrix $G$ of a code $\Gamma(L, g_1 g)$ randomly chosen in $\mathcal{G}$
  - a $n \times n$ permutation matrix $P$,
  - a $k \times k$ non-singular matrix $S$.

*Public Key* To the difference of the original scheme it consists in two parts

  - the product $G' = SGP$,
  - the way to generate a $t$-tower decodable set $\mathcal{E}$.

*Encryption* Let $x$ be the $k$-bit message that has to be transmitted. The sender chooses randomly a word $e$ in $\mathcal{E}$, then sends $x' = xG' + e$.

*Decryption* The receiver first computes $x'P^{-1} = xSG + eP^{-1}$.

Since $e$ is in the $t$-tower decodable set $\mathcal{E}$, from Definition 1 there is a linear combination $\sum_{i=0}^{m/s-1} \epsilon_i \sigma^i(e)$ of weight less than the error correcting capability $t$ of $\Gamma(L, g)$.

The receiver computes

$$\sum_{i=0}^{m/s-1} \epsilon_i \sigma^i(x'P^{-1}) = \sum_{i=0}^{m/s-1} \epsilon_i \sigma^i(xSG) + \sum_{i=0}^{m/s-1} \epsilon_i \sigma^i(eP^{-1})$$

Note that $xSG$ is a word in the code $\Gamma(L, g_1g)$. However, by construction, $\Gamma(L, g_1g)$ is a subcode of $\Gamma(L, g)$. Therefore we can consider that $xSG$ is a word in $\Gamma(L, g)$. Moreover, since $\sigma$ is in the automorphism group $\Gamma(L, g)$ by construction, $\sum_{i=0}^{m/s-1} \epsilon_i \sigma^i(mSG)$ is also a codeword of $\Gamma(L, g)$. Since $P^{-1}$ is a permutation we have

$$\sum_{i=0}^{m/s-1} \epsilon_i \sigma^i(eP^{-1}) = \left(\sum_{i=0}^{m/s-1} \epsilon_i \sigma^i(e)\right) \cdot P^{-1}$$

which is a decodable pattern in $\Gamma(L, g)$. The receiver gets thus the vector $E = \left(\sum_{i=0}^{m/s-1} \epsilon_i \sigma^i(e)\right)$ of weight less than $t$. $E$ can thus be recovered by applying the decoding algorithm of $\Gamma(L, g)$. The knowledge of $E$ provides a unique way to find $e$.

*Complexity of the Scheme* The complexity of the encryption is exactly the same as in the original system, since consisting in matricial products and picking up a random vector.

The decryption requires additional operations. However, the cost strongly depends on the structure of the $t$-tower decodable set $\mathcal{E}$.

*Conditions on $\mathcal{E}$* From a cryptological point of view, the $t$-tower decodable set must satisfy the following conditions:

1. $\mathcal{E}$ has to be a set of words of weight larger than the error-correcting capability of the code. This conditions strengthens the system against decoding attacks,
2. $\mathcal{E}$ has to be large enough to avoid enumeration. Namely, if an exhaustive search on the possible error-words were feasible the initial message $x$ would be easily recovered,
3. the way to generate $\mathcal{E}$ must be public, and must not reveal information that could help an attacker.

*Importance of the Hiding Polynomial $g_1$* We introduced the concept of hiding polynomial $g_1$ to satisfy the third condition on $\mathcal{E}$. If we used for $\mathcal{G}$ the family of irreducible Goppa codes with generating polynomial over $\mathbf{F}_{2^s}$, by applying the support splitting algorithm to the public key $G'$ any attacker would be able to recover $\sigma$. Then one could apply linear transformations of the Frobenius automorphism and reduce the problem of finding the error vector $e$ to the problem of finding the vector $E$ of lower weight.

The codes $\Gamma(L, g_1g)$ are subcodes of the codes $\Gamma(L, g)$ with a large structure. The introduction of the hiding polynomial scrambles the structure of the code rendering the automorphism group of $\Gamma(L, g_1g)$ trivial. Moreover, the hiding polynomial $g_1$ can be published since its knowledge does not give any exploitable information.

# 4   Extension of Degree 5

In the previous section we introduced the theoretical concept of tower decodability and how to use it in cryptography. In practice however, it is uneasy to build $t$-tower decodable sets satisfying the cryptological requirements. Therefore we focus on the example of extensions of degree 5. They not only turned out to be suitable from a cryptological viewpoint but they also intervene in the original parameters of the system. When using such $t$-tower decodable sets we show that, without increasing the size of the public-key, the security of the modified system is increased.

## 4.1   Construction of a $t$-Tower Decodable Set

We consider the field extension $\mathbf{F}_{2^{5s}}$ of $\mathbf{F}_{2^s}$, and the corresponding Frobenius automorphism $\sigma : z \mapsto z^{2^s}$. Since 5 is prime, the orbits of the elements of $\mathbf{F}_{2^{5s}}$ have size 5 except the orbits of the elements of $\mathbf{F}_{2^s}$ of size 1. Hence there are exactly $N_5 = (2^{5s} - 2^s)/5$ orbits of size 5. Let $L = (\alpha_1, \dots, \alpha_n)$ be a labeling of the field $\mathbf{F}_{2^{5s}}$. From now on, we suppose that any word of length $n$ is labeled by $L$.

The action of the Frobenius automorphism $\sigma$ on $e$ corresponds exactly to the action of the automorphism on the coordinates of $e$: if $e = (e_{\alpha_1}, \dots, e_{\alpha_n})$ then $\sigma(e) = \left(e_{\sigma(\alpha_1)}, \dots, e_{\sigma(\alpha_n)}\right)$.

We define a $t$-tower decodable set with respect to the Frobenius automorphism as follows,

**Definition 2.** *Let $\mathcal{E}$ be the set of all the possible words of length $n = 2^{5s}$ constructed this way:*

1. *one chooses randomly $p$ orbits out of the $N_5$ orbits of size 5 in the generating vector $L$, where $p$ satisfies $p = \lfloor t/2 \rfloor$*
2. *puts randomly 3 bits on every chosen orbit.*
3. *puts the coordinates to zero on the remaining positions.*

The set $\mathcal{E}$ contains words of weight $3p = 3\lfloor t/2 \rfloor$. The construction of $\mathcal{E}$ relies on the knowledge of the position of the orbits in the labeling $L$ of the field.

**Proposition 3.** *Let $\mathcal{E}$ be the set of words previously defined, we have*

1. *the cardinality of $\mathcal{E}$ is $10^p \cdot \binom{N_5}{p}$,*
2. *$\mathcal{E}$ is $t$-tower decodable.*

*Proof.* There are $\binom{N_5}{p}$ possibilities in choosing $p$ orbits of size 5 out of $N_5$. Once these orbits have been chosen, there are $\binom{5}{3} = 10$ possibilities in choosing three bits out of 5, proving the first assertion.

Let $e = (e_{\alpha_1}, \dots, e_{\alpha_n})$ be a word in the set $\mathcal{E}$. We reorder the labeling $L$ of the support in such a way that $e$ is written

$$e = (e_1, e_2, \dots, e_{N_5}, \underbrace{0, \dots, 0}_{\mathbf{F}_{2^s}})$$

where the $e_i = (e_{\alpha_i}, e_{\sigma(\alpha_i)}, e_{\sigma^2(\alpha_i)}, e_{\sigma^3(\alpha_i)}, e_{\sigma^4(\alpha_i)})$ denote the subvectors of length 5 of $e$ labeled by the orbit corresponding to the element $\alpha_i$. By construction of $\mathcal{E}$, the $e_i$ have either weight 0 or weight 3.

After the reordering, the action of the Frobenius automorphism $\sigma$ on the word $e$ becomes a combination of cyclic shifts on the vectors $e_i$. Therefore all 5-bit patterns of weight 3 can be divided into two classes $f_1$ and $f_2$ up to the Frobenius shifting equivalence:

| Type 1 | Type 2 |
|---|---|
| $f_1 = (11100)$ | $f_2 = (11010)$ |
| $\sigma(f_1) = (01110)$ | $\sigma(f_2) = (01101)$ |
| $\sigma^2(f_1) = (00111)$ | $\sigma^2(f_2) = (10110)$ |
| $\sigma^3(f_1) = (10011)$ | $\sigma^3(f_2) = (01011)$ |
| $\sigma^4(f_1) = (11001)$ | $\sigma^4(f_2) = (10101)$ |

The patterns $f_1$ and $f_2$ play dual roles. There exist linear combinations of $f_1$, $f_2$ and of their Frobenius images that enables one to reduce the weight of one pattern from 3 to 1 preserving the weight of the other. The average weight of the pattern is thus decreased. Namely we have

$$f_1 + \sigma(f_1) + \sigma^2(f_1) = (10101), \quad f_2 + \sigma(f_2) + \sigma^2(f_2) = (00001)$$

and

$$f_1 + \sigma^2(f_1) + \sigma^3(f_1) = (01000), \quad f_2 + \sigma^2(f_2) + \sigma^3(f_2) = (00111)$$

Whenever one has the image $f + \sigma(f) + \sigma^2(f)$ or $f + \sigma^2(f) + \sigma^3(f)$ of a pattern $f$ of weight 3 , $f$ is recoverable in a unique way:

1. from the weight of the obtained pattern one gets the type of the pattern, either type 1 or type 2,
2. from the positions of the bits on the obtained pattern, one gets the original one.

To prove that $\mathcal{E}$ is $t$-tower decodable, it is sufficient to prove that one of the linear combinations $e + \sigma(e) + \sigma^2(e)$ and $e + \sigma^2(e) + \sigma^3(e)$ has weight less than $t$. Suppose now that $e$ is made of $p_1$ patterns of type 1 and $p_2$ patterns of 2. Then $e + \sigma(e) + \sigma^2(e)$ has weight $3p_1 + p_2$, and $e + \sigma^2(e) + \sigma^3(e)$ has weight $p_1 + 3p_2$. Since by construction $p = p_1 + p_2 = \lfloor t/2 \rfloor$, we have

$$2(p_1 + p_2) \leq t$$

This implies in particular that at least either $3p_1 + p_2$ or $p_1 + 3p_2$ is less than $t$. Hence at least one of the images of $e$ by the previous combinations has weight less than $t$.

For instance if $e + \sigma^2(e) + \sigma^3(e)$ has weight less than $t$ then by using the property of one-to-one correspondence between the patterns of weight 3 and their image by this transformation, the word $e$ can be recovered entirely.

Thus $\mathcal{E}$ is $t$-tower decodable.    □

*Remark 1.* The optimal parameters for $\mathcal{E}$ are $2p = t$. In this case each word in $\mathcal{E}$ has weight $3t/2$. With this method one can decode up to one half beyond the error-correcting capability $t$.

*McEliece Parameters* The set considered is the set of irreducible polynomials over $\mathbf{F}_{2^{10}}$ and error-correcting capability 50. Since $m = 10$, we have $s = 2$. The number of orbits of size 5 is 204. By taking the parameter $p = 25$, then the set $\mathcal{E}$ generated is composed of $2^{188}$ words of weight 75. Still it is negligible compared to the $2^{284}$ patterns of weight 50 but remains large enough to avoid enumeration.

## 4.2   Application to the Cryptosystem

Section 3.3 was dedicated to the modification of McEliece system by using the general properties of $t$-tower decodability to strengthen the system against decoding attacks. In this section we apply this modification with the $t$-tower decodable sets previously defined over extensions of degree 5. In particular we show that it is possible to publish how to generate $\mathcal{E}$ without giving the possibility for an attacker to reduce the complexity of the attacks on the system.

### Parameters of the System

*Family of Goppa Codes* As a hiding polynomial we take an irreducible polynomial $g_1$ of degree 2 over $\mathbf{F}_{2^{5s}}$. Let $L$ be a labeling of the field $\mathbf{F}_{2^{5s}}$, we consider the family $\mathcal{G}$ of Goppa codes $\Gamma(L, g_1 g)$ where $g$ has degree $t$ and coefficients over $\mathbf{F}_{2^s}$.

*Private Key* It consists in 3 parts:

  − a $k \times n$-generating matrix $G$ of a code picked up randomly in $\mathcal{G}$,
  − a $n \times n$-permutation matrix $P$,
  − a non-singular $k \times k$-matrix $S$.

*Public Key*

  1. the matrix $G' = SGP$,
  2. the positions of the $N_5$ orbits of cardinality 5 in the generating vector $L$.

Note that if the positions of the orbits are in some way canonical the size of the public-key can be made as low as the size of $G'$.

**Encryption-Decryption** Since the positions of the orbits are public, the sender can generate the set $\mathcal{E}$ of $t$-tower decodable words described in the previous section.

*Encryption* Let $x$ be the $k$-bit plaintext one has to transmit, the sender chooses randomly a word $e$ in $\mathcal{E}$: he first picks up $\lfloor t/2 \rfloor$ orbits out of the $N_5$ possible and puts randomly 3 bits on each orbit. The corresponding ciphertext is $x' = xG' + e$.

*Decryption* The receiver computes $x'P^{-1} = xSG + eP^{-1}$. Since permuting the coordinates does not change the structure of the automorphism group, we can consider that $eP^{-1}$ is still in $\mathcal{E}$. It was shown in 3 that $E$ is $t$-tower decodable, therefore by applying the right linear combinations of the powers of the Frobenius automorphism, the receiver first recovers $eP^{-1}$, then recovers $x$.

To evaluate the relative cost of the procedure compared to the original scheme, we have to separate it into different steps.

1. First one has to compute the two linear combinations $x'_1 = x' + \sigma(x') + \sigma^2(x')$ and $x'_2 = x' + \sigma^2(x') + \sigma^3(x')$. Let $s_n$ be the cost of computing the action of $\sigma$ on a vector of length $n$, and let $a_n$ be the cost of xoring two words of length $n$. Overall the cost is $3(s_n + a_n)$. The action of $\sigma$ is the product of cyclic shifts thus we neglect the cost of this step compared to the complexity of the decoding part.
2. Decoding part: let $A_t = 3mnt + 4m^2t^2$ be the cost (given in 2.1) of decoding one word corrupted by a $t$-bit error-vector. In the original system the decoding part costs exactly $A_t$ operations. In this modification the cost is at most $2A_t$ and can be greatly reduced. Namely we first try to decode $x'_1$ and only if the decoding fails then we decode $x'_2$. Thus the additive cost of the procedure to recover $E$ from $x'_1$ or $x'_2$ is on average at most $1/2A_t$.
3. The cost to recover $e$ from $E$ is a few times the cost of running over the $n$ positions of the word, so it can be neglected compared to the cost of the decoding procedure.

Thus if $D$ is the cost of the decryption in the original scheme, and $D_1$ is the cost in time of the decryption in the modified scheme we have

$$D_1 \leq D + 1/2A_t$$

By taking the original parameters – $n = 1024$, $t = 50$ – the number of binary operations per information bit for decryption becomes: $W^D/k = 7521.5$, which remains much smaller than the 738 112.5 operations per information bit required for the RSA-1024 decryption.

The memory cost is identical in both schemes.

**Security of the System** In the conception of the scheme the positions of the orbits of size 5 in the support of the code are public stuff. It does not jeopardize the scheme since it does not provide a potential attacker with exploitable information. Given indeed this information it seems difficult to recover additional properties enabling to recover the Frobenius automorphism. This would imply that given the public code one could build a larger code from which we only know the non-ordered orbits through its automorphism group.

By considering the McEliece parameters we show that this system provides a better security against decoding attacks than the original scheme.

*McEliece Parameters* If $\mathcal{G}$ is the set $\Gamma(L, g_1g)$ where $L$ is a labeling of $\mathbf{F}_{2^{10}}$ and $g$ runs over the polynomials of degree 50 over $\mathbf{F}_{2^2}$ then

- the size of $\mathcal{G}$ is approximately $2^{95}$,
- the size of the public key is of the same order as in the original system. $\Gamma(L, g_1 g)$ being a subcode of $\Gamma(L, g)$ the size a generating matrix for $\Gamma(L, g_1 g)$ will be slightly smaller than the size of a generating matrix for $\Gamma(L, g)$,
- $\mathcal{E}$ is a family of 50-tower decodable codewords of weight 75 and has cardinality $2^{188}$. This is very few compared to the set of patterns of length 1024 and weight 50 having cardinality $2^{284}$ that are decodable, but still remains largely out of range for the computers.

  In that case applying the best algorithm for decoding attack [CS98] gives roughly $2^{91}$ binary operations compared to the $2^{64}$ involved for breaking the original system.

# 5    Conclusion

In the paper we showed how to use the automorphism group of Goppa codes to increase the security of the McEliece system against decoding attacks. This approach can be easily transferred to its Niederreiter type version, the security of which is the same. Of course the specific structure we require from the family of Goppa codes enables any attacker to greatly reduce the complexity of a structural attack compared to the cost of a structural attack on the original version. However, in the example developed above concerning the extensions of degree 5 the size of the family of codes to enumerate remains largely beyond the capabilities of the computers. The security is thus the result of a trade-off between the two kinds of attacks.

Such an approach can be generalized to any finite field extension with characteristic 2. Still, in that case the problem is to find $t$-tower decodable sets satisfying the simple cryptographical constraints such as being a large set of large-weight words. The ideal would be to find a decodable set whose words have weight larger than half of the code-length. Decoding attacks would be then completely obsolete, and as a consequence, the main problematic factor which is the large size of the public-key would vanish. Gabidulin, Paramonov and Tretjakov proposed such a cryptosystem based on error-correcting codes [GPT91] with very nice properties. This system is unbreakable with a decoding attack and has a very low key size (less than 10kbits). Unfortunately the codes in the key space have so much structure that in its first version it was efficiently broken by K. Gibson [Gib95].

# References

[Can96]    Anne Canteaut. *Attaques de cryptosystèmes à mots de poids faible et construction de fonctions t-résilientes*. PhD thesis, Université Paris-VI, 1996.

[CS98]    A. Canteaut and N. Sendrier. Cryptanalysis of the original McEliece cryptosystem. In Kazuo Ohta and Dingyi Pei, editors, *Advances in Cryptology - ASIACRYPT'98*, number 1514 in LNCS, pages 187–199, 1998.

[Gib91]     J. K. Gibson. Equivalent Goppa codes and trapdoors to McEliece's public key cryptosystem. In D. W. Davies, editor, *Advances in Cryptology - EUROCRYPT'91*, number 547 in LNCS, pages 517–521. Springer-Verlag, 1991.

[Gib95]     J. K. Gibson. Severely Denting the Gabidulin Version of the McEliece Public Key Cryptosystem. *Designs, Codes and Cryptography*, 6:37–45, 1995.

[Gop70]     V. D. Goppa. A new class of linear error-correcting codes. *Problemy Peredachi Informatsii*, 6(3):207–212, 1970.

[GPT91]     E .M. Gabidulin, A. V. Paramonov, and O. V. Tretjakov. Ideals over a non-commutative ring and their application in cryptology. *LNCS*, 573:482 – 489, 1991.

[LDW94]   Y. X. Li, R. H. Deng, and X. M. Wang. On the equivalence of McEliece's and Niederreiter's public-key cryptosystems. *IEEE Transactions Information Theory*, 40(1):271–273, 1994.

[LS98]      P. Loidreau and N. Sendrier. Some weak keys in McEliece public-key cryptosystem. In *IEEE International Symposium on Information Theory, ISIT'98, Boston*, page 382, 1998.

[McE78]    R. J. McEliece. A public-key cryptosystem based on algebraic coding theory. Technical report, Jet Propulsion Lab. DSN Progress Report, 1978.

[MS77]      F. J. MacWilliams and N. J. A. Sloane. *The Theory of Error–Correcting Codes*. North Holland, 1977.

[Nie86]     H. Niederreiter. Knapsack-type cryptosystems and algebraic coding theory. *Problems of Control and Information Theory*, 15(2):159 – 166, 1986.

[Pat75]     N. J. Patterson. The algebraic decoding of GOPPA codes. *IEEE Transactions Information Theory*, 21:203–207, 1975.

[Sen98]     N. Sendrier. On the concatenated structure of a linear code. *AAECC*, 9(3):221–242, 1998.

[Sen99]     Nicolas Sendrier. The Support Splitting Algorithm. Technical Report 3637, INRIA, March 1999. http://www.inria.fr/RRRT/RR-3637.html.

[SS92]      V. M. Sidel'nikov and S. O. Shestakov. On cryptosystems based on generalized REED-SOLOMON codes. *Discrete Mathematics*, 4(3):57–63, 1992. in russian.

# Password-Authenticated Key Exchange Based on RSA

Philip MacKenzie[1], Sarvar Patel[1], and Ram Swaminathan[2]

[1] Bell Laboratories, Lucent Technologies,
{philmac, sarvar}@lucent.com
[2] Hewlett-Packard Research Laboratories,
swaram@godel.hpl.hp.com

**Abstract.** There have been many proposals in recent years for password-authenticated key exchange protocols. Many of these have been shown to be insecure, and the only ones that seemed likely to be proven secure (against active adversaries who may attempt to perform off-line dictionary attacks against the password) were based on the *Diffie-Hellman* problem. In fact, some protocols based on Diffie-Hellman have been recently proven secure in the random-oracle model. We examine how to design a provably-secure password-authenticated key exchange protocol based on *RSA*. We first look at the OKE and protected-OKE protocols (both RSA-based) and show that they are insecure. Then we show how to modify the OKE protocol to obtain a password-authenticated key exchange protocol that can be proven secure (in the random oracle model). The resulting protocol is very practical; in fact the basic protocol requires about the same amount of computation as the Diffie-Hellman-based protocols or the well-known `ssh` protocol.

## 1 Introduction

Consider the following scenario: Alice and Bob share a short secret (say, a 4 digit PIN number or a 6 character password) that they wish to use to identify and authenticate each other over an insecure network (say, the Internet). They do not carry any other information with them. Of course, neither wants to reveal the secret to the other until the other has revealed his/her own knowledge of the secret. In fact, neither wants to reveal anything that could be used to verify the secret (such as a one-way function applied to the secret) since the secret can then be found by anyone using a dictionary attack (by simply iterating through the relatively small number of possible secrets, applying the one-way function to each of them, and comparing each result to the transmitted value). So how do Alice and Bob authenticate themselves? In general, Alice and Bob will want to not only authenticate themselves, but set up a secure channel between themselves. For this they need a cryptographically strong shared *session key*. So a variation of the question above would be: how do Alice and Bob bootstrap a short secret into a secure strong secret?

This problem, which we call *password-authenticated key exchange*, was first proposed in Bellovin and Merritt [BM92]. In that paper, the *Encrypted Key*

T. Okamoto (Ed.): ASIACRYPT 2000, LNCS 1976, pp. 599–613, 2000.

*Exchange (EKE)* protocol was proposed as a solution. The problem has since been studied extensively [BM93, GLNS93, Gon95, Jab96, Jab97, Luc97, STW95, Wu98], but only two recent papers [BPR00, BMP00] present protocols along with proofs of security, and in fact, many of the previously-proposed protocols have been shown to be insecure [Ble99, Pat97]. Both of the protocols that were proven secure were based on Diffie-Hellman. Specifically, [BPR00] developed a clean and elegant protocol based on EKE and proved its security based on Computational Diffie-Hellman (CDH), using the random oracle and ideal symmetric encryption function assumptions. The protocol in [BMP00] is similar, but with the proof of security based on Decisional Diffie-Hellman (DDH), using only the random oracle assumption.

## 1.1 Overview of Our Results

We study password-authenticated key exchange protocols based on RSA. We first look at the OKE (Open Key Exchange) and protected-OKE protocols of Lucks [Luc97], since they are the first ones that were based on RSA and were claimed to have proofs of security. We show that in fact they are insecure. Then we show how to modify the OKE protocol to obtain a protocol that we prove to be secure. This new protocol requires only 4 moves, and only one public-key operation (i.e., a modular exponentiation) per side (either encryption or decryption). Thus it is efficient enough to be used in practice, e.g., for securing remote user access to a server, and is roughly as efficient as the other Diffie-Hellman-based protocols or the ssh protocol, all of which require two exponentiations per side.[1]

In this scenario, it is actually useful for the server to store only some verification information for the password (such as a one-way function applied to the password) but not the password itself. This provides resilience to server compromise, meaning that an adversary that compromises the server and steals the password information is still not able to impersonate a user, unless the adversary actually performs a dictionary attack on the verification information. We show how to extend our protocol to provide some resilience to server compromise, but due to space limitations, we omit the full proof of security for this extended protocol.

The proposals presented in this paper have been presented in informal settings under the names SNAPI (Secure Network Authentication with Password Information) and SNAPI-X. To avoid confusion, we will continue to use those names here.

## 1.2 Security Model and Definitions

What does it mean for a password-authenticated key exchange protocol to be secure? Informally, it means that the probability that an adversary can successfully authenticate itself is at most negligibly more than that of an adversary

---

[1] It is difficult to do more than rough comparisons, since modulus size and exponent size may vary among the different protocols.

who runs a trivial attack of simply iteratively guessing passwords and running the authentication protocol (i.e., attempting to login). In SNAPI, we specifically show that if the adversary can do non-negligibly better than this trivial attack, then one can break RSA [RSA78]. We use the random-oracle model [BR93a] for our proofs. While a protocol having a security proof in the random-oracle model is certainly less desirable than a protocol having a proof in the standard model (using standard cryptographic assumptions) [CGH98], it is certainly preferable over a protocol which lacks any proof. Other techniques proven secure in the random-oracle model include Optimal Asymmetric Encryption Padding [BR94] (used in PKCS #1 v. 2 [Not99]) and Provably Secure Signatures [BR96].

For our proofs we use the security model for password-authenticated key exchange from [BMP00], in which the adversary totally controls the network, a la [BR93b], and which is based on the multi-party simulatability paradigm as described in [Bea91, BCK98, Sho99]. In this paradigm, security is defined using an ideal system, which describes the service (of key exchange) that is to be provided, and a real system, which describes the world in which the protocol participants and adversaries work. The ideal system should be defined such that an "ideal world adversary" cannot (by definition) break the security. Then, intuitively, a proof of security would show that anything an adversary can do in the real system can also be done in the ideal system, and thus it would follow that the protocol is secure in the real system.

Although it is *not* a password-only protocol, we do point out that the (one-way) authentication protocol given in Halevi and Krawczyk [HK98] is the first password-based authentication protocol to be formally proven secure, with standard security assumptions. The proof methods in this paper are significantly influenced by their techniques. Boyarsky [Boy99] has recently discussed enhancements to the protocol of Halevi and Krawczyk to make it secure in the multi-user scenario.

We note that basic shared-secret authentication protocols (e.g., [BR93b]) are not secure when the parties share *short secrets*. However, there is a similarity between basic authentication and password-based authentication: both seem to be very difficult to get correct, and many protocols have been published for both, which have subsequently been broken. This is precisely the reason why we emphasize *provable security* in this paper.

# 2    Attack on the RSA Open Key Exchange Protocol

Interest in developing RSA-based password-authenticated key exchange protocols has been strong [Luc97, RCW98] ever since Bellovin and Merritt first described the RSA-EKE protocol [BM92], their RSA version of the Encrypted Key Exchange protocol. However, the use of RSA in password-only protocols has proven to be quite tricky. Many of the RSA-based password-authenticated key exchange protocols have been shown to be insecure [Ble99, Pat97]. A different approach from the EKE protocols was used by Lucks [Luc97] to propose an RSA-based protocol which has so far resisted attacks. In this section, we present an efficient attack against this RSA-based protocol. In later sections, we modify the basic protocol in [Luc97] to obtain an RSA-based password-authenticated key exchange protocol that can be proven secure.

Lucks presented two protocols: the Open Key Exchange (OKE) protocol and the protected-OKE protocol, both described in terms of generic encryption functions with certain properties. When instantiated with RSA, the OKE protocol has a problem, as noted by Lucks, which can allow an attacker to recover the password. Hence, Lucks modified the basic OKE protocol to create the protected-OKE protocol that was supposed to be secure when the encryption function was instantiated with RSA. We present the basic steps of the RSA versions of the OKE and protected-OKE protocols along with our attack. Details of OKE and protected-OKE can be found in [Luc97]. We first describe OKE:

**Step A** Alice and Bob agree on a common secret $\pi$. Alice in advance generates an RSA key pair $((e, N), (d, N))$.

**Step B** Alice chooses a random $m$ and sends $(e, N)$ and $m$ to Bob.

**Step C** Bob chooses a random $\mu$ and then a random $a$ from $\mathbb{Z}_N^*$, computes $p = H(e|N|m|\mu|\pi)$ and $q = E(a) \diamond p$ and sends $\mu$ and $q$ to Alice. $H$ is a random function with range $\mathbb{Z}_N^*$, $E$ is the RSA encryption function and $\diamond$ is the RSA multiplication operation.

**Step D** Alice computes $p$ like Bob, and recovers $a$ by performing an RSA decryption of $q \diamond p^{-1}$.

The remaining authentication and key generation steps are omitted because they are not needed for the attack. Lucks noted that there is a problem with this scheme because when an attacker can choose $(e, N)$ such that function $E$ may not be invertible, $E(a)$ would not be uniformly distributed. Hence some information about $p$ will be leaked which would be useful in ruling out candidate values for $p$ and $\pi$. This lead Lucks to propose protected OKE which changes step C to step C′ below:

**Step C′** Bob chooses $a \in_R \mathbb{Z}_N^*$, but instead of $\mu$, Bob chooses 2 values $\mu_{-1}$, $\mu_0 \in_R \mathbb{Z}_N^*$. Bob uses $\mu_i = E(\mu_{i-2} \diamond H'(\mu_{i-1}))$ to compute $\mu_1, \mu_2, ..., \mu_K$ where $H'$ is another random function mapping to $\mathbb{Z}_N^*$. The value $p = H(e|n|m|\mu_{-1}| \mu_0|\pi)$ is computed, and $q = E(a) \diamond p$ is sent to Alice along with the last two values $\mu_{K-1}$ and $\mu_K$.

Lucks reasoned that if $E$ is not invertible then there are at least two choices for $\mu_{K-2}$, and then for every choice of $\mu_{K-2}$ there are two choices for $\mu_{K-3}$ and so on. Thus we expect $2^K$ choices for $\mu_{-1}$. For suitably large values of K, say 80, it would be infeasible for the adversary to evaluate all possible $p$; so the information leaked about $p$ from $q = E(a) \diamond p$ is of no use to the adversary. Unfortunately, RSA protected OKE has a weakness and we present an example attack:

**Step 1** The attacker picks $e$ and $N$ such that $e$ is 3, $N$ is a large prime, and $3|N-1$ and then sends $m$ and $(e, N)$ to Bob.

**Step 2** Unwittingly, Bob calculates $p = H(e|N|m|\mu_{-1}|\mu_0|\pi)$ and sends out $q$, $\mu_{K-1}$ and $\mu_K$.

**Step 3** We will now uniquely recover values up to $\mu_2$ and $\mu_1$ using some basic results from number theory by showing how to recover $\mu_{i-2}$ from $\mu_{i-1}$ and $\mu_i$. We make a note that we can efficiently find $dth$ roots if $d|N-1$ [BS96]. We thus decrypt $\mu_i = E(\mu_{i-2} \diamond H'(\mu_{i-1}))$ by solving for the three cubic roots of $\mu_i$. Then we multiply each root with $(H'(\mu_{i-1}))^{-1}$ to get three possible solutions for $\mu_{i-2}$. Of the three possible solutions, only one will be a cubic residue. We know apriori that the correct value for $\mu_{i-2}$ will be a cubic residue because $\mu_{i-2}$ was formed by encrypting (i.e., cubing): $\mu_{i-2} = E(\mu_{i-4} \diamond H'(\mu_{i-3}))$. Of the three possible values, we can identify the cubic residue because $\mu_{i-2}^{\frac{N-1}{3}} \equiv 1 \bmod N$. We continue to recover the rest of the $\mu_i$ values until $\mu_1$.

**Step 4** $\mu_0$ and $\mu_{-1}$ are random values and thus cannot be uniquely recovered using Step 3. There will be three possible values for $\mu_0$ and for each $\mu_0$ value there will be three possible values for $\mu_{-1}$. Hence there are nine possible $(\mu_0, \mu_{-1})$ pairs. We will now try to eliminate some candidate passwords from the list of possible password for Bob. If the password is guessed correctly and the $(\mu_0, \mu_{-1})$ pair is correct then solving for $E(a)$ from $q = E(a) \diamond p$ will result in an $E(a)$ which is a cubic residue. Conversely, if the solved $E(a)$ is not a cubic residue, assuming $(\mu_0, \mu_{-1})$ pair is correct, then we know the password guess is incorrect. We do not know the correct $(\mu_0, \mu_{-1})$ pair, however, if for all 9 pairs the 9 possible solutions for $E(a)$ turn out not to be cubic residues then we can eliminate this password guess.

This will happen with a significant probability and thus we can eliminate a significant portion of the possible passwords. The probability for a given password that a $(\mu_0, \mu_{-1})$ pair will be such that the result will be non-cubic residue is equivalent to a random number being a non-cubic residue which is $\frac{2}{3}$. The probability that all 9 $(\mu_0, \mu_{-1})$ pairs result in non-cubic residues is $\left(\frac{2}{3}\right)^9$ which is about 2.5%.

**Step 5** We repeat the above procedure (Step 1 - Step 4) eliminating a constant fraction of the remaining passwords in each run, until only one password remains.

It may be tempting to propose blocking this attack by checking for primality of $N$ and rejecting the session if $N$ is prime. Although we have described the example attack using a prime $N$ to keep the presentation simple, we could have done the same steps using a composite $N = pq$ and using the chinese remainder theorem where necessary; we omit the details. The above attack can efficiently discover a user's password after a small number of sessions. One can also try to reduce the probability of the attack's success by requiring $e$ to have only large factors. However, this may still allow some leakage and does not rule out other attacks. Ad hoc countermeasures are not very satisfactory in password-based protocols because every avenue of information leakage has to be blocked. Details matter.

## 3   Model

For our proofs, we use the model defined in [BMP00], which extends the formal notion of security for key exchange protocols from Shoup [Sho99] to password-authenticated key exchange. We assume the adversary totally controls the network, a la [BR93b].

Briefly, this model is defined using an ideal key exchange system, and a real system in which the protocol participants and adversaries work. The ideal system will be secure by definition, and the idea is to show that anything an adversary can do to our protocol in the real system can also be done in the ideal system, and thus it would follow that the protocol is secure in the real system.

### 3.1   Ideal System

We assume there is a set of (honest) *users*, indexed $i = 1, 2, \ldots$. Each user $i$ may have several *instances* $j = 1, 2, \ldots$. Then $(i, j)$ refers to a given *user instance*. A user instance $(i, j)$ is told the identity of its partner, i.e., the user it is supposed to connect to (or receive a connection from). An instance is also told its *role* in the session, i.e., whether it is going to *open* itself for connection, or whether it is going to *connect* to another instance.

There is also an *adversary* that may perform certain operations, and a *ring master* that handles these operations by generating certain random variables and enforcing certain global consistency constraints. Some operations result in a record being placed in a *transcript*.

The ring master keeps track of session keys $\{K_{ij}\}$ that are set up among user instances (as will be explained below, the key of an instance is set when that instance starts a session). In addition, the ring master has access to a random bit string $R$ of some agreed-upon length (this string is not revealed to the adversary). We will refer to $R$ as *the environment*. The purpose of the environment is to model information shared by users in higher-level protocols.

We will denote a password shared between users $A$ and $B$ as $\pi[A, B]$.

The adversary may perform the following operations: (1) *initialize user* operation with a new user number $i$ and a new identifier $ID_i$ as parameters; (2) *set password* with a new user number $i$, a new identifier $ID'$, and a password $\pi$ as parameters (modeling the adversary creating his own account); (3) *initialize user instance* with parameters including a user instance $(i, j)$, its role, and a user identifier denoting the partner with whom it wants to connect; (4) *terminate user instance* with a user instance $(i, j)$ as a parameter; (5) *test instance password* with a user instance $(i, j)$ and a password guess $\pi$ as parameters (this query can only be asked once per instance and models the adversary guessing a password and attempting to authenticate herself); (6) *start session* with a user instance $(i, j)$ as a parameter (modeling the user instance successfully connecting to its partner and establishing a random session key; (7) *application* with a function $f$ as parameter, and returning the function $f$ applied to the environment and any session keys that have been established (modeling leakage of session key information in a real protocol through the use of the key in, for example, encryptions of messages); (8) *implementation*, with a comment as parameter (modeling real world queries that are not needed in the ideal world).

For an adversary $\mathcal{A}^*$, *IdealWorld*($\mathcal{A}^*$) is the random variable denoting the transcript of the adversary's operations.

For a detailed description of the syntax and semantics of the above operations, see [BMP00].

## 3.2   Real System

In the real system, users and user instances are denoted as in the ideal system. User instances are defined as state machines with implicit access to the user's $ID$, $PID$, and password (i.e., user instance $(i, j)$ is given access to $\pi[ID_i, PID_{ij}]$). User instances also have access to private random inputs (i.e., they may be randomized). A user instance starts in some initial state, and may transform its state only when it receives a message. At that point it updates its state, generates a response message, and reports its status, either *continue*, *accept*, or *reject*, with the following meanings:

- *continue*: the user instance is prepared to receive another message.
- *accept*: the user instance (say $(i, j)$) is finished and has generated a session key $K_{ij}$.
- *reject*: the user instance is finished, but has not generated a session key.

The adversary may perform the following types of operations: (1) *initialize user* operation as in the ideal system; (2) *set password* operation as in the ideal system; (3) *initialize user instance* as in the ideal system; (4) *deliver message* with an input message $m$ and a user instance $(i, j)$ as parameters, and returning the message output from $(i, j)$ upon receiving $m$; (5) *random oracle* with the random oracle index $i$ and input value $x$ as parameters, and returning the result of applying random oracle $H_i$ to $x$; (6) *application* as in the ideal system.

For an adversary $\mathcal{A}$, *RealWorld*($\mathcal{A}$) denotes the transcript of the adversary's operations.

Again, details of these operations can be found in [BMP00].

## 3.3   Definition of Security

Our definition of security is the same as the one in [Sho99] for key exchange. It requires

1. **completeness**: for any real world adversary that faithfully delivers messages between two user instances with complimentary roles and identities, both user instances accept; and
2. **simulatability**: for every efficient real world adversary $\mathcal{A}$, there exists an efficient ideal world adversary $\mathcal{A}^*$ such that *RealWorld*($\mathcal{A}$) and *IdealWorld*($\mathcal{A}^*$) are computationally indistinguishable.

## 4    SNAPI

In this section we will start by presenting the definition of RSA and giving a standard and well-accepted version of the RSA security assumption.[2] Then we will present the SNAPI and SNAPI-X protocols.

First we give some preliminary definitions. Let $k$ and $\ell$ denote our security parameters, where $k$ is the "main" security parameter and can be thought of as a general security parameter for hash functions and secret keys (say 128 or 160 bits), and $\ell > k$ can be thought of as a security parameter for RSA or discrete-log-type public keys (say 1024 bits). Let $\{0,1\}^*$ denote the set of finite binary strings and $\{0,1\}^n$ the set of binary strings of length $n$. Let "|" denote the concatenation of bit strings in $\{0,1\}^*$. A real-valued function $\epsilon(n)$ is *negligible* if for every $c > 0$, there exists a $n_c > 0$ such that $\epsilon(n) < 1/n^c$ for all $n > n_c$.

The RSA encryption scheme is generally defined as follows: Let key generator $GE$ define a family of RSA functions to be $(e, d, N) \leftarrow GE(1^\ell)$ such that $N = PQ$, where $P$ and $Q$ are prime numbers. Then, the public key is the pair $(e, N)$ where $\gcd(e, \phi(N)) = 1$ and the order of the group $\phi(N) = (P - 1) \cdot (Q - 1)$. The encryption function $E : \mathbb{Z}_N^* \to \mathbb{Z}_N^*$ is defined by $E(x) \equiv x^e \bmod N$ and the decryption function $D : \mathbb{Z}_N^* \to \mathbb{Z}_N^*$ is $D(x) \equiv x^d \bmod N$, where the secret exponent $d$ is chosen such that $ed \equiv 1 \bmod \phi(N)$.

The choice of $P$, $Q$, and $e$ is generally left to the implementation, although it is recommended that $P$ and $Q$ be random large primes with about the same bit length (about $\ell/2$ for security parameter $\ell$) [IEE98], and for efficiency $e$ is often chosen to be a small prime and with a small number of ones in its binary representation, such as 3, 17, or 65537.

For the security of SNAPI, we make *explicit* requirements on the generation of $P$, $Q$, and $e$, which are well within the scope of the general RSA security recommendations. Specifically, we require that $GE(1^\ell)$ chooses two random primes $P$ and $Q$ from the range $\{2^{\ell/2-1}, \ldots, 2^{\ell/2}\}$ (for convenience, we assume $\ell$ is a multiple of 2). This implies that $2^{\ell-2} \leq N \leq 2^\ell$. We also require that $e$ be a prime in the range $\{2^\ell + 1, \ldots, 2^{\ell+1}\}$. Note that this guarantees that $\gcd(e, \phi(N)) = 1$. For efficiency in our protocol, a standard value of $e$ for a given security parameter $\ell$ could be chosen beforehand. This would eliminate the need for a primality test by Bob. (An alternative requirement on $e$ would be that $e$ is a prime, $e \geq \sqrt{N}$ and $(N \bmod e) \nmid N$, since this can be checked in (probabilistic) polynomial time, and also implies that $\gcd(e, \phi(N)) = 1$ [Len84].)

Given these requirements on $GE$, we use the following assumption on RSA:

**RSA Security Assumption:** Let $\ell$ be the security parameter. Let key generator $GE$ define a family of RSA functions (i.e., $(e, d, N) \leftarrow GE(1^\ell)$). For any probabilistic polynomial-time algorithm $A$, $\Pr[u^e \equiv w \bmod N : (e, d, N) \leftarrow GE(1^\ell); w \in_R \{0,1\}^\ell; u \leftarrow A(1^\ell, w, e, N)]$ is negligible.

---

[2]   The security of the SNAPI protocol can actually be proven under a slightly more general security assumption. Details are omitted.

## 4.1   The Protocol

Before the SNAPI protocol starts, two players agree on a common password $\pi \in P$. Let $A$ and $B$ be the identities of the two players, with $A$ playing the role of Alice, and $B$ playing the role of Bob. From this point on, we will refer to $A$ as Alice and $B$ as Bob, except when we must explicitly use their identities.

We assume that Alice has chosen an RSA key pair $((e, N), (d, N))$. In general, Alice would most likely use the same key pair in many sessions, although for perfect forward secrecy, Alice would need to choose a new pair in each session. That is, if Adv discovers the decryption key then Adv can determine all session keys obtained in sessions using that key pair. Obviously, some tradeoffs of security versus efficiency could be performed. Alternatively, the two parties could obtain perfect forward secrecy by computing the session key with a Diffie-Hellman key exchange [DH76] using, for instance, the $m$ and $\mu$ values. This, however, would require the Diffie-Hellman assumption for security, along with the RSA assumption. For simplicity, we will assume Alice uses the same encryption/decryption pair for each session, although if Adv impersonates Alice, Adv could use a different one.

Define hash functions $h, h', h'' : \{0, 1\}^* \rightarrow \{0, 1\}^k$ and $H : \{0, 1\}^* \rightarrow \{0, 1\}^\eta$ (where $\eta \geq \ell + k$). We will assume that $h$, $h'$, $h''$, and $H$ are independent random functions. Let $S_N = \{p : p \leq 2^\eta - (2^\eta \bmod N) \text{ and } \gcd(p, N) = 1\}$.

The protocol is shown in Figure 1. Alice and Bob exchange random values, and Alice also gives her public key to Bob. They both compute hashes of all of these values, plus the password. Then Bob encrypts a random value $a$, multiplies it by the hash, and sends it to Alice. Alice can divide the received value by the hash and decrypt the result (using her private key) to obtain $a$. This value $a$ can be used as a "long" secret for authentication. The idea of why this works is that even if Bob computes hashes corresponding to other passwords, Bob cannot find another value $a'$ whose encryption times the other hash would equal the value sent to Alice, since Bob does not have the private key.[3]

**Theorem 1** *The SNAPI protocol is a secure password-authenticated key exchange protocol under the RSA assumption and the random oracle model.*

Proof in appendix.

## 5   SNAPI-X

We now present a protocol for password-only authentication that is "weakly" resilient to server compromise.[4]

Let $g$ be a *generator* of a cyclic group $\Omega$ of size $\omega$ superpolynomial in $k$ in which the Diffie-Hellman problem is hard. In the SNAPI-X protocol, we assume

---

[3] Naturally, *proving* that this is the case is much more difficult.

[4] By "weakly," we mean that our protocol can be proven secure assuming that once the adversary has corrupted the server it does not actually impersonate the server to a real client, perhaps because it is unable to do network address spoofing.

---

Alice ($A$) Step 1: Choose $m \in_R \{0,1\}^k$, and send $(A, m, (N, e))$ to Bob.
Bob ($B$) Step 2: If $m \notin \{0,1\}^k$, $N \notin [2^{\ell-2}, 2^\ell]$, $e \notin (2^\ell, 2^{\ell+1}]$,
  or $e$ is not prime, then reject,
  Else
  1. Choose $\mu \in_R \{0,1\}^k$ and $a \in_R \mathbb{Z}_N^*$.
  2. Compute $p = H(N|e|m|\mu|A|B|\pi)$.
  3. If $p \notin S_N$ then set $q = a$,
    Else set $q \equiv pa^e \bmod N$.
  4. Send $(\mu, q)$ to Alice.
Alice ($A$) Step 3: If $\mu \notin \{0,1\}^k$ or $\gcd(q, N) \neq 1$, then reject,
  Else
  1. Compute $p = H(N|e|m|\mu|A|B|\pi)$.
  2. If $p \notin S_N$ then reject,
    Else, set $a \equiv (q/p)^d \bmod N$ and
    send $r = h(N|e|m|\mu|A|B|q|a)$ to Bob.
Bob ($B$) Step 4: If $p \notin S_N$ or $r \neq h(N|e|m|\mu|A|B|q|a)$, then reject,
  Else, send $t = h'(N|e|m|\mu|A|B|q|a)$ to Alice,
  set $K = h''(N|e|m|\mu|A|B|q|a)$, and accept.
Alice ($A$) Step 5: If $t \neq h'(N|e|m|\mu|A|B|q|a)$, then reject,
  Else set $K = h''(N|e|m|\mu|A|B|q|a)$ and accept.

---

**Fig. 1.** SNAPI Protocol

there is an initialization in which a client with identity $B$ (whom we will refer to as Bob) chooses a password $\pi \in P$, computes $x = H'(A|B|\pi)$ (where $A$ is the identity of the server, whom we will refer to as Alice) and sends Alice $X = g^x$, which we call the *password verifier*. Alice generates an RSA key pair $((e, N), (d, N))$. After the initialization Bob only needs to remember $\pi$. As in SNAPI, we assume that Alice has chosen an RSA key pair $((e, N), (d, N))$.

Define hash functions $h, h', h'' : \{0,1\}^* \to \{0,1\}^k$ and $H, H' : \{0,1\}^* \to \{0,1\}^\eta$ (where $\eta \geq \ell + k$). We will assume that $h$, $h'$, $h''$, $H$, and $H'$ are independent random functions. Let $S_N = \{p : p \leq 2^\eta - (2^\eta \bmod N) \text{ and } \gcd(p, N) = 1\}$.

The protocol is shown in Figure 2. Alice and Bob exchange random values, and Alice also gives her public key to Bob. They both compute hashes of all of these values, plus the password verifier. Then Bob encrypts a random value $a$, multiplies it by the hash, and sends it to Alice. Alice can divide the received value by the hash and decrypt the result (using her private key) to obtain $a$. This value $a$ can be used as a "long" secret for authentication. Also, to verify that Bob knows the password and not just the password verifier, a type of "Diffie-Hellman" exchange is used. Alice generates her Diffie-Hellman values randomly, and Bob uses the password verifier along with its discrete log as his value. The secret Diffie-Hellman value can thus be computed by both parties and included in the authentication value sent by Bob. Due to space restrictions, we omit the discussion of the security model and proof, and simply state our theorem.

Bob $(B)$ Step 0: Send $B$ to Alice.

Alice $(A)$ Step 1: Retrieve $X$ from password file for $B$.

Choose $m \in_R \{0,1\}^k$, and send $(A, m, (N, e))$ to Bob.

Bob $(B)$ Step 2: If $m \notin \{0,1\}^k$, $N \notin [2^{\ell-2}, 2^\ell]$, $e \notin (2^\ell, 2^{\ell+1}]$,

or $e$ is not prime, then reject,

Else

1. Set $x = H'(A|B|\pi)$.
2. Choose $\mu \in_R \{0,1\}^k$ and $a \in_R \mathbb{Z}_N^*$.
3. Compute $p = H(N|e|m|\mu|A|B|g^x)$.
4. If $p \notin S_N$ then set $q = a$, Else set $q \equiv pa^e \bmod N$.
5. Send $(\mu, q)$ to Alice.

Alice $(A)$ Step 3: If $\mu \notin \{0,1\}^k$ or $\gcd(q, N) \neq 1$, then reject, Else

1. Compute $p = H(N|e|m|\mu|A|B|X)$.
2. If $p \notin S_N$ then reject, Else,
   (a) Set $a \equiv (q/p)^d \bmod N$
   (b) Choose $\gamma \in \mathbb{Z}_\omega$.
   (c) Set $r = h(N|e|m|\mu|A|B|q|a)$ and $y = g^\gamma$.
   (d) Send $(r, y)$ to Bob.

Bob $(B)$ Step 4: If $p \notin S_N$ or $r \neq h(N|e|m|\mu|A|B|q|a)$, then reject,

Else,

1. Send $t = h'(N|e|m|\mu|A|B|q|a|y^x)$ to Alice.
2. Set $K = h''(N|e|m|\mu|A|B|q|a)$, and accept.

Alice $(A)$ Step 5: If $t \neq h'(N|e|m|\mu|A|B|q|a|X^\gamma)$, then reject,

Else set $K = h''(N|e|m|\mu|A|B|q|a)$ and accept.

**Fig. 2.** SNAPI-X Protocol

**Theorem 2** *The SNAPI-X protocol is a secure password-only authentication and key exchange protocol with weak resilience to server compromise, in the random oracle model under the RSA assumption and assuming the hardness of Decision Diffie-Hellman.*

# Acknowledgements

We thank Daniel Bleichenbacher for helpful and stimulating discussions, and for showing us how our protocol can remain secure with a shorter RSA public exponent $e$ (our alternative requirement for $e$ in Section 4). We also thank Victor Boyko for his thorough reading of the paper and helpful comments.

# References

[BCK98]  M. Bellare, R. Canetti, and H. Krawczyk. A modular approach to the design and analysis of authentication and key exchange protocols. In STOC'98 [STO98], pages 419–428.

[Bea91]  Donald Beaver. Secure multiparty protocols and zero-knowledge proof systems tolerating a faulty minority. *Journal of Cryptology*, 4(2):75–122, 1991.

[Ble99]     D. Bleichenbacher, 1999. Personal Communication.

[BM92]      S. M. Bellovin and M. Merritt. Encrypted key exchange: Password-based protocols secure against dictionary attacks. In *Proceedings of the IEEE Symposium on Research in Security and Privacy*, pages 72–84, 1992.

[BM93]      S. M. Bellovin and M. Merritt. Augumented encrypted key exchange: A password-based protocol secure against dictionary attacks and password file compromise. In CCS'93 [CCS93], pages 244–250.

[BMP00]     V. Boyko, P. MacKenzie, and S. Patel. Provably-secure password authentication and key exchange using Diffie-Hellman. In EUROCRYPT2000 [EUR00].

[Boy99]     M. Boyarsky. Public-key cryptography and password protocols: The multiuser case. In CCS'99 [CCS99], pages 63–72.

[BPR00]     M. Bellare, D. Pointcheval, and P. Rogaway. Authenticated key exchange secure against dictionary attacks. In EUROCRYPT2000 [EUR00].

[BR93a]     M. Bellare and P. Rogaway. Random oracles are practical: A paradigm for designing efficient protocols. In CCS'93 [CCS93], pages 62–73.

[BR93b]     M. Bellare and P. Rogaway. Entity authentication and key distribution. In *CRYPTO '93, LNCS* vol. 773, pages 232–249. Springer-Verlag, August 1993.

[BR94]      Mihir Bellare and Phillip Rogaway. Optimal asymmetric encryption. In *EUROCRYPT 94, LNCS* vol. 950, pages 92–111. Springer-Verlag, May 1994.

[BR96]      M. Bellare and P. Rogaway. The exact security of digital signatures—how to sign with RSA and Rabin. In *EUROCRYPT 96*, pages 399–416, 1996.

[BS96]      E. Bach and J. Shallit. *Algorithmic Number Theory: Volume 1 Efficient Algorithms*. The MIT Press, Cambridge, Massachusetts, 1996.

[CCS93]     *First ACM Conference on Computer and Communications Security*, 1993.

[CCS99]     *Sixth ACM Conference on Computer and Communications Security*, 1999.

[CGH98]     R. Canetti, O. Goldreich, and S. Halevi. The random oracle methodology, revisited. In STOC'98 [STO98], pages 209–218.

[DH76]      W. Diffie and M. Hellman. New directions in cryptography. *IEEE Trans. Info. Theory*, 22(6):644–654, 1976.

[EUR00]     *Advances in Cryptology—EUROCRYPT '2000, LNCS* vol. 1807. Springer-Verlag, 14–18 May 2000.

[GLNS93]    L. Gong, T. M. A. Lomas, R. M. Needham, and J. H. Saltzer. Protecting poorly chosen secrets from guessing attacks. *IEEE Journal on Selected Areas in Communications*, 11(5):648–656, June 1993.

[Gon95]     L. Gong. Optimal authentication protocols resistant to password guessing attacks. In *Proc. 8th IEEE Computer Security Foundations Workshop*, pages 24–29, 1995.

[HK98]      S. Halevi and H. Krawczyk. Public-key cryptography and password protocols. In *Proceedings of the Fifth Annual Conference on Computer and Communications Security*, pages 122–131, 1998.

[IEE98]     IEEE P1363 Annex D/Editorial Contribution 1c: Standard specifications for public-key cryptography, June 1998.

[Jab]       D. Jablon. Integrity sciences web site. http://www.IntegritySciences.com.

[Jab96]     D. Jablon. Strong password-only authenticated key exchange. *ACM Computer Communication Review, ACM SIGCOMM*, 26(5):5–20, 1996.

[Jab97]     D. Jablon. Extended password key exchange protocols immune to dictionary attack. In *WETICE'97 Workshop on Enterprise Security*, 1997.

[Len84]    H. W. Lenstra. Divisors in residue classes. *Mathematics of Computation*, 42:331–340, 1984.

[Luc97]    Stephan Lucks. Open key exchange: How to defeat dictionary attacks without encrypting public keys. In *Proc. Workshop on Security Protocols*, 1997.

[MPS]      P. MacKenzie, S. Patel, and R. Swaminathan. Password-authenticated key exchange based on rsa. full version.

[Not99]    RSA Laboratories Technical Note. PKCS #1, version 2, RSA encryption standard. http://www.rsa.com/rsalabs/pubs/PKCS/, 1999.

[Pat97]    S. Patel. Number theoretic attacks on secure password schemes. In *Proc. IEEE Symposium on Research in Security and Privacy*, pages 236–247, 1997.

[RCW98]    M. Roe, B. Christianson, and D. Wheeler. Secure sessions from weak secrets. Technical report, Univ. of Cambridge and Univ. of Hertfordshire, 1998.

[RSA78]    R. Rivest, A. Shamir and L. Adleman. A method for obtaining digital signature and public key cryptosystems. *Comm. of the ACM*, 21:120–126, 1978.

[Sho99]    V. Shoup. On formal models for secure key exchange. IBM Research Report RZ 3121, April 1999.

[STO98]    *Thirtieth ACM Symposium on Theory of Computing*, May 1998.

[STW95]    M. Steiner, G. Tsudik, and M. Waidner. Refinement and extension of encrypted key exchange. *ACM Operating System Review*, 29:22–30, 1995.

[Wu98]     T. Wu. The secure remote password protocol. In *Proc. 1998 Internet Society Network and Distributed System Security Symposium*, pages 97–111, 1998.

[Wu99]     T. Wu. A real world analysis of kerberos password security. In *1999 Internet Society Network and Distributed System Security Symposium*, 1999.

# A    Security of the SNAPI Protocol

The completeness requirement follows directly by inspection. Here we prove that the simulatability requirement holds. The basic technique is essentially that of Shoup [Sho99]. The idea is to create an ideal world adversary $\mathcal{A}^*$ by running the real world adversary $\mathcal{A}$ against a simulated real system, which is built on top of the underlying ideal system. In particular, $\mathcal{A}^*$ (i.e., the simulator combined with $\mathcal{A}$) will behave in the ideal world just like $\mathcal{A}$ behaves in the real world, except that idealized session keys will be used in the real world simulation instead of the actual session keys computed in the real system.

Thus our proof consists of constructing a simulator (that is built on top of an ideal system) for a real system so that the transcript of an adversary attacking the simulator is computationally indistinguishable from the transcript of an adversary attacking the real system. Due to space restrictions we are only able to sketch the simulation. Details may be found in the full version of the paper [MPS]. The difficult part of the simulation is to answer queries to user instances and random oracles that are consistent with the ideal world, but without a priori knowing the passwords.

First we deal with the random oracle queries. Note that the user IDs and nonces allow the simulator to know which conversations they correspond to.

The simulator always answers an $H$ query with the encryption of a known value, which helps in the later simulation.

For an $h$ query, the simulator is able to test whether this corresponds to a password test by encrypting the $a$ value in the query, and then for each $H$ query corresponding to the same conversation, multiplying the encryption of $a$ by the result of that query and testing if the result equals the $q$ value sent in the conversation. If for any $H$ query this test is positive, the simulator must make a test instance password query to the ringmaster in the ideal world. Naturally we must show that the simulator never makes a test instance password query unless the adversary is actively involved in the conversation, and in that case the simulator makes at most one test instance password query. (We sketch the proofs of those below.) If there is no password being tested, or if there is a password being tested and it is incorrect, the simulator simply responds with a random bit string. Otherwise, the simulator responds with a bit string consistent with previous values of the protocol.

For $h'$ and $h''$ queries, the simulator simply responds with random bit strings, and these will be indistinguishable (details omitted).

Now we deal with user instance queries. In general, they are handled as in the actual protocol, except that the $q$ value is set to a random encryption (not multiplied by the result of an $H$ query, since the password is not known to the simulator), and the authentication values $r$ and $t$ are generated randomly, except when a password test by the adversary is detected (by examing the random oracle queries). If necessary, the simulator makes a test instance password query to the ideal world ringmaster, and responds accordingly. If the simulator detects a matching conversation, i.e., an incoming authentication value was sent by a valid partner using the same nonces, then the simulator accepts the authentication value (even though it cannot actually check it since the simulator does not know the password).

To prove that the adversary never forces the simulator to make a test instance password query for a matching conversation, we assume that the adversary does and break the RSA assumption as follows. We take a challenge RSA key and ciphertext and guess the user $A$ involved in the offending conversation. The simulator sends the challenge RSA key when simulating user $A$, and for any user instance in a conversation with $A$ sends $q$ equal to a random encryption multiplied by the challenge ciphertext. Then for any random oracle query that tests a password, one can compute the decryption of the ciphertext (using the fact that the output of the $H$ oracle is a value whose decryption is known to the simulator).

To prove that the adversary never forces the simulator to make two test instance password queries for a non-matched conversation with an "Alice" user instance, we assume that the adversary does and break the RSA assumption as follows. We take a challenge RSA key and ciphertext and guess the user $A$ involved in the offending conversation. The simulator sends the challenge RSA key when simulating user $A$, and for any $H$ query involving user $A$, flips a coin to decide whether to set the output to the encryption of a known value, or the encryption of a known value multiplied by the challenge ciphertext. If two $h$ or $h'$ queries are made along with two $H$ queries such that two password tests must be performed, then these correspond to the same $q$ value sent to user $A$ and thus can be related by an equation which allows one to solve for the decryption of the challenge ciphertext, as long as exactly one of the $H$ query outputs included the challenge ciphertext. This happens with probability $\frac{1}{2}$.

# Round-Efficient Conference Key Agreement Protocols with Provable Security*

Wen-Guey Tzeng and Zhi-Jia Tzeng

Department of Computer and Information Science
National Chiao Tung University
Hsinchu, Taiwan 30050
{tzeng,zjtzeng}@cis.nctu.edu.tw

**Abstract.** A conference key protocol allows a group of participants to establish a secret communication (conference) key so that all their communications thereafter are protected by the key. In this paper we consider the distributed conference key (conference key agreement) protocol. We present two round-efficient conference key agreement protocols, which achieve the optimum in terms of the number of rounds. Our protocols are secure against both passive and active adversaries under the random oracle model. They release no useful information to passive adversaries and achieve fault tolerance against *any coalition* of malicious participants. We achieve the optimal round by transferring an interactive proof system to a non-interactive version, while preserving its security capability.

## 1 Introduction

A conference key protocol allows a group of participants to establish a secret communication (conference) key so that all their communications thereafter are protected by the key. In this paper we consider the distributed conference key (conference key agreement) protocol under the broadcast channel model in which sent messages are guaranteed to be received intact. Nevertheless, the attacker can inject false messages.

For security, we consider both active and passive adversaries. A passive adversary, eavesdropper, tries to learn information by listening to the communication of the participants. There are two types of active adversaries: impersonators and malicious participants. An impersonator tries to impersonate as a legal participant. A malicious participant tries to disrupt conference key establishment among honest participants.

Our protocols focus on round efficiency. We would like to have a conference key agreement protocol by which the participants exchange messages with as few rounds as possible even when active adversaries are present. In this paper we present two round-efficient conference key agreement protocols that achieve

---

* Research supported in part by the National Science Council grant NSC-89-2213-E-009-180 and by the Ministry of Education grant 89-E-FA04-1-4, Taiwan, ROC.

T. Okamoto (Ed.): ASIACRYPT 2000, LNCS 1976, pp. 614–627, 2000.

the optimum in terms of the number of rounds, that is, they use only one round even in the worst scenario. After each participant sends messages to and receives messages from other participants, they go on to compute the conference key no matter whether the attack of active adversaries occurs. Our protocols are secure against both passive and active adversaries under the random oracle model. They release no useful information to passive adversaries and achieve fault tolerance against *any* *coalition* of malicious participants. We achieve the optimal round by transferring an interactive proof system to a non-interactive version, while preserving its security capability.

## 1.1 Related Work

Computing a conference key among a set of participants is a special case of secure multiparty computation in which a group of people, who each possesses a private input $k_i$, computes a function $f(k_1, k_2, \cdots)$ securely [2]. Therefore, it is possible to have a secure conference key agreement protocol by the generic construction for secure multiparty computation. Nevertheless, it is an overkill. Furthermore, there are some distinct features for the conference key agreement protocol. First, a cheater's goal in conference key agreement is to disrupt conference key establishment among the set of honest participants, which is quite different from that in secure multiparty computation. Second, since a cheater's secret is not a necessity in conference key agreement, the cheater can be simply excluded when detected. On the other hand, in secure multiparty computation when a cheater is found, the cheater's secret $x_i$, which is shared into others, is recovered by honest participants so that evaluation can proceed.

There have been intensive research on conference key protocols. Conference key distribution protocols (with a chairman) have been studied in [3,9,10,19]. Pre-distributed conference key protocols have been studied in [4,5,22]. And conference key agreement protocols have been studied in [17,19,20,27,29,28]. Information-theoretically secure conference key protocols have been studied in [5,12]. Most proposed protocols except [18,28] do not have the capability of fault-tolerance so that a malicious participant can easily mislead other participants to compute different conference keys so that the honest participants cannot confer correctly.

Burmester and Desmedt [7] proposed a round-efficient (two-round) protocol (Protocol 3) with $f(k_1, k_2, \ldots, k_n) = g^{k_1 k_2 + k_2 k_3 + \cdots + k_n k_1} \mod p$. In the modified Protocol 7 (authenticated key distribution), they used an interactive proof for authenticating sent messages to show that the protocol is secure against impersonators. However, both protocols cannot withstand the attack of malicious participants. The fault-tolerant conference key agreement protocol of Klein et al. [18] is quite inefficient and its security is not rigidly proved. In [28], when malicious participants are detected, the protocol restarts for the remained participants. It can may be that a participant behaves maliciously in a new round and thus the protocol has to restart again. So, the protocols have to run $O(m)$ times for $m$ malicious participants in the worst case. This may be inefficient since the number of rounds may entail main communication cost,

## 2  Preliminaries

A user in a conference key system is a probabilistic polynomial-time Turing machine. Each user $U_i$ has a secret key $x_i$ and a corresponding public key $y_i$. The system has a public directory of recording the system's public parameters and each user's public key that can be accessed by every one. All users are connected by a broadcast network such that the messages sent on the network cannot be altered, blocked or delayed. For simplicity, we assume that the network is synchronous, that is, for a given phase of a round, all users send their messages to other recipients (or receive messages from others senders) simultaneously. No private channel exists between users. A group of users who wants to establish a conference key is called the *set of participants*.

We consider three types of adversaries. They are all probabilistic polynomial-time Turing machines. An *eavesdropper*, who is not a participant, listens to the broadcast channel and tries to learn the conference key established by the honest participants. An *impersonator*, who is an outsider, tries to impersonate as a legal participant. A *malicious participant*, who is a participant, tries to disrupt establishment of a common conference key among the honest participants. A malicious participant mainly sends "malicious" messages to fool an honest participant to believe that he has computed the same conference key as that of other honest participants, while he does not indeed. We do not care about the possibility that two or more cheating participants collaborate and result in one of them or other malicious participants not being able to compute the key. For example, a malicious participant $U_i$ sends "malicious" messages, but all honest participants compute the same key. Another malicious participant $U_j$, though receiving an incorrect key, still claims that he has had received the correct key. We tolerate this case since this type of collaboration between malicious $U_i$ and $U_j$ do no harm to the honest participants. We do not restrict the number of malicious participants in a conference.

A conference key agreement protocol should meet the following requirements:

- Authentication: an outsider cannot impersonate as a legal participant.
- Correctness: the set of honest participants who follow the protocol computes a common conference key.
- Fairness: the conference key should be determined unbiasedly by all honest participants together.
- Fault tolerance: no coalition of malicious participants can spoil the conference by making honest participants compute different conference keys.
- Privacy: an eavesdropper can not get any information about the conference key established by the honest participants.

We consider two types of communication cost:

- Message efficiency: the total number of messages sent by the participants for completing the protocol. This includes the extra messages for dealing with malicious participants.

– Round efficiency: the total number of rounds executed by the participant for completing the protocol. This includes the extra rounds for dealing with the malicious participants.

In security analysis we use the random oracle model [1], which assumes that a cryptographically strong (collision-free) hash function is a random function. Although this is only a security argument [8], it is a suitable paradigm for analyzing our first protocol.

## 3   Basic Techniques

We use the following setting for the system and users throughout the rest of the paper. The system has public parameters:

– $p$: a large prime number that is $2q + 1$, where $q$ is a large prime also.
– $g$: a generator for the subgroup $G_q$ of all quadratic residues in $Z_p^*$.

Each user $U_i$ has two parameters:

– Private parameter $x_i$: a number in $Z_q^* - \{1\}$.
– Public parameter $y_i = g^{x_i} \bmod p$. Since $q$ is a prime number, $y_i$ is a generator for $G_q$.

Let $x \in_R S$ denote that $x$ is chosen from the set $S$ uniformly and independently and $[a..b]$ denote the set of numbers in between $a$ and $b$, where $a \leq b$. In order to simplify presentation, we omit the or complexity measure $n$ from the related parameters, unless necessary. For example, when we say a probability $\epsilon$ is negligible, we mean that for any positive constant $c$, $\epsilon = \epsilon(n) < 1/n^c$ for large enough $n$. A probability $\delta$ is overwhelming if $\delta = 1 - \epsilon$ for some negligible probability $\epsilon$.

The discrete logarithm (DL) problem is to compute $x \equiv \log_g y \pmod{p}$ from given $(y, g, p)$, where $p = 2q + 1$, $g$ is a generator of $G_q$ and $y \in_R G_q$. The decisional Diffie-Hellman (DDH) problem is to distinguish the distributions

$$(g_1, g_2, g_1^r \bmod p, g_2^r \bmod p) \text{ and } (g_1, g_2, u_1, u_2)$$

with a non-negligible probability, where $g_1$ and $g_2$ are generators of $G_q$, $r \in_R Z_q$ and $u_1, u_2 \in_R G_q$. We assume that the DL and DDH problems are computationally infeasible. They are called the DL assumption (DLA) and the DDH assumption (DDHA). In particular, any probabilistic polynomial-time algorithm cannot solve even a non-negligible fraction of input of the DL problem.

The main building block of our conference key agreement protocols is a protocol of sending a (random) secret to the other participants such that any one can verify that all participants receive the same secret. We call this as the protocol for a publicly verifiable secret (PVS). Let $t$ be the security parameter of the system. If participant $U_i$ wants to send the secret (subkey) $g^{k_i} \bmod p$ to all other participants in a publicly verifiable way, it broadcasts

$$u_{i,j} = y_j^{k_i} \bmod p, 1 \leq j \leq n,$$

where $k_i \in_R Z_q$. Another participant $U_j$ can obtain the shared secret $g^{k_i} \bmod p$ with $U_i$ by computing $(u_{i,j})^{x_j^{-1}} \bmod p$. The PVS proof system shows that

- $\log_{y_1} u_{i,1} \equiv \log_{y_2} u_{i,2} \equiv \cdots \equiv \log_{y_n} u_{i,n} \pmod{p}$, and
- $U_i$ knows that the exponent $k_i = \log_{y_j} u_{i,j} \bmod p$, $1 \le j \le n$.

with error probability $1/2^t$. We can make the error probability inverse exponentially small by repeating the system for a polynomial number of times. The PVS proof system is:

1. $P \to V$: $b_j = y_j^r \bmod p$, $1 \le j \le n$, where $r \in_R Z_q$;
2. $V \to P$: $c \in_R [0..2^t - 1]$;
3. $P \to V$: $w = r - ck_i \bmod q$;
4. $V$ checks whether $b_j = y_j^w \cdot u_{i,j}^c \bmod p$, $1 \le j \le n$.

**Theorem 1.** *Assume the DLA. The PVS proof system above is complete, sound and zero-knowledge.*

*Proof.* The completeness property can be verified easily. For soundness, if a probabilistic polynomial-time adversary $A$ can impersonate $P$ with a non-negligible probability $\epsilon$, the verifier $V$ and $A$ together can solve the discrete logarithm problem with an overwhelming probability. Since the probability $\epsilon$ is non-negligible, one can use $A$ to generate two responses $w_1 = r - c_1 k_i \bmod q$ and $w_2 = r - c_2 k_i \bmod q$ for the same commitment $b_j$'s and two different challenges $c_1$ and $c_2$. One can compute $U_i$'s secret key $k_i = (w_1 - w_2)(c_2 - c_1)^{-1} \bmod q$. Furthermore, if the prover does not know $k_i$, he can pass a challenge by the verifier $V$ with the probability of $1/2^t$.

To simulate the view of a verifier $V^*$, the simulator $S$ first selects $c \in_R [0..2^t-1]$ and $w \in_R Z_q$ and computes $b_j = y_j^w \cdot u_{i,j}^c \bmod p$, $1 \le j \le n$. $S$ then simulates $V^*(b_1, b_2, \ldots, b_n)$ to get $c'$. If $c = c'$, then $S$ outputs $(b_1, b_2, \ldots, b_n, c, w)$. Otherwise, $S$ resets $V^*$ to its original state before this round of simulation and starts the next round of simulation. The output of $S$ and the view of $V^*$ are statistically indistinguishable. $\square$

We need the proof system to be non-interactive. By the standard technique [14], we replace $V$ with a cryptographically strong (collision-resistant) hash function $\mathcal{H}$ for generating the challenge $c$. In the non-interactive paradigm, the interactive version of a proof system need only be zero-knowledge for the honest verifier. Our PVS proof system is honest-verifier zero-knowledge even when $c \in Z_q$. Therefore, we choose $\mathcal{H} : \{0,1\}^* \to \{0,1\}^{\lceil \log q \rceil}$.

The message $(c, w)$ sent by $U_i$ for non-interactive PVS satisfies

$$c = \mathcal{H}(g||y_1|| \cdots ||y_n||u_{i,1}|| \cdots ||u_{i,n}||y_1^w u_{i,1}^c|| \cdots ||y_n^w u_{i,n}^c)$$

where $||$ is the concatenation operator of strings. $U_i$ can compute $(c, w)$ by choosing $r \in_R Z_q$, computing $c = \mathcal{H}(g||y_1|| \cdots ||y_n||u_{i,1}|| \cdots ||u_{i,n}||y_1^r|| \cdots ||y_n^r)$, and setting $w = r - ck_i$. We shall use $\mathrm{NIPVS}(g, y_1, y_2, \ldots, y_n, u_{i,1}, u_{i,2}, \ldots, u_{i,n})$

to denote the non-interactive proof system described above. By verifying $\log_{y_1} u_{i,1} \equiv \log_{y_2} u_{i,2} \equiv \cdots \equiv \log_{y_n} u_{i,n}$, one can be assured that all participants receive the same secret value $g^{k_i}$. The proof system releases no useful information assuming the DLA and the random oracle model.

The integral APVS (PVS with authentication) proof system achieves public verification of a secret and authentication of an identity simultaneously. For APVS, the participant $U_i$ broadcasts

$$u_{i,j} = y_j^{k_i} \bmod p, 1 \leq j \leq n,$$

to other participants, where $k_i \in_R Z_q$. Given the broadcast messages, the APVS proof system is to show that

- $\log_{y_1} u_{i,1} \equiv \log_{y_2} u_{i,2} \equiv \cdots \equiv \log_{y_n} u_{i,n} \pmod{p}$,
- $U_i$ knows that the exponent $k_i = \log_{y_j} u_{i,j} \bmod p$, $1 \leq j \leq n$, and
- $U_i$ knows that the secret $x_i = \log_g y_i \bmod p$.

The APVS proof system with input $(g, y_1, y_2, \ldots, y_n, u_{i,1}, u_{i,2}, \ldots, u_{i,n})$ is:

1. $P \to V$: $b_j = y_j^{r_1} g^{r_2} \bmod p, 1 \leq j \leq n$;
2. $V \to P$: $c \in [0..2^t - 1]$;
3. $P \to V$: $w_1 = r_1 - ck_i \bmod q, w_2 = r_2 - cx_i \bmod q$;
4. $V$ checks $b_j = y_j^{w_1} g^{w_2} (y_i u_{i,j})^c \bmod p, 1 \leq j \leq n$.

**Theorem 2.** *Assume the DLA. The APVS proof system is complete, sound and zero-knowledge.*

*Proof.* The completeness and soundness properties are easily checked.

For zero-knowledge, the simulator $S$ simulates $P$'s interaction with any verifier $V^*$. $S$ randomly selects $c \in_R [0..2^t - 1]$, and $w_1, w_2 \in_R Z_q$ and computes $b_j = y_j^{w_1} g^{w_2} (y_i u_{i,j})^c \bmod p, 1 \leq j \leq n$. $S$ then simulates $V^*(b_1, b_2, \ldots, b_n)$ to get $c'$. If $c = c'$, then $S$ outputs $(b_1, b_2, \ldots, b_n, c, w_1, w_2)$. Otherwise, $S$ resets $V^*$ to its original state before this round of simulation. We can see that the output of $S$ and $V^*$'s view with $P$ are statistically indistinguishable. $\square$

Again, we can make the proof system non-interactive by using a cryptographically strong hash function $\mathcal{H}$ in place of $V$. Let

$$\mathrm{NIAPVS}(g, y_1, y_2, \ldots, y_n, u_{i,1}, u_{i,2}, \ldots, u_{i,n}) = (w_1, w_2, c)$$

denote the non-interactive APVS proof system such that

$$c = \mathcal{H}(g||y_1|| \cdots ||y_n||u_{i,1}|| \cdots ||u_{i,n}||y_1^{w_1} g^{w_2} (y_i u_{i,1})^c|| \cdots ||y_n^{w_1} g^{w_2} (y_i u_{i,n})^c),$$

where $c, w_1, w_2 \in_R Z_q$.

We now present two proofs for the Diffie-Hellman and equality properties, respectively. The proof system DH for the Diffie-Hellman property is to show that an input has the form $(g, u, v, z) = (g, g^a, g^b, g^{ab})$ and the prover knows $a$ and $b$. The proof system EQ for the equality property is to show that an input has the form $(g, u, y, v) = (g, g^a, y, y^a)$ and the prover knows $a$.

The DH proof system is as follows.

1. $P \rightarrow V$: $a_1 = g^{r_1}, a_2 = u^{r_1}, b_1 = g^{r_2}, b_2 = v^{r_2}$, where $r_1, r_2 \in_R Z_q$;
2. $V \rightarrow P$: $c \in_R [0..2^t - 1]$;
3. $P \rightarrow V$: $w_1 = r_1 + bc \bmod q, w_2 = r_2 + ac \bmod q$;
4. $V$ checks $g^{w_1} = a_1 v^c, u^{w_1} = a_2(z)^c, g^{w_2} = b_1 u^c, v^{w_2} = b_2(z)^c$.

**Theorem 3.** *The DH proof system is complete, sound and zero-knowledge.*

*Proof.* The system's completeness follows easily. For soundness, if an adversary, who does not know $a$ and $b$, can impersonate the prover with a non-negligible probability, it can answer two different challenges $c$ and $c'$ of the same commitment $(a_1, a_2, b_1, b_2)$, corresponding to $r_1$ and $r_2$, from the verifier. Let the adversary give the answers $(w_1, w_2)$ and $(w'_1, w'_2)$. We can compute $b = (w_1 - w'_1)/(c - c')^{-1} \bmod q$ and $a = (w_2 - w'_2)/(c - c')^{-1} \bmod q$.

For zero-knowledge, the simulator $S$ simulates $P$'s interaction with any verifier $V^*$. $S$ randomly selects $c \in_R [0..2^t - 1]$, and $w_1, w_2 \in_R Z_q$ and computes $a_1 = g^{w_1}/v^c, a_2 = u^{w_2}/z^c, b_1 = g^{w_2}/u^c$, and $b_2 = v^{w_2}/z^c$. $S$ then simulates $V^*(a_1, a_2, b_1, b_2)$ to get $c'$. If $c = c'$, $S$ outputs $(a_1, a_2, b_1, b_2, c, w_1, w_2)$; otherwise, $S$ resets $V^*$ to its original state before this round of simulation. We can see that the output of $S$ and $V^*$'s view with $P$ are statistically indistinguishable. $\square$

The EQ proof system is:

1. $P \rightarrow V$: $b_1 = g^r \bmod p, b_2 = y^r \bmod p$, where $r \in_R Z_q$;
2. $V \rightarrow P$: $c \in_R [0..2^t - 1]$;
3. $P \rightarrow V$: $w = r - ca \bmod q$;
4. $V$ checks $b_1 = g^w u^c \bmod p$ and $b_2 = y^w v^c \bmod p$.

**Theorem 4.** *The EQ proof system is complete, sound and zero-knowledge [11].*

We use NIDH and NIEQ to denote the non-interactive versions of the DH and EQ proof systems, respectively.

# 4   Our Round-Efficient Protocols

We present two round-efficient conference key protocols and show their security. The first one uses the PVS protocol for verifying the sender's subkey and digital signature for sender's identity. The second protocol uses the integral APVS for both subkey verification and identity authentication.

Since our protocols are non-interactive, we need use a session token ST, which is new for each conference session, in the hash functions to prevent the replay attack. Thus, in our protocols each collision-free hash function $\mathcal{H}(\cdot)$ is computed as $\mathcal{H}(ST, \cdot)$.

## 4.1   Protocol CONF-1

The protocol starts with that an initiator calls for a conference for a set $\mathcal{U}$ of participants and sets the session token ST. Without loss of generality, let $\mathcal{U} = \{U_1, U_2, \dots, U_n\}$ be the initial participant set. Each participant $U_i$, $1 \leq i \leq n$, knows $\mathcal{U}$. Let $H$ be a collision-resistant hashing function, which is used in the modified ElGamal signature scheme. In the protocol, each participant $U_i$ first selects a random number $k_i$ and computes his subkey $g^{k_i} \bmod p$. This subkey is conveyed to the other participants by sending $u_{i,j} = y_j^{k_i} \bmod p, 1 \leq j \leq n$. $U_i$ sends NIPVS$(g, y_1, y_2, \dots, y_n, u_{i,1}, u_{i,2}, \dots, u_{i,n})$ for convincing other participants that all other participants receive the same subkey. $U_i$ also sends the signature $(r_i, s_i)$ of his subkey for authentication. After receiving messages from other participants, $U_i$ checks whether the participant $U_j, j \neq i$, sends the correct messages and authenticates $U_j$'s identity. If not, $U_i$ excludes $U_j$ from the set of honest participants. Then, $U_i$ computes the conference key according to the set of honest participants. Our protocol is as follows.

1. **Message sending:** each participant $U_i$ does the following:
   (a) Randomly select $k_i, R_i \in Z_q$.
   (b) Compute and broadcast $u_{i,j} = y_j^{k_i} \bmod p$, $1 \leq j \leq n$, NIPVS$(g,$ $y_1,$ $y_2,$ $\dots,$ $y_n,$ $u_{i,1},$ $u_{i,2},$ $\dots,$ $u_{i,n})$, $r_i = g^{R_i} \bmod p$ and $s_i = R_i^{-1}(H(ST, r_i, g^{k_i}) - r_i x_i) \bmod q$.
2. **Conference key computing:** each participant $U_i$ does the following:
   (a) Fault detection and exclusion: for each $j \neq i$,
       - Compute $z_j = (u_{j,i})^{x_i^{-1}} \bmod p$ and verify whether $(r_j, s_j)$ is the signature of $z_j$.
       - Verify NIPVS$(g, y_1, y_2, \dots, y_n, u_{j,1}, u_{j,2}, \dots, u_{j,n})$.
       If both checkings are correct, add $U_j$ to its honest participant set $\mathcal{U}_i$.
   (b) Compute the conference key: assume that $U_i$'s honest participant set $\mathcal{U}_i$ is $\{U_{i_1}, U_{i_2}, \dots, U_{i_m}\}$. $U_i$ computes the conference key

$$K = (u_{i_1,i} u_{i_2,i} \cdots u_{i_m,i})^{x_i^{-1}} \bmod p$$
$$= g^{k_{i_1} + k_{i_2} + \cdots + k_{i_m}} \bmod p.$$

Note that only legal participants can verify whether $(r_i, s_i)$ is the signature of the subkey $z_i$. This property is crucial to the proof of releasing no useful information in the random oracle model.

## 4.2   Security Analysis of CONF-1

We now show security of protocol CONF-1 on authentication, correctness, fairness, fault tolerance against malicious participants, and releasing no useful information.

We first show that all honest participants who follow the protocol compute the same conference key. The conference key is determined by all honest participant unbiasedly.

**Theorem 5 (Fault tolerance, correctness and fairness).** *All honest participants who follow the protocol compute a common conference key with an overwhelming probability no matter how many participants are malicious. Furthermore, the common conference key is determined by the honest participants unbiasedly.*

*Proof.* For fault tolerance, we show two things. First, any malicious participant $U_i$ who tries to cheat another participant $U_j$ to accept a different subkey will be excluded by all honest participants. Second, any honest participant will not be excluded by any other honest participant.

Since we assume the broadcast channel, every participant receives the same messages. If a malicious participant $U_i$ sends $(y_1, y_2, \ldots, y_n, u_{i,1}, u_{i,2}, \ldots, u_{i,n})$ such that not all $\log_{y_j} u_{i,j}, 1 \leq j \leq n$, are equal, the probability that he can construct $\mathrm{NIPVS}(y_1, y_2, \ldots, y_n, u_{i,1}, u_{i,2}, \ldots, u_{i,n})$ is at most $T/q$, which is negligible, where $T$ is $U_i$'s runtime. Thus, all honest participants will exclude the malicious participant with an overwhelming probability. We can easily check that an honest participant who follows the protocol shall be accepted by other honest participants as "honest". Therefore, each honest participant computes the same honest participant set with an overwhelming probability.

For correctness, since each honest participant $U_i$ computes the same participant set, $U_i$ uses his private key $x_i$ to compute the subkeys $z_j = g^{k_j} \bmod p$ of all honest participants. Thus, they compute the same conference key with an overwhelming probability.

For fairness, since the common conference key is $g^{k_1 + k_2 + \cdots + k_n} \bmod p$, it is unbiased if any of $k_i, 1 \leq i \leq n$, is selected over $Z_q$ uniformly and independently. Therefore, no participants can bias the conference key as long as one of the honest participants behaves properly.                                                     □

In our protocol, we let each participant sign his broadcast subkey by the modified ElGamal signature scheme, which is existentially unforgeable against the chosen ciphertext attack under the random oracle model [23]. No outsider can impersonate as a legal participant with a non-negligible probability under the chosen-ciphertext attack in the random oracle model.

**Theorem 6 (Authentication).** *Assume the random oracle model. If an outsider $A$ can impersonate as a legal participant $U_i$ to $V$ with a non-negligible probability, $A$ and $V$ together can extract $U_i$'s secret $x_i$ from $A$ with an overwhelming probability.*

*Proof.* Since the modified ElGamal signature scheme is secure against existential forgery under the chosen ciphertext attack, successful impersonation in the interactive system with a non-negligible probability would lead to computing $U_i$'s secret $x_i$ with an overwhelming probability. The use of session token ST makes our non-interactive protocol secure against the replay attack. Note that without special care, the replay attack is inevitable in non-interactive systems. Therefore, our protocol is authenticated under the random oracle model.       □

Since we replace the verifier's challenge with a cryptographically strong hash function, the protocol is not zero-knowledge. But, it releases no useful information that can be used in the protocol under the random oracle model.

**Theorem 7 (No useful information leakage).** *Assume the DLA and the random oracle model. Protocol* CONF-1 *releases no useful information that can be used in the protocol.*

*Proof.* Even though the PVS protocol is complete, sound and honest verifier zero-knowledge, we cannot claim that our protocol release no useful information directly since $U_i$ sends a signature $(r_i, s_i)$ in addition. The simulator $S$ should handle this case. To simulate $U_i$'s output, $1 \leq i \leq n$, $S$ selects $k_i \in Z_q$ randomly and computes $u_{i,j} = y_j^{k_i} \bmod p$, $1 \leq j \leq n$, and NIPVS$(g, y_1, y_2, \ldots, y_n, u_{i,1}, u_{i,2}, \ldots, u_{i,n})$. $S$ then computes a forged signature $(r_i', s_i')$ of $U_i$ for the hash value $h$, i.e., $r_i' = g^a y_i^b \bmod p$, $s_i' = -r_i' b^{-1} \bmod q$ and $h = -r_i ab^{-1} \bmod p$ for $a \in_R Z_q$ and $b \in_R Z_q^*$. Since $H$ is assumed to be a random function under the random oracle model, we let $H(ST, r_i', g^{k_i}) = h$. Finally, $S$ outputs $(u_{i,1}, u_{i,2}, \ldots, u_{i,n}, \text{NIPVS}(g, y_1, y_2, \ldots, y_n, u_{i,1}, u_{i,2}, \ldots, u_{i,n}), r_i', s_i')$, together with a partial description of the random oracle $H$, i.e., setting $H(ST, r_i', g^{k_i}) = h$.

We now compare the output distributions of $U_i$ and $S$. For the output of $U_i$, since $h$ is random, $(r_i, s_i)$ is independent of $(u_{i,1}, u_{i,2}, \ldots, u_{i,n}, \text{NIPVS}(g, y_1, y_2, \ldots, y_n, u_{i,1}, u_{i,2}, \ldots, u_{i,n}))$ and uniformly distributed over $G_q \times Z_q$ that satisfies $g^h \equiv y^{r_i} r_i^{s_i} \pmod p$. For the output of S, the distribution of $(u_{i,1}, u_{i,2}, \ldots, u_{i,n}, \text{NIPVS}(g, y_1, y_2, \ldots, y_n, u_{i,1}, u_{i,2}, \ldots, u_{i,n}))$ is the same as that of $U_i$. The distribution of $(r_i', s_i')$ is also uniformly distributed over $G_q \times Z_q$ that satisfies $g^h \equiv y^{r_i'} r_i'^{s_i'} \pmod p$ since $a$ and $b$ are randomly chosen to fit the equation. Thus, the output distribution of $S$ is equal to that of $U_i$ under the random oracle model. Therefore, our protocol releases no useful information under the random oracle model. $\square$

### 4.3   Protocol CONF-2

In this protocol, identity authentication is achieved by NIAPVS. The protocol is as follows.

1. **Message sending:** each participant $U_i$ does the following:
   (a) Randomly select $k_i \in Z_q$.
   (b) Compute and broadcast $u_{i,j} = y_j^{k_i} \bmod p$, $1 \leq j \leq n$, NIAPVS$(g, y_1, y_2, \ldots, y_n, u_{i,1}, u_{i,2}, \ldots, u_{i,n})$.
2. **Conference key computing:** each participant $U_i$ does the following:
   (a) Fault detection and exclusion: for each $j \neq i$, Verify NIAPVS$(g, y_1, y_2, \ldots, y_n, u_{j,1}, u_{j,2}, \ldots, u_{j,n})$. If the verification holds, add $U_j$ to its honest participant set $\mathcal{U}_i$.

(b) Compute the conference key: assume that $U_i$'s honest participant set $\mathcal{U}_i$ is $\{U_{i_1}, U_{i_2}, \ldots, U_{i_m}\}$. $U_i$ computes the conference key

$$K = (u_{i_1,i} u_{i_2,i} \cdots u_{i_m,i})^{x_i^{-1}} \bmod p$$
$$= g^{k_{i_1} + k_{i_2} + \cdots + k_{i_m}} \bmod p.$$

## 4.4   Security Analysis of CONF-2

The security analysis of NIAPVS-based CONF-2 is similar to that of NIPVS-based CONF-1.

**Theorem 8.** *Assume the DLA and the random oracle model. Protocol CONF-2 is correct, fair, fault-tolerant, and authenticated, and releases no useful information that can be used in the protocol.*

*Proof.* The only difference between CONF-1 and CONF-2 is that CONF-1 uses digital signature to authenticate participants, while CONF-2 uses NIAPVS to authenticate participants. Since Theorem 2 shows that participant's identity is authenticated, CONF-2 meets the security requirements.     □

## 5   A Message-Efficient Protocol

The main protocol in [7] is message-efficient. If all the participants are honest, they broadcast $O(n)$ messages totally. But, the protocol is not fault tolerant, that is, it cannot withstand the attack of malicious participants. We can apply the technique of publicly verifiable secrets to obtain a message-efficient, but not round-efficient, conference key agreement protocol that meets the security requirements. The protocols's message complexity is $O(n)$ in the best case and $O(n^2)$ in the worst case. It seems that there is no easy way to augment the protocol to be both message- and round-efficient. The modified protocol is as follows.

1. **Message sending:** each participant $U_i$ does the following:
   (a) Randomly select $k_i \in Z_q$.
   (b) Compute and broadcast $z_i = g^{k_i x_i} \bmod p$, $t_i = g^{k_i} \bmod p$ and $NIDH(g, y_i, t_i, z_i)$.
2. **Message sending and fault detection:** each participant $U_i$ does the following:
   (a) For each $j, j \neq i$, check whether $NIDH(g, y_j, t_j, z_j)$ is valid. If yes, add $U_j$ to his honest participant set $\mathcal{U}_i$.
   (b) Let $\mathcal{U}_i = \{U_1, U_2, \ldots, U_m\}$ and $Z_i = z_{(i+1 \bmod m)} / z_{(i-1 \bmod m)} \bmod p$. Compute and broadcast $Y_i = Z_i^{k_i x_i} \bmod p$ and $NIEQ(g, z_i, Z_i, Y_i)$.
3. **Conference key computing:** each participant $U_i$ does the following:
   (a) Fault detection and exclusion: for each $j \neq i$, validate $NIEQ(g, z_j, Z_j, Y_j)$. If the validation does not hold, remove $U_j$ from its honest participant set $\mathcal{U}_i$ and *restart* the protocol with the new honest participant set.

(b) Compute the conference key: assume that $U_i$'s honest participant set $\mathcal{U}_i$ is $\{U_1, U_2, \ldots, U_m\}$. $U_i$ computes the conference key

$$K = (z_{i-1})^{mk_i x_i} Y_i^{m-1} Y_{i+1}^{m-2} \cdots Y_m^{i-1} Y_1^{i-2} Y_2^{i-3} \cdots Y_{i-2} \bmod p$$
$$= g^{x_1 x_2 k_1 k_2 + x_2 x_3 k_2 k_3 + \cdots + x_n x_1 k_n k_1} \bmod p.$$

**Theorem 9.** *Assume the DDHA and the random oracle model. The protocol above is correct and secure with authentication, fault tolerance, and leaking no useful information.*

*Proof.* The correctness follows from [7], while the security follows from the previous two round-efficient conference key agreement protocols.     □

## 6   Conclusion

We have presented two round-efficient conference key agreement protocols. The protocols meet the security requirements: authentication, correctness, fairness, fault tolerance (robustness) and privacy. Their message complexity is $O(n^2)$ for $n$ participants. We also modified Burmester and Desmedt's protocol so that it can withstand the attack of malicious participants.

It would be interesting to find a round-efficient protocol that meets all security requirements and has $O(n)$ message complexity.

## References

1. M. Bellare, P. Rogaway, "Random oracles are practical: a paradigm for designing efficient protocols", Proceedings of the First ACM Conference on Computer and Communications Security, pp.62-73, 1993.
2. M. Ben-Or, S. Goldwasser, A. Wigderson, "Completeness Theorems for Non-Cryptographic Fault-Tolerant Distributed Computation", Proceedings of the 20th ACM Symposium on the Theory of Computing, pp.1-10, 1988.
3. S. Berkovits, "How to Broadcast a Secret", Proceedings of Advances in Cryptology - Eurocrypt '91, Lecture Notes in Computer Science 547, Springer-Verlag, pp.535-541, 1991.
4. R. Blom, "An Optimal Class of Symmetric Key Generation Systems", Proceedings of Advances in Cryptology - Eurocrypt '84, Lecture Notes in Computer Science 196, Springer-Verlag, pp.335-338, 1984.
5. C. Blundo, A.D. Santis, A. Herzberg, S. Kutten, U. Vaccaro, M. Yung, "Perfectly-Secure Key Distribution for Dynamic Conferences", Proceedings of Advances in Cryptology - Crypto '92, Lecture Notes in Computer Science 740, Springer-Verlag, pp.471-486, 1992.
6. D. Boneh, R. Venkatesan, "Hardness of Computing the Most Significant Bits of Secret Keys in Diffie-Hellman and Related Problems", Proceedings of Advances in Cryptology - Crypto '96, Lecture Notes in Computer Science 1109, Springer-Verlag, pp.129-142, 1996.

7. M. Burmester, Y. Desmedt, "A Secure and Efficient Conference Key Distribution System", Proceedings of Advances in Cryptology - Eurocrypt '94, Lecture Notes in Computer Science 950, Springer-Verlag, pp.275-286, 1994.

8. R. Canetti, O. Goldreich, S. Halevi, "The Random Oracle Methodology Revisited", Proceedings of the 30th STOC, pp.209-218, 1998.

9. C.C. Chang, C.H. Lin, "How to Converse Securely in a Conference", Proceedings of IEEE 30th Annual International Carnahan Conference, pp.42-45, 1996.

10. C.C. Chang, T.C. Wu, C.P. Chen, "The Design of a Conference Key Distribution System", Proceedings of Advances in Cryptology - Auscrypt '92, Lecture Notes in Computer Science 718, Springer-Verlag, pp.459-466, 1992.

11. D. Chaum, T.P. Pedersen, "Wallet DataBases with Observers", Proceedings of Advances in Cryptography - Crypto'92, pp.90-105, 1992.

12. Y. Desmedt, V. Viswandathan, "Unconditionally secure dynamic conference distribution", IEEE International Symposium on Information Theory 98, pp.383, 1998.

13. W. Diffie, P.C. van Oorschot, M.J. Weiner, "Authentication and Authenticated Key Exchanges", Design, Codes and Cryptography Vol. 2, pp.107-125, 1992.

14. U. Feige, A. Fiat, A. Shamir, "Zero-Knowledge Proof of Identity", Journal of Cryptology Vol. 1, pp.77-94, 1988.

15. O. Goldreich, H. Krawczyk, "On the Composition of Zero-Knowledge Proof Systems", ICALP 90, Lecture Notes in Computer Science 443, pp.268-282, Springer-Verlag, 1990.

16. T. Hwang, J.L. Chen, "Identity-Based Conference Key Broadcast Systems", Proceedings of IEE Computers and Digital Techniques, Vol. 141, No. 1, pp.57-60, 1994.

17. I. Ingemarsson, D.T. Tang, C.K. Wong, "A Conference Key Distribution System", IEEE Transactions on Information Theory, Vol. IT-28, No. 5, pp.714-720, 1982.

18. B. Klein, M. Otten, T. Beth, "Conference Key Distribution Protocols in Distributed Systems", Proceedings of Codes and Ciphers-Cryptography and Coding IV, IMA, pp.225-242, 1995.

19. K. Koyama, "Secure Conference Key Distribution Schemes for Conspiracy Attack", Proceedings of Advances in Cryptology - Eurocrypt '92, Lecture Notes in Computer Science 658, Springer-Verlag, pp.449-453, 1992.

20. K. Koyama, K. Ohta, "Identity-Based Conference Key Distribution Systems", Proceedings of Advances in Cryptology - Crypto '87, Lecture Notes in Computer Science 293, Springer-Verlag, pp.175-184, 1987.

21. K. Koyama, K. Ohta, "Security of Improved Identity-Based Conference Key Distribuitoin Systems", Proceedings of Advances in Cryptology - Eurocrypt '88, Lecture Notes in Computer Science 330, Springer-Verlag, pp.11-19, 1988.

22. T. Matsumoto, H. Imai, "On the Key Predistribution System: A Practical Solution to the Key Distribution Problem", Proceedings of Advances in Cryptology - '87, Lecture Notes in Computer Science 293, Springer-Verlag, pp.185-193, 1987.

23. D. Pointcheval, J. Stern. "Security proofs for signatue schemes", Proceedings of Advances in Cryptology - Eurocrypt '96, Lecture Notes in Computer Science 1070, Springer-Verlag, pp.387-398, 1996.

24. R.A. Rueppel, P.C. Van Oorschot, "Modern Key Agreement Techniques", Computer Communications, 1994.

25. A. Shimbo, S.I. Kawamura, "Cryptanalysis of Several Conference Key Distribution Schemes", Proceedings of Advances in Cryptology - Asiacrypt '91, Lecture Notes in Computer Science 739, Springer-Verlag, pp.265-276, 1991.

26. V. Shoup, "Lower Bounds for Discrete Logarithms and Related Problems", Proceedings of Advances in Cryptology - Eurocrypt '97, Lecture Notes in Computer Science 1233, Springer-Verlag, pp.256-266, 1997.
27. D.G. Steer, L. Strawczynski, W. Diffie, M. Wiener, "A Secure Audio Teleconference System", Proceedings of Advances in Cryptology - Crypto '88, Lecture Notes in Computer Science 409, Springer-Verlag, pp.520-528, 1988.
28. W.G. Tzeng, "A Practical and Secure Fault-tolerant Conference-key Agreement Protocol", Proceedings of Public Key Cryptography - PKC 2000, Lecture Notes in Computer Science 1751, Springer-Verlag, pp.1-13, 2000.
29. T.C. Wu, "Conference Key Distribution System with User Anonymity Based on Algebraic Approach", Proceedings of IEE Computers and Digital Techniques, Vol. 144, No 2, pp.145-148, 1997.
30. Y. Yacobi, "Attack on the Koyama-Ohta Identity Based Key Distribution Scheme", Proceedings of Advances in Cryptology - Crypto '87, Lecture Notes in Computer Science 293, Springer-Verlag, pp429-433, 1987.

# Author Index

# Lecture Notes in Computer Science

For information about Vols. 1–1882
please contact your bookseller or Springer-Verlag